MW00846590

Architecture and Mathematics from Antiquity to the Future

Kim Williams • Michael J. Ostwald
Editors

Architecture and Mathematics from Antiquity to the Future

Volume I: Antiquity to the 1500s

Editors
Kim Williams
Kim Williams Books
Torino, Italy

Michael J. Ostwald
School of Architecture and Built Environment
The University of Newcastle
Callaghan, Australia

Cover photo: The Pyramids of Akhet Khufu, Khafre and Menkaura,
Giza, Egypt. Photograph © Giulio Magli, reproduced by permission.

ISBN 978-3-319-00136-4 ISBN 978-3-319-00137-1 (eBook)
DOI 10.1007/978-3-319-00137-1
Springer Cham Heidelberg New York Dordrecht London

Library of Congress Control Number: 2014958157

Printed on acid-free paper

Springer International Publisher is part of Springer Science+Business Media
(www.birkhauser-science.com)

Preface to Architecture and Mathematics from Antiquity to the Future

In June of 1996, in his keynote address at the conference 'Nexus'96: Relationships Between Architecture and Mathematics', the founding international conference of what would become an international community for research in a new interdisciplinary field, eminent engineer Mario Salvadori asked, '[c]an there be any relationship between architecture and mathematics?' Over the next 18 years, the Nexus community came together for a series of bi-yearly conferences in Italy, Portugal, Mexico, Turkey and the USA to examine, debate and celebrate the relationships that exist between architecture and mathematics. The conferences were hosted in locations where important historic connections had been proposed between architecture and mathematics: in Europe these locations include Fucecchio (1996), Mantua (1998), Ferrara (2000), Óbidos (2002), Genoa (2006), Porto (2010) and Milan (2012). Further afield, conferences were held in Mexico City (2004), San Diego (2008) and Ankara (2014). Conference venues were chosen to permit participants to visit local sites of historic importance for architecture and mathematics in post-conference workshops, such as Pompeii and Herculaneum in 1996, the villas of Palladio in 1998 and Teotihuacan in 2004. The speakers at these events include some of the most influential people in architecture, art, mathematics and engineering. Lionel March, Robert Tavenor, Alberto Pérez-Gómez, Marco Frascari, Michele Emmer, Leonard Eaton and Mario Salvadori, amongst many other luminaries, have all presented at the Nexus conferences and taken part in round-table discussions, forums and visits to some of the great architecture of these regions.

The first Nexus conference was actually conceived out of the frustration caused by the difficulty of finding a venue for publishing interdisciplinary research: papers in architecture and mathematics were seen as too mathematical for architectural journals, but not mathematical enough for mathematics journals. At best, such research was viewed as a curiosity, too far from the mainstream to garner much interest. Because there was no single journal that encouraged such research, when authors were fortunate enough to have an article accepted, publications were scattered, and authors seldom knew about the work of others examining similar topics. The Internet was in its infancy at that time, leaving far-flung scholars to work in isolation. One journal, *The Mathematical Intelligencer*, and its particularly open-minded editor-in-chief, Chandler Davis, had accepted papers by three of the participants at the first conference, Kim Williams, Benno Artmann and Heinz Götze, who subsequently began to correspond. The 23 people who met in 1996 at the first conference knew of each other's work by word of mouth: friends sending their work to friends. But already by the second conference, 2 years later, the growing group felt the need for a publishing venue, and it was decided to found the *Nexus Network Journal—Nexus*, from the name chosen for the first conference to represent the idea of interweaving ideas from two disciplines, and *Network*, to describe the group of people whose acquaintances and collaborations were continuing to expand. The first issue of the journal, with Kim Williams as editor-in-chief, was introduced online in 1999, was added to at trimester intervals of the course of that year and was produced in print at its end. The journal continued in that way for its first 2 years, but by volume 3 in 2001, submissions had grown so much that it was published in two issues per year, until with volume 9 in 2010, it grew to three issues per year.

Across 15 volumes, 35 issues and over 500 refereed papers, the international reputation and impact of the journal have grown considerably. Now published jointly in the Birkhäuser programme of Springer-Basel and Kim Williams Books of Torino, Italy, the journal is highly respected and has a growing readership. Beginning with volume 16 in 2014, the *NNJ* will be overseen jointly by the editors of these present two volumes.

Foreseen along with the conferences was the publication of the proceedings. The series entitled 'Nexus: Architecture and Mathematics' comprised seven volumes from the first seven conferences. At the beginning, the conference books were seen as separate from the journal. This changed with the eighth conference, when speakers voiced the desire to see their papers published in the *NNJ*, which was by that time mature and esteemed. Thus, since 2010, papers presented in the Nexus conferences have been published in special issues of the journal and are available online. However, the research presented at the early conferences was only available in a series of limited edition books. With many of these being out of print there has been growing pressure to make the most highly cited works from the early years of the Nexus conferences available. Rather than simply republishing selected works in the order in which they were written, such was the scope of these early Nexus publications that an alternative proposition presented itself.

We, the editors, have assembled almost a hundred papers from the early years of the Nexus conferences, and arranged them both thematically and chronologically to trace key moments in the history and theory of architecture and mathematics, from antiquity to the present day, along with predictions for the future. These chapters describe over 60 major buildings and architectural works, consider more than twenty major theories of geometry and design and cover themes and ideas arising from five continents and spanning over four millenia.

Having said this, the present two-volume work does not pretend to be a comprehensive encyclopaedia of the history and theory of every facet of the relationship between architecture and mathematics. Being works by more than one hundred authors with backgrounds in not only architecture and mathematics but also engineering, physics, chemistry, philosophy, music and more, there is a rich diversity of approaches to the topic, along with some insightful synergies and informative disagreements. All of the chapters have undergone minor editorial revisions, including, in some cases, updated bibliographies. In a few cases authors have chosen to make more substantial revisions, to bring their chapters up to date, or direct the reader to advances that are currently occurring in their areas. In addition to this, we have provided an overview chapter for each volume (Chap. 1 in vol. I and Chap. 48 in vol. II), to frame the sequence and structure of the whole as well as a chapter entitled 'Mathematics *in*, *of* and *for* Architecture: A Framework of Types' (Chap. 3) which seeks to classify, and thereby make more accessible, the myriad connections proposed across this work.

Each of the chapters in the present work have became crucial landmarks in the scholarly landscape of architecture and mathematics. Some represent pioneering research, the first studies of the relationships between architecture and mathematics in a specific period, or in the oeuvre of a given architect. They serve as both points of departure for new voyages of discovery and as destinations for people entering unfamiliar terrain. For the novice researcher these works provide a grounding for their explorations, and for seasoned scholars these chapters offer a critical record of the efforts of fellow travellers. We, the editors, hope that through this two-volume work, these chapters can continue to inspire and guide future generations.

We wish to thank Maria Roberts, Valentina Filemio and Marco Giorgio Bevilacqua for assistance with editing and proofing, and Michael Dawes for support with image preparation and research assistance. We thank all authors for permission to reuse their material, and for their help in updating texts and references. Finally, we thank Anna Mätzener, Editor for Mathematics and History of Science, and Thomas Hempfling, Executive Editor for Mathematics, Birkhäuser, for their support of the Nexus conferences and the *Nexus Network Journal* throughout the years, and especially for their support of this present work.

Torino, Italy — Kim Williams
Newcastle, Australia — Michael J. Ostwald
January 2015

Contents for Volume I

Contents for Volume II

Contributors to Volume I

Benno Artmann (1933–2010)

Niels Bandholm Kloster, Ringkoebing, Denmark

Javier Barrallo School of Architecture, The University of the Basque Country (UPV-EHU), San Sebastian, Spain

Maria Teresa Bartoli Dipartimento di Architettura, Università degli studi di Firenze, Florence, Italy

Avril Behan Dublin Institute of Technology, College of Engineering & Built Environment, Dublin, Ireland

Gerardo Burkle-Elizondo Centro Interinstitucional de Investigaciones en Arte y Humanidades, Universidad Autónoma de Zacatecas, Zacatecas Zac, Mexico

Paul A. Calter Vermont Technical College, Randolph Center, VT, USA

Luisa Rossi Costa Department of Mathematics, Politecnico di Milano, Milan, Italy

Salvatore di Pasquale (1931–2004)

Yvonne Dold-Samplonius Ruprecht-Karls-Universität Heidelberg, Institute of Mathematics, Heidelberg, Germany

Sylvie Duvernoy Politecnico di Milano, Milan, Italy

Holger Falter Hochschule für angewandte Wissenschaften Coburg, Coburg, Germany

Marco Frascari (1945–2013)

Heinz Götze (1912–2001)

Paulus Gerdes (1952–2014)

Gulzar Haider School of Architecture, Beaconhouse National University, Lahore, Pakistan

István Hargittai Materials Structure and Modeling Research Group, Budapest University of Technology and Economics, Budapest, Hungary

Magdolna Hargittai Materials Structure and Modeling Research Group, Budapest University of Technology and Economics, Budapest, Hungary

Silvia L. Harmsen St. Leon-Rot, Germany

George Gheverghese Joseph University of Manchester, Manchester, UK

Jay Kappraff Department of Mathematics, New Jersey Institute of Technology, University Heights, Newark, NY, USA

Izumi Kuroishi School of Cultural and Creative Studies, Aoyama Gakuin University, Shibuya-ku, Tokyo, Japan

Andrew I-kang Li Tokyo, Japan

Elena Marchetti Department of Mathematics, Politecnico di Milano, Milan, Italy

Rafael Ramírez Melendez Pompeu Fabra University, Barcelona, Spain

Rachel Moss Department of the History of Art and Architecture, Trinity College, Dublin 2, Ireland

Muhammad Moussa HHCP Architects, Orlando, FL, USA

Vini Nathan College of Architecture, Design and Construction, Auburn University, Auburn, AL, USA

Michael J. Ostwald School of Architecture and Built Environment, University of Newcastle, Callaghan, NSW, Australia

Alpay Özdural (1944–2003)

Alfonso Ramírez Ponce E 21, M XII Educación, Coyoacán, México, D.F.

Graham Pont Balmain, NSW, Australia

Mark A. Reynolds Academy of Art University, San Francisco

Clara Silvia Roero Department of Mathematics "Giuseppe Peano", Università di Torino, Turin(Torino), Italy

Nicoletta Sala Accademia di Architettura, Università della Svizzera italiana, Mendrisio, Switzerland

Peter Saltzman Berkeley, CA, USA

Mario Salvadori (1907–1997)

Santiago Sanchez-Beitia School of Architecture, The University of the Basque Country (UPV-EHU), San Sebastian, Spain

William D. Sapp San Bernardino National Forest, San Bernardino, CA, USA

Peter Schneider College of Architecture and Planning, University of Colorado, Denver, USA

David Speiser Université catholique de Louvain, Louvain-La-Neuve, Belgium

Gert Sperling Vellmar, Germany

Rudolf H.W. Stichel Fachbereich Architektur, Technische Universität Darmstadt, Darmstadt, Germany

Helge Svenshon Fachbereich Architektur, Technische Universität Darmstadt, Darmstadt, Germany

Robert Tavernor Tavernor Consultancy, Pimlico, London, UK

Ricardo David Valdez-Cepeda Centro Regional Universitario Centro Norte Apdo, Universidad Autónoma Chapingo, Zacatecas Zac, Mexico

Livio G. Volpi Ghirardini Accademia Nazionale Virgiliana, Mantova, Italy

Stephen R. Wassell Department of Mathematics and Computer Science, Sweet Briar College, Sweet Briar, VA, USA

Carol Martin Watts The College of Architecture, Planning & Design, Kansas State University, Manhattan, KS, USA

Kim Williams Kim Williams Books, Turin (Torino), Italy

Mark Wilson Jones Department of Architecture and Civil Engineering, University of Bath, Bath, UK

Michael R. Ytterberg BLT Architects, Philadelphia, PA, USA

Radoslav Zuk School of Architecture, McGill University, Montreal, QC, Canada

Chapter 1
Relationships Between Architecture and Mathematics

Michael J. Ostwald and Kim Williams

architecture |ˈärkiˌtek ch ər| *noun*
The art or practice of designing and constructing buildings. The style in which a building is designed or constructed, esp. with regard to a specific period, place, or culture.

mathematics |maθ(ə)ˈmatiks| *plural noun*
The abstract science of number, quantity, and space. Mathematics may be studied in its own right (pure mathematics), or as it is applied to other disciplines such as physics and engineering (applied mathematics).

Introduction

What is the nature of the relationship between architects and mathematicians or between architecture and mathematics? As they are commonly understood these two groups seem to have few obvious connections. The word 'architecture' is used to describe either the practice of creating buildings or a particular class of constructed—architectural—objects. In contrast, the word 'mathematics' denotes a domain of pure or applied knowledge that is associated with the study or use of abstract objects such as numbers and shapes or forms. Professions, like architecture, tend to be isolated and controlled, restricting membership to experts who have been awarded particular qualifications and have fulfilled certain criteria (Fournier 2000).

M.J. Ostwald
School of Architecture and Built Environment, University of Newcastle, Callaghan, New South Wales 2308, Australia
e-mail: michael.ostwald@newcastle.edu.au

K. Williams (✉)
Kim Williams Books, Corso Regina Margherita, 72, 10153 Turin (Torino), Italy
e-mail: kwb@kimwilliamsbooks.com

K. Williams and M.J. Ostwald (eds.), *Architecture and Mathematics from Antiquity to the Future*, DOI 10.1007/978-3-319-00137-1_1,

Instead, disciplines, like mathematics, are ways of grouping and identifying bodies of knowledge and expertise that are both pertinent within the discipline and which might also be applied in other fields (Klein 1996). However, despite these apparent differences, the distance between these two—between a profession and a discipline and between an object (of design) and a subject (of study)—is far less than many would assume. In order to make sense of the contemporary view which considers architecture and mathematics as dissimilar pursuits, it is useful to trace a brief history of the growth of disciplines and professions over time; a history which, most importantly, reminds us that in the past these two were very closely connected.

In the world of ancient Greece, learned professionals were trained in one of three fields: religion, ethics, or the human condition. While some distinction was made between the ways students were educated in each of these fields, a sound foundation in both philosophy and mathematics was considered a necessity (Marrou 1956). At the same time, the acts of creating a chair, designing a house or constructing a ballista were all tasks that could be undertaken by artisans. Such skilled workers were trained in the practical processes of making (stone-cutting, woodwork) along with a range of knowledge domains including both geometry and metrology (Kostof 1977a).

Later, in Roman times, Vitruvius tells us that the architect was neither a scientist nor a craftsman, but one who had sufficient knowledge of a range of scientific fields—first and foremost geometry, but also history, philosophy, music, medicine, law and astronomy—to be able to oversee the work of all other disciplines (Vitruvius 2009: 5).

A similar pattern of relationships is found in medieval Europe where the first universities recognised graduates in law, divinity and medicine. In the medieval university students were expected to complete preparatory studies on the *trivium*—grammar, logic and rhetoric—along with studies in arithmetic, geometry and aesthetics (Janin 2008). Seated in rows of carved pews, in high-roofed tapestry-lined halls, these students were able to directly observe the geometric mysteries of space and form. But the very halls they inhabited during these studies had been created by master builders and teams of craftsmen and labourers with a fund of geometric knowledge. Evidence of this knowledge is found in Villard de Honnecourt's *Livre de Portraiture*, where he wrote, "...in this book ... you are able to find the technique of representation as the discipline of geometry requires and instructs" (Barnes 2009: 35). The most senior of the master builders, often the direct descendants of artisan families, were given the title 'architect' or 'engineer' (Kostof 1977b). These master builders were trained in both the arts and sciences of design and construction. Furthermore, in both the Classical and Medieval eras, the mathematical disciplines provided a critical foundation for construction (Fitchen 1961; James 1981). This is why, for much of this era, there was little distinction between professions and disciplines, or between architects and mathematicians. This close and productive relationship between architecture and mathematics was to continue for several centuries before reaching its most visible apogee during the Renaissance.

During the Renaissance the Medieval educational foundation found in the *trivium* was expanded to encompass a second tier of four arts (the *quadrivium*) that typically included studies of arithmetic, geometry, music and astronomy. The archetypical 'Renaissance man' was expected to have both a broad and comprehensive knowledge across these seven subjects, along with their potential application in at least one, and possibly more, of the following pursuits; science, art, medicine and architecture. Leon Battista Alberti was one such polymath; an author, artist, poet and linguist who had also mastered optics, perspective and cryptography (Williams et al. 2010). Today, Alberti is best known as an architect, because it was in the application of these ideas to design, in both written and built work, that he achieved his most enduring success. For a similar reason Christopher Wren is also regarded as one of the world's great architects, even though he originally distinguished himself in astronomy, physics and mathematics and only began to design buildings when he was already a respected scientist (Bennett 1982). Robert Hooke is another example of the natural philosopher-cum-architect. Even the great Isaac Newton demonstrated an amateur, but informed, interest in architectural theory. Such cases suggest that for centuries architecture and mathematics were closely related, and equally respected, areas of inquiry.

However, despite the examples of Wren and others, the seventeenth century was more generally marked by the rise in power of guilds and colleges who sought to define and preserve their members' interests (Melton 2001). For example, it was around this time that the Freemasons formulated a series of rules of membership and practice that sought to protect the knowledge and skills of the stonemasons. Amongst the earliest articles of Freemasonry is a set of practical and symbolic rules showing the essential relationship between architecture and geometry (Berman 2012). Thus, while the tradition of the 'Renaissance man' was still being valorised in the eighteenth century, the guilds and technical colleges remained the driving force which gradually separated professions from disciplines and, inadvertently, increased specialisation began to distance architecture from mathematics (Clarke 1994).

Over the ensuing 200 years, in parallel with the emergence of new technology and the need for more focussed trades and skills, disciplines and professions became increasingly specialised and their roles began to change (Duffy and Hutton 1998). For example, established in 1794 the French *École Polytechnique* was at the forefront of approaches to training a new, elite class of technocrats. Embracing a scientific disposition, the curriculum included issues of aesthetic perception, positivism and rationalism. In part because of this educational focus, this era marked a growing separation between architects and engineers (Picon 1992). The French Enlightenment was also one of the last periods wherein architects, still trained in descriptive geometry, directly contributed to mathematical knowledge. In particular, the discipline of stereotomy was developed largely by architects to allow stone blocks to be cut and assembled in complex forms (Warren 1875).

By the early years of the twentieth century architectural education was split between an atelier-based model, which traced its origins to the École des

Beaux-Arts, and the technical college model, which acknowledged a lineage to the apprenticeship system (Cuff 1991; Crinson and Lubbock 1994). The atelier system, modelled on fine art practice, included geometry in its core curriculum but the role of arithmetic was less apparent (Draper 1977). The technical college system included both science and mathematics in its syllabus along with more extensive applications of geometry. Significantly though, despite the apparent differences between these systems, both effectively positioned mathematics as a secondary discipline which merely served to buttress the education of architects (Boyer and Mitgang 1996). In a comparable way, the discipline of mathematics divided itself into a 'pure' and an 'applied' strand, with the former being regarded as the path for specialists, and the latter for those who sought to engage more directly with other fields (Davis et al. 1995). This pattern was repeated around the world with the combination of increasing specialisation and the desire for professional recognition gradually separating and isolating different knowledge domains and disciplines from each other (Fournier 2000).

Such was the compartmentalisation of knowledge that occurred in the early part of the twentieth century that British scientist and novelist Charles Percy Snow famously criticised the rise of two distinct and separate cultures—science and humanities—each seemingly unaware of the basic values and lessons of the other (Snow 1998). Snow's observations, derived from his identification of the growing separation between disciplinary groups, were both widely reported and criticised (Carafoli et al. 2009). Certainly it was becoming harder for a person to be qualified in two or more fields and the era of the peripatetic scholar was effectively over. This was true even within the discipline of mathematics itself; Henri Poincaré was considered the "last universal mathematician". Yet, the growing accessibility of knowledge, no longer protected by professional guilds or enshrined in the lore of esoteric societies, meant that rather than fostering the divide between two distinct cultures, there were potentially a multitude of secondary connections to be made between different groups, each creating new sub-cultures (Nicolescu 2002). However, such new transdisciplinary groups face a twofold problem: visibility and recognition (Doucet and Janssens 2011). In the first instance, while important connections exist between fields, like architecture and mathematics, they are often rendered invisible by contemporary educational practices and the legal implications of professional ethics (Sokolowski 1991). In the second, sustained research must take place before such hidden associations can be recognised, investigated and celebrated. This is especially the case in contemporary society where these same two problems of visibility and recognition continue to hinder our capacity to engage with transdisciplinary knowledge.

Today, the extent to which architects are formally trained in mathematics is probably lower than in any previous period in history (Ostwald and Williams 2008). The degree to which mathematicians directly engage with building design and construction is at an equally low ebb. This is unfortunately true of the general public as well, since no formal architectural education at all is offered in public schools, while at least of minimum of mathematics is taught. Yet, this situation is not entirely as it seems. Advances in computing have placed mathematical techniques and processes at the fingertips of every young architect in the world, providing a means of using complex geometry which was previously unavailable (Szalapaj 2005; Littlefield 2008; Ostwald 2012). In the last decade buildings have been produced which model non-linear dynamical systems, are covered in fractals or aperiodic tiles, are roofed in complex membranes and are optimised for energy performance and wind-load using Boussinesq or Bernoulli equations (Ostwald 2006; Burry and Burry 2012). All of these developments rely on advances in mathematics yet, paradoxically, the distance between the architectural profession and the mathematical disciplines has seemingly never been greater. The problem is, as stated previously, that a myriad of connections continue to exist between architecture and mathematics, just as they have done for several millennium, but in order to understand and appreciate these connections—to perceive and recognise both their historical and theoretical significance—academics and professionals must be willing to engage in transdisciplinary scholarship. This is where the present work has an important role to play.

The relationship between architecture and mathematics is most visible and recognisable when members of these two groups cross the divide and productively work with concepts, themes and topics that have been developed in the other field. Thus, for architects to talk with authority about geometry or arithmetic requires both a willingness and a capacity to traverse disciplinary boundaries. Similarly, a mathematician venturing into the realm of architectural history, theory and design must engage with a field that has its own language and traditions. The present, two-volume, edited collection represents the work of some 100 authors who have made such important transdisciplinary incursions. They have investigated the complex interplay of connections between architecture and mathematics and have engaged with bodies of disciplinary and professional knowledge that are often separated in contemporary society. These authors have worked to reconnect two fields that were once closely reliant on each other and, in doing so, blaze a path so that other scientists, scholars, professionals and gifted amateurs may follow.

At the start of this chapter two questions were raised. The first asked: "What is the nature of the relationship that exists between architects and mathematicians?" The short answer is that the two share a common intellectual heritage and similar values and concerns. Both work with a highly structured system of symbols that support each other hierarchically to achieve an edifice. Mathematicians, in fact, speak quite often of "the mathematical edifice". Here is Bertrand Russell on the study of mathematics:

> The discovery that all mathematics follows inevitably from a small collection of fundamental laws is one which immeasurably enhances the intellectual beauty of the whole; to those who have been oppressed by the fragmentary and incomplete nature of most existing chains of deduction this discovery comes with all the overwhelming force of a revelation; like a palace emerging from the autumn mist as the traveller ascends an Italian hillside, the stately storeys of the mathematical edifice appear in their due order and proportion, with a new perfection in every part (Russell 2009: 67).

The second question was more general and inclusive: "How are the subjects of architecture and mathematics connected?" The answer to this involves a consideration of several thousand years of history, along with an investigation of philosophy, number and shape, construction and material science. Issues of representation, meaning, religion, culture and ethics are also pertinent to this question. Thus, to begin to approach this larger topic, the following section of this chapter describes the structure of the work, which alternates between historical and theoretical themes.

The Structure of the Work

The chapters that comprise these present volumes have been arranged following a predominantly chronological approach. We chose this convention because it best represents the way architecture and mathematics—but especially architecture—has developed. The architecture of any given historical period is usefully seen in relation to the one that preceded it, perhaps as a natural development or outgrowth, but also potentially as a rebellion against it. Often the underlying reason for a change in style is a shift in thinking, or in knowledge or technique. Thus we have interspersed the historically grouped chapters with groups relating to ideas and theory. We believe in this way it is possible for ideas and concepts, along with built works and processes, to juxtapose and illuminate each other.

The relationship between architecture and mathematics is both longstanding and complex, with the two being bound together in a multitude of practical, representational and contingent ways. On a practical level, the design and construction process for a building relies on mathematics for measurements, timelines, weights and structural calculations (Salvadori 1968; Swallow et al. 2004). In a different way, architects have used numbers and shapes to represent—through symbolic, metaphoric or semiotic means—a broad range of themes that are socially and culturally significant (Rowe 1947; Preziosi 1979; Evans 1995). A parallel tradition in architecture uses geometry and other branches of mathematics to analyse the way designers approach form (Stiny 1975), to investigate spatial hierarchies (Hillier 1995) or measure visual and phenomenal properties (Benedikt 1979). All of these examples involve different types of connections between architecture and mathematics. Furthermore, throughout history there have been various periods characterised by particularly dominant types of relationships between architecture and mathematics.

In the ancient Greek and Roman worlds architects repeatedly used combinations of shapes and numbers to evoke spiritual or cosmic themes. For example, the Roman architect Vitruvius suggested that because the human body possesses distinct geometric proportions, an architecture that is produced in accordance with those proportions represents a microcosm of the divine universe (Rykwert 1996). In parallel with this symbolic application of geometry, the challenges of supporting construction and trade throughout a rapidly expanding empire led engineers and architects to develop a standard system of measurement (Kostof 1977a). In contrast, during the Renaissance systems of proportion based on the musical ratios were the focus of much aesthetic debate. With the rise of Baroque architecture, a broader geometrical vocabulary including ovals and ellipses was used to reinforce the combined spiritual and experiential power of space (Norberg-Schulz 1971). In the eighteenth century French Enlightenment architects designed buildings in the shape of monumental Phileban solids (spheres, cones, and cubes) in an attempt to encourage contemplation of higher order scientific or philosophical principles (Vidler 2006). However, by the nineteenth century widespread concerns with health standards and overpopulation encouraged architects to embrace industrial techniques, along with the new ways of managing production scheduling and understanding material tolerances and limits. In our own day, algorithmic and computational approaches to architecture have embedded mathematics in every line of every CAD model or BIM file, and in every form or shape evolved or generated in the computer (Szalapaj 2005). While measurement remains of practical importance in contemporary architecture, it is no longer as critical as it was in the ancient world. Similarly, while several Renaissance treatises suggest that design could be parametrically determined, the power of computing has allowed this once peripheral notion to become increasingly important. Thus, throughout history the relationship between architecture and mathematics has shifted and changed, with some key moments in this relationship being embodied in a single building of a period or the work of a particular architect, but at other times being only understood retrospectively through the efforts of historians and theorists.

The structure of the present book reflects this interweaving of the theory and history of architecture and mathematics. The first volume commences, in Part I, with a set of chapters that are concerned with overarching theories connecting architecture and mathematics, before progressing, in Part II, to the examination of a particular period in architectural history: the ancient world prior to 1000 AD. This transition from theory to history and back again is repeated for the remainder of the two volumes. For example, in Part III, a series of chapters focus on the historic importance of mathematics in creating systems of measurement and structural stability. Thereafter, returning to the historic time line, in Part IV, architectural examples drawn from 1100 AD to 1400 AD (from Medieval to Romanesque) are analysed. This alternating structure, which knits a historical chronology of buildings and architects together with theories that were of relevance to the era, is repeated throughout the two volumes. The present work can therefore be read either historically or thematically, but it is only in the combination of the two—through viewing the complete fabric and not just its warp or weft—that the wealth and profundity of connections between architecture and mathematics can be grasped.

The theory strand that connects the two volumes—Parts I, III, V, VII, X and XI—examines the significance of essential concepts including measurement, proportion, symmetry and representation. Theories and applications of tiling—both periodic and aperiodic—fractals and scanning technology are examined in the penultimate theory section. In the final of these parts, computational and parametric theories are described along with the philosophical implications of these developments in the context of architectural history. While this is the shorter of the two strands that connect the two volumes of the present work, it frames the historical work, providing a foundation for ideas developed throughout the chapters and a context for the changing role of mathematics in architecture.

The history strand comprises the majority of the two volumes—Parts II, IV, VI, VIII and IX. It commences with Neolithic and Copper age construction, before progressing through examples of ancient Egyptian, Greek, Roman and Mayan architecture. The Renaissance and Baroque eras are covered in the intermediate sections along with Islamic, Christian and Ottoman design. Architects and theorists whose works are considered in this section include such giants as Michelangelo Buonarroti, Andrea Palladio, Francesco Borromini and Christopher Wren, along with lesser-known figures such as Juan Bautista Villalpando and Antonio Rodrigues. The final section in this strand is focussed on the twentieth century and the first decade of the twenty-first century. It considers examples of Modern, Organic, Postmodern and Computer-Generated architecture. Works and texts by Frank Lloyd Wright, Marcel Breuer, Le Corbusier, Louis Kahn and Oscar Niemeyer, amongst many others, are considered in this part.

In what follows, we provide an overview of the first six parts of the present work which collectively make up Volume I. A separate introduction and overview is provided at the start of Volume II (Chap. 48) covering material in that work and offering a framework for understanding why certain eras offer a multitude of examples of connections between architecture and mathematics, while others have a relative paucity.

Part I: Connections Between Architecture and Mathematics

The first part of the present volume comprises a series of chapters which consider a range of themes pertaining to the overarching relationship between architecture and mathematics. In 'Can there be any relationships between Mathematics and Architecture?' (Chap. 2), the eminent engineer and mathematician Mario Salvadori offers a provocative reading of the relationship between pure mathematics and what he calls 'concrete real architecture'. Salvadori, in a personal address to the reader, illustrates pure mathematics using examples of Euclidean and non-Euclidean geometries, and concludes that the beauty of mathematics is that 'it is totally free, it is abstract'. In contrast, architecture must be constructed and it is because of this property that, in a mathematical sense, the two are incomparable. But then, Salvadori reverses his own argument to demonstrate that the connections between mathematics and architecture are so many that 'if mathematics had not been invented, architects would have had to invent it themselves.'

In Chap. 3, "Mathematics *in*, *of* and *for* Architecture" we editors of this publication, Michael Ostwald and Kim Williams, begin to formulate an answer to Salvadori's question; 'can there be any relationships between mathematics and architecture?' Starting with a review of the two famous origin myths of architecture, then though a consideration of more contemporary modes of architecture practice, the chapter identifies three types of relationships. The first type, covers geometric or numeric properties that are intentionally designed *in*to or demonstrated *in* architecture. The second type encompasses analytical methods for quantifying or determining various properties *of* architecture. The final includes practical tools *for* the support of architectural design, construction, generation and conservation.

In 'Relationships between History of Mathematics and History of Art' (Chap. 4), which is similarly broad in its scope, Clara Silvia Roero examines the relationship between the history of art in general and the history of mathematics. Her chapter proposes three levels of connection between art and mathematics—a surface or substrate level, the conscious or unconscious application of principles, and finally a higher-level demonstration of knowledge.

While the close relationship between architecture and mathematics was especially prevalent during the Renaissance, in 'Art and Mathematics Before the Quattrocento: A Context for Understanding Renaissance Architecture' (Chap. 5), Stephen Wassell demonstrates that this achievement should be understood by considering the development of mathematics in previous eras. Starting with Neolithic speculative geometry, and then progressing through mathematical principles developed in Ancient Greece and Rome, Wassell proposes that the practical and metaphysical foundation for Renaissance architecture can be attributed to much earlier developments in both architectural and mathematical knowledge.

In 'The Influence of Mathematics on the Development of Structural Form' (Chap. 6), Holger Falter examines the role played by mathematics in the development of structural form. By tracing examples from different eras, Falter argues that not only is mathematics (in the form of structural mechanics) essential to every design, but that each era developed different architectural structures in response to changing cultural conceptions and mathematical developments. While Falter acknowledges that geometry often served important symbolic functions in architecture, he argues that the structural and material properties cannot be forgotten.

Part II: Architecture from 2000 BC to 1000 AD

Histories of architecture typically suggest that it was around 9000 BC, in the early Mesolithic era, that the first consciously designed structures were built. Prior to that time natural features in the landscape, including caves, hollows and rock overhangs, provided rudimentary shelter from both the environment and predators. However, in the Mesolithic era small communities began to create simple, but often extensive, structures from naturally weathered stone that could be moved and arranged to create earthen mounds or barrows. By selectively covering these mounds in branches and soil, and inhabiting the voids between the larger rocks, these structures became the equivalent of artificial caves. The remains of a large number of these barrows have been excavated throughout the last century, with some evidently serving community or ceremonial purposes, while others acted as burial mounds (Spikins 2008).

In the early Neolithic era, dating from around 5000 BC, a number of communities in different parts of the world developed a capacity to create simple masonry structures (Hofmann and Smyth 2013). Some of the most extensive of these featured large, walled enclosures made of crude mud bricks which were sometimes finished in a type of plaster. For example, in Southern Anatolia in Turkey the Chalcolithic (Copper Age) people of Çatalhöyük constructed a dense cellular architecture, wherein a network of broadly orthogonal masonry walls was roofed with timber shingles. At around the same time, wattle and daub structures with thatched roofs over timber beams began appearing across both Eastern and Western Europe. In Germany, England and Ireland more elaborate timber structures, including long houses, tombs and cisterns were also being constructed (Cruickshank 1996).

In parallel with the latter part of the European Neolithic period, in North Africa, the higher temperatures allowed for the development of large-scale techniques for casting and drying mud-bricks. It was this advance which lead to the rise of Sumerian architecture (Kramer 1963). Using clay-based masonry the Mesopotamian culture developed methods that allowed them to stack layers of bricks, typically without mortar, to produce buttressed walls and stepped pyramids, or ziggurats. Supplemented with a limited amount of timber and occasionally clad in coloured stone, these buildings were large in scale, but the friable nature of the masonry core of these structures rendered them impermanent.

Just as architecture developed throughout the Mesolithic, Neolithic and Sumerian eras, so too did mathematics. Simple attempts to record the passing of time were common in early agrarian cultures. The need to plant and harvest crops at appropriate times led to the need to measure and predict the seasons and thus the first calendars were developed. The desire to trade with neighbouring lands was responsible for the Sumerians developing simple weights and measures. However, the first great developments in both architecture and mathematics can be traced to the Ancient Egyptians. Responsible for a variation of the decimal system, for surveying and the astronomical calendar, the Egyptian civilisation developed both architecture and mathematics to a new level of refinement (Rossi 2004).

Spanning broadly from around 3000 BC to 500 AD, Egypt provided one of the most enduring civilisations of the ancient world. In this era, however, Egypt was not alone in making both mathematical and architectural advances; both the Ancient Indian Vedic texts and Ancient Mayan (Mesoamerican) hieroglyphs reveal similar levels of advancement, with the latter inventing a base-20 (*vigesimal* rather than *decimal*, or base-10) system of counting. It is this age, when the Egyptian, Mayan and Indian cultures developed systems of counting, measuring and geometry, which is the focus of the opening four chapters in Part II of the present work. The majority of the latter chapters in Part II address the architecture and mathematics of the Ancient Greek and Roman civilisations.

In 'Old Shoes, New Feet and the Puzzle of the First Square in Ancient Egyptian Architecture' (Chap. 7), Peter Schneider examines the origins of the first square, one of the most primitive and recurring geometric figures in architecture. Schneider finds evidence of a simple, yet sophisticated technique for constructing the square, in the ancient Egyptian measures of the 20 digit *remen* and the 28 digit *cubit*, suggesting that this allowed architects to construct a primal square which could be readily replicated. In 'Geometric and Complex Analyses of Maya Architecture: Some Examples' (Chap. 8), Gerardo Burkle-Elizondo, Nicoletta Sala and Ricardo David Valdez-Cepeda undertake a detailed geometric analysis of 26 pyramids constructed by the Mayan civilization. Using fractal analysis they examine the relationship between the formal properties of these structures and their cosmological purpose. Again with regard to Mesoamerica and the stepped pyramids of the first century AD, in 'A New Geometric Analysis of the Plan of the Teotihuacan Complex in Mexico' (Chap. 9), Mark Reynolds offers a geometric analysis of the plan of the Teotihuacan complex in Mexico. Rather than considering the significance of these structures in an astronomical or archaeological sense, Reynolds uses plan overlays to examine the organization of these culturally significant buildings.

George Gherveghese Joseph proposes a different cultural relationship between architecture and mathematics in 'The Geometry of Vedic Altars' (Chap. 10). Ancient Indian systems of weights and measures can be traced in the construction methods and forms of a series of sacrificial altars (*vedi*). Through an analysis of these altars, Joseph demonstrates that while the origin of Indian mathematics has often been linked to largely theological or symbolic purposes, the degree to which this culture was able to solve practical problems in geometry has been underestimated.

Site planning is the focus of Graham Pont's chapter, 'Inauguration: Ritual Planning in Ancient Greece and Italy' (Chap. 11). Drawing on a theory of planning by polar coordinates, Pont considers 29 ancient sites, noting the evidence for a system of site planning which involves the division of sites using visual arcs. Such a ritual system, based on visual relations as defined from distinct points in space, might explain why many Greek towns and sites have complex angled plans, which have resisted standard geometric analysis techniques.

In 'The Geometry of the Master Plan of Roman Florence and its Surroundings' (Chap. 12), Carol Martin Watts describes the Roman town planning practice of laying out major streets in accordance with cardinal points. Thereafter she offers a possible explanation for the planning of both the city and countryside of Florence, which is in accordance with Roman practices, relies on a range of clear geometric processes and responds to symbolic concerns about *genius loci*. Continuing the focus on Roman architecture, in 'Architecture and Mathematics in Roman Amphitheaters' (Chap. 13), Sylvie Duvernoy examines a particular elliptical shape used almost exclusively for the design of amphitheatres. The geometric basis for Roman amphitheatres has been the subject of a body of specialised research, leading to debates about the difference between oval and elliptical forms. Starting with amphitheatres from the late Republic and early Roman Empires, Duvernoy traces a series of simple geometric relations. Then, using measured surveys of the amphitheatres of Pompeii, Roselle, and Veleia, she analyses the nature of the geometric curves which provide the basis for each. Geometry in Roman architecture is also the topic of Carol Martin Watts's second chapter in Part II, 'The Square and the Roman House: Architecture and Decoration at Pompeii and Herculaneum' (Chap. 14). Through an analysis of the Roman house and its decoration, Watts uncovers several geometric systems and operations (including a variation of the 'sacred cut') in the proportional relationships found in the overall site planning, spatial organisation, and decorations and tiling of the house. The penultimate chapter in this part continues the close analysis of geometry in Roman architecture. In 'The Quadrivium in the Pantheon of Rome' (Chap. 15), Gert Sperling examines the Pantheon in Rome, reviewing several famous interpretations of its form, and partially rejecting the cosmological connections that have been made between the building's dome and the heavens. Taking into account a survey of the metrical dimensions of the complex, Sperling proposes that the geometry of the Pantheon not only reflects the heavens, but is also derived from the mathematical knowledge of the heavens recorded in the ancient *quadrivium*.

In '"Systems of Monads" in the Hagia Sophia: Neo-Platonic Mathematics in the Architecture of Late Antiquity' (Chap. 16), Helge Svenshon and Rudolf Stichel examine the relationship between mathematics and design in one of the most iconic buildings of the first millennium. Hagia Sophia or *Ayasofya* in Istanbul (formerly Constantinople), was built in the fifth century for the Byzantine emperor Justinian. Famous for its enormous dome and the majesty of its interior, significantly, the original designers commissioned for Hagia Sophia were not architects, but were rather a scientist and a mathematician; respectively, Isidore of Miletus and Anthemius of Tralles.

Part III: Theories of Measurement and Structure

The eight chapters in Part III are centred on theories of measurement and the role of mathematics in the construction and structuring of architecture. In 'Measure, Metre, Irony: Reuniting Pure Mathematics with Architecture' (Chap. 17), Robert Tavernor takes as his starting point the argument that systems of measurement are representative of the values of the culture that defined them. Thus, in the ancient world the measures and proportions of the human, male, body were often used for theological, rather than practical reasons. Tavernor illuminates the fundamental irony that the systems of measurement developed and used by architects and mathematicians are rarely so universal as they seem. In 'Facade Measurement by Trigonometry' (Chap. 18), Paul Calter examines a different way of measuring the built environment, derived from surveying and optimised for the consideration of historic buildings and ruins.

Mark Wilson Jones describes his motivation for writing 'Ancient Architecture and Mathematics: Methodology and the Doric Temple' (Chap. 19) as a desire to bring greater rigour to the measurement of ancient Greek and Roman architecture. In particular Jones is critical of proportional studies of façades and plans that are neither precise nor objective enough. Jones offers seven criteria for evaluation of the efficacy of a particular set of measures and their interpretation, each of which is illustrated with specific examples from the ancient world.

The act of measuring buildings is also the focus of 'Calculation of Arches and Domes in Fifteenth-Century Samarkand' (Chap. 20) by Yvonne Dold Samplonius. This chapter features a reading of the great mathematician al-Kashi's work, *Key of Arithmetic*, an important text about the measurement of buildings. The techniques developed by al-Kashi were used to determine the surface area of vaults and domes in order to support the practical construction (and estimates of costs and materials) of architectural work. The mathematics of domes and vaults is also the subject of 'Curves of Clay: Mexican Brick Vaults and Domes' (Chap. 21) by Alfonso Ramírez Ponce and Rafael Ramírez Melendez. The authors, father and son, analyse the properties of traditional Mexican brick vaults and domes that were constructed using brick stacking techniques that required no scaffolding or centering.

The structural properties of architecture are the primary topic of 'Mathematics and Structural Repair of Gothic Structures' (Chap. 22) by Javier Barrallo and Santiago Sanchez-Beitia. Founded in the proposition that architectural education should include an introduction to mathematics, Barrallo describes processes for the restoration and maintenance of Gothic buildings. While Barrallo's chapter is concerned with the European tradition, in 'Mathematics of Carpentry in Historic Japanese Architecture' (Chap. 23) Izumi Kuroishi describes how Heinouchi Masaomi, a nineteenth-century master carpenter and mathematician, developed a theory of *Kikujutu* (stereotomy) and an application of the Japanese mathematical system, *Wasan*, to carpentry. Kuroishi's research into Masaomi's theory demonstrates both the technical and cultural significance of mathematics in the construction process.

The final two chapters in this section continue the examination of different cultural traditions of geometric construction and measurement. In 'On Some Geometrical and Architectural Ideas from African Art and Craft' (Chap. 24), Mozambique mathematician Paulus Gerdes suggests that the practice of weaving large baskets or surfaces represents the early stages of architectural development in Africa. Gerdes's work undertakes a close analysis of geometric patterns and construction methods found in woven mats, which were often used to line or decorate traditional structures. Continuing the cross-cultural survey, in 'Design, Construction and Measurement in the Inka Empire' (Chap. 25) William Sapp commences by noting that the lack of consistency in Inkan architectural features (doors, niches, windows) has lead to the suggestion that there were no clear mathematical rules for generating architectural dimensions and proportions. Sapp partially refutes this position demonstrating how the use of plumb bobs and a proportional system, both known to have been in use by the Inka civilisations, can explain the way their architecture was laid out.

Part IV: Architecture from 1100 AD to 1400 AD

The twelfth century AD saw both the end of the late Romanesque architecture and the rise of the first Gothic buildings that characterised so much European design at that time. For the next 300 years developments in Early English and Muscovite architecture occurred in parallel with more advanced Gothic brick and stone structures until, around the beginning of the fifteenth century, the first Renaissance buildings began to be completed. Part IV of the present volume covers this 300-year period, which broadly coincides with the rise and fall of the Gothic. However, of the 11 chapters in Part IV only four are about Gothic architecture and only five are set in central Europe. The remainder of Part IV features two chapters about developments in the northern parts of the continent (Ireland and the Baltic Sea), two chapters about developments in Asia (India and China) and two, North Africa (Egypt and Persia).

In 'Vastu Purusha Mandala' (Chap. 26) Vini Nathan describes the geometric diagram, part of the *Vastu Shastra* doctrine, which was developed in medieval India to provide a set of rules for translating cosmic or theological ideas and values into architectural form. Nathan initially describes the way in which the *Vastu purusha* mandala was used in India for the planning of buildings and towns. Thereafter she argues that the geometric logic of the mandala had a much greater influence on the cultural milieu of India than previously thought. While ostensibly serving a function similar to the Indian *Vastu Shastra*, the *Yingzao fashi*, was a twelfth-century Chinese manual of building and construction standards and techniques. In 'Algorithmic Architecture in Twelfth-Century China: The *Yingzao Fashi*' (Chap. 27), this historic rule-based system is used by Andrew I-kang Li as the basis for a shape grammar; a set of rules which, collectively, provide an algorithmic basis for understanding, and potentially replicating, a particular approach to design.

The way the *Vastu purusha* mandala operated in Indian culture to mediate between the heavens and the earth has clear parallels to the use of stereographic projections of the Heavenly Sphere in parts of Europe. In 'The Celestial Key: Heaven Projected on Earth' (Chap. 28) by Niels Bandholm, the placement and proportions of 15 medieval churches (constructed between 1150 and 1250) on the island of Bornholm in the Baltic Sea are examined, and potentially explained, using stereographic projection. Late Medieval architecture remains the focus of the following two chapters. In 'Friedrich II and the Love of Geometry' (Chap. 29), Heinz Götze revisits the Castel del Monte, describing its geometry and tracing several precedents for its form in navigational charts, wind stars and mosaics. The Holy Roman Emperor Friedrich II built the Castel del Monte in the thirteenth century in southern Italy. This small castle features a particular nested octagonal plan structure, which has fascinated architects and mathematicians for many hundreds of years. In 'Metrology and Proportion in the Ecclesiastical Architecture of Medieval Ireland' (Chap. 30), Avril Behan and Rachel Moss examine the extent to which a close, empirical analysis of medieval Irish window tracery can illuminate the backgrounds, training and work practices of the masons who constructed them. This study offers a different way of thinking about both architecture and measurement, not for symbolic purposes, but for supporting historical and archival research into construction.

Window tracery is also the topic of 'The Cloisters of Hauterive' (Chap. 31) by Benno Artmann, although his focus is on Gothic church windows in the Cistercian monastery of Hauterive near Fribourg, Switzerland. Gothic tracery is constructed geometrically from an elaborate combination of circular arcs and straight lines. Artmann uncovers a series of complex geometric patterns and rules which represent a departure from the early Gothic tracery.

The next pair of chapters is concerned with the architecture and geometry of the Islamic world. The first of these commences with a reading of the twelfth-century Persian text *On interlocks of similar or complementary figures*, an important treatise of the era which describes 68 geometric constructions for use by artisans. In 'The Use of Cubic Equations in Islamic Art and Architecture' (Chap. 32), Alpay Özdural uncovers a series of techniques that were historically used by artisans to solve complex geometric relations. Such rich geometric relationships, which are found in North African and Islamic decoration, are also present in the geometric order of the Sultan Hassan Floor in Cairo. In 'Explicit and Implicit Geometric Orders in Mamluk Floors: Secrets of the Sultan Hassan Floor in Cairo' (Chap. 33), Gulzar Haider and Muhammad Moussa use a detailed measured survey of the largest of these Mamluk floors to describe a computer-assisted analysis of the geometric orders in the design.

The final three chapters in this part are all focussed on the architecture of Italy. The first of these is about the Palazzo della Signoria in Florence, the second Milan Cathedral and the last the Baptistery and the Campanile of Pisa. In 'The Sequence of Fibonacci and the Palazzo della Signoria in Florence' (Chap. 34), Maria Teresa Bartoli describes the significance of the Fibonacci sequence (a sequence of numbers wherein each is the sum of the two preceding values) and which converges to ϕ, the golden section, and traces its geometric presence (in the form of the Fibonacci rectangle, which is also an approximation of the golden section) in the Palazzo della Signoria. Through this analysis Bartoli demonstrates the extent to which the Fibonacci sequence, when expressed as a set of geometric proportions, was significant in society at that time. In 'What Geometries in Milan Cathedral?' (Chap. 35) Elena Marchetti and Luisa Rossi Costa show a similar point of convergence on the use of geometric proportional systems in architecture. Milan cathedral, with its late Gothic structure and early Renaissance façade, was completed in the fourteenth century. Marchetti and Rossi Costa examine the cathedral, seeking evidence for any of the common proportional systems including the golden mean and other 'metallic' numbers. In 'The Symmetries of the Baptistery and the Leaning Tower of Pisa' (Chap. 36), physicist David Speiser describes and demonstrates the 15-fold and 30-fold symmetry of the penta-decagonal campanile (the 'Leaning Tower') at Pisa, before examining the even more striking 12-fold symmetry in the plan of the Baptistery at Pisa.

Part V: Theories of Proportion, Symmetry, Periodicity

The Renaissance saw the rise of a particular fascination with relationships between parts, involving theories of proportionality, symmetry and periodicity, many of which were either explained or rationalised using musical or harmonic notions. Part V contains five chapters, the first two of which are concerned with the use of music as a means of connecting architecture and mathematics. In 'Musical Proportions at the Basis of Systems of Architectural Proportions both Ancient and Modern' (Chap. 37), Jay Kappraff provides an overview of the way in which musical proportions have been used to shape architectural form. A recurring theme throughout this time was the conviction that the application of certain ratios or proportions would endow a design with both an overarching sense of unity as well as a distinct harmony between its component parts. Through this process Kappraff identifies three significant proportional systems—Alberti's musical ratios, the Roman application of the 'sacred cut' and Le Corbusier's Modulor—all of which have a similar purpose, but use ratios in different ways. In 'From Renaissance Musical Proportions to Polytonality in Twentieth Century Architecture' (Chap. 38), Radoslav Zuk describes a related proportional system derived from consonant musical intervals and traces its evolution. Zuk argues that for such proportional relationships between architecture and music to be meaningful, they must incorporate the three-dimensional properties of architecture, as well as the more common two-dimensional relationships found in plans and elevations.

The following pair of chapters shift the focus away from proportions and ratios and towards issues of symmetry (the mirrored or translated version of a shape or form) and periodicity (the repetition of a shape or form at set intervals). In 'Quasi-Periodicity in Islamic Geometric Design' (Chap. 39), Peter Saltzman undertakes a detailed analysis of Islamic geometry and tiling, reviewing both published research and important examples. He concludes by noting the presence of complex quasi-periodicity within tiling fragments in fifteenth-century Iranian decoration and architecture. More abstract geometric notions also inform 'The Universality of the Symmetry Concept' (Chap. 40), István Hargittai and Magdolna Hargittai's chapter about the apparent universality of symmetrical form. The authors, both chemists, commence by describing the ubiquity of symmetry, before arguing that a better understanding and appreciation of different types of symmetry may assist the development of trans-disciplinary knowledge.

The final chapter in Part V contains one of the better-known counterarguments to the practice of seeking hidden geometric traces, constructions and proportions in historic buildings. In 'Contra Divinam Proportione' (Chap. 41), Marco Frascari and Livio Volpi Ghirardini argue that the presence and importance of the golden mean in historic architecture has been much exaggerated. They propose that such is the allure and simplicity of the golden mean that it has been uncritically adopted to explain a growing number of forms which neither closely conform to its geometry nor have any theoretical affinity with it. By considering the way architecture was historically constructed, Frascari and Volpi Ghirardini demonstrate the potential fallacy at the heart of many interpretations of historic buildings.

Part VI: Architecture from 1400 AD to 1500 AD

The fifteenth century saw the waning of the Gothic era, the growth in importance of Renaissance architecture, and the first stages of the Tudor style in England. The first four chapters in Part VI are about Leon Battista Alberti, his architecture and theory. The following two chapters also have a focus on the Italian Renaissance while the final chapter examines the geometry of *Muqarnas* vaulting.

Alberti's Sant'Andrea in Mantua is widely regarded as one of his most refined and complete works. In 'Alberti's Sant'Andrea and the Etruscan Proportion' (Chap. 42), Michael Ytterberg uncovers an unusual proportional system in the completed building and demonstrates that this ratio can be accounted for by considering the Etruscan architectural tradition. Livio Volpi Ghirardini's chapter, 'The Numberable Architecture of Leon Battista Alberti as a Universal Sign of Order and Harmony' (Chap. 43), is also concerned with Alberti's architecture, but with a particular emphasis on signs of harmony and number symbolism. Considering both Sant' Andrea and San Sebastiano in Mantua, Volpi Ghirardini investigates two conflicting interpretations of Alberti's proportional geometry. Volpi Ghirardini concludes by observing that Alberti relied on finite, but not musical, ratios and progressions, which produced 'numerically proportionate triads, which are themselves proportionately interrelated.' In 'Leon Battista Alberti and the Art of Building' (Chap. 44), Salvatore di Pasquale examines Alberti's defence of the use of models, noting that while Alberti apparently viewed the design concept as a fixed or inviolate proposition, he was also aware of the importance of material and structural properties, using models to test basic principles. The next pair of chapters are centred on the city of Florence in Tuscany. The first of these analyses the geometric design of a single tombslab, while the second considers the geometric composition of an entire church. Kim Williams deciphers a geometric code, in 'Verrocchio's Tombslab for Cosimo de' Medici: Designing with a Mathematical Vocabulary' (Chap. 45). Credited to Florentine sculptor Andrea del Verrocchio, the tombslab in the basilica of San Lorenzo features a set of complex geometrical forms and proportions. Williams (one of the editors of the present volume) analyses Cosimo de' Medici's tombslab and then compares its design with that of three pavements which were completed at a similar time: the Sistine Chapel, the Medici Chapel in the Palazzo Medici and the Chapel of the Cardinal of Portugal in S. Miniato al Monte. Williams's conclusion stresses the way geometry was used to reinforce humanity's symbolic centrality in the cosmos. In 'A New Geometric Analysis of the Pazzi Chapel in Santa Croce' (Chap. 46), Mark Reynolds undertakes a geometric analysis of a single work by Brunelleschi. Reynolds, whose analysis of the Teotihuacan complex was featured in an earlier chapter, uses tracing techniques, overlaid on plans, elevations and sections, to analyse architecture. While informed by the interpretations of past scholars, Reynolds's approach applies a range of geometric constructions (like the 'sacred cut' or the *vesica pisces*) to measured drawings to seek an underlying order in the architecture. Here he develops evidence to support several common interpretations of the Pazzi Chapel's construction using divided squares and circles, along with the suggestion that the altar space in the building may be constructed around the golden Section.

A *muqarnas* vault is one of the characteristic architectural forms found across North Africa and in Arabic and Islamic architecture. While examples of *muqarnas* vaulting have been traced to the tenth century, it was during the fifteenth century that some of the most complex were produced, and, not coincidentally, when the mathematician Ghiyath al-Din al Kashi developed a technique for measuring their surface area. In 'Muqarnas, Construction and Reconstruction' (Chap. 47), Yvonne Dold-Samplonius and Silvia Harmsen describe their development of a database of *muqarnas* dimensions and constructions to allow for the mapping of styles, regions and timeframes using subtle developments that occurred in the geometry and form of the vault.

Conclusion

Across 47 chapters, authored by 53 scholars of architecture, mathematics, engineering and philosophy, the present volume contains examinations of key theories, buildings and treatises which, between 2000 BC and 1500 AD, evidence the crucial relationships between architecture and mathematics. In the theory strand, overarching ideas have been introduced and analysed, while in the history strand, precise examples have been considered in great detail, often testing past ideas or proposing new ways of viewing famous buildings. Volume II, featuring a similar number of chapters and with a commensurately wide range of authors, spans the history and theory of architecture from 1500 AD to the present day.

Biography Michael J. Ostwald is Professor and Dean of Architecture at the University of Newcastle (Australia) and a visiting Professor at RMIT University. He has previously been a Professorial Research Fellow at Victoria University Wellington, an Australian Research Council (ARC) Future Fellow at Newcastle and a visiting fellow at UCLA and MIT. He has a PhD in architectural history and theory and a DSc in design mathematics and computing. He completed post-doctoral research on baroque geometry at the CCA (Montreal) and at the Loeb Archives (Harvard). He is Co-Editor-in-Chief of the *Nexus Network Journal* and on the editorial boards of *ARQ* and *Architectural Theory Review*. He has authored more than 300 scholarly publications including 20 books and his architectural designs have been published and exhibited internationally.

Kim Williams was a practicing architect before moving to Italy and dedicating her attention to studies in architecture and mathematics. She is the founder of the conference series "Nexus: Relationships between Architecture and Mathematics" and the founder and Co-Editor-in-Chief of the *Nexus Network Journal*. She has written extensively on architecture and mathematics for the past 20 years. Her latest publication, with Stephen Wassell and Lionel March, is *The Mathematical Works of Leon Battista Alberti* (Basel: Birkhäuser, 2011).

References

BARNES, Carl F. 2009. *The Portfolio of Villard de Honnecourt: A New Critical Edition and Color Facsimile*. Farnham, Surrey, UK and Burlington, VT, USA: Ashgate.

BENEDIKT, Michael L. 1979. To Take Hold of Space: Isovists and Isovist View Fields. *Environment and Planning B: Planning and Design* **6**, 1: 47–65.

BENNETT, J. A. 1982. *The Mathematical Science of Christopher Wren*. Cambridge: Cambridge University Press.

BERMAN, Ric. 2012. *The Foundations of Modern Freemasonry*. Eastbourne: Sussex Academic Press.

BOYER, Ernest L, and Lee D. MITGANG. 1996. *Building Community: A New Future for Architecture Education and Practice*. New Jersey: The Carnegie Foundation for the Advancement of Teaching.

BURRY, Jane and Mark BURRY. 2012. *The New Mathematics of Architecture*. London: Thames and Hudson.

CARAFOLI, Ernesto, Gian Antonio DANIELI and Giuseppe O. LONGO, eds. 2009. *The Two Cultures: Shared Problems*. Milan: Springer.

CLARKE, David. 1994. *The Architecture of Alienation: The Political Economy of Professional Education*. New Brunswick and London: Transaction Publishers.

CRINSON, Mark, and Jules LUBBOCK. 1994. *Architecture: Art or Profession?* Manchester: Manchester University Press.

CRUICKSHANK, Dan, ed. 1996. *Sir Bannister Fletcher's History of Architecture*. 20th edn. Oxford: The Architectural Press.

CUFF, Dana. 1991. *Architecture: The Story of Practice*. Cambridge and London: MIT Press.

DAVIS, Philip J., Reuben HERSH and Elena Anne MARCHISOTTO. 1995. *The Mathematical Experience*. Basel: Birkhauser.

DOUCET, Isabelle and Nel JANSSENS. 2011. Transdisciplinarity, the Hybridisation of Knowledge Production and Space-Related Research. Pp. 1–15 in *Transdisciplinary Knowledge Production in Architecture and Urbanism: Towards Hybrid Modes of Inquiry*. Isabelle Doucet and Nel Janssens, eds. Heidelberg, New York: Springer.

DRAPER, Joan. 1977. The Ecole des Beaux-Arts and the Architectural Profession in the United States: The Case of John Galen Howard. Pp. 209–237 in *The Architect: Chapters in the History of the Profession*. Spiro Kostof, ed. New York: Oxford University Press.

DUFFY, Francis and Les HUTTON. 1998. *Architectural Knowledge: The Idea of a Profession*. New York: Routledge.

EVANS, Robin. 1995. *The Projective Cast: Architecture and its Three Geometries*. Cambridge, Massachusetts: MIT Press.

FITCHEN, John. 1961. *The Construction of Gothic Cathedrals: A Study of Medieval Vault Erection*. Chicago: University of Chicago Press.

FOURNIER, Valerie. 2000. Boundary Work and the (un)Making of the Professions. Pp. 67–86 in *Professionalism, Boundaries and the Workplace*. Nigel Malin, ed. New York: Routledge.

Hillier, B. 1995. *Space is the Machine,* Cambridge: Cambridge University Press.

Hofmann, Daniela, and Jessica Smyth, eds. 2013. *Tracking the Neolithic House in Europe: Sedentism, Architecture and Practice.* New York: Springer.

James, John. 1981. *The Contractors of Chartres.* London: Croom Helm.

Janin, Hunt. 2008. *The University in Medieval Life, 1179–1499.* Jefferson North Carolina: Macfarland and Company.

Klein, Julie T. 1996. *Crossing Boundaries: Knowledge, Disciplinarities, and Interdisciplinarities.* Charlottesville: University of Virginia Press.

Kostof, Spiro. 1977a. The Practice of Architecture in the Ancient World: Egypt and Greece. Pp. 3–27 in *The Architect: Chapters in the History of the Profession.* Spiro Kostof, ed. New York: Oxford University Press.

——. 1977b. The Architect in the Middle Ages, East and West. Pp. 59–95 in *The Architect: Chapters in the History of the Profession.* Spiro Kostof, ed. New York: Oxford University Press.

Kramer, Samuel. 1963. *The Sumerians: Their History, Culture and Character.* Chicago: The University of Chicago Press.

Littlefield, David. 2008. *Space Craft: Developments in Architectural Computing.* London: RIBA Publications.

Marrou, Henri Irénée. 1956. *A History of Education in Antiquity.* Madison: University of Wisconsin.

Melton, James Van Horn. 2001. *The Rise of the Public in Enlightenment Europe.* Cambridge: Cambridge University Press.

Nicolescu, Basarab. 2002. *Manifesto of Transdisciplinarity.* New York: SUNY (State University of New York Press).

Norberg-Schulz, Christian. 1971. *Baroque Architecture.* New York: Harry N. Abrams.

Ostwald, Michael J. 2006. *The Architecture of the New Baroque: A Comparative Study of the Historic and New Baroque Movements in Architecture.* Singapore: Global Arts.

——. 2012. Systems and Enablers: Modelling the Impact of Contemporary Computational Methods and Technologies on the Design Process. Pp. 1–17 in Ning Gu and Xiangyu Wang, eds. *Computational Design Methods and Technologies: Applications in CAD, CAM and CAE Education.* Pennsylvania: IGI Global.

Ostwald, Michael J. and Anthony Williams. 2008. *Understanding Architectural Education In Australasia. Volume 1: An Analysis of Architecture Schools, Programs, Academics and Students.* Sydney: Australian Learning and Teaching Council and Carrick Institute for Learning and Teaching in Higher Education.

Picon, Antoine, 1992. *French Architects and Engineers in the Age of Enlightenment.* Cambridge: Cambridge University Press.

Preziosi, Donald. 1979. *Architecture, Language and Meaning.* The Hague: Mouton.

Rossi, Corinna. 2004. *Architecture and Mathematics in Ancient Egypt.* Cambridge: Cambridge University Press.

Rowe, Colin. 1947. *The Mathematics of the Ideal Villa and other Essays.* Cambridge, Massachusetts: MIT Press.

Russell, Bertrand. 2009. *Mysticism and Logic and Other Essays (1918).* Ithaca, NY: Cornell University Library.

Rykwert, Joseph, 1996. *The Dancing Column: On Order of Architecture.* Cambridge, Massachusetts: MIT Press.

Salvadori, Mario. 1968. *Mathematics In Architecture.* New Jersey: Prentice Hall.

Snow, C. P. 1998. *The Two Cultures* (1959). Cambridge: Cambridge University Press.

Sokolowski, Robert. 1991. The Fiduciary Relationship and the Nature of Professions. Pp. 23–44 in Edmund D. Pellegrino, Robert M. Veatch and John P. Langan, eds. *Ethics, Trust and the Professions: Philosophical and Cultural Aspects.* Georgetown: Georgetown University Press.

Spikins, Penny. 2008. Mesolithic Europe: Glimpses of Another World. Pp. 1–17 in Geoff Bailey and Penny Spikins, eds. *Mesolithic Europe.* Cambridge: Cambridge University Press.

Stiny, G. 1975. *Pictorial and Formal Aspects of Shapes and Shape Grammars*. Basel: Birkhauser.
Swallow, P., D. Watt, and R. Ashton. 2004. *Measurement and Recording of Historic Buildings*. Shaftsbury: Donhead.
Szalapaj, P. 2005. *Contemporary Architecture and the Digital Design Process*. Oxford: Architectural Press.
Vidler, A. 2006. *Claude-Nicolas Ledoux: Architecture and Utopia in the Era of the French Revolution*. Basel: Birkhäuser.
Vitruvius. 2009. *On Architecture*. Richard Schofield, trans. London: Penguin Books.
Warren, S. Edward. 1875. *Stereotomy: Problems in stone cutting*. New York: John Wiley and Son.
Williams, Kim, Lionel March and Stephen R. Wassell, eds. 2010. *The Mathematical Works of Leon Battista Alberti*. Basel: Birkhauser.

Part I
Mathematics in Architecture

Chapter 2
Can There Be Any Relationships Between Mathematics and Architecture?

Mario Salvadori

My dear friend Kim,
Ladies and Gentlemen of the Congress,
Signore e Signori,
Señores y Señoras
Mes Dames et Messieurs,
Meine Damen und Herren,

I am greatly honoured to be asked to present a lecture at this congress on the theme of Mathematics and Architecture, because I happen to be a mathematician and because, although academically untrained in the difficult discipline of architecture, it has been my good luck to collaborate as a structural engineer in the creation of architectural buildings of all types and all over the world with architects like Gropius, Breuer, Saarinen and many others, during the 30 years I spent in a well known architectural engineering office in the United States.

Let me say that, as I prepared myself for today's demanding assignment, I ran immediately into a basic question. Since there is not just *one* mathematics, but many, and there is not *one* architecture but many, *which* mathematics should I discuss in relationship to *which* architecture?

I am afraid that, to clear these doubts, I will have to give the architects in this hall some idea of how many mathematics there have been, there are and there will be in

Mario Salvadori (1907–1997).

First published as: Mario Salvadori, "Can There Be Any Relationships Between Mathematics and Architecture?", pp. 9–13 in *Nexus I: Architecture and Mathematics*, ed. Kim Williams, Fucecchio (Florence): Edizioni dell'Erba, 1996.

Editors' note: The text that follows is a transcript of the keynote address at Nexus'96: Relationships Between Architecture and Mathematics, the inaugural conference of what would become the Nexus series.

K. Williams and M.J. Ostwald (eds.), *Architecture and Mathematics from Antiquity to the Future*, DOI 10.1007/978-3-319-00137-1_2,

the future, and to the mathematicians sitting next to them how many architectures there have been, there are and there will be in the future. But, in order not to bore both groups of specialists, I will try to be brief and simple.

To start with, we all know that there is the mathematics we use to go to market, but that kind of mathematics is an obvious practical derivation from a more complex field of higher mathematics called *number theory*. Then there is the kind of mathematics used by all technologists, which is extremely useful to all of them but has little to do with the *real* mathematics. Finally there is the mathematics used by the great scientists, that has allowed them to better understand the universe, of which we are such a minimal part, and amazes all of us poor mortals with the astonishing results it allows us to obtain. Yet, even *this* is not the highest level of mathematics, which is usually called *pure mathematics*.

The essential character of *pure* mathematics is its purity and derives from the fact that it is the fruit of our minds and has no relationship *whatsoever* to what some people call "reality". Just as we may *think* of a green cow, although there are no green cows in our world, mathematics, *pure* mathematics, is the purest product of our mind and has nothing to do with nature or the constructs of man.

Let me illustrate what I mean by a simple example. You are all familiar with the geometry of Euclid that asserts how from a point *outside* a given line one can draw a *single* line *parallel* to the given line. When discussing Euclidian geometry I like to ask one of my young students: "And how long are the lines Euclid is talking about?" He or she usually answers: "They last forever", or: "They are infinite." Upon hearing this statement I like to remind the student that we both live on earth and that, if we keep going on for ever, we will describe a *circle* and *not* a straight line. The student becomes confused and I then explain that the concepts of point and of line used by Euclid are purely *abstract* concepts and have nothing to do with the earth on which we live, because straight lines cannot exist on a round earth. And to convince my students of the total abstraction of mathematics I mention that towards the end of the last century one Russian and one Hungarian mathematician invented a new geometry in which not one but *two* lines could be drawn parallel to a given line and, as if this were not enough, in 1907 the German mathematician Riemann invented a geometry in which an *infinite* number of lines can be drawn parallel to a given line.

I seem to hear some of our architects say: "Well, if mathematicians like to play with abstractions and have a good time with them, let them. But *we* are interested in 'reality'" (whatever this word means to them). To which I answer that if Riemann had not invented the Riemannian geometry Einstein could not have invented his general theory of relativity, which has solved a number of mysteries unsolvable by the apparently "more real" Euclidian geometry.

In one word, mathematics is a fruit of the human spirit, like poetry, and, like poetry, it is one of the infinitely beautiful fruits of the human spirit because it is totally free, it is abstract.

Let me now touch upon some characteristics of architecture. I know that there are in this room many, highly knowledgeable, ladies and gentlemen who have investigated the architectures of the past, those of the present and even tried to guess those of the future. They know the *theory* of architecture better than many famous architects and I have the greatest respect for their knowledge. But I must *unequivocally* state that they are *not* real architects because the basic characteristic of architecture is that it is among the most *concrete* of all human endeavours and, may I add, that (in my humble opinion) it is also one of the most demanding human endeavours, if not *the* most demanding. I know by experience that architecture is much more difficult than mathematics, whose freedom of invention is only bound by the needs of logic, while architecture is bound by innumerable laws, opinions, traditions and, above all, by the whim of man.

What I am trying to say is that no architecture exists unless it is *architected,* that is, *concretely* built without any reference to *pure ideas,* but constrained by the laws of nature and, above all, I repeat, by the whim of man, the most unsatisfied animal of the animal kingdom.

I am therefore asking myself, and all of you: "How can there be a relationship between the totally abstract real mathematics and the totally concrete real architecture?"

This is where I seem to hear a murmur in this hall that says: "But, come on, you have forgotten geometry! If you don't wish to consider other possible relationships, how can you ignore the importance of geometry in architecture?"

Ladies and gentlemen, here again I must state that there are at least two aspects of geometry: one that is obviously of interest to the architects and of no interest to the mathematicians, and one that does just the opposite. The architect may be interested in a geometrical shape called a triangle, but the mathematician doesn't care about *the shape* of the triangle. What excites him is that *whatever the shape of the triangle* the sum of its three angles adds always to 180°. And he is not even interested in the shape of a so called *right triangle,* but only in the fact that the sum of the square of its two sides, *whatever the shape and dimension of the right triangle,* always adds up to the square of its longest side, its hypotenuse! And (may I add in parentheses) that it took over 300 years of very hard work by the greatest mathematicians in the world to prove that, calling a and b the sides of a right triangle and c its hypotenuse, by Pythagoras theorem:

$$a^2 + b^2 = c^2,$$

but that there will never, *never* be a right triangle *with integer sides* for which:

$$a^3 + b^3 = c^3,$$

or any other exponent *larger than two.* And let me finally add that all the mathematicians in the world rejoice that one of them, just about a year ago, succeeded, after 7 years of concentration on this unique problem, to solve the so-called *Fermat's last theorem,* where everybody else had failed in the preceding 300 years.

Ladies and Gentlemen, having "proved" that to look for relationships between as abstract a science as mathematics and as concrete an art as architecture is theoretically inconceivable, allow me now to take off my mathematical hat and put on my engineering hat. As soon as I change hats, I realize that all my disquisitions on the impossibility of relating mathematics and architecture vanish and, as a technologist, I must agree with you that the relationships between mathematics and architecture are so many and so important that, if mathematics had not been invented, architects would have had to invent it themselves.

But there are so many architects and investigators of architecture in this hall that I would not dare to address a topic they are more than capable and eager to present themselves.

I therefore apologize for my long initial statement, but with an explanation. I love mathematics and architecture equally, but find architecture so difficult that, as the lazy man I am. I prefer to limit my activity in this field to helping the architects. And I confess that, were it not for my technological knowledge, I would not dare touch the sublime beauty and the scary difficulties of architecture, pacifying my big ego with the thought that the architects of today perhaps could not architect without the contributions of great engineers like Pier Luigi Nervi and so many more of us, even if we are not of Nervi's calibre.

Thank you very much for your courteous patience.

Biography Mario Salvadori (1907–1997) earned doctoral degrees in both civil engineering and mathematics from the University of Rome in 1930 and 1933 respectively. An outspoken critic of Fascist regime, he left Italy in 1938 for New York at the recommendation of his teacher and friend, Enrico Fermi. After the war, he took up teaching at Columbia University, where he would become a professor in 1959 in the School of Architecture, Planning and Preservation; he taught at Columbia for 50 years. From 1954 to 1960, Salvadori worked as a consultant and then principal at the engineering firm Weidlinger Associates. He was a partner until 1991, when he became honorary chairman. As a structural engineer, Salvadori became known for the design of thin concrete shells. As he reached retirement age, he began volunteering to work with under-privileged minority students from inner-city New York public schools. He is the author of numerous books, including *Mathematics in Architecture* (1968), *Why Buildings Stand Up* (1980) and *Why Buildings Fall Down* (1992).

Further Reading

Salvadori, Mario. 1986. *Structure in Architecture: The Building of Buildings*. New Jersey: Prentice Hall

——. 2002. *Why Buildings Stand Up: The Strength of Architecture*. New York: W. W. Norton & Company.

——. 2000. *The Art of Construction: Projects and Principles for Beginning Engineers and Architects*. Chicago: Chicago Review Press.

Levy, Matthys and Salvadori, Mario. 1994. *Why Buildings Fall Down: How Structures Fail*. New York: W. W. Norton & Company.

Chapter 3
Mathematics *in*, *of* and *for* Architecture: A Framework of Types

Michael J. Ostwald and Kim Williams

aetiology |ˌaētēˈäləjēl *noun*
The investigation or attribution of the cause or reason for something, often expressed in terms of historical or mythical explanation.

teleology |ˌtelēˈäləjē | *noun*
The explanation of phenomena by the purpose they serve rather than by postulated causes.

Introduction

The frontispiece of the thirteenth century *Bible Moralisee* conserved in Vienna portrays a Christ-like figure leaning over a primordial world and using a pair of compasses to measure and inscribe its limits (Fig. 3.1). Titled 'God as architect of the world', it depicts the use of a mathematical instrument to determine the functional, symbolic and aesthetic properties of the universe. The pair of compasses is a symbol of all of the possible ways in which mathematics is used to support design. Such symbols are useful for reinforcing the simple message that the creative impulse relies on mathematics to translate a concept into reality. At the same time, however, this symbolism masks the fact that the relationships between architecture and mathematics are both richer and more diverse than the sign implies. The purpose of the present chapter is to look behind the symbol of the

M.J. Ostwald (✉)
School of Architecture and Built Environment, University of Newcastle, Callaghan, New South Wales 2308, Australia
e-mail: michael.ostwald@newcastle.edu.au

K. Williams
Kim Williams Books, Corso Regina Margherita, 72, 10153 Turin (Torino), Italy
e-mail: kwb@kimwilliamsbooks.com

K. Williams and M.J. Ostwald (eds.), *Architecture and Mathematics from Antiquity to the Future*, DOI 10.1007/978-3-319-00137-1_3,

Fig. 3.1 'God as architect of the world'. *Bible Moralisée*, Paris (ca. 1220–1230) (Image: Osterreichische Nationalbibliothek, Vienna, Codex Vindobonensis 2554, fol. Iv. Reproduced by permission)

pair of compasses and to begin to identify the different ways in which mathematics is used in architecture.

The *Bible Moralisee* was an illuminated manuscript in the medieval tradition that used images to communicate important biblical themes. The illuminations were evocative visual counterparts to the myths, beliefs, parables and morality tales, originally transmitted orally, that sought to educate people about the world. The conflation of God as both architect and geometer in the frontispiece is especially noteworthy because it communicates mathematics' fundamental contribution as intermediary between the creative impulse and the product of that divine vision (Kline 2001). What is often forgotten in this reading of the frontispiece is that the analogy not only communicates something about God's power and wisdom, but also about the accepted role and skills of the architect. The allegorical effectiveness of this image relies on the viewer being aware that architects use mathematics to create structure. This message is reinforced by the representation of God stepping through a timber portal, with one foot resting in the quotidian world of the designer or artisan as user of geometry, and the other transcending this as maker of the universe (Husband 2009). The frontispiece of the *Bible Moralisee* is a culturally-coded representation of the vital bond that exists between architecture and mathematics. Yet, while it presents this relationship as both natural and necessary, it says nothing about the connection itself.

A common question in architectural scholarship asks why architects use mathematics (Kappraff 1990; Rossi 2004; Goldberger 2009). Despite multiple answers being offered (Scruton 1983; Evans 1995), the majority of such responses have served a rhetorical purpose, providing the impetus for a personal manifesto or theory (Salingaros 2006). For example, Mario Salvadori (2014) asks, '[c]an there be any relationship between architecture and mathematics?', and after considering several responses, concludes that architecture simply cannot exist without mathematics. Salvadori's answer, like many of the others that have been offered, is eminently reasonable but it does not provide a holistic insight into the different ways architects use mathematics.

Here we will identify some of the types of applications of mathematics that conventionally occur in architecture, drawing on historic and contemporary myths and models to propose a framework for classifying the ways architects use numbers and geometry. We commence by examining connections between architecture and mathematics first from a *causal* or mythopoeic perspective, and second from an *effects*-based viewpoint. Here, the causes and effects are disconnected, each informing and shaping the framework, but unable to be directly correlated through that mechanism. This discontinuity is unavoidable because relationships between architecture and mathematics are not predicated on a singular need, desire or process; they serve a multiplicity of different and sometimes conflicting agendas. Cause and effect cannot be perfectly aligned under such conditions, but there are ways of investigating the two that are informative and useful for this purpose.

The study of the cause or genesis of an occurrence is called *aetiology*. This approach to understanding the origin of an idea or relationship is often undertaken through an investigation of the founding myths of a discipline. The present chapter commences by examining the classic Western myths of the first building—the primitive hut of the ancients—and the first architect, Daedalus. The purpose of this strategy is to reveal the presence of mathematics within the earliest accounts of architecture. Such myths distil a series of ideas in such a way that their essential message is retained while other peripheral issues are excised (Kirk 1975). A study of myths reveals the values, superstitions and beliefs that are the historic cornerstone of a discipline (Bettelheim 1978). The myths of the primitive hut and Daedalus are crucial indicators of architectural attitudes towards geometry, pattern and metrology and they resonate with other canonical value structures including the Vitruvian triad of *firmitas*, *utilitas*, and *venustas*; terms that were aptly translated by Sir Henry Wotton as *firmness*, *commodity* and *delight* (Kostof 1977; Johnson 1994).[1]

Whereas aetiology supports the consideration of causes without effects, the examination of effects without causes is called *teleology*. A teleological investigation of a relationship seeks to comprehend it in terms of its outcome and without reference to its source. In the second major section of this chapter a more modern myth—the collectively accepted model of the design process—is reviewed to reveal the breadth and depth of uses of mathematics in more recent times. This model has endured for many hundreds of years, embedded as it is in the practices of the architectural discipline through pedagogical, fiduciary and curatorial mechanisms such that, despite countless practical changes, the primary creative systems continue to be conceptualised in this way (Miller 1995; Ostwald 2012).

Combining both the aetiological and the teleological readings of the relationships between architecture and mathematics allows us to propose a framework of types. Three purposive agendas are at the core of this framework: the use of knowledge *for* supporting the design process, the desire to embed knowledge *in* an aesthetic construct, and the application *of* knowledge through design analysis. Within this framework 13 different types of mathematical applications in architecture are identified. These are: logic; measurement; surveying; modularity; performance and prediction; generation; aesthetics; symbolism and semiotics; phenomenality and rationalism; inspiration; surface articulation; analysis and informatics.

The framework proposed in this chapter is not intended to provide a definitive epistemology; rather, its purpose is more akin to a genealogist's study of kinship and consanguinity. It investigates the natural mathematical relations or bloodlines that have historically sustained architecture. Furthermore, the goal of this chapter is not to explain why these different applications of mathematics occur in architecture, but to provide a mechanism for recording the different types of applications and for understanding them holistically, as either occurring in a particular stage of the design process, or in support of a specific architectural quality. Through this dual aetiological and teleological process the breadth of

[1] "Well building hath three Conditions. *Commoditie*, *Firmenes* and *Delight*" (Wotton 1624: 1).

approaches, applications and techniques—all symbolically represented by the pair of compasses in the frontispiece of the *Bible Moralisee*—is revealed.

Myths of Architecture: An Aetiology

A common practice in the historiography of many disciplines is to link the origins of ideas to specific incidents, either real or imagined. For example, in 1665, while convalescing at the family home in Lincolnshire, Isaac Newton observed an apple falling from a tree. In his later life he would recount this event, describing it as the catalyst for his formulation of a universal theory of gravitation (Hall 1999). The story of Newton and the apple has since become one of the enduring myths of modern science. However, despite being allegedly based on real events, the term 'myth' is appropriate here because there are multiple conflicting versions of Newton's account (Brewster 1835). Indeed, five decades passed between the windfall occurring and Newton describing its significance. Newton actually invested several decades of his life in detailed research into the topic of gravity but when called upon in his later life to explain the genesis of his work, he repeated variations of this account of the falling apple. An aetiological perspective of this event is not concerned with its historical veracity but with the reason Newton chose to present his work in this way, emphasising the manner in which it uses an everyday occurrence to evoke the presence of a universal system of physical laws (Berkun 2010).

Every discipline has an equivalent origin myth, a tale that serves to elucidate and authorise a set of actions or values. In Western mythology the two great origin myths of architecture are both, as is typical of the genre, largely apocryphal. This is why they should only be read as a post-rationalised or figurative explanation of why certain acts should continue or particular relationships are important. The two origin myths of Western architecture describe the construction of the first building and the skills of the first architect. Whether one can be said to precede the other is a point of minor contention, but the myth of the first building, the archetypal primitive hut, is deliberately composed without the presence of an architect and so it is the first that is considered here.

Joseph Rykwert (1981) argues that throughout history architects have returned to the idea of the first house, the primitive hut of the ancients, whenever they have sought to make sense of the purpose of architecture. According to Rykwert, an interest in the primitive hut has been a constant throughout history: '[it] seems to have been displayed by practically all peoples at all times, and the meaning given to this elaborate figure does not appear to have shifted much from place to place, from time to time' (Rykwert 1981: 183). The myth of the primitive hut provides a philosophical foundation for understanding, questioning or reinvigorating architecture. Alberti, Laugier, Perrault, Viollet-le-Duc, Ruskin, Le Corbusier and Wright have each studied the primitive hut in its various incarnations (Harries 1993; Vogt 1998). Whether they have attempted to find its site, reconstruct its form, or study its construction, they have been drawn to seek inspiration from its imagined properties (Mitias 1999). Rykwert maintains that the primitive hut provides a 'point of reference for all speculation on the essentials of building' (Rykwert 1981: 183) including the relationship between architecture and systems of knowledge (like mathematics). The earliest extant version of this myth, from which most others can be traced, is found in Vitruvius's *De Architectura.*

Marcus Vitruvius Pollio, writing around the time of the Emperor Augustus in the first century B.C., provides an imagined account of a primitive race of men who 'were born like the wild beasts, [and lived in] woods, caves, and groves' (Vitruvius 1914: 38). During a storm, the branches of some trees near the tribe's cave 'caught fire, and so the inhabitants of the place were put to flight, being terrified by the furious flame' (38). After the storm had subsided, they gathered around the flames and learnt to sustain them, and the fire in turn kept the tribe safe from predators. To maintain both the fire and the community that had formed around it, a shelter had to be constructed. This compulsion to create a structure in a specific location, rather than to inhabit an existing cave or hollow, was to be the impetus for the first building:

> At first they set up forked stakes connected by twigs and covered these walls with mud. Others made walls of lumps of dried mud, covering them with reeds and leaves to keep out the rain and the heat. Finding that such roofs could not stand the rain during the storms of winter, they built them with peaks daubed with mud, the roofs sloping and projecting so as to carry off the rain water (Vitruvius 1914: 39).

A woodcut illustration in the 1521 edition of Vitruvius by Cesare Cesariano depicts a large fire surrounded by a primitive tribe. In the foreground people are gathering branches to feed the flames, while in the background, glimpsed through the smoke-haze, the branches of the living trees can be seen entwined together, suggesting a pitched or woven-roofed form. A second woodcut by Cesariano—much like subsequent ones from later editions of *De Architectura* and those in *Vitruvius Teutsch*—is less allegorical in its intent, displaying a more literal representation of the first hut. In that woodcut, rows of evenly spaced, vertically-arrayed tree trunks each end in a forked bough, which creates a natural cradle for a horizontal timber spar to connect the columns and create an edge to the roof. Between these columnar trunks with their forked pinnacles, smaller branches have been woven (Fig. 3.2). The regularly spaced, if roughly hewn, rafters and beams are also plaited together, creating an alternating surface of branches and grass, woven as if 'in imitation of the nests of swallows' (Vitruvius 1914: 38).

The architecture of the Vitruvian primitive hut is founded, initially at least, on the crafts of weaving or plaiting; the regular interleaving of elements forms a reinforced surface which is also a recurring geometric pattern. Starting with living branches and leaves, in groves or bowers, and then including loose grass and partially dressed timber, woven structures formed the basis for tents, screens and simple roofs. The first structures were created using felled trees as columns, arrayed in such a way that their forked joints created natural supports, and sized and spaced to achieve a consistent wall. These timber frames were the basis for

Fig. 3.2 The primitive hut according to Cesariano's edition of Vitruvius. Image: Cesariano (1521: Bk. II, ch. 1, p. XXXI v

subsequent woven and layered enclosures. Vitruvius states that variations of these techniques can be seen in the primitive dwellings of many cultures, including the Colchians of Pontus (near present day Georgia on the Black Sea). The Colchians would commence by laying.

> ... down entire trees flat on the ground to the right and the left, leaving between them a space to suit the length of the trees, and then place above these another pair of trees, resting on the ends of the former and at right angles with them. These four trees enclose the space for the dwelling. Then upon these they place sticks of timber, one after the other on the four sides, crossing each other at the angles, and so, proceeding with their walls of trees laid perpendicularly above the lowest, they build up high towers. The interstices, which are left on account of the thickness of the building material, are stopped up with chips and mud.

> As for the roofs, by cutting away the ends of the crossbeams and making them converge gradually as they lay them across, they bring them up to the top from the four sides in the shape of a pyramid (Vitruvius 1914: 39).

In the Colchian hut, stacked logs, carefully sized, spaced and cut to measure, create both structure and enclosure. The form of this dwelling is square in plan with a pyramid-shaped roof. In the various examples of the primitive hut the importance of measurement (typically relative to other elements in a building), structural stability (intuitively or empirically determined), geometry (for the creation of symmetrical and stable forms in three dimensions) and pattern (in the construction and expression of woven forms) are all reinforced.

A second architectural aetiology is found in Greek mythology where Daedalus, the father of Icarus, is characterized as the first architect. Daedalus was an Athenian craftsman who is credited with the design of Ariadne's dancing floor and the Labyrinth at Knossos. Whether he was a real person or an amalgam of several different designers is unknown. Homer, Euripides and Ovid describe his actions in poetic terms, dwelling on his invention of animated statues, the golden thread of Ariadne and the waxed and feathered wings of Icarus that famously melted, sending Daedalus's son plummeting to his death. In contrast, Pliny the Elder treats Daedalus as a historic figure, with a known parentage and birthplace.

In one of the earliest references to Daedalus, Homer's epic poem the *Illiad* (written in the seventh or eighth century B.C.) describes a 'cunningly wrought dancing-floor like unto that which in wide Cnosus Daedalus fashioned of old for fair-tressed Ariadne' (Homer 1924: 590). Produced in 415 B.C., Euripides' play *Hecuba* refers to Daedalus's almost godlike power to give life to inanimate objects. Aristotle, in Book I of *Politics* (ca. 330 B.C.), presents Daedalus as a legendary sculptor and Plato in Book III of *Laws*, refers to the great inventions of Daedalus. In Ovid's *Metamorphoses* (ca. 8 A.D.) Daedalus is described as 'an architect of wonderful ability' who 'built with intricate design' (Ovid 1922: 152). In 78 A.D., Pliny the Elder's *Naturalis Historia* commends Daedalus on being the 'first person who worked in wood' (Pliny 1893: 226). Pliny states, 'it was [Daedalus] who invented the saw, the axe, the plummet, the gimlet, glue, and isinglass' (1893: 226). Notwithstanding the obvious fallacy of Pliny's statement (axes existed long before Daedalus is thought to have been born), Horace (Horatius Flaccus), Virgil (Virgili Maronis), William Shakespeare and John Ruskin, amongst many others from antiquity to modern times, have portrayed Daedalus as a master sculptor, inventor and architect.

In mythology Daedalus's most famous work is the Labyrinth at Knossos. Ovid's account of the origins of the maze commences with the unnatural birth of the bull-headed man, the Minotaur. King Minos, seeking to imprison the Minotaur, commissioned Daedalus to design and construct a maze:

> This he planned of mazey wanderings that deceived the eyes, and labyrinthic passages involved. So sports the clear Maeander, in the fields of Phrygia winding doubtful; back and forth it meets itself, until the wandering stream fatigued, impedes its wearied waters' flow; from source to sea, from sea to source involved. So Daedalus contrived innumerous paths, and windings vague, so intricate that he, the architect, hardly could retrace his steps (Ovid 1922: 152).

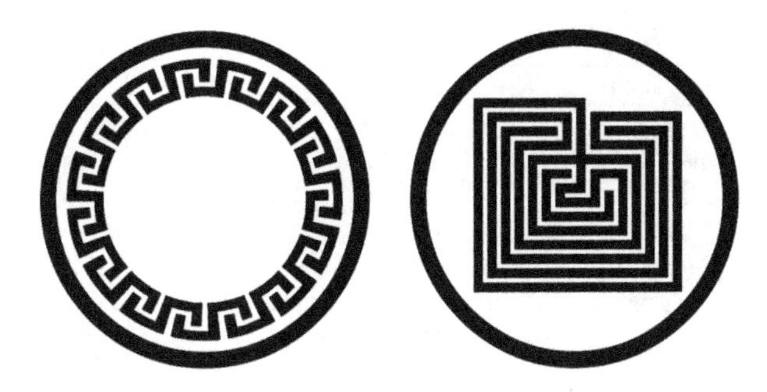

Fig. 3.3 Representation of the Labyrinth at Knossos reconstructed from silver coins (ca. 400 B.C.). Image: Michael Dawes

Thus imprisoned, the Minotaur had to be appeased with the lives of Athenian youths and maidens. This sacrificial rite continued for many years until Theseus, guided through the maze by Ariadne's golden thread, slew the Minotaur and escaped the Labyrinth (Castleden 1990).

Opinion is divided over whether Daedalus built Ariadne's *choros*—an intricate dancing floor—before or after Theseus's escape from the Labyrinth (Ovid 1922; Nichols 1995). Part of the confusion relates to the language used to describe the two designs. Indra Kagis McEwen (1993) has demonstrated that several of Daedalus's inventions share a common etymology. That is, the Greek words used to describe the act of dancing, a patterned dancing floor and a maze are all related to the concepts of weaving or animation. Noting this connection, Kerenyi (1976) gave Ariadne the title 'Mistress of the Labyrinth', a reference not only to the Minotaur's maze, but also to the elaborate formal structure of Ariadne's dance and to its divine or transcendent aspiration. Kern (2000) suggests that the geometry of the dancing floor, itself a symbol of the ritual and possibly erotic conjoining of two bodies, was repeated at larger scale in the Labyrinth, which explains why the two share the same geometric pattern and language. For these reasons, in Greek mythology Daedalus's claim to the title architect is not a result of his ability to oversee the construction of a building, but rather of his capacity to weave geometry into space and form which has both symbolic and phenomenal significance.

In the examples of the *choros* and the Labyrinth, geometry is placed in the service of design in three broad ways, each of which is aligned to one of the Vitruvian triad of architectural qualities. First, it delineates and structures space (firmness): both the dancing floor and the maze are geometrically defined and controlled. Second, it fulfils a program function (commodity): the maze is a geometric structure with a distinct spatial function—to disorientate, restrain, or beguile visitors. The function of the *choros* was to enable the 'crane dance', a tightly constrained marriage ritual. Finally, geometry provides a decorative motif (delight): the geometric weave of the *choros* and the maze has since appeared on coins, reliefs, pottery and in wood carvings (Fig. 3.3).

Homer notes that the Daedalic geometric weave is found in the decoration on Achilles' shield, and Ruskin traces the aesthetic and moral importance of 'Daedalic Right Line' in Gothic architecture (Moore and Ostwald 1997). Like the application of measurement, structure and pattern in the primitive hut, the presence of geometric function, foundation and fascination in the work of Daedalus

Table 3.1 Mathematical applications in the foundation myths of architecture

Application		Myth		
Type	General definition	Primitive hut	Daedalus	Instance
Measurement	The use of mathematics to record and communicate dimensional information	■	□	Sourcing or modifying materials to achieve consistent, relative dimensions
Surveying	The use of mathematics to derive and translate locational or site-related measures	■	□	Information relating to the position and relative spacing of columns and the efficient sourcing or transportation of materials
Performance and prediction	The use of mathematics to inform decisions about structural, acoustic, environmental, visual and related physical properties	■	□	Empirically or intuitively derived estimates of the size of structural members for stability and endurance
Surface articulation	The use of mathematics to achieve an efficient or controlled coverage of a defined plane	■	■	Empirically or intuitively derived methods for achieving a waterproof, or wind-proof woven or thatched surface. The use of geometry to achieve an intricate, patterned surface covering
Generation	The use of algorithms or rules to evolve or parameterise aspects of a design	⃠	■	The Labyrinth is a mathematical construct with a distinct set of geometric and spatial parameters
Inspiration	The use of mathematics as influence, motivation or animation	⃠	■	Both the form of the dancing floor and the Labyrinth are geometric mazes
Aesthetics	The use of mathematics to achieve a particular appearance or visual effect	□	■	The woven path of Ariadne conforms to a pre-determined symmetrical field, within which separate circular and orthogonal patterns reinforce the overall structure
Symbolism and semiotics	The use of mathematics to represent or communicate something about a building	□	■	The geometric decoration of Achilles shield (likened to Daedalus's dance floor) is intended to communicate both a connection to Ariadne and to the heavens
Phenomenality and rationalism	The use of mathematics to evoke a connection by way of the senses or the mind	⃠	■	The geometric path on Ariadne's dance floor evokes and enables a particular physical and sensual ritual—the 'crane dance'

(continued)

Table 3.1 (continued)

Application		Myth		
Type	General definition	Primitive hut	Daedalus	Instance
Logic	The reasoned or disciplined application of knowledge	□	□	Underpinning the majority of the applications of mathematics found in the two myths is the presence of a reasoned and consistent use of information

Key: ■ = application explicit, □ = application inferred, ⃠ = application absent

reinforces the early, mythopoeically delineated set of relationships between architecture and mathematics.

When the two foundation myths are viewed together, they present complementary visions of the role of architecture and of the architect (Table 3.1).

The primitive hut stresses the importance of construction, structure and utility, while the work of Daedalus emphasises aesthetic, inspirational and phenomenal applications. Furthermore, despite their differing emphases, both myths contain references to a larger set of pragmatic and poetic applications. For example, a crucial function of the primitive hut is to shelter a community, both physically and spiritually. Social and cultural concerns are present in this myth, even if its brevity curtails them. The primitive huts described by Vitruvius also possess symmetrical cross sections and plans, something that is especially significant when viewed in the context of the larger body of his theory which uses geometry to evoke divine relations. Similarly, in the Daedalus myth, technical skills are praised along with the ability to work with particular materials. His capacity to measure and survey is also assumed as a basic prerequisite skill of his craft. In addition, while not explicitly stated in either myth, there is an implication that underlying all of the basic actions and decisions is a capacity to think logically and consistently. Thus, the correct size for a rafter in the primitive hut was not calculated, it was determined either empirically (by loading different size beams until structural failure occurred) or intuitively (by using a knowledge of the size of rafters that had worked in the past). For this reason, and despite identifying nine different rudimentary applications of mathematics in these myths, the central role of logic, the tenth type, cannot be ignored.

Finally, it is possible to conceptualize each of these ten types of applications as serving at least one of the core qualities of architecture. For example, if we accept the Vitruvian triad then performance-related applications of mathematics may be associated with firmness and aesthetic applications are related to delight. However, some other types, like measurement, can be mapped to two categories—firmness and commodity—while surface articulation and logic can potentially be used to fulfil parts of all three Vitruvian qualities (Fig. 3.4).

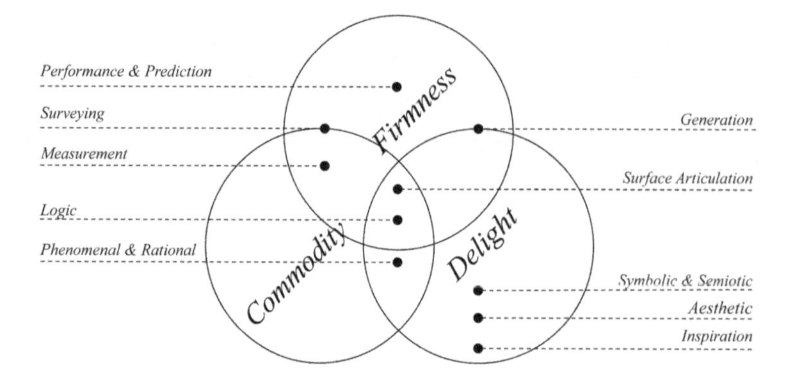

Fig. 3.4 Conceptual mapping of application types against the Vitruvian triad (definitions in Table 3.1)

Models of Architecture: A Teleology

The focus of this section shifts from historic myths to contemporary models. Just as aetiology and teleology have a close, but inverse relationship, so too do myths and models. Ian Barbour defines a myth as an archetypal event that reinforces a pattern of behaviour in society. Myths 'integrate the community around common memories and common goals'; they 'are neither true nor false; they are useful fictions which fulfil these important social functions' (Barbour 1974: 3). The modern counterpart of the myth is the model or paradigm. For Barbour, the model is 'a symbolic representation of selected aspects of the behaviour of a complex system' (3). Whereas a myth describes the world, a model is an 'imaginative tool for ordering experience' (3). The critical difference between myths and models is that the myth derives a universal message from a specific event (thereby relating the particular to the general), while the model starts with a universal system from which a specific response is derived (progressing from the general to the particular) (Coupe 2009). Thus, despite the way models are positioned in contemporary discourse as encapsulating a global truth, they have innate fictional, imagined or conceptual properties that are similar to those of myth. It is also often assumed that the model is more cogently founded in reason, observation or data, but the myth too, represents a body of received wisdom.

The primary role of the architect has historically been, and remains to the present day, the visualization of a design and the communication of this intent, in such a way as to support the construction of a building. The same is also true if the architect's purpose is to refurbish an existing structure, to design a landscape for a park or create a new urban space. While the tools and technologies available to architects have changed over many centuries, the conceptual process of designing and executing a building has remained a surprisingly durable one. There are many subtle variations of this model of the design process, although the majority are conceptualised as an iterative or staged sequence with occasional recursive loops. This model of the design process as a system is found in educational settings

(Pressman 1993; Anderson 2011) and there is evidence that it is used by professionals (Rowe 1987; Lawson 2005; Pressman 2012).[2] Two of the more common variations of the model are framed around cognitive and contractual processes. The cognitive variation commences with problem definition, analysis and synthesis stages, prior to conceptual and schematic diagramming, and finally solution proposition, testing (the recursive loop) and realisation (Pressman 2012). The more contractual or practical variation commences with client briefing, conceptual design, schematic design, developed and detail design, and construction. Several of these steps allow for a limited return to the previous stage to revise or correct any errors which have occurred in the process or to take account of any revisions—to the brief, budget or site conditions—which require a more substantial redesign. More nuanced variations of this contractual model note that there are parallel approval and review processes and that design often continues throughout the construction period and through to post-occupancy evaluation and optimisation. The cognitive variation of the model continues to cycle through the same stages, but with each subsequent series the focus is on a smaller sub-problem within the larger design. Although there are differences between these variations, they both describe a simplified and universal vision of the role of the architect in society. This model, and especially the contractual or practical variation, is useful for identifying the various ways in which architecture uses mathematics.

A necessary precursor to the design process is the production of a design brief, a document which defines the practical and functional limits of a project. The brief typically comprises a list of functional zones, along with information about the scale, critical dimensions and performance criteria. For example, a brief might state that a particular house requires a living room which is at least 15 m^2 in floor area, with a minimum ceiling height of 3.5 m, and with a south-facing wall which is mostly (between 5 and 8 m^2) glass, at least 30 % of which is operable. These measures or conditions are a numeric reflection of the need to accommodate a certain size of social gathering in a space that doesn't feel vertically constrained, is illuminated with natural light, and allows for some natural ventilation.[3]

[2] A common and reasonable concern that has been raised with the standard design process model is that design is not necessarily a linear or systematic process. Design is often characterised as an 'ill-defined' or 'wicked' problem (Brown et al. 2010). Design problems, unlike many mathematical ones, rarely have a single ideal solution. Instead, design involves handling a range of challenges that are described by scientists and engineers as either 'non-trivial' or 'sub-optimal'. Design involves balanced compromise between issues, some of which may be described with great rigour (like structural stability and material strength) while others cannot (like the symbolic power of a building, or the message its iconography communicates to society). This is why the design process model, which may be appropriate for simple or formulaic buildings, is much less useful for more complex building types.

[3] For some complex building types, a much higher level of performance is specified in the architectural brief including lighting levels, acoustic reverberation times and structural bearing capacities. In the last few decades it has also become common for technically advanced buildings, like hospitals, to rely on a relative performance brief. For example, a client might state that a new oncology centre for Rome must function at least as well as the recently completed oncology centre

Once the brief is defined, then the architect engages in a process of parametrically-informed idea generation, wherein he or she seeks to derive a solution to the constraints and opportunities of a brief and a site. This so-called 'conceptual design' stage draws on the architect's ability to manage multiple, sometimes conflicting requirements, simultaneously juggling both relative spatial issues (like the relationship between a living room, a dining room and a kitchen) and absolute ones (like the orientation of the site and the address or access to the building). These interconnecting performance parameters may often be solved in a larger number of alternative spatial configurations and thus the architect must be guided by a vision or set of values, often embodied in a *parti* or organising principle, which assists in determining which conceptual design variations to present to a client. The vision or inspiration for a design remains ever-present throughout the remainder of the project, but its core aesthetic, poetic or representational agenda is typically delineated at this stage, along with possible strategies for achieving this vision. Furthermore, the architect's core values become evident at this point, including the factors driving their design aspiration, from ecological to social, technical and poetic values. Many of these factors involve geometry in an aesthetic, symbolic, semiotic or inspirational role.

Whereas in the concept design stage spatial and contextual relations are described in a topological manner (that is, through connections and relations rather than absolute dimensions), in the schematic design stage, the concept and *parti* of a design are given scale and dimensionality, in accordance with the original brief, along with relative proportions. The first sense of structure and three-dimensional massing (width, depth, height, bulk) is typically tested at this stage, along with an early sense of fenestration and materiality. A preliminary estimate of the cost of the design, typically based on 'square metre' or 'floor rates', is also calculated to determine if the client's brief and budget are viable. For particular building types, the schematic design stage can also include simple modelling and simulation of performance requirements, like the volume of indirect natural light in an art gallery, clear sight lines in a theatre or overshadowing caused by a tall building.

Once the schematic design has been approved, the next stage requires the refinement of its principles. Depending of the building type, the developed design stage can commence with extensive testing and modelling of design variations to optimise important factors (light, security, efficiency, environmental impact) and with each refinement the spatial program evolves while seeking to maintain the topological and geographic relations agreed with the client in the previous stages, but which are now forced to change in response to more detailed design considerations. As the design is finalised, its overarching dimensions and properties are delineated and cost estimates made prior to seeking approval to commence construction.

---

in Sydney, but accommodate a 25 % growth in treatment capacity. Such a brief involves both the measuring of the properties of the reference structure and then the interpretation and interpolation of these performance criteria into the new design with increased capacity.

Prior to construction commencing, construction documentation must be produced to translate the design into a system which allows for multiple contractors to undertake the work. At this time, structural engineers complete and certify their designs of columns, beams and bracing, mechanical engineers design services and equipment, and other specialist consultants use a range of mathematical and computational approaches and techniques to determine and specify systems which can be installed during the construction process (Ambrose and Tripeny 2012). Some of the sub-contractors involved in this stage can include pre-fabrication and curtain wall consultants, professionals who extract information from architectural drawings and models to quantify the time it will take to manufacture components, the implications for tolerances and batching (storing pre-fabricated elements prior to construction) and site handling. Meanwhile, the architect often coordinates all of these activities, defining the dimensions and limits for each part of the building.

Several variations of the design process model end with a 'post occupancy' stage, in which the completed building is analysed to optimise or assess its performance. Increasingly, theories and techniques have been developed which can be applied to support improved social interaction, wayfinding and security in buildings, amongst other factors. Such mathematical techniques are useful for refurbishment and improvement and also for scholarly analysis. With the rise of global information and positioning systems, data developed from a building may also be applied to much larger models of suburbs or cities (Hilton 2007). Whereas architectural analysis is typically focussed on extracting information from a building so as to better understand its properties, the field of spatial or urban informatics combines information from multiple buildings, transport networks and infrastructure systems to analyse larger regions (Foth 2009).

If the complete set of applications listed in this section are categorised, a set of thirteen types is identified, each of which can be cross-referenced to the stages in the design process model in which they are likely to occur (Table 3.2).

Some application types are concentrated (but not exclusively present) in certain stages. For example, applications of mathematics associated with modularity are typically less important early in the design process, but become more significant in the detail design and construction stages. Modularity may well be a consideration in earlier stages in the design of particular building types, or for architects whose theories rely on systematised construction, but it is more likely to be used in the detail design stage (Kroll 1986). Similarly, the use of mathematics to generate the form of a design is something that is most likely to occur in the concept and schematic stages, and is often indirectly evolved from the brief itself. In certain projects such generative or parametric techniques might continue to be important in the detail design stage as well, but this is less common. Aesthetic and phenomenal considerations are likely to be more prominent in these same early stages (concept and schematic design) and play a lesser, supporting role, in later parts of the process. Viewed in this way, the different types of mathematical applications in

Table 3.2 Mathematical applications in the traditional model of the design process

Application		Stage in the design process model						
Type	General definition	Brief	Concept	Schematic	Developed	Detail	Const.	Post Occ.
Logic	The reasoned or disciplined application of knowledge	■	■	■	■	■	■	■
Measurement	The use of mathematics to record and communicate dimensional information	■	⊘	□	■	■	■	⊘
Surveying	The use of mathematics to derive and translate locational or site-related measures	■	□	□	■	□	■	⊘
Modularity	The use of mathematics for achieving coordination and consistency within a larger system	⊘	□	□	■	■	■	⊘
Performance and prediction	The use of mathematics to inform decisions about structural, acoustic, visual, environmental and related physical properties	⊘	⊘	□	■	■	□	⊘
Surface articulation	The use of mathematics to achieve an efficient or controlled coverage of a defined plane	⊘	⊘	□	■	■	⊘	⊘
Analysis	The use of mathematics to better understand the properties of a design	⊘	⊘	□	■	□	⊘	■

(continued)

Table 3.2 (continued)

Application		Stage in the design process model						
Type	General definition	Brief	Concept	Schematic	Developed	Detail	Const.	Post Occ.
Informatics	The use of mathematics to visualise or characterise architectural, urban and regional spatial and formal properties.	⃠	■	□	⃠	⃠	⃠	■
Generation	The use of algorithms or rules to evolve or parameterise aspects of a design	□	■	■	⃠	⃠	⃠	⃠
Inspiration	The use of mathematics as influence, motivation or animation	⃠	■	■	□	⃠	⃠	⃠
Aesthetics	The use mathematics to achieve a particular appearance or visual effect	⃠	■	■	□	⃠	⃠	⃠
Symbolism and semiotics	The use of mathematics to represent or communicate something about a building	⃠	■	■	⃠	⃠	⃠	⃠
Phenomenality and rationalism	The use of mathematics to evoke a connection by way of the senses or the mind	⃠	■	■	□	⃠	⃠	⃠

Key: ■ = common application, □ = less common application, ⃠ = rare application

the design process can be understood in terms of their shifting potential at various times in the project, rather than as a set of absolute values. This means, for example, that while deriving inspiration from mathematics is something that usually happens early in a process (Hahn 2012), it can also, in certain circumstances, be useful for final detail design decisions (Jencks 1985).

The complete set of application types found in the contemporary design process also include variations of those identified in the review of foundation myths. However, whereas the first group could be readily conceptualised as serving core architectural values, as exemplified in the Vitruvian triad, the second, larger group have taken on a more directed quality. That is, the categories define types by their use or application for different purposes. Thus, while it is possible to map the more extensive set of contemporary types to the categories of firmness, commodity and delight, this is less useful for the applications found in modern architectural practice. For this reason, the following section considers an alternative, triadic framework.

A Framework of Types

There are three overarching categories in the proposed framework for classifying the types of applications of mathematics found in architecture. These categories distinguish between mathematics that is used *for* the support of the design and construction process, that which is visible *in* the design product, and finally mathematics which is a property *of* the design itself. The first of the three categories could be thought of as encompassing all the factors conventionally considered under the Vitruvian rubric 'firmness', as well as some of those associated with the more functional dimensions of 'commodity'. This first part of the framework, mathematics *for* architecture, is related to, amongst other things, stability, function and environmental performance. The second category, mathematics *in* architecture closely correlates to the classic Vitruvian quality, 'delight', and generally comprises aesthetic, sensual or intellectual properties. The final category in the framework has no clear parallel in Vitruvius, although some of the derived properties of 'commodity' in the sense of usefulness or function may resonate with its purpose. The mathematics *of* architecture is concerned with reasoning and analysis about spatial and formal relations present in a design. Analysis, as a stand-alone activity, was uncommon in ancient times but it has since become increasingly important. Collectively the three purposive categories—*for*, *in* and *of*—provide an indicative way of classifying application types (Fig. 3.5).

Importantly though, while the majority of the 13 applications identified in this chapter are aligned to one of the three categories, a few potentially cross between them depending on their purpose or application (Table 3.3).

Furthermore, logic, as a foundation or core value for any reasoned practice, is a member of all three sets, although it has been listed here as part of the analytical

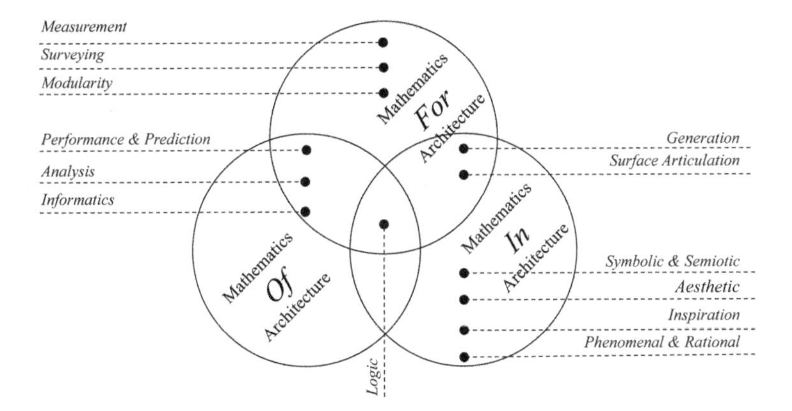

Fig. 3.5 Conceptual mapping of application types against the proposed framework (definitions in Table 3.2)

group of applications. Each of the three categories is described in more detail in what follows.

The first category in the framework—mathematics *for* architecture—includes practical techniques and tools that support architectural design, construction and conservation. These applications occur as a natural part of the design process informing decision-making about functional issues. They enable the development of intrinsic properties of a building that are critical for its stability, environmental performance and programmatic function, but are not necessarily expressed or visually apparent in its final form. This means that the mathematical application itself—for example, the calculation of the depth of a beam or the level of insulation required in a wall—is completed as a precursor to construction commencing. The outcomes of these calculations—the stability of the beam or the wall's capacity to mitigate heat—shape the ongoing function of the building, but the application itself is over and its result is implicit in the finished building rather than explicit. For example, surveying is a critical mathematical technique at certain stages in the design process, but once these stages are completed the active role of mathematics in the process is over. This first category is the most extensive in its list of types, but also the most temporal, because most of these only occur briefly and as part of the design process.

The specific application types that are exclusively allied with this first category include measurement, surveying and modularity. Measurement is associated with the use of mathematics to record and communicate dimensional information, whereas surveying applies and develops such measures in the context of a specific site. Applications of measurement, like those of logic, are so ubiquitous that they are rarely acknowledged in architecture, whereas site-specific surveying techniques have become more specialised over time. Modularity describes those practices that support coordination and consistency in the construction processes.

Table 3.3 A framework of types: mathematics, *in*, *of* and *for* architecture

Primary category	Definition	Type	Examples
Mathematics *for* Architecture	Practical or functional tools or techniques for the support of architectural design, construction and conservation	Measurement—the use of mathematics to record and communicate dimensional information	Theories/Techniques: relative measures, (cubits, rods and rulers), decimal and non-decimal systems, absolute measures (metric)
		Surveying—the use of mathematics to derive and translate locational or site-related measures	Theories/Techniques: levelling, traversing, triangulation and topography, laser scanning, GPS, GIF
		Modularity—the use of mathematics for achieving coordination and consistency within a larger design system	People: Moshe Safdie, Lucien Kroll, Richard Meier Buildings: Habitat (Montreal), Medical Faculty, Louvain University (Brussels), Sainsbury Centre (Norwich)
		Performance and Prediction—the use of mathematics to inform decisions about structural, acoustic, visual, environmental and related physical properties	People: Ove Arup, Renzo Piano, Norman Foster, Future Systems Buildings: Menil Collection (Houston), London City Hall (London)
		Generation—the use of algorithms or rules to evolve or parameterise aspects of a design	People: George Stiny, William Mitchell, Patrik Schumacher Buildings: Water Cube (Beijing), British Museum Atrium (London) Theories: Parametric design, generative design, shape grammar
Mathematics *in* Architecture	Geometric or numeric properties which are demonstrated, visible or sensible in architecture	Aesthetic—the use of mathematics to achieve a particular appearance or visual effect	People: Brunelleschi, Andrea Palladio, Le Corbusier, Hans Van Der Laan. Buildings: Parthenon (Athens), Chartres Cathedral (Chartres), Notre Dame Cathedral, (Paris), Unité d'Habitation (Marseilles) Styles: Medieval, Gothic, Palladian, Renaissance, Mannerism Theories: Golden Section, The Modulor, The Plastic Number

(continued)

Table 3.3 (continued)

Primary category	Definition	Type	Examples
		Symbolic and Semiotic—the use of mathematics to represent or communicate something about a building	People: Leon Battista Alberti, John Ruskin, Robert Venturi, Charles Correa Buildings: Hagia Sophia (Istanbul), Vana Venturi House (Philadelphia), Jawahar Kala Kendra (Jaipur) Styles: Medieval, Gothic, Postmodern
		Phenomenal and Rational—the use of mathematics to evoke a connection by way of the senses or the mind	People: Walter Burley Griffon, Richard Neutra, Louis Kahn, Stephen Holl Buildings: Salk Institute (La Jolla), Casa Del Fascio (Como), Simmons Hall (Cambridge, Mass.) Styles: Rationalism, Organic Modernism, Regionalism
		Inspirational—a use of mathematics as influence, motivation or animation	People: Oscar Niemeyer, Zaha Hadid, Peter Eisenman Buildings: National Congress of Brazil (Brasília), Sydney Opera House (Sydney) Styles: Modernism, Deconstructivism, Generative
		Surface articulation—the use of mathematics to achieve an efficient or controlled coverage of a defined plane	People: Antonio Gaudí, Lab, ARM Buildings: Park Güell (Barcelona), Federation Square (Melbourne), Storey Hall (Melbourne) Theories: Tessellations and Tilings, Aperiodic and Quasi-periodic Tiling
Mathematics *of* Architecture	Logical and analytical methods for quantifying or determining various properties of architecture	Analysis—a use of mathematics to better understand the properties of a design	People: Christopher Alexander, Bill Hillier, Lionel March Theories: Space Syntax, Fractal Analysis, Graph Theory, Fuzzy Theor

(continued)

Table 3.3 (continued)

Primary category	Definition	Type	Examples
		Informatics—the use of mathematics to visualise or characterise architectural, urban and regional spatial and formal properties	People: John James, Michael Benedikt, Michael Batty Theories: Tochymetry, Isovists, GIS mapping
		Logic—the reasoned or disciplined application of knowledge	Theories: Inductive, abductive and deductive reasoning; computational and heuristic reasoning

In a sense, these are specialised routines for relative or rule-based measurement. The remaining application types associated with this primary category can also be found in other secondary categories as well. They include the use of mathematics to predict and optimise the performance of a design, the use of rules or parameters to generate and evaluate design alternatives, and applications of plane-filling geometry (tiling, weaving and patterns). Specific examples of mathematical applications in this category include static and dynamic load calculations for structural stability, computational thermo-fluid dynamics for determining wind load, and thermal conductivity formulas for estimating human comfort levels.

The second of the three categories—mathematics *in* architecture—encompasses geometric or numeric properties that are intentionally designed into and are demonstrated in the form and materiality of a building. This category includes those applications of mathematics that are visible in the architecture, but are not necessarily products of its structure, construction or other functional or performance-related factors. Therefore, this category encompasses applications which augment or supplant those associated with the basic needs for stability and shelter. They could be described as being extrinsic factors because they are integral to the expression of a building, whereas those in the previous category were intrinsic to the function of the building. The type of mathematics that is found *in* architecture is expressed in ways that can be seen, sensed or read in the completed building. It includes the use of mathematics as inspiration for a design, the use of numbers and geometry to perform symbolic or semiotic functions, and properties that can be sensed either intellectually (aesthetic properties) or sensually (phenomenological properties). It is possible, and indeed likely, that the structure of a building will play a role in the expression of mathematics in architecture, but this is not necessarily a factor of the practical performance of that structure (its load-bearing or bracing capacity), but of the meaning or message it conveys visually or perceptually. Thus, while acknowledging that the meaning of symbols changes over time and that inspiration and phenomena cannot be consistently transmitted from a building to a viewer, this category could be understood as pertaining to those mathematical properties of architecture which are enduring or continue to operate after the building is complete. Some specific examples of applications in this category include proportional systems (like the golden mean and the Modulor), symbols which use geometry or number to communicate religious or cultural ideas (the Star of David), rationalist applications of Phileban solids in architecture, and phenomenological uses of geometry to evoke connections to nature.

The final category—mathematics *of* architecture—comprises analytical methods and approaches that are used for quantifying or determining various properties of a completed building or its context. These are mathematically-derived properties, rather than innate ones. They are the by-products of other design decisions which can be understood or modelled mathematically. The types of applications found in this category are focussed on the analysis of information about buildings and cities, for the purpose of understanding or optimising some aspect of a design. This category could also be understood as relating to those mathematical properties of a building that are only apparent when the building is subjected to a methodical investigation using approaches which are not otherwise intrinsic in the design. Examples of this category include space syntax and fractal analysis techniques, isovist analysis and spatial cognition and urban spatial informatics.

The set of types which make up this framework represents a compromise between accuracy and usefulness. At one extreme, it is possible to group almost all of the applications into just two or three categories that broadly correspond to the three overarching groups that are present in the final framework presented here. But, as the examples, tables and diagrams demonstrate, there are multiple overlaps between the three which undermine their utility. At a much finer-grained level, a notably larger list of types was originally identified which separated out multiple specific applications of mathematics, almost a third of which were used for structural and environmental calculations. However, these mathematical and computational approaches have changed over time and with increasing processing power, techniques which were impractical a decade ago, are now in common use. What has not changed is the core intent of all of these applications of mathematics—to ensure that the performance of a part of a building meets a given standard. Thus, a more extensive list of applications was merged into a single type: performance and prediction.

A different challenge was present in the topics of measurement, surveying and modularity. It could be argued that the first two are the same and that the third is simply a specialised application of rule-based measurement. However, measurement, like logic, is part of the base language of architectural design, whereas surveying is, to extend the metaphor, a separate dialect with a specific purpose and application. Modularity is a more contingent type because it is also potentially related to the aesthetic consideration of proportions. Nevertheless, in the pre-fabrication process and as part of a design approach, a separate tradition has developed around the topic of modularity to such an extent that it is worthy of separation from the other types.

The decision to merge the symbolic and semiotic uses into a single type was made to overcome the lack of distinction between them in many applications. For the first of these, some historic uses of symbolic geometry are clear in their application, while in much postmodern architecture, numbers are used as signs and with an understanding of their semiotic and linguistic properties. Nevertheless, for the majority of cases such a segregation is irrelevant because the design is simply called upon to communicate an idea using one technique or another. A similar logic was behind the decision to merge the phenomenal and the rational into a single type. Proponents of phenomenological design tend to deny or understate the role of the mind in responding to architecture and conversely, supporters of rationalist design tend to consider the senses a debased extension of the mind which distracts it from higher thoughts. Despite these differences, both phenomenal and rational approaches rely on geometry and form to elicit either a physical reaction or a mental one. It is their common desire for provocation that binds them together, much as it is impulse to communicate that led to symbolism and semiotics being similarly grouped.

Conclusion

In the thirteenth century, the pair of compasses in the hands of 'God the architect' symbolised the complete set of tools and devices used by designers to translate a vision into reality. The central message—that mathematics serves to translate the imagined into the physical—was reinforced by God the architect's stance, framed by a constructed portal and poised midway between the heavens and the earth. The pair of compasses encapsulates the many different types of applications of mathematics in architecture, with the majority present, in some rudimentary way at least, in even the earliest myths of this discipline. The more extensive set of application types in use today shares a clear lineage to these ancestral cases. The specific formulas used by architects and engineers may have changed, and, amongst other things, their capacity to work with non-orthogonal geometries has also improved, but the fundamental purpose of the application of mathematics in architecture has endured throughout history.

Biography Michael J. Ostwald is Professor and Dean of Architecture at the University of Newcastle (Australia) and a visiting Professor at RMIT University. He has previously been a Professorial Research Fellow at Victoria University Wellington, an Australian Research Council (ARC) Future Fellow at Newcastle and a visiting fellow at UCLA and MIT. He has a PhD in architectural history and theory and a DSc in design mathematics and computing. He completed post-doctoral research on baroque geometry at the CCA (Montreal) and at the Loeb Archives (Harvard). He is Co-Editor-in-Chief of the *Nexus Network Journal* and on the editorial boards of *ARQ* and *Architectural Theory Review*. He has authored more than 300 scholarly publications including 20 books and his architectural designs have been published and exhibited internationally.

Kim Williams was a practicing architect before moving to Italy and dedicating her attention to studies in architecture and mathematics. She is the founder of the conference series "Nexus: Relationships between Architecture and Mathematics" and the founder and Co-Editor-in-Chief of the *Nexus Network Journal*. She has written extensively on architecture and mathematics for the past 20 years. Her latest publication, with Stephen Wassell and Lionel March, is *The Mathematical Works of Leon Battista Alberti* (Basel: Birkhäuser, 2011).

References

AMBROSE, James, and Patrick TRIPENY. 2012. *Building Structures*. Hoboken, New Jersey: John Wiley.

ANDERSON, Jane. 2011. *Basics Architecture 03: Architectural Design*. Lausanne: AVA.

BARBOUR, Ian G. 1974. *Myths, Models and Paradigms*. New York: Harper Collins.

BERKUN, Scott. 2010. *The Myths of Innovation*. Sebastopol, California: O'Reilly.

BETTELHEIM, Bruno. 1978. *The Uses of Enchantment: The Meaning and Importance of Fairy Tales*. New York: Penguin Books.

BREWSTER, David. 1835. *The Life of Sir Isaac Newton*. New York: Harper and Brothers.

BROWN, Valerie A., John A. HARRIS, and Jacqueline RUSSELL. 2010. *Tackling Wicked Problems: Through the Transdisciplinary Imagination*. London: Routledge.

CASTLEDEN, Rodney. 1990. *Knossos Labyrinth*. London: Routledge.

CESARIANO, Cesare. 1521. *Di Lucio Vitruvio Pollione de architectura libri dece traducti de latino in vulgare affigurati*. Como: G. da Ponte.

COUPE, Laurence. 2009. *Myth*. London: Routledge.

EVANS, Robin. 1995. *The Projective Cast: Architecture and its Three Geometries*. Cambridge, Massachusetts: MIT Press.

FOTH, Marcus. 2009. *Handbook of Research on Urban Informatics: The Practice and Promise of the Real-time City*. Pennsylvania: IGI Global.

GOLDBERGER, Paul. 2009. *Why Architecture Matters*. New Haven: Yale University Press.

HAHN, Alexander. 2012. *Mathematical Excursions to the World's Great Buildings*. Princeton, New Jersey: Princeton University Press.

HALL, A. Rupert. 1999. *Isaac Newton: Eighteenth Century Perspectives*. Oxford: Oxford University Press.

HARRIES, Karsten. 1993. Thoughts on a Non-Arbitrary Architecture. Pp. 41–60 in David Seamon, ed. *Dwelling, Seeing and Designing: Toward a Phenomenological Ecology*. Albany, New York: State University of New York Press.

HILTON, Brian N., ed. 2007. *Emerging Spatial Information Systems and Applications*. London: Idea.

HOMER. 1924. *The Iliad Volume II.* A.T. Murray, trans. Cambridge, Massachusetts: Harvard University Press.

HUSBAND, Timothy B. 2009. *The Art of Illumination: The Limbourg Brothers and the Belles Heures of Jean de France and Duc de Berry.* New York: The Metropolitan Museum of Art.

JENCKS, Charles, 1985. *Towards a Symbolic Architecture.* New York: Rizzoli.

JOHNSON, Paul-Alan. 1994. *The Theory of Architecture: Concepts, Themes and Practices.* New York: Van Nostrand Reinhold.

KAPPRAFF, Jay. 1990. *Connections: The Geometric Bridge Between Art and Science.* New York: McGraw-Hill.

KERENYI, Carl. 1976. *Dionysos: Archetypal Image of Indestructible Life.* Ralph Manheim, trans. Princeton: Princeton University Press.

KERN, Hermann. 2000. *Through the Labyrinth: Designs and Meanings over 5000 Years.* New York: Prestel.

KIRK, Geoffrey Stephen. 1975. *Myth: Its Meaning and Functions in Ancient and Other Cultures.* London: Cambridge University Press.

KLINE, Naomi Reed. 2001. *Maps of Medieval Thought: The Hereford Paradigm.* Suffolk: Boydell Press.

KOSTOF, Spiro, ed. 1977. *The Architect: Chapters in the History of the Profession.* New York: Oxford University Press.

KROLL, Lucien. 1986. *The Architecture of Complexity.* Peter Blundell Jones, trans. London: Batsford.

LAWSON, Bryan. 2005. *How Designers Think: The Design Process Demystified.* Burlington, Massachusetts: Elsevier.

MCEWEN, Indra Kagis. 1993. *Socrates' Ancestor: An Essay on Architectural Beginnings.* Cambridge, Massachusetts: MIT Press.

MILLER, Sam. F. 1995. *Design Process: A Primer for Architectural and Interior Design.* New York: Van Nostrand Reinhold.

MITIAS, Michael H. 1999. Is Architecture an Art of Representation? Pp. 59–80 in Michael. H. Mitias, ed. *Architecture and Civilisation.* The Netherlands: Editions Rudopi.

MOORE, R. John, and Michael J. OSTWALD. 1997. Choral Dance: Ruskin and Dædalus. *Assemblage.* 32 (1997): 88–107.

NICHOLS, Nina da Vinci. 1995. *Ariadne's Lives.* London: Associated University Presses.

OSTWALD, Michael J. 2012. Systems and Enablers: Modelling the Impact of Contemporary Computational Methods and Technologies on the Design Process. Pp. 1–17 in Ning Gu and Xiangyu Wang, eds. *Computational Design Methods and Technologies: Applications in CAD, CAM and CAE Education.* Pennsylvania: IGI Global.

OVID. 1922. *Metamorphoses.* Brookes More, trans. Boston: Cornhill Publishing Co.

PLINY THE ELDER. 1893. *The Natural History, Volume II.* John Bostock and Henry Thomas Riley, trans. London: George Bell and Sons.

PRESSMAN, Andrew. 1993. *Architecture 101: A Guide to the Design Studio.* London: Wiley.

——. 2012. *Designing Architecture: The Elements of Process.* London: Routledge.

ROSSI, Corinna. 2004. *Architecture and Mathematics in Ancient Egypt.* Cambridge: Cambridge University Press.

ROWE, Peter G. 1987. *Design Thinking.* Cambridge, Massachusetts: MIT Press.

RYKWERT, Joseph. 1981. *On Adam's House in Paradise: The Idea of the Primitive Hut in Architectural History.* Cambridge, Massachusetts: MIT Press.

SALINGAROS, Nikos A. 2006. *A Theory Of Architecture.* Solingen: Umbau Verlag.

SALVADORI, Mario. 2014. Can There Be Any Relationships Between Mathematics and Architecture? Chap. 2, Pp. 25–29 in this present volume.

SCRUTON, Roger. 1983. *The Aesthetic Understanding: Essays in the Philosophy of Art and Culture.* London: Methuen.

Vitruvius. 1914. *The Ten Books on Architecture*. Morris Hicky Morgan, trans. Cambridge, Massachusetts: Harvard University Press.
Vogt, Adolf Max. 1998. *Le Corbusier, the Noble Savage: Toward an Archaeology of Modernism*. Cambridge, Massachusetts: MIT Press.
Wotton, Henry. 1624. *The Elements of Architecture*. London: Iohn Bill.

Chapter 4
Relationships Between History of Mathematics and History of Art

Clara Silvia Roero

During the course of centuries mathematics has interacted in many ways with culture and human activities, and among these a place of privilege has been reserved for art and architecture. Numerous artists, architects and historians of mathematics have made these relationships evident, such as Piero della Francesca, Leonardo, Albrecht Dürer, Maurits Cornelis Escher, Le Corbusier, Felix Klein, G. David Birkhoff, Andreas Speiser and Federigo Enriques, to mention only the most celebrated.

In this chapter I will show several examples of the existence of three levels of interaction between mathematics and art: the presence of a mathematical substrate in various archaeological and artistic relics from antiquity, the conscious or unconscious application by artists of mathematical principles whose theories had not yet been fully developed, and finally the relationship established by some mathematicians with artists and art theorists that permitted an awareness and acquisition of mathematical knowledge and rules that were then applied to artistic creations. The development of these three levels of interactions between mathematics and art can be a valid aid to the creation of a unified vision of the history of culture of peoples and civilizations, indicating various kinds of influence: technical-practical, theoretical-scientific, mystical-sacred, principles and customs, etc.

Indeed, in the wake of a long-term historiographic approach, new research perspectives have emerged recently that have been favourably received by art historians and critics. In particular, I wish to refer to some of the studies of Tullio Viola (1904–1985), who in the latter years of his life was partial to interdisciplinary

First published as: Clara Silvia Roero, "Relationships between History of Mathematics and History of Art". Pp. 105–110 in *Nexus VI: Architecture and Mathematics,* Sylvie Duvernoy and Orietta Pedemonte, eds. Turin: Kim Williams Books, 2006.

C.S. Roero (✉)
Department of Mathematics "Giuseppe Peano", Università di Torino, Via Carlo Alberto 10, 10123 Turin(Torino), Italy
e-mail: clarasilvia.roero@unito.it

K. Williams and M.J. Ostwald (eds.), *Architecture and Mathematics from Antiquity to the Future*, DOI 10.1007/978-3-319-00137-1_4,

investigations connected to archaeology, art and technology, especially of the most remote antiquity.

In 1984 Viola wrote:

> Every geometric property found in a figurative work is in some way and some measure an index of geometric sensitivity that cannot but manifest itself or be manifest in the works that we are used to calling 'scientific'. The 'geometry of the deep', which physicist Wolfgang Pauli calls "*Urintuition*"[1] is like the submerged mass of the iceberg, of which we can see and know only the tip that rises above the surface. The rational systemizations of geometric theories has their psychological, and therefore historical, roots in man's irrationality. Constructions logically demonstrated are based on spontaneous, unconscious intuitions, of which aesthetic feeling is a trustworthy guide (Viola 1986: 314).

Under Viola's guidance, since 1979 Livia Giacardi and I have studied a Sumerian game in which a serpent that bites its own tail moves with extraordinary regularity through a certain number of compartments, and if during the course there is a succession of natural numbers, there emerges a magic square, that is a square in which the sum of the elements of each row, column and diagonal is constant (Giacardi and Roero 1979) (Fig. 4.1).

Through an examination of other relics (with intertwined serpents, ornaments of the checkerboard of the royal tombs of Ur (Fig. 4.2) and polygonal disks with stepped sides) we arrived at interesting topological properties of such interlaced motifs and to the formulation of hypotheses as to their geometrical and arithmetical-magical meanings.[2]

Each checkerboard, however formed, can be covered in one, and only one, way, by intertwined serpents, regardless of the orientation of the intertwining, of the exchange of "underpassages" with "overpassages" in the crossing of the "doors of communication" between one compartment and the next, and of the collocation of the heads of the serpents along the corresponding paths. For rectangular checkerboards, in which the number of rows and columns are prime numbers, the braid is always composed of a single serpent, while in square checkerboards of n rows and n columns, there are exactly n serpents. In all cases, the methods of numbering of the compartments can be described so as to obtain numeric tables that present magic properties. This is an example of a mathematical substrate in an artistic creation.

In 1980 Viola studied the problem of the passage from the contemplation of ideal geometric figures of primitive man to that of the rational geometry in the work of

[1] *Jedes Verstehen ist ein lanwieriger Prozess, der lange vor der rationalen Formulierbarkeit des Bewusstseinsinhaltes durch Prozesse im Unbewussten eingeleitet wird: auf der vorbewussten Stufe der Erkenntnis sind an Stelle von klaren Begriffen Bilder mit starkem emotionalem Gehalt vorhanden, die nicht gedacht, sondern gleichsam malend geschant werden* (Every process of mental comprehension is of long duration. Much before the possibility of the conscious formulation of its content, it takes the form by means of an unconscious process. At the level of pre-consciousness, in place of clear concepts are present images of a strong emotional content. These are not only thought of, but are looked at as though painted) (Pauli 1961: 91).

[2] Viola and I presented two papers on this theme at the eleventh Congress of the Union of Italian Mathematicians in Palermo; see Giacardi et al. (1979, 1980).

Fig. 4.1

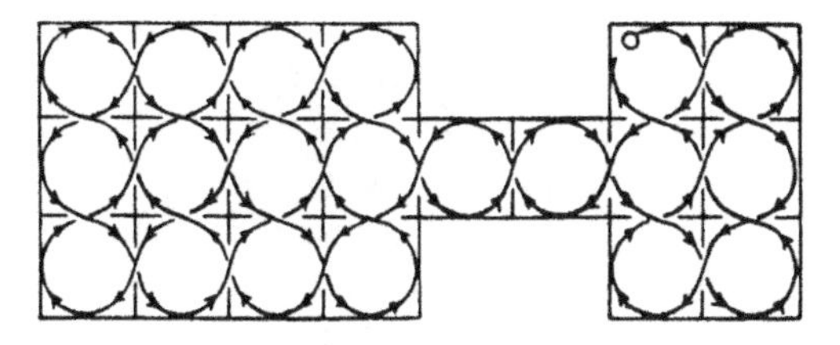

Fig. 4.2

Fig. 4.3

Thales of Mileto (Rizzi and Viola 1980). In order to demonstrate this, Viola chose an ornament in gold leaf that was found in Iran and is datable to the second millennium B.C., representing the head of a ram whose long horns curl in the form of a volute, each of which has the forms of four ovals that are noticeably elliptical (Figs. 4.3 and 4.4).

Having performed an accurate graphical analysis, Viola underlined the extraordinary approximation of an ellipse of the external oval of one of the horns and concluded:

> It seems evident that the Iranian artist who created this jewel let himself be guided by exclusively aesthetic requirements to reproduce a geometric figure (the ellipse) that he contemplated in his own mind at the very moment that he was working. But simultaneous and complementary contemplation and artistic creation are not in themselves sufficient to permit the birth of the mathematical concept: for this, contemplation had to be enriched by rational needs, and this effectively occurs a millennium and a half later, in a faraway land, by another people, in an extremely complex, refined and philosophically profound cultural context... [T]he geometry of Thales was not yet a rational geometry, in the way in which we think of that, ... in that, the contemplation of figures was no longer exclusively of an aesthetic nature but was already enriched by the attempt at deductive justification, going in search of the 'reason' behind certain properties (Rizzi and Viola 1980).

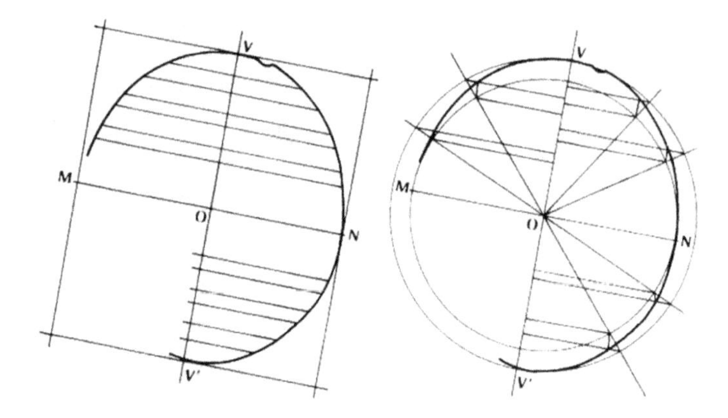

Fig. 4.4

Fig. 4.5

In the same year, Viola suggested that I study a bas-relief of a metope of the Parthenon, the results of which enjoyed a considerable degree of success in the international arena (Roero 1981, 1982; Ragghianti 1982). It was in fact possible to demonstrate how a great artist such as Phideas intuited and visualized admirably in his frieze the physical principle of equilibrium, two centuries before that would be explicitly formulated by Archimedes (Figs. 4.5 and 4.6).

The rearing horse with only one hoof touching the ground is in static equilibrium. Analogous research was undertaken by Viola in collaboration with Maria Teresa Navale and Silvia Mazzoni (Manzoni and Navale 1980; Manzoni et al. 1980), in which they were able to identify the possible construction technique of the tunnel of the island of Samos in the fourth century B.C. by Eupalino of Megara with the aid of triangulation (Manzoni et al. 1980, 1985).

Fig. 4.6

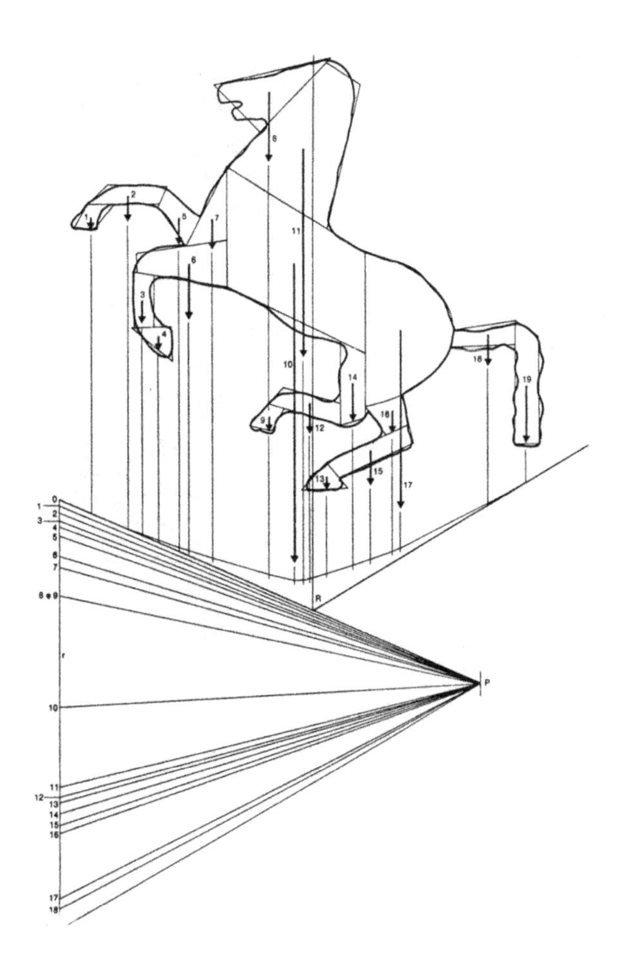

Together with Navale, Viola was then able to define the geometric form of the profile of the Narmer Palette, a celebrated Egyptian masterpiece in slate of 3000 B. C., demonstrating that it is an excellent approximation of the catenary curve (Navale and Viola 1985, 1986) (Fig. 4.7).

They also found very interesting results in a long and complex mathematical-historical analysis of some Ionic volutes of temples in Greece and Italy (Navale and Viola 1980, 1982). Together with Silvio Curto, then director of Turin's Egyptian Museum, Viola studied the measurements of some Egyptian colossi (Curto and Viola 1980) and conjectured as to the construction of the pyramid of Cheops, without however arriving at definitive conclusions. His historic approach was further stimulated by the research of Lina Mancini Proia and Marta Menghini on the evolution of the shape of cupolas in churches, from circular to ovals and only in the seventeenth century to a form that was decidedly elliptical (Mancini Proia and Menghini 1984). These authors

Fig. 4.7

maintain, for valid reasons, that the three Roman architects of the Baroque Gianlorenzo Bernini, Giacomo Berrettini da Cortona, and above all, Francesco Borromini, to whom we owe the first elliptical cupola, that of S. Carlo alle Quattro Fontane were inspired by astronomical research and the fascinating findings of Galileo.

The ellipse of the Iranian ram, the rearing horse of the Parthenon frieze and the curve of the Narmer Palette are all examples of conscious or unconscious application of mathematical principles whose formal theories would be fully developed only much later.

Finally, to illustrate the third level of interaction and exchange of knowledge between mathematicians and artists we can recall the historic studies conducted on the proportional models to represent the beauty of the human body in classical Greece and the Renaissance (Roero 1999, 2000), on the geometry of the fixed compass from the Medieval to the 1900s (Roero 2006), and the symmetry of Guarino Guarini (Roero 2005).

Translated from the Italian by Kim Williams

Biography Clara Silvia Roero is full professor of History of Mathematics at the University of Torino. She is a member of the editorial board of several journals, including *Bollettino di Storia delle Scienze Matematiche, Revue d'histoire des mathematiques, Lettera Matematica Pristem,* and *Il Maurolico.* She is on the Scientific Committee for the collected scientific papers of the mathematicians and physicists of the Bernoulli family, for the National Edition of R. G. Boscovich's works, and for the papers of M. G. Agnesi. She was President of the Italian Society of History of Mathematics (SISM) from 2000 to 2008, and a member of the International Commission of History of Mathematics. She is currently Director of the Torino Research Group on History of Mathematics.

References

Curto, S. and Viola, T. 1980. Per un computo ponderale di alcuni manufatti litici egizi. *Atti dell'Accademia delle Scienze di Torino*, Cl. Scienze Morali, **114**: pp. 155-171. Turin: Accademia delle Scienze.

Giacardi, L. and Roero, C. S. 1979. *La matematica delle civiltà arcaiche*, Turin: Stampatori.

Giacardi, L., Roero, C. S. and Viola, T. 1979. Ipotesi sull'esistenza di una matematica magico-sacrale presso gli antichi Sumeri. Pp. 143-160 in *Arithmos-Arrythmos, Skizzen aus der Wissenschaftsgeschichte, Festschrift für J.O. Fleckenstein zum 65 Geburtstag*. Munich: Minerva Publikation.

Giacardi, L., Roero, C. S. and Viola, T. 1980. Proposte d'interpretazione di alcuni reperti sumerici risalenti al III millennio a.C. (Ipotesi sull'esistenza di una matematica magico-sacrale). *Quaderni di Matematica della Università di Torino*, **10**: pp. 1-64. Turin: Università degli Studi di Torino.

Mancini Proia, L. and Menghini, M. 1984. Dalle ovali policentriche alle ellissi, nell'architettura barocca (Una possibile derivazione concettuale dall'astronomia). *Atti dell'Accademia delle Scienze di Torino*, Cl. Sci. Fis. Mat. Nat., **118**: pp. 325-338. Turin: Accademia delle Scienze.

Manzoni, S. and Navale, M. T. Osservazioni del presentatore della nota di, su talune proprietà matematiche di un bassorilievo di Fidia. *Atti dell'Accademia delle Scienze di Torino* **114**, 1980, pp. 466-467. Turin: Accademia delle Scienze.

Manzoni, S., Navale, M. T. and Viola, T. 1980. Problemi geometrici applicati alle tecniche costruttive e rappresentative, con particolare riguardo al tunnel di Samo (Ipotesi di triangolazione topografica nel VI sec. a.C.). *Quaderni della Scuola di Disegno*, **80.3**: pp. 1-66. Turin: Università degli Studi di Torino.

———. 1985. Imago et mensura mundi. Pp. 505-514 in *Atti del IX Congresso Internazionale di storia della cartografia*. Rome: Istituto dell'Enciclopedia Italiana.

Navale, M. T. and Viola, T. 1980. Le volute joniche nei capitelli della Grecia classica: saggio di un'analisi strutturale oggettiva. *Atti dell'Accademia delle Scienze di Torino*, **114**: pp. 303-317. Turin: Accademia delle Scienze.

Navale, M. T. and Viola, T. 1982. Nuove ricerche sulle volute dei capitelli jonici della Grecia classica. *Atti e Rassegna tecnica della Società degli Ingegneri e degli Architetti in Torino*, **36**, 9: pp. 489-512. Turin: Società degli Ingegneri e degli Architetti.

———. 1985. Il profilo della Tavolozza di Narmer. *Atti dell'Accademia delle Scienze di Torino*, Cl. Scienze Morali, **119**: pp. 87- 94. Turin: Accademia delle Scienze.

———. 1986. *Memorie dell'Accademia delle Scienze di Torino*, Cl. Sci. Mor., **5-10**, pp. 3-29. Turin: Accademia delle Scienze.

Pauli, W. 1961. *Aufsätze und Vorträge über Physik und Erkentnis Theorie*. Braunschweig: Verlag Vieweg.

RAGGHIANTI, L. 1982. Recenti scoperte a proposito di arte e geometria. Quel cavallo di Fidia . . . *La Nazione*. Florence, 2 September 1982, p. 3.

RIZZI, B. and VIOLA, T. 1980. Dalla contemplazione ideale delle figure geometriche nell'uomo primitivo a quella della geometria razionale attraverso l'opera di Talete di Mileto. *Atti dell'Accademia delle Scienze di Torino*, **114**,: pp. 355-363. Turin: Accademia delle Scienze.

ROERO, C. S. 1981. La statica dall'arte alla scienza. *Le Scienze*, **150**: pp. 88-97. Roma: Gruppo Editoriale l'Espresso.

———. 1982. Statik zwischen Kunst und Wissenschaft. *Spektrum der Wissenschaft*, pp. 68-77. Stuttgart: Verlagsgruppe Georg von Holtzbrinck.

———. 1999. Mean, Proportion and Symmetry in Greek and Renaissance Art, Symmetry: Culture and Science. In *The Quarterly of the International Society for the Interdisciplinary Study of Symmetry* (ISIS-Symmetry). Special Issue: G. Darvas, ed. *Chapters from the History of Symmetry*. Vol. **9, 1-2**, pp. 17-47.

———. 2000. Media, proporzione e simmetria nella matematica e nell'arte, da Policleto a Dürer. Pp. 40-59 in L. Giacardi and C. S. Roero, eds. *Conferenze e Seminari 1999-2000*. Turin: Associazione Subalpina Mathesis.

———. 2005. Les symétries admirables de Guarino Guarini. In *Symétries, Contribution au sé minaire de Han-sur-Lesse*. Pp. 425-442 in P. Radelet de Grave, ed. *Réminiscences*, **7**. Turnhout: Brepols.

———. 2006. Il compasso nella geometria e nell'arte. In L. Giacardi and C.S. Roero, eds. *Matematica Arte e Tecnica nella storia. Atti del Convegno in memoria di T. Viola*. Foligno: Edizioni dell'Arquata.

VIOLA, T. 1986. Aspetti e problemi della matematica antica (Alcune proposte di un nuovo indirizzo di ricerca). Pp. 314-315 in *Atti del Convegno Storia degli studi sui fondamenti della matematica e connessi sviluppi interdisciplinari*, Vol. **1**. Rome: Tip. Luciani.

Chapter 5
Art and Mathematics Before the Quattrocento: A Context for Understanding Renaissance Architecture

Stephen R. Wassell

Introduction

In his classic *Architectural Principles in the Age of Humanism*, Rudolf Wittkower convincingly argues that an understanding of the roots of Renaissance architecture designed by masters such as Alberti and Palladio can be developed only by appreciating the relationships between architecture, music and mathematics as seen through the eyes of Renaissance architects and theorists (Wittkower 1998). Crucial to developing this appreciation is the ability approach the world of knowledge as Renaissance scholars would have, without inherently accepting the artificial division of this world into arts and sciences—and the compartmentalized disciplines within each.

Lionel March suggests "... the Renaissance might be called the era of conspicuous erudition in which patrons, scholars, and artists displayed their breadth of classical learning in various works and commissions" (March 1998: xii). The foundation of learning upon which artists of the Renaissance built was constructed through a determined search for reason in aesthetics, logic in beauty and rational explanations to intangible phenomena, a search involving at least implicit use of mathematics.

This is a vast topic, and I wish to state three restrictions from the outset. The focus is almost solely on Western cultures; it is restricted to literature in English; the emphasis is on recent literature. I have chosen selected topics in order to exhibit the innate human desire to rationalize aesthetics.

First published as: Stephen R. Wassell, "Art and Mathematics before the Quattrocento: A Context for Understanding Renaissance Architecture", pp. 157–168 in *Nexus III: Architecture and Mathematics,* ed. Kim Williams, Ospedaletto (Pisa): Pacini Editore, 2000.

S.R. Wassell (✉)
Department of Mathematics and Computer Science, Sweet Briar College, Sweet Briar, VA 24595, USA
e-mail: wassell@sbc.edu

K. Williams and M.J. Ostwald (eds.), *Architecture and Mathematics from Antiquity to the Future*, DOI 10.1007/978-3-319-00137-1_5,

Neolithic Speculation

When humans were able to intellectualize, but before recorded history, they developed modest skills concerning geometry and applied them to speculation. Unfortunately modern investigators can only speculate on those! An example of reasonable speculation is Tons Brunés' *The Secrets of Ancient Geometry and Its Use* (1967: vol 1, 19–108). The author develops a series of geometric constructions based on the role of the circle in the perceptible universe. He takes the natural steps of dividing the circle into four quadrants using a cross (thus marking the centre) and inscribing and circumscribing squares within and around the circle (Fig. 5.1). Brunés suggests that an early primitive observer would deduce the fact that the inscribed square has exactly half the area of the circumscribed square. The side–length of the inscribed square, which is equal to half of the diagonal-length of the circumscribed square, thus may have been of interest.

This is the basis for the principal and most well known of Brunés' constructions, the Sacred Cut (Fig. 5.2). Brunés makes a quite compelling argument concerning the importance of this construction in ancient geometry, based on the assumption that, at some point, a prehistoric observer must have chosen to work with a square of side 10, circumscribed about a circle of radius 5 (these numbers being natural choices for physiological reasons). If a compass point—or more probably, one end of a piece of twine—was placed in one corner of the square, and the other end of the "compass" at the centre of the circle, then the distance from the corner of the square to where the new arc cuts its side would measure 7 units (the exact measure is $5\sqrt{2}$, or about 7.07, but 7 is merely 1 % off from the exact). Thus, the inscribed square of Fig. 5.1 would have a perimeter of $4 \times 7 = 28$ units, magically reflecting the cycles of the moon. Brunés suggests that this may very well have been one of the reasons for the prominence of the number 7 in early writings, such as the Old Testament.

There is evidence that the Sacred Cut has been used as a foundation for architectural designs. Brunés' myriad examples include the Great Pyramid of Khufu, the Parthenon, and the Pantheon (Brunés 1967: vol 1, 123–147; vol 1, 301–310; vol 2, 38–56)[1]; more recent analyses involve the layout of a Roman housing complex in Ostia, and the Baptistery of San Giovanni in Florence (Watts and Watts 1986; Williams 1994). The Sacred Cut is perhaps the first component in the long history of using *ad quadratum* relationships in architectural design.[2]

Another important geometric construction that dates to prehistoric times is the *vesica piscis* or *mandorla* (literally, "fish bladder" or "almond"), formed by the intersection of two circles whose circumferences pass through each other's centres. Ubiquitous in Christian art, it has been called the "shape in architecture that

[1] The validity of Brunés analysis suffers somewhat from his exuberance.

[2] Design *ad quadratum* (by the square) and *ad triangulum* (by the equilateral triangle) are extensively discussed in March, *Architectonics of Humanism*; other examples are referenced below in the context of the Middle Ages.

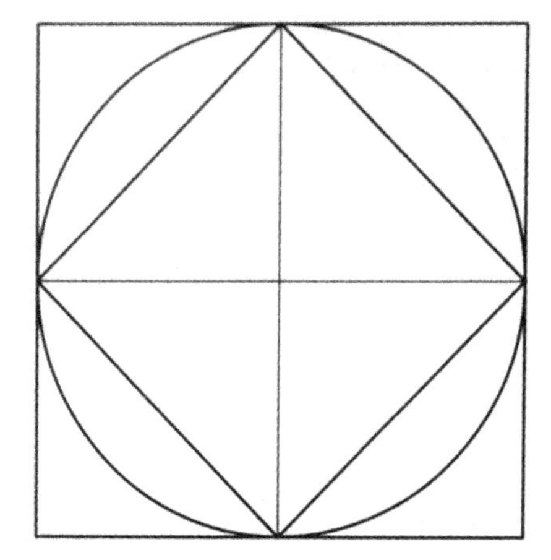

Fig. 5.1 An elementary geometric construction. Drawing: author

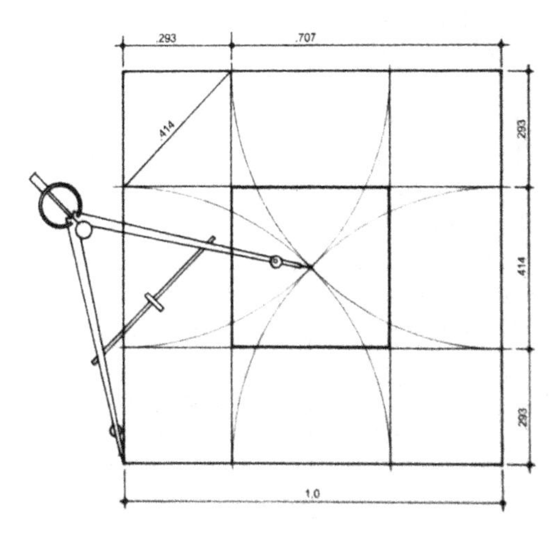

Fig. 5.2 The construction of the Sacred Cut. Drawing: author

symbolizes life, that represents the materialization of the spirit," (Hale 1994:76) and "the central diagram of Sacred Geometry for the Christian mysticism of the Middle Ages" (Lawlor 1982: 31).

> The light and energy emanating from Jesus in paintings of the Transfigurations are often manifest as a vesica piscis. Paintings showing a vision of the Virgin Mary also commonly employ this device, particularly at the moment of her Assumption (Speake 1994: 150; see also Schiller 1971 for numerous examples).

The top half of a *vesica piscis* may have been the original inspiration for the pointed arch.

I would like to speculate on a possible origin of the *vesica piscis* that seems so natural that it is surprising to me that nobody else has suggested it. Our prehistoric relatives associated the equilateral triangle with the number 3, the square with the

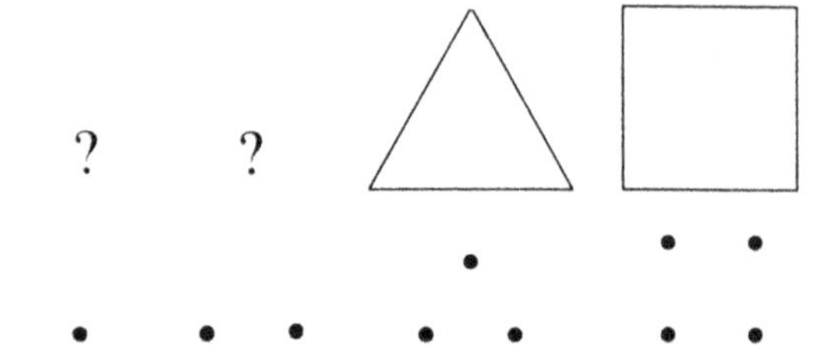

Fig. 5.3 The association of numbers 1 through 4 with geometric shapes. Drawing: author

number 4, thc regular pentagon with the number 5 and so forth (Fig. 5.3).[3] What shapes were associated with the numbers 1 and 2? One possible answer is that "neither monad, nor dyad exhibit [that] shape which is first to be found in the triad's triangularity, the copula of the monad and dyad" (March 1998: 32).

There is reason to believe, however, that the circle was associated with unity (Lawlor 1982: 12). While the triangle has three sides, the square four and so forth, the circle has one continuous (curved) side. Any shape with only two sides must necessarily involve at least one curve as well, since a polygon—the sides of which must be straight line segments—must have at least three sides. A semicircle would be a possible choice, albeit asymmetrical. Might the most suitable, most perfect, shape for the number 2 be found in the *vesica piscis*? It is constructed from two circles, i.e., two monads (if indeed the circle was associated with the number 1). The *vesica piscis* is also a natural bridge between the numbers 1 and 3, because the equilateral triangle results from its construction (Fig. 5.4); note that the first proposition of Euclid's *Elements* is based on this construction (Euclid 1956).

The megalithic stone circles of the British Isles provide evidence that neolithic cultures did use and appreciate simple geometric forms. The solar orientation of the giant circular structure at Stonehenge is unmistakable; its builders may have used it as an astronomical guide. The existence of the altar in the circle's middle provides evidence of ceremonial use. Perhaps its circular shape was as important aesthetically for ritual as computationally for astronomy. Benno Artmann relates that the aesthetics of pure geometric forms were appreciated by these cultures:

> About 390 neolithic carved stone balls of fist size dating from before or about 2000 B.C.E. have been found in Scotland. All of the five regular solids appear in these decorations, the dodecahedron on one specimen in the Museum of Edinburgh (Artmann 1999: 300–301).

While the exact usage of the megalithic circular structures may never be known, we may speculate that neolithic humans appreciated simple geometry for its own sake.

[3] I am well aware of the association of the point with 1, the line with 2, the equilateral triangle with 3, and the tetrahedron with 4; see, e.g., Christopher Butler, *Number Symbolism,* London: Routledge & Kegan Paul, 1970, pp 3–4. This association was undoubtedly quite attractive to classical scholars.

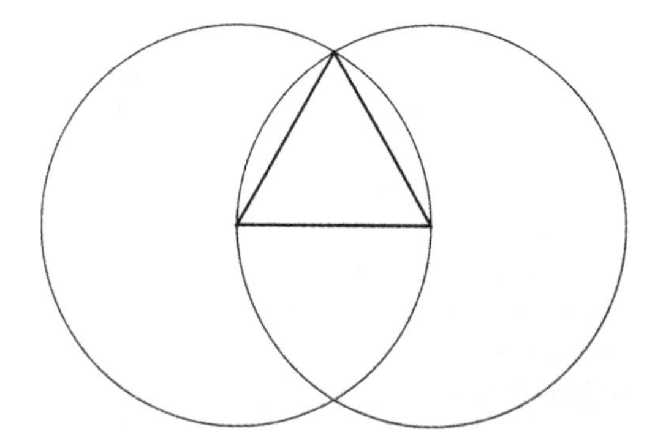

Fig. 5.4 An equilateral triangle constructed in the top half of a vesica piscis. Drawing: author

Beginnings of History in the Middle East

What mathematics is inherent in the sizable architectural bequest left to us by Egypt? The Great Pyramid of Khufu is the largest and best known of the famous collection of pyramids. Its proportions have been the subject of intense scrutiny. Brunés hypothesizes that the Sacred Cut was central to its design; others postulate the use of the Golden Section and ë.[4] What is certain about the pyramids is the use of pure geometric form to make a statement of grandeur. As in Stonehenge, this shows the inherent human fascination with simple geometry, especially on a grand scale.

In art and architectural decoration, figural and natural representations such as statues of humans or animals, or palm, papyrus and lotus plant forms, not themselves inherently mathematical, were often governed by mathematical principles. The prevalence of symmetry in Sumerian, Egyptian, Babylonian and Persian art exhibits an early rational approach to aesthetics (Weyl 1952: 8). Moreover, the rules of proportion by which the artists designed these biomorphic elements, especially human forms, shows once again the desire for reasoned guidelines in artistic endeavours. Looking at the evolution of rules of proportion from Egyptian times through the Renaissance, Erwin Panofsky has describes two types of proportional systems, "objective" and "technical" (Panofsky 1982: 55–107). An objective system defines the normal or ideal proportions that the artist must strive to capture, through whatever means possible; a technical system provides rules of construction which dictate what proportions are actually laid out on canvas or in stone.

Panofsky maintains that Egyptian art is the only example where both types of systems were used simultaneously. The possible conflicts between the two did not

[4] See e.g., Peter Tompkins, *Secrets of the Great Pyramid*, New York: Harper & Row, 1971, 189ff; John Michell, *The New View over Atlantis,* New York: Harper Collins, 1982, 144 ff.; and Mark Reynolds, "A Comparative Geometric Analysis of the Heights and Bases of the Great Pyramid of Khufu and the Pyramid of the Sun at Teotihuacan", *Nexus Network Journal*, vol. 1 (Florence: Edizioni Cadmo, 2000). The last of these contains not only a good review but also some fresh ideas on the subject.

arise essentially because of the simplicity of Egyptian art, which removed the need to depict oblique views, movement or foreshortening. The Egyptian artist needed to know the proportional relationships of humans and other animals, only in full frontal or profile view (or ground plan view, for sculpture such as sphinxes); these "objective" proportions were directly applied in a purely "technical" way. Panofsky cites direct evidence that this was achieved by means of a grid of equally sized squares, not merely to aid in transferring subject to canvas or stone but actually to construct the representation from the start. Rules of proportion became specifications for where the key points of the body landed on the standard grid: an excellent example of the innate human desire to create rational guidelines for beauty.

The First True Mathematicians

Of Greek art, in contrast to Egyptian art, Panofsky writes,

> Classical Greek art took into account the shifting of the dimensions as a result of organic movement; the foreshortening resulting from the process of vision; and the necessity of correcting, in certain instances, the optical impression of the beholder by 'eurhythmic' adjustments (Panofsky 1982: 62–63).[5]

Greek artists strove to ascertain "objective" proportions, and then applied them without being constrained to any specific "technical" system. This carried over into architecture as well, with the use of such optical refinements as column entasis and curvature of stylobates and architraves.

How were the correct objective proportions determined? Panofsky addresses this by translating a passage from Galen's *Placita Hippocratis et Platonis*.

> Chrysippus ... holds that beauty does not consist in the elements but in the harmonious proportions of the parts, the proportion of one finger to the other, of all the fingers to the rest of the hand, of the rest of the hand to the wrist, of these to the forearm, of the forearm to the whole arm, in fine, of all parts to all others, as it is written in the canon of Polyclitus (Panofsky 1982: 64).

He later states,

> With the sole exception of Plotinus and his followers, classical aesthetics identified the principle of beauty with the consonance of the parts with each other and the whole (Panofsky 1982: 68).[6]

The classical theory of proportion was intended to capture beauty, not just aid in the correct construction of art. Again, we have an example of the human search for reason in aesthetics.

---

[5] Panofsky cites a passage from Plato's *Sophistes* that directly supports these ideas.

[6] The reader who has studied the writings of Alberti or Palladio surely recognizes these ideas!

Studies of proportion (by which authors often mean "ratio" rather than an equality of ratios) were central to early Greek mathematics as well. Pythagoras is generally credited for discovering that string lengths related by certain commensurate ratios, when plucked, produce sounds that resonate with each other. The Pythagorean school also discovered, to their dismay, the irrationality of $\sqrt{2}$. This implies, for example, that the length of the diagonal of a 5 by 5 square is not 7, as our neolithic ancestors believed. For ease of calculation (and perhaps to maintain tradition), Greek mathematicians approximated irrational numbers with ratios.[7] This may be the basis for the practice of using rational approximations for important irrational numbers (e.g. 7/5, 10/7, or 17/12 for $\sqrt{2}$, and 7/4, 19/11, or 26/15 for $\sqrt{3}$) in art and architecture (March 1998). One major milestone in mathematics occurred when Archimedes calculated the true value of π to be between 3–10/71 and 3–10/70 (Artmann 1999: 272).[8]

Since Greek mathematicians were wed to the use of rational numbers, they were extremely interested in the general theory of proportions. The number of results concerning proportions in Euclid's *Elements* testifies to the importance of the subject in Greek mathematical history. The basic question is, under what conditions on s, t, u and v will the equality $s: t = u: v$ hold? Interesting relationships develop when the same variable occupies more than one slot, such as in the proportion $a: b = b: c$. This implies that b is the geometric mean of the two extremes a and c; Greek mathematicians were well acquainted with the arithmetic and the harmonic means as well. These three means, and several others, were discovered by Greek mathematicians to be generated by proportions involving only three variables; for example, the harmonic mean of a and c is given by b using the proportion $(b - a): (c - b) = a: b$.[9]

The Golden Section was known in early Greece as well. Called "the extreme and mean ratio," it is the result of a proportion involving only two variables: $(a + b): b = b: a$. In his *Elements*, Euclid says: "A straight line is said to have been cut in extreme and mean ratio when, as the whole line is to the greater segment, so is the greater to the less" (Artmann 1999: 104). Because the number of references to this construction is actually quite limited (Fowler 1987: 87), this tends to cast doubt on the plethora of analyses that purport to show the ubiquity of the Golden Section in Greek art and architecture.[10] Although the debate continues on both sides, many

---

[7] For an excellent treatment of this subject, see D.M. Fowler, *The Mathematics of Plato's Academy: A New Reconstruction*, New York: Oxford University Press, 1987.

[8] For different views of Archimedes' techniques, see D. H. Fowler, *The Mathematics of Plato's Academy*, New York: Oxford University Press, p. 31 ff. and Tobias Dantzig, *The Bequest of the Greeks*, New York: Greenwood Press, 1969, pp. 152–157.

[9] See Stephen Wassell, "Rediscovering (and Renaming) a Family of Means," Pp. 58–65 in *The Mathematical Intelligencer* **24**, 2 (2002), for a review of the rich history of means.

[10] Probably the most famous example is Jay Hambidge, *The Parthenon and Other Greek Temples: Their Dynamic Symmetry*, New Haven, Connecticut.: Yale University Press, 1924; see also Fredrik Macody Lund, *Ad Quadratum: A Study of the Geometrical Bases of Classic & Medieval Religious Architecture*, London: B.T. Batsford, 1921, vols. 1 and 2.

respected modern historians fairly adamantly reject the notion that the Golden Section was a driving force in art and architecture before the second millennium (Artmann 1999: 305; March 1998: 102; Frascari and Volpi Ghirardini 2015).

Not only Greek mathematics, but Greek metaphysics as well was highly influential to later cultures in regards to the relationships between art and mathematics. The association of gender to numbers, the infatuation with perfect numbers, triangular numbers, square numbers, etc., and the general imbuing of numbers with specific qualities was an integral part of Pythagorean "mathematics" (March 1998). Plato, who maintained that mathematics is a crucial tool in uncovering universal truths, played a significant role in inspiring others to study mathematics; his metaphysical use of mathematics continued the Pythagorean tradition (Fowler 1987: e.g. 106–108).

An example is Plato's assignment of the five Platonic solids to the four elements (fire, air, water, and earth) plus the universe. As cited above, these five regular polyhedra were known to ancient humans (Euclid showed that there were only five Platonic solids in Book XIII of his *Elements*). Plato hypothesized in *Timaeus* the divine association between the elements and the Platonic solids. For example, "*The first will be the simplest*, the tetrahedron, *which is the original element and seed of fire*" (Emmer 1993: 215–220 - Plato's words are shown in italics). Plato assigns the octahedron to air, the icosahedron to water, and the cube to earth; leaving the dodecahedron, "*which God used in the delineation of the Universe*" (Emmer 1993: 216). That Kepler, some two millennia later, would attempt to assign the Platonic solids to the planets, even though he was much better aware of the facts of planetary motion, testifies to the desire to rationalize the unknown through a combination of reason, tradition and aesthetics.

The Masters of Engineering

While classical Greek mathematicians generally had a strong distaste for the application of mathematics (except on metaphysical topics), the Romans held just the opposite view. They viewed mathematics not as an end unto itself but as a means of engineering. This would mean that rigorous mathematics (such as Euclid's *Elements*, for example) would be all but abandoned in classical Rome. The role of mathematics in architecture, however, was heightened by this non-theoretical, applied focus.

Roman architects greatly expanded the geometric palette upon which subsequent cultures would draw. Their widespread use of the arch system was central to this expansion. Influenced by the modest use of arches by Egyptians, Mesopotamians, and Greeks, of vaults by the Etruscans and of concrete by the Greeks, the Romans applied their engineering prowess to raise the use of the arch, vault, and dome to an art form (Trachtenberg and Hyman 1986: 116). Moreover, the use and appreciation of the arch system turned into a love affair with circular and semicircular forms in general, and this often manifested itself in plan as well as elevation. The Greek amphitheatre and *tholos* were sources of inspiration, of course, but the Romans freed the circle and semicircle to be applied as desired. Hadrian's Villa provides a wonderful example of the geometric freedom that evolved during the Roman era.

Of course the Romans were very careful to uphold established tradition when it came to the canons of architectural form. Nowhere is this more evident than in the treatise of Marcus Vitruvius Pollio. In his *De architectura*, Vitruvius carefully describes the origins of the three principal orders as derived from Greek temples, specifying the correct dimensions and proportions of the columns, capitals, entablatures, intercolumniations, doorways, etc., as well as the same elements from the Tuscan order as derived from Etruscan architecture. After detailing these mathematical rules, Vitruvius states:

> The laws which should govern the design of temples built in the Doric, Ionic, and Corinthian styles, have now, so far as I could arrive at them, been set forth according to what may be called the accepted methods (Vitruvius 1960: iv, vi, 120).

His goal is to quantify the beauty of the architecture passed down to the Romans. Vitruvius proceeds to detail the correct proportions for the length, width, and height of rooms using "symmetrical proportions," "so that the beholder may feel no doubt of the eurhythmy of its effect" (Vitruvius 1960: vi, iii, 175, 178).[11] The specifications governing design aesthetics found in *De architectura* became the authoritative source for centuries of great architects.

Roman influence on Western art is most prevalent in architecture. Their combination of the arch, vault, and dome system with the accepted orders of architecture became the standard practice of later cultures. With their engineering prowess, they were able to shape space elegantly; they taught later cultures to focus on the void rather than the mass of the building. Their strong adherence to symmetry in organizing spaces is another important aspect of their architectural heritage. Much of the success of Roman architecture depended on their rational approach to design aesthetics.

The Middle Ages

> At the borders between the classical and the medieval worlds we find St. Augustine and Boethius. Both of these transmitted onwards the more Pythagorean aspects of the philosophy of proportion, for they dealt with it chiefly in the context of musical theory (Eco 1986: 29–30).

Augustine introduced the concept of Christianity-as-philosophy, and in many ways medieval scholarship constituted an attempt to reconcile classical philosophy, essentially considered pagan, with the increasingly powerful church doctrine. The attention that Augustine paid to number symbolism sanctioned its continuation by

[11] For an illuminating discussion of Vitrivius's use of the words *proportio*, *symmetria*, and *eurhythmia*, see footnote 19 in Panofsy, *Meaning in the Visual Arts*, pp. 68–69.

his successors. Medieval number symbolism was heavily influenced both by biblical themes and by Pythagorean philosophy (Hopper 1938; see also March 1998: 115 for examples of early number symbolism in Judaism).

Boethius strove to keep classical knowledge alive, including the acceptance of the mathematical underpinnings of beauty, especially regarding the role of proportion in musical theory. His role in securing the prominence in medieval education of the *trivium* and *quadrivium* ensured the influence of mathematical thought in medieval philosophy. Umberto Eco points out in *Art and Beauty in the Middle Ages*,

> All of the medieval treatises on the figurative arts reveal an ambition to raise them to the same mathematical level as music. In these treatises, mathematical conceptions are translated into canons of practice and rules of composition, usually detached from the matrix of cosmology and philosophy, though united to them nevertheless by subterranean currents of taste and preference (Eco 1986: 41).[12]

Returning to Panofsky's analyses, in contrast to the Egyptian marriage of the "technical" and "objective" proportions, or the Greek stress only on "objective" goals, the medieval theory of proportions was based exclusively on "technical" proportions:

> Where the Egyptian method had been constructional, and that of classical antiquity anthropometric, that of the Middle Ages may be described as schematic (Panofsky 1982: 73).

This is seen in the "three-circle scheme" of Byzantine and Byzantinizing art (Fig. 5.5) or the Gothic system illustrated by French architect Villard de Honnecourt (Panofsky 1982: 84, 86–87).

Other mathematical aspects of medieval art include the common use of geometric elements, including symmetry, in triptychs and other altarpieces, in textile art, and in Romanesque sculpture (Bouleau 1963: 50). An important trend in medieval painting was the attention paid to frame design—both the overall frame and the framing of individual subjects within the painting—and geometry and symmetry were crucial in this regard. Moreover, to place the subjects within the frame(s), medallions—geometric constructions, often based on inscribing a regular polygon within a circle—were used to govern the layout of the key elements of the subject.

Surely the most impressive form of medieval art is found in its architecture, and mathematics was integral to its design and construction. Byzantine architecture, with its influences from the East, provided new geometric elements such as the centralized plan, the pointed arch and the use of pendentives, which would have significant effects on subsequent eras. The evolution of cathedral design from the Romanesque through Gothic periods showed an increasing dependence on geometry, including *ad quadratum* and *ad triangulum* design of plan, elevation,

---

[12] Examples include the *Painter's Manual of Mount Athos*, and Cennino Cennini's *Il Libro dell'Art*; see also Panofsky, *Meaning in the Visual Arts*, 74 ff.

Fig. 5.5 The "three-circle scheme" of Byzantine and Byzantinizing art. Drawing: author, after (Panofsky 1982)

and in details such as tracery (Lund 1921; Krier 1988; Artmann 2015: 15–26; Bispham 1996). That geometry was central to cathedral design is without doubt, given the existence of stonemason's marks—based on intricate geometric constructions—engraved on stones of various Gothic masterpieces (Fig. 5.6) (see Bouleau 1963 for numerous examples). Even castle design was based on strong geometric influences (Götze 2015: 62–80).

The Dark Ages were just that for pure mathematics in Western cultures. Progress was slowly being made in quantitative reasoning, and the conflict between this and qualitative reasoning characterized the medieval psyche. By the beginning of the Renaissance,

> The West was making up its mind (most of its mind, at least) to treat the universe in terms of quanta uniform in one or more characteristics, quanta that are often thought of as arranged in lines, squares, circles, and other symmetrical forms: music staffs, platoons, ledger columns, planetary orbits (Crosby 1997: 10–11).

This did not all occur at once. Indeed, the modest gains in mathematics and science in the West, which were partially aided by influences from the Near East, through Leonardo Pisano (Fibonacci) in no small part, had to come in spite of church and royal dogma. While inventions such as water mills and windmills depended on increasing sophistication with levels, gears, and wheels, the spiritual, even mystical, approach of medieval leadership hindered faster progress (Crosby 1997: 52–53).

Conclusions

During the Middle Ages, the qualitative view of neo-Platonic metaphysics slowly gave way to a more quantitative view of reality that would eventually allow science to progress at a steady pace. This transition was slow, and Renaissance scholars

Fig. 5.6 Stonemason's mark from Strasburg Cathedral. Drawing: author, after (Bouleau 1963)

were still heavily influenced by long-held ideas on number symbolism and sacred geometry. As did their predecessors, Renaissance artists sought to incorporate meaning into their design by using a rational approach towards aesthetics. Some of the most prominent theorists—Barbaro, Pacioli, and Dürer –focused their efforts largely on concerns of geometry and proportion, often doing so within the context of the ideas described above. Indeed, this was completely natural before the divorce of arts and sciences in the Age of Reason! It is crucial, therefore, to be able to put oneself in this same context if one is truly to understand Renaissance architecture.

Acknowledgments This research has been partially supported by grant #9846 from the Graham Foundation for Advanced Studies in the Fine Arts, as well as by the Sweet Briar College Faculty Grants program.

Biography Stephen R. Wassell received a B.S. in architecture in 1984, a Ph.D. in mathematics (mathematical physics) in 1990, and an M.C.S. in computer science in 1999, all from the University of Virginia. He is a Professor of Mathematical Sciences at Sweet Briar College, where he joined the faculty in 1990. Steve's primary research focus is on the relationships between architecture and mathematics. He has co-authored various books, one with Kim Williams entitled *On Ratio and Proportion* (a translation and commentary of Silvio Belli, *Della proportione et proportionalità*); one with Branko Mitrović entitled *Andrea Palladio: Villa Cornaro in Piombino Dese,* and another with Kim Williams and Lionel March entitled *The Mathematical Treatises of Leon Battista Alberi.* Steve's overall aim is to explore and extol the mathematics of beauty and the beauty of mathematics.

References

ARTMANN, Benno. 1999. *Euclid: The Creation of Mathematics*. New York: Springer.

———. 2015. The Cloisters of Hauterive. Pp 453–466 in Kim Williams and Michael J. Ostwald eds. *Architecture and Mathematics from Antiquity to the Future: Volume I Antiquity to the 1500s*. Cham: Springer International Publishing.

BISPHAM. Michael. 1996. *Platonic Geometric Atomism in Medieval Design*. Originally sourced from http://www.fupro.com/plat/index.htm. Archived copy available at http://www.luckymojo.com/atomism.html

BOULEAU, Charles. 1963. *The Painter's Secret Geometry: A Study of Composition in Art*. New York: Harcourt, Brace & World.

BRUNÉS, Tons. 1967. *The Secrets of Ancient Geometry and Its Use*. Copenhagen: Rhodos.

BUTLER, Christopher. 1970. *Number Symbolism*. London: Routledge & Kegan Paul.

CROSBY, Alfred W. 1997. *The Measure of Reality: Quantification and Western Society, 1250–1600*. Cambridge: Cambridge University.

DANTZIG, Tobias. 1969. *The Bequest of the Greeks*. New York: Greenwood Press.

ECO, Umberto. 1986. *Art and Beauty in the Middle Ages*. Hugh Bredin trans. New Haven, Connecticut: Yale University Press.

EMMER, Michele. 1993. Art and Mathematics: The Platonic Solids. Pp. 215–220 in *The Visual Mind: Art and Mathematics*. Cambridge, Massachusetts: MIT Press.

EUCLID. 1956. *The Thirteen Books of the Elements*, trans. Sir Thomas L. Heath. New York: Dover.

FOWLER, D.M. 1987. *The Mathematics of Plato's Academy: A New Reconstruction*. New York: Oxford University Press.

FRASCARI, Marco and LIVIO VOLPI GHIRARDINI. 2015. Contra Divinam Proportionem. Pp 619–626. in Kim Williams and Michael J. Ostwald eds. *Architecture and Mathematics from Antiquity to the Future: Volume I Antiquity to the 1500s*. Cham: Springer International Publishing.

GÖTZE, Heinz. 2015. Friedrich II and the Love of Geometry. Pp 423–436. in Kim Williams and Michael J. Ostwald eds. *Architecture and Mathematics from Antiquity to the Future: Volume I Antiquity to the 1500s*. Cham: Springer International Publishing.

HALE, Jonathan. 1994. *The Old Way of Seeing*. New York: Houghton Mifflin.

HAMBIDGE, Jay. 1924. *The Parthenon and Other Greek Temples: Their Dynamic Symmetry*. New Haven, Connecticut: Yale University Press.

HOPPER, Vincent Foster. 1938. *Medieval Number Symbolism: Its Sources, Meaning, and Influence on Thought and Expression*. New York: Columbia University Press.

KRIER, Rob. 1988. *Architectural Composition*. New York, Rizzoli.

LAWLOR, Robert. 1982. *Sacred Geometry: Philosophy and Practice*. London: Thames and Hudson.

LUND, Fredrik Macody. 1921. *Ad Quadratum: A Study of the Geometrical Bases of Classic and Medieval Religious Architecture*. London: B.T. Batsford.

MARCH, Lionel. 1998. *Architectonics of Humanism: Essays on Number in Architecture*. London: Academy Editions.

MICHELL, John. 1982. *The New View over Atlantis*. New York: Harper Collins.

PANOFSKY, Erwin. 1982. The History of the Theory of Human Proportions as a Reflection of the History of Styles. *Meaning in the Visual Arts*. Chicago: University of Chicago Press.

REYNOLDS, Mark. 2000. A Comparative Geometric Analysis of the Heights and Bases of the Great Pyramid of Khufu and the Pyramid of the Sun at Teotihuacan. *Nexus Network Journal*, vol. 1. Florence: Edizioni Cadmo.

SCHILLER, Gertrude. 1971. *Iconography of Christian Art*. Greenwich, Connecticut: New York Graphic Society.

SPEAKE, Jennifer. 1994. *The Dent Dictionary of Symbols in Christian Art*. London: J.M. Dent.

TOMPKINS, Peter. 1971. *Secrets of the Great Pyramid*. New York: Harper & Row.

TRACHTENBERG, Marvin and Isabelle HYMAN, 1986. *Architecture: From Prehistory to Post–Modernism*. Englewood Cliffs, New Jersey: Prentice Hall.

VITRUVIUS, Marcus (Pollio). 1960. *De architectura (The Ten Books on Architecture)*. Morris Hicky Morgan, trans. New York: Dover.

WASSELL, Stephen. 2002. Rediscovering (and Renaming) a Family of Means. *The Mathematical Intelligencer* **24**, 2 (2002): 58–65.

WATTS, Donald J. and Carol Martin WATTS. 1986. A Roman Apartment Complex. *Scientific American*. **255**, 6 (1986): 132–140.

WEYL, Hermann. 1952. *Symmetry*. Princeton, New Jersey: Princeton University Press.

WILLIAMS, Kim. 1994. The Sacred Cut Revisited: The Pavement of the Baptistery of San Giovanni, Florence. *The Mathematical Intelligencer*, **16**, 2 (1994): 18–24.

WITTKOWER, Rudolf. 1998. *Architectural Principles in the Age of Humanism*. Chichester, West Sussex: Academy Editions.

Chapter 6
The Influence of Mathematics on the Development of Structural Form

Holger Falter

Introduction

To analyse the influence of mathematics on architecture from antiquity up to today, the difference between the criteria which have always been the basis for the design process must be recognized. Such criteria are the geometry, the structure's function, the loadbearing behaviour of the structural elements and of the structure, the manufacturing technique and the choice of materials, the lighting as well as the interior and exterior decoration (Falter and Mirabella 1996). Lately, ecological and economical aspects have been added.

Although all these terms are well known, each era assigned different meanings to them. While modern man understands the term "function" as purely meeting the primary task, i.e. a bridge as a means to cross a valley, or walls and a roof as protection from external influences, the definitions of past eras far exceeded this understanding. A structure not only had to be of material usefulness—similar to today's idea of function—but also had to have a psychologically beneficial and intellectually fruitful influence.

The religious structures before the Industrial Revolution were to form the connection between man and divinity. Geometry was an important tool, as was the choice of forms and proportions that relate the structural elements to each other. By creating a pleasant atmosphere through harmonic proportions, they were to have a psychologically beneficial influence. Form and proportion combined with the symbolism of measurement and figure were to remind the visitor of the spiritual background. However, the choice of the geometry and the measurements was

First published as: Holger Falter, "The Influence of Mathematics on the Development of Structural Form", pp. 51–64 in *Nexus II: Architecture and Mathematics,* ed. Kim Williams, Fucecchio (Florence): Edizioni dell'Erba, 1998.

H. Falter (✉)
Hochschule für angewandte Wissenschaften Coburg, Postfach 1652, 96406 Coburg, Germany
e-mail: falter@hs-coburg.de

K. Williams and M.J. Ostwald (eds.), *Architecture and Mathematics from Antiquity to the Future*, DOI 10.1007/978-3-319-00137-1_6,

limited by the technically feasible and the loadbearing behaviour of the selected structural elements and the chosen materials, which in turn are closely connected to the choice of manufacturing procedure. The manufacturing procedure also dictated the form of the structural elements and consequently the geometry and the overall appearance of the structure.

Mathematics combined with technology and mechanics is a part of every design criterion. Each era applied this very differently to the structures, based on the respective cultural understanding.

Pythagoras' Intellectual Function of Mathematics and Its Effects on Structures

The origin of the numeral's meaning may certainly not be found in Pythagoreanism, but it provides an essential basis for comprehending some developments in structural history. The important Pythagorean aspect for structural engineering is the connection between the qualitative—the sound—and the quantitative—the chord of the monochord. Pythagoras transferred the audible (the qualitative) to numbers (the quantitative). However, this also means that the object, the monochord, attains a psychic and spiritual value through the number expressed by the sounds and intervals (Kaiser 1991). The chord's measurement is reflected by the value of the sound and vice versa. Thus, following the harmonic theory, a structure may express a spiritual meaning. Furthermore, material objects provided man with an entrance to the spiritual body of thought and to the world of ideas. Through the harmonic design of the form, Pythagoreanism attempted to create the connection with intellectuality.

Up to the period of the Renaissance master builders endeavoured to shape the structures according to various harmonic rules in order to manifest the qualitative within the form. However, what was soon lost was the flip-side of this relationship: recognizing ideas through the means of the form.

With Pythagoras and Plato, natural numbers have symbolic meaning based on the observation of nature and the ideas about the origin of the world and mankind. A symbol stands for something which cannot be described with words. Plato formed the term "idea" and connected it to numbers. By recognizing the symbolic meaning of the numbers, man should be able to experience these ideas. Plato views the symbol as a way to make ideas accessible; materializing these numbers in a form so people may recognize them. The number, hidden in the harmonic theory or the geometrical form, made ideas accessible. Plato, Euclid and Vitruvius describe such geometrical forms. The square, the cross and the circle especially affect structural engineering and consequently the development of structures.

In all cultures with an architectural history, the square and the cube were assigned to the material or the element "earth" and therefore the number 4. Plato assigned the cube, consisting of 6 squares, to the element "earth." Vitruvius considers the number 4 to be man's number since, with outstretched arms, his width equals his height, thus marking the height and width of a ideal square (Vitruvius 1960: III i, paragraph iii 72–73). The Middle Ages distinguished between four elements: 4 major winds, 4 seasons and 4 continents (Eco 1991). Not only the square but also the cross represented Fig. 6.4 and therefore the material and earth-bound aspect throughout several pre-Christian cultures. Aside from the square and the cross, the circle and the sphere are significant. As an infinite line, the circle may symbolize anything infinite: time, eternity, infinity. The circle symbolizes the periodicity of human life and the laws of nature. In architecture the square represents exoteric human existence on earth, while the circle represents eternity, intellectuality.

From ancient times to the period of the Renaissance, the circle, the square and the cross were the basic elements of the design process. The Pantheon in Rome is an impressive example for the use of the circle and the sphere (Fig. 6.1).

In elevation, a perfect sphere may be embedded in the interior, which is circular in plan. The circle determined the plan and the interior vertical wall as well as the hemisphere on top. The loadbearing structure had to absorb the dome's shear. This resulted in a 6-m-thick rotunda guided to the outside across the dome's base, creating the Pantheon's depressed outer shape.

The Pantheon shows how a construction must fulfil the desire for a certain interior shape. The interior shape and the spiritual function of the structure are predominant. The structure is only the necessary means to accomplish this function.

All three basic shapes—the cross, the square and the circle—are used in the Hagia Sophia (Fig. 6.2).

The church's outline is almost square. The interior is a cross with a longitudinal axis emphasized by the support-free nave. The circular dome rises over the crossing. The structure must accomplish the transfer from the square crossing, the symbol for everything earthly, to the circular base of the dome, representing everything spiritual. The dome's form and crown were chosen to have the most favourable influence on the loadbearing sub-structure (Heinle and Schlaich 1996: 30–32).

These examples portray the desire for a certain basic form of a structure which is especially prevalent in the interior. The structure had to meet these requirements and substantially influenced the exterior form. Loadbearing behaviour and material characteristics further limited the choices.

The use of the forms described above continued throughout the Middle Ages. In the Byzantine Empire cross-plan churches similar to the Hagia Sophia were erected over a square plan with a cross-shaped interior and a central dome over the crossing. The naves and aisles of Gothic and Romanesque basilicas are a sequence of square bays dictating the naves' width and length.

Especially in the Middle Ages, but also in the Renaissance, the harmonic theory was applied to pursue the spiritual and psychological function described above. Examples are the Gothic cathedrals or the structures by the famous master-builder Palladio.

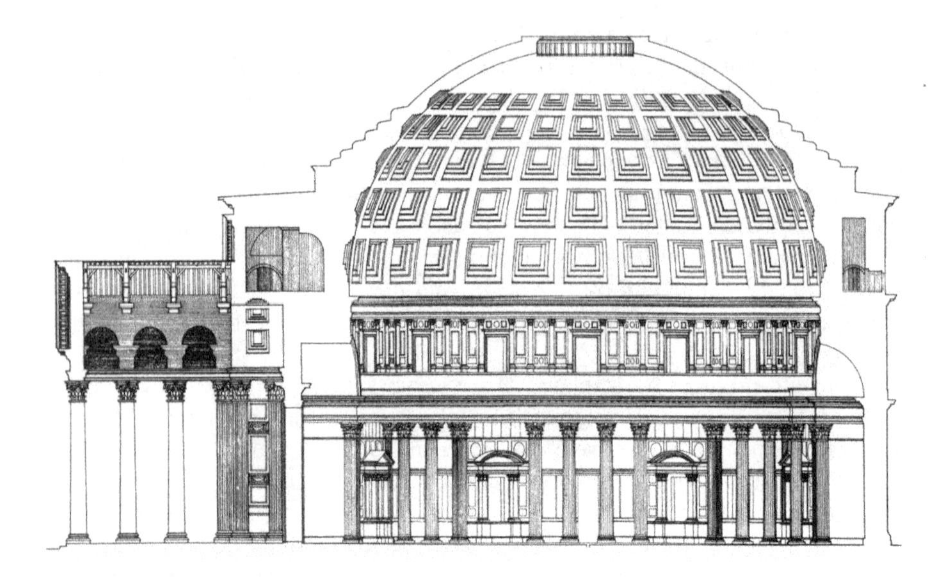

Fig. 6.1 Cross-section of the Pantheon in Rome. Drawing: author

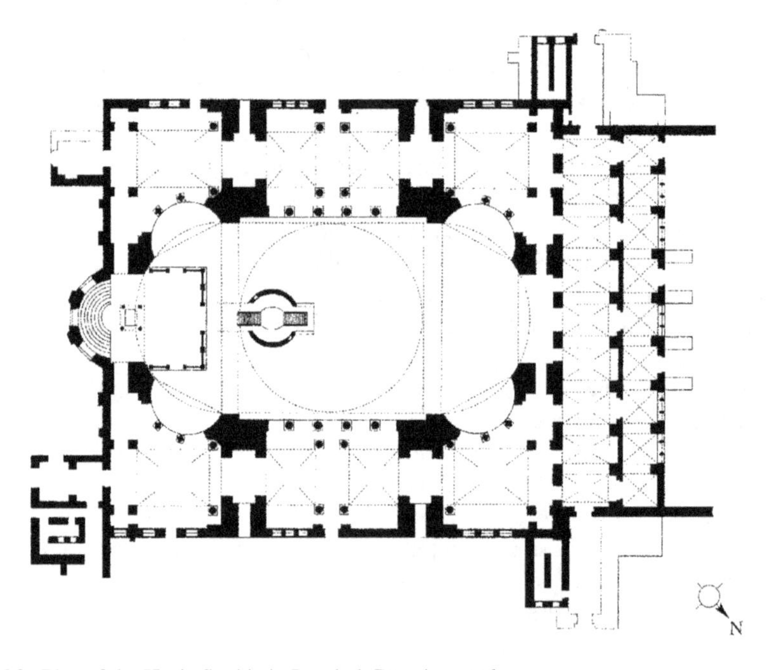

Fig. 6.2 Plan of the Hagia Sophia in Istanbul. Drawing: author

Visually Comprehending the Mechanics to the Empirical Rule

In order to realize their ideas of form, the master-builders had to use structural elements such as columns, arches, hemispheres, walls and piers without being able to mathematically describe their mechanical behaviour. One reason was the fact that the development of mechanics was not directly dependent on the natural sciences because research was aimed at enlightenment and not on practical application. Mechanics could only be experienced visually. Deformations, cracks and other damage, even up to the collapse, could only hint at the mechanical behaviour and facilitated conclusions in view of the necessary dimensioning of a certain structural element.

For example, the relations between span, vertex-height, thickness of the arch and of the column could be experienced during the construction of vaults by "trial and error." This knowledge about the mechanical interrelations remained a well-kept secret among the master-builders up to the period of the Renaissance. The increase in scientific activity during the Age of Humanism created empirical rules of thumb, i.e., Alberti who related by means of numerical ratios the thickness of the arch and of the column to the span of stone arch bridges (Alberti 1965) (Fig. 6.3). This development replaces Pythagoras' mathematics, based originally solely on the spiritual meaning of numbers and the musical harmonies, with a purely practical functional handling of the numbers.

From the Qualitative Assessment to the Quantitative Dimensioning of the Structure

Although the huge domes of the Renaissance refer to the harmonic theory in the choice of proportions, the sheer size of these domes demanded an increasing attention to the influence of the loadbearing behaviour on the design and the construction. Brunelleschi, in view of the large loads, deviated from the antique examples and placed a steep lantern on his dome in Florence. He reduced the mass of the dome by using a two-tier dome.

Shortly after the completion of Santa Maria del Fiore and long before the construction of the dome of St. Peter's in Rome, the first cracks appeared in the dome in Florence. In the same period, Leonardo da Vinci studied crack formation in walls, the behaviour of vaults and the loadbearing behaviour of domes. In the *Codex Arundel* (1500–1506) Leonardo explained the causes for cracks in domes. He realized that the joints will open just like an orange whose peel is cut into meridional strips and pressed at the top (di Teodoro 1989: 33–34). This led to the discovery that an excessive single load at the top of a dome will result in meridional cracks. Leonardo concluded that a dome with a large single load would have to be as steep as possible and its base would have to be supported at the sides (Fig. 6.4).

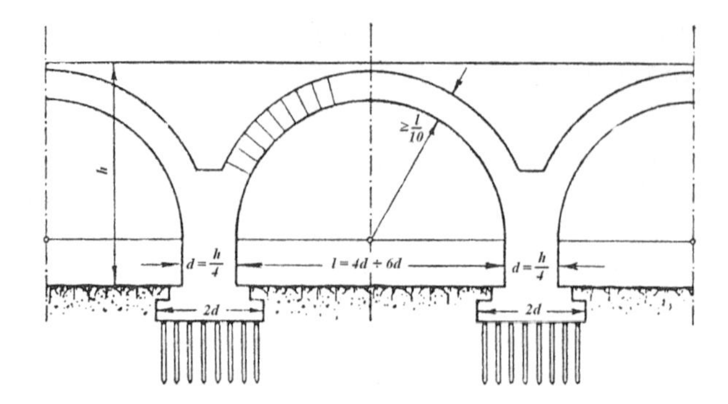

Fig. 6.3 Arch bridge according to Leon Battista Alberti. Image: author, after (Straub 1992: 129, Fig. 32)

Fig. 6.4 Study of a steep dome with a large single load by Leonardo da Vinci, *Codex Trivulziano,* fol 20v

Leonardo's work proves that the humanistic holistic account in writing and in drawings took the place of scientific proof and mathematical formulation.

In 1675, Robert Hooke published the first theory on form-finding recognizing an analogy between the form of the chain and the pressure line of an arch. According to his diary, Hooke was in close contact with Sir Christopher Wren, the architect and builder of St. Paul's Cathedral. Therefore it may be assumed that Hooke inspired Wren to form the supporting dome shell in the form of the inverse geometry of a relevantly loaded sagging cable/rope. The supporting shell is conical at the bottom with a slight external belly and levels off just below the lantern, thus corresponding to the load introduced by the heavy lantern and the weight of the exterior light shell (Heinle and Schlaich 1996: 119–119).

The supporting shell's profile proves that Wren understood the theory of the pressure line and drew the correct conclusions from the damages to the shallow dome of St. Peter in Rome (Dorn and Mark 1981). But this design method was still based on experiments and could not yet be mathematically described. Leonardo da Vinci's purely intellectual connection between the necessity to transfer dead load and the dome's form became qualitatively depictable and comprehensible by experiment through Hooke's realizations. Apparently the result of this design process failed to correspond to the traditional ideas of a large cathedral and therefore the structure itself was encased internally and externally, thus separating the structurally determined form from the desired, visual architectural form.

The dome of St. Peter's in Rome is a steep dome designed by Michelangelo and altered by the master-builders Giacomo della Porta and Domenico Fontana. As with the dome in Florence, the choice of structure was heavily influenced by the most favourable load transfer. Three peripheral tie beams of iron were to assist in absorbing the dome's shear. Despite these conclusive measures, cracks were reported as early as in 1631. In 1742, three mathematicians, Tomaso Le Seur, Francesco Jacquier and Ruggero Boscovich, were requested to provide an expert opinion as to the cause for these alarming cracks in the dome. This analysis was published in 1743 (Fig. 6.5).

The authors explored new ground and produced history's first statical proof, resulting in the dimensioning of a structural element. They attempted to determine the horizontal shear and prove that the three peripheral tie beams were inadequate to absorb this force by applying a method probably most closely resembling today's statical principle of virtual translation.

In 1743, Giovanni Poleni was also requested to investigate the dome's condition (Fig. 6.6). His results were published in 1748. Following Hooke's findings, Poleni used a sagging chain loaded with balls and attempted to draw conclusions about the dome's behaviour. From the shape assumed by the freely sagging chain, he deduced that the lantern was too heavy for the profile of the vault. To avoid further cracks, he concurred with the three mathematicians and proposed reinforcing the dome with additional peripheral tie beams.

The two methods described above were totally different: the three mathematicians with their mathematically determined method and Poleni with his solely qualitative descriptive method in the tradition of Leonardo and Hooke. In consequence, Poleni heavily criticized the work of his colleagues; he writes, *Buonarroti non sapeva di Mathematica, e poi sempre seppe architettare la Cupola … Perchè appunto ho grandissima stima di questa Scienza, altremente me ne dispiace il suo abuso* (Michelangelo was unfamiliar with mathematics and could still build the dome. … Mathematics is an esteemed science but was abused in this case) (Straub 1992).

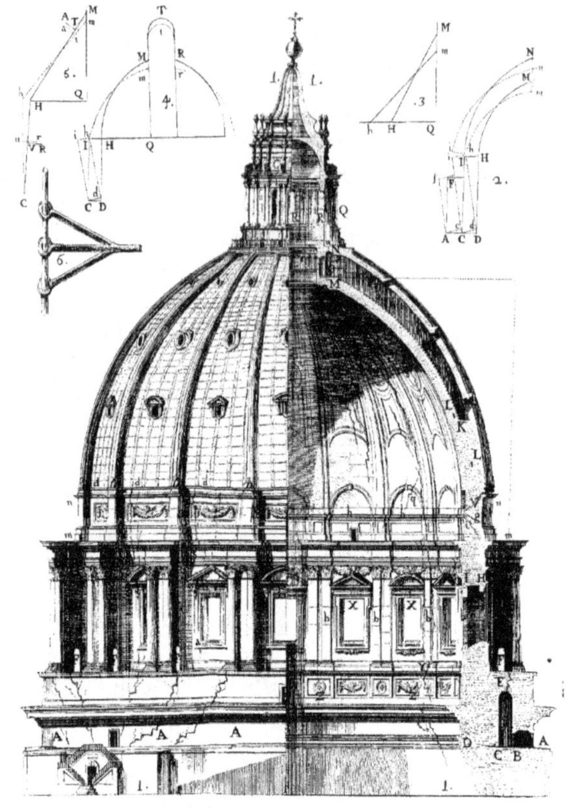

Fig. 6.5 Copperplate illustration from Le Seur, Jacquier and Boscovich, *Parere di tre mattematici sopra i danni che si sono trovati nella Cupola di S. Pietro, 1742*. Next to the dome with the sketched damages the plate shows in the upper right-hand corner the graphic drawing, including the cracks as joints

The Mathematical Calculation of Structures

The effects of the French Revolution and the Industrial Revolution influenced the political climate of the nineteenth century. Liberalism, strengthened by the French Revolution's spirit of enlightenment, aimed at the rule of rationalism and a free economic system. The tremendous progress in natural science and technology and the achievements of the Industrial Revolution resulted in an optimistic belief in man's almost unlimited ability to shape the world.

Auguste Comte's (1798–1857) positivism, founded on Newton's work, based mankind's progress on raising the understanding to the positive level (i.e. the scientific level), the highest possible for the human mind (Comte 1994). In this scientific state an observation leads to a rule, the search for the ultimate cause is abandoned and the interest is aimed towards the existing facts. To mathematics is assigned the highest possible positivistic degree because it is ascertainable by pure reason and does not require interpretation. The meaning of positive is the veritable and the utilitarian; consequently, this implies the separation of theory and practical application.

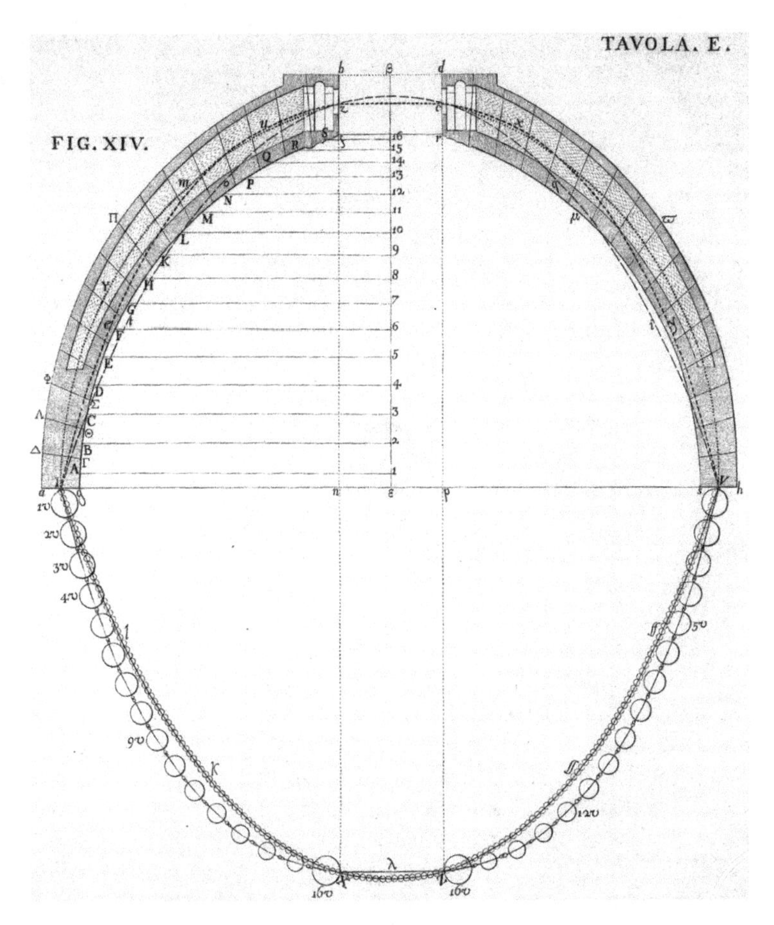

Fig. 6.6 Reconstruction of St. Peter's pressure line by Giovanni Poleni, from *Memorie istoriche della gran cupola del tempio vaticano,* 1748

Structural engineers gratefully accepted the theoretical knowledge of natural science, translating it into practical applications. Mathematics, as far as structural engineering was concerned, took two different routes. The new material iron was inextricably linked with this development by concentrating forces in linear loadbearing members. It is not difficult to see that the mechanics of such structures were more easily described mathematically than those of the earlier, massive masonry structures. At first this new material was applied only in stone bridge construction by replacing the massive stones with parallel connected iron bars. Although technological development had already surpassed this level, it was only very slowly applied to structures. Therefore, it is impossible to relate the development in mathematics directly to the structures. However, some relations are to be explained.

At the beginning of the nineteenth century, the development of mathematical methods went in two major directions: the so-called "graphical statics" and the "analytical statics". Based on the geometry by Sebastiano Serlio, Albrecht Dürer, Girard Desargues and Rene Descartes, surveyor Gaspard de Monge (1746–1818) and Jean Victor Poncelet (1788–1867) developed projective geometry. In Germany Karl Georg Staudt refined the scientific principles of projective geometry, thus creating the basis for graphic statics (Kurrer 1994: 79–86). The Swiss Karl Culmann (1821–1881), the Italian Luigi Cremona (1830–1903) and the German Wilhelm Ritter (1847–1906) applied the knowledge of graphic statics to develop a complex graphic statics facilitating the analysis and dimensioning of complicated three-dimensional structures. The method of graphic statics created the possibility to visually conceive the flow of forces, thus uniting the design and construction processes. The graphic method was especially suited for statically determined structures consisting of linearly hinged iron bars; these prerequisites characterized the "industrial architecture" with its roofs and bridges.

The shape of the Pauli-girder (fishbelly girder), named after its inventor August von Pauli (1802–1883), keeps the forces constant across the entire length of the upper and bottom chord (Fig. 6.7). In the trusses developed a few years later (1867) by Johann Wilhelm Schwedler (1823–1894) the diagonals are only tension-stressed (Straub 1992). The loadbearing behaviour and the mathematical method used for the calculation determined the design in each of these three cases.

While the Pauli and Schwedler girders were used in bridge construction, the three-hinge truss-arch is the loadbearing element in the final quarter of the nineteenth century. This three-hinge truss-arch separated the wall from the roof. Rooms were created by adding arches and longitudinally connecting them with purlins and glass roofs. For over three decades this was the basic structure of numerous train-stations in England and Germany and of a multitude of industrial sheds.

By the end of the nineteenth century graphic statics reached its peak and the gradual acceptance of analytical methods already caused its slow demise. The new material ferroconcrete especially contributed to the favoured use of analytical methods in statical calculations by the end of the nineteenth century. The Frenchmen Navier (1785–1836), Cauchy (1789–1857) and Clapeyron (1799–1864), major advocates of this method, treated the theory of mechanics and strength in a strictly mathematical way, completely rejecting graphic descriptions. This might explain why their methods were not practically applied but consequently pursued further. The breakthrough came with Müller-Breslau's (1851–1925) flexibility matrix method published in 1904, the first major procedure to calculate statically indeterminate member systems (Kurrer 1987: 1–8).

Tension and compression strength were characteristics of ferroconcrete, and it could be shaped almost at will. Forming hinges was simple with iron, but difficult with ferroconcrete. It created statically indeterminate frames continuing through several spans, a difficulty only surmountable by using analytical methods. This was the foundation for constructing the first major story frames, especially in the United States.

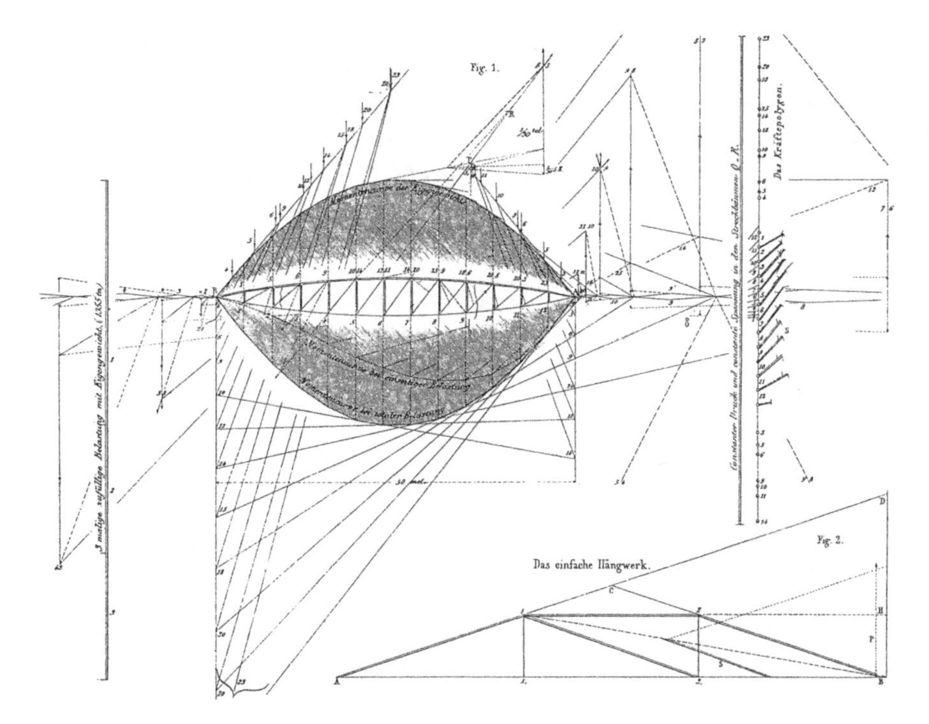

Fig. 6.7 Static calculation of the "Pauli girder" with the graphic method by Karl Culmann. Image: (Culmann 1866: Tafel 18)

This change in paradigm, from graphic to analytical statics, caused the loss of the analogy between design, calculation and construction, threatening the flow-of-forces-oriented design of structures as dictated by the graphic procedure. Only a limited number of statically indeterminates could be solved with these new procedures, which is probably the reason for the mostly simple structures causing as little calculating effort as possible. The structures by Pier Luigi Nervi or Maillard's bridges are examples which were partly calculated using the graphic method.

Computer-Oriented Mathematical Procedures Facilitate "Any" Structure

The Finite Element Method created a new fundamental development in the 1970s. Although its basic method was well-known as the translation method, it gained significance as a calculation procedure only through efficient computers. Almost any structure could be calculated using this method. At the onset of this rapid development its application had to be reduced to reasonable measure. The method of finite elements was only partially applied to flow-of-forces-oriented structural designs, resulting in beautiful, constructively sophisticated structures such for the 1972 Olympic Games in Munich. However, even more efficient computers could deal with almost any problem occurring in structural engineering. Though complex designs could be executed such as membrane or cable structures, the constructive aspect of the design process might be neglected at the same time, resulting in the construction of structurally inferior designs based on calculation results. Design, calculation and construction are drifting further apart and very often the structures are proof of it. This "the-sky-is-the-limit-calculating" creates an unprecedented variety of different structures. But this freedom necessitates an increased consideration for the design and its effects on man and the environment.

Conclusions

There is a direct link between the function of mathematics in the story of its culture and architecture. Pythagoras and Plato attributed primarily symbolic and spiritual meaning to numbers. Consequently antique structures were designed to meet the spiritual requirement, to connect man with divinity. The end of the Renaissance brought the application of mathematics to the dimensioning of structural elements. Until then there were only rules of thumb, expressed by numbers taking the visually conceived mechanics into consideration. In the Renaissance the task of mathematics consisted of contributing to the structure's pleasant form determined by the dimensions designed with aid of the harmonic theory. Although the loadbearing behaviour gained increasing consideration in the design, it was more an intuitive than a mathematical procedure. The simultaneous development in mathematics and natural science found no practical application in structural engineering at first. At the same time mathematics was completely losing its spiritual background and the harmonic theory fell into oblivion. The Industrial Revolution and the emergence of the material iron resulted in the practical application of mathematics in structural engineering by facilitating the description of the structure's mechanical behaviour and the dimensioning of structural elements. But the necessary idealization of the structures leads to a replacement rather than to a description of the actual structure. Today the method of finite elements combined with the computer results in an improved comprehension of the structure. But not only reasonably optimized structures are created, but also "everything possible" might be realized. These examples show that the application of the new method requires intensive reflection about the possibilities and the demands of designing.

Biography Holger Falter studied structural engineering at the University of Stuttgart. After completing his *Diplom* degree, he continued his studies and at the University of Stuttgart and the Politecnico di Milan in Italy. In 1998 he received his Ph.D. in Engineering at the University of Stuttgart. He then joined Arup and worked as a structural engineer on large international projects in their offices in Berlin, London and Dublin until 2010. He has published several articles on the conceptual design of structures and other related topics. Between 2007 and 2011 he taught as a part-time lecturer at University College Dublin. In 2010 he was appointed Professor at the school of design of Coburg University of Applied Sciences and Arts.

References

ALBERTI, Leon Battista. 1965. *Ten Books on Architecture*. London: Alec Tiranti.

COMTE, Auguste. 1994. Geist des Positivismus. *Philosophische Bibliothek*. Hamburg: Felix Meiner.

CULMANN, Karl. 1866. *Die graphische Statik*. Zurich: Von Meyer & Zeller.

DI TEODORO, F.P. 1989. Cracks and Chains. Leonardo and the Cracks on Brunelleschi's Dome. In C.A. Brebbia, ed. *Structural Repair and Maintenance of Historical Buildings*. Basel: Birkhäuser.

DORN, Harold and Robert MARK. 1981. The Architecture of Christopher Wren. *Scientific American* July 1981: 160-173.

ECO, Umberto. 1991. *Kunst und Schönheit im Mittelalter*. Münich: Carl Hanser.

FALTER, Holger and Giulio Mirabella ROBERTI. 1996. Conceptual Design of Ancient Vaulted Structures. *Proceedings of the IASS-Symposium*. Stuttgart.

HEINLE, Erwin and Jörg SCHLAICH. 1996. *Kuppeln aller Zeiten - aller Kulturen*. Stuttgart: Deutsche Verlagsanstalt.

KAISER, Hans. 1991. Von der Erfahrung der Quantität und Qualität. In Inge von Wedemaier, ed. *Pythagoras*. Ahlerstedt: Param.

KURRER, Karl-Eugen. 1987. Beitrag zur Entwicklungsgeschichte des Kraftgrößenverfahrens. *Bautechnik*. **64** 1 (January 1987): 1–8.

———. 1994. Von der Grafischen Statik zur Graphostatik. *Dresdener Beiträge zur Geschichte der Technikwissenschaften*. **23** 1.

STRAUB, Hans. 1992. *Die Geschichte des Bauingenieurs*. Basel: Birkhäuser.

VITRUVIUS. 1960. *The Ten Books on Architecture*. Trans. Morris Hicky Morgan. New York: Dover Publications.

Part II
From 2000 B.C. to 1000 A.D.

Chapter 7
Old Shoes, New Feet, and the Puzzle of the First Square in Ancient Egyptian Architecture

Peter Schneider

Introduction

One day, a little under 4,000 years ago, in the city of On—Heliopolis—in ancient Egypt, the twelfth dynasty pharaoh Kheperkare, Sesostris I, called together his court. In the presence of the "companions of the palace", he commissioned the process of building his pyramid, established its design, and set its basic geometry. The text known as the *Building Inscription of Sesostris I* records the steps Kheperkare followed in commissioning his "house of eternity". It moves through four clear and ritualized stages:

Kheperkare[1] first appears before his court and speaks, announcing his plan to build his pyramid:

> Behold, my majesty plans a work contemplates an act of value. For the future I will make a monument, I will construct a great house... The hill is my name, the lake is my memory... He who builds himself does not know oblivion, for his name is still pronounced.[2]

First published as: Peter Schneider, "The Puzzle of the First Square in Ancient Egyptian Architecture", pp. 207–221 in *Nexus IV: Architecture and Mathematics,* Kim Williams and Jose Francisco Rodrigues, eds. Fucecchio (Florence): Kim Williams Books, 2002.

[1] The pharaoh Kheperkare is also known by the names Sesostris I, Senwosret I, and Senusert I. He lived and reigned in the twelfth dynasty, between 1971 and 1927 B.C. His building inscription is preserved on a leather roll from the eighteenth dynasty, and is one remarkable for its completeness, and its presentation of the formal act of commissioning a building. The inscription is long, and much of its language is honorific. The excerpts that follow have been extracted from the text to give an idea of the process that was followed. Three different translations on the Kheperkare text were consulted. They are from Breasted (1988: Vol. I. Inscriptions 501–506), Lichtheim (1975: Vol. I, 116–117), Parkinson (1991, 40–43).

[2] The tradition of giving unique names to the individual pyramids had a long history in ancient Egypt. The first recorded pyramid names date to the hills of the pharaohs of the fourth dynasty, in

P. Schneider (✉)
College of Architecture and Planning, University of Colorado Denver, USA
e-mail: peter.schneider@ucdenver.edu; psch@ecentral.com

K. Williams and M.J. Ostwald (eds.), *Architecture and Mathematics from Antiquity to the Future*, DOI 10.1007/978-3-319-00137-1_7,

The court rejoices in the royal plan, delighting in its brilliance and significance. They respond:

> Discernment is in your words. Wisdom supports you. What you plan will come about. It will serve as your image, and will establish you for all eternity.

Kheperkare then instructs his architect, the overseer of all works of the king, to plan for and begin the great project.

> Your counsel completes all of the works of my desires; you act according to my wish. Skill and cunning belong to you... Now is the time to act...do according to your design.

Kheperkare finally participates in the ancient ritual called "stretching-the-cord", consecrating the place for the new monument, and establishing its proper orientation and form: "The king appeared in his plumed crown, with all of the people following him. The chief lector priest and scribe of the god's books stretched the cord. The cord was released, laid in the ground, made to be his monument. The king ordered work to begin. Joined together were upper and lower Egypt."

Stretching the Cord

The building inscription of Sesostris I is one of an extensive range of texts that belong to the genre of the building inscription in ancient Egyptian literature. Within that genre, it is both the most extensive, and the most complete. The story it tells explains and expands on a range of other texts, some much older and others much later, that describe the ritualized practice through which the great monuments that mark the ancient Egyptian landscape came into being. It also sheds light—although it is only a glimmer—on the structure of one particular part of that practice: the archaic ritual of stretching-the-cord.

The practice known as stretching-the-cord finds it origins in Egypt's ancient and mythic past. Its constant use is documented in a range of texts that span Egypt's long 3,000-year history: from the annals of the Archaic Period, through the records of the Old, Middle and New Kingdoms, to the chronicles of the Late Period.[3] While

---

2600 B.C. Kheperkare's pyramid was eventually given its own unique version of his name: "Kheperkare, most favored; Sesostris, who gazes out over the two lands".

[3] The Palermo stone records the fragmentary and cryptic annals of the first through the early fifth dynasties. They are translated in Breasted (1988). Inscriptions 76 through 166. The inscriptions for the first and second dynasties mention the stretching-the-cord ceremony five times. Those inscriptions are usually preceded by another inscription, recording another parallel ceremony that had to do with the design and commissioning of a monument—a "house". The paired inscriptions typically read: "Design of the house called Shelter-of-the Gods", and then, in the following year, "Stretching-the-cord for the house called Shelter-of-the Gods by the prophet of Seshat". The paired design-of-the-house and stretching-the-cord rituals that are mentioned in the ancient annals are echoed in the ritual followed by Kheperkare in establishing his monument: the first three parts of the ritual deal with the process of its "design", the fourth with its establishment and construction.

there can be no doubt that the ceremony was an essential part of the foundation and construction of all important buildings in ancient Egyptian times, there is no fuller description of the nature of the ceremony than the brief comment recorded in Kheperkare's inscription: "the chief lector priest and scribe of the god's books stretched the cord. The rope was released, laid in the ground, and made to be the monument."

Just what the set of practices that constituted that ceremony might actually have been, and just what was embodied in and instanced through the ritual of stretching that cord and laying it in the ground, is one of the puzzles that have intrigued Egyptologists for over 150 years. While the details that surround the act of stretching the cord are shrouded in the deep mists of the past, there is a general acceptance of four facts that are established by and through its physical consequences: the monuments themselves.

The first of these is the fact that the stretched cord inevitably established the primary axial orientation of the monument: its alignment in the spaces of both the physical and metaphysical landscapes of ancient Egyptian culture. The second is the fact is that the stretched cord also inevitably established a right angle.[4] The third is the fact that the stretched cord—in one way or another—established and set out the primary geometric figure—most often the square—that controlled the eventual form of the building with a very high degree of precision and accuracy.[5] The fourth and final fact is that the ritual—again in some way or another—established the measure and proportions from and through which the consequent geometries of the building or complex of buildings could be derived: from and through which the cord, laid in the ground, could be made to be the monument.

In spite of a general agreement that recognizes these four consequences of the ritual of stretching-the-cord, the puzzle itself remains unanswered: Just what was that ritual? How was it used to establish the right angle? How did it result in the description of that first square with such precision? How could it control the form, organization, design and proportions of the monuments so effectively, and so predictably? These are the major questions this chapter sets out to answer.

---

[4] Most Egyptologists agree that the method for the construction of that right angle was separate and distinct from that of the 3:4:5 or Pythagorean triangle. There is no contemporary evidence that the ancient Egyptians were aware of the method of constructing the right angle using an early version of the Pythagorean triangle. Gillings (1972: 242) cites conclusions by a range of scholars who dealt with Egyptian mathematics that make that very clear. The proto-Pythagorean triangle with its 3:4:5 relationships was a Babylonian invention, and was evidently passed on to the Greeks through their contacts with the 'Persians' in the archaic period of Greek history. The Babylonian origins of the Pythagorean theorem are discussed in Neugebauer (1957).

[5] In the case of a pyramid, applying an established and well-known ratio called the *seked* to the dimension of its base derived its height. That same formula also set the gradient of the pyramid's surfaces. See Gillings (1972: 185–187) for a discussion of the application of the *seked*. An interesting fact about the *seked:* the triangle that is the result of the ratios established by the Rhind Mathematical Papyrus is a 'proto-Pythagorean'right-angled triangle with sides of 21:28:35 digits.

To do so, it moves into a future distant from Kheperkare's time to encounter variants of the ritual's tradition: a tradition than involves both a simple geometric method and a set of measures that embody the intrinsic geometries inherent in the ritual's basic figure: the square. After encountering traces of the tradition in a number of different contexts, forms, and times, the narrative returns to Kheperkare's time—the time of that ritual geometry's archaic origins—to confront the puzzle of the cord's unanswered questions and to sketch out the framework within which one answer to these questions can be framed.[6]

Setting Out Muckross Abbey

Sometime between the years of 1440 and 1450 CE, near a place called Muckross in County Kerry, Ireland, a small group of Franciscan Friars set about building their new friary. It was eventually called Muckross Abbey.[7] One of their first actions was set out the building on its site using a simple and straightforward method that established and fixed the form and geometry for the new abbey.[8]

The method that the friars used to set out their abbey was clearly one that was well known, and absolutely familiar. It was a process that, for its time, was both sophisticated and elegant. The controlling figure from which the whole abbey's organization is derived is its cloister: its first and original square.[9] Once the first

[6] The documents known as the *Regius Poem* and the *Cooke Manuscript* date from the late-fourteenth and mid-fifteenth centuries, although the oral tradition they capture is certainly much earlier. They are published in Knoop and Jones (1938). The manuscripts document the points and charges that regulated the conduct of the lodges of masons and their gatherings in the late-medieval period. They also record the traditional histories of the art of geometry and the craft of masonry as these were known to the late-medieval masons. There are two histories—one called the long and the other the short—recorded in the *Cooke Manuscript*, and one in the *Regius Poem*. All of the histories locate the origins of the "true" geometry in ancient Egypt, in the time of the first pharaohs and the invention of monumental stone building.

[7] The history, geometry and characteristics of Muckross Abbey are discussed in Stall (1990).

[8] Eric Fernie observes, "since the majority of pre-modern architectural designs in the western world appear to have been laid out geometrically, one can if one wishes restrict oneself to coming to terms with the diagrams of the argument" (Fernie 1990: 230). He also writes: "one proportion appears to have been overwhelmingly more popular that any other in the design of [medieval] buildings, namely the ratio of the side of a square to its diagonal, which is one to the square root of two..." Roger Stall's examination of the organization of Muckross Abbey is an exercise in coming to terms with the diagram its particular argument, which is informed by its reliance on that 'popular proportion' of $1{:}\ \sqrt{2}$.

[9] They began by setting out the friary's cloister: a great square measuring 48 units by 48 units. Building on the geometry of that original square, they used the measure of its diagonals to establish a set of subsidiary figures, and a second pair of squares. They did so by describing a series of arcs using the corners of the original square as centers, and its diagonals as radii. Having established the second pair of squares, they repeated the process, using the measure generated by their diagonals to describe another set of arcs, which in turn established more figures and a third square. They used the arc of the diagonal of that square one final time to establish the location for the east wall of the choir.

square had been marked out and accurately described, the subsequent manipulations that allowed the form, proportion, organization, and design of the building to emerge were quite predictable, and entirely consequential.[10]

The friar's successive operations on the figure of the original square of the cloister extended its geometries through a series of measured transformations using a variant of the geometric technique for partitioning a square known as the "sacred cut". That technique exploits the relationship that exists between the measures of a square's side and its diagonal to produce a sequence of measures and an ordered disposition of its various parts and components that are always related in a ratio of $1:\sqrt{2}$. The final organization of Muckross Abbey—its final "sacred cut" geometry—emerged and evolved inevitably out of these carefully-managed manipulations and operations on that prime figure: that first and archetypal square (Fig. 7.1).

First Make a Square...

At almost the same time that those Franciscan friars were setting out their first square and building their abbey at Muckross in County Kerry, three German master craftsmen—Mathes Roriczer, Hanns Schmuttermayer, and Lorenz Lechler—were practicing their art and craft in the south of Germany. In the closing years of the fifteenth century or the early years of the sixteenth, at almost the same time that construction on the abbey was finally being completed, each mason published one or more brief but important booklets of 'instructions' explaining some aspect of the art of geometry as it related to the craft of masonry and its methods and practices.[11]

In each case, after a brief preface and dedication, the text of their instructions and teachings begins with an almost identical phrase. Mathes Roriczer says: "begin by making a square". Hanss Schmuttermayer exhorts: "first make a square". Lorenz Lechler insists: "first draw a square". There are no instructions in any of the

[10] There are many examples of plan and sectional organizations in history that are the outcomes of the process of partitioning a square using the measure known as the "sacred cut". The same kind of manipulations and operations occur again and again in the smaller late-medieval Franciscan and Cistercian friaries in Ireland, and in the general organization of many late-medieval monastic complexes on the continent. Muckross Abbey is an interesting example of this process of establishing $1:\sqrt{2}$ sequences geometrically, in that its geometry is the outcome of a process of successively enlarging the original square through the manipulation of its diagonals, rather than the accepted and common practice of successively reducing the square through the manipulation of the midpoints of its diagonal structure. The quantitative $1:\sqrt{2}$ sequence that results at Muckross through the manipulation of the sides and diagonals of its figures belongs the 12:17:24:34:48:64:96:128 series.

[11] The booklets are: Roriczer (1486, 1488a, b), Schmuttermayer ca. (1489), Lechler (1516). Translations of Roriczer's booklets and that of Schmuttermayer appear in Shelby (1977). A translation of selected excerpts from Lorenz Lechler's unpublished manuscript appear in Shelby (1971).

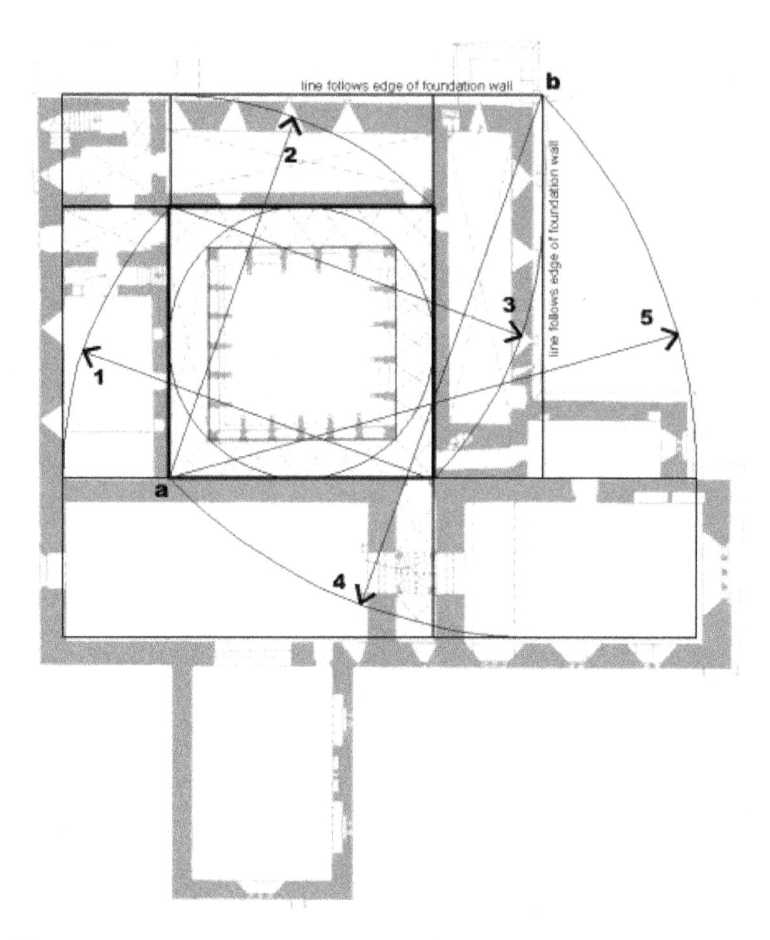

Fig. 7.1 The basic geometric system used to set out Muckross Abbey. The successive arcs establishing the major geometric figures are numbered consecutively, and the apex of each of the major squares are indicated by the letters a, b and c. Drawing: author

booklets that tell one exactly how to begin: that tell one how to make that first square.[12] In each method, the first square is evidently a familiar thing: made, drawn, and described. Its construction must have been so well known, such an essential part of the knowledge of the craft, that it needed no explanation. In each of the methods, that familiar first square was the condition precedent to all of the operations that

[12] Roriczer does set out a method for making a "true square" in his *Geometria deutsch* (Roriczer 1488a; Shelby 1977: 114). Mathes's method for constructing a right angle is a variant of the well-known theorem that asserts that the angle described in any semi-circle is a right angle. That theorem was evidently one of several mathematical propositions "borrowed" by the Ionian Greeks in the archaic period from the Egyptians. Robert Hahn mentions this "borrowing" in Hahn (2001: 57–59).

would follow. One made the first square, the prime figure, to make any and all other operations on its hidden geometries possible, and visible.

In each method, the first square—that prime figure—is carefully operated on using the process of *ad quadratum*: the process of defining successively smaller squares by connecting the mid-points of the original—and the subsequent—figure's sides.

> Mark the midpoints of each side... and draw lines between these points to make a second square through the first...Do this again to make a third square inside the second (Shelby 1977: 84–85).[13]

Mathes Roriczer's method requires three iterations of *ad quadratum*. Schmuttermayer's instructions call for eight. Lechler needs just two.

In each case, once the first square had been precisely made, drawn and described, the subsequent quadratic manipulations let the form, organization and structure of a pinnacle, a gable or a spire emerge quite predictably and consequentially by virtue of the qualities and properties that inhere in the prime figure. Their final form is inevitable, a product of their geometric process, just as the form of Muckross Abbey exists as a consequential product of its geometric process.

Roriczer's method differs significantly from those of Schmuttermayer and Lechler in the absolute reliance it places on geometry as the sole means through which a pinnacle and a gable are to be properly constructed. Mathes Roriczer says so in his dedication: "[this method] arises out of the fundamentals of geometry through manipulation of the dividers" (Shelby 1977: 83). Roriczer's method is a precise sequence involving over 250 subsequent operations of the dividers once the first square has been described.

In contrast, Schmuttermayer's and Lechler's methods use the initial geometric operations to establish a set of measures that are then applied as the controlling dimensions that guide the design of a pinnacle, a gable, a mullion, or even a building. Both 'take' the measures out of the original square using the process of *ad quadratum*, and each measure is consequently related to the next in the series in the ratio of $1{:}\sqrt{2}$. These measures are identified as the *old* and the *young* in both methods: the *old and young shoes* in Schmuttermayer's; the *old and young measures* in Lechler's.

To explain his method, Schmuttermayer writes:

> If you wish to draw a pinnacle and a gable, the first make a square, however large you wish. In that square make eight squares smaller and smaller, so that each fits into the other on the diagonal. Then set the eight squares beside one another...The first [square] is called the Old Shoe, the next is called the New Shoe...Out of these eight squares and their measure come all the settings-out of the pinnacle, the gable, and all measured work (Shelby 1977: 128–129).[14]

[13] Conflation of Mathes Roriczer's initial instructions for setting out a pinnacle.

[14] The other six squares, taken out of the first figure, are expressed as fractions—as halves, thirds, quarters, sixths or eights—of either the old or the new "shoes".

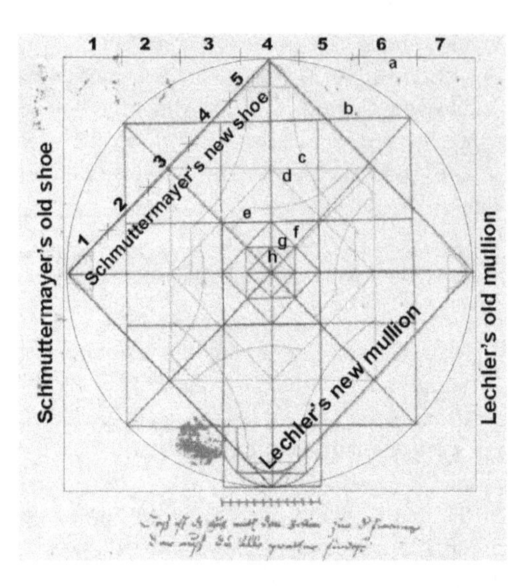

Fig. 7.2 Composite of Lechler's drawings for deriving the old and new measures, with the letters a to h indicating the rotated *ad quadratum* squares. Drawing: author

Lechler derives just two measures using his method. He does so by rotating two squares of the same size over one another, and then rotating another smaller square inside the first. The vertical diagonal of the larger square is the "old" measure, the horizontal diagonal of the smaller the "young". He writes:

> draw three squares through one another; in this manner you will find the length and the breadth that are the correct measures out of which can be taken all of the measures that one puts to use. . .through the device of [making] two squares within a square you will find the two correct measures, the Old and the Young (Shelby 1971: 150–152, and Fig. 9).

Having described a geometric method for deriving the equivalent of Schmuttermayer's old and new 'shoes,' Lorenz Lechler then goes on to describe another way—an alternative method—for finding the correct "old" and "young" measures (Fig. 7.2). He writes:

> Take one a part [side] of the (first) square and divide it into seven parts. That is the correct old measure for all buildings. If you then wish to make a young measure, which one uses often, then take two parts from the seven parts, leaving five parts. Those five parts are the young measure. So, the old measure has seven parts, and the young measure five, and the young is taken out of the old (Shelby 1971: 147–154).

Lechler's twofold instructions are intriguing, and highly significant. In the first place, they link and couple two distinct and different methods that derive the $1{:}\ \sqrt{2}$ ratio out of a square: one method based in geometry, the other in number and measure.[15] Lechler's young-to-old ratio of 5:7 is the first whole—and therefore

[15] The use of the coupled methods at Muckross—the geometric in the successive $\sqrt{2}$ manipulations of the original and subsequent squares, and the numeric in the use of the 12:17:24:34:48 $\sqrt{2}$ series—supports a reading of Lechler's instructions that clearly shows that these two methods and practices of dealing with the puzzle of $1{:}\sqrt{2}$ were familiar and customary in the late-medieval period.

rational—number approximation of the irrational ratio of 1: √2. It begins the √2 series 5:7:10:14:20:28. The twofold instructions also confirm both the importance and validity of Kossmann's discovery—and Paul Frankl's later corroboration of that discovery—that these two measures, five and seven, constituted a "great unit" and a "lesser unit" that were used rather extensively in the medieval period to set out buildings and complexes of buildings.[16] They finally explain the existence and significance of the repeated measures of five and seven that appear cryptically in various drawings in the mid-thirteenth century *Sketchbook* of Villard de Honnecourt, written some 250 years before Lechler began his manuscript.[17]

The method of *ad quadratum* and its parallel and coupled method of using known sets of numbers and measures that established close approximations of the 1: √2 ratio were clearly a part of the secrets and mysteries medieval apprentices were taught as they became adept in the mason's craft (Frankl 1945: 46–51 and 57–60).[18] They were fundamental parts of the tradition and its art: methods that been obviously established by the "old ones" and passed down through geometry's long and rich tradition.[19]

---

[16] Kossmann's discovery is mentioned by Frankl (1945: pp. 49–51 and note 14). Kossmann's observations of the existence of these measures were evidently supported by other, parallel discoveries by Steiglitz and Jenner; see Frankl (1945: note 14). Frankl himself discusses the repeated appearance of the five and seven measures, and their derivatives of 10, 14, 20, 28, 40, 56, as these were used in the mathematician Stornaloco's 1391 'geometry' for the Cathedral in Milan; see Frankl (1945: p. 55).

[17] The most significant examples of these measures of five and seven occur on sheets 39 and 40 of the folio, sometimes attributed to Magister 2. One particular figure shows a triangle whose short side has been 'cut' into five divisions. If one increases the length that side by two of those divisions, the new seven-unit length matches the length of its longer side. There are also two sketches on those sheets that show quite unequivocally that Villard was familiar with both the method of *ad quadratum*, and that of the "sacred cut". Frankl discusses the *ad quadratum* sketch in "Secret," (1945: 57). The diagram showing Villard's familiarity with the principles of the "sacred cut" geometry—which has not to my knowledge been previously identified and discussed—is a square inside of which are drawn two intersecting lines that connect the points on the square's opposite sides marked by the arcs of the "sacred cut": the arc described by a circle with its center on the square's corner, and a radius of one half of the square's diagonal. A final, curious coincidence occurs in Villard's drawings of the various mason's squares in the *Sketchbook*. Almost all of them have been drawn so that the ratio between their shorter and longer sides is—given the limitations of deriving exact measures from the old drawings—5:7 or, in modern terms, 1:√2.

[18] Frankl clearly established the process of *ad quadratum*—and by inference the parallel method of the 'sacred cut'—as the "secret" of the medieval masons. What Lechler's manuscript reveals is that there were in fact two secrets: one geometric, the other numeric. Both exist in parallel: as a twinned pair that holds and reveals the "secret" of the masons. A close reading of the well-known Vitruvian text that describes the squaring of the square—Book IX, Introduction—shows Vitruvius using both 10 and 14, the second pair in the 1:√2 series 5:7:10:14:20:28, to show how one can only approximate the irrational products of 1:√2 arithmetically. The other common numeric √2 sequences were 12:17:24:34:48:68; and 29:41:58:82:116:164.

[19] In their individual booklets, Roriczer, Schmuttermayer and Lechler each mention that their knowledge has been inherited from "the old ones who knew this art", from "the old ones among us, who invented this art that has its origins in the level, the square, the triangle, the divider and the

The Square in Classical Rome and Archaic Greece

Each of these well-established methods originates in and depends on the particular attributes of the true square. Each is, through this, absolutely dependent of the existence of an original square as the primary, archetypal figure. Any attempt to uncover the origins of the coupled methods inevitably confronts an enduring puzzle: the puzzle of the first square. The evidences of the geometric version of the methods—and the puzzle—can be easily traced back through Vitruvius and Plato's *Meno* to the emerging intellectual culture of archaic Greece in Ionia in the seventh century B.C. (Frankl 1945: 57; Hahn 2001: 56–60).[20]

The origins of the numeric method can be traced back—with a little more difficulty—through Cetius Faventius and Vitruvius to a point a great deal further back in time than that. It can be followed back to an archaic point in space and time where an original square was evidently imagined, described and constructed in ancient Egypt using a method that relied on the existence of two very ancient measures that embodied the crucial $1{:}\sqrt{2}$ ratio in what was perhaps an original and archetypal form.

M. Cetius Faventius is known for his third century commentaries on Vitruvius, and for the use Palladius made of those commentaries in writing the *De Re Rustica*. Faventius's commentaries are brief: glosses to the Vitruvian texts. They were collated as a compendium in the edition of the *Ten Books* prepared by Valentin Rose in 1867, and were published after the Vitruvian text in a section titled *De Diversis Fabricis Architectonicae*. In the next-to-last chapter of his commentaries, Chap. 28, Faventius describes with great precision the method one should use to make a 'perfect' square. He writes:

> Since the square is an invention with many uses, one without which nothing useful can be constructed, here is how you can make a proper square: take three rods of equal thickness, two each two feet long, the other two feet and ten inches long. Point their ends, and then join them at the points to form a triangle. When you have the triangle, you have made a proper square (Plommer 1973: 80–81, 109).

If the art of making the square was a familiar part of the technical knowledge of the medieval architect, it was evidently not quite so familiar in the case of the architects of Faventius's time. Faventius's short passage gives explicit instructions on just how to do that: how to make a proper square. It is the first text we have encountered that does so. One takes two measures, one of 24 units, and the other of

measure", and from "our old fathers, who used this method". While it is evident that they were referring to their own fathers and older colleagues, it is also clear from a careful study of the texts themselves that they were also referring to a much older set of "old ones": a set whose history is evidently contemporaneous with the emergence of the art of geometry and the craft of masonry in ancient Egypt. See note 7 above.

[20] Proclus writes: "Thales came to Miletus from Egypt, and brought geometry to Greece. His pupil Anaximander, according to the Suda, 'was the first to discover the square, bringing it into Greece and making known the basis of geometry'."

34 and, by doubling and combining them in a particular way, constructs both a right angle and a half-square that are *proper, expert and perfect.*[21]

Faventius's instruction is simple, clear and entirely memorable. It is also compelling in its brevity and clarity. Faventius' use of those two special measures—24 and 34—allowed any Roman architect or craftsman to make a right angle and half-square simply, easily and perfectly. Lechler's two *correct measures for all buildings*—five and seven, young and old—*the one taken out of the other,* would have let the architects, craftsmen and even friars of his time to do exactly the same thing, using this method inherited from "old ones who knew this art".[22]

The Remen and the Cubit

As one looks at the history of measure in ancient Egypt, one finds there a "new" measure that, like Lechler's, is also taken directly out of the "old". It is a 20-digit measure called the *rmn* (remen), and is derived directly out of the 28 digit measure known as the great or royal cubit.[23] The remen's numeric and dimensional relationship to the cubit is identical to the $1{:}\sqrt{2}$ ratio that characterizes the coupled 10:14 unit ratio mentioned by Vitruvius, the 24:34 ratio used by Faventius, Stornaloco's 10:14 series at Milan, the paired 5:7 ratio derived by Lechler, and even the 34:48 ratio used by the Franciscan friars at Muckross.

The existence of these two ancient Egyptian measures—the 20-digit remen and the 28-digit cubit—and their clear relationship to these other measures that are a part of architecture's history solve in an elegant and convincing way the puzzle of both stretching-the-cord and the construction of the first square in ancient Egypt. Each of these those ratios, as we have seen, establish the correct and proper

[21] Faventius's familiarity with the 12:17:24:34:48:68 $\sqrt{2}$ series is echoed by Vitruvius's apparent familiarity with the 5:7:10:14:20:28 series, as he mentions it in his introduction to his Book IX. The persistent appearance of measures from these two series—and measures from the 29:41:58:82:116:164 series—that occur in Roman houses and other architectural settings clearly shows that familiarity with systems of measure based on the $1{:}\ \sqrt{2}$ ratio were an intrinsic part of the customary, practical knowledge of the time. It is also interesting that two of the important measures used by the Romans *agrimensores* are directly related to the 5:7 series: the *passus* and double-*passus* of 5 and 10 ft. respectively, and the *actus* of 120 ft.

[22] It is perhaps no coincidence that the prime measures for Muckross Abbey—the inner and outer squares of its cloister—are 24 and 34 feet, or shoes, and that the young or smaller measure must have 'been taken out of the other' in just the way that Lechler describes in his method—by an *ad quadratum* rotation.

[23] The existence of the remen as a unit of measure in Ancient Egypt is well documented in a range of contemporary papyri and inscriptions. It is also documented in the examples of a range of measuring rods and staffs, which are clearly marked with both the cubit and remen measures.

measures from and through which one may inevitably derive the perfect right angle, and the perfect square.[24] Simply, easily, precisely, predictably and conclusively.

Faventius did not "invent" his simple method for describing the proper square. It was clearly an inherited part of the technological culture of his time, as it was in the times of Vitruvius and Plato, and Anaximander and Thales. It had originated with "the old ones who knew this art", "who used this method". Faventius, in framing his instruction, was repeating a teaching that had been handed down through the generations: from craftsman to craftsman, architect to architect. That older teaching may well have been framed just a little differently, using different words, and different tools:

> Here is how you make a perfect square. Take three cords, two twenty-four digits long, the third, thirty-four. Now join the cords end to end to form a triangle. When you have the triangle, you have made a proper square.

What these words describe is a simple, clear, effective and efficient technique using two cords with very particular characteristics to construct a square. This same technique, when it was used with two cords using other, different sets of measures—the one knotted in remen and the other knotted in cubits—allowed the ancient Egyptian "architect" to stretch the cord and construct the perfect "first" square from which all subsequent squares and proportions were derived with an amazingly high degree of geometric accuracy. Simply, easily, precisely, predictably and conclusively (Fig. 7.3).[25]

The instructions for the ritual called stretching-the-cord were well known to the chief lector priest and scribe of the god's books mentioned in Kheperkare's inscription. The god's book was where those instructions had been recorded for time immemorial: at the time of first occurrences in the dawn of Egypt's history. They had been established in the archaic time of the first architects, the "old ones who invented this art that has its origins in the square and the measure".

---

[24] Richard Gillings mentions another unit, the *double-remen* of 40 units, which is the diagonal of a square with a side of one cubit. He explains the usefulness of the remen, the cubit and the double-remen: "Doubling was standard technique in Egyptian arithmetic... in measuring land the relations between the double remen, the cubit, and the remen enabled areas (whether squares, rectangles, triangles, circles or other shapes) to be doubled and halved merely by changing the units of measurement [from double-remen to cubit to remen]... a square on the side of the double-remen is double the area of a square on the cubit, while a square on the side of the remen is half a square on the cubit" (Gillings 1972: 208–209).

[25] Most surviving paintings and drawings of Egyptian 'surveyors' show the surveyors as having two cords: the one stretched out and being used to measure a line, the other coiled and carried over the shoulder. It is clear why the paintings show the two cords, given Richard Gillings's explanation of the role of the two cords in note 25 above. Most contemporary measuring rods and staffs had marks that established the remen and cubit measures. A simple technique for constructing a right angle and/or square using a one-cubit rod would have been to mark out a line one cubit long and then, using the remen measure as a radius, describing the intersection of two arcs centered on the ends of the cubit-line. Joining those ends points to the arc's intersection would have described a true right angle. Repeating the process again would have led to the construction of a full square. It is a deceptively simple technique, yet it is highly accurate, very efficient, and remarkably effective.

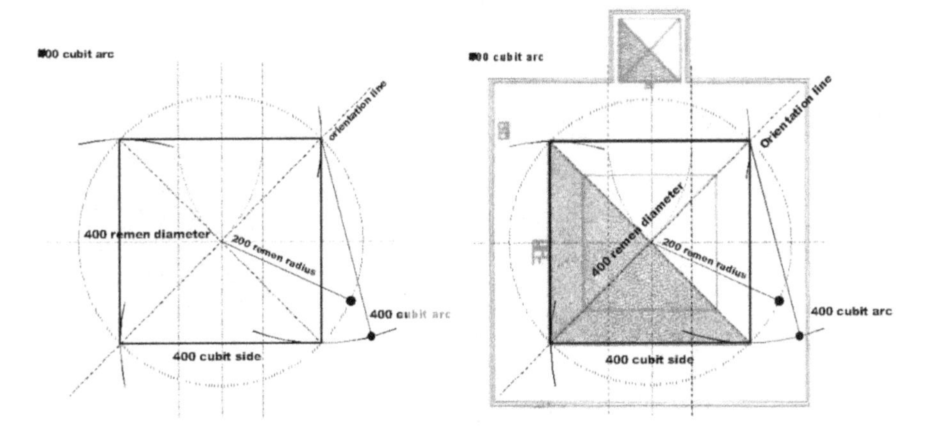

Fig. 7.3 The method for constructing a perfect square using the remen and cubit cords. The composite illustration shows the geometry superimposed on the plan of a typical old kingdom pyramid complex. Drawing: author

Those ancient instructions might, if we expanded on and extended the basic method preserved in Faventius's instructions, have read something like this (Fig. 7.4):

> Stretch a cord so that it makes a line pointing to the place of the north star. Now, take a remen cord and on that line mark out a circle whose radius is four hundred remen. Next, mark out on the line of the circle the place of the north star. Take a cubit cord. Holding one end of the cord on the mark of the north star, make a mark where its four-hundredth cubit knot touches the circle again. Do that again, and again, and again. Draw lines between the four marks you have made on the circle's edge, and you will have made a true and perfect square.[26] Release the cords, lay them in the ground, and make them to be the monument.[27]

[26] The measures used to establish the circle's diameter, and then to section its circumference using four chords related to its diameter in the ratio of $1{:}\sqrt{2}$, inevitably lead to the construction of a figure having four right angles, and four equal sides. It is the simplest, clearest and most direct method of constructing a square using cords, and chords. It seems as such to meet the requirements of the logical principle known as Occam's Razor which holds that it is vain to do with more what might be done with less. Put in another way: Occam's Razor asserts that the simplest, clearest explanation is always the best, and most often the most correct. The method described in this hypothetical reconstruction inevitably finds its historic origins in the measures and methods invented and used by the first geometers—the first measurers of the earth—who stretched the cords for the many houses built in Egypt's archaic period.

[27] This is one method using the two cords that may have been used to set out the pyramids and other Egyptian buildings. There are others, if one follows Faventius's original method, in which figures can be set out starting with a side, a central axis, or a diagonal. Each, however, depends on the existence and use of the paired cords, one knotted in remen or double-remen, the other in the great cubit. The use of the method of the great circle is certainly more elegant, and also more appropriate given the structure of the ancient Egyptian cosmologies and creation myths.

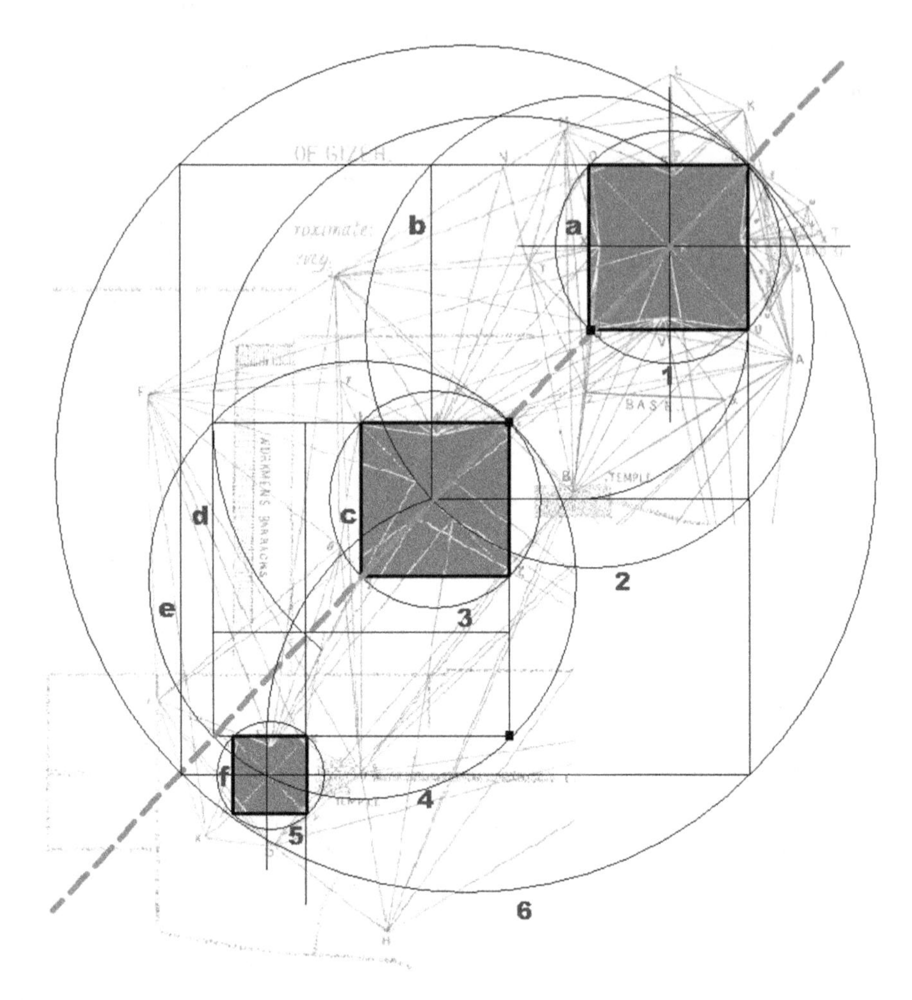

Fig. 7.4 The numbers show the successive circles used to set out the pyramids, the letters the successive squares inscribed in those circles, the roman numerals the three arcs that establish centers and edges. The diagonal orientation axis is the line used to establish the true north/south orientation for the complex. Drawing: author

Conclusion

And so, the pharaoh Kheperkare appeared in his plumed crown, with all of the people following him. The chief lector priest and scribe of the god's books stretched the cords. The cords were released, laid in the ground, made to be his monument. Kheperkare ordered work to begin: "Construct the great house called Kheperkare, most favored; Sesostris, who gazes out over the two lands," he might have said, 'its hill is my name, its lake my memory." Joined together were upper and lower Egypt, and also joined together by an enduring tradition were the architects and geometers of Kheperkare's times, those of Anaximander's, Vitruvius's and Roriczer's times, and those all-but-forgotten friars of Muckross Abbey. Their works, separated by the millennia, all emerged and evolved inevitably out of their customary and measured operations on a prime figure: a first and archetypal square derived though the use of a technique that might more properly be called stretching-the-cords.

Biography Peter Schneider's scholarly and research interests have been focused on the history of the architect: on the architect's mind, methods and manners as these have occurred in history, and in the way that the interactions between these three forces have shaped the discipline's traditions. He, like Louis Kahn, loves beginnings, and so his particular focus has been on exploring and understanding the forces that shaped the origin of the architectural tradition, and on the persistent influence those original ideas have had in shaping and forming the persona of the architect. His writings on the history of the architect and the architect's methods and practices have been widely published, and he has lectured on the topic extensively. He is currently exploring and writing about the connections that exist between the myths of architecture's archaic origins and the traditions embedded in its first, archetypal artifacts.

References

Breasted, J. 1988. *Ancient Records of Egypt*, vol. I. Irvine: Michael Sanders.

Fernie, E. 1990. A Beginner's Guide to the Study of Architectural Proportions. Pp. 229-237 in E. Fernie & P. Crossley, eds. *Medieval Architecture and Its Intellectual Context. Studies in Honour of Peter Kidson*. London: Hambledon Press.

Frankl, P. 1945. The Secret of the Masons. *Art Bulletin* **27**: 46-60.

Gillings, R.J. 1972. *Mathematics in the Time of the Pharaohs*. Cambridge: MIT Press.

Hahn, R. 2001. *Anaximander and the Architects*. Albany: SUNY Press.

Knoop, D. and Jones, G.P. 1938. *The Two Earliest Masonic MSS*. Manchester: Manchester University Press.

Lechler, L. 1516. *Unterweisung*. Heidelberg: unpublished manuscript.

Lichtheim, M. 1975. *Ancient Egyptian Literature*. Vol. I. Berkeley: University of California Press.

Neugebauer, O. 1957. *The Exact Sciences In Antiquity*. Providence, Brown University Press.

Parkinson, R. B. 1991. *Voices From Ancient Egypt*. Norman: University of Oklahoma Press.

Plommer, H. 1973. *Vitruvius and Later Roman Building Manuals*. Cambridge: Cambridge University Press.

Roriczer, M. 1486. *Büchelin von Der Finialen Gerechtigkeit*. Regensburg.

———. 1488a ca.. *Geometria Deutsch*. Regensburg.

———.1488b ca.. *Wimpergbüchelin*. Regensburg.

Schmuttermayer, H. 1489. *Finalenbüchelin*. Nürnberg.

Shelby, L.R. 1971. Medieval Masons' Templates. *Journal of the Society of Architectural Historians* **30**, 2 (May): 140-54.

Shelby, L.R. 1977. *Gothic Design Techniques: The Fifteenth Century Design Booklets of Mathes Roriczer and Hanns Schmuttermayer*. Carbondale: Southern Illinois University Press.

Stall, R. 1990. Gaelic Friars and Gothic Design. Pp. 191-202 in E. Fernie & P. Crossley, eds. *Medieval Architecture and Its Intellectual Context. Studies in Honour of Peter Kidson*. London: Hambledon Press.

Chapter 8
Geometric and Complex Analyses of Maya Architecture: Some Examples

Gerardo Burkle-Elizondo, Nicoletta Sala, and Ricardo David Valdez-Cepeda

Introduction

Few written documents about the works of art, sculpture, and architecture from Mesoamerica have survived. Even so, there is no doubt that these are all products of talented artists, because the execution often exhibits a great precision, showing a definite mathematical knowledge of this extraordinary civilization. Analysis seems to show that the ancient Mesoamerican architects and artists developed geometrical concepts and used them in their works, for example, to orient their buildings with a relationship with geomancy and alignment with the equinox, as described by Aveni (1997), Hartung (1980) and Broda (1991). Mesoamerican cultural life existed for nearly 2,000 years in Mexico, Guatemala, and Honduras, and the main civilizations that developed were the Maya and the Aztec cultures, which reached two golden periods, the first one around 650 AD, and the second one about 800 years later.

First published as: Gerardo Burkle-Elizondo, Nicoletta Sala and Ricardo David Valdez-Cepeda, "Geometric and Complex Analyses of Maya Architecture: Some Examples" pp. 57–68 in *Nexus V: Architecture and Mathematics,* Kim Williams and Francisco Delgado Cepeda, eds. Fucecchio (Florence): Kim Williams Books, 2004.

G. Burkle-Elizondo (✉)
Centro Interinstitucional de Investigaciones en Arte y Humanidades, Universidad Autónoma de Zacatecas, Unidad de Postgrado II, 98060 Zacatecas Zac, Mexico
e-mail: burklecaos@hotmail.com

N. Sala
Accademia di Architettura, Università della Svizzera italiana, Largo Bernasconi, 6850 Mendrisio, Switzerland
e-mail: nicoletta.sala@usi.ch

R.D. Valdez-Cepeda
Centro Regional Universitario Centro Norte Apdo, Universidad Autónoma Chapingo, Postal 196, 98001 Zacatecas Zac, Mexico
e-mail: vacrida@hotmail.com

K. Williams and M.J. Ostwald (eds.), *Architecture and Mathematics from Antiquity to the Future*, DOI 10.1007/978-3-319-00137-1_8,

Cosmological systems are viewed as structures together with the ideological apparatus of the culture. To understand the complex structure of the architecture and of development of the pyramids of Mesoamerica, we first need to realize that this was a universe that combined the viewpoints of archaeology, anthropology, and the history of their religion and politics. Pyramids had a religious function related with the myth and the ritual expressions, traditions and ideology. These buildings were established as specific sacred spaces in order that the people could experience a powerful sacred event. The relationship between death, art, and the architecture is also evident (Matos Moctezuma 1981). Many ritual ceremonies take place on the top of the mountains or pyramids or on the platform-like steps. A pyramid was fundamentally a ceremonial building that represented a pattern of the cosmological organization and the center of the world (Freidel et al. 1999).

The role of the major cities in the organization of the pre-Hispanic societies is well known, and has a relationship with the idea of authority as a complex power. In their artistic works, architecture, and urbanism, the idea of harmony was very important as a nexus between ideology and politics. For example, it is in myth that the Mayas envisaged an end when creation would return to its beginnings. In their calendar this is a great cyclical concept moving through space-time in which the gods, trapped within the stars and the sun, travel ceaselessly in a daily cycle from the upper to the underworld, beginning life again and again. Their cities, their ceremonies, and their sacrifices, as reproductions of the cyclic cosmological design, were permanent witnesses to the celestial order. By means of their pyramids—like the center of the world, represented geometrically by a flat cross-section and the cardinal points—and a ritual, they were able to rise to heaven or go down the steps to hell (López-Austin 1994).

History, Pyramids, Styles and Geometry

In the ancient Mesoamerican religion, a pyramid was a sacred symbolic mountain (Broda 2001). These geometries in Mesoamerica might be deemed arbitrary, except for the finding by some authors of units of measure, and because it is obvious that dynamics of the artistic composition and style are constant, as for example, in the extraordinarily precise drawings of the Palenque panels in Chiapas, Mexico, which Linda Schele and others have shown to have a geometrically structured composition, involving standard measures and the use of mathematical canons of proportion, as the primary tools inherent to a cosmic concept (Schele and Freidel 1999; Lhuillier 1992). Further study is needed to understand better the relationships between observation of the sky, nature, and symbolism in relation to their artistic and architectural expressions. At present, we know that the Mesoamericans had a cosmic diagram that mediated between their faith and the material world, which allowed them to imagine a universe of multiple planes in a vertical distribution, with thirteen levels above the *Tamoanchan*, earth or upper world, and nine below *Tlalocan*, underworld, the deepest level of which was the *Mictlan* (López-Austin

1994). All these were very important, because the Mesoamerican people symbolized and ordered their entire cosmic vision so as to integrate with it the space where the pyramids and the cities were built. This whole process takes place in time as a vehicle for men to return through material creation back to his maker, to rebirth. This was the mechanics of the temple; this was the power of the pyramids. Often the pyramids were crowned by a temple, sometimes connected with the feathered serpent, symbol of instinct and life. In this way, Mayan architecture and art reflect the Mayan character: serene, formal, but at the same time, sober and magnificent in its stylistic development.

Mayan Architectural Styles

Mayan buildings reflect a characteristic attitude toward nature, especially in the landscape forms imitating the mountains, in which pyramids are seen as sacred mountains and their forms conceived in relation to the power of the divinity. Each temple is a sacred description and has a complementary relationship with a god, symbolizing, for example, the water from the sky, and the earth, which focused primarily on the sacred mountain as a place of ceremonial operation between human ritual and their cosmic vision.

Therefore, in all these sacred sites the relationship is fundamentally the same, but with different styles within the context of the general landscape pattern, for example, with different kinds of anthropomorphic figures, the forms of the bases and platforms of the pyramids, the stairs, and the temples that crown the top like caves. At different periods of time and at different places in Mexico, Guatemala, and Honduras we see special characters from the human concept of an ideal universe, and a way to create a connection with the world of the gods. It is important to remember that the kind of political government that was used widely in Mesoamerica was the city-state.

From this point of view, the city styles in the Mayan culture are the following:

- *El Petén*: slim, high pyramids with blocks composed by a mixture of a small slope (*talud*), a molding, and a big slope, as at Tikal and Uaxactun.
- *Cuenca del Río Motagua*: plinths of vertical slopes, stairs with hieroglyphic texts and decoration at the friezes, walls and cresting, and with altars and stelae at the base, as at Copán and Quirigua.
- *Costa del Pacífico*: early influence of the Olmeca culture and after from the center of Mexico, characteristically by vertical slopes, vertical *alfardas*, twin temples at the top, and few decoration, as at Cahyup and Zacaleu.
- *Cuenca del Río Usumacinta*: temples with a hall with a vaulted ceiling and a small sanctuary inside, thick walls, inclined slopes, high cresting, and plentiful decoration as at Palenque, Yaxchilán, and Bonampak.

- *Río Bec*: steeply inclined pyramids with plinths and stairs with rather a decorative function, with a simulated temple at the top and cresting like a hideous mask, as at Xpuhil, Río Bec and Hormiguero.
- *Costa de Quintana Roo*: with Toltec influence but with distinct characteristics such as internal spaces, temples and columns, flat ceilings, cornices simulating fastenings, and friezes with niches, as at Tulun, El Meco, and El Rey.
- *Chenes*: façades with plentiful decoration, serpentine elements and motifs making up a big mask of the god Chaac, the mouth of which is the entrance to the temple; a small base and no columns, as at Hochob, Chicanná and Tabasqueño.
- *Puuc*: temples of a palatial aspect with many rooms, smooth walls decorated with plentiful geometric forms, big serpentine forms, with small columns and gods, without cresting and with arches, as at Uxmal, Labná and Kabah.
- *Maya-Tolteca*: with influences from the highlands of Mexico, for example, erecting the main temple to the god Kukulcan (or Quetzalcoatl), the use of board slope (*tablero-talud*) style at the platforms of the pyramids, the presence of serpentine columns, Atlantis statues, a surplus of warriors, ceremonial skull platform, *Chaac Mool* sculptures and circular temples, as at Chichen Itza and Mayapan.

The structure of a pyramid reflects dimensions that refer to numbers representing the mass, the length, the weight and even time. Numbers without dimensions refer to trigonometry and exponents of power laws, etc. It seems that they employed in their reckoning systems numbers without dimensions to refer to constant number series, constant in terms of computation (Mora-Echeverría 1984).

We think that they were able to use theoretical mathematics as well, and symbolic representations of numbers and their behavior even in computations, for example in the use of astronomical stones and some pyramids (Garcés Contreras 1995). Like good sky watchers they were, they found numbers relating to exponents of power that reflect the orbits of the planets around the sun (Aveni 1997; Garcés Contreras 1995; Trejo 1994). The symmetries, like patterns in the calendars, codices, artistic works, urban designs, and architecture, suggest that number was an intentional property (Thompson and Eric 1950). The only historical evidence that indicates that ancient pyramidal structures were related to one another is the fact that they all existed in the framework of their civilizations, but it is not difficult to imagine that almost identical design diagrams were used by different cultures, and that the underlying reasons for building in a pyramidal form were the same. Each segment and line are geometrically determinate with a function to join the relationships of the different meanings of individual or several elements without a hierarchical scheme, except the whole structure has no center, which leaves the imagination free to develop any possibility.

The use and the boundaries of linear dimensions remain to be resolved; for example, their role in determining an area or volume, especially in very complicated forms, is still not understood. Very often, the structures appear to involve symmetrical quadrilaterals; as Margarita Martínez del Sobral explained, these squares and rectangles are formal units (Martínez del Sobral 2000). She shows, as have others, that the Mesoamerican architects utilized geometry and mathematics consciously (De La Fuente 1984). This geometry seems to be somehow abstract, especially considering that it comes from the abstraction of mythology joined with the possibility of imagining the geometrical order taken from the sky and nature and made concrete in structure, platforms, and elevations. It appears that in order to conceive these monumental structures it was necessary to think in the same monumental way, looking at the buildings that came before as very big multi-structural geometrical forms and a cosmos themselves within a landscape containing this spectacular geometry (Harleston 1974). These stunning buildings have complex symbolic meanings that can be studied from different approaches—as art, philosophy, anthropology, history, psychology, and, of course, mathematics and geometry (Euclidean and fractal). Pyramidal structures mean or represent something—a sacred mountain, a calendar, a cave, the place of the gods and the rituals dedicated to them—within a total universe (Broda and Báez 2001). The answer may be found in their scientific knowledge, related always to the symbolic world, and their cosmic vision.

Characterization of Mesoamerican Pyramids by Different Fractal Geometry Models

The pyramids and their urban contexts function together to confer an aesthetic value to open spaces. These physical objects have also logical value as structures of perception and coherence with the space, form, and historical moment. Geometrically a pyramid maintains a balance with other structures, and the overlapping of their platforms gives symmetry to the subjective perception that causes a strong aesthetic power (Mangino Tazzer 1996). The geometric mix of vertical and horizontal axes gives not only the idea of massive volume but a wonderful equilibrium that contains at once both the rhythm and the movement of the monumental scale proportions. It may be that the fractal geometry will permit us to access a different way to study this architecture with a non-integer dimension, helping us to understand better the practical reason underlying their forms (Burkle-Elizondo et al. 2003).

Methods of Study

A pyramid is composed of a varying number of platforms of different dimensions. We have to analyze the structure as a whole, but on the other hand, we must also see these buildings as boundaries and try to study their individual, sequential, segments in order to achieve an understanding of the distinct aspects of the correlations describing the fractality. Geometric studies of monumental art and architecture have been done looking the pyramids as Platonic bodies (cubes or tetrahedrons) and Euclidean structures (rectangles, squares, triangles, and circles). But the massive, dense, pyramids are irregular and complex buildings.

The aim of our research is to discover the patterns, designs, and forms of the complex geometry that appears to have a specific kind information encoded within them. We have attempted to decipher the possible interconnections of different reckoning systems. To do this, we carried out three different procedures of analysis. The first studies the structures as series, from the point of view of areas against volumes. In the second, we visualize the pyramid as volume interpolated with its empty complementary mould. The third calculates the fractal dimension of a large number of pyramids using the box counting method, which indicates the roughness of an object or fluctuations of height over length. In the past, we have found that Mesoamerican artworks, sculptures and architecture have fractal dimensions (Burkle-Elizondo and Valdéz Cepeda 2001).

Procedure 1: Addition of the Areas Interpolated with Addition of Volumes To understand this model, we must first break down the pyramid according to its platforms so that many segments can be studied by individual mean values (Fig. 8.1). The series proceeds in an inverse direction, that is, the correlation of the areas of the platforms from the top to the base is interpolated with the volumes from the base to the top of these segments. In this group we studied 16 pyramids. The general average Fractal Dimension (Dv) was 1.236 ± 0.108 with $r^2 = 0.918$.

Procedure 2: Ratio of the Volume to the Empty Complement To get a logical structure it is necessary to add an extra, imaginary, platform at the top of the pyramid in order to form a mould, making a complex model. The extra rectangle added to the top of a figure in Mesoamerican geometry is called the *ome* rectangle, which in native theology means *Omeyocan* or "house of Ometéotl god", used by them in design (Martínez del Sobral 2000: 54). This series was constructed by taking into consideration an imaginary parallelepiped that included the extra platform at the top as a whole, thus: one variable was estimated by adding volumes occupied by platforms; the other one was calculated by subtracting from the whole volume that was occupied by the building. Using this model we find that:

$$\sum_{i=1}^{n} \boldsymbol{VP}_i \propto \left(\boldsymbol{VT} - \sum_{i=1}^{n} \boldsymbol{VP}_i \right)^D$$

VP_i is the volume of the platforms and VT is the total volume of the parallelepiped that contains the pyramid.

Fourteen pyramids were included for the analysis in this group. The general average fractal dimension (Dv) of this group was 1.312 ± 0.179 with $r^2 = 0.874$. Figure 8.1 shows the theoretical image of the second model with the extra platform forming the mould. Figure 8.2 illustrates the variogram that belongs to "Tikal I" pyramid.

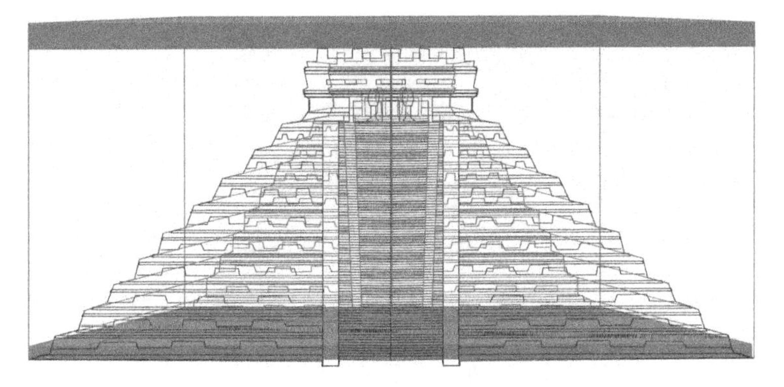

Fig. 8.1 Theoretical image of the second model. Rendering: Authors

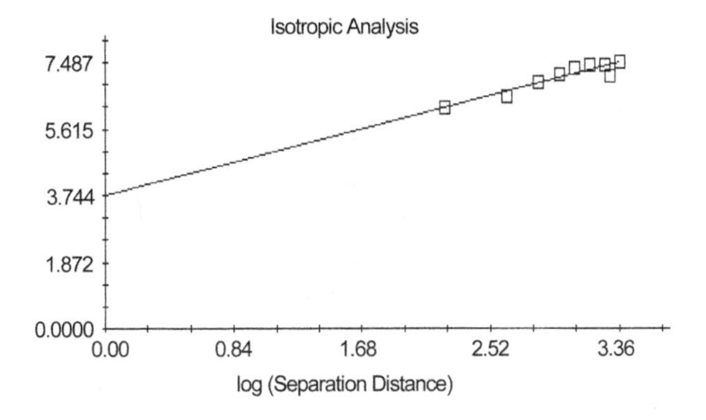

Fig. 8.2 Variogram of Tikal I pyramid, from the second model

In order to calculate the extra platform or base of the mould, we used an arithmetic progression from the base rectangle, adding square units, an approach that was recommended by Martínez del Sobral (2000).

Procedure 3: Fractal Dimension Twenty-six images from pyramids of different Mesoamerican cultures were scanned and saved as bitmap files in a computer. Then, in order to calculate Box, Information and Mass Dimension, and their intercepts on log–log plots, the images were analyzed with the program Benoit® (version 1.3: Fractal Analysis System, manufactured by TruSoft International Inc.). The Box Dimension is defined as the exponent Db in the relationship:

$$N(d) \approx 1/d^{Db}$$

where N(d) is the number of boxes of linear size d (the number of pixels in this study) that is necessary to cover a data set of points distributed in a two dimensional plane.

The Information Dimension assigns weights to the boxes in such a way that boxes containing a greater number of points count more than boxes with fewer points. Considering a set of points evenly distributed on the two dimensional plane, we have:

$$N(d) = 1/d^2$$

we assume that $m_i = d^2$, so the we have the equation:

$$I(d) \approx -N(d)\left[d^2 \log(d^2)\right] \approx -{}^1/d^2 = \left[2d^2 \log(d)\right] = -2\log(d)$$

where for the set of points composing a smooth line, we would find:

$$I(d) \approx -\log(d).$$

Therefore we can define the Information Dimension D_i as:

$$I(d) \approx D_i \log(d).$$

In the Mass Dimension we have a draw of a circle of radius *r* on the data set of points distributed in a two-dimensional plane, and count the number of points in the set that are inside the circle as M(r). If there are M points in the whole set, we can define the "Mass" *m*(*r*) in the circle of radius *r* as:

$$m(r) = \frac{M(r)}{M}$$

We can also determine the Mass Dimension D_M as the exponent in the relationship:

$$m(r) \approx r^{D_M}$$

We obtain the measurement of the mass *m*(*r*) of circles of increasing radius starting from the center of the set and plot the logarithm of *m*(*r*) versus the logarithm of *r*. If the set is a fractal, the plot will follow a straight line with a positive slope equal to D_M, approach that is the best suited to objects that follow some radial symmetry.

The total averages of this group were for Box Dimension $Db = 1.931 \pm 0.010$, Information Dimension $Di = 1.941 \pm 0.00017$ and Mass Dimension $D_M = 1.959 \pm 0.042$.

Figure 8.3 shows a log–log plot graph of "El Castillo de Kukulcán" pyramid, and Fig. 8.4 illustrates the image of the same temple. In Table 8.1 we can see the particular results of the values for the Fractal Dimension that we obtained in each of the three procedures.[1]

[1] The measurements are approximations of the actual, and the measurements and plans of reference for our calculations were taken from (Marquina 1990).

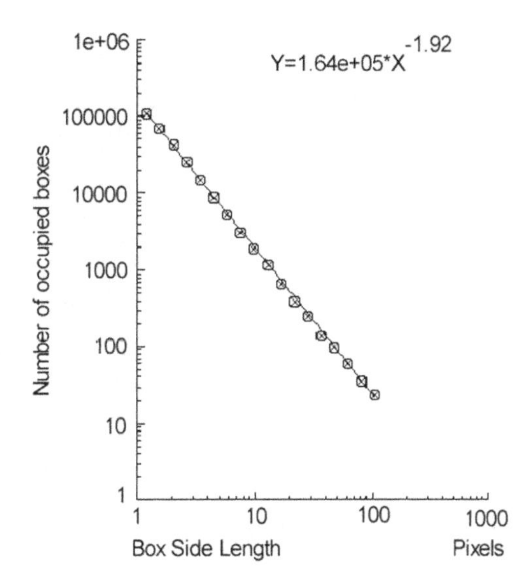

Fig. 8.3 Log–log plot for "El Castillo de Kukulcan" (Box dimension of 1.920). It can be appreciated as a *straight line*

Fig. 8.4 "El Castillo de Kukulcan" pyramid, upper view. Rendering: Authors

Conclusions

It would be speculative to conclude that comparison between our research and the existing pyramids proves that the builders of these pyramids conceived their models using the same structures that we have presented in this work. However, we think that this system of mathematical computation, which works well for these

Table 8.1 Procedures and methods to study the fractal dimension of Mesoamerican pyramids

	Fractal dimension method Df (Benoit)			Dimension Dv from procedures 1 and 2		
Pyramids	Box dimension (Db)	Information dimension (Di)	Mass dimension (D_m)	Volume – Empty complement (Dv)	Σ Areas – Σ Volumes (Dv)	Segments
Temple V Tikal	1.940	1.937	1.9540	1.223	1.4950	14
	±0.016	±0.003	±0.086	±0.206	±0.339	
				$r^2 = 0.835$	$r^2 = 0.735$	
Temple I Tikal	1.925	1.951	1.995	1.286	1.204	12
	±0.019	±0.0009	±0.173	±0.264	±0.037	
				$r^2 = 0.772$	$r^2 = 0.993$	
Temple III Tikal	1.928	1.924	2.016	1.341	1.113	12
	±0.014	±0.002	±0.020	±0.176	±0.028	
				$r^2 = 0.879$	$r^2 = 0.996$	
Pyramid Uaxactun E-VII	1.941	1.950	1.938	1.171	1.193	8
	±0.005	±0.001	±0.013	±0.122	±0.069	
				$r^2 = 0.929$	$r^2 = 0.977$	
Temple of Inscriptions, Palenque	1.922	1.938	1.936	1.117	1.157	12
	±0.011	±0.001	±0.011	±0.079	±0.038	
				$r^2 = 0.966$	$r^2 = 0.992$	
Temple of the Sun, Palenque	1.933	1.936	1.925	1.181	1.310	14
	±0.007	±0.002	±0.180	±0.114	±0.078	
				$r^2 = 0.938$	$r^2 = 0.976$	
Main Temple, Cempoala	1.932	1.947	1.917	1.221	1.100	20
	±0.005	±0.002	±0.043	±0.113	±0.029	
				$r^2 = 0.936$	$r^2 = 0.994$	
Xpujil (side tower only)	1.933	1.939	1.948	1.238	1.113	13
	±0.013	±0.005	±0.007	±0.118	±0.026	
				$r^2 = 0.940$	$r^2 = 0.996$	
Yaxchilan Structure 30	1.934	1.955	1.973	1.277	1.317	13
	±0.020	±0.001	±0.035	±0.197	±0.081	
				$r^2 = 0.857$	$r^2 = 0.974$	
Pyramid Monte Albán Building M	1.934	1.934	1.932	1.146	1.266	10
	±0.010	±0.001	±0.015	±0.106	±0.099	
				$r^2 = 0.944$	$r^2 = 0.959$	
Pyramide of Edzná Campeche	1.938	1.955	1.960		1.101	6
	±0.010	±0.001	±0.066		±0.030	
					$r^2 = 0.997$	
Pyramid 364 Nichos Tajin	1.926	1.910	1.927	1.100		8
	±0.007	±0.002	±0.003	±0.166		
				$r^2 = 0.879$		
Pyramid Calixtlahuaca Adoratorio Ehecatl	1.924	1.945	1.948	1.224		8
	±0.008	±0.002	±0.025	±0.166		
				$r^2 = 0.872$		
Pyramid of Cholula	1.941	1.964	2.001	1.172		9
	±0.003	±0.001	±0.053	±0.167		
				$r^2 = 0.875$		

(continued)

Table 8.1 (continued)

	Fractal dimension method Df (Benoit)			Dimension Dv from procedures 1 and 2		
Pyramids	Box dimension (Db)	Information dimension (Di)	Mass dimension (D_m)	Volume − Empty complement (Dv)	Σ Areas − Σ Volumes (Dv)	Segments
Temple IV Tikal	1.940	1.957	1.944	1.212	1.100	12
	±0.011	±0.0008	±0.013	±179	±0.013	
				$r^2 = 0.868$	$r^2 = 0.999$	
"The Castle" Kukulcan Chichen Itza	1.920	1.904	1.909	1.210	1.190	10
	±0.009	±0.003	± 0.038	±0.173	±0.075	
				$r^2 = 0.875$	$r^2 = 0.973$	
Temple I Tancah	1.935	1.958	1.945		1.131	14
	±0.022	±0.0009	±0.006		±0.031	
					$r^2 = 0.994$	
Great Palace Tower, Palenque	1.935	1.945	1.948		1.639	9
	±0.012	±0.002	±0.007		±0.373	
					$r^2 = 0.506$	
Temple of Tlahuizcan-pantecuhtli, Tula	1.942	1.941	1.937	1.488	1.346	11
	± 0.006	± 0.002	±0.005	±0.294	±0.388	
				$r^2 = 785$	$r^2 = 0.632$	
Pyramid of Quet-zalcoatl Teotihuacan	1.937	1.946	1.952			
	±0.006	±0.002	±0.010			
Observatory, Chichen Itza	1.927	1.943	1.894			
	±0.009	±0.003	±0.010			
Temple of the Sun Teotihuacan	1.923	1.913	2.000			
	±0.004	±0.003	±0.014			
Temple of the Magician, Uxmal	1.908	1.911	2.085			
	±0.006	±0.0005	±0.124			
Temple of the Descending God, Tulum	1.929	1.950	2.085			
	±0.006	±0.0005	±0.126			
Pyramid Huichapa building C	1.937	1.967	1.952	1.158		12
	±0.020	±0.001	±0.006	±0.107		
				$r^2 = 0.944$		
Hall of Columns, Mitla	1.928	1.949	1.902	1.236		21
	±0.004	±0.001	±0.004	±0.116		
				$r^2 = 0.942$		
Las Chimeneas, Cempoala				1.349 ± 0.268		10
				$r^2 = 0.760$		

buildings, reflects the presence of significant numbers and their fractal expressions out of a pure randomness. This leads us to consider the possibility that Mesoamerican architects conceived spatial representations before building a pyramid or other of their monumental works. Given the ideological elements, computational mathematics and a complex reckoning system had to exist in order for the Mesoamerican architects to imagine and build their cities and temples, and to express as well a correspondence between astronomical data and patterns and these massive forms. Our first two procedures show the existence of a fractality in series of different kind of pyramids and the relation between the parameters. In the third procedure, we have found bigger fractal dimensions than in the first two, because the third procedure measures the roughness of an object as a whole, and is completely different from measuring the geometric relations of particular scalar properties and studying the power function that describes the fractality of the pyramids. We presume that the first two procedures have generated a very similar fractal dimension because they are not complex. Our findings lead us to believe in the possibility that, when they designed their buildings, the architects were thinking on the basis of the concept of movement. Keeping in mind their vision of the cosmos, this complex mathematical design system could include the idea how the universe works.

Biography Gerardo Burkle Elizondo: Human Medicine, 1973–1978, School of Medicine, Universidad La Salle, México D.F. Residence in Internal Medicine, 1979–1982, Instituto Nacional de La Nutrición, México D.F. Master in Philosophy, 1989–1992. Universidad Autónoma de Zacatecas, México. Doctorate PhD in History, 1999–2004. Universidad Autónoma de Zacatecas, México. Several publications about Fractal and Chaos Theory applied to Mesoamerican Art and Architecture, from 2001 to 2004. Author of a book about Chaos Theory and Medicine and Psychology published by the University of Zacatecas, 1999.

Nicoletta Sala received her degree in physics, applied cybernetics, at State University of Milan, Italy. PhD in Communication Science at Università della Svizzera italiana of Lugano, Switzerland. Postgraduates (each 2 years) in "Didactics of the communication and multimedia technologies" and "Journalism and mass media". She teaches Mathematics Thought and Computer Graphics and New Media at the Academy of Architecture of Mendrisio, Switzerland. She is co-editor of the *Chaos and Complexity Letters International Journal of Dynamical System* (Nova Science, New York). She studies the interconnection between mathematics, fractal geometry and architecture. She has written 14 mathematics and information technology textbooks and several scientific papers dedicated to the complexity and fractal geometry in arts and architecture.

Ricardo Valdez-Cepeda: Engineer in Agronomy. Master and Doctor PhD in Agronomy. He applied the fractal geometry in different research fields. He works at the Universidad Autónoma Chapingo (Mexico).

References

AVENI, A. F. 1997. *Observadores del Cielo en el México Antiguo*. México City: Fondo de Cultura Económica–CONACULTA.

BRODA, J. and BÁEZ, J. F. 2001. La etnografía de la fiesta de la Santa Cruz: una perspectiva histórica. Pp. 165-238 in *Cosmovisión, ritual e identidad de los pueblos indígenas de México*. México City: Fondo de Cultura Económica–CONACULTA.

Broda, J. 1991. *Arqueoastronomía y Etnoastronomía en Mesoamérica*. México City: Universidad Nacional Autónoma de México.

———. 2001. Stanislaw Iwaniszewski and Raúl Carlos Aranda Monroy. *La Montaña en el paisaje ritual*. México City: CONACULTA. INAH. UNAM. UAP.

Burkle-Elizondo, G. and Valdéz Cepeda, R. 2001. Do The Artistic and Architectural Works Have Fractal Dimension? Pp. 431-432 in M. N. Miroslav, ed. *Emergent Nature. Patterns, Growth and Scaling in the Sciences*. Singapore: World Scientific.

Burkle-Elizondo, G., Valdéz Cepeda, R. and Sala, N. 2003. Complexity in the Mesoamerican Artistic and Architectural Works. P. 18 in Proceedings of International Nonlinear Sciences Conference: Research and Applications in the Life Sciences, H. Haken, ed. Vienna: INSC.

De La Fuente, B. 1984. *Los Hombres de Piedra*. México City: Escultura Olmeca. UNAM.

Freidel, D., Schele, L. and Parker, J. 1999. *El Cosmos Maya*. México City: Fondo de Cultura Económica–CONACULTA.

Garcés Contreras, G. 1995. *Pensamiento Matemático y Astronómico en el México Precolombino*. México City: Instituto Politécnico Nacional.

Harleston, H. Jr. 1974. A Mathematical Analysis of Teotihuacan. *Proceedings of XLI International Congress of Americanists*. México City: Comision de Publicacion de las Actas y Memorias.

Hartung, H. 1980. Arquitectura y planificación entre los antiguos mayas: posibilidades y limitaciones para los estudios astronómicos. Pp. 145-167 in A. Aveni, ed. *Astronomía en la América Antigua*. México City: Siglo XXI.

Lhuillier, A. L. 1992. *El Templo de las Inscripciones: Palenque*. México City: Fondo de Cultura Económica–CONACULTA.

López-Austin, A. 1994. *Tamoanchan y Tlalocan*. México City: Fondo de Cultura Económica–CONACULTA.

Mangino Tazzer, A. 1996. *Arquitectura Mesoamericana*. México City: Editorial Trillas.

Marquina, I. 1990. *Arquitectura Prehispánica*. México City: Instituto Nacional de Antropología e Historia.

Martínez del Sobral, M. 2000. *Geometría Mesoamericana*. México City: Fondo de Cultura Económica–CONACULTA.

Matos Moctezuma, E. 1981. The Great Temple of Tenochtitlan: Model of Aztec Cosmovisión. Pp. 71-86 in E. P. Benson, ed. *Mesoamerican Sites and World Views*, Washington D.C.: Dumbarton Oaks.

Mora-Echeverría, J. I. 1984. Prácticas y Conceptos Prehispánicos sobre espacio y tiempo: a propósito del origen del calendario ritual mesoamericano. *Boletín de Antropología Americana*, **9** (July): pp. 5- 46. México City: Instituto Panamericano de Geografía e Histoira.

Schele, L. and Freidel, D. 1999. *Una Selva de Reyes. La Asombrosa Historia de los Antiguos Mayas*. México: Fondo de Cultura Económica–CONACULTA.

Thompson, J. and Eric, S. 1950. *Maya Hieroglyphic Writing*. Carnegie Institute of Washington, pub. 589. Washington D.C.: Carnegie Institute of Washington.

Trejo, G. J. 1994. *Arqueoastronomía en la América Antigua*. México City: conacyt Ed. Equipo Sirius S.A.

Chapter 9
A New Geometric Analysis of the Teotihuacan Complex

Mark A. Reynolds

When anyone died, they used to say of him that he was now teotl, meaning to say he had died in order to become spirit or god.

Bernardino de Sahagun, sixteenth cen.

Introduction

San Juan Teotihuacan, which dates anywhere from 1500 and 1000 BC, was the largest Mesoamerican city in antiquity, yet there are very few documents, drawings, and stories that tell the story of this remarkable city. Little is known regarding the identity of the city's builders and inhabitants, whether the city plan was carried out continuously or sporadically, and what time frame encompassed the building and completion of the urban plan whose ruins we see today.

Finding reliable data concerning the geometric plans of the city is made more problematic because of the great amount of deterioration at the site. Furthermore, it is difficult to know with certainty which parts of the structures are original and which parts have been altered and/or rebuilt.[1] Additionally, although the complex appears to be laid out on a square grid, almost everything at the site is skewed from true right angles,[2] and precise geometric alignments within the architectural elements are virtually nonexistent. This creates technical problems when searching for geometric

First published as: Mark Reynolds, "A New Geometric Analysis of the Plan of the Teotihuacan Complex in Mexico", pp. 155–171 in *Nexus V: Architecture and Mathematics*, Kim Williams and Francisco Delgado Cepeda, eds. Fucecchio (Florence): Kim Williams Books.

[1] This is especially true considering the abuse and neglect of the site by Leopoldo Bartres.

[2] It could be that the builders' main concern was to just have the neighborhood communities placed informally around the ceremonial complex rather than developing a strict right-angled grid. We just don't know.

M.A. Reynolds (✉)
Academy of Art University, San Francisco, CA, USA
e-mail: marart@pacbell.net

K. Williams and M.J. Ostwald (eds.), *Architecture and Mathematics from Antiquity to the Future*, DOI 10.1007/978-3-319-00137-1_9,

systems based on the square and the rectangle, the standard shapes of choice when analyzing most ancient and classical architecture. Still, I was able to overcome many of these difficulties, and to find a good deal of information in my analysis of the city.

I used the excellent survey maps from *The Teotihuacan Map* by Millon et al. (1976), compiled from both land and aerial surveys, perhaps the best extant plan document. I also referred to information on the Ciudadela area of the complex by Drucker (1974).

The Analysis of the Overall Complex

In order to establish a geometric gridwork with which to develop the analysis, I used the north/south axis of the Avenue of the Dead. Essential for my analysis, and shown by line OQ in Fig. 9.1, this axis is called "Teotihuacan North/South" and is about 15.25° east of True, or Astronomic, North. Next I constructed KM perpendicular to OQ, which runs along the northern side of the Pyramid of the Moon.

This line ends at points K and M, at the retaining walls of the two flanking areas, called Group 5 and Group 5_1, to the east and west of the pyramid. It is believed that the Avenue was constructed to unite the area that was originally just to the Northwest of this wall, which was composed of several small villages, with the area of the Pyramid and Plaza of the Moon. As Kostof writes in *The City Shaped:*

> ...In...Teotihuacan, the administrative powers invested in the religious complex were sufficient to substitute a formal orthogonality (i.e., The Avenue of the Dead) for the pattern of villages that originally occupied the site (Kostof 1999: 35).

From this information, I focused on the area of the Pyramid and Plaza of the Moon as a key to the development of the analysis: if any plan had been originally laid out, its roots would be found here.

After several attempts at determining a working length that appeared to fit well in the area of the Pyramid of the Moon and the Plaza of the Moon, and could also conceivably have been used as a reference or boundary line by the builders, line KM was defined as 1.[3] My decision was reinforced when I found that KM is very nearly equal to both the east/west axis of the platform of the Pyramid of the Sun and the east/west measurement of the platform for the Ciudadela, both of major significance in the complex. This measure is also the distance from the center of the bridge along the Avenue of the Dead that crosses over the Rio San Juan to the east/west axis of the Temple of Quetzalcoatl in the Ciudadela.

By defining KM as 1, the north/south length of the complex, from the Northern side of the Pyramid of the Moon to the Southern retaining wall of the Ciudadela, can be defined as three √3 rectangles—AKMZ, UAZV, and GUVN—plus a reciprocal √3 rectangle, SGNT, or, a little more than 5.75:1. Another way to lay off this long

[3] Millon was the first to realize, and document, the very large area that the city encompassed, and that the extent of the city wasn't determined only by the structures along the edges of the Avenue.

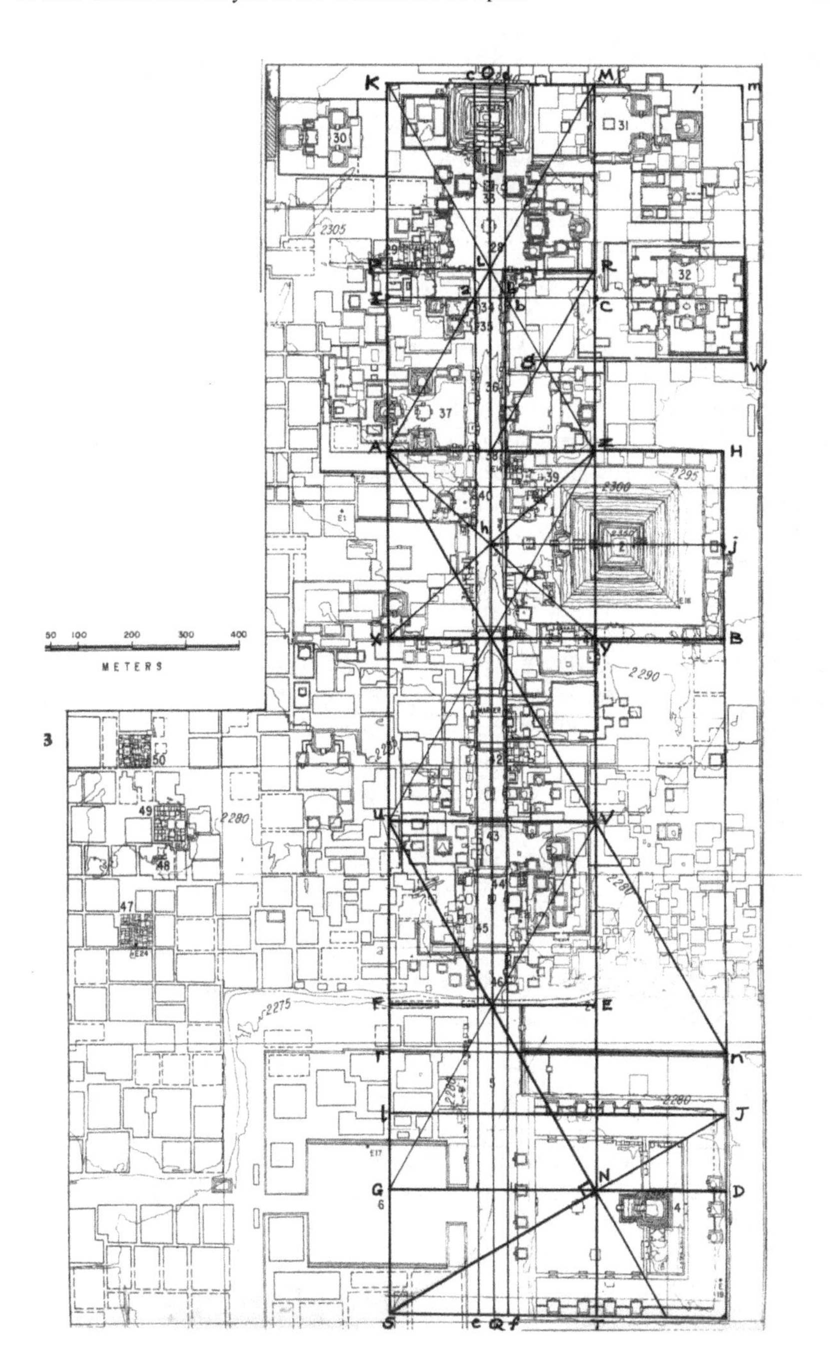

Fig. 9.1 Drawing: author

space is to use the Rectangle of the Equilateral Triangle (which we will call 0.866 from now on for brevity), PKMR, which is half of a √3 rectangle. This particular ratio, 1:0.866, clearly defines the area of the Plaza of the Moon and in locating the east/west axis of the Pyramid of the Sun. There would then be six of these rectangles, plus the reciprocal √3. In either case, the numbers 3 and 6 are present, with a system of 0.866 rectangles and the square being identified.

Also in Fig. 9.1, KLM is an equilateral triangle. By extension of lines KL and ML to meet the base of square IKMC at points a and b, the sides of the Avenue of the Dead, from the Plaza of the Moon to the Rio San Juan, are indicated. This width, ab, is also the width of the base of the steps on the southern face of the Pyramid of the Moon. The 0.866 rectangle cleanly marks off the perimeter of the Plaza of the Moon, and abuts the Northern side of the Temple of Agriculture compound on the Western side of the Avenue of the Dead. Point g, where two half-diagonals intersect in the second 0.866 rectangle, APRZ, defines the southern retaining wall of the Xala Compound, along line gW.

There are few right angles to be found in any of the major structures in the entire complex.[4] One of the exceptions is the Pyramid of the Sun. By establishing line jh perpendicular to the Avenue of the Dead, OQ, from the center of the square base of the pyramid, I found that jh intersects OQ at point h, which is the center of the second 0.866 rectangle, XAZY. This is the third rectangle down from the Pyramid of the Moon. The base of this rectangle, XY, can be extended to point B, the southeastern corner of the Pyramid of the Sun area.

In Fig. 9.2, KM was used to construct square PKMR, from which three irrational rectangles based on the square were constructed: √2 rectangle PKGL; √3 rectangle PKWE; and golden section rectangle PKNH. I did this in order to determine the placement of various architectural elements that were built on the eastern side of the Avenue of the Dead, and to see if these ratios could be applied. Although there are significant buildings and features on the western side, such as the Temple of Agriculture, the Plaza of the Columns, and the Great Compound, the most significant features of the complex, for my interests, are on the eastern side.

Of other significance in Fig. 9.2 is:

- A north/south line, Nh, which can be drawn from the golden section rectangle, PKNH, tangent to the Easternmost sides of the Pyramid of the Sun and the Ciudadela, gh.
- A north/south line, Gm, which can be drawn from the √2 rectangle, PKGL, tangent to the eastern base of the Pyramid of the Sun, bx, and the inner eastern platform of the Ciudadela, tu.
- The western base of the Pyramid of the Sun, ef, which can be generated by a fractional sixth point, v, of the 0.866 rectangle, VCDw, with point e tangent to this rectangle's diagonal, VD.

[4] Dr. Bruce Drewitt was kind enough to send me eight blueprints of various rooms and floor plans in some of the apartments/compounds in the complex done during the Teotihuacan Mapping Project, and no right angles are to be found, even in the smallest of these architectural plans.

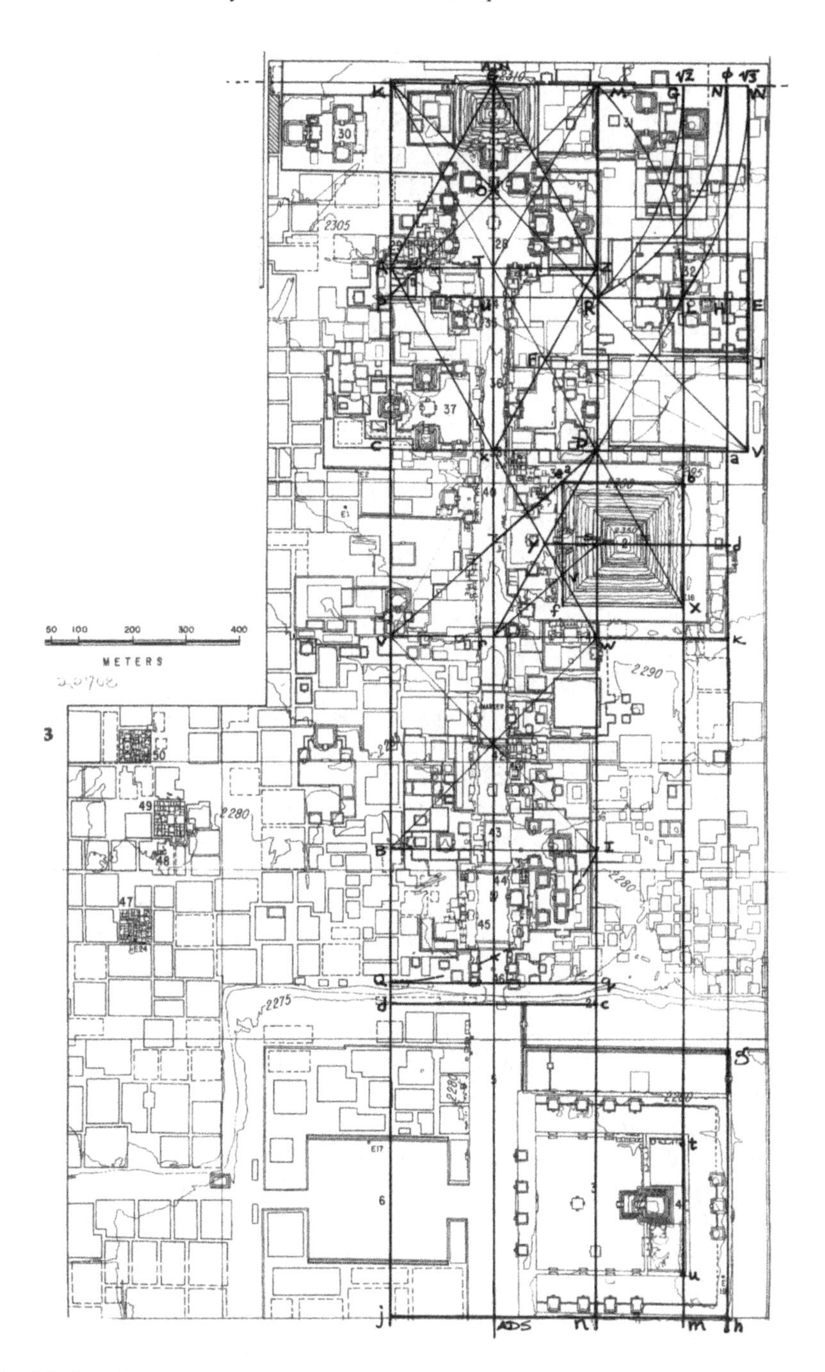

Fig. 9.2 Drawing: author

- A north/south line, WV, which can be drawn from √3 rectangle PKWE, giving the Eastern side of the Xala Compound.
- Square VBWI has been drawn immediately below the third 0.866 rectangle, VCDW, which is directly to the side of the Pyramid of the Sun. From VBWI, golden section rectangle QVWq and the √3 rectangle dVWc were constructed. The Rio San Juan passes directly through the difference in the heights of the two rectangles, dQ and cq.[5]

By using the golden section rectangle in Fig. 9.2, we have the golden section rectangle, AKPR, in Fig. 9.3. Extending sides AK and PR to the southernmost part of the complex, that is, the southern edge of the platform of the Ciudadela, at points S and W, we can draw square SEFW. The width of golden section rectangle KP could also be considered as a basic unit, in that most of the important elements are enclosed within the width of this unit.

Given this unit, then, STUW is a double square, with TW its diagonal. TU defines the inner southern edge of the platform of the Pyramid of the Sun. By the application of the construction of the golden section from the double square, we find the golden section of ST at point H. HB then becomes the northernmost wall before the Rio San Juan. SCVW is a reciprocal golden section rectangle whose long side, HB, defines the northernmost edge of the Ciudadela.

In Fig. 9.4, we have a detail of the area of the Pyramids of the Moon and Sun. AKMZ is the square base of the Pyramid of the Sun, from which golden section rectangle RPMZ has been generated. The left side of rectangle AKMZ is almost tangent to the Millon's survey line of the Avenue of the Dead. Millon's axis does not divide all of the architectural elements along the Avenue of the Dead equally on either side of it. This can be seen in the difference in the north/south axial bisectors of the Pyramid of the Moon and the Plaza of the Moon, which simply do not coincide. Still, Millon's axis is the best possible average axial line. Within a small margin, side RP could be considered to be congruent to the central axis of the Avenue of the Dead.

Also in Fig. 9.4, square UVWX is equal to square AKMZ, and fits well around the Plaza of the Moon. Its northern edge is tangent to the platform at the top of the stairs of the pyramid, and is also tangent to the easternmost and westernmost platforms in the plaza. The center of the square, Q, marks the center of the platform in the center of the plaza.

Also of note in Fig. 9.4 is the reciprocal golden section rectangle, TSGN, which is generated by double square AKGN in master square AKMZ The left side of this rectangle, TS, aligns with the right edge of master square UVWX around the plaza.

Additionally, the sides of rectangle EFMZ are in a 2:3 ratio. When extended vertically to pass through the square of the Plaza of the Moon, UVWX, side EF will generate the reciprocal golden section rectangle, UVLH, which is equal to RPKA in size and ratio. EF becomes the eastern side of the Avenue of the Dead, and is tangent to the eastern edge of the steps of the Pyramid of the Moon.

---

[5] When the city was being developed, the builders altered the course of the river so that it would run where we see it in Millon's survey.

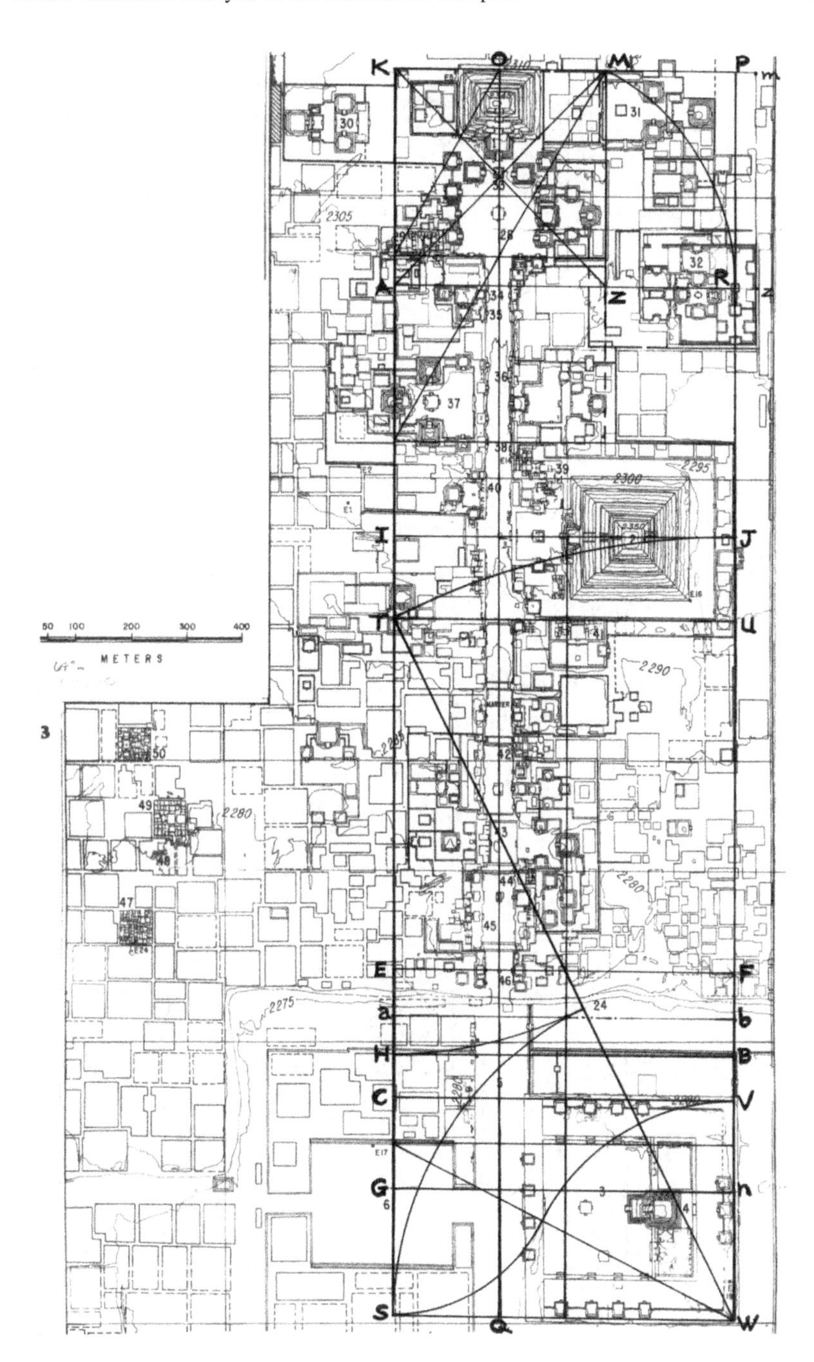

Fig. 9.3 Drawing: author

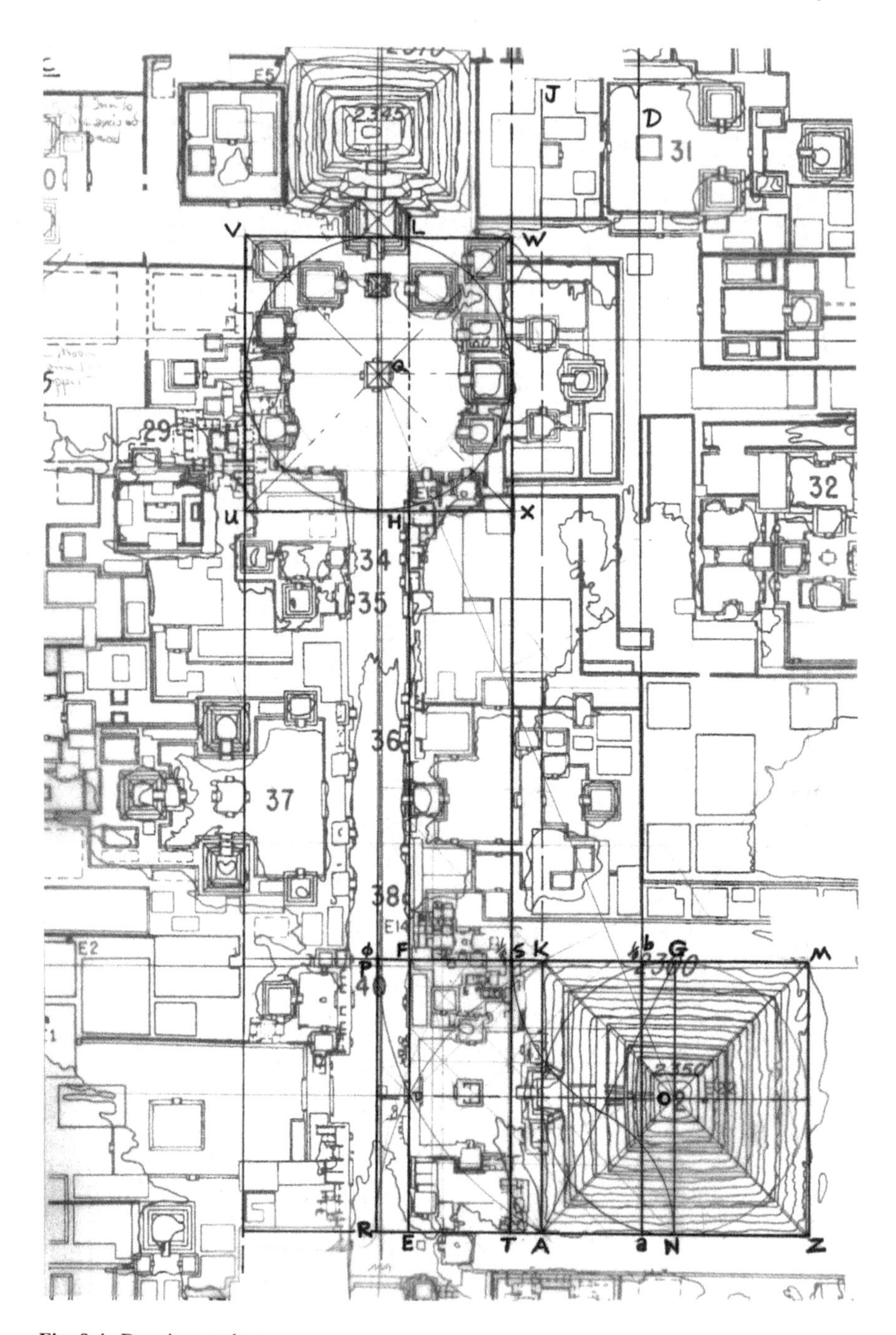

Fig. 9.4 Drawing: author

The Analysis of the Northern Half of the Complex

The primary focus of this part of the analysis is on the Pyramid of the Moon and the Plaza of the Moon. Although Group 5 and Group 5_1 have been utilized, and although there is much more that could be investigated, such as the Temple of Agriculture and the Plaza of the Columns, a complete analysis is beyond the scope of the present chapter.

Initial indications are that the entire area of the Pyramid of the Moon and its Plaza conform to the 0.866 rectangle, and to the configurations generated by the combination of *ad quadratum* and *ad triangulum* constructions. Figure 9.5 shows the basic master grid that incorporates the equilateral triangle and the square, with square PAZR and 0.866 rectangle KAZM. ADN/ADS is the axis of the Avenue of the Dead, with SY as the vertical bisector to the square and the equilateral triangle in the 0.866. SY will, as mentioned above, at times be adjusted slightly, but generally is an acceptable axis. Additionally, we do not know with certainty which foundations are original, and this adds to the difficulties in maintaining bilateral symmetry along SY.

In other words, the "curbs" of the avenue, CG and XN, do not define the exact edges of the architectural elements along the avenue.

Points e and f along the horizontal midline QH, and points t and r along the diagonals of the square where they intersect arcs ZTP and ATR for the *ad triangulum* construction, divide the width, QH, into a relationship where JD and EF provide the caesurae for the base of the Pyramid of the Moon and both Groups 5 and 5_1. Also, caesura JD, extended down to L, defines the two flanking platforms at the entrance to the Plaza of the Columns, at points a and b. Other architectural elements are also tangent to LJ and EW and indicate that it is quite possible that the layout of the complex could be based on two relationships: 1) between the square and the equilateral triangle as found in the rectangle, PAZM; 2) simple fractional parts of the original unit of 1 referred to previously. I will return to this a little later.

In Fig. 9.5, sides AT and ZT of the equilateral triangle have been extended to points G and N on the base of square PR, in order to see if the width of the Avenue of the Dead could be ascertained. Although this appears close, I found, in Fig. 9.6, that by using the half-diagonals of the square, KY and MY, and their intersection, at G and N, with the base of the 0.866 rectangle, PR, that G and N may provide the other reasonable, and perhaps more precise, solution to establishing the width of the avenue. In each case, the approximations are very close.

The width of GN, defined by vertical lines VT and WU, in addition to determining the width of the avenue, also approximates the base of the steps on the southern face of the Pyramid of the Moon, and the edges of the two northernmost platforms of the "ring" of platforms in the center of the plaza. The western edge of the top level of the pyramid falls along VT. The base of rectangle PKMR, along PR, clearly marks the southern end of the Plaza of the Moon, and G and N define the corners of the avenue as it enters the plaza.

Figure 9.7 shows 0.866 rectangle KAMZ divided into thirds, which provides a reasonable scenario for the geometry of the plaza area and the pyramid. Half of the

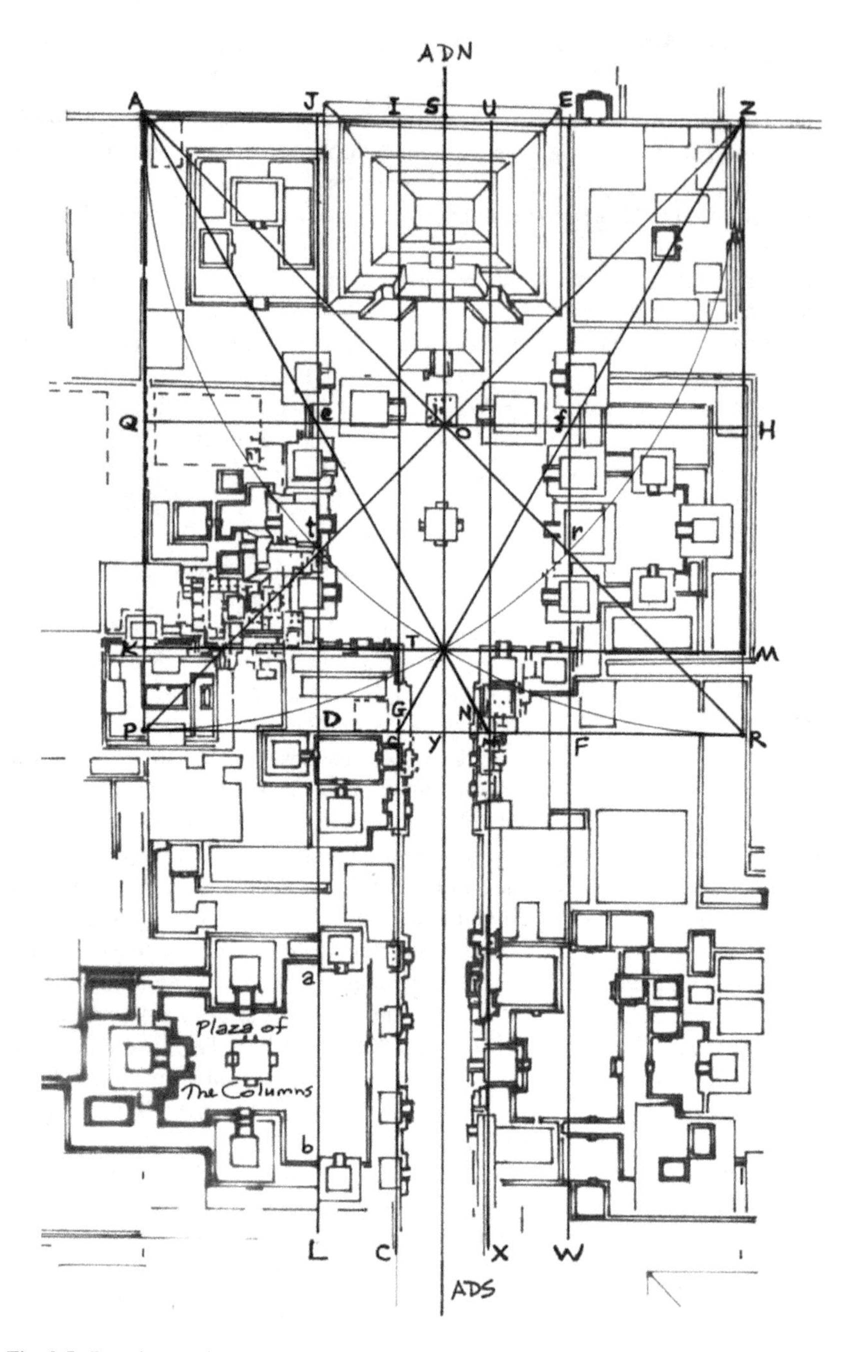

Fig. 9.5 Drawing: author

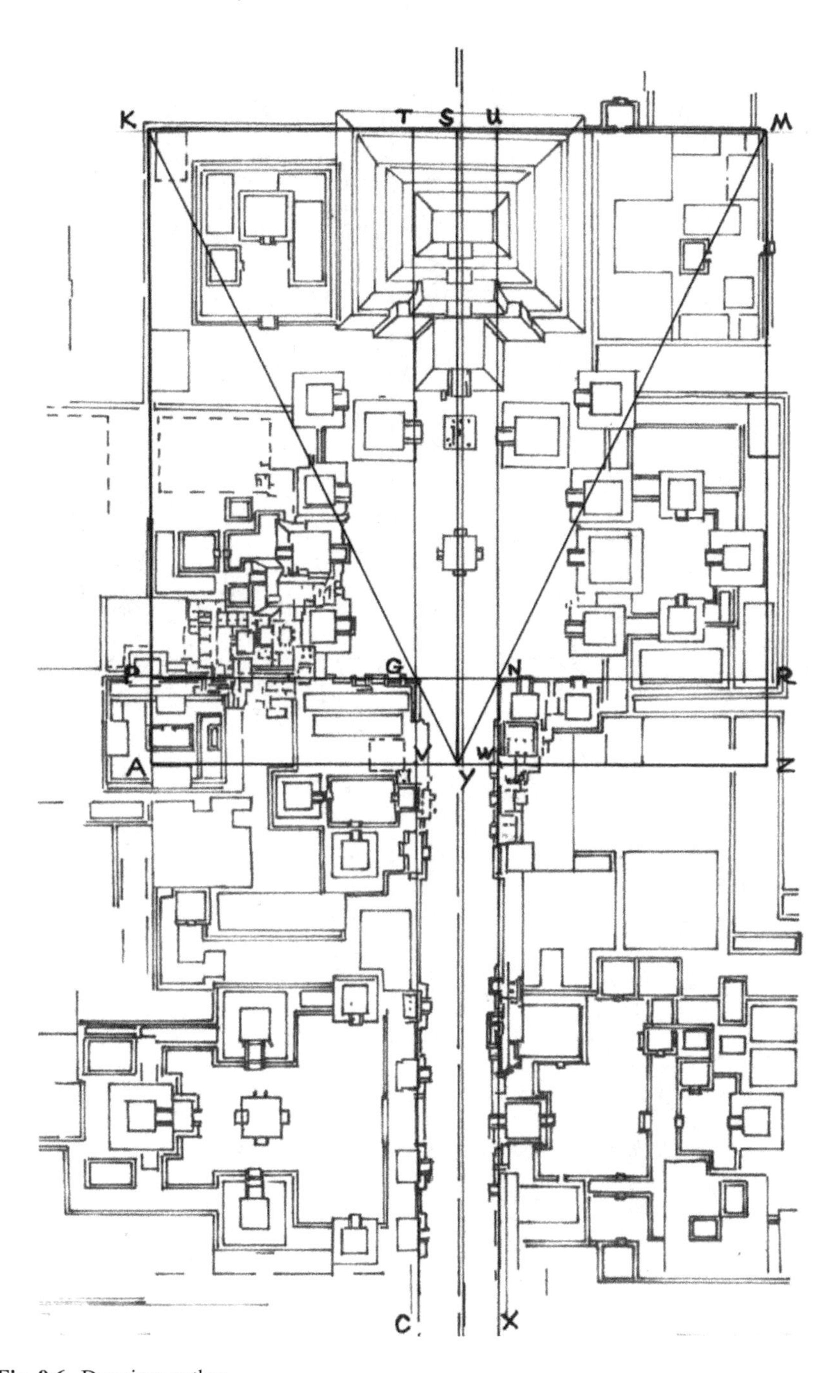

Fig. 9.6 Drawing: author

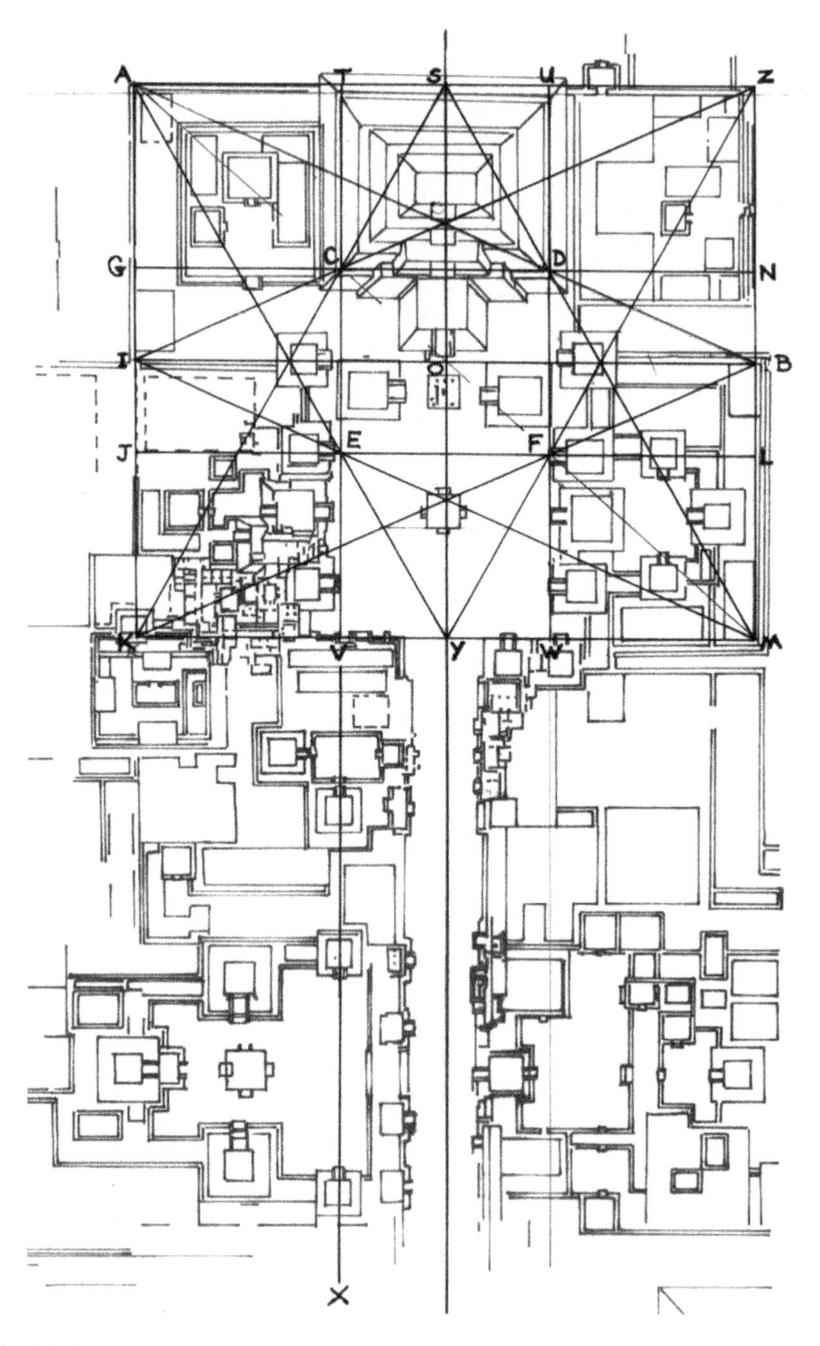

Fig. 9.7 Drawing: author

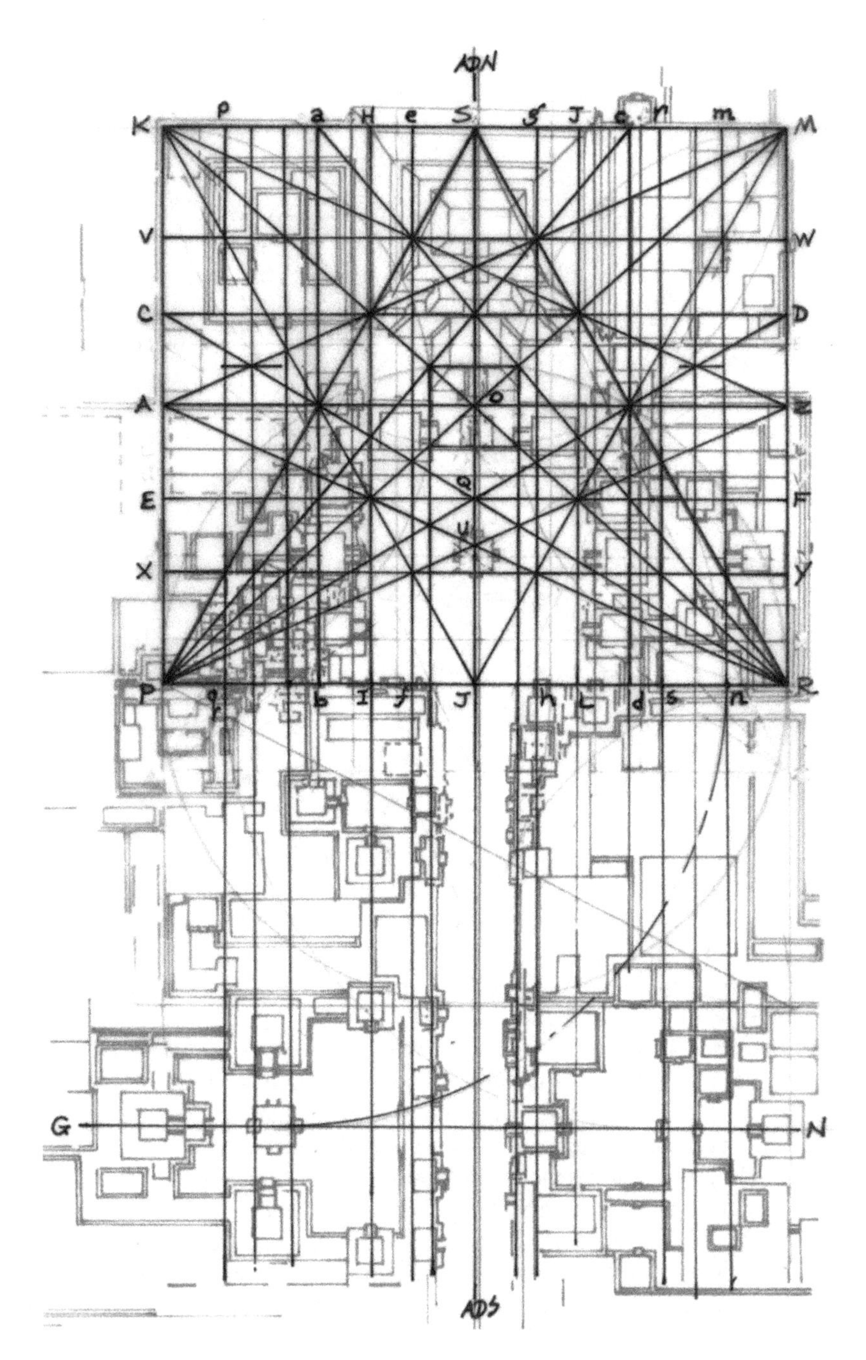

Fig. 9.8 Drawing: author

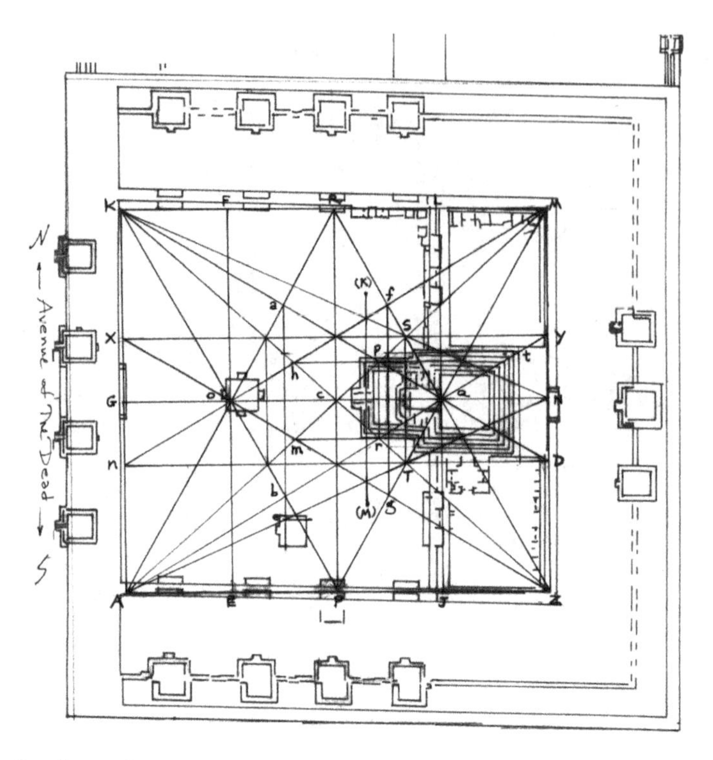

Fig. 9.9 Drawing: author

0.866 rectangle, IB, provide the caesura for the walls that divide the area in half as well as the axial bisectors for the two northernmost buildings in this circular arrangement of buildings in the plaza. This half is very nearly tangent to the base of the steps on the southern face of the pyramid. Although not all elements are precisely tangent to all the grid lines, it can still be seen that the plan follows the general pattern of the construction.

Note that the thirds caesura, VT, extended down to X, provides the axial bisector for the two flanking buildings in the Plaza of the Columns mentioned above.

In Fig. 9.8, the same 0.866 rectangle has been divided into halves, thirds, fourths, fifths, and tenths. These fractional parts were used to further develop the layout of the plaza. They were also extended southwards to see what alignments might occur. Although there seem to be an infinity of elements, it can be observed that the sides or central axes of most of the elements align with these fractional lines. Also, observe that GN passes through the Plaza of the Columns as an axial bisector. GN is 4/5 of the Unit 1, KM used as the standard in this analysis. Further studies are needed on this fractional parts approach.

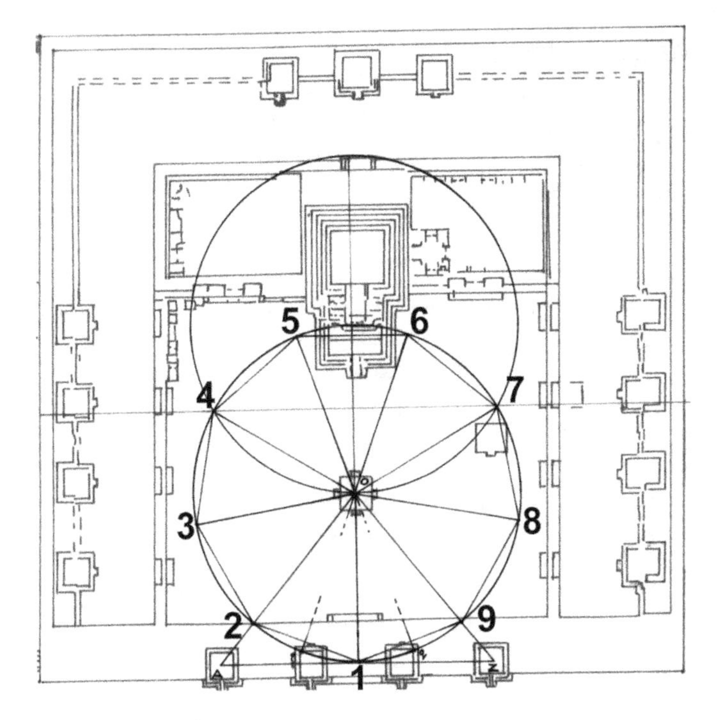

Fig. 9.10 Drawing: author

The Southern Half of the Complex

In keeping with the possibility that the square and the equilateral triangle were used in the northern part of the complex, I considered it likely that those same systems are a part of the Ciudadela in the southern part of the complex. I did not spend much time on the Great Compound or other features in this area because of the scope of this initial analysis. I can say that the Great Compound is in a ratio of 5:6, and there are some general relationships with the Ciudadela, but nothing of major significance.

The Ciudadela presented me with the same difficulties as other parts of the complex: it is not a rectangle, but is an almost rectangular parallelogram.[6] It is still possible that its plan was originally rectangular and was then adjusted as specific celestial observations became known, if this was, in fact, its original function.[7] I proceeded as if it were originally intended as rectangular, applying several ratios—the 0.866 rectangle, the enneagon, or nine-sided figure, and the golden

[6] Drucker believes this to be a result of the Ciudadela being an observatory for the study of Venus.

[7] Prof. Drewitt made the comment to me that it seems people are quick to surmise that much of the architecture in Mesoamerica was devoted to cosmological/astronomical observations, when, in fact, this may not be the case.

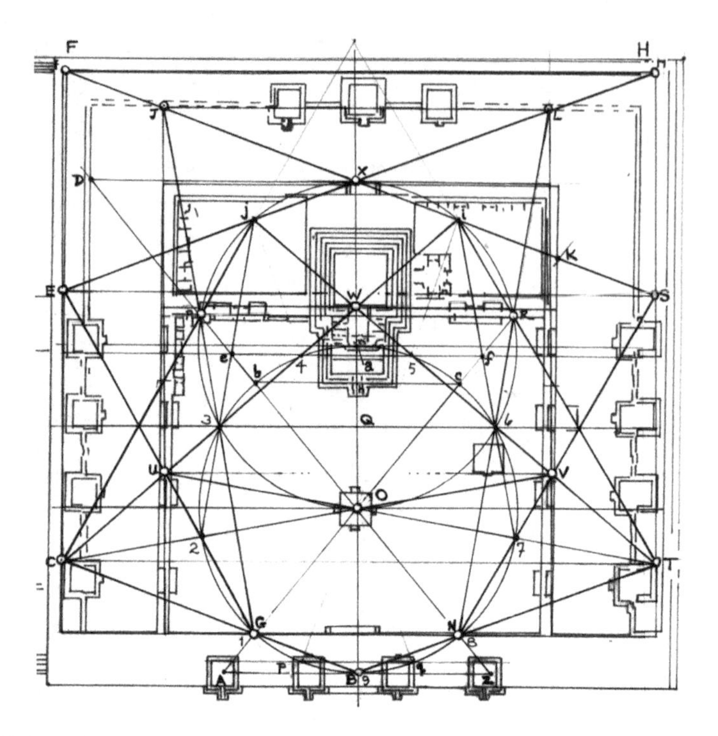

Fig. 9.11 Drawing: author

section rectangle—to the various areas that could be considered to be the total area of the plan of the Ciudadela. When the walls are aligned to right angles, the inner space does in fact fit into the 0.866 rectangle, as seen in Fig. 9.9.

AKMZ is a 0.866 rectangle, and ARZ and KPM are equilateral triangles. Their intersections, at O and Q, define the Western edges of the two center platforms, one of which is on the top of the Temple of the Feathered Serpent. Point Q also determines the line JL, which defines walls and platform changes in the compound. Points h, m, p, and r give the width of the Western portion of the temple. The asymmetry is noticeable in the rectangle, yet, at the same time, one senses the underpinnings of a rectangular grid. This is the most common feature found throughout the site.

During my long explorations on this portion of the complex, I also found an extraordinary relationship involving the enneagram, or nine-pointed star. Figures 9.9 and 9.10 show this construction. Having noticed the four platforms along the Avenue of the Dead and the central platform in the middle of the compound, I decided investigate by drawing triangle AOZ using the approximate centers of all the platforms. I had drawn the *vesica piscis* to get a perpendicular at Q, and noticed that the second circle was almost tangent with the eastern edge of the compound. I found that angle AOZ was 80°, which meant that, when bisected, there were two 40° angles, creating the enneagon within the circle. Also convincing was

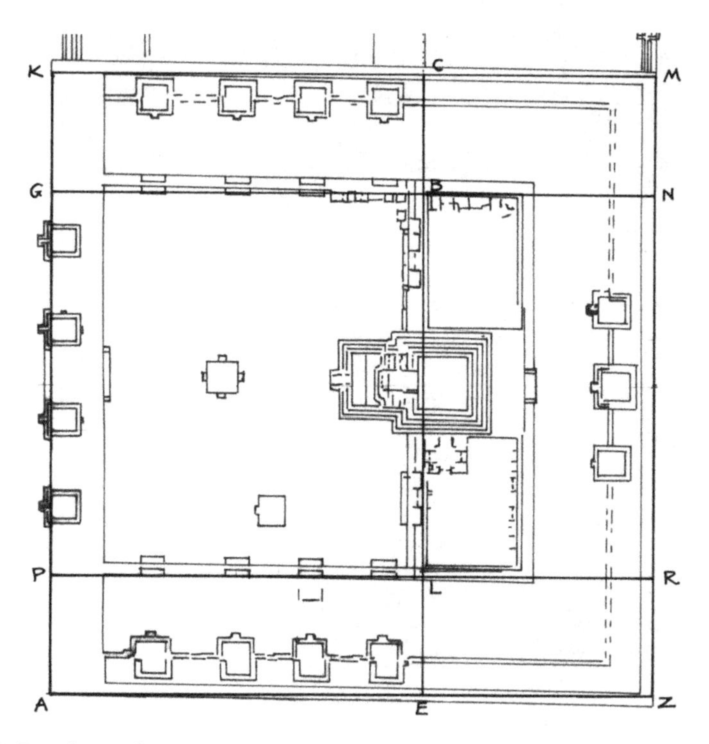

Fig. 9.12 Drawing: author

the line parallel to the Avenue of the Dead that passes through points 2 and 9, which lies exactly on the far western edge of the compound.

Figure 9.11 shows that all of the walls in the compound, as well as many of the architectural features, are tangent to the apexes of two overlapping enneagrams—for example, points C, E, J, L, U, V, S, and T—generated from the two enneagons in the circles O and a. For clarity, I did not draw the entire star.

Before leaving the Ciudadela, I also looked at a slightly larger area that included the additional outside walls, and was pleased to find master square AKMZ, as seen in Fig. 9.12, which led me to explore the use of the golden section rectangle within the overall area. Rectangles PGNR, AKCE, and LBNR are golden section rectangles, and work quite well in defining the space. In defining the total area in front of the temple, square PGBL further supports this view.

Figure 9.13 shows the diagonals of the two master golden section rectangles, PGNR and AKCE. The diagonal systems, KE, AC, RB-V, and NL-W, further define the architecture in a most convincing manner.

In Fig. 9.14, long side PR of golden section rectangle PGNR was rotated to intersect GN at point Y, generating $\sqrt{\phi}$ rectangle PGYW and intersecting the eastern wall at YW, behind the temple. Continuing the rotation to point E on the wall facing the Avenue of the Dead defines the northernmost wall of the compound, EF, which separates the compound from the Rio San Juan.

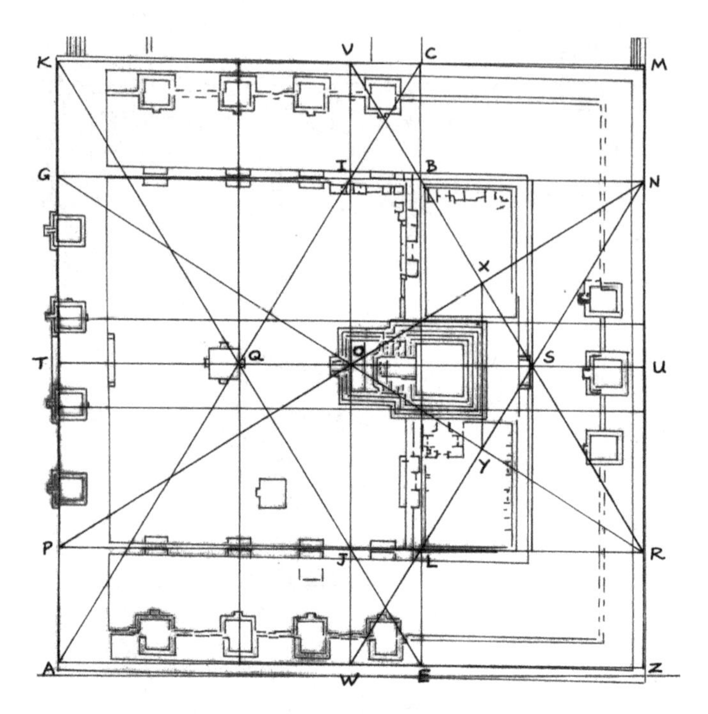

Fig. 9.13 Drawing: author

Figure 9.15 further supports the evidence for the golden ratio in the Ciudadela. PGNR is a golden section rectangle in square AKMR. LBNR is a reciprocal golden section rectangle. Q is the center of square PGBL, and OO is the center of square YXNR. Caesurae pq and no are the golden sections of rectangle PGNR.

Caesurae ab and cd are fourths. In Fig. 9.15, it will be seen that almost every aspect of the Ciudadela plan fits into this golden section grid.

Conclusion

As mentioned in the Introduction, the plan analysis of San Juan Teotihuacan presented notable difficulties. This may explain why so little has been written on the geometric and analytical aspects of the site, other than the survey work done by the Millon team and some earlier documentation done in the early 1900s, especially by Manuel Gamio. Thus little of an authoritative nature has been established.[8] It appears that the builders had a working knowledge of the circle, square, the

[8] With this, it must be stated for the record that all the professors I spoke with were quick to dismiss the work of Hugh Harleston as being unscientific and without merit. Harleston's writings, presented in Tompkins (1976), and his claim of a "Teotihuacan Unit" precisely equalling the

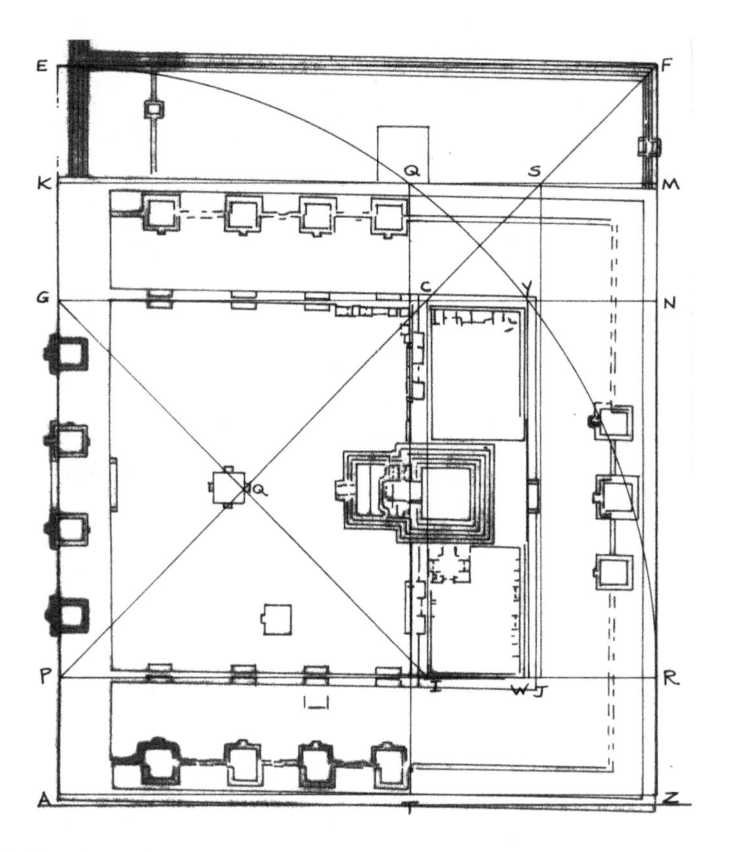

Fig. 9.14 Drawing: author

equilateral triangle, and various rectangular systems, especially those generated from the square (the golden section) and the equilateral triangle (the 0.866). There is also some evidence of fractional parts being applied to the overall plan.

Although there is no conclusive evidence that the Pyramids of the Moon and Sun were contemporaneously, it appears that there are some relationships regarding the square base of the Pyramid of the Sun and the Plaza of the Moon, which sits between the two pyramids. I believe that the Pyramids of the Moon and the Sun may contain all the numerical and geometric elements that were used in the construction of the city.

Further studies need to be done on these relationships by building on the work of Millon et al. (1976) who established a standard unit of measure of 82.3 cm (variations run from 80 to 83 cm), or a little less than a yard (32.15″), found throughout the site.

The 0.866 rectangle appears to have been used in a number of ways, especially as it may have been applied to both the areas of the Plaza of the Moon and the inner

twelfth root of two generated doubts in my mind as well. It is highly doubtful that the Teotihuacanos were aware of this scientific, mathematical, knowledge.

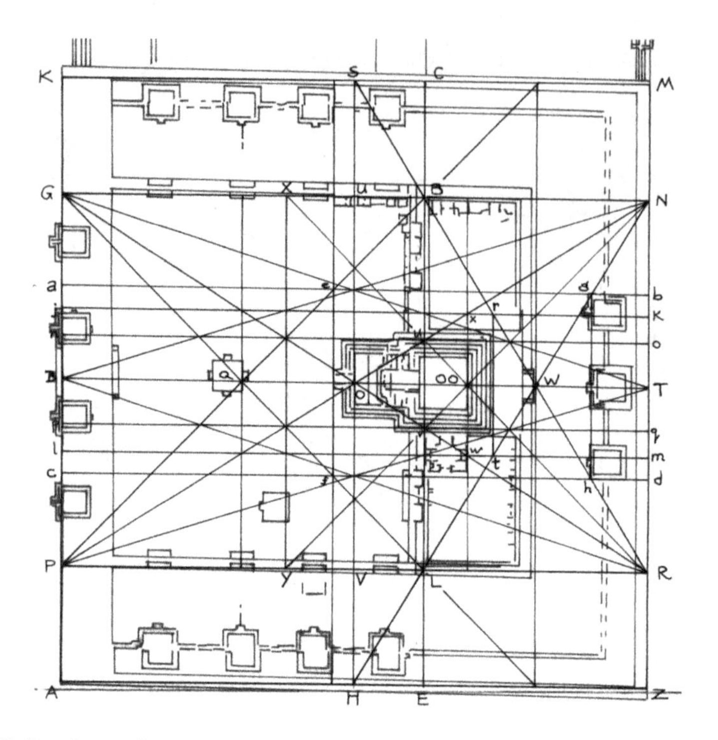

Fig. 9.15 Drawing: author

compound of the Ciudadela, effectively tying together the beginning and end of the significant temple and palace architecture in the complex.

As my analysis is one of the few attempts at a detailed study of the geometric layout of Teotihuacan, I had little on which to base it, but when I focused on what is believed to be the earliest part of the city, the Pyramid and Plaza of the Moon complex, the analysis began to develop some direction. Unfortunately, I have not yet done any measurements *in situ*. However, new ground has been broken with my examination of the Millon map.

We need to know more about the builders' geometrics skills and knowledge at the time when Teotihuacan was begun. Were there influences from Egypt, Africa, or European and Greco/Roman sources? Did the ancient Mesoamericans know about geometry as it was known and used elsewhere? If we understood more about the Teotihuacanos' practical and philosophical views regarding geometry, specifically as an ordering and compositional tool, we would have a better chance of solving the mysteries of the city. The search is a worthy one, for we may find another connection regarding our common cultural heritage as dwellers on this globe, and perhaps, also find the links that geometry can provide for us in the realization of our common ancient history.

Acknowledgments I am grateful to Rene Millon and George Cowgill, with whom I corresponded by telephone and e-mail, and to Bruce Drewitt, with whom I corresponded by post and e-mail.

Biography Mark Reynolds is a visual artist who works primarily in drawing, printmaking and mixed media. He received his Bachelor's and Master's Degrees in Art and Art Education at Towson University. He was awarded the Andelot Fellowship to do post-graduate work in drawing and printmaking at the University of Delaware. For years Mr. Reynolds has been at work on an extensive body of drawings, paintings and prints that incorporate and explore the ancient science of sacred, or contemplative, geometry. He is widely exhibited, showing his work in group competitions and one person shows, especially in California. Mark's work is in corporate, public, and private collections. A born teacher, Mr. Reynolds teaches sacred geometry, linear perspective, drawing, and printmaking to both graduate and undergraduate students in various departments at the Academy of Art University in San Francisco, California. Additionally, Reynolds is a geometer, and his specialities in this field include doing geometric analyses of architecture, paintings, and design.

References

DRUCKER, R. D. 1974. Renovating a Reconstruction: The Ciudadela at Teotihuacan, Mexico. Construction Sequence, Layout, and Possible Uses for the Structure. Ph.D. thesis. Rochester: University of Rochester, Anthropology Department.

KOSTOF, S. 1999. *The City Shaped: Urban Patterns and Meanings Through History*. New York: Thames & Hudson.

MILLON, R., COWGILL, G. and DREWITT, R. B. 1976. *The Teotihuacan Map*. Austin: The University of Texas Press.

TOMPKINS, P. 1976. *Mysteries of the Mexican Pyramids*. New York: Harper and Row.

Chapter 10
Geometry of Vedic Altars

George Gheverghese Joseph

Introduction

The earliest material evidence of Indian mathematics is found among the ruins of the Harappa civilization, which dates back to the start of the third millennium before the Christian Era. Archaeological finds show an elaborate system of weights and measures. Plumb-bobs of uniform size and weight found throughout the vast area of this culture conform to two series (binary and decimal) and their combinations, in the ratio of 1, 2, 4, 8, 16, 32, 64 and 10, 20, 40, 160, 200, 300, 640, 1,600, 6,400, 8,000, and 12,800. Until recently, equivalent weights formed the basis of an elaborate system of barter in certain parts of India, with conversion rates almost identical to some of the above ratios.

Scales and instruments for measuring length have been found at major urban centres of this civilization, such as Mohenjo-Daro, Harappa and Lothal. The Mohenjo-Daro scale is a fragment of shell 66.2 mm long, with nine carefully sawn, equally spaced parallel lines, on average 6.7056 mm apart. The accuracy of the graduation is remarkably high, with a mean error of only 0.075 mm. One of the lines is marked by a hollow circle, and the sixth line from the circle is shown by a large circular dot. The distance between the two markers is 1.32 in. (33.5 mm) and is known as the 'Indus inch'.[1]

First published as: George Gherveghese Joseph, "The Geometry of Vedic Altars", pp. 97–113 in *Nexus: Architecture and Mathematics*, ed. Kim Williams, Fucecchio (Florence): Edizioni dell'Erba, 1996.

[1] There are a number of interesting links between this unit of length (if indeed that is what it was) and others found elsewhere. A Sumerian shushi is exactly half an Indus inch, which supports other archaeological evidence of contacts between the two ancient civilizations. In northwest India, a traditional yard known as the gaz was in use from very early times. In the sixteenth century, the Mughal Emperor Akbar attempted unsuccessfully to have the gaz adopted as a standard measure in

G.G. Joseph (✉)
University of Manchester, Manchester, UK
e-mail: george.joseph@manchester.ac.uk

K. Williams and M.J. Ostwald (eds.), *Architecture and Mathematics from Antiquity to the Future*, DOI 10.1007/978-3-319-00137-1_10,

A notable feature of the Harappa culture was its extensive use of kiln-fired bricks for building and flood control. This bricks were exceptionally well baked and of high quality, and could still be used provided care is taken in removing them in the first place. They contain no straw or other binding material. While 15 different sizes of Harappan bricks have been identified, the standard ratio of the three dimensions—the length, breadth and thickness—is close to 4:2:1. Even today this is considered the optimal ratio for efficient bonding.

In the absence of any written evidence (the Harappa script has not yet been deciphered), these bricks may serve as the only "link" over a period of 1,500 years between the Harappa civilization and the beginning of the Vedic period of Indian history—a link, as it were, between the "frozen" geometry from archaeology and the first appearance of written geometry in the form of a surveyor's guide to constructing Vedic brick altars.

Geometry of the Vedic Age

An examination of the earliest written record of geometry in India involves a study of the *Sulbasutras*,[2] conservatively dated around 800–500 BC, although knowledge from earlier times is incorporated as well.[3] The *Sulbasutras* contain instructions for the construction of sacrificial altars (*vedi*) and the location of sacred fires (*agni*),

his kingdom. The gaz, which is 33 in. (or 5,840 mm) by our measurement, equals 25 Indus inches. Furthermore, the gaz is only a fraction (0.36 in.) longer than the megalithic yard, a measure that seems to have been prevalent in northwest Europe around the second millennium BC. This has led to the conjecture that a decimal scale of measurement, originating somewhere in Western Asia, spread widely as far as Britain, ancient Egypt and the Indus Valley (Mackie 1977).

[2] Three of the more mathematically important *Sulbasutras* were the ones recorded by Baudhyana, Apastamba and Katyayana. Little is known about these *sulhakaras* (i.e., authors of *Sulbusutras)* except that they were not just scribes but also priest-craftsmen performing a multitude of tasks including design, construction and maintenance of sacrificial altars (Thibaut 1875, Sen and Bag 1983).

[3] Chronologically, this period of Indian astronomy and mathematics should be taken to commence from when the Vedic hymns began to be composed, which some date as going back to 1500 BC. Certain issues regarding early Indian chronology have unfortunately become tug-of-war between those Westerners who see themselves as the guardians and promoters of impartial scholarship and invariably adopt conservative dating, and certain Indians who make excessive claims of antiquity for the early sources of Indian mathematics and astronomy. The tunnel vision of both groups make the task of incorporating recent discoveries in archaeology, necessitating drastic revisions the conservative dating of the Vedic period, more difficult. What this evidence would indicate is that earlier versions of both *Sathapatha Brahmana* and *Sulbasutras* should be placed about a 1,000 years earlier than the conservative dates attributed to these texts. For further details on recent evidence from archaeology, see Frawley (1991) and Kak (1987, 1993). A thorny question on the history of early Indian mathematics is how much importance should be given to oral evidence compared to written texts. Ignoring the oral evidence and regretting the paucity of written evidence has led to a fissure in the ranks of historians of Indian mathematics, generating more heat than light even in current discussions of the subject.

which, in order to be effective instruments of worship and sacrifice, had to conform to clearly laid-down requirements about their orientation, shapes and size. There were two main types of ritual: worship at home and communal worship. Square and circular altars were sufficient for household rituals, while more elaborate altars, of shapes which were combinations of these and other basic figures, were required for public worship.

The composers of the *Sulbasutras* made it clear that their work was not original but was to be found in earlier texts, notably the *Samhitas* and the *Brahmanas*, of which the most relevant extant text, the *Sathapatha Brahmana*, is at least 3,000 years old.[4] Despite its obscurities and archaic character, this text contains a valuable discussion of the technical aspects of altar construction.

An important section of the *Sathapatha Brahmana* deals with the construction of altars to carry out a 12-day *Agnicayana* ("piling up of *Agni*") ceremony. The ceremony often took place in an area containing two sections (Fig. 10.1):

i. The *Mahavedi* (Great Altar): Shaped as an isosceles trapezium, the two parallel sides of this structure were constructed so that the larger side measured 30 prakramas on the west and the smaller side 24 prakramas on the east, with the altitude of the trapezium being 36 prakramas.[5] Contained within the *Mahavedi* was a falcon-shaped brick altar (*Vakrapraksa-syena*) representing time.[6] Since many of the interesting results in Vedic geometry arose from the construction of this altar, it will be discussed in the next section.
ii. To the west of the *Mahavedi* was a smaller rectangular area called *Pracinavamsa* in which, at specified positions, were three fire altars called *Garhapatya* (of circular shape symbolising the earth), *Dakshinagni* (of semicircular shape representing space) and *Ahavaniya* (a square indicating

[4] The literature includes, four Vedic *Samhitas* (i.e., *Rigveda, Yajurveda, Samayeda and Atharavdveda*) in their various recensions, being the collection and presentation in a classified form of a large number of Vedic hymns; a set of elucidatory literature called *Brahmanas* of which the *Sathapatha Brahmana* is the most important for our purposes; a set of philosophical treatises called *Upanishads*; and six *Vedangas,* written for the purpose of instilling the correct methods of recitation of the *Vedas* and performing Vedic rituals, of which two, the *Jyotisa* and the *Kalpa*, are particularly important, since the first contains early knowledge of astronomy and the last contains the *Sulbasutras*.

[5] The measures used in the *Sathapatha Brahmana* were the same as in the *Sulbasutras*. The important units of measurement were:

1 *pada* = 15 *angulas*,

1 *prakrama* = 2 *padas*,

1 *purusha* = 4 *prakramas* = 120 *angulas*.

A *prakarama* is about 0.5 m.

[6] In Vedic mythology, time was represented by the metaphor of a bird. The year was divided into six seasons, with the head of the bird being the *vasant,* the body being both *hemanta and sisira* the wings being *sarad* and *grishma* and the tail being *varsha*.

Fig. 10.1 Drawing: author

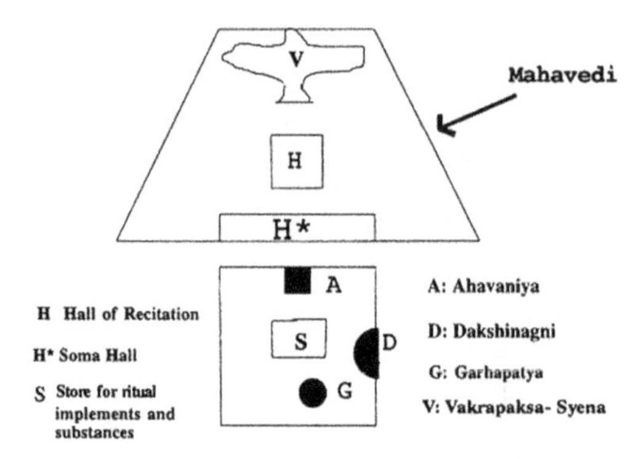

the sky). In the *Sulbasutras*, there is the suggestion that the areas of the three fire altars were equivalent and equalled 1 square purusha.[7]

There were other structures contained in the *Mahavedi* and *Pracinavamsa* (H, H* and S in Fig. 10.1) of functional and ritual significance, but yielding little of mathematical interest.

The instructions for the design of the *Mahavedi* provide an insight into the practical nature of the texts of the period. The *Apastamba Sulbasutra* (V.2) contains the following instructions, a longer version of the original cryptic instruction:

> To a cord of length 36 *prakramas*, add 18 *prakramas*. Make two marks on the cord, one at 12 *prakramas* and the other at 15 *prakramas* from the western end. Tie the ends [of the cord] to pegs on the ends of the East–west [*prsthya*] line of length 36 *prakramas*. Take the cord by the mark at 15 *prakramas* and stretch it to the south and mark the point with a peg. Do the same to the north. These are respectively the south west and the north west corners of the *Mahavedi*. Untie the ends of the cord from the East–west line and retie the end that was fastened previously to the peg on the east end to the west end and vice versa. Repeat the previous procedure but using the mark at 12 *prakramas* to obtain the south east and north east corners of the *Mahavedi*.

In Fig. 10.2, the length of the extended cord is 36 + 18 = 54 *prakramas*. From the other end, the 12th mark is half of the smaller parallel side while the 15th mark forms the base (AB) of a right-angled triangle (ABC), with its hypotenuse (BC) being the remainder of the cord (i.e., 36 + 3 = 39 *prakramas*) and the other side being the East–west line (AC) which measures 36 *prakramas*.

Apastamba gave other rational right-angled triangles that would satisfy the measurements required by the *Mahavedi*. These are the Pythagorean triples (3, 4, 5) multiplied by 4 or 5; (12, 5, 13) multiplied by 3; (15, 8, 17) and (12, 35, 37). All these triples may have been chosen initially to ensure that at least one side was of the same

[7] Seidenberg (1983, pp. 113–116) contains an interesting discussion of the ambiguities in the Vedic texts relating to equivalences of area as well as the philosophical underpinnings of such a requirement.

Fig. 10.2 Drawing: author

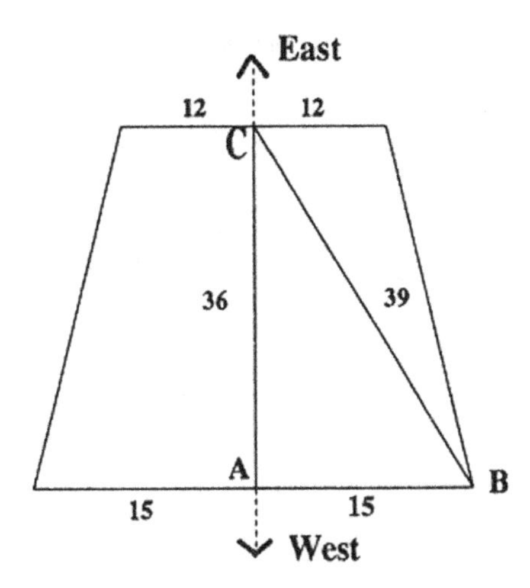

length as one side of the *Mahavedi*. In the same text appears a modification to the method of designing the *Mahavedi* that is suitable for any length of a given side.[8]

Another problem that led to some interesting mathematics related to the precise distances and relative positions of the three fire altars, *Garhapatya*, *Dakshinagni and Ahavaniya* contained in the *Pracinavamsa*, shown in Fig. 10.1. The general requirement was: *Dakshinagni* should lie south of the line joining the other two fire altars and its distance from the *Garhapatya* should be one third the distance between the other two fire altars.

The *Baudhyana Sulbasutra* contains three different versions of how this could be achieved. To quote the relevant passages, as given in Datta (1932, pp. 203–205), with some modifications for sake of clarity:

(1) With the third part of the length [i.e. the distance between Garhapatya (G) and Ahavaniya (A)], construct three squares closely following one another [from west towards the east]. Garhapatya is at the northwestern corner of the western square; Dakshinagni (D) is at its southeastern corner; and Ahavaniya at the northeastern corner of the eastern square (Baudhyana *Sulbasutra* 1.67).
(2) Divide the distance between the Garhapatya and Ahavaniya into five or six [equal] parts; add [to it] a sixth or seventh part; Divide [a cord as long as] the whole increased length into three parts and mark the end of the two parts from the eastern end [of the cord]. Fasten the two ends of the cord [to two] pegs at either end of the distance between the Garhapatya and Ahavaniya, stretch it towards the south, having taken it to the mark and fix a peg at the point reached This is the position of the Dakshinagni (Baudhyana *Sulbasutra* 1.68).

[8] Apastamba's procedure may be interpreted as follows: Let the cord placed on the East-west line be x units in length. If the length of the cord is extended by half the original length ($x + 1/2x$), and a mark is made at a distance of $5x/12$ from one end, then remaining part of the cord is $13x/12$. If we now tie the cord to the ends of the East-west line, and stretch it up to the mark, we get a right-angled triangle who sides are x, $5x/12$ and $13x/12$. This relationship will hold for any integral value of x.

Fig. 10.3 Drawing: author

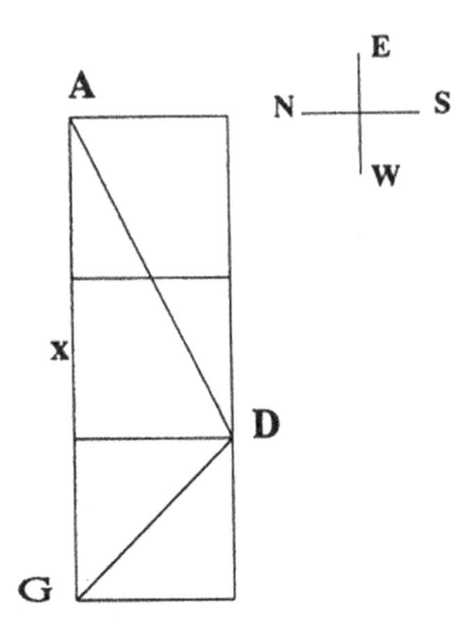

(3) Increase the measure [between the Garhapatya and Ahavaniya] by a fifth; divide [a cord of that length] into five parts and make a mark at the end of two parts from the western end [of the cord] after fastening the two ends to the east–west line. Stretch the cord towards the south having taken it to the mark and fix a peg at the point reached. This is the position of the *Dakshinagni* (*Baudhyana Sulbasutra* 1.69).

The figure constructed by Datta on the basis of these instructions, is shown in Fig. 10.3. He obtained various estimates of the relative distances between the fire altars.

Thus, if $AG = x$, it is easily shown that:

$$AD = \sqrt{\frac{5x}{3}} \text{ and } GD = \sqrt{\frac{2x}{3}} \text{ from (1);}$$

$$AD = \frac{7}{9}x \text{ or } AD = \frac{16}{21}x \text{ and } GD = \frac{7}{18}x \text{ or } GD = \frac{8}{21}x \text{ from (2);}$$

$$AD = \frac{18}{25}x \text{ and } GD = \frac{12}{25}x \text{ from (3).}$$

Moreover, if one assumes that the relative positions of all three fire altars are the same irrespective of the rule used, then the estimates for $\sqrt{5}$ and $\sqrt{2}$ are:

$\sqrt{5} = 2.333, 2.286, 2.169$; $\sqrt{2} = 1.166, 1.143, 1.44$

None of these estimates are accurate approximations, the best being only correct to the first place of decimals. These rules were essentially practical "rules of thumb" that an early surveyor might use, without mathematical considerations being predominant. However, this does not mean that considerations of accuracy did

Fig. 10.4 The first layer of a *Vakrapaksa-syena* altar: the wings are each made from 60 bricks of type a, and the body, head and tail from 50 type b, 6 type c, and 24 type d bricks. Each subsequent layer was laid out using different patterns of bricks with the total number of bricks equalling 200. Drawing: author

not occur in early Indian geometry. For example, to calculate the square root of 2, the following instructions are given both by Apastamba (1.6) and Katyayana (2.13) who came after Baudhyana: "Increase the measure by its third and this third by its own fourth less the thirty-fourth part of that fourth. This is the value of a special quantity in excess [which needs to be deducted]".

If we take one unit as the dimension of the side of a square, the above formula gives the approximate length of its diagonal as:

$$\sqrt{2} = 1 + \frac{1}{3} + \frac{1}{12} - \frac{1}{408} = 1.4142156\ldots.$$

The true value is 1.414213...

The *Sulbasutras* contain no clue as to the manner in which this accurate approximation was arrived at over 2,500 years ago. A number of theories or explanations have been proposed. Of these, a plausible one is that of Datta (1932), based on the "dissection and reassembly" principle, and discussed in Joseph (1992, pp. 234–36).

The Geometry of the Falcon-Shaped Altar

One of the most elaborate of the public altars (also found in the *Mahavedi* constructed for the *Agnicayana* ceremony) was shaped like a giant falcon just about to take flight (Fig. 10.4). It was believed that offering a sacrifice on such an altar would enable the soul of a supplicant to be conveyed directly to heaven by a falcon.

Most falcon-shaped altars were constructed with five layers of 200 bricks,[9] with each of these constructions reaching the height of the knee. For special occasions 10, 15, and, improbably, up to a maximum of 95 layers of bricks were used in their construction. The top layer of the basic altar was constrained to an area of 7.5 square *purushas*.[10] A *purusha* was defined as the height of a man with his arms stretched above him, say 2 m, which would give the altar an area measure of approximately 30 m^2. For the second layer from the top, the prescription was that 1 square *purusha* should be added, so that the total area would be 8.5 square *purushas*.[11] Similarly, each successive layer area was increased by 1 square *purusha*, so that in the rather exceptional (or more likely hypothetical) case of the 94th successive increase of 1 square *purusha*, the area of the base of this huge construction would be 101.5 square *purushas* or about 400 m^2![12]

---

[9] Two different types of bricks were used in altar construction. There were ordinary bricks (*lokamprina*) and special (*yajushmatt*) bricks, each of which was consecrated and then marked for purposes of identification. The bricks varied by size and shape and were used in different combinations in constructing different layers of the altars. Thus for example, the first, third and fifth layers of a falcon-shaped altar with six-tipped wings was made of 38 squares, 58 rectangles (of two sizes) and 104 triangles (of two sizes); the second and fourth layers were constructed from 11 squares, 88 rectangles (of two sizes) and 101 triangles (of six sizes and five shapes). Different configurations of these basic figures were used in construction of falcon-shaped altars with five-tipped wings, with different rituals being performed on different altars. Staal (1978) provides a detailed description of the construction of these altars and their accompanying rituals, with one of the most recent ones involving a five-tipped falcon-shaped altasr being performed in Kerala, South India, in 1990.

[10] Apart from minor variations, the body of the top layer of the falcon-shaped altar was 4 square *purushas*. The wings and tail were 1 square *purusha* each plus the wing increased by 1/5 of a square *purusha* each and the tail by 1/10 of a square *purusha* so that the image would more closely approximate the shape of a falcon. Thus the total area of the top layer is: $4 + (2 \times 1.2) + 1.1 = 7.5$ square *purushas* or approximately 30 m^2.

[11] In Katyayana *Sulbasutra* (5.4) appears the following instruction: "For the purpose of adding a square *purusha* [to the original falcon-shaped altar], construct a square equivalent to the original altar together with the wings and tail, add to it a square of one purusha. Divide the sum [i.e., the resulting square] into fifteen parts and combine two of these into a square. This will be the [new] unit of square purusha [for the construction of the enlarged figure]".

[12] The instructions given in *Sathapatha Brahmana* (X.2.3.11–14) for constructing a falcon-shaped altar consisting of 95 layers of bricks may be interpreted as follows:

Area of the body $= 56 + (12/7)(56)$

Area of two wings $= 2(14) + (3/7)(14) + (1/5)(1/7)(3)(14)$

Area of tail $= 14 + (3/7)(14) + (1/10)(1/7)(3)(14)$

The total area is about 116 square *purushas*, which is an over-estimate of the actual 101.5 square *purushas*, arising in part from a rounding-off error resulting from taking 14 rather than $(13 + 8/15)$. *Baudhyana Sulbasutra* contains an explanation of how the estimate of the total area was obtained. Expressed in modem notation:

Let the new unit after the mth augmentation be x.

Then

$x^2 = 1 + (2m/15)$ where m runs from 1 to 94.

For $m = 94$,

$x^2 = 13 + 8/15$.

Clearly, if in the construction of these altars the builders had to conform to certain basic shapes and prescribed areas or perimeters, two geometrical problems would soon arise:

(1) the problem of finding a square equal in area to two or more given squares;
(2) the problem of converting other shapes (for example, a circle or a trapezium or a rectangle) into a square of equal area or vice versa.

The constructions were probably achieved through a judicious combination of concrete geometry (the principle of dissection and reassembly),[13] and the application of ingenious algorithms, including the so-called Pythagorean theorem.

In the *Katyayana Sulbasutra* (named after one of the authors) appears the following proposition: "The cord [stretched along the length] of the diagonal of a rectangle makes an [area] which the vertical and horizontal sides make together" (2.11).

Using this version of the Pythagorean theorem, the *Sulbasutras* showed how to construct both a square equal to the sum of two given squares and a square equal to the difference of two given squares. Other constructions, including the

---

So we see that 14, the estimate used, is a rounding-up of this number. The use of the more accurate figure gives the calculated total area as 110 square *purusha*.

[13] The essence of this method involves two commonsense assumptions:

i. Both the area of a plane figure and the volume of a solid remain the same under rigid translation to another place.
ii. If a plane figure or solid is cut into several sections, the sum of the areas or volumes of the sections is equal to the area or volume of the original figure or solid. For example, the following sizes and shapes of bricks used to construct one of the layers of a falcon-shaped altar can be "dissected and reassembled" from a square:

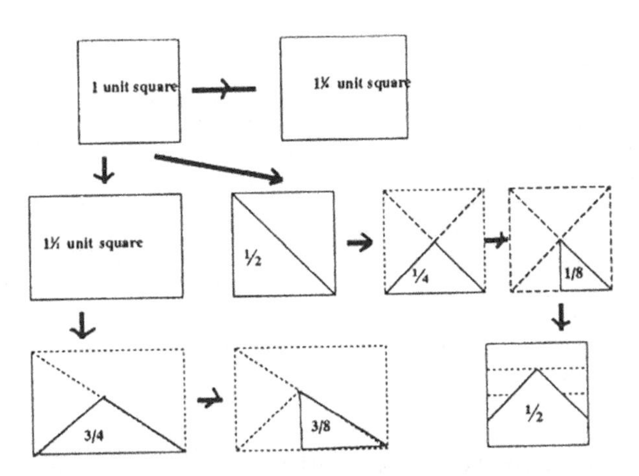

The reasoning behind this approach was very different from that behind Euclidean geometry, but the method was often just as effective, as shown in the Indian (and Chinese) "proofs" of the Pythagorean theorem. For further details, see Joseph (1992, 1994).

transformation of a rectangle (square) to a square (rectangle) of equal area and of square (circle) to a circle (square) of approximately equal area, were part of the repertoire. In geometrical terms, the constructions "doubling the square" and "squaring the circle" lead naturally to the devising of algorithms for the square root of 2 and other numbers in the first case, and the discovery of the inexact nature of the relationship between circumference of the circle and the diagonal (diameter) of the square in the second case.[14]

Apart from equivalences through area, the Vedic texts contain equivalences established between phenomena through numbers. The starting point is the centrality of the number 360 in the Vedic calendar and philosophy. Parallels are drawn between human anatomy and planetary motions. Thus, the *Caraka Samhita*, an early medical text, reckons the total number of *asthis* (or bones, teeth, nails, hard cartilages) in the human body to be 360, obtained from considering the 308 bones of the newborn babe (before they fuse into a smaller number of 206 in the adult), 32 teeth and 20 nails where each of these *asthis* are associated with each day of the year. The parallel between the nominal year (360 days) and man (*purusha*) *is* carried further in *Sathapatha Brahmana* where the basic falcon-shaped altar of 7.5 square *purushas* or 108,000 square *angulas* (1 *purusha* = 120 *angulas*) is linked to 10 nominal years or 108,000 *muhurtas* (1 *muhurta* = 48 min).[15] It is interesting in this context that a total of 10,800 ordinary (or *lokamprina*) bricks were used in the construction of the three fire altars found in the *Pracinavamsa*, the same as the number of *muhurtas* in a nominal year.[16]

Kak (1993) has argued that the concept of equivalence is of central importance in interpreting Vedic astronomical knowledge, so that in the design of altars an astronomical code was present which required deciphering. For example, the circular *Garhapatya* fire altar, which symbolised earth or the womb, was constructed with 21 ordinary bricks and had an area of 1 square *purusha*.[17] If the basic falcon-shaped altar having an area of 7.5 square *purusha* corresponded to

---

[14] There are interesting similarities and differences between the geometry of Greece and that of Vedic India. Both were used in the construction of sacred altars for ritual purposes; both had to solve the fundamental practical problem of how to construct a square equal to the area of a given rectangle. However, an important difference which shaped the way geometry developed in the two cultures was the Greek concentration on volume, notably the problem of "doubling the cube", while in India the principal questions involved the area of altars: the circular, the square, the trapezoid and combinations of these shapes. For a discussion of these constructions and the mathematics underlying them, see Joseph (1992, pp. 228–236; 1993, pp. 6–11; 1994, p. 184–189).

[15] Various rituals required the day-time and night-time to be divided into 2, 3, 4, 5 and 15 equal parts. In the 15-fold division, each part was a *muhurta*, which would be equivalent to 1/15 of (12 × 60), or 48 min.

[16] A number of other parallels based on the equivalence of numbers is found the Vedic literature of the period. For further examples, see Kak (1993).

[17] The choice of 21 is supposedly symbolic. It is the sum of 12 months, 5 seasons, 3 worlds and the sun; or the three sets of *rishis* (or planets); or the sum of five elements (earth, water, fire, air, space), five breaths (*prana, apana, vyana, udana, samana*), five organs of cognition (*jnanednriyas*), five organs of action (*karmendriyas*) and the inner ear (*antakarana*).

360 days, then 1 square purusha would be equivalent to 48 days. The augmentation of the basic falcon-shaped altar by 1 square purusha at a time was to be seen as a correction to make the altar correspond closer to an actual year (366 days or 372 *tithis* or lunar days). A *nakshatra* year was taken as the number of *nakshatras* (27) multiplied by the number of months (12) which would give 324 days (or *tithis*). An additional 48 *tithis* as a correction was needed to get an actual year. However, this would mean an excess of 0.93761 *tithi* every year, since the number of *tithis* in a solar year is 371.06239. Thus, by constructing a falcon-shaped altar of 95 layers symbolising a 95-year cycle, with each augmentation being 1 square *purusha* (or 48 *tithis*), starting from 7.5 square *purushas* to 101.5 square purushas at the end of that cycle, the practice of a major adjustment every 95 years to the calendar by 90 *tithis* (or 3 lunar months) made sense. Such a correction implied that the length of the solar year is: $372 - (90/95) = 371.05263$ *tithis* which corresponds to 365.24675 days. Comparing this value to the present-day estimate for the tropical year of 365.25636 days, we are struck by the accuracy of estimates which are at least 3,000 old.

Conclusion

There is a view that Indian mathematics originated in the service of religion. The proponents of this view have sought their main support in the complexity of motives behind the recording of the *Sulbasutras*. Since time immemorial, they argue, the needs of religion have determined not only the character of Indian social and political institutions, but also the development of her scientific, knowledge. Astronomy was developed to help determine the auspicious day and hour for performing sacrifices. The 49 verses of *Jyotisutras* (the *Vedanga* containing astronomical information) gave procedures for calculating the time and position of the Sun and Moon in various *nakshatras* (signs of the zodiac). A strong reason in Vedic India for the study of phonetics and. grammar was to ensure perfect accuracy in pronouncing every syllable in a prayer or sacrificial chant. So too, the construction of altars and the location of sacred fires, as we have seen, had to conform to clearly laid-down instructions about their shapes and areas if they were to be effective instruments of sacrifice.

However, there is a danger that the magico-religious beliefs surrounding the Vedic rituals may be overly emphasized when considering the origins of Indian mathematics. We have seen the crucial role played by the *Agnicayana* ceremony in generating geometrical concepts and techniques found in the *Sulbasutras*. The rituals associated with the construction of fire altars may be looked at from two standpoints. The first is from the standpoint of the beliefs connecting the shapes of altars with the specific desires to be fulfilled by their use in the sacrifices. The second is that of technology pure and simple: How exactly were the altars constructed to conform to specific shapes, specified size and by using a specific numbers and types of bricks?

It is clear that the geometry originating in the *Sulbasutras* had little to do with the first standpoint. Thus, for example, whether a falcon-shaped altar ensured for the sacrificer heaven or the annihilation of enemies was totally irrelevant to the problem of constructing it to conform to certain size and shape. As a matter of fact, these problems would be the same if somebody wanted a structure to be built in the garden for ornamental purposes. In other words, the geometry developed in the *Sulbasutras* was aimed at solving technological problems involved in constructing brick structures. It is this geometry that is being studied by historians of mathematics.[18]

Once the *Sulbsutras* are seen as primarily manuals for technicians, the questions then arise of where and when the practical knowledge relating to bricks and brick technology were acquired. References to bricks are conspicuous by their absence from the most sacred and earliest of Vedic literature, the *Rigveda Samhita*. When they do make an appearance in a recension (*Tattiriya Samhita*) of a later *Veda*, the *Yajurveda Samhita*, bricks are viewed as marvellous and mysterious entities.[19] In the same text, there are exhortations that "tiles or potsherds" from the ruined Harappa cities should be gathered for ritual purposes. It is, therefore, likely that the priests were acquainted with the burnt bricks from the same sites and would in course of time invest them with magico-religious properties.

In one of the last recensions to *Yajurveda* appears the *Sathapatha Brahmana*, which, as mentioned earlier, contains a description of the *Agnicayana*. The magico-religious elements of this ritual are accompanied by a short discourse on the construction of brick altars of various shapes and sizes. While the discussion lacks the geometrical sophistication of the *Sulbasutras*, it is clear that the knowledge of brick technology which was abundantly evident in the Harappa culture was slowly percolating into the Vedic rituals, to become the most critical element of Vedic constructions.

Where, then are we to look for the origins of geometry in India? The common view is that the *Sulbasutras* are the source. However, one hypothesis is that if the geometry embodied in the *Sulbasutra* texts is to be viewed as the outcome of a long and sophisticated tradition of brick technology, this geometry must have come into being when there was in fact an advanced form of brick technology with a long tradition behind it. This, in other words, would mean that whatever may be the time

[18] There are indications in the texts that the authors of the *Sulbasutras* were aware of this distinction, for often one comes across expressions, "thus we are told", "such are our instructions", etc. The implication is that these instructions (say, on the sacrificial efficacy of different shaped altars or the astronomical codes to be adhered to) were not particularly relevant to the main purposes of the texts. These instructions were simply taken for granted, while the texts themselves paid exclusive attention to the technique of executing them. In fact, the texts are exemplars of how exact science may grow directly out of applications.

[19] Consider the following passage from the *Taittiriya Samhita* (iv.4.11): "May these bricks, O Agni, be milch cows for me, one, and a thousand, and a million, and ten million, and a hundred million, and a thousand million, and ten thousand million,.,,; may these bricks, O Agni, be for me milkers of desires named the glorious yonder in yon world".

of the actual codification of the *Sulbasutras*, their contents come down from a different period. That must have been a period of flourishing brick technology. Only one period answering to all this is known in ancient Indian history, and that is the period of the Harappa civilisation mentioned in the introduction. The presumption, in short, is that geometry which was eventually codified in the *Sulbasutras* could have come down from the Harappan period. If this presumption is correct, the first and earliest of the discontinuity in the chronology of Indian mathematics has been filled with the assistance of bricks.

Biography George Gheverghese Joseph received his PhD and MA from the University of Manchester. In October 2000, he was called to the Bar of the Middle Temple, London. He is currently Honorary Reader, School of Education, University of Manchester. He has travelled widely, holding university appointments in East and Central Africa, Singapore, Papua New Guinea and New Zealand as well as a Royal Society Visiting Fellowship (twice) in India during which he gave lectures at several universities. A third edition of his classic (greatly revised and expanded) *"The Crest of the Peacock: Non-European Roots of Mathematics" published by Princeton University Press appeared in 2011, following his book on Kerala School of Mathematics and Astronomy,* "The Passage to Infinity". Other major books include: *Women at Work* (Philip Allan, Oxford, 1983), *Multicultural Mathematics: Teaching Mathematics from a Global Perspective* (Oxford University Press, 1993); *George Joseph: Life and Times of a Kerala Christian Nationalist* (Orient Longman, 2003).

References

Datta, B. 1932. *The Science of the Sulba*. Calcutta: University Press.

Frawley, D. 1991. *Gods, Sages and Kings*. Salt Lake City: Passage Press.

Joseph, G. G. 1992. *The Crest of the Peacock: Non-European Roots of Mathematics*, 3rd printing. London: Penguin. (A third and greatly expanded third edition was published in 2011 by Princeton University Press)

———. 1993. What is a Square Root: A study of geometrical representation in different mathematical traditions. Pp. 3–14 in *Proceedings of the 1993 Annual Meeting of CM.E.S.G.*, M. Quigley, ed. Calgary: University of Calgary.

———. 1994. Different Ways of Knowing: Contrasting Styles of Argument in India and the West. Pp. 183–198 in *Selected Lectures from the 7th International Conference on Mathematical Education*. Sainte-Foy, Quebec: Les Presses de l'Universite Laval.

Kak, S. C. 1987. On the astronomy in ancient India. *Indian Journal of History of Science* **22**: 205–221.

———. 1993. Astronomy of the Vedic Altars. *Vistas in Astronomy* **36**: 117–140.

Mackie, E. W. 1977. *The Megalithic Builders*. Oxford: Phaidon.

Seidenberg, A. 1983. The geometry of the Vedic Rituals. Pp. 95–126 in *Agni, The Vedic Ritual of the Fire Altar*, vol. II, F. Staal, ed. Berkeley: Asian Humanities Press.

SEN, S. N. and A. K. BAG. 1983. *The Sulbasutras.* New Delhi: Indian National Science Academy.
STAAL, F. 1978. "The Ignorant Brahmin of the Agnicayana," Annals of the Bhandarkar Oriental Research Institute, Diamond Jubilee Number, pp. 337–348.
THIBAUT, G. 1875. On the Sulbasutra. *Journal of the Asiatic Society of Bengal* **44**, 3: 227–275.

Chapter 11
Inauguration: Ritual Planning in Ancient Greece and Italy

Graham Pont

Like so many visitors to Greece, I was immediately overwhelmed by the beauty of the old temples and their harmonious relationship with each other and the surrounding landscape. It was only later that I realised how such powerful impressions are repeatedly conveyed by a range of building types, in various states of disintegration, ruin and reconstruction, and in widely differing settings, from coastal to mountainous. It took even longer to notice a strange anomaly: many of these sacred sites seem to have been laid out with almost complete disregard of orientation, axial alignment and geometrical planning; yet the buildings themselves are masterpieces of geometrical symmetry and axial planning.

It was more than 30 years later that I came across the doctoral thesis of Constantinos A. Doxiadis (1913–1975) which offers an ingenious and original explanation of the apparently random planning of the Greek temple sites. First published in German in 1937, his work has become widely known in the English translation, *Architectural Space in Ancient Greece* (Doxiadis 1972), which includes revisions, corrections, additional photographs and supplementary references to the literature.

Doxiadis maintains that, despite appearances, the ancient temple sites (and some secular spaces as well) were laid out rationally not by the system of axial or Cartesian coordinates used at the drawing board, but according to a 'natural' system of polar coordinates established on the actual building site (Doxiadis 1972: 4–5). Polar coordinates are measures taken from a pole or fixed point in this case, the position of the human planner standing at the site which precisely locate another point or object in the surrounding landscape. That point is located by two numbers: the first gives the distance **r** of that point from the pole or viewer's

First published as: Graham Pont, "Inauguration: Ritual Planning in Ancient Greece and Italy", pp. 93–104 in *Nexus VI: Architecture and Mathematics,* Sylvie Duvernoy and Orietta Pedemonte, eds. Turin: Kim Williams Books, 2006.

G. Pont (✉)
54 Birchgrove Road, Balmain, NSW 2041, Australia
e-mail: graham_pont@hotmail.com

K. Williams and M.J. Ostwald (eds.), *Architecture and Mathematics from Antiquity to the Future*, DOI 10.1007/978-3-319-00137-1_11,

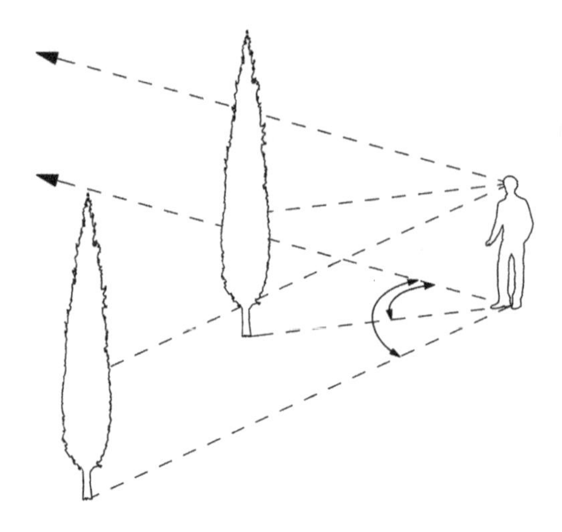

Fig. 11.1 Establishing polar coordinates. Image: Michael Dawes, after Doxiadis (1972: 4)

position; the second gives the position of that same point on the circumference of a circle whose centre is the pole and whose radius is **r**. This second number gives the size of the angle (**x** degrees) between any two 'directed' radii or sightlines from the viewing point to the points being viewed (Fig. 11.1).

The position of the planner's viewpoint is critical to this system: for the Greek planners it was usually located at the propylon of an acropolis or some similarly important entrance to the site (Figs. 11.2 and 11.3).

Standing at this point, the planner could make coordinated sightings to the perceived horizon, thus creating radii of an enclosing circle centred on the human viewer. These radii extend either to the outermost point where the sky appears to meet the earth or to some nearer object that interrupts the long view. Thus the position of a temple in the perceived landscape can be accurately defined by the angle between the two radii that touch on the visible vertical extremities of the temple (the edges of the cornice or stylobate) and the distance of these points from the viewer (**rx**). Similarly, one can plot the relative position of other temples and monuments on the site, as viewed from the same vantage point.

To test his hypothesis, Doxiadis examined 29 ancient sites, of which only eight were reasonably intact or 'authoritatively reconstructed' at the time of the research. Despite the limitations of his sample, Doxiadis's results are highly significant and, indeed, quite astonishing. Most significant is his conclusion that urban sites of the archaic, classic and hellenistic periods were developed through the employment of a 'uniform system in the disposition of buildings in space that was based on principles of human cognition' (Doxiadis 1972: 3).

For 400 years or more, the determining principle of sacred and at least some secular planning in Greece seems to have been anthropocentric[1]: Doxiadis found

[1] Doxiadis cites the famous dictum of the fifth-century sophist Protagoras of Abdera: 'Man is the measure of all things'. This certainly seems relevant, even though Doxiadis's system is found in sixth-century sites.

Fig. 11.2 Reconstruction of the Acropolis after 450 BC: Perspective view from Point A (see Fig. 11.3). Image: Michael Dawes, after Doxiadis (1972: 35)

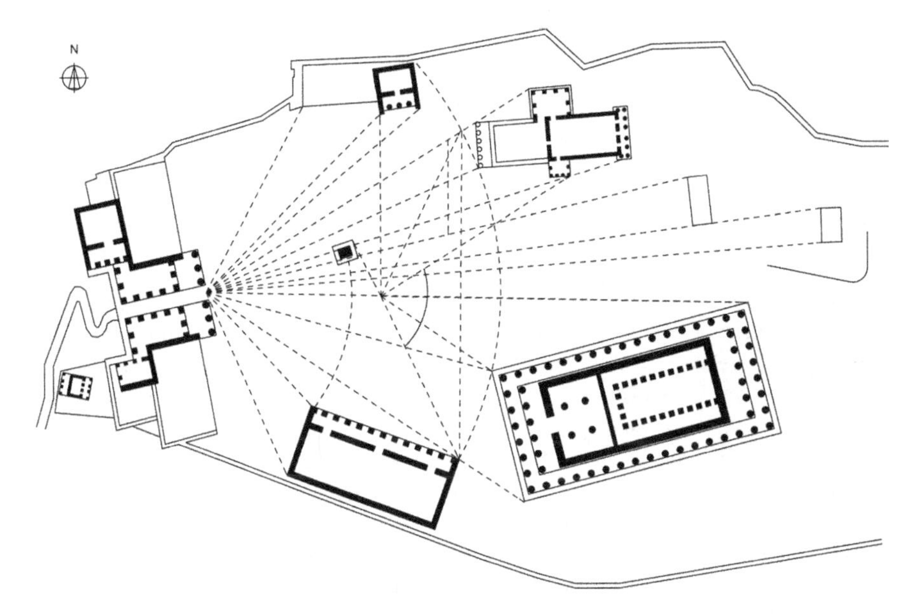

Fig. 11.3 Athens, Acropolis after 450 BC: Plan by C.A., showing Point A at eastern entrance of the Propylaea and radiating sightlines to other buildings. Image: Michael Dawes, after Doxiadis (1972: 37)

that, with few exceptions, temple site development and redevelopment were based on the human view of the scene, as taken from the 'first and most important position from which the whole site could be observed' (Doxiadis 1972: 5). This polar point was usually the principal entrance, at the end of the traditional approach or sacred way. The crucial viewpoint lies 'where the mathematical axis of the propylon intersects the line of its innermost step (i.e., the final step before one entered the sanctuary) at the height of approximately 5′7″, the eye level of a man of average height' (Doxiadis 1972: 5).

Even though the precise location of the original viewing point is sometimes debatable (Stevens 1940) and some of Doxiadis's sightlines might have been obstructed by lesser monuments (Scully 1962: 5), there remains a considerable body of evidence indicating that the Greek buildings were positioned by canonic angles of vision and the distances from the viewer (the distances being apparently based on geometric ratios derived from the angles of vision). Important buildings, such as the Parthenon and the Erechtheum, were placed so that the observer could enjoy a three-quarter view of the entire edifice. If this was not possible, a building could be completely hidden from view but not partially concealed. Buildings were positioned so as to relate to features of the surrounding landscape and create a 'unified composition'. Ritually important views were left open to the countryside and usually oriented east or west, or in a direction determined by the sacred way or local cult traditions (Doxiadis 1972: 3–5).

In these principles, Doxiadis concluded, lay the secret of ancient Greek planning, its sense of human scale and its power of 'satisfying man and uplifting his spirit as he entered a public space whether it was a precinct sacred to the gods with its temples and votive columns or the agora with its stoas and statuary' (Doxiadis 1972: ix).

The 10- and 12-Part Systems

Doxiadis's most surprising results came from a mathematical analysis of the angles of vision, which revealed that the laying out of these sites was determined by two distinct geometries: generally speaking, the sites of buildings in the Doric style of architecture were organised on a 12-part division of the 360° horizon circle, whereas those of the Ionic style were mostly based on a 10-part division of the horizon circle. Typical angles of vision in the former are 30°, 60° and 90° and, in the latter, 18°, 36° and 72°. Doxiadis also observed that the Doric 12-part layout always involved a path or sacred way that was left open and unobstructed so that the views from the temple site could extend out into the surrounding landscape, whereas the Ionic 10-part layout had closed views (or the impression of closure) and the path was 'wholly incorporated within the layout' (Doxiadis 1972: 8).[2]

In both systems the visible horizon was viewed as a continuous line intersected by the radii of the canonic angles of vision at mathematically precise points the vertical edges of temples, altars, statues and other constructions, all carefully positioned so as to appear in the field of vision as part of a regular rhythmic pattern that united the built with the natural environment. Thus the main aim of the planners was 'to bring the outlines of the buildings into harmony with the lines of the landscape' (Doxiadis 1972: 8); but the system also applies to enclosed spaces, such as peristyles and courtyards viewed from one or more points of entry (Doxiadis 1972: xviff, 52, 58–61, 66, etc.) and even to sites which are orthogonally planned (Doxiadis 1972: 66, 138, 147).

The planning technique revealed by Doxiadis appears to be a form of *architectural scenography*, the mathematical organisation of the perceived

[2] Was this remarkable difference of layout associated with gender stereotypes? The Greeks considered Doric to be masculine, and Ionic feminine.

environment in the manner of large-scale perspective scene-painting: this art was developed by the fifth-century painter Agartharcus of Samos and further investigated by the philosophers Democritus and Anaxagoras. According to Vitruvius, all three wrote books on the subject (Elderkin 1912). A well-known example of architectural scene-painting is the irregular placement of the door and windows in the western wings of the Propylaea, so that the openings might appear to the viewer coming up the sacred way as being symmetrically placed between the columns of the porch (Elderkin 1912).

When the original viewing point can be identified, Doxiadis's theory is precise and empirically testable; when the viewing point is unknown, canonic angles of vision can sometimes help to locate that viewpoint. The Propylaea not only opens eastward into the temple precinct but also looks out west over the sacred way from the Agora. If this prospect is surveyed from another very likely viewpoint at the intersection of the entry axis and the western edge of the stylobate we find that the view of the temple to Athena Nike on the left falls exactly within a 30° angle of vision: another application, presumably, of the 12-part system in the Periclean reconstruction of the Acropolis (Fig. 11.4).[3]

According to Doxiadis, the earliest 10-part system dates from c.550 BC and the earliest 12-part from 530 BC, and both systems were employed by the Greeks until the second century BC. But he also found evidence of the 10-part system being used by the Romans at Palmyra, during the first century AD (Doxiadis 1972: 21–2, 29). It should be noted that both systems apply to the ground plans of temple sites and the horizontal disposition of their monuments. In summarising his results Doxiadis has included details of temple proportions (Doxiadis 1972: 9–14), but he does not examine the possibility that polar coordinates might also have been used in the vertical plane, to control the apparent height of buildings and their perceived relationships with each other and the surrounding landscape (Stevens 1940: 5).

If the Romans did employ the Greek planning system in their colonies, they might also have used it back home. Furthermore, that system could have been introduced to Italy through the establishment of Greek colonies, especially in the south (Magna Graecia) where there were numerous settlements from the eighth century BC onwards. Rome itself has a very irregular plan[4]; and so has Hadrian's Villa (c. 118–134 AD), which was designed by the Emperor himself an architect and well-travelled connoisseur of Greek culture. Since the Villa was packed with

[3] When the other principal buildings are viewed from the eastern end of the Propylaea, the angles of vision are all 30° (Doxiadis 1972: 32); except for the angle between the west porch of the Erechtheum and the nearest corner of the Parthenon, which is 36° (a tenth part of 360). This change to the Ionic ten-part system may have been an imperial gesture towards reconciliation of the western and eastern systems just as the Periclean rebuilding included two Ionic temples as well as incorporating Ionic columns in the Doric Propylaea. See Scully (1991: 83–84).

[4] Rykwert (1988: 72, 106ff). Note especially the reconstructed perspective view of the Capitol at the Roman colony of Cosa (Rykwert 1988: 122): the temples are not all orthogonally aligned and the view is very similar to some of Doxiadis's irregular Greek sites e.g., (Doxiadis 1972: 35, 81, 87, 89).

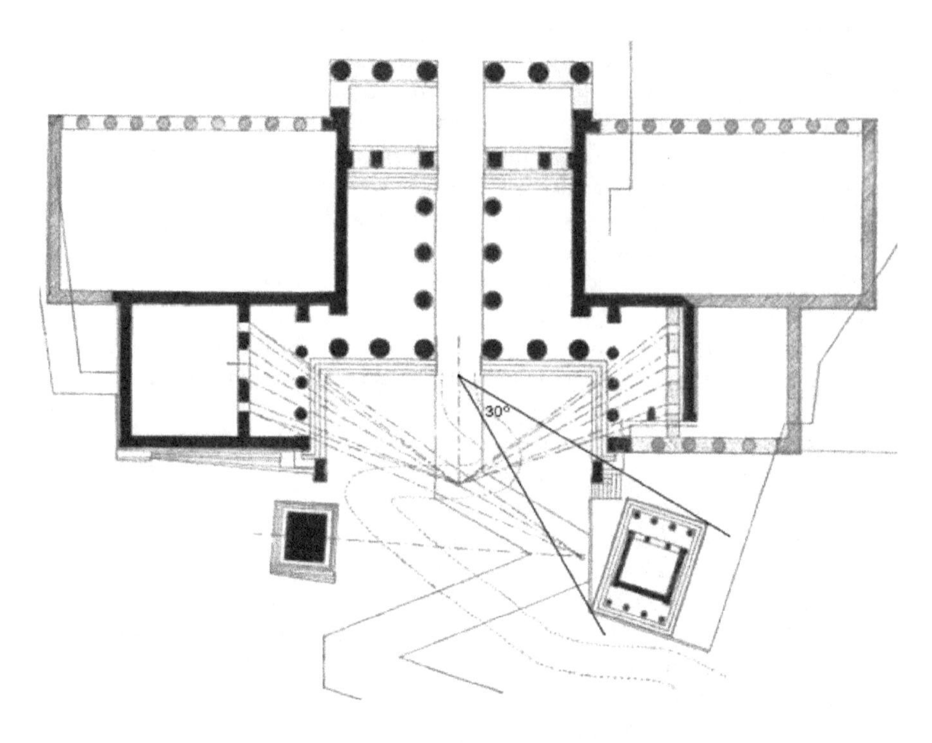

Fig. 11.4 Athens, Propylaea: Plan showing conjectural approaches, viewpoint and sightlines to porches. The temple of Athena Nike appears to have been positioned so that it would be viewed from the western entrance within a 30° angle of vision. Image: Elderkin (1912: 12)

allusions to Greek architecture and landscape, Hadrian may well have organised the scenery and its principal viewpoints according to the old Greek system.

After reviewing the very limited evidence of early Greek ideas on planning, Doxiadis concluded that the method of planning by polar coordinates was what Aristotle meant by the 'traditional system' (*archaioteros tropos*), as opposed to the 'new system' of grid-iron or chequer-board planning frequently but wrongly attributed to Hippodamus of Miletus (fifth century BC) (Doxiadis 1972: 20). Doxiadis apparently did not notice that *tropos* was also used as a technical term in Greek music theory to mean 'tuning' (as a synonym of *tonos* and *harmonia*) (Michaelides 1978). This striking coincidence of musical and architectural terminology could be another significant clue to the lost theory of Greek planning.

Reconstructing the Etruscan Rite

No one, to my knowledge, has attempted to extend Doxiadis's analysis to ancient Greek or Roman town plans of Italy. His work has been entirely ignored by Joseph Rykwert in three editions of an influential book, *The Idea of a Town*[5]; yet Rykwert's inquiry arose from a similar dissatisfaction with the modern city, its chaos, alienation and lack of focus. His primary subject, however, is Italy and particularly Rome itself; but his investigation of ancient foundation rituals and city planning allows for the possibility of analogous Greek procedures, of which almost nothing is known (Rykwert 1988: 86–88).

Early Roman planning borrowed much from the Etruscans, particularly in the rituals of divination and inauguration conducted to determine a propitious time and place for the creation of a new city. These were known as the 'Etruscan Rite' or 'Discipline'. The foundation rite was the responsibility of a state official, an augur, who conducted the procedure on a prominence which gave him a clear view of the surrounding countryside. Here he sought the will of the gods in various omens, including the flight of birds. An essential part of the ritual was the formal delineation of a 'templum':

> Temples and wild lands be mine in this manner, up to where I have named them with my tongue in proper fashion.
>
> Of whatever kind that truthful tree is, which I consider that I have mentioned, temple and wild land be mine to that point on the left.
>
> Of whatever kind that truthful tree is which I consider that I have mentioned, temple and wild land be mine to that point on the right.
>
> Between these points, temples and wild lands be mine for direction, for viewing, and for interpreting, and just as I have felt assured that I have mentioned them in proper fashion (Kent 1958: II, 275).

In commenting on Varro's record of this incantation, Rykwert cites the story of a famous Etruscan augur who drew a templum on the ground with his augural staff (*lituus*) to mark the site (or centre) of a new city. Rykwert then goes on to show how the ritual of inauguration proceeded with the division of the site into four quarters with the east-west *decumanus* and the north-south *cardo*, which, he assumes, would become the principal streets of a chequer-board town-plan (Rykwert 1988: 89ff).

It is important to note here, however, that Rykwert has interpreted the Etruscan ritual formula as if it was used *solely* as a means of establishing a cardinally oriented and orthogonally planned city.[6] But his extensive survey of the ancient world has identified only one Etruscan city, Marzabotto, with this idealised form and no more than half a dozen other early sites which exhibit that same form (Rykwert 1988: 42, 72ff, 78ff, 85ff, 194). *So Rykwert's version of the foundation*

---

[5] Joseph Rykwert in three editions of an influential book, *The Idea of a Town* (1976; revised 1988; reissued 2002).

[6] Rykwert's argument is further weakened by his admission that 'orthogonal planning... is not immediately dependent on the Etruscan or any other related rite...' (Rykwert 1988: 72).

scenario leaves all the irregular Greek, Etruscan and Roman city plans completely unaccounted for.[7] Furthermore, since Rykwert also admits that 'we have no guide to tell us how the ancients laid out the public buildings and temples in relation to the plan of the town' (Rykwert 1988: 57), his ignoring of Doxiadis's theory which might well have solved the mystery is all the more puzzling.[8]

Doxiadis's theory is immediately brought to mind by Varro's comment that 'in making this temple, it is evident that the trees are set as boundaries, and that within them the regions are set where the eyes are to view...' (Kent 1958: II, 275).[9] The augur, it seems, established coordinates emanating from where he stood (with his staff, presumably, as the originating pole) to nominate objects which marked the limits of a *templum* or sacred site. Note also that the augur's gaze or sighting (*conspicio*) takes in 'temples and wild lands' the open and continuous view of the horizon that seems to have been the basis of early Greek planning too (Rowland and Howe 1999: 152 ff). This interpretation is confirmed by Varro's quotation from the poet Naevius:

> Where land's semicircle lies,
> Fenced by the azure vault.

'Of this temple, the four quarters are named thus: the left quarter, to the east; the right quarter, to the west; the front quarter, to the south; the back quarter, to the north.' (Kent 1958: II, 275). Varro's comment on these verses leaves little doubt that *templum* here refers to the 180° view bounded by the horizon and that the complete circle of 360° was ritually divided into four equal parts of 90°. However, when Varro asserts that the *templum* 'ought to be fenced in uninterruptedly and have not more than one entrance' (Kent 1958: II, 281), he is evidently referring to the augur's viewing position (the *auguraculum*) (Rowland and Howe 1999: 154, Fig. 12). This curious requirement has no apparent relevance to Rykwert's scenario of orientation and orthogonal planning but some such restriction would have been absolutely necessary in Doxiadis's system to ensure that the visitor arrives at the principal viewing point of a sacred site and surveys the scene appropriately.

The sectioning of the visible field into four quadrants is central to the rite of inauguration, but the quadrant can be equally subdivided by either of Doxiadis's 12- and 10-part systems; so Rykwert's highly conjectural reconstruction of the foundation ritual could easily be modified to accommodate Doxiadis's system.

---

[7] In other words, Rykwert's reconstruction of the Etruscan rite is not supported, to any significant degree, by the available archaeological evidence. However, his fifth chapter, 'The Parallels', confirms that he will not abandon his fundamental urban paradigm, the rectilinear grid: this turns out to be a Procrustean bed which simply cannot accommodate the irregular sites so neatly explained by Doxiadis.

[8] Space does not permit a review of the mixed reception accorded to Doxiadis's theory. For a rare sympathetic (yet not uncritical) response, see Scully (1962: 5, 1991: 68–69).

[9] Since ancient times, trees have been revered as sacred sites and boundary marks. Cf. the Australian colloquialism 'beyond the Black Stump' (meaning the remote outback or inland).

Whether the Greek system of polar coordinates was actually employed in Italy (where there were numerous early cities and temple sites laid out irregularly) remains an open question; but our inquiry has at least established the possibility that the 'traditional' Greek system of planning was based on ritual practices analogous to the 'Etruscan Discipline'.[10]

Temple as Macrocosm and Microcosm

Templum is a systematically ambiguous term. Varro distinguishes three levels of templa, the celestial, the earthly and the subterranean a conception deriving from archaic shamanism (Kent 1958: II, 271). *Templum* was probably derived from the same root as the Greek *temenos*, meaning 'a space cut off' (that is, reserved as a sacred enclosure); though *templum* can also refer to a celestial form or pattern.

The delineation of a heavenly *templum* and the marking out of its earthly imitation or symbol on the ground was an essential part of ritual inauguration; but this process did not necessarily involve actual building (Kent 1958: II, 281; Rykwert 1988: 100)[11]: the ritual primarily consisted in 'cutting off' an earthly space (creating a *temenos* or microcosm) according to some celestial or macrocosmic template and so the ritual could have been older than the art of monumental building. Varro's augural incantation clearly derives from the pre-literate era when boundaries, land-use and ownership were ritually sanctioned and remembered in song and dance.[12]

All human knowledge was once preserved in song and dance by a global culture that saw the motions of the Sun, Moon and stars as the dance of animate beings. As our distant ancestors slowly came to recognise the cycles of the months, seasons and years they began to imitate them more or less exactly in ritual formation and religious liturgy. To them the world was not only cyclic but visibly circular: from the blue vault of the sky and the sublime procession of the stars to the ever-encompassing horizon. Their view of the world was regularly displayed and remembered in astral or round dances and other circular symbolism; and, when they finally came to build permanent houses and temples, the first seem to have been mostly circular too.

[10] On possible connections between early Greek and Etruscan planning, see Rykwert (1988: 85–88, 195).

[11] Thus 'templum' usually denotes the demarcation and limits of space; and, by extension, a sacred building erected therein; but 'templum' was sometimes also used to denote a 'cut off' portion of time. Cf. Rykwert's suggestive inference that, according to Roman law, 'the sunlit day is the equivalent in time to the space of the *templum*'.

[12] The old British custom of 'beating the bounds' was clearly a ritual means of preserving the memory of parish boundaries. The ancient Romans observed a similar rite on 23 February in the festival of Terminalis (god of boundaries).

From this archaic 'musical' culture, we have inherited the circle as the prime model of the world, as a type and symbol of perfect harmony and a fundamental tool for measuring and allocating both space and time. Much of the conceptual richness of that world-view is encapsulated in the ancient 'tem-' vocabulary: Tempe, temper, tempera, temperament, temperamental, temperance, temperate, temperature, tempest, tempestuous, template, temple, tempo, temporal, temporary, temporize and, apparently, tempt, temptation, etc. To this we could add derivatives such as contemplate, distemper, esteem, estimate and extemporise, along with words of variant spellings, such as tamper, tense and terminus, as well as several others that have fallen into disuse.[13]

These terms are all linked by the core meaning of 'section cut off'; that is, measure: *precise measure*, as in template, terminus and temperature; *due measure*, as in temper, temperament, temperance; even *undue measure*, as in tempestuous, temperamental, intemperate etc. But how did this vocabulary extend to the sacred *temenos* and the augur's *templum*? And what do these have in common with *tempus* (time[14])?

Having mastered the understanding and imitation of the circle, our ancestors hit upon the regular or rhythmic division of the circle at first, presumably, in myth, ritual, dance and seasonal celebration. That sense of rhythm, of the circular pattern and its measured divisions became the basis of the primitive 'musical' world-view which finally gave rise to a sophisticated mathematical cosmology, the 'Harmony of the Spheres'. According to the archaic world-view possibly thousands of years older than Pythagoras the circular forms and cyclic movements of the heavenly *templum* were perfectly tuned (tempered) and therefore ideal models (templates) for the harmonious organisation of earthly space (the *temenos* and its defining *termini*) and the precise measurement (cutting off) of time. In essence, the *temenos-templum* is a circular space ritually cut off and dedicated (among other things) to the sectioning of space and time: hence temples are commonly built on (or made into) elevated sites suitable for terrestrial and celestial observations.

Thus, from remote but uncertain origins, we have inherited the *perceived* circle and its regular divisions as a paradigm for conceptualising and measuring space and time. In the heavenly system we discern three great circles. The **Ecliptic** is the apparent orbit of the Sun; the **Zodiac** is the broader celestial circle which includes the observed motions of the Sun, the Moon and the Planets (their circles have long been divided by the 12-part system but we also know of zodiacs based on the decimal system). The third great circle is the terrestrial **Equator** which encircles the Earth equidistant from the poles: it is also divided by the 12-part system.

[13] The classic study of the harmonic world-view and its pervasive vocabulary is Spitzer (1963). A similar study of the harmonic vocabulary in Asiatic languages would doubtless yield similar results.

[14] The English term is derived, not from the *tem* root, but a (related?) Indo-European root *di-mon* whose base *da* also means 'cut off'.

This last circle is in the same plane as the astronomical Equator whose plane is perpendicular to the Earth's axis. This circle is essential to the modern art of time keeping, as it enables the equating of the hours, the division of both day and night into 12 equal parts. The ancient art of time reckoning employed the circle and the same duodecimal divisions for the notional year of 360 days from which we developed our sexagesimal division of hours, minutes and seconds. There were decimal calendars too as our names for the last 4 months testify but, while the decimal system had obvious advantages in some areas, it was of limited use in making and measuring music. With the duodecimal system, however, it was possible to develop a more comprehensive system of musical time (the metres of song and dance[15]); to explore the intricacies of harmonic space with monochord circles divided into 12 parts[16]; and, most importantly, to develop a precise analogy between the perceived circles of heaven and earth and a common mathematical system that brought both cosmic orders into rational harmony.

Conclusion

Viewed against this vast panorama, the Etruscan augur's ritual surveying no longer appears far removed from the mathematical scenography of the Greek temple planner. Both, no doubt, were derived from sacred arts (Rykwert 1988: 60); and both appear to have addressed the surrounding landscape as a visual *templum*, surveying the chosen site from a precisely defined vantage point, sectioning the horizon circle with natural or made-made land-marks or *termini* (mountains, rocks, trees, boundary stones, and, finally, buildings and other monuments) all with the aim of creating a harmonious microcosm in imitation of the heavenly macrocosm. While there is no documentary record of ritual surveying in ancient Greece (Rykwert 1988: 80), Doxiadis has identified precise archaeological evidence indicating that the methods used by the Greek surveyor-planners did not differ in principle from the Etruscan Discipline. This conclusion, I suggest, is supported by the close philological connection between *temenos* and *templum* and their common origin and/or long association in the prehistoric musical world-view.

The musical or harmonic world-view and its fundamental numbers have played a major role in the arts, techniques and sciences since ancient times: the archetypal harmonic number 360, for instance, was central to the mensuration of the Assyrians, Egyptians, Hebrews and Indians and the name for that number in Assyrian and Hebrew also meant 'earth' and 'horizon' (cf. Latin *orbis*) (McClain 1976: 33 ff. 1978: 26 ff. Oppert 1887: 87).

[15] Duodecimal forms of music include the 12-bar Blues and the Jig, Tarantella and Siciliana in 12/8. The chromatic octave scale is divided into 12 semitones.

[16] See e.g. McClain (1978: 4, 11, etc.). The rich history and prehistory of the numbers 10 and 12 have been greatly illuminated by this book and its wide-ranging predecessor (McClain 1976).

Our synopsis leaves outstanding a problem that is too large to resolve here: how, why, when and where did the decimal and duodecimal systems arise and come to coexist? My tentative answer would start from something like this: the decimal or 10-finger system belongs more to the ordinary public world, the market square, the agora; whereas the duodecimal system belongs more to the esoteric priestly world, the temple and the acropolis. However, as Vitruvius points out, the ancient Greeks regarded both 10 and 12 as perfect numbers (Rowland and Howe 1999: 47–48) and, while this fact alone would suffice to explain the canonic use of those numbers in temple planning, it does not account for the interesting regional distribution of the 12- and 10-part systems or explain the significance of their combined use, as at the Acropolis of Athens. Vitruvius's brief discussion of the two number systems concludes that they arose from a common origin in the size, shape and symmetries of the human figure and that both systems were adopted in the mathematics of temple design.

This philosophically suggestive section comes from the introduction to his third book, on Temples; and, while there is no indication here that the numbers he discusses might extend beyond the temple to the *temenos*, the argument leaves no doubt that Doxiadis's 12- and 10-part systems of temple layout are wholly in accordance with what Vitruvius conceives to be the essential mathematics of Greek sacred architecture (Doxiadis 1972: 15, 17–18).

And, whatever their system of measurement, the early Etruscan, Greek and Roman planners do seem to have shared the essentials of an ancient ritual art (Rykwert 1988: 195)[17] one that was based, not on the square grid, but on a circular world-view whose dim origins immeasurably antedate the chequered culture of cities.

Biography Graham Pont, a specialist in interdisciplinary studies, taught in the General Education programme at the University of New South Wales for 30 years, where he introduced the world's first undergraduate courses in Gastronomy. He was a founding convener of the Symposium of Australian Gastronomy (1984) and co-editor of *Landmarks of Australian Gastronomy* (1988). His last appointment was a visiting professorship in the School of Science and Technology Studies, UNSW. Trained in philosophy, his principal research area has been history and philosophy of music but his interests have also extended to environmental studies,

[17] While I have proposed a common anthropocentric basis to ancient Greek and Italian planning, I doubt that the methodologies and conclusions of Doxiadis and Rykwert could be entirely reconciled. Doxiadis has offered a precise, quantitative and empirically testable hypothesis which is supported by hard evidence and remains open to further testing. Rykwert has offered a less precise theory, with some supporting evidence of cardinally-oriented and orthogonally-planned cities. But he has no explanation for the far more numerous early cities that are irregularly planned and non-aligned: that is, which exhibit characteristics already explained by Doxiadis at least in some Greek and Graeco-Roman cities. To account for these would require a radical revision of Rykwert's foundation scenario; whereas Doxiadis has already shown that his theory extends to orthogonally planned sites.

landscape, history of gardening, philosophy of technology, bio-acoustics and wine history. In 2000 he published the results of the first major computer analysis of Handel's music and he is completing a biography of Australia's first composer and musicologist, Isaac Nathan (1792–1864).

References

Doxiadis, C. A. 1972. *Architectural Space in Ancient Greece*. Cambridge: Mass. & London: MIT Press

Elderkin, G. W. 1912. *Problems in Periclean Building*. Princeton NJ: Princeton University Press.

Kent, R. G., ed. 1958. *Varro on the Latin Language with an English translation*. London: William Heinemann Ltd; Cambridge: Mass: Harvard University Press.

McClain, E. G. 1976. *The Myth of Invariance*. New York: Nicolas Hays Ltd.

——. 1978. *The Pythagorean Plato; Prelude to the Song Itself*. Stony Brook NY: Nicolas Hays Ltd.

Michaelides, S. 1978. *The Music of Ancient Greece: An Encyclopaedia*. London: Faber & Faber.

Oppert, M. J. 1887. *L'Étalon des mesures assyriennes, fixé par les textes cunéiformes*. Paris: Imprimerie Nationale.

Rowland, I. D. and howe, T. N. 1999. *Vitruvius; Ten Books on Architecture*. Cambridge: Cambridge University Press.

Rykwert, J. 1988. *The Idea of a Town. The Anthropology of Urban Form in Rome, Italy and the Ancient World*. Cambridge, Mass. & London: MIT Press.

Scully, V. 1962. *The Earth, the Temple, and the Gods; Greek Sacred Architecture*. New Haven and London: Yale University Press.

——. 1991. *Architecture: The Natural and the Manmade*. New York: St. Martin's Press.

Spitzer, L. 1963. *Classical and Christian Ideas of World Harmony*. Baltimore: Johns Hopkins Press.

Stevens, G. P. 1940. *The Setting of the Periclean Parthenon. Hesperia Supplement III*. Athens: American School of Classical Studies.

Chapter 12
The Geometry of the Master Plan of Roman Florence and Its Surroundings

Carol Martin Watts

The Romans brought a remarkably consistent approach to all scales of design, using similar principles to order everything from pavement mosaics to the layout of cities and entire regions. This consistent approach to all scales of design is a reflection of their comprehensive and unified world view. This chapter will look at the larger end of this spectrum.[1] Florence, the Roman Florentia, was founded as a new town to settle military veterans. Like many cities founded by the Romans, it was oriented to the cardinal points. The major streets, a north-south *cardo* and east-west *decumanus*, met in the centre of the rectangular walled town. A gate was located at each of the four intersections of major street and city wall. A grid of secondary streets divided the city into blocks. Even today, looking at an aerial view or a plan of the city, this orderly central portion of Florence is evident.[2]

First published as: Carol Martin Watts, "The Geometry of the Master Plan of Roman Florence and its Surroundings", pp. 169–181 in *Nexus III: Architecture and Mathematics,* ed. Kim Williams, Ospedaletto (Pisa): Pacini Editore, 2000.

[1] This chapter is the outgrowth of joint research with Donald J. Watts on the Roman layers under the Duomo of Florence. We presented this ongoing work at the Nexus'98 conference in Mantua, Italy, "Roman Code", and the 1998 Bridges Conference, Winfield, Kansas, "Traces of the Geometrical Ordering of Roman Florence". I wish to acknowledge the important contributions of Donald J. Watts to the hypothesis presented in this chapter, as well as his work on the illustrations. What follows is based on maps at 1:5,000 and 1:25,000 published in 1996 by the Istituto Geografico Militare, Florence, based on satellite imagery. Given the scale involved, it was impossible to personally take measurements.

[2] Little remains of Roman Florence, but recently there has been a revived interest in understanding the city's origins. A good recent source is *Alle Origini di Firenze, dalla Preistoria alla Citta Romana* (Capecchi 1996), the catalogue for an exhibition at the Museo Firenze Come'era. Although the two main streets and the location of city gates followed typical Roman typology, the city wall of Florentia appears to have deviated from the ideal rectangle in the southeast corner, to respond to topography of the site near the river, as revealed by excavations below the Piazza

C.M. Watts (✉)
The College of Architecture, Planning and Design, Kansas State University, Seaton Hall 21, Manhattan, KS 66506-2902, USA
e-mail: cmwatts@k-state.edu

K. Williams and M.J. Ostwald (eds.), *Architecture and Mathematics from Antiquity to the Future*, DOI 10.1007/978-3-319-00137-1_12,

It was standard Roman practice, when founding a new colony, to survey the countryside and assign plots of land to the settlers.[3] The crossing of two major lines, the *cardo maximus* and *decumanus maximus*, was established for the region, as with the layout of a city. The land was then subdivided into a grid, using lines parallel to the regional cardo and decumanus, generally into square plots 20 *actus* on a side. An actus was 120 Roman feet.[4] Every fifth division was particularly important, and surveyed more precisely to keep the system accurate. Each *century*, as these 20 actus squares were known, was further subdivided into farm plots. This process of organizing the countryside is commonly known as *centuriation.* Evidence of such grids upon the countryside can still be seen, particularly in aerial photography, throughout what had been the Roman Empire. The surveyors made maps of each colony, with a copy (in bronze) sent to the land registry office in Rome, and another kept locally. There was a standard notation system used to describe the location of each allotment of land, and subsequent transfers of property. Centuriation was a practical way to ensure equitable distribution of land and recordkeeping of land ownership, and relied on the skills of the surveyors whose responsibility it was.

Traces of the Roman centuriation of the region surrounding Florence reveal a 20 actus square grid. Usually cities and centuriation have the same orientation, and often the same centre point, but at Florence the centuriation was oriented so that the regional decumanus, running from northwest to southeast, traversed the length of the long valley in which the colony was situated.[5] Oriented to the cardinal points, the city of Florentia was in a rotated position relative to the surrounding centuriation (Fig. 12.1) The two orientations are off by a seemingly random almost 31°. Some scholars have thought that this meant they were established at different times, but more recently it has been assumed that the regional centuriation and the city of Florentia were established at the same time, about 41 BC, and that for practical reasons different orientations were used. The Roman colony was located very close to the river Arno, in a flat and swampy plain. The nearest town at the time of its founding was the Etruscan site of Fiesole, on a hill to the northeast. Unlike many other Roman towns founded along pre-existing or concurrently built major roads, Florence seems to have been located some distance from roads. The

della Signoria. In the late nineteenth century the urban renewal of a large area around the present Piazza della Repubblica brought to light many Roman remains, but did not allow time for systematic excavation. They were recorded by the architect Corinto Corinti, whose drawings and notebooks are catalogued in (Orefice 1986). Today one can visit the south gate of the Roman city, see the outline of the city wall on the east marked in the street at one point, and visit the crypt of Santa Reparata below the Duomo, where a sequence of churches were built over Roman housing. A Roman house excavated in the late 19th century can also be glimpsed below the Baptistery of San Giovanni.

[3] Dilke (1971) is a comprehensive work on what is known of Roman surveying techniques and practices.

[4] The Roman foot (as used in Florence) was 0.295 m = 1 Roman foot.

[5] For the centuriation around Florence see Instituto Geografico Militare (2003, PL 27–28).

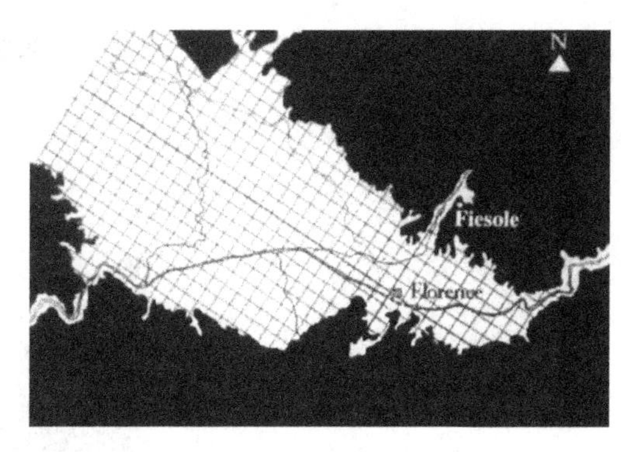

Fig. 12.1 Roman centuriation surrounding Florence Drawing: author

Via Cassia, a major road, appears to have been originally located some distance away, in the hills rather than the valley (Hardie 1965: 122–140).[6]

According to Hardie, the location of Florentia near the eastern end of the regional colonial centuriation was due to numerous practical reasons.

> The site of Florence was well chosen to exploit both the centuriated plain and the land on the south bank. It was also near, though not on, the Via Cassia. The site also had some natural defensible features being beside the Arno and also at the junction of the Arno with the Torrente Mugnone that could serve as a moat for the eastern side of the city. Lastly, the Arno was apparently navigatible up to the location of the city (Hardie 1965: 134).

A number of scholars have pointed out that there is one point of intersection between the city and the centuriation (Fanelli 1997: 1–5; Hardie 1965: 132). This is immediately outside the western gate of the city, which is also the *umbilicus*, or origin point, where the regional cardo intersects at right angles the regional decumanus.

It is my hypothesis that the angle between the two orientations was not random, and that the relationship between the two systems goes far beyond overlapping at a single point. Rather, there was a deliberate, sophisticated geometrical process linking landscape and urban settlement.

In order to understand the following hypothesis of the process of laying out the city and the region, it is important to note that surveyors' practice entailed the use of geometrical knowledge from surveying texts as well as the utilization of drawing equipment during the execution of the survey. While most of the process described involves the actual laying out of points and lines upon the site, a few of would have been done as drawn geometric constructs by the surveyor in determining the steps of the physical layout. Although there is no way to know exactly the sequence of steps or thinking which went behind them, the explanation which follows is a hypothesis which explains the observable phenomena in light of known Roman practices.

[6] The usual explanation for the differing orientations has been that the centuriation, for pragmatic reasons, parallels the long, narrow valley and is thus a result of topography, while the city follows established preferences to orient to the cardinal points, for cosmological reasons.

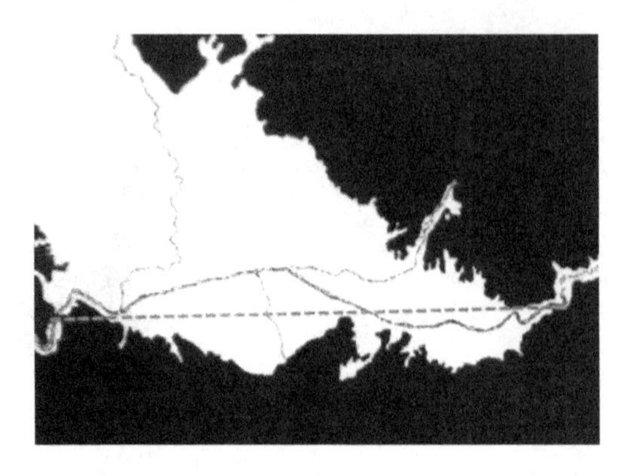

Fig. 12.2 Urban decumanus marking congruent paths of river and sun from east to west Drawing: author

A Hypothesis for the Process of Laying Out the City and Countryside of Florentia

A political decision was made to found a colony in the general area near Fiesole, to utilize the uncultivated plain of the Arno River. The next decision would have been just where to put the city and how to begin to organize the landscape. A look at the pre-existing landscape in the area, as the Romans found it can help us understand their thinking. Figure 12.2 shows the valley in question (in white) in which hills (in black) enclose a flood plain for the Arno River and several tributaries. The topography suggests some natural axes. The exact location of the colony is defined by the intersection of two of these axes.

The first of these is shown by the dashed line in Fig. 12.2. A due east-west line can be observed to connect the point of the eastern entry of the Arno River into the valley with the point of departure of the Arno out of the valley. This inherent east-west line became the alignment of the decumanus of the city. It thus located the city at this particular point near, but not on, the river, rather than in some other position further north or south. Although the city contrasts in orientation to the grid of the region, its order is based on the congruence of the path of the sun and the natural east-west axis of the river.[7]

The second natural axis is northeast to southwest and establishes the line of the regional cardo of the centuriation (Fig. 12.3). This line extends into small valleys on the north and south edges of the Arno valley. The valley to the north is just below the Etruscan town of Fiesole, where the river Mugnone, and probably a pre-existing road crossing the Appenines, came through the mountains. The valley to the south was the point where the Etruscan road from Volterra probably entered the Arno valley.[8]

[7] According to Dilke (1971: 89), the east–west *decumanus maximus* was typically the first line established in the Roman surveying process. The augur typically faced east.

[8] This road may have crossed the river near the site of the city. One theory for the location of Florence at this point has been that the Arno was navigable up to this point, or that it was a natural crossing point for the pre-existing road system, at or near the site of the present-day Ponte Vecchio.

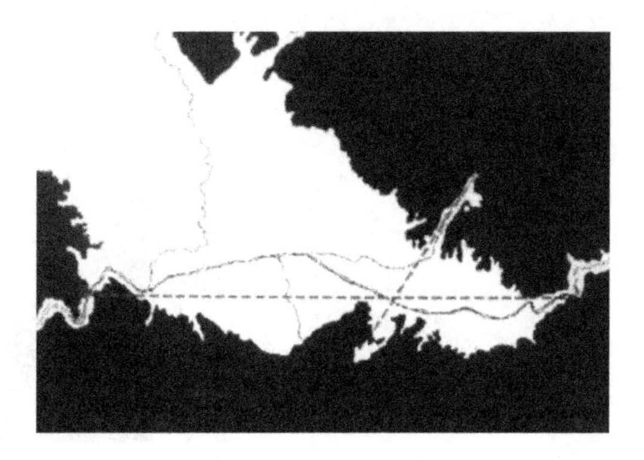

Fig. 12.3 Regional cardo. Drawing: author

The crossing of these two lines (the urban decumanus and the regional cardo) establishes an origin point 0,0 and forms the kernel of geometric order for anchoring both city and region to the cosmos.

In the Roman procedures for founding a city as described by Rykwert (1988), the priest known as the auger was the first and most important official involved. A congruence was established between the heavens above and the future settlement in view before his eyes. The vault of heaven, depicted as a cross within the sky, was conceptually mapped upon the landscape through the waving of the auger's special cane. City founding therefore began with establishing a specific conceptual and spatial relationship between a specific primary feature of the site, as revealed to the auger, and the cosmos. In the mind of the auger, the urban decumanus was not his creation but rather already pre-existent and made apparent to him through his devotion to religious ritual. Proceeding from this revelation, the duty of the auger, and his servant the surveyor, was to make manifest the order of the place.

Through the use of a circle drawn upon the ground with its centre marked by a sundial gnomon, a solar east-west orientation line is drawn as a cord cutting through the circle (Dilke 1971: 57–58). This cord of the circle is bisected by a line coming from the centre of the circle and this bisector defines the north-south orientation line. Figure 12.4 shows the addition of the north-south line which follows from the east-west urban decumanus.

The approximate angle between the two orientations was thus based on the topography of the region as well as the cardinal points. The surveyors must have then chosen the specific angle based on the closest angle having certain known geometrical and numerical properties, probably by consulting a handbook. The rotational relationship between the regional and urban grid at Florence can be defined by a simple right triangle having a three to five proportion of its legs. Such a proportion results in a rotation of 30°, 58 min or about 31°.[9]

[9] The angle is described as 30° 58′ or about 31° by Hardie (1965: 132). The fact that this is an angle based on a 3:5 triangle was pointed out by John Peterson, in personal correspondence instigated after viewing his Internet site. In an unpublished paper presented at the Theoretical Roman Archaeology conference in 1998 he discussed meaning in the geometry of Roman planned

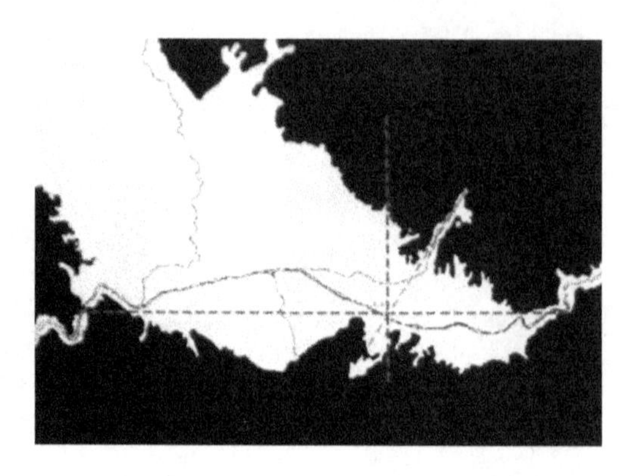

Fig. 12.4 North-south line. Drawing: author

Figure 12.5 shows the establishment of two large 3:5 triangles with the sides along the regional cardo as one century or 2,400 Roman feet (20 actus × 120 ft). The other sides are 1,440 and 2,800 Roman feet. These two triangles establish three points (including 0,0) in order to accurately site this line. The same pair of opposing triangles are plotted to locate the regional decumanus, at right angles to the regional cardo, and to mark the one century distance along this axis (Fig. 12.6). These triangles mark the first century division in each direction from the origin point. The surveying proceeded outward from this point to establish the grid of farmland. These triangles also create four squares one century on a side, or one large square two centuries on a side (Fig. 12.7).

Through the study of the Roman use of geometrical ordering systems at many scales of their built environment, I have observed their persistent use of regulating squares (Watts 1996; Watts and Watts 1992, 1986). The four squares of the centuriation grid can be seen as a single 40 actus square template (or regulating square) within which the specific design of the city will unfold.

As described above, the original conception of the city and colony called for the orientation of the city to remain true to the cardinal points. Therefore, in the process of developing the design of the specific city plan, a regulating square was needed that fell within the larger 40 actus centuriation template (Fig. 12.8) and was also oriented to the cardinal points, thus ensuring mathematical commensurability and geometric unity of both city and colony. The points of tangency where the urban regulating square touches the sides of the 40 actus centuriation square lie 5 actus from each of the midpoints of the centuriation square, or at the quarter points of each 20 actus square (Fig. 12.9) (it was common procedure to subdivide the century into quarter points to facilitate allotment of farmland). These points of tangency can also be established through the use of the 3:5 triangle, as indicated on the diagram in Fig. 12.9, which, as we have seen, is the rotational relationship between the centuriation and city orientations. Whereas the entire urban regulating square would have been part of the surveyor's geometric diagram, only a portion of the

landscapes, including Florence. Peterson identifies a number of whole number ratios which can be expressed as triangles which he finds used in Roman surveying.

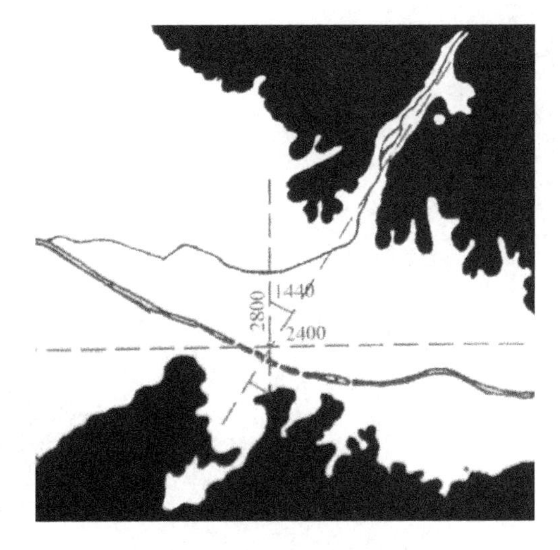

Fig. 12.5 Triangles in the ratio 3:5 define precise rotation between centuriation and city. Drawing: author

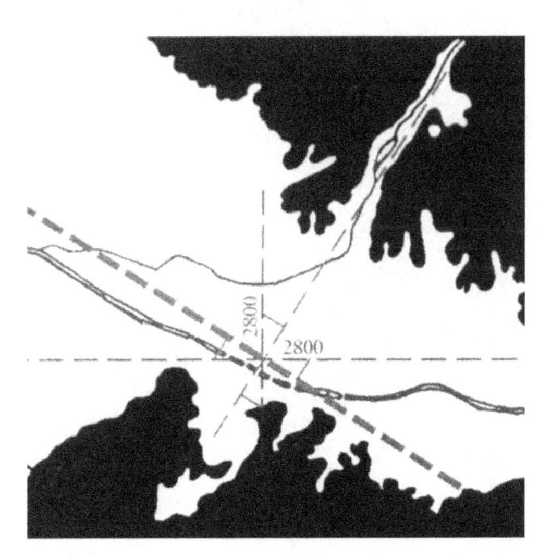

Fig. 12.6 Establishment of regional decumanus using 3:5 triangles. Drawing: author

eastern side of the square served as an actual component of the city. Using the points 5 actus from the midpoints of the 40 actus centuriation square, the eastern side of the urban regulating square could be located upon the site. This edge of the regulating square defines the eastern wall of the city.

The Sacred Cut is a geometric procedure defined by Brunés (1967). As shown in Fig. 12.10, arcs centred at the lower right and upper right corners of the urban regulating square, passing through the centre of the square (equal in length to half the diagonal of the square), intersect the right side of the square as shown. These arcs determine the north and south limits of the city of Florentia. The sacred cuts yield a dimension of 12.076 actus for this distance. It appears likely that by

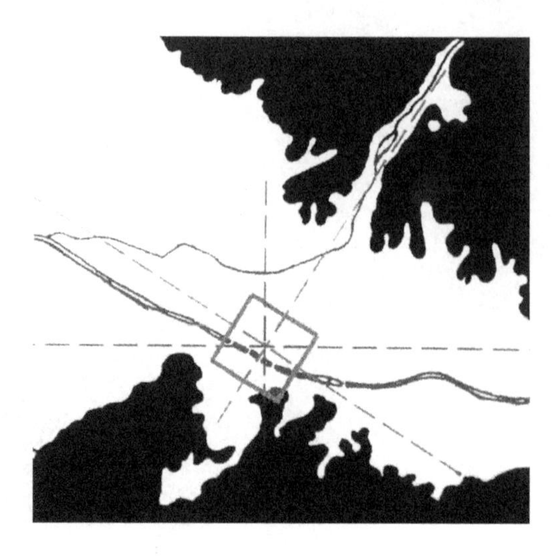

Fig. 12.7 Four century square (40 actus per side). Drawing: author

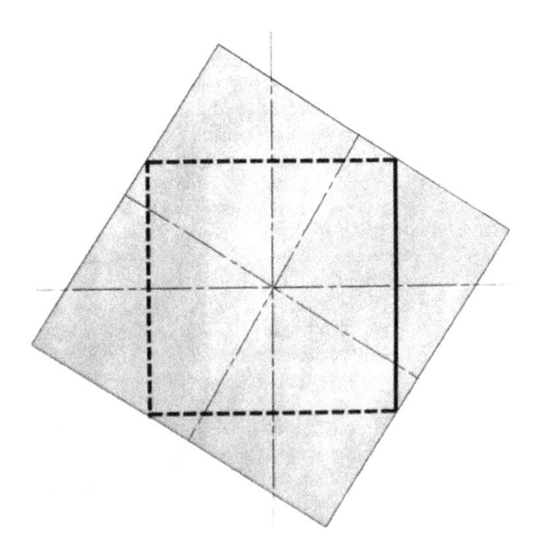

Fig. 12.8 Urban regulating square within four century square. Drawing: author

rounding this off to an even 12 actus, or 6 actus either side of the urban decumanus, it was possible to continue laying out the city using whole numbers. This distance could have been measured either side of the urban decumanus, as shown by the solid arcs in Fig. 12.10.[10]

Figure 12.11 shows two alternatives for establishing the position of the north and south walls of the city. The Sacred Cut could have been repeated to the west, and

[10] Continuing the Sacred Cut arcs creates a lozenge or petal-like form, and the complete Sacred Cut operation would create four such petals, perhaps giving rise to the name for the city, Florentia. Such a flower-like form was also common in Roman mosaic patterns.

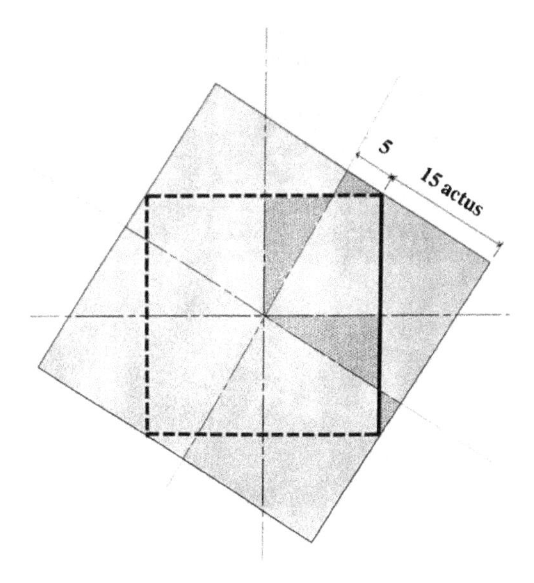

Fig. 12.9 Establishment of urban regulating square. Drawing: author

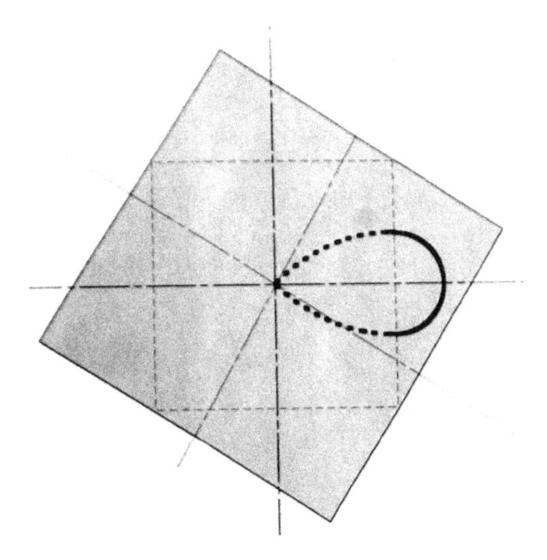

Fig. 12.10 Establishment of length of east city wall using Sacred Cut. Drawing: author

the points joined to position the edges of the city, or, knowing that the Sacred Cut was a nominal 12 actus, the surveyors could have measured 6 actus to the north and south of the decumanus, both from the 0,0 point and where the urban decumanus intersected the eastern edge of the urban regulating square, and connected these points. Since making the Sacred Cut on the site would have crossed the river, it is likely that measuring 6 actus was more expedient. The discrepancy between these two methods is 9 Roman feet, or 4.5 ft at each city wall. This could have allowed for the wall thickness, with one system measuring to the outside of the wall and the

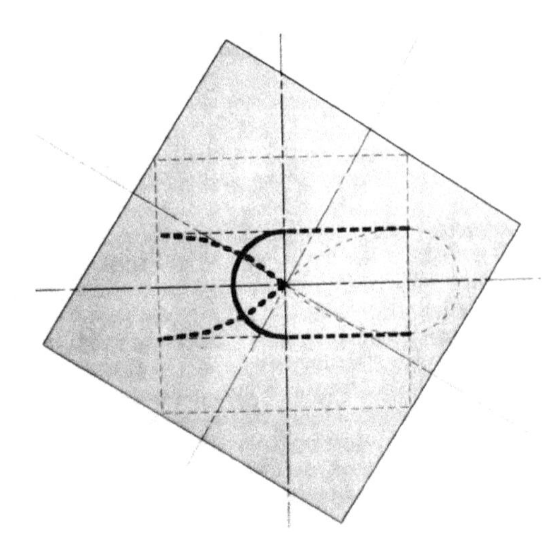

Fig. 12.11 Establishment of north and south city walls using Sacred Cut. Drawing: author

other to the interior of the wall.[11] The next steps locate the position of the urban cardo, the major north-south street (Fig. 12.12). The rectangle of the city defined thus far is subdivided, using once again a 3:5 triangle. Another way to look at this is as the drawing of a line parallel to the centuriation orientation, 7 actus in length, crossing the urban decumanus, and continuing another 7 actus. It then turns due north for 6 actus. The lengths of this line $(7 + 7 + 6) = 20$ actus, or one century. This procedure locates the centre point of the city, and thus locates the urban cardo. Continuing the procedure (Fig. 12.13) locates the western edge of the city, 12 Roman feet east of the 0,0 point. It has long been noted that the origin point of the centuriation was slightly outside the western gate of the city. One explanation for this has been the presence of a stream which was diverted to flow just outside the city wall, crossed at a bridge at this point.[12]

Conclusion

The preceding steps indicate a logical way in which the colony of Florentia, both city and countryside, could have been laid out, consistent with what is known of Roman practices and with the observable traces of the Roman settlement. Relatively simple geometrical operations integrate the city with its landscape, and with the cosmos as well. The landscape of the region, linked with the cosmos

[11] A similar accommodation occurs between dimensions commonly used for Roman house regulating dimensions and block sizes. Another parallel is with pavement patterns, where a border zone mediates different dimensions and accommodates irregularities between the outer wall of a room and its mosaic pattern.

[12] Archaeological evidence of the western gate has not been located.

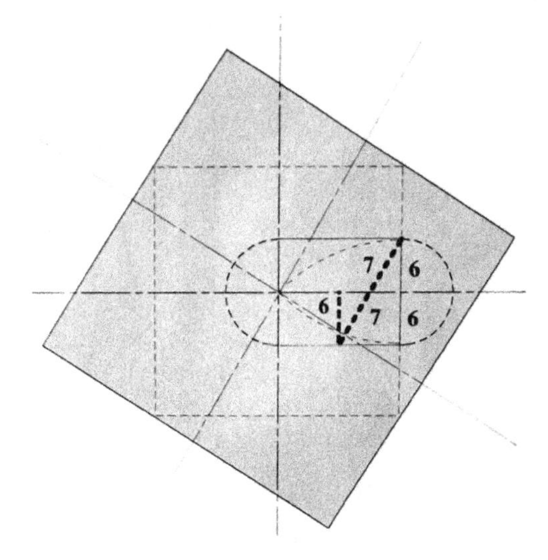

Fig. 12.12 Location of urban cardo and south gate. Drawing: author

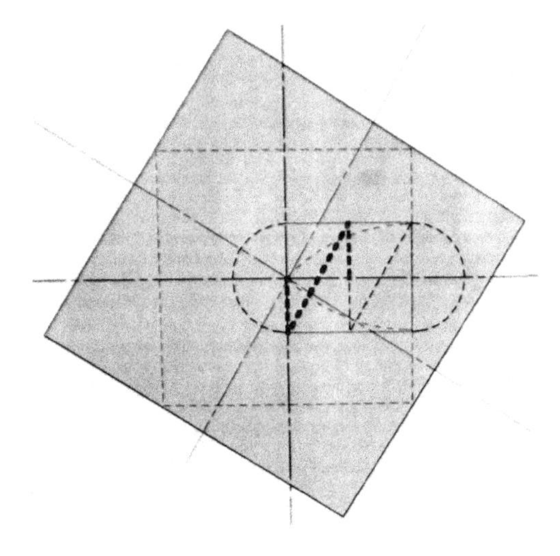

Fig. 12.13 Location of western city edge. Drawing: author

through the path of the sun, provided the initial positioning of the city. A geometry based on the 3:5 triangle establishes the precise rotational relationship between the two levels of scale, the region and the city. A conceptual square mediating between the two orientations, and its Sacred Cut, establishes the limits of the city to the east, north and south. The further use of the 3:5 triangle locates the centre of the city, and its cardo, as well as its western edge. What at first glance may seem to be a pragmatic adaptation to the site conditions is shown to be a rigorous geometric relationship, predicated on using a meaningful set of whole numbers, in an affirmation of the *genius loci* of Florentia.

Biography Carol Martin Watts is an architectural historian and faculty member in the Department of Architecture at Kansas State University, Manhattan, Kansas. She received her Ph D from the University of Texas at Austin, writing her dissertation on "A Pattern Language for Houses at Pompeii, Herculaneum and Ostia." She has published several articles, together with Donald J. Watts, on geometry in Roman architecture, painting and mosaics, including "A Roman Apartment Complex" in *Scientific American. Her current research interests include vernacular Italian architecture and the impact of Roman development on subsequent layers in urban environments. She is a past member of the editorial board of the* Nexus Network Journal.

References

Brunés, Tons. 1967. *The Secrets of Ancient Geometry and its Use*. 2 vols. Copenhagen: Rhodos.

Capecchi, Gabriella. 1996. *Alle Origini di Firenze, dalla Preistoria alla Citta Romana*. Firenze: Edizioni Polistampa.

Dilke, O.A.W. 1971. *The Roman Land Surveyors*. New York: Barnes and Noble.

Fanelli, Giovanani. 1997. *Le Citta Nella Storia d'Italia, Firenze*. 7th edn. Rome: Editori Laterza.

Hardie, Colin. 1965. The Origin and Plan of Roman Florence. *Journal of Roman Studies*. 55: 122–140.

Instituto Geografico Militare. 2003. *Atlante Aereofotografico delle Sedi Umane in Italia*, part 3. Firenze: Instituto Geografico Militare.

Orefice, Gabriella. 1986. *Rilievi e memorie dell'antico centro di Firenze 1885–1895*. Firenze: Alinea Editrice.

Rykwert, Joseph. 1988. *The Idea of a Town, the Anthropology of Urban Form in Rome, Italy and the Ancient World*. Cambridge: MIT Press.

Watts, C.M. 1996. The Square and the Roman House: Architecture and Decoration at Pompeii and Herculaneum. Pp. 167–181 in Kim Williams, ed. *Nexus: Architecture and Mathematics*. Fucecchio, Florence: Edizioni dell'Erba

Watts, Donald J. and C.M. watts. 1986. A Roman Apartment Complex. *Scientific American*, **255** (6): 132–139.

——. 1992. The Role of Monuments in the Geometrical Ordering of the Roman Master Plan of Gerasa. *Journal of the Society of Architectural Historians*. **51**(3): 306–314.

Chapter 13
Architecture and Mathematics in Roman Amphitheatres

Sylvie Duvernoy

Early Relationships Between Mathematics and Graphics

When we refer to the relationship between mathematics and architecture, we often make the assumption that geometry existed before architecture, and that the former always provided the latter with a solid background, a sort of database of shapes, figures and proportional systems among which designers could pick, at any given time, the solution that best answered their needs. This idea was given currency by architectural treatises from the Renaissance on, which all began with a summary of whatever contemporary available geometric knowledge might prove useful to the architect before addressing the subject of the design itself. However, if we consider geometry as a science that grew simultaneously with artistic culture, acting as a motor as well as a recipient of research progression and cultural evolution, its relationship with architecture appears more interesting and reveals traces of reciprocal influences.

In ancient times, the true value of a number, today represented by an abstract Arab symbol, was physically drawn and represented by an appropriate quantity of dots, or by a line of appropriate length. The research methodology for arithmetic and geometry consisted in the drawing of numbers. The first graphic demonstrations in arithmetic are due to Pythagoras, the eldest of the brilliant dynasty of Greek mathematicians, who lived approximately between 560 and 480 B.C. By arranging dots (or pebbles called *calculi* by the Romans) according to

First published as: Sylvie Duvernoy, "Architecture and Mathematics in Roman Amphitheaters", pp. 81–93 in *Nexus IV: Architecture and Mathematics,* eds. Kim Williams and Jose Francisco Rodrigues, Fucecchio (Florence): Kim Williams Books, 2002.

S. Duvernoy (✉)
Politecnico di Milano, Milan, Italy
e-mail: syld@kimwilliamsbooks.com

K. Williams and M.J. Ostwald (eds.), *Architecture and Mathematics from Antiquity to the Future*, DOI 10.1007/978-3-319-00137-1_13,

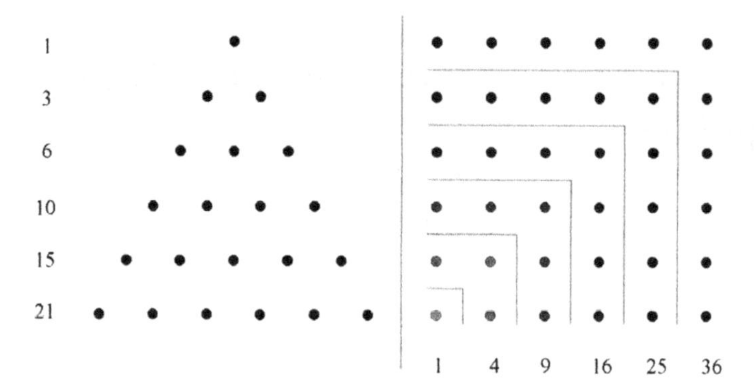

Fig. 13.1 Pythagoras's *triangular* and *square numbers*. The sum of two successive *triangular numbers* is always equal to a *square number*. The sum of the n first odd numbers is equal to n^2. Drawing: Author

specific geometric patterns, Pythagoras classified numbers in different categories named after their shape. By means of their graphic image, he proved that "triangular numbers", "square numbers", "pentagonal numbers" all had particular properties, and could be related to one another by specific arithmetical laws (Fig. 13.1).

For other purposes, numbers could be represented by lines of appropriate length. In this case, adding two numbers meant joining the two respective lines, thus obtaining the length corresponding to the sum. Multiplying two numbers meant drawing a rectangle whose sides were equal to the numbers in question: the area would be the graphic visualization of the result of the calculation. Multiplying a number by itself lead to drawing a square, and we still today talk about "squaring a number", or about "square roots". Consequently, drawing geometric figures was a valid method for solving "quadratic" equations (Fig. 13.2). In addition to being a calculation methodology, graphic representation is also a powerful means of communication. Thus drawing acts simultaneously as a research tool and as a persuasive demonstration.

By the time Archimedes was studying conics, between 280 and 212 B.C. three big problems were still unsolved: the duplication of the cube, the quadrature of the circle and the trisection of the angle. As often happened in ancient Greece, the first problem was supported by a legend. We are told that the people of Delos, struck by a severe plague, asked the oracle of Apollo how to calm the gods' wrath. The answer was that they had only to build a new altar to Apollo twice as big as the existing one, which was of a cubic form. The locals immediately built a new altar, whose edge was twice as long as the previous one... but the plague did not stop. The problem of the duplication of the cube could not be solved as long as the construction of plane geometric figures remained the only research methodology. We now know that this method is limited to solving "quadratic" equations, but not "cubic" ones.

We know as well that the second of the three problems, the quadrature of the circle, could not be solved... by any means. Nevertheless, many Greek

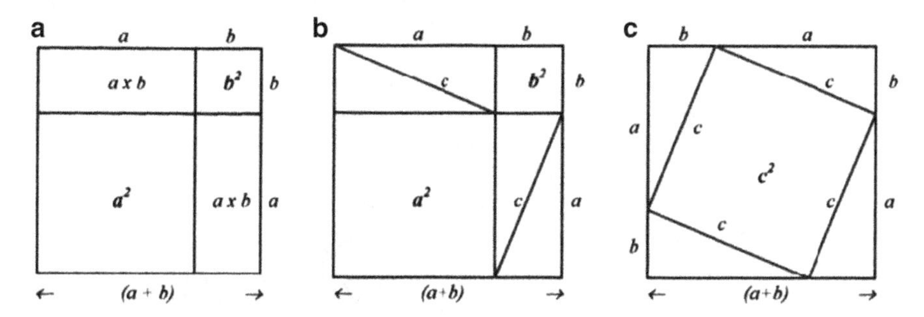

Fig. 13.2 Graphic demonstration of a quadratic equation $(a + b)^2 = a^2 + 2ab + b^2$, and of the Pythagorean theorem of triangles, $c^2 = a^2 + b^2$. Drawing: Author

mathematicians tried to find a solution, and among them, the results of Archimedes's efforts were to influence all further generations. The idea was to find a squared figure—a polygon—the area of which would be equal to the area of the circle. It sought to determine the true value of the number π. By comparing the areas of two polygons of 96 sides, one inscribed inside a circle and the other one circumscribing the same circle, Archimedes reached a most precise approximation of π. He claimed that it lay in the narrow interval between 3 + 10/71 and 3 + 1/7. This result was of fundamental importance for all further calculations. Since 3 + 1/7 suffices for the practical purposes of metrical geometry, it became the value commonly adopted for π from then on, and the ratio that best expressed this incommensurable quantity was 22/7.

The last problem, the trisection of the angle, turned out to be the easiest, and we have at our disposal several solutions proposed by different scientists. Among those, Archimedes's graphic appears rather complicated to our modern minds trained to the use of the goniometer, but it provides excellent evidence of the fact that the only methodology for solving arithmetic and geometric queries around the end of the third century B.C. was still the graphic approach. The Babylonian division of the circle into 360° had not yet been imported into western culture, and trigonometry was not yet common knowledge. Greek mathematicians still relied on the construction of plane figures for their studies and research, and used ungraduated straightedges and compasses as their only graphic tools.

This need to sketch for research purposes was also familiar to architects, who used to depend on the manipulation of the same professional tools for the solution of technical and aesthetic queries. In his treaty, Vitruvius insists strongly on the superiority of drawing over calculation in the process of researching the harmonious proportions of the architectural object. In the Preface to Book IX, quoting Plato and reporting his famous theorem on the duplication of the square, he demonstrates that when a certain number is required that "[n]obody can discover ... by calculation", accurate drawing is always possible (Vitruvius 2009: 243]. Vitruvius's schemes for the design of Greek and Roman theaters provide the first historical evidence of a close relationship between mathematics and architecture. The schemes (Fig. 13.3) are clearly drawn with a ruler and a compass only, and the

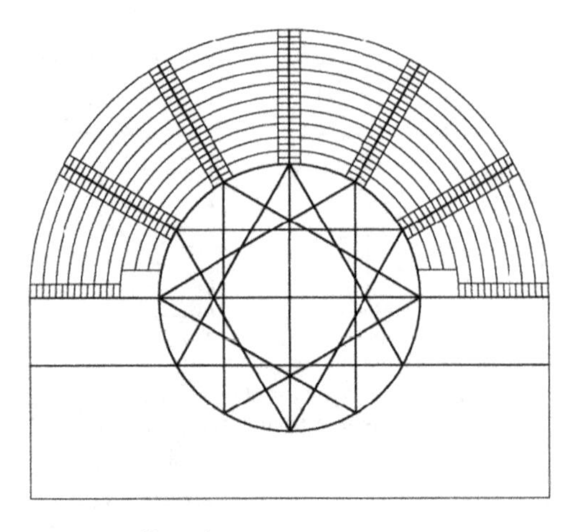

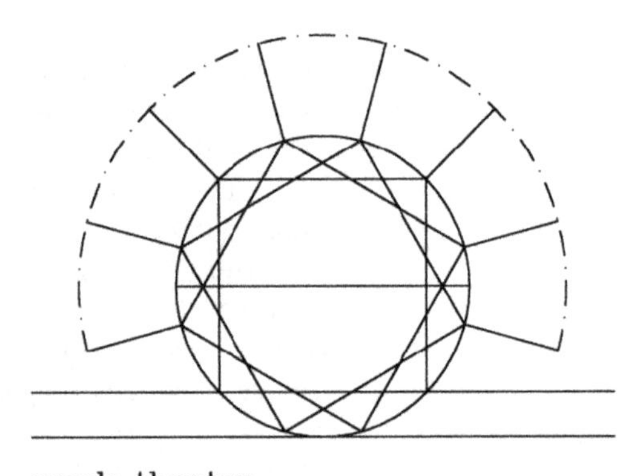

Fig. 13.3 Geometric diagrams for Roman and Greek theater design according to Vitruvius, *De architectura libri decem*. Drawing: Author

radial division of the tiered seating area called the *cavea* (the trisection of the right angle of the half-*cavea*) is obtained thanks to the inscription of several polygons (four equilateral triangles or three rotated squares) in the circle of the *orchestra*, the space between the stage and the first rows of seats.

Geometry and Architecture in the Roman Amphitheatre

Evidence of dynamic interaction between geometric research and architectural design can be found in the Roman world, around late Republican times and early Empire. In those years a new type of building appears in Roman culture: the amphitheater. Romans had a number of different buildings for shows and plays: ballet, lyrics and music were performed in the theaters and odeons; stadiums were built to host athletic games; circuses were used for horse races. All these building types, as well as the spectacles they housed, came from the Greek heritage. Amphitheaters, however, are a Roman invention as they were designed to host shows of gladiator fights (*munera* and *venationes*), nonexistent in ancient Greece or in oriental cultures. Amphitheaters first appeared in Campania at the beginning of the second century B.C. They were immediately characterized by a closed elliptic shape that had never been previously adopted in architectural design, and would never be used for other purposes in Roman culture.

The end of the third century B.C. corresponds to the acme of the Golden Age of Greek mathematics, whose last heroes were Archimedes (d. in 212 B.C. in Syracuse), Eratosthenes (d. 192 B.C.), Apollonius of Perga (d. 190 B.C.) and Nicomedes. Archimedes and Apollonius are famous for having devoted much effort to the study of conics and conoids. Thus, the first apparition of elliptic-shaped buildings in southern Italy no earlier than the beginning of the second century B.C. puts us in front of a triple coincidence: historical, geographical and morphologic. We may therefore consider that these simultaneities and similarities are not due to chance, but that the necessity of designing a new building type provided theoretical mathematics with a successful field of direct and immediate experimentation.

Are amphitheaters elliptic or oval? The question has already been widely debated. The ellipse is a natural curve: it is the shape of the shadow of a sphere or a disk on a plane; it is the path of the planets orbiting in the sky around the sun. On the other hand, the oval is a closed polycentric curve, a graphic composition that can be either egg-shaped or elliptic-shaped. Ellipses and ovals have different mathematical properties. The tracing of an elliptic curve relies on the prior determination of the length of its mean axes and, on occasion, the position of its two focal points. The construction of an oval consists in joining at least three or four arcs of different dimensions the centers of which can be arranged according to a variety of patterns. Nevertheless any ellipse can be closely approximated by a specific oval. The difference between the two curves of identical axes would only be conceptual and therefore irrelevant as far as visual perception is concerned.

The Amphitheatre of Pompeii

In order to understand how the new typological pattern for amphitheaters was first drawn, and how it was transformed into a variety of archetypal models during its historical evolution, inquiries must be made on early samples, buildings that were erected in the late Republic or early Empire. These are much less sophisticated monuments than the late Flavian ones, but they testify to the first attempts to apply a new geometric scheme to architectural design.

The most interesting building, from this point of view, is the one in Pompeii (Fig. 13.4). Rather awkward and ungraceful from the outside, this monument nevertheless shows an elegant and well-organized *cavea*. It is the oldest among the surviving buildings, and the first one in chronological order. Unveiling its geometric pattern means discovering the starting point of the design research.

Fig. 13.4 The amphitheater in Pompeii. Photo: Author

The construction of Pompeii's amphitheater involved partly excavating and partly raising the natural ground. The level of the arena lies underneath the exterior ground level, while the solid peripheral wall (partly formed by the city wall itself) contains the upper part of the *cavea*. The stone seats used to lie directly on the ground, pre-modeled to provide a suitable slope running all around the central void of the arena.

Only very accurate measurements could lead to reliable conclusions in the search for the mathematical identity of the curves. The survey of the monument was conducted according to the "polar methodology", by means of a single electronic theodolite, approximately located in the center of the arena. The reflecting prisma, target of the theodolite telescope, was moved, at regular intervals, along the arena wall, until the whole perimeter had been measured. The same thing was done along the back podium line, and along the upper wall pierced by forty gates, that encloses the *summa cavea*, the uppermost part of the tiered seating area. Thus a significant amount of data was collected for each of the three reasonably complete curves of the monument: the inner, the outer, and an intermediate.

The polar coordinates so obtained were transformed into coordinates of a Cartesian grid having the minor and major diameters of the curves as x- and y-axes with their point of intersection (the exact center of the arena) as its origin. The plot of the survey reveals three series of a hundred dots, approximating each of the curves. In order to understand their real mathematical identity, a comparison was made between the coordinates of the measured points, and those of points lying either on an ellipse or on an oval curve of four centers, algebraically generated, having the same diameters as the actual building. It appeared that the arena wall and the back podium wall could unequivocally be considered ellipses.

Tracing an oval curve for the arena with minor and major axes of 10 and 19 units respectively[1] would have meant that two out of the four centers would have been high in the middle of the slope of the raised ground, much above the level of the arena itself, while tracing ellipses using "the gardener's method" (two poles planted in the ground, and a rope) meant working on a horizontal surface, and dealing with two "centers" only: the two focal points of the curve located symmetrically on its major axis, inside its perimeter. Even the foci of the outer perimeter of the corridor are inside the arena.[2] The layout of the outer (and upper) perimeter of the *cavea* is the most imprecise, and measures show a wide range of irregularity.

The design of amphitheaters involves two of the classic problems of ancient mathematics: the quadrature of the circle and the trisection of the angle. The upper wall closing Pompeii's *cavea* is pierced by 40 gates, and was thus approximated by a polygon of 40 sides. From each gate, stairs used to come down the steps of the *cavea*, dividing it into *cunei* (wedges) of more or less identical size. But the ever-changing curvature of the elliptic ring interferes with its regular division, and the stairs do not converge towards the foci of the central ellipse but rather to independent points that are of no use for radial geometry; neither do they converge to the center of the arena, the crossing point of its axes, nor to points that could act as the centers of an oval curve.

The layout of Pompeii is rather awkward and irregular. It shows a number of imprecisions and adjustments that are typical of "work in progress". Perfection is not yet achieved in this graphic/arithmetic pattern, but the search for it is obvious. The errors and incoherencies of the diagram point out the problems that the later architects would have to solve.

The Amphitheatres of Roselle and Veleia

Other inquiries for geometric patterns were done upon two other amphitheaters: Roselle and Veleia (Duvernoy 2000). They were both built a little later than Pompeii, presumably in the first half of the first century A.D. They therefore belong to the first generation of monuments, and are typical small, provincial amphitheaters of the Republican period. Respectively situated in the Regio VII (Etruria) and VIII (Emilia) of the Empire, they appear very similar when considering their estimated date of construction, size, and simple building type. Like Pompeii they were partly dug, and partly erected above the natural ground, and few rows of seats were arranged directly upon the modeled slope, forming a *cavea*

[1] The architectural design unit in Pompeii's amphitheater is equal to 12 ft of approximately 29.25 cm.

[2] This hidden underground curve could not be measured during the survey of the monument. Nevertheless, its position is known by the gates that lead to the corridor. Its elliptic shape is the only assumption included in the geometric pattern proposed here, while all other conclusions come from accurate measurements.

of rather small width. No underground areas were present, and only in Roselle do we find four small *carceres* opening directly on the arena, located under the first steps of the *cavea*. The design of such open-air structures, lacking any sort of decorated façades, mostly consists in the drawing of concentric curves, with very few radial divisions. Not much is preserved of the Veleia amphitheater, whereas the ruins in Roselle are a little more complete and in better shape. In both cases the study focused mainly on the curved wall enclosing the arena, the best-preserved part of the buildings.

The surveys were conducted according to the same methodology used in Pompeii, and the analysis of the data followed the same procedure.

Both in Roselle and in Veleia, it appeared that the curve was a polycentric one, composed of four arcs of circles, the four centers of which were symmetrically arranged on the two orthogonal mean axes, although in different ways in each case.

In Roselle the centers are located on the vertices of a square inscribed inside the arena, its diagonal corresponding to the minor axis (Fig. 13.5a). The numbers used for the choice of the dimensions are particularly interesting. The square connecting the centers of the curve has sides of five modules.[3] Therefore the minor axis of the arena, equal to the diagonal of the square and to the radius of the widest arcs, is of seven modules. The number seven acts simultaneously as an approximation of $5 \times \sqrt{2}$, and as a factor in the formula approximating $\pi = 3 + 1/7$.[4] The major axis is of 11 modules, and is related to the minor one by a ratio of $\pi/2$. The designer of Roselle's amphitheater obviously worked out an arithmetic answer to the architectural query, cleverly applying Archimedes's numbers in the dimensions of his project.

In Veleia (Fig. 13.5b) we can observe a game that is more graphic than arithmetic, even though numbers are accurately chosen here too. The centers of the oval curves lie on the vertices of a rhombus, formed by two equilateral triangles, inscribed inside the arena, the height of which is equal to its minor axis. The proportions of this basic figure for locating the centers lead to the drawing of a less elongated oval ring than in Roselle. Here, the search may have included more exacting aesthetic demands. The specific oval curve drawn by two circles circumscribing the central rhombus, which runs exactly along the middle of the width of the *cavea*, would be later called "the perfect oval" by the Italian architects of the Renaissance.

Ellipses are clearly totally absent from the layouts of Roselle and Veleia, even though the two amphitheaters are only slightly later than Pompeii. The reason for the preference for polycentric diagrams, when ellipses and ovals are so aesthetically equivalent, must be sought in the realm of ease and practical necessities. Even though early Greek mathematicians only had straightedges and compasses as their tools, we have evidence that later on, other instruments had to be invented in order

[3] The architectural design module in Roselle corresponds to 12 ft of 29.38 cm.

[4] A circle having a radius of seven (or multiple of seven) will allow simple calculations for perimeter or area, and the results will be expressed through round numbers only.

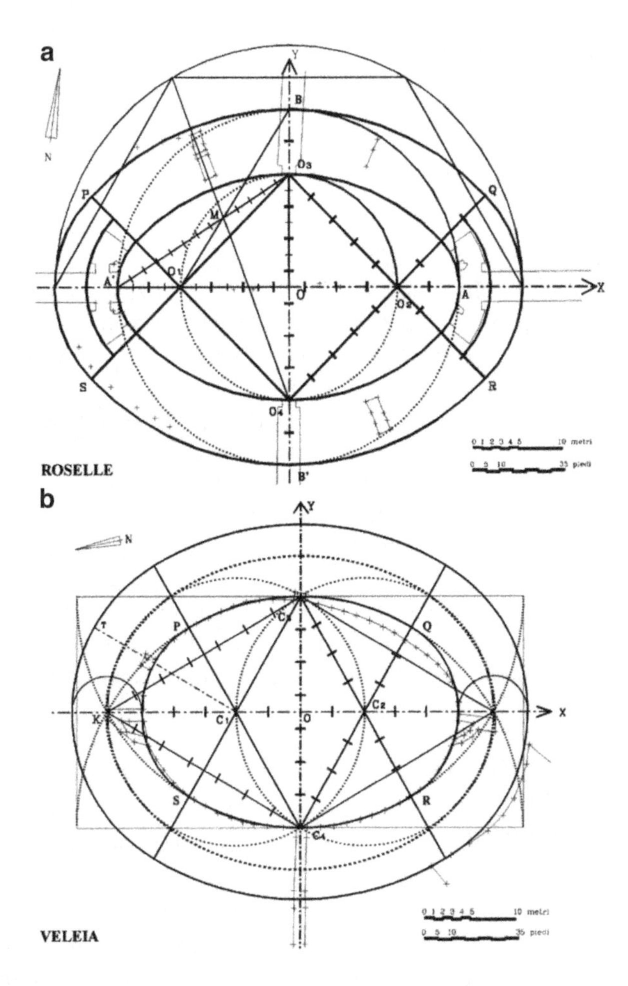

Fig. 13.5 Geometric patterns of the amphitheaters of Roselle (**a**) and Veleia (**b**). Drawing: Author

to allow progress in science. We know for sure that Nicomedes built himself a graphic tool in order to be able to draw the concoid curves that he was studying.[5] This special sort of compass, made of two wooden pieces, one sliding in the other, is conceptually very similar to the well-known elliptic compass, whose first appearance in Western culture cannot be determined with certainty (Fig. 13.6).

This mechanical instrument is made of two orthogonal and fixed rulers that allow a third one to rotate above them. On the mobile ruler two pivots and a graphic point can slide and be fixed wherever necessary. To draw an ellipse having axes equal to A and B, the first pivot must be fixed at a distance of $A/2$ from the pencil, and the second one at a distance of $B/2$. The interval between the two pivots is thus equal to the difference ($A/2 - B/2$). Moving the pencil only, without sliding the pivots, the instrument makes it possible to draw "parallel" ellipses, regardless of the

[5] An approximate image of this tool is published in Loria (1914).

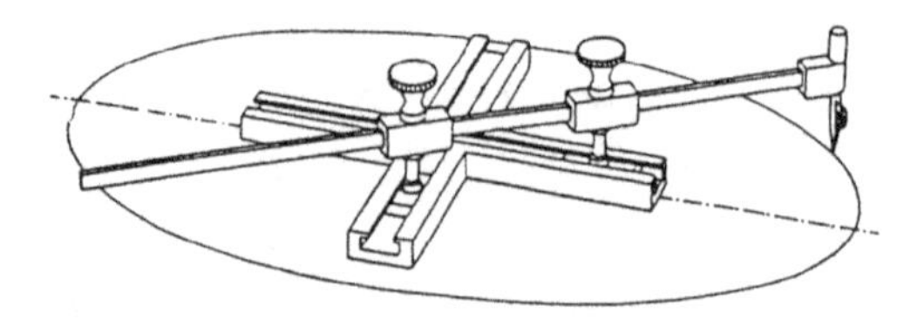

Fig. 13.6 Ancient elliptic compass. Image: Author, after Catalano (1990)

position of their foci. A similar tool may have been used by Pompeii's architect in his preparatory sketches. Thc three main central ellipses (arena, back podium and outer wall of the "underground" corridor) have diameters whose difference is constantly equal to three modules.

If the late Greek mathematicians and Roman architects were able to draw precise ellipses on their tablets by means of appropriate tools, the "gardener's method" remained the only available one to trace an ellipse in full scale on the ground at the building site. While the tracing of a single curve is as simple as laying out a circle, it is much more difficult to set out a series of parallel ellipses than it is to set out several concentric circles. Parallel ellipses are not homofocal, so the foci have to be re-positioned every time. The oval is an excellent graphic imitation of the ellipse, and it offers a number of interesting advantages, all leading to simplification and ease as soon as problems of parallelism, concentricity, radial division and quantity calculation arise.

Conclusion

Architecture and geometry interact when new morphologies or building types appear. Architecture has such qualitative and quantitative demands that it not only brings theoretical geometry into practical application, but also leads it to the establishment of new theories. The creation of a new shape always requires extensions of previous knowledge, as it opens new queries and forces one to work out new diagrams and models. Architectural design by means of the drawing tool (and construction practicalities) surely represented an opportunity for geometry to grow and define new theorems and laws.

Oval diagrams became archetypal theoretical models, and were cited in the later architectural treatises of the Renaissance. In those books, ovals are always compared to ellipses: the former acting as basic theoretical knowledge for design, the latter being the "vulgar" builders' practice. Among all authors, Serlio is the most exhaustive about oval geometry. In the first book of his treaty "*Di Geometria*", he gives examples of graphics for bridge or vault design, based on the drawing of an elliptic curve. But the world "ellipse" is never mentioned. We are told about this particular curve "of lesser height than the half circle" which "really pleases the eye"... Masons trace it with a rope, whereas architects draw it by points with the help of inscribed and circumscribing circles. This curve, he claims, "*is similar to some oval forms drawn with the compass*". He then proposes four possible ways to draw oval shapes, based on two sorts of different patterns and different proportions. The second and fourth examples strongly recall the diagrams of the amphitheaters of Roselle and Veleia.

By Serlio's times, the ellipse was thus considered to be an imitation of the oval, whereas the historical evolution of geometrical architectural diagrams involving these sorts of curves, from antiquity to Baroque, seems to suggest that the oval replaced the ellipse without compromising aesthetics. The move from classical antiquity to the Renaissance marked an inversion between theory and practice, and a decisive shift from the beauty of the mysterious ellipse to the grace of the perfect oval.

Biography Sylvie Duvernoy is an architect, graduated from Paris University in 1982. She was awarded the Italian degree of *Dottore di Ricerca* in 1998. After having taught architectural drawing for several years at the engineering and architecture faculties of Florence University, she recently began teaching at the Politecnico di Milano. Her research mainly focuses on the reciprocal influences between graphic mathematics and architecture. The results of her studies were published and communicated in several international meetings and journals. In addition to research and teaching, she has always maintained a private professional practice. She is the author of *Elementi di disegni, 12 lezioni di disegno di Architettura* (Florence, Le Lettere, 2011).

References

CATALANO, G.M. 1990. Il compasso conico – Uno strumento per tracciare qualsiasi conica con moto continuo. *Disegnare* **1** (October 1990).

DUVERNOY, Sylvie. 2000. Due anfiteatri repubblicani: Roselle e Veleia. *Disegnare* **20/21** (June-December 2000).

LORIA, Gino. 1914. *Scienze esatte nell'antica Grecia*. Milan: Hoepli.

VITRUVIUS POLLIO. 2009. *On Architecture*. Richard Schofield, trans. London: Penguin Classics.

Further Reading

BARTOLI, Maria Teresa. 1998. *Le ragioni geometriche del segno architettonico*. Florence: Alinea.

DUVERNOY, Sylvie and Paul L. ROSIN. 2015. The Compass, the Ruler, and the Computer. Pp 525–540 in Kim Williams and Michael J. Ostwald eds. *Architecture and Mathematics from Antiquity to the Future: Volume II the 1500s to the Future*. Cham: Springer International Publishing.

GOLVIN, Jean Claude. 1988. *L'amphithéâtre romain, essai sur la teéorisation de sa forme et de sa fonction*. Paris: Centre Pierre.

WILSON JONES, Mark. 1993. Designing amphitheaters. *Mitteilungen des deutschen archäologischen Instituts-Römische* **100** (1993): 391-441.

ZERLENGA, Ornella. 1996. Il tracciamento delle «forme ovali» nella trattatistica del XVI secolo. La pratica del filo e del compasso. *XY* **27/28**.

Chapter 14
The Square and the Roman House: Architecture and Decoration at Pompeii and Herculaneum

Carol Martin Watts

Introduction

The *domus* is the ancient Roman single-family urban house type. It was known from descriptions in the treatise by Vitruvius, the late first century BC Roman architect, even before the re-discovery of Pompeii and Herculaneum in the middle of the eighteenth century. These towns on the Bay of Naples were destroyed in 79 AD by the eruption of the volcano Vesuvius. Excavation of these sites (still ongoing) has provided the opportunity to study Roman provincial towns frozen at a moment in time. Houses, mostly of the domus type, occupied the majority of each town. They varied in size, housing families of all social classes, but followed the same traditions of organization and decoration. Examples from Pompeii and Herculaneum range from the second century BC to the destruction of the cities in 79 AD, and over these three centuries there was relatively little change in the building type. The origins of the domus are obscure, but it had developed over several centuries before the earliest examples known at Pompeii.

The Domus

The term domus was used by the Romans to refer to a single-family townhouse organized around a central space, the *atrium,* (see Fig. 14.1). The domus was an

First published as: Carol Martin Watts, "The Square and the Roman House: Architecture and Decoration at Pompeii and Herculaneum", pp. 167–181 in *Nexus I: Architecture and Mathematics*, ed. Kim Williams, Fucecchio (Florence): Edizioni dell'Erba, 1996.

C.M. Watts (✉)
The College of Architecture, Planning & Design, Kansas State University, Seaton Hall 211, Manhattan, KS 66506-2902, USA
e-mail: cmwatts@k-state.edu

K. Williams and M.J. Ostwald (eds.), *Architecture and Mathematics from Antiquity to the Future*, DOI 10.1007/978-3-319-00137-1_14,

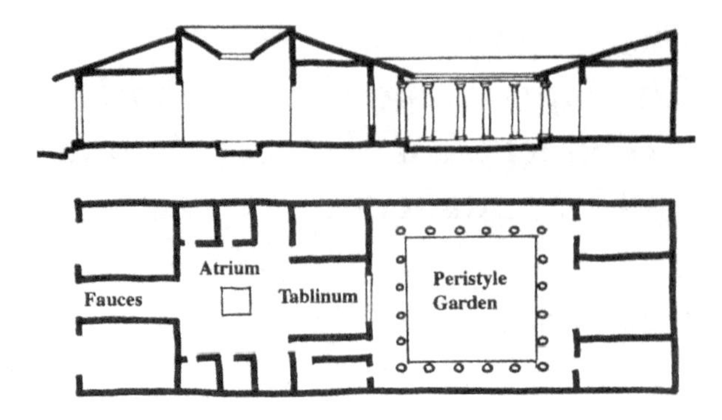

Fig. 14.1 Plan and section of typical domus. Image: author

inwardly-oriented house, with few openings to the street, and generally surrounded by other buildings except on the street facade. Originally only one story in height, an upper floor was sometimes added. The most characteristic feature of the domus was the large central space, the atrium, which usually had an opening in the roof, providing light into the centre of the house. The *fauces* was the name for the entry corridor leading from the street to the atrium. The *tablinum* or major reception room stood at the rear of the atrium, and opened onto a walled garden or a *peristyle*. The peristyle was a garden surrounded by colonnaded porticos, adapted from the Greeks.

These spaces, specialized by function, formed the canonical core of the house, with their relationship to each other rigidly prescribed. Ideally they were disposed around an axis of symmetry, although this was often a visual axis rather than a strict geometric axis, giving the experience of symmetry to the visitor entering the more public spaces of the house. In addition to the major centre of the house (the atrium), secondary centres organized parts of larger houses. These secondary centres included the peristyle area, service areas, and suites of rooms. Other rooms found in the domus in variable locations included *triclinia* (dining rooms), *cubicula* (bedrooms,) reception rooms, kitchens and latrines.

Besides the traditional organization of the spaces of the domus, there were traditions associated with the decoration of interior surfaces. These reflect more rapid changes in style or fashion, but still follow deeply-held traditions. The decoration of surfaces served many purposes, including the expression of the status and interests of the owner, the reinforcement of architectural features and the differentiation of functions of spaces. Walls and ceilings covered with brightly coloured fresco paintings, and floors with patterns, borders, and demarcations were all an integral part of the architecture of the house. Within these decorated surfaces, there were special "pictures", often small reproductions of famous Greek works, executed in fresco in the centre of walls, or in coloured mosaic in the centre of floors. These were the focal points of the decoration, but the "background" was also important in contributing to the architecture of each space and the relationship of one space to another.

It was while studying the relationships of these decorative ensembles (the way the floor, ceiling, and wall surfaces related to each other and to the house as a whole), that I discovered an underlying geometry which appears to operate at all scales within the Roman house. Several well-preserved case study houses were inventoried in detail, supplemented by less extensive documentation of a large number of other houses at both sites.[1] In the process, measurements were taken, plans sketched, and the composition of painted walls and mosaic pavements recorded. Analysis of these measurements, and of published drawings by others, revealed a repeated pattern of proportions and dimensional relationships.

The Square

Analysis of houses at Pompeii and Herculaneum suggests that two simple geometric systems, both based on the square, underlie the design of the Roman house at all scales.[2] These geometric systems explain the proportional relationships which are found in the overall shape of the site and its organization and subdivision, the relationship of volumes of space, and of planes such as walls and floors, throughout the house.

Although Vitruvius does not specifically discuss the geometric systems to which I refer, he does mention the importance of geometry to the Roman architect.

> Geometry, also, is of much assistance in architecture, and in particular it teaches us the use of the rule and compasses, by which especially we acquire readiness in making plans for buildings in their grounds, and rightly apply the square, the level, and the plummet. ... difficult questions involving symmetry are solved by means of geometrical theories and methods (Vitruvius 1960: I, 1, 4).

The two geometric systems, both based on the square, which I found in the domus, can be constructed using only compass and straight edge. They could have been laid out full scale on the site using chains or ropes and stakes. They both enable one to design a plan or other composition in which the parts are all commensurate and proportionally related, at many levels of scale. The square and circle are closely related in this geometry.

Vitruvius emphasized the relationship of both circle and square to the human body with the well-known "Vitruvian man" inscribed within a circle and a square.

> Therefore, since nature has designed the human body so that its members are duly proportioned to the frame as a whole, it appears that the ancients had good reason for

[1] Case study houses included the houses of the Labyrinth, L. Ceius Secundus, Tragic Poet, M. Lucretius Pronto, Vettii, Faun and Sallustius at Pompeii, and the houses of the Carbonized Furniture, Samnite, Tuscan Colonnade, Wooden Partition, and Bicentenary at Herculaneum. For the larger study of which the geometric analysis forms a part, see Watts (1987).

[2] Similar geometric systems were also found in apartments and apartment complexes (*insulae*) at Ostia Antica, near Rome. See Watts and Watts (1987); see also Watts and Watts (1986).

their rule, that in perfect buildings the different members must be in exact symmetrical relations to the whole general scheme (Vitruvius 1960: III, 1, 4).

By "symmetrical relations" Vitruvius is referring to commensurate relationships rather than bilateral symmetry.

One geometric system frequently used in the domus is the square root of 2 progression, often known as *ad quadratum*. Figure 14.2, a photographic detail of a marble pavement in a Roman house, illustrates this system. In this series of squares the side of each square is equal to the diagonal of the next smaller square. Each square thus relates to the next by a root 2 proportion. The area of each successive square is either halved or doubled. This progression can be constructed by joining the midpoint of the sides of a square, thus creating the next smaller square, or alternatively by constructing a square tangent to the comers of the original square and at 45° to it.

Vitruvius discusses a similar system, attributing it to Plato (Vitruvius 1960: IX, intro 4). He explains how one can use geometry, rather than arithmetic, to double a square by making the diagonal of the original square equal the side of the new square. Vitruvius also mentions a root 2 proportion as one alternative for determining the proportions of the atrium of a domus, "by using the width to describe a square figure with equal sides, drawing a diagonal line in this square, and giving the atrium the length of this diagonal line" (Vitruvius 1960: VI, 3, 1). Unfortunately we have no other Roman texts which explain how a house or other building was designed or laid out on the site, but we do have the evidence of the actual buildings to give us further insight into how the Romans used geometry.

A related geometric system also based on the square, shown in Fig. 14.3, is known as the "sacred cut" (Brunès 1967). Starting with a given square and its diagonals, arcs are drawn with a compass centred at each corner of the square, with a radius equal to half the diagonal, going through the centre point of the square. Each of these arcs intersect two adjoining sides of the square, dividing each side into three segments, with the centre larger than the side segments. Connecting the points where the sides are cut by the arcs gives the nine-part grid shown in Fig 14.3. The large central square of the grid will be referred to as the sacred cut square of the original square. The smaller squares in the corners of the grid have sides equal to half the diagonal of the sacred cut square. The rectangles within the grid have sides in the ratio of one to the square root of 2. This geometrical construct can be extended to smaller scales (by repeating the process within the sacred cut square), and to larger scales. Figure 14.3 shows the construction of the next larger square, of which the original square is the sacred cut.

Brunès called this geometric system the sacred cut because it combines the circle with the square. The length of the arc of the sacred cut is within 0.6 % of the length of the diagonal of half the reference square. Brunès argues that this level of accuracy was sufficient for ancient builders to believe that with straight edge and compass they were able to construct a circle with the same perimeter as a given square and vice versa. The "squaring of the circle" is of great importance to cosmological geometry because the circle represented the spirit or unknowable

Fig. 14.2 Photograph of marble pavement from a Roman house using *ad quadratum* geometry. Photo: author

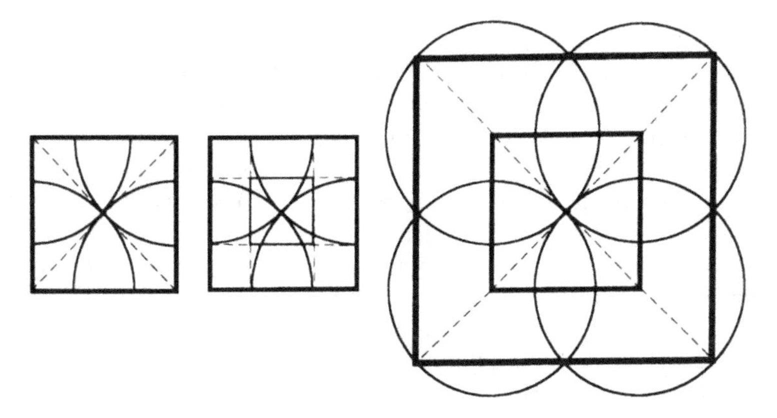

Fig. 14.3 Diagrams of the sacred cut: *left*, *arcs* equal to half the diagonal intersect the *regulating square*. *Centre*, the nine part grid created by joining the points where the *arcs* intersect the *square*. *Right*, the generation of the next *larger square*, of which the *original square* is the sacred cut. Image: author

features of the universe, while the square represented the comprehensible world. The expression of a near-equality between the circle and square thus represented a way in which the unknown could be expressed by the known (Lawlor 1982: 74–79).

Both the sacred cut and the root 2 progression or ad quadratum would have had pragmatic reasons for their use by the Romans, as an easy way to ensure proportional relationships among the parts of a house without the need to do mathematical calculations. They could have been used in drawings for the house as well as on the site during construction to lay out the building and its parts, and to guide the various workmen involved in the construction process, including craftsmen responsible for painted and mosaic decoration. But beyond this pragmatic application, such geometry also seems to have had a strong symbolic meaning, tying all parts of the domestic environment together into a totality which expressed a view of the world, and ultimately relating the house to its place within the universe.

Both ad quadratum and the sacred cut are systems which emphasize a centre through concentric reiterations of the same geometric operation. The emphasis on centre is important in the domus at a variety of scales. The Roman world view, placing themselves in the centre of the universe, permeates their design at all scales, including the organization of the empire, the layout of cities, and the design of most building types.

Geometric Organization of the House Site

Herculaneum was laid out with a regular grid of streets and blocks, and Pompeii, while more irregular, also used grids of different sizes as the town expanded. These generally rectangular blocks were subdivided into building lots. There is variety in the size of the blocks and the size and proportion of the house lots, but all of the examples which I have studied have lot proportions which can be expressed in terms of squares. The placement of major elements of the house, such as the atrium, peristyle, tablinum, and other major spaces appear to be related to a regulating square and its sacred cut and/or ad quadratum. At Pompeii and Herculaneum the houses use a limited number of proportions and of regulating square sizes. The most common proportion of width to depth was 1:3, followed by 1:2. There were a few occurrences of 1:2½ and 1:1. These can be expressed as squares as shown in Fig. 14.4. In some cases the entire depth of the house could be divided into an exact number of squares, and the diagonal of these squares was the width of the house.

It is likely that the laying out of the house was conceived in terms of squares, rather than numerical ratios, because further subdivisions of the house relate to the geometry of such squares. The organization of the house site would first have been conceived in a simple drawing, and then laid out full-scale on the actual site. The "regulating square" would be established on the site, usually using the given site width, and its sides and diagonals indicated with ropes or string stretched between stakes, or even lines inscribed in the ground. The sacred cut could be determined using arcs of rope stretched from the corners of the square to the centre (marked by the crossing of diagonals), then rotated to intersect the sides of the square. The sacred cut grid could be established by connecting these points, and these lines used to position major walls of the house. A similar process could also work for the ad quadratum series, and both appear to have been used in most houses. Such a method would allow the plan to be laid out without the use of dimensions, although dimensions in whole numbers of feet frequently occur in house plans, as the size of regulating squares were often such to allow a series of dimensions close to whole numbers. An important result of such a design method is to ensure commensurable proportions throughout the house.

House	Diagram	Number of Squares	Size of Regulating Square
Labyrinth, Pompeii	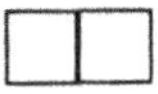	2	116 Oscan ft.
L. Ceius Secundus, Pompeii	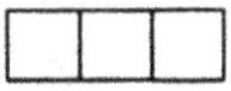	3	34 Oscan ft.
Tragic Poet, Pompeii	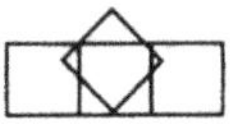	3 width=diag.	34 Roman ft.
M. Lucretius Fronto, Pompeii	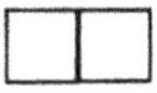 or 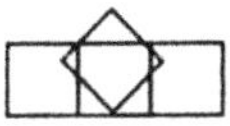	2 or 3 with width=diag.	58 Oscan ft. or 41 Oscan ft.
Vettii, Pompeii		1	116 Oscan ft.
Faun, Pompeii	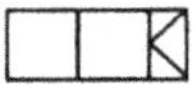	2 1/2	116 Oscan ft.
Sallustius, Pompeii		1	116 Oscan ft.
Carbonized Furniture, Herculaneum	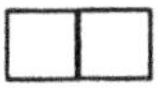 or 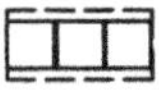	2 or 3	41 or 24 Roman ft.
Samnite, Herculaneum	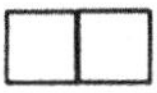	2 (to original lot line)	41 Oscan ft.
Tuscan Colonnade, Herculaneum	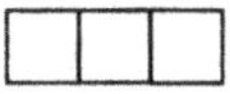	3	41 Oscan ft.
Wooden Partition, Herculaneum	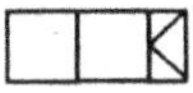	2 1/2	58 Oscan ft.
Bicentenary, Herculaneum	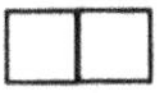	2	70 Oscan ft.

Fig. 14.4 Table of proportions and *regulating square size* for case study houses at Pompeii and Herculaneum

The regulating square of a house can be seen in the basic layout of the plan, as a geometric overlay, but it also provides whole number dimensions in either Oscan feet or Roman feet which occur throughout the house.[3] There are two common and interrelated numerical series frequently found in dimensions of Roman houses. The diagram in Fig. 14.5 illustrates a series using 12, 17, 29, 41, and 99 derived from the sacred cut, as well as the ad quadratum series of 29, 41, 58, and 82. These series

[3] The Oscan foot, used in the earliest houses at Pompeii and Herculaneum, is 0.275 m = 1 ft. The Roman foot in use at these sites is 0.297 m = 1 ft. See Mau (1982, p. 280).

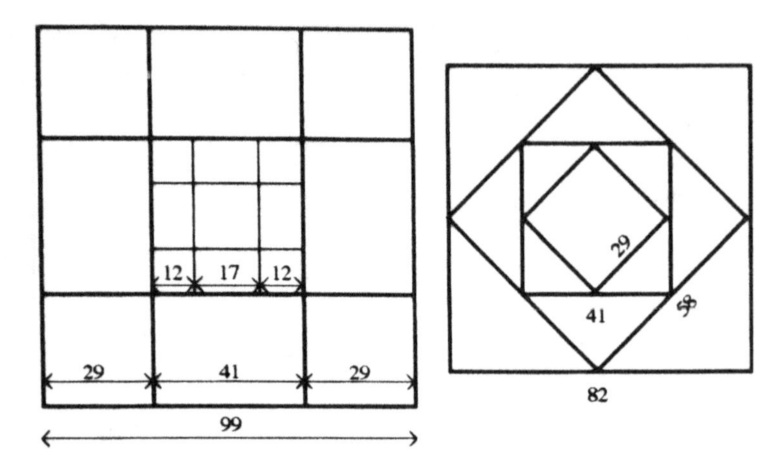

Fig. 14.5 Diagram of related sacred cut and ad quadratum numerical series. Image: author

have 29 and 41 in common. The particular numbers found in this sacred cut series are those occurring in the Principle of Alternation, the Pythagorean method of finding successive approximations of the square root of 2.[4]

Figure 14.6 illustrates the use of dimensions derived from the regulating square in the House of the Tuscan Colonnade, Herculaneum. The site is three squares deep, with each square 41 Oscan ft. The sacred cut of this 41 ft. square gives the dimensions 17 and 12; 17 plus 12 equals 29, part of the ad quadratum series from 41. The diagonal of 41 is 58. The plan indicates where these dimensions are found within the house. The width of the atrium is very close to 29 Oscan feet. The width of the tablinum is 17 Oscan ft. The peristyle with its porticoes is 41 ft. square, and the diagonal of the peristyle colonnade is also 41. The peristyle width within the colonnades is 29 ft. Other spaces use dimensions of 17, 12, 7 and 5 Oscan ft. Other houses surveyed show similar use of a set of geometrically related dimensions, particularly for the major spaces (later additions, alterations, or service areas may not conform to the system used in the core of the house). Although best documented in plan, from what evidence survives of full height walls it appears that similar dimensions also were used in three dimensions, so that the spaces of a house were proportionally related volumetrically as well as in plan.

Within individual spaces, similar geometric relationships can also be seen. Many spaces, particularly the more important rooms of the house, are cubes, or rectangular spaces with two square walls. Figure 14.7 illustrates common room proportions. Vitruvius devotes chapter 3 of book VI to the proportions of the principal rooms of a house. He discusses the appropriate proportions in plan and height for the atrium, alae, tablinum, peristyle, triclinia, and *oeci* (reception rooms). Most of these proportions are based on the width of the room. He is concerned both with the proportions of one room to another, and with proportions of the individual

[4] Lawlor gives a geometric demonstration of this principle in (1982, pp. 38–43).

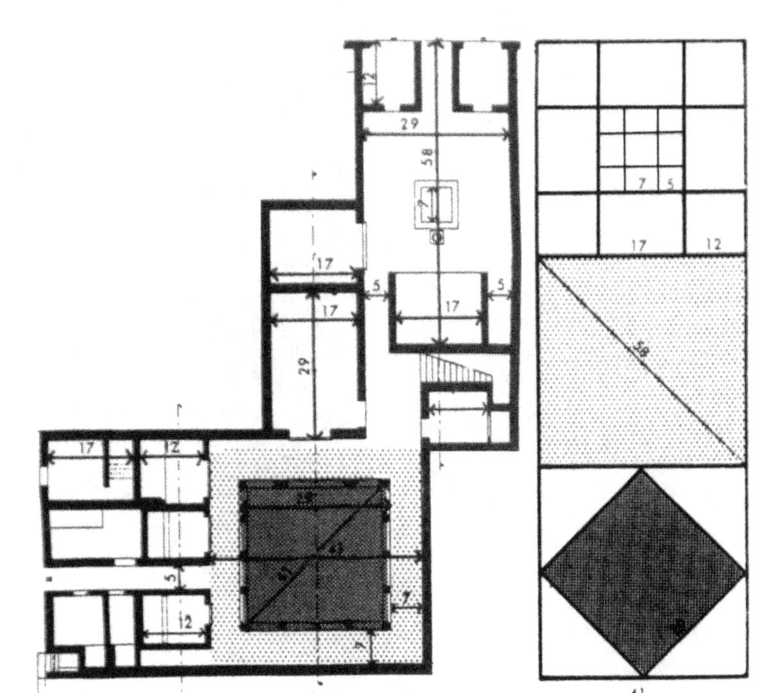

Fig. 14.6 The *regulating square* of the House of the Tuscan Colonnade, Herculaneum. Image: author

space. Studies of actual examples of the domus at Pompeii have shown only some houses following Vitruvian proportion (Weiskittel 1979: 25–38). More examples appear to use the geometry discussed here.

A common proportion in plan for rectangular rooms is one to the square root of 2. Vitruvius mentions this as one of three ways to proportion an atrium (Vitruvius 1960: VI, 3, 3). If the short side of such a space is a square in elevation, then the length of the long side equals the diagonal of the square of the short side. The ceiling line coincides with the sacred cut line of the regulating square of the long side (Fig. 14.7). The same proportions in plan may have a different volume, with the walls of the long sides equal to a square. In this case the short walls are taller than a square. Their height is equal to the diagonal of their regulating square (or a square plus the central sacred cut square). An example of this proportion is a triclinium in the House of L. Ceius Secundus at Pompeii. The room measures 12 by 17 Oscan ft, and appears to be very close to 17 ft high. The tablinum of the House of the Tuscan Colonnade at Herculaneum has the same dimensions in plan, 12 by 17 Oscan ft, but is a variation on the theme in height (Fig. 14.8). The long wall is the same height to the top of the vault as it is wide, thus creating a conceptual square. The flat ceiling on either side of the vault creates two short walls that are of a different proportion than in the previous example. Here, the smaller wall can be seen as a square plus the smaller dimension created by the sacred cut of that square. The floor of this room

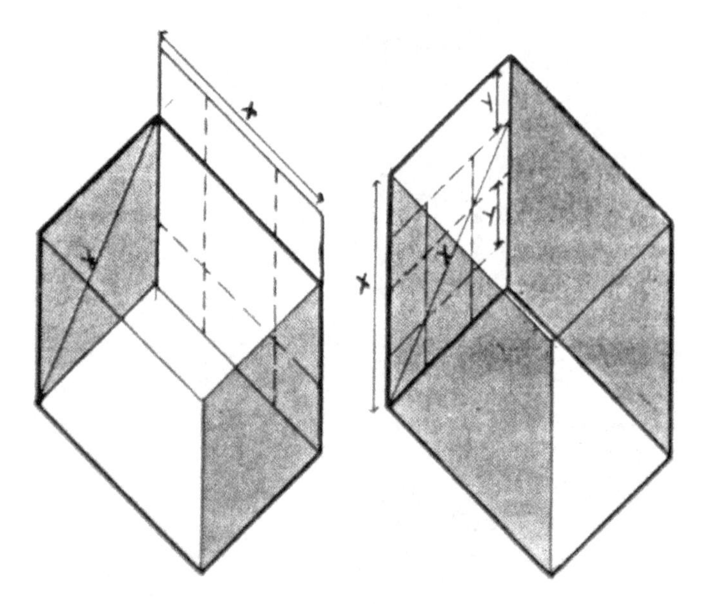

Fig. 14.7 Diagrams of the common Roman room proportions. Image: author

consists of an over-all white mosaic with a black border. The border creates a rectangle related to wall dimensions as shown on the diagram. The window placement and size also relate to the regulating squares.

The same geometrical systems appear to operate in most cases at the smallest scale, that of individual planes (walls, ceilings and floors) and their decoration. A regulating square (the width in the case of walls) appears to determine the organization of the composition and dimensions of parts. Figure 14.9 shows one common motif used in different types of pavements as well as ceiling ornament in Roman houses, and the diagram illustrates how this pattern is based on both the sacred cut and the ad quadratum. This geometry permeates the Roman domestic environment at all levels of scale, at times quite obviously expressed (as in Fig. 14.2) and at other times not evident without analysis.

Conclusions

The house was of great importance to the Romans. The domus was one of their earliest building types and remained the archetype for other buildings. As such, it embodies the basic characteristics of Roman design, including an emphasis on a centre, organizing axes, the differentiation of spaces, and the integration of architecture and ornament. The geometry of the square also appears to be an integral part of the house. Further study may reveal its operation in other more monumental, Roman buildings. For its symbolic as well as pragmatic utility, the

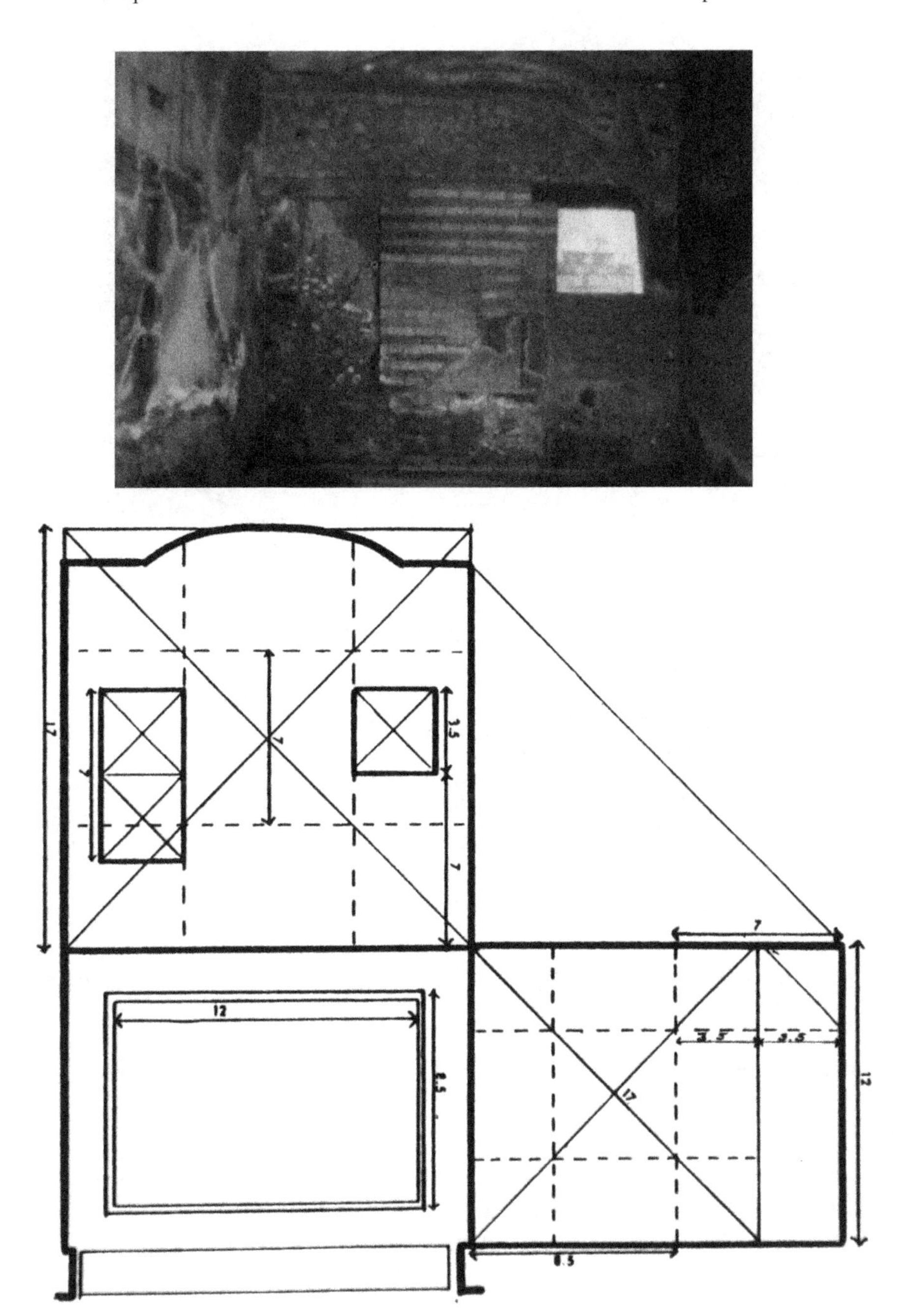

Fig. 14.8 The tablinum of the House of the Tuscan Colonnade, Herculaneum—photograph and diagram showing the relationship of wall and floor planes and the dimensioning based on the sacred cut. Image and photo: author

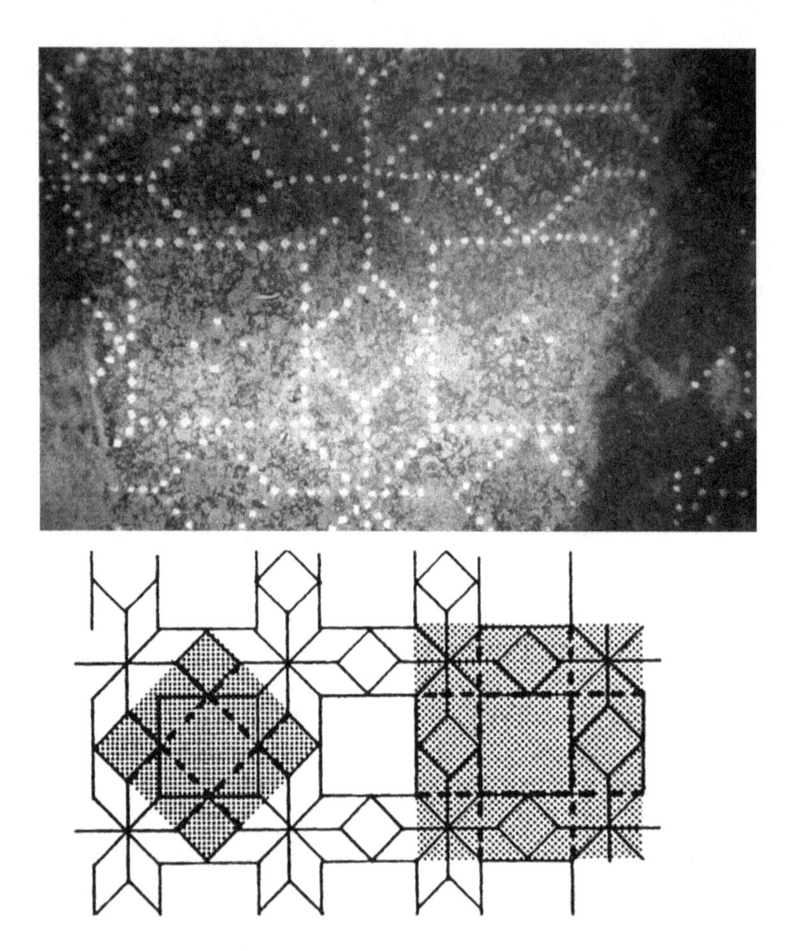

Fig. 14.9 Photograph of a floor pattern from Pompeii, and analysis of this pattern, showing both the sacred cut and ad quadratum. Image and photo: author

square was the basis for the conception and construction of the domus, as we know it from the well-preserved sites of Pompeii and Herculaneum.

Biography Carol Martin Watts is an architectural historian and faculty member in the Department of Architecture at Kansas State University, Manhattan, Kansas. She received her Ph.D. from the University of Texas at Austin, writing her dissertation on "A Pattern Language for Houses at Pompeii, Herculaneum and Ostia". She has published several articles, together with Donald J. Watts, on geometry in Roman architecture, painting and mosaics, including "A Roman Apartment Complex" in *Scientific American*. Her current research interests include vernacular Italian architecture and the impact of Roman development on subsequent layers in urban environments. She is a past member of the editorial board of the *Nexus Network Journal.*

References

BRUNÈS, Tons. 1967. *The Secrets of Ancient Geometry and its Use.* 2 vols. Copenhagen: Rhodos.

LAWLOR, Robert. 1982. *Sacred Geometry, Philosophy and Practice.* New York: Crossroads.

MAU, August. 1902. *Pompeii, its Life and Art.* Francis W. Kelsey trans. (reprinted 1982) New Rochelle, New York: Caratzas Bros.

VITRUVIUS. 1960. *The Ten Books on Architecture.* Morris Hicky Morgan trans. New York: Dover Publications.

WATTS, Donald J. and Carol Martin WATTS. 1986. A Roman Apartment Complex. *Scientific American* **255**, 6 (December 1986): 132–139.

WATTS, Carol Martin. 1987. A Pattern Language for Houses at Pompeii, Herculaneum and Ostia. Ph.D. Diss. The University of Texas at Austin.

WATTS, Carol Martin and Donald J. WATTS. 1987. Geometrical Ordering of the Garden Houses at Ostia. *Journal of the Society of Architectural Historians* **46**, 3 (September 1987): 265–276.

WEISKITTEL, S. Ford. 1979. Vitruvius and Domestic Architecture at Pompeii. In *Pompeii and the Vesuvian Landscape.* Washington: American Institute of Archaeology and Smithsonian Institution.

Chapter 15
The "Quadrivium" in the Pantheon of Rome

Gert Sperling

The Mathematic Concept of the Pantheon Complex

The Pantheon complex has been the object of countless interpretations. There is no certainty as to how and why it was created and what it is meant to express, because there are no documents concerning the identity of the architect, the exact datas of conception, its origin and its function. Since ancient times we find vague references to its symbolic function: according to Cassius Dio, it resembles the heavens. But the cosmological interpretations do not take into consideration the real metrical dimensions of the whole complex nor the relation between its numbers, shapes, forms and proportions. Even the modules are identified very differently, so that it is difficult to compare the various analyses (De Fine Licht 1966; Brunés 1967: 38). This present analysis is based on a 1989 survey of the rotunda by Marco Pelletti (1989: 10–18).[1]

There is an unusually high precision in the construction of the cylinder and the cupola of the rotunda, as well as in the division of the cupola into 28 coffered segments necessitating the realization of irrational angles (12.857...°) (Martines 1989). The curved distances of the upper zone, derived from the factorization of 28 as 4×7 are identical to pilaster-distances of the lower zone, derived from the factorization of 32 as 4×8. The square modules of the pavement are integrated into the architectural design of the *cella* (rotunda) through the derivation of their

First published as: Gert Sperling, "The "Quadrivium" in the Pantheon of Rome", pp. 127–142 in *Nexus II: Architecture and Mathematics*, ed. Kim Williams, Fucecchio (Florence): Edizioni dell'Erba, 1998.

[1] For a description of the mathematics involved in this system, see Calter (2014).

G. Sperling (✉)
Wilhelm-Busch-Str. 5, 34246 Vellmar, Germany
e-mail: sperling.dehair@t-online.de

K. Williams and M.J. Ostwald (eds.), *Architecture and Mathematics from Antiquity to the Future*, DOI 10.1007/978-3-319-00137-1_15,

dimensions from the interior dimensions of the Pantheon; the pavement design "surgically" meets the perimeter of the rotunda.[2]

This ensemble of geometrical forms and their dimensions reflects the fusion of opposite elements in order to achieve harmony. There is a fusion of integer numbers. There is also a fusion of irrational numbers, represented by geometrical shapes, such as π, $\sqrt{2}$, the Pythagorean theorem $c^2 = a^2 + b^2$, the Golden Mean (the Fibonacci series), the Sacred Cut and the *ad quadratum*. Represented also are the three "classical" mathematical problems: the doubling of the cube, the trisection of an angle, and the squaring of a circle.

The main module of the rotunda is 147 Roman ft (1 Roman ft = 0.2956 m).[3] The module may be thought of as $3 \times 7 \times 7$, measured from the inner edge of opposite columns. The number also relates to smaller modules determined by the height of the column capitals, the column diameters, the length of the column shafts, the diameter of the *opaion* (oculus) and formal elements of the former *attica*. The fact that the main module is found there links the Pantheon to many sacred buildings with a Greek tradition (Jacobson 1986). My own measurements and calculations have identified the fusion of the Roman foot (0.2956 m) with the Roman palm (the width of a hand, equal to 0.2173 m) in many of the Pantheon's architectural details, so that the building may be interpreted using two anthropomorphical measures (ratio 34:25) as well as the geodetical meter: the radius of the main module, 73.5 ft, equals 100 palmi, while the measured radius of the cupola 22.039 m equals 7π m!

The Golden Mean has been identified in the distance of the five rows of coffers, and is related to data and angles of the positions of the sun between the equinoxes and the summer solstice in ancient times (114 AC) (Alvegård 1987: 17). I believe this fact explains the supposed uncentred position (winter solstice) of the *Arcus Pietatis Trajani* in the forecourt, a point of view neglected by Fine Licht (1966: 295, note 43).

Two Roman "blueprints" were found in the travertine pavement in front of the Augustus-Mausoleum, the larger one of them deciphered as a full-scale representation of the gable of the *pronaos* of the Pantheon (Haselberger 1995a; b). The angle of the gable shown is exactly 24°, referred to by Vitruvius as the ecliptic angle of the polygonal with 15 angles (Vitruvius 1991: IX, 439–445), while the angle of the actual gable is smaller, perhaps taken from the real ecliptic of those times (23°41′) and realized using geometrical gnomon theory and empirical measurements of the angle. The edge blocks of the gable could have been prepared as 24° in a first moment, then adapted and fitted into the shape of the real given ecliptic-angle during the construction phase, so that the ancient sun dictated the integration of the *pronaos* into the rotunda as well as, I believe, the arrangement of the whole complex. The second gable above is a kind of "legend" of the gnomon theory, projected onto the wall of the intermediate block.

[2] For discussions of the Pantheon pavement, see Williams (1997a, b).

[3] This value was measured and calculated by Pelletti (1989). Most Imperial buildings are based on a foot of 0.29476 m, the 100th part of Cestius pyramid.

While the rotunda represents the sphere of the sky, the convex pavement, some 28 cm higher in the centre that at the perimeter, is intended as an image of the earth's surface (representing, according to Pelletti one-seventh of the radius of the earth) and originally supposed to be a real section of a sphere. The projection of the rays of the sun on the pavement at the summer solstice indicates knowledge about the real size of the earth and the Roman Empire. The east-west axis of the rotunda represents the northern Tropic of Cancer (23°40′), the centre of the beam of light on 21 June marks Rome's latitude (41.88°). The rotunda is a circle with a radius equivalent to the distance of the northern tropic to the equator, touching the Canary Islands in the west and the border to the Parthers in the east: hence, the northern part of the rotunda is an image of the Roman Empire under Trajan.

Haselberger has traced the date of construction to the time of Trajan by analysing the substance of the walls, and maintains that the rotunda was already half built in 115 A.C. Specific mathematical features of the whole building and its relation to the Markets and the Column of Trajan indicate Apollodorus of Damascus, a Greek-educated intellectual and famous architect, as the author of the Pantheon (Heilmeyer 1975).

Scholars such as Giangiacomo Martines take the Neopythagorean roots of the Pantheon seriously, interpreting the architecture as an integrated visualization of the Greek mathematically-conceptualized theory of the cosmos, which consisted of an amalgamation of cosmological, geodetical and anthropomorphical dimensions. To generate harmony, the laws of arithmetic, geometry, astronomy and musical-proportions are fused. The Pantheon can be considered an architectural image of the Pythagorean cosmos, a "living organism" with a mathematically proportioning "soul" and unchanging, "eternal," consonant-symphonic ratios. It "resembles the heavens", but is a resemblance based on mathematical knowledge, a summary of the ancient *Quadrivium* (Munxelhaus 1976: 41).

The Fundamental Role of Arithmetic and the "Perfect Numbers"

The Pantheon is arithmetically related to the "perfect number" 28 (4 × 7), 7 being a "holy" number in ancient symbolism and dedicated to Athena (Minerva) and Zeus (Jupiter) (Nicomachus of Gerasa 1926; Stahl 1971; Capella 1977: 281). It is also related in an important way to the planetary system and the ages of man. Twenty-eight is the second member of a small family of really rare numbers, which Nicomachus praised in his *Introduction to Arithmetic* as "perfect." They are the sum of their factors (this is the basis of harmony) and there is only one number within the unit:

6	= 1 x 2 x 3	= 1 + 2 + 3 = 6
28	= 1 x 2 x 2 x 7	= 1 + 2 + 4 + 7 + 14 = 28
496	= 1 x 2 x 2 x 2 x 2 x 31	= 1 + 2 + 4 + 8 + 16 + 31 + 62 + 124 + 248
8128	= 1 x 2 x 2 x 2 x 2 x 2 x 2 x 127	

Twenty-eight is emphasized in the Pantheon in each of the four quadrivial disciplines: arithmetic, geometry, astronomy and music. Further, there are remarkable relationships to the $7 \times 7 = 49$ beneath the old Pythagorean dogmatical meaning of the 10 (*tetraktys*) and its powers.

According to early Pythagorean philosophy, the whole of reality was a subdivision of the unit (*logos* = 1 = the whole of being) into fragmented and incessantly changing parts. Behind this instable process of materialization, there is a constant system of numbers, which represents the "true," everlasting aspect of all being in its forms, sizes, quantities, qualities, colours, movements and so on. Harmony is the mathematical power to fuse what consists of or represents the opposite elements of being. And it is the same with numbers, because their characteristics can also be different and sometimes opposite (for example: odd and even); harmony is needed to put them together. According to Nicomachus, the numbers 1 and 2 are not real "numbers", they are both dualistically understood as opposite elements or sides of the unit. The dualism has to be overcome by mathematically fusing the opposite parts of worldly phenomena, represented by the numbers, forms, shapes, proportions and the laws of movement.

Arithmetic being the fundament of the *quadrivium*, I have discerned many arithmetic relationships within the formal arrangement of the pavement, which is determined by a grid of 11×11 squares cut by the perimeter of the rotunda:

1. When a square is constructed within the Pantheon so that its corners touch the interior perimeter, it contains $7 \times 7 = 49$ square units within the paving grid. Further, the total number of uncut (whole) paving squares is divisible in three ways: 4 lateral rows of 7 = 28; 3 rows of 11 = 33; 4 longitudinal rows of 7 = 28.
2. The sum of the odd numbers (circle-in-square panels) is 45; the sum of the even numbers (square-in-square panels) is 44. Taking 1 and 2 together as the unit (according to Nicomachus), there are 22 divisible odd numbers, 22 indivisible prime numbers, and 44 even numbers. That means a proportion of 88 (total):44 (even):22 (odd: prime or divisible) = 2:1; this is the octave-interval in the Pythagorean music system, a queen of harmony.
3. Integers also are able to generate geometrical forms; because of this the stereotypical change of these circles and squares in the panels can be read as a diagram of qualities of the so-called "number-families." The main difference of

"odd and even" generates the *diagonals* (binary code); on the diagonals you find the odd and even *square numbers*, beginning with 1 in the middle (centre of the sixth row in both directions). The perfect numbers 6, 10, 28 are elements of the *triangular* numbers, 6 consisting of 3 units, 10 of 4 units (*tetraktys*), 28 of 7 units on each side of the triangle. The Platonic solids are based on triangles.

Other arithmetic relationships are found in the inner column shafts, which are 8 diameters and 7 capital-heights high, a fusing of 8 and 7 according to the upper order (7) and the lower order (8). If the circle-in-square panels of the pavement had been originally of the same size, they would have had a diameter of 8 ft, while the squares have 7 ft, the greater ones 10 ft, again a ratio of $1{:}\sqrt{2}$.The column shafts equal the inner width of the oculus. The length of the whole Pantheon complex equals nine times the radius of the rotunda; subtracting 1 radius for the south building and the rotunda wall, there remain 8:2 radii or 4:1 diameters from the south apse to the north entrance. This is the double-octave system derived from an old model of sphere harmony; it is a distance of $21 \times 28 = 588$ ft. Including the apse, there are 600 ft, the 224th part (8×28) of the earth's circumference according to Eratosthenes (2 B.C.), in metres instead of kilometres ($\times 10^{-3}$). Nicomachus' arithmetical row of numbers contains 28, 56, 112, 224, a row of geometrical means. These 600 ft are equivalent to 100 average heights of man according to Vitruvius (a man's height = 6 Roman ft) and this corresponds geodetically to the division of the first three perfect numbers: 496 m:28 m = 17,714 m = 6×10 ft = 10 Vitruvian man-heights. The fusing by arithmetical numbers of geodetical and anthropomorphical qualities and musical consonant proportions is obvious.

The Symmetry of Cylinder and Sphere, Their Irrational Divisibility and the Inheritance of Archimedes

That the creator of the world should use irrational numbers as well as and instead of integer numbers and "ratios" was a shock to the Pythagoreans. Archimedes wished that his greatest discovery, the symmetry between cylinder and sphere, be inscribed in his grave-stone: the surfaces of cylinder (*tambour*) and sphere with the same radius have a ratio of 1:1, the *volumina* of 3:2; the irrational number π is involved, approximated as 3.14 using his "method of exhaustion." The ratios correspond to the musical intervals of a prime and the fifth. The rotunda is the fusion of a cylinder with a (semi)sphere, ideally combined with a cube; this is reflected by the two-dimensional squares and circles of the pavement (Williams 1997b: 5). The cylinder inside is half of the height of the rotunda, an interval of the octave (1:2), outside it is higher due to reasons of statics, but still with a difference of $\sqrt{2}$ to the inner height of the rotunda: 104–147 ft. The *pronaos* also is influenced by $\sqrt{2}$: the square cutting the exedrae walls of the rotunda has a diagonal that fits exactly into the depth of the pronaos (Williams 1997b: 5).

The geometrical elements of the Pantheon and their dimensions are closely related to the Mausoleum of Augustus: the circle of the *tumulus* has a diameter of 104 (8 × 13) ft; a square circumscribing this circle has a diagonal of 147 (104 × √2) ft, equal to the diameter of the rotunda. The pavement of the rotunda is divided exactly in 8 × 13 × √2 ft with reference to the centre points of the grid bands (Williams 1997b: 5). Both the Mausoleum and the Pantheon introduce the factor √2, symbolizing "an ovation to Augustus" (Andreae 1973: 527–529).

The volumes of cubes with sides measuring 104 ft and 147 ft have a ratio of 10:28 (*tetraktys:* perfect number); the volume of the rotunda is approximately 100 times the cube based on the length of the column shafts (Geertman 1980: 203–229).

The cupola is divided in 28 segments with 5 rows of coffers, a division impossible to construct with compass and straightedge. This division was cited first in Rome in the era of Barock (St. Maria in Campitelli). The trisection (as well as the division in seven parts) of a right angle is possible only using the *trisettrice* of Hippias of Elis, or the *conchoide* of Archimedes.

The squaring of the circle is referred to in the dimensioning of the infrastructure of the rotunda wall: the size of the pavement square is equal to the circle bordering the semi-circular niches inside the wall. This circle provides the basis for a consistent geometrical interpretation of the whole complex using derivations of √2. The doubling of the cube relates to the inner and the outer square of the rotunda wall, being identical to the squares obtained by performing the Sacred Cut.

The "Gnomon-Theory" and the Sun as Architectural Constructor

Considering the fact that in 9 B.C. Facundus Novus, in order to pronounce the cosmological relationship between the birthday of the first princeps (23 September, autumn equinox) and his generation at the winter solstice, exactly projected the Augustean *Horologium* in relation to the main architectonic lines of the Ara Pacis in its further position and the Mausoleum of Augustus (Buchner 1982) using the positions of the sun, it is plausible to look for connections between the gnomon theory and the construction of the Pantheon.

It is easy to demonstrate that the gnomon theory played a principal role in the planning and construction of the rotunda, the intermediate block and the *pronaos*, as well as determining the dimensions of the forecourt: historical astronomical data indicates an ecliptic angle of 23°40′ around the year 114 A.C. The gnomon theory referred to by Vitruvius has to be centred in the middle of the oculus (Vitruvius 1991: IX). The latitude of Rome, 41.88° appears in the angle formed by the cornice and the connection of the upper niches to the centre of the oculus. A plumbline

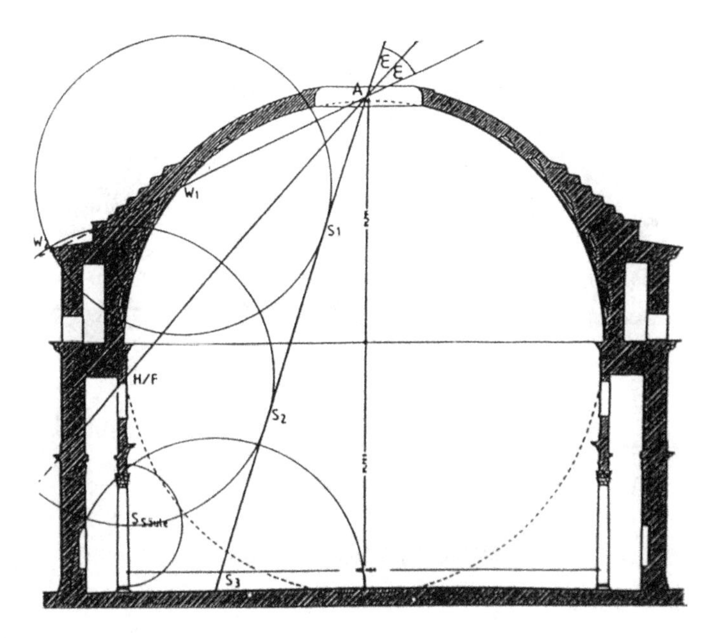

Fig. 15.1 Drawing: author

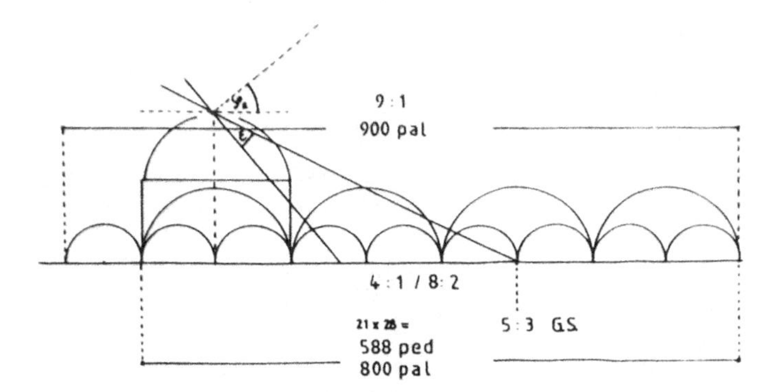

Fig. 15.2 Drawing: author

along this line gives the position of the sun at the equinox, which exactly halves the depth of the *pronaos*. Along the equinox line you can reconstruct the location of sun rays during the summer and winter solstices: evidently the architectonic frames of the building are cut by a circle with the same radius, given by the three sun positions (Fig. 15.1). The point where the ray of the winter solstice meets the forecourt is exactly given by the doubling of the rotunda's diameter, the main module of building (Fig. 15.2).

The triangle formed by the sunrays of the equinox and the point where the sunrays of the summer solstice cut a plumbline dropped through the middle of the rotunda is an exact configuration of the position, angle and form of the upper gable

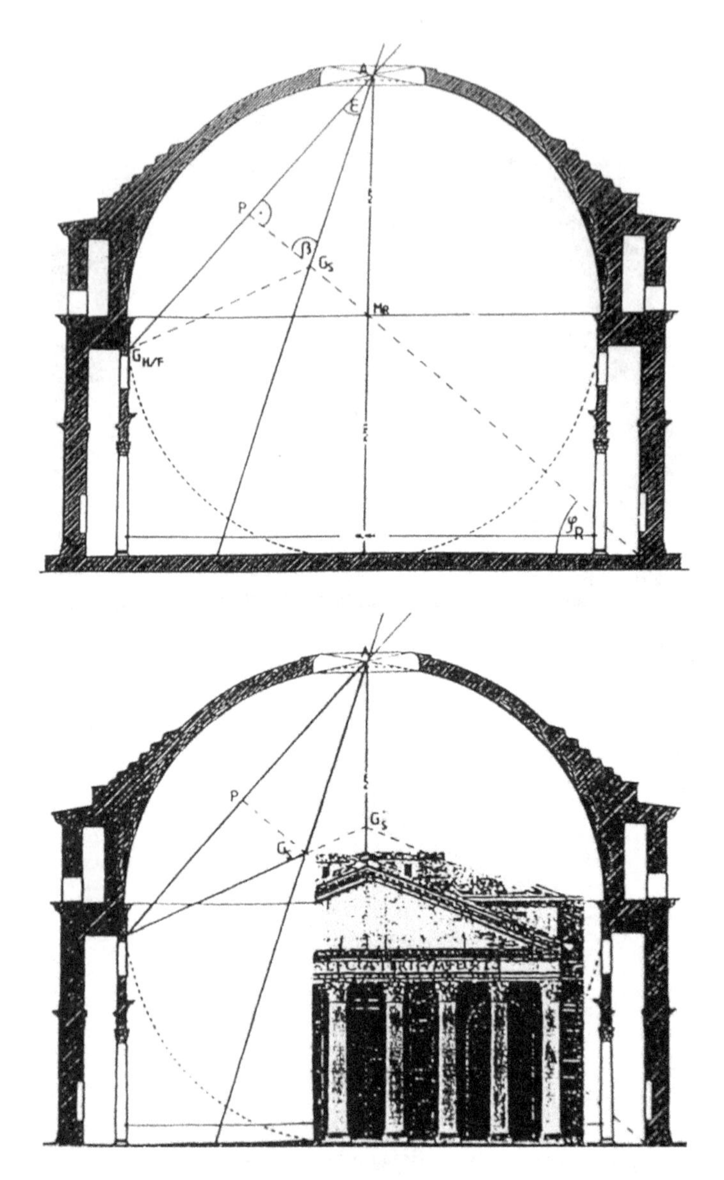

Fig. 15.3 Drawing: author

upon the intermediate block. The base line is identical with the cornice inside the rotunda, its centre point is the navel of the inscribed *Vitruvian Man*, whose dimensions relate to the height of the inner capitals by the factor 28 times the irrational value of the Golden Mean (Fig. 15.3) (Alvegård 1987: 8). The head of the Roman eagle also falls on the navel (Fine Licht 1966: 46) (Fig. 15.4), thus the building's proportions, contrived with the infallibility of scientific knowledge, symbolize the cosmic legitimacy of Roman imperialism as an image of the

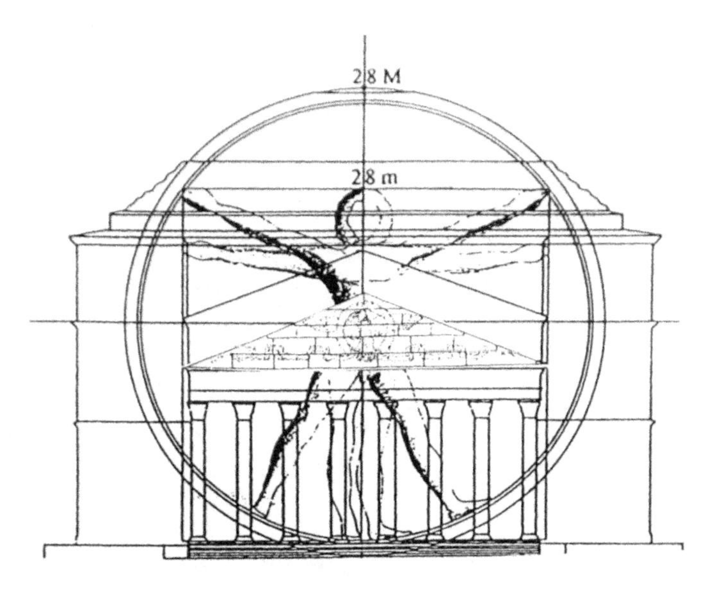

Fig. 15.4 Drawing: author

cosmological *logos* and its reason, harmony and beauty. The *pronaos* was intentionally contracted with the given height, angles and proportions of the columns. Their arrangement is a fusing of eustyle and systyle to create the harmonic concept of Hermogenes of Alabander (Haselberger 1995b: 303). According to the hypothesis of Davies et al. (1987), a *pronaos* equal in height to the intermediate block would have destroyed the importance of the sun positions as significant lines to integrate rotunda, intermediate block, pronaos and forecourt.

It is evident that in the rotunda the gnomon theory is theoretically revolutionized: the centre point of the gnomon model is the motionless point of the geometrical arrangement; the top of a gnomon, or the globe on top of an obelisk, represents the motionless earth in the middle of the geocentric view of the cosmos. The changing length and position of the shadow is thought of as a *reflection of the sun's movement*. In the rotunda, the motionless point of the gnomon model is represented by the middle of the oculus, projecting the sunlight as a cylindric beam into the darkness of the rotunda. It is, therefore, an image of the sun and consequently, within the theory, the motionless point. If the sun is thought of as motionless and situated in the centre of the cosmos, the movement of the light-shape inside the rotunda is a *reflection of the two movements of the earth*: the axial rotation (day and night) and the orbit through the zodiac on the inclined plane of the eclipse (year). The rotunda is a document of the first Greek heliocentrical hypothesis, inaugurated by Herakleides Pontikos, Aristarchus of Samos, and Seleukos of Seleukia, both cited by Plutarch in *Quaestiones Platonicae* shortly before the construction of the Pantheon (Van der Waerden 1988: 148–155) (Fig. 15.5).

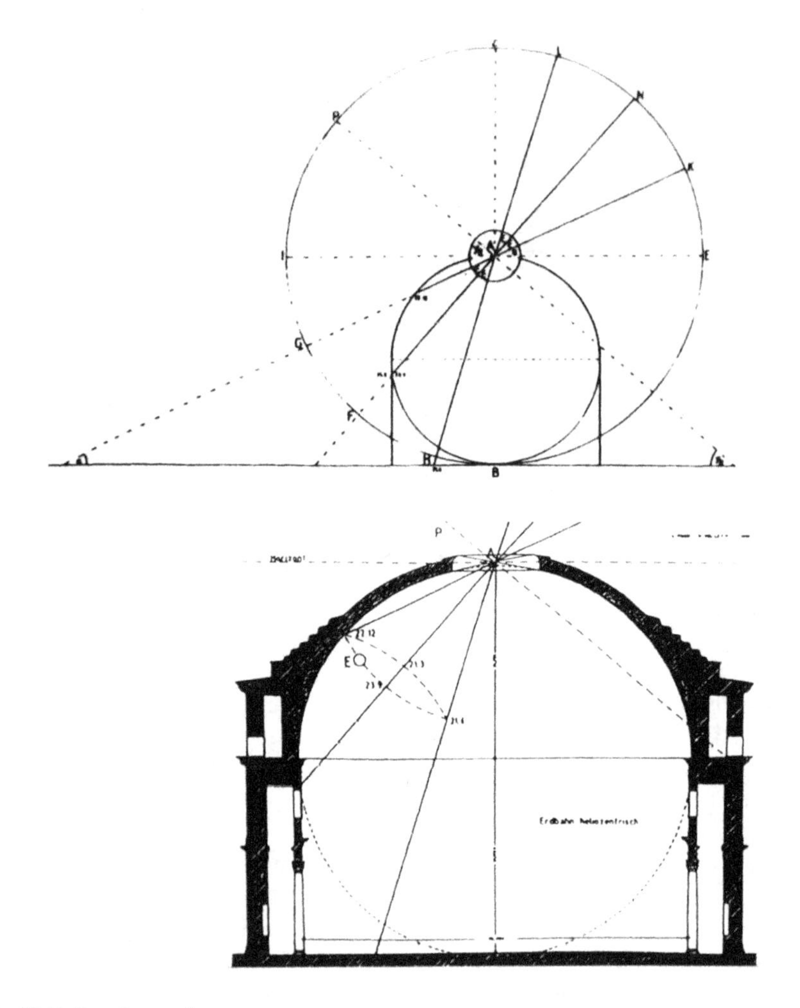

Fig. 15.5 Drawing: author

The Ancient Dogma of "Sphere Harmony" and the Interval System in the Pantheon

By studying the mathematical laws of the consonant or dissonant intervals of tones on a monochord, the Pythagoreans developed a harmonic system of a double-octave. Unlike the single octave, it can be divided exactly in the middle; division of the single octave into two parts can only result in a fifth and a fourth. F. Zaminer (1984) has reconstructed the double-octave system from the earliest Pythagorean ideas of "sphere harmony", a "symphonic" system of consonant harmony derived from the movements of the planets and the intervals between them. Taking the earth-moon relationship as the basic tone and assigning to it the number 1, the Pythagorean tradition constructed the following proportional steps:

first octave:

earth	-	moon	=	1	
earth	-	mercury	=	1 1/3	(4:3 = fourth)
earth	-	venus	=	1 1/2	(3:2 = fifth)
earth	-	sun	=	2 (middle)	(2:1 = octave)

second octave:

earth	-	sun	=	1	
earth	-	mars	=	1 1/3	(4:3 = fourth)
earth	-	jupiter	=	1 1/2	(3:1 = fifth above octave)
earth	-	saturn	=	2	(4:1 = double-octave)

This system includes the distinction between *inner* ("in front of the sun") and *outer* ("behind the sun") planets as one important condition of the heliocentric hypothesis. The sun is the centre of the interval system, and the movements of the planets are calculated in relation to the position of the earth. If an image of the cosmos had to be built, it was important to use these proportions in the structure and in the architectonic arrangement of the building and its design elements.

On first examination, the double-octave system may be discerned in the well-known structure of the rotunda: the cornice divides the rotunda in 2:1, and the perimeter wall itself is divided horizontally in 2:1 by the first ledge above the columns and pilasters. Taking into consideration the movement of the sunrays passing through the *opaion*, these proportions can be identified as an astronomical proportioning of the main horizontal lines inside the rotunda, as described by Ptolemy with his model of the so-called *helikon* (Munxelhaus 1976: 123). Together with the extended *tetraktys* of Nicomachus of Gerasa (12:9:8:6) (Munxelhaus 1976: 23), it is possible to demonstrate a harmonic arrangement of the movement of the light in correspondence to the horizontal lines: the sunrays will be divided by the cornice in 12:6 (2:1/4:2, Gnomon-Modell (heliozentrisch) octave/double-octave); the base of the *attica* creates the proportion 12:8 (3:2, fifth), and the ledge above the columns 12:9 (4:3, fourth). This proportion is the same on each day of the year with regards to the rays of the sun at high noon; the dividing points replicate the interval system derived from the sphere-harmony of planets (Fig. 15.6).

It is not possible to identify these proportions in the decorative elements of the rotunda wall in either the horizontal or the vertical distances of significant architectural lines. They are hidden by mathematical laws: the significant lines of the proportions are neither vertical nor horizontal, but are instead the diagonals of the different rectangles inside the *attica*. Relating vertical and horizontal distances to these diagonals reveals the traditional proportions of the Pythagorean system, but only in combination with $\sqrt{(a^2 + b^2)}$, the "law of Pythagoras." Thus, the "sacred proportions" of the soul of the cosmos are incorporated into the structure of the building by mathematical transformations.

The length of the whole complex can be reconstructed using these proportions, because they significantly agree with astronomical distances, arithmetical relations

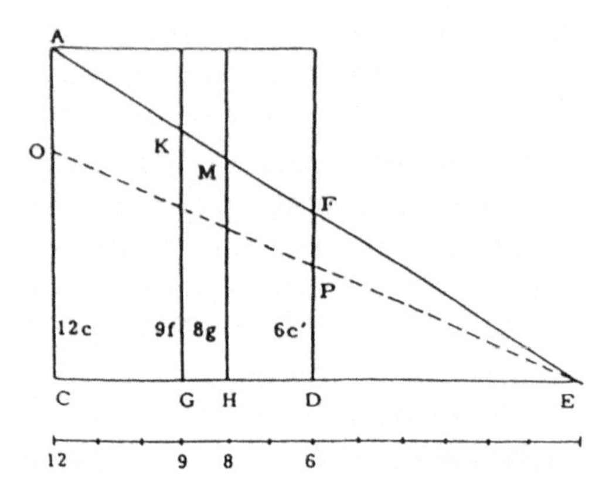

Fig. 15.6 Drawing: author

to the number 28, the radius of the main circle of 147 ft and the Golden Mean proportion. The rays of the sun during the winter solstice striking the forecourt result in exactly the doubled diameter as a main module of the rotunda; this gives the supposed position of the *Arcus Pietatis Traiani*. If the double-octave system is applied, this would be the midpoint (representing the sun), so that another 2 diameters due north would have been the entrance of the forecourt; these 4 diameters correspond to $3 \times 7 \times 28$ ft. The distance from the centrepoints inside the rotunda to the supposed entrance corresponds to 700 palms (152.1 m), or exactly the 109th part of the distance between earth and sun in the *aphel* (at the summer solstice).[4]

Biography Now retired, former Presbyterian minister Gert Sperling has researched topics concerning the Pantheon for over 30 years. He is well known from his many reports on the subject. His most complete work is his book, *Das Pantheon in Rom: Abbild und Mass des Kosmos (Munich/Neuried: ars una Verlag, 1999).*

References

Alvegård, L. 1987. *The Pantheon Metrological System*. Billdal.

Andreae, B. 1973. *Römische Kunst*. Freiburg.

Brunés, Tons. 1967. *The Secrets of Ancient Geometry and Its Uses*. Vol. 2. Copenhagen: Rhodos International Science Publishers.

Buchner, E. 1982. *Die Sonnenuhr des Augustus, Sonderdruck aus RM 1976 und 1980 und Nachwort über die Ausgrabung 1980–81*. Mainz.

[4] For a full explanation of the relationship between the Pantheon complex and the *quadrivium*, see Sperling (1998).

CALTER, Paul. 2014. Façade Measurement by Trigonometry. Chap. 18, pp. 261–269 in this present volume.

CAPELLA, Martianus. 1977. *The Marriage of Philology and Mercury*. Vol. II. William Harris Stahl and Richard Johnson with E.L. Burge, trans. New York: Columbia University Press.

DAVIES, P., D. HEMSOLL and M.W. JONES. 1987. The Pantheon: Triumph of Rome or Triumph of Compromise? *Art History* (June 1987).

DE FINE LICHT, Kjeld. 1966. *The Rotunda in Rome. A study of Hadrian's Pantheon*. Jutland Archaeological Society Series, viii. Copenhagen: Gyldenalske Goghandel, Norkisk Forlag.

GEERTMAN, Herman. 1980. AEDIFICIUM CELEBERRIMUM: Studio Sulla Geometria del Pantheon. *Bulletin Antieke Beschaving* **55** (1980): 203–229.

HASELBERGER, L. 1995a. Deciphering a Roman Blueprint. *Scientific American* (June 1995).

———. 1995b. *Romische Mitteilungen: Ein Grundriß der Vorhalle des Pantheon – Die Werkrisse vor dem Augustus-Mausoleum*. Mainz: des Dt. Arch. Instituts.

HEILMEYER, W. 1975. Apollodorus von Damaskus, der Architekt des Pantheon. *Jahrbuch des Deutschen Archäologischen Instituts* **90** (1975).

JACOBSON, D.M. 1986. Hadrianic Architecture and Geometry. *American Journal of Archeology* **90** (1986).

MARTINES, Giangiacomo. 1989. Argomenti di Geometria Antica a Proposito della Cupola del Pantheon. *Quaderni dell'Istituto di Storia dell'Architettura* **13** (1989).

MUNXELHAUS, B. 1976. *Pythagoras Musicus: Zur Rezeption der Pythagoreischen Musiktheorie als Quadrivialer Wissenschaft im Lateinischen Mittelalter*. Bonn-Bad: Godesberg.

NICOMACHUS of Gerasa. 1926. *Introduction to Arithmetic*. Martin Luther D'Ooge, trans. (with studies in Greek arithmetic by F.E.Robbins and L.C. Karpinski). New York, London: Macmillan.

PELLETTI, Marco. 1989. Note al Rilievo del Pantheon. *Quaderni dell'Istituto di Storia dell'Architettura* **13** (1989): 10–18.

SPERLING, Gert. 1998. *Abbild und Maß des Kosmos*. Munich: Ars Una Verlag.

STAHL, William Harris. 1971. *The Quadrivium of Martianus Capella: Latin Traditions in the Mathematical Sciences 50 B.C.–A.D. 1250*. Martianus Capella and the Seven Liberal Arts Vol. I. New York: Columbia University Press.

VAN DER WAERDEN, B.L. 1988. *Die Astronomie der Griechen, Eine Einführung*. Darmstadt: Wissenschaftliche Buchgesellschaft.

VITRUVIUS. 1991. *De Architectura Libri Decem*. Curt Fensterbusch ed. Darmstadt: IX, 439–445.

WILLIAMS, Kim. 1997a. *Italian Pavements: Patterns in Space*. Houston: Anchorage Press.

———. 1997b. II Pantheon e la Creazione dell'Universo. *Lettera Matematica Pristem* **24** (June 1997): 4–9.

ZAMINER, F. 1984. Hypate, Mese und Nete im frühgriechischen Denken: Ein altes musikterminologisches Problem in neuem Licht. *Archiv für Musikwissenschaft* **XLI**, 1 (1984).

Chapter 16
"Systems of Monads" in the Hagia Sophia: Neo-Platonic Mathematics in the Architecture of Late Antiquity

Helge Svenshon and Rudolf H.W. Stichel

Introduction

The Hagia Sophia in Istanbul, built between 532 and 537 in the time of Emperor Justinian, is universally acknowledged as one of the few examples of outstanding architecture in the world (Fig. 16.1).[1]

An ongoing stream of descriptions continues in trying to explain this fascinating "wonder of space". Exuberant praise of the building is to be found as early as Procopius (*Aedificia* I.1.9 (1940)), the historian of Justinian, who especially and exceptionally highlighted the extraordinary "splendour and harmony in the measures" of the temple.

Despite various and extensive scientific discussion the design of Hagia Sophia still escapes complete and sufficient understanding. Anthemius of Tralleis and Isidorus of Milet, the two architects of the Great Church, were among the best mathematicians and engineers of their time. Thus it seems natural to search for latent mathematics hidden in the extent of the building, to determine the substantial structures of mathematical relevance and finally to develop an explanatory model for the design and for the meaning of this spectacular architecture.

First published as: Rudolf H. W. Stichel and Helge Svenshon, "Systems of Monads as Design Principle in the Hagia Sophia: Neo-Platonic Mathematics in the Architecture of Late Antiquity", pp. 111–120 in *Nexus VI: Architecture and Mathematics,* Sylvie Duvernoy and Orietta Pedemonte, eds. Turin: Kim Williams Books, 2006.

[1] For a general description of the Hagia Sophia, see (Mainstone 1988; Kähler 1967).

H. Svenshon (✉) • R.H.W. Stichel
Fachbereich Architektur, Technische Universität Darmstadt, El-Lissitzky-Str.1, 64287 Darmstadt, Germany
e-mail: svenshon@gta.tu-darmstadt.de; stichel@klarch.tu-darmstadt.de

K. Williams and M.J. Ostwald (eds.), *Architecture and Mathematics from Antiquity to the Future*, DOI 10.1007/978-3-319-00137-1_16,

Fig. 16.1 Hagia Sophia, interior. Photo: Authors

According to general consensus, one has to start from a barely visible square, which can be found in the floor plan between the four large main piers. With a side length of almost exactly 31 m, it is obviously planned with a high degree of accuracy. Such a distance can usually and at first sight quite convincingly be translated to exactly 100 ft. Following this hypothesis the diagonal between the piers results in an irrational number of $100 \times \sqrt{2}$ (= 1.4142135...) and consequently leads to a large number of further irrational building dimensions caused by their geometrical dependence on the basic square with which it would have been hardly possible to plan, draw and mark out this building precisely (Fig. 16.2).

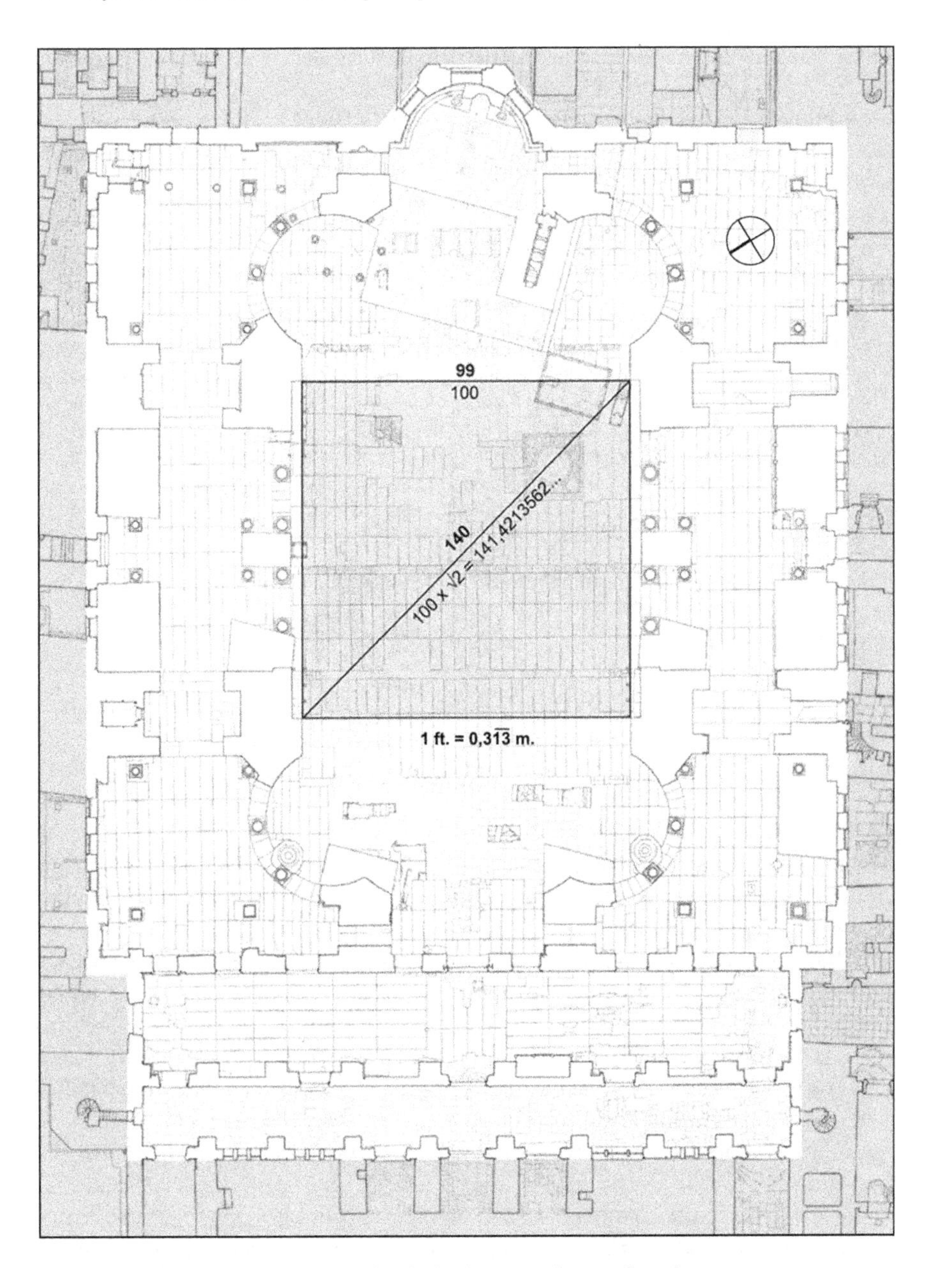

Fig. 16.2 Hagia Sophia, floor plan with the basic square. Image: Svenshon

However, the problem can be solved by a surprisingly simple process employing a mathematical 'trick'. If one assumes the length of the square side to be 99 ft. instead of 100, the diagonal can be calculated at almost precisely 140 ft.; the margin of error is now only within a few thousandths of a foot and thus remains irrelevant for the building procedure (Fig. 16.2).

A proposition for this idea can be found in early Greek mathematics: a series of so-called side and diagonal numbers. Theon of Smyrna, in his mathematical explanations of Plato's works (Hiller 2009: 43, 5–8) was the first to formulate it as a rule. Some 300 years later Proclus Diadochus's commentary on Plato's *Republic* (Kroll 1901: II, 24f.) summarised it within a precise mathematical formula:

> The Pythagoreans and Plato thought to say that, the side being expressible, the diagonal is not absolutely expressible, but, in the squares whose sides they are, (the square of the diagonal) is either less by a unit or more by a unit than the double ratio which the diagonal ought to make: more, as for instance is 9 than 4, less as for instance 49 than 25. The Pythagoreans put forward the following kind of elegant theorem of this, about the diagonals and sides, that when the diagonal receives the side of which it is diagonal it becomes a side, while the side, added to itself and receiving in addition its own diagonal, becomes a diagonal. And this is demonstrated by lines through the things in the second (book) of Elements by him [Euclid] (Fowler 1990: 101) (Fig. 16.3).

Starting with a square with a side of 1 unit (= *monas*) and according to the rule

$$s_1, s_2 = s_1 + d_1, s_3 = s_2 + d_2, \ldots \text{and } d_1, d_2 = 2s_1 + d_1, d_3 = 2s_2 + d_2, \ldots,$$

a series of squares is developed. Its ratio between side and diagonal delivers at every further step of that series a more and more precise approximation to $\sqrt{2}$, because the difference between the square of the diagonals and the double square of the sides alternates solely between +1 and 1, i.e., in every case exactly one unit (Heller 1956: 4). The unit, however, does not by itself determine the elegant result of this rule, but, "similar to a sperm, that carries within itself all the attributes of the future life seminal, is by itself capable to produce this ratio between diagonal and side in a square" (Heller 1965: 335f).

Theon of Smyrna justifies this with unmistakable clearness: "Therefore since the unit, according to the supreme generative principle, is the starting-point of all the figures, so also in the unit will be found the ratio of the diameter to the side." It is those properties "which give harmony to the figures" (Thomas 1980: 133ff.).

Thus it is not surprising that the simple practicality of the calculation, combined with the hinted cosmological claim, promotes the side and diagonal numbers an important component of Greek applied mathematics and geometry (*logistike* and *geodaisia*).

Especially in the writings of Heron of Alexandria, under whose name mathematical and engineering handbooks were edited between the first centuries A.D. and the Byzantine middle ages, numerous examples can be found that document the application of this method of calculation for handling integer approximations of $\sqrt{2}$ (Cantor 1907: 381; Meissner 1999: 140f; Smily 1944: 18–26).

Depending on the needs and the demand for accuracy these values were varied but never deviated from the side and diagonal numerical series and their derivatives.

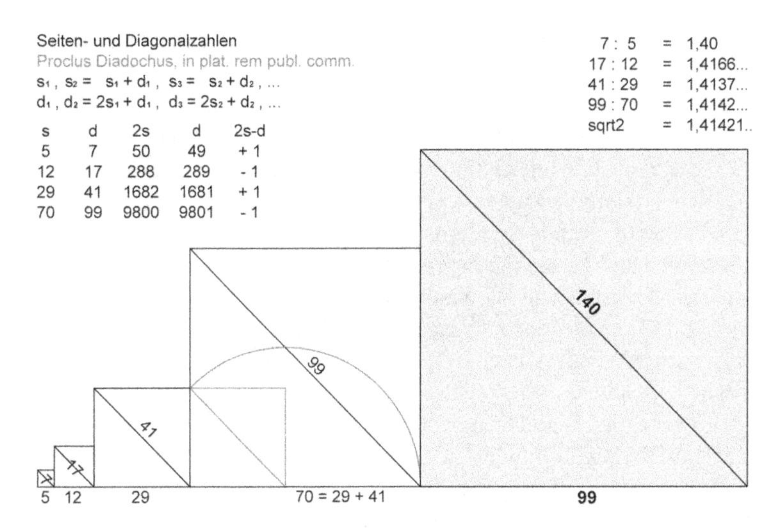

Fig. 16.3 System of side and diagonal numbers. Image: Svenshon, after (Heller 1965: 334)

Consequently Didymus of Alexandria uses for the surface calculations of a square block of wood with the side measurements of 10 ft. a diagonal measuring "approximately 14 1/7" (=99/7), which corresponds with the decuple of the very accurate approximation of √2 provided in the sixth square of our series (Bruins 1964: I, 125: II, 84; III, 178).

In contrast, Heron in his *Metrika* (Schöne 1903: III, 57f), an extensive and probably original treatise about the methodical visualisation of surface and volume calculations, utilizes the slightly more inaccurate values of the third and fourth square, the 7:5 and the 17:12 ratio respectively, the latter a more familiar approximation for the classical antiquity and the middle ages. Those also appear as exemplifications in the formulas of Theon and Proclus.

But the practical advantages of this calculation method were known outside Greek mathematics as well: In the cuneiform texts of ancient Babylon and especially in the so-called coefficient list of Susa, which was written in the early second millennium B.C., the ratio 17:12 is used as a√2 constant (in the Babylonian sexagesimal system in the form of 1:25 = 60/60 + 25/60 = 85/60 = 17/12) (Bruins and Rutten 1961: 25ff pl. 4/5). Derivations that can only be imaginable with knowledge of the square series, can be found on the ancient Babylonian cuneiform tablets Plimpton 322 (Neugebauer and Sachs 1945: 38ff, 130; Robson 2001) and VAT 8512 (Neugebauer 1935: 341f, pl. 27, 52; Høyrup 2002: 234). Moreover, Lennart Berggren's synopsis of ancient and medieval approximations of irrational numbers (Berggren 2002) shows that the use of single values from the side and diagonal numerical series is verified in nearly all antique cultures.

This exceptional phenomenon is easily understood, as approximations for √2 can be demonstrated geometrically in a simple fashion. Theon's remark that the side and diagonal numbers in particular give "harmony to the figures" has also a strong pragmatic connotation.

The exact numerical proportion, which can be calculated with the previously mentioned algorithm, result in the side measurement of a regular octagon, after the choice of the appropriate initial dimensions (s/d numbers). The evidence for this is again supplied by Heron's *Metrika* (Schöne 1903: III, 57f). For the area calculation of a regular octagon with a side length of 10, two right-angled triangles with the rational sides of 5 and 12 and the hypotenuse of 13 are constructed (Fig. 16.4). Next, Heron divides the longer side, which is also the perpendicular bisector of the octagon side, in two sections measuring 5 and 7, whereby an isosceles, orthogonal triangle with catheti measuring 5 is formed over the half side of the octagon. The hypotenuse corresponds approximately to the larger section of the first triangle's longer cathetus and is logically numeralised with the rational value of 7 in order to accomplish the following calculations of the octagon's surface area on the solid basis of the side and diagonal numbers. The efforts of the ancient geodesists and logisticians to be able to perform their computations on the foundations provided by expressible numbers, interlinked through a super-ordinate system, hardly become clearer than in this example.

The fact that these are not isolated cases of special knowledge, which only a few experts would be able to draw upon, but rather are broadly received common knowledge is proven by a letter of Gregorius of Nyssa written in late fourth century A.D. It was his request for administrative assistance to Amphilochius of Iconium for the building of an octagonal church (Teske 1997: 85–88). Besides the detailed description of the structure, precise measurements are mentioned in this letter, which were used in order to calculate the costs for the building materials as exactly as possible. The particular data provided by Gregorius allow the reconstruction of a four-winged structure with an octagonal nucleus, whose dimension between axes (octagon side = 10 cubits, circumcircle radius = 13 cubits, internal radius = 12 cubits) correspond exactly to the ones found in Heron's *Metrika* (Fig. 16.5).

It is known that the bishop's friend and colleague, Gregorius of Nazianz, had included Heron among those he considered the three key figures in Greek mathematics (Cantor 1875: 12), along with Ptolemaeus and Euclid. Therefore it is only natural to assume that Gregorius of Nyssa was acquainted with Heron's handbooks.

This source illustrates how interwoven the building processes and mathematical thought of late antiquity were. It also demonstrates how the theoretical and philosophical fundamentals of scientific mathematics were introduced to architecture through the application-friendly disciplines of *geodaisia* and *logistike*.

Anthemius and Isidorus, the architects of Hagia Sophia, were mathematicians and engineers, trained theoretically as well as practically. It is a fact that both of them re-edited older treatises but also wrote scientific handbooks of their own. The imperial request to plan the Great Church put them before a task for which no adequate experience existed. Therefore it is even more probable that for the conceptual work they resorted to well-tried basic principles from ancient handbooks of mathematics, which at the same time they possibly also amended.

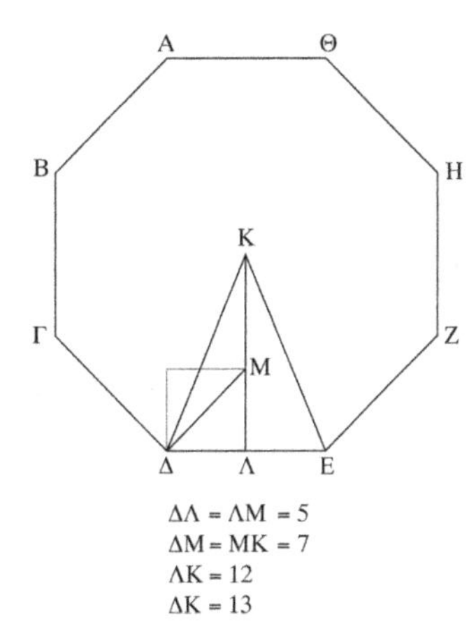

Fig. 16.4 Heron's octagon construction. Image: Author after (Schöne 1903: III, 57f)

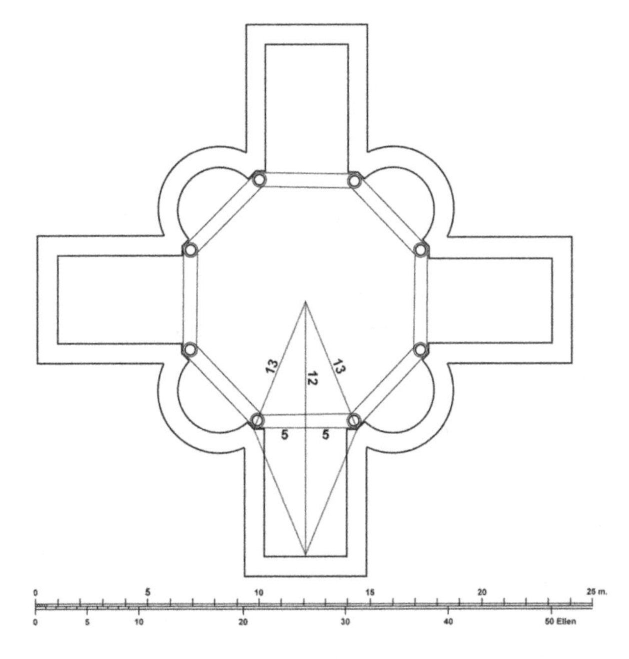

Fig. 16.5 The octagonal church according to Gregorius of Nyssa. Image: Author, after (Restle 1979: pl. 58)

If one also considers the central role that the described octagon with its rational side measurements plays in ancient surveying (Cantor 1907: I 559), it encourages the assumption that this figure was used for the building in a greater scale as well.

Constructing an octagon from two congruent squares with the side and diagonal lengths of 99 ft. and of 140 ft. respectively by rotating one to the other by 45° results in the division of the squares' sides in nearly rational segments measuring 29 ft./ 41 ft./29 ft. (= 99 ft.), which correspond to the side and diagonal numerical proportion of the fifth square. Precisely those numbers are now used constitutively for further construction of the church's floor plan.

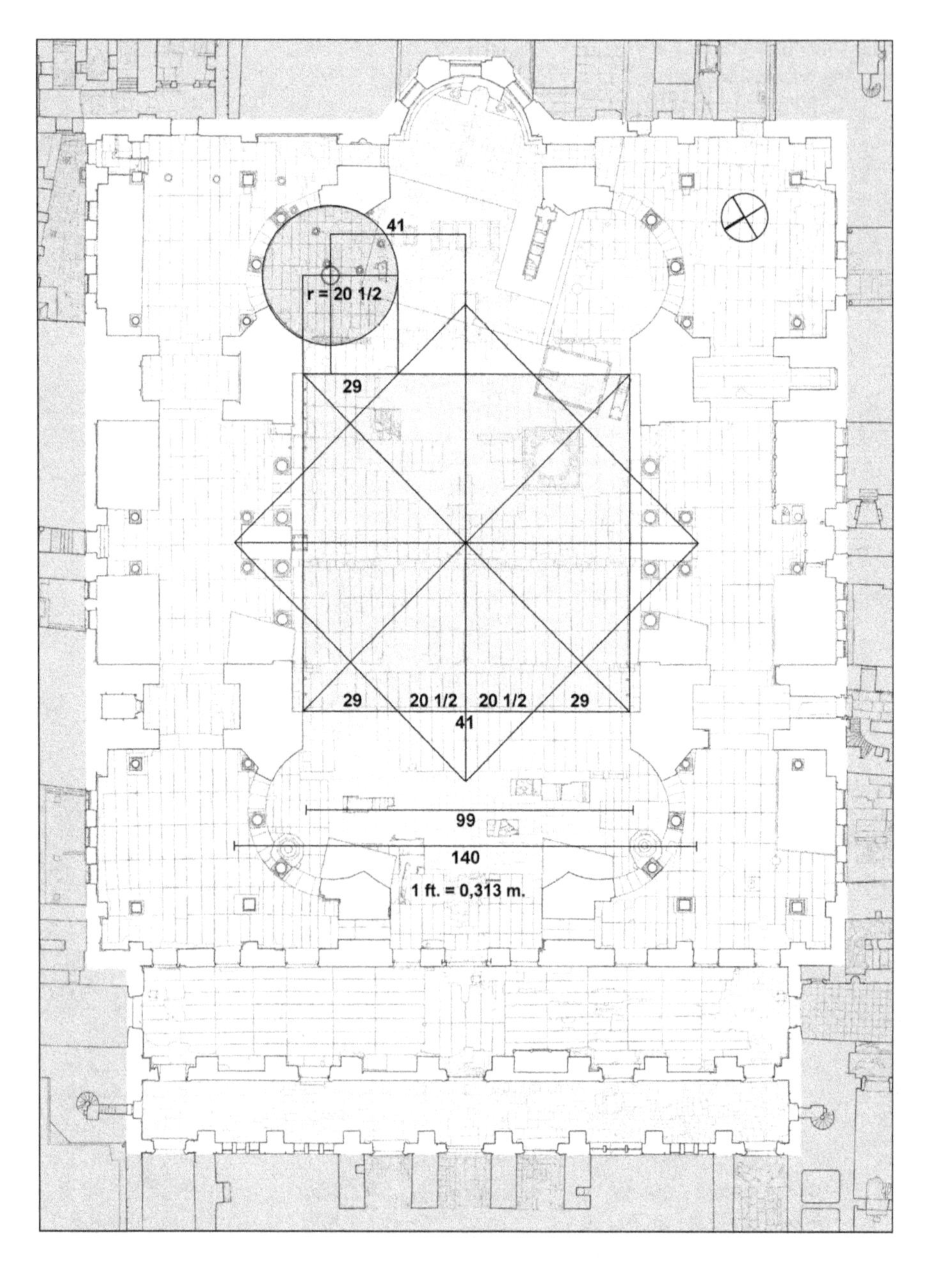

Fig. 16.6 Hagia Sophia, construction of the conchs. Image: Author, after (Svenshon 2003)

The concept of the concha circles can serve as an example, because they can be derived from the basic geometric figure in a few steps. If one completes the lines resulting at the north-eastern main square sides to further squares with the side measurements of 29 and 41 ft. the intersection point of the two squares also creates the centre of the concha circles that thus obtain their final size with an exact radius of 20 1/2 ft. (= 41/2 ft.) (Fig. 16.6). Following this pattern, the orthogonal structure of the floor plan can be developed simply and using only rational measurements, which nearly without exception follow or descend from the side and diagonal numerical series.

Given, however, that the Hagia Sophia is a centralized structure dominated by a dome, a new aspect of calculating with irrational numbers is introduced. In a similar way as with the square side and diagonal, the diameter of the circle also stands in an irrational ratio to the circumference, defined by the number ð (=3.14159...). However, for the planning of the dome, expressible numbers were essential to make the execution of both exact and conveyable calculations possible, directly at the building site.

Greek mathematics offers a practical formula for this purpose as well, which is compatible with the side and diagonal numbers. Archimedes positions the approximation for ð between 3 1/7 and 3 10/71. Even though this value is very accurate, it is logistically very inconvenient. Therefore Heron suggested the use of the less exact value of 3 1/7 (22/7), which is still completely feasible for all circle-related calculations (Schöne 1903: III, 67). In the many exemplary calculations that follow, the fact that the radius and diameter numbers are only slightly varied stands out. The number 7 comes into use most frequently, followed by divisors or multiples of this value. This is because the number 7 in the denominator can be easily reduced in the circumference formula $U = 2 \times ð \times r$ and thus produce an integer value for the circumference. In his *Geodaesia*, Heron of Byzantium used exactly those examples for the measurement of circles (Sullivan 2000: 130ff) (Fig. 16.7):

$$70 \times 22/7 = 220(= 140/2 \times 22/7)$$
$$210 \times 22/7 = 660(= 2 \times 105 \times 22)$$

These values can be rediscovered in the dimensions of Hagia Sophia. The diagonal of the main square measures exactly 140 ft. and defines the circle on which the great pendentives rest. The diameter of the dome is 105 ft. (3/4 × 140) and so further divisors can be derived that are also compatible with the system of the side and diagonal numbers.

As shown in the plan in Fig. 16.8, the principal architectural measurements of Justinian's Hagia Sophia may be expressed by integer numbers. They all are related by well-defined arithmetical and (also) geometrical proportions. Therefore one might speak of a "system of monads", though perhaps not in exactly the same sense as this phrase is used in ancient mathematical philosophy, e.g., in Domninus of Larissa in the fifth century A.D. (Boissanade 1832: IV.413) or even earlier in Iamblichus (Pistelli 1894: 10, 11) in the fourth century A.D., who is allegedly citing Thales of Miletus from the sixth century B.C. (not cited in Diels and Kranz 1951/ 1952: I§11).

This "system of monads" found in the geometry of the Hagia Sophia was well known in Classical Antiquity and widely discussed not only by mathematicians, but

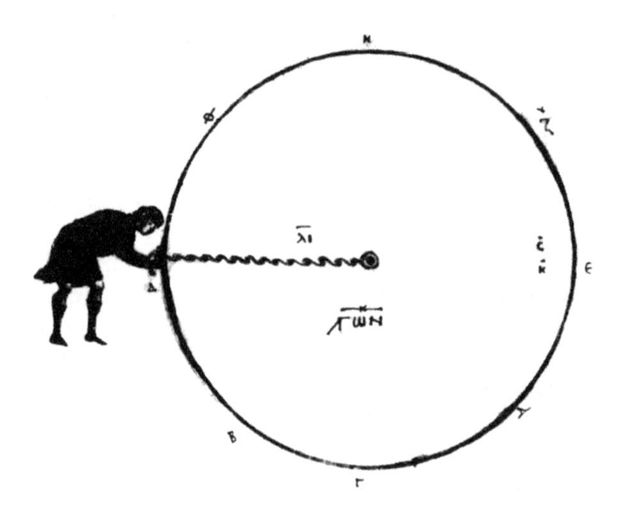

Fig. 16.7 Heron's circle measurement. Image: Sullivan (2000: pl. 36)

also in philosophical circles, as the creation of the world was understood as a geometrical problem. As a *locus classicus* one may cite Plato's *Timaeus*:

> . . . all . . . kinds were without proportion and number . . . in such a condition as we should expect for anything when deity is absent from it. Such being their nature at the time when the ordering of the universe was taken in hand, the god then began by giving them a distinct configuration by means of shapes and numbers . . . with the greatest possible perfection (53a-b; trans. Crawford 1959).

In a paragraph following this quote, Plato declares the isosceles right-angled triangle the very first of all shapes used by the demiurge.

When the neo-Platonic philosophers Theon of Smyrna, second century A.D. (Hiller 2009: 43, 5–8) and Proclus Diadochus, late fifth century A.D. (Kroll 1901: II, 24f), discuss at length the mathematical phenomenon of the numbers of the sides and the diagonal of the square, they are giving no more than an explanation of Plato's theories. In the *Theologumena Arithmeticae* of Pseudo-Iamblichus, a work of ca. 300 A.D., the *Diametriká* are again cited in connection with the creation of the world as in Theon's work combined with other groups of integer numbers through which all is determined *spermatikôs* (through the sperm) (De Falco 1922: 79).

In the fourth chapter of *Introduction to Arithmetic* by Nicomachus of Gerasa, a philosopher of Roman imperial times, one might find the same perception, but expressed with a higher degree of verbal complexity:

> . . . arithmetic existed before all the others in the mind of the creating God like some universal and exemplary plan, relying upon which as a design and archetypal example the creator of the universe sets in order his material creations and makes them attain to their proper ends . . . (D'Ooge et al. 1926: 813).

This text was broadly received, especially in the sixth century A.D. Almost the exact same words can be read in the commentaries of John Philoponus (Hoche 1866: 10) and of Asklepius of Tralleis (Tarán 1969: 30), both relying on the lessons

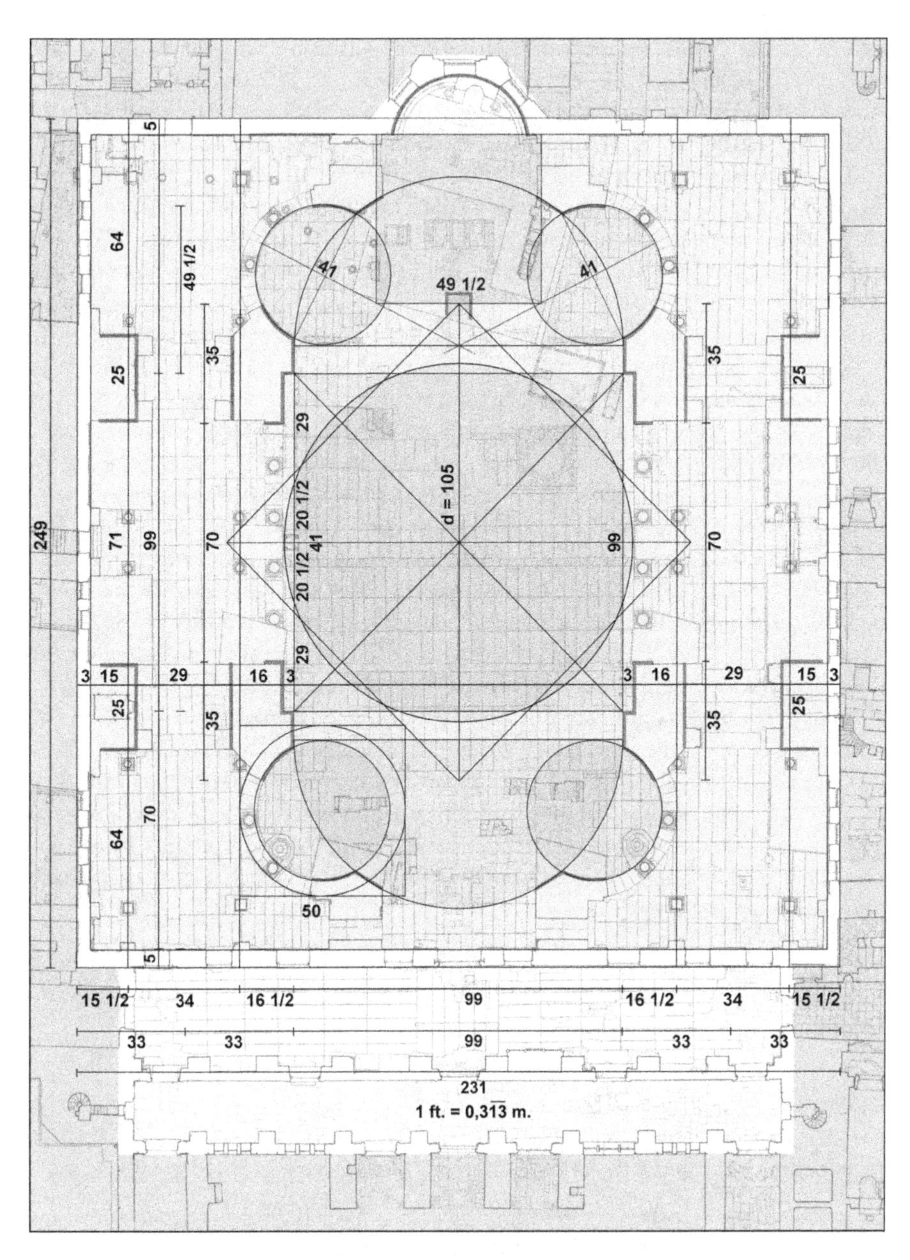

Fig. 16.8 Hagia Sophia, floor plan with the most important measurements. Image: Author, after (Svenshon 2003)

of Ammonius, son of Hermeias, a scholar of Proclus Diadochus, teaching in Alexandria. It is also repeated in the Latin translation of Nicomachus by Boethius (Friedlein 1867: 10, 10–15), the well-known statesman executed in the times of Theodorich. The same thought is expressed in poetical form in his *consolatio philosophiae*: "You steer the world by reason everlasting, Creator of heaven and earth ... ... You bind the elements by numbers ..." (O'Donnell 1984: 3, 9). According to the text that proceeds this impressive poem, it is to be understood as a prayer to the 'father of all things', the God of Plato's *Timaeus*.

The architects of Justinian's Hagia Sophia may have heard of this interpretation, especially because they were in friendly contact with Eutocius of Ascalon, a former pupil of Ammonius. Unfortunately it cannot be proven that Anthemius of Tralleis and Isidorus of Miletus were indeed conscious of the philosophical discussion while designing the church. However, if in their time and age someone would have proclaimed, "As God when creating the kosmos, Anthemius and Isidorus in building Hagia Sophia are using only integer numbers", no doubt the meaning of this proclamation would be comprehended by all. As a matter of fact Procopius of Caesarea, historian of Justinian, speaks of God's assistance in conceiving the Hagia Sophia: "Whenever one goes to this church to pray, one understands immediately that this work has been fashioned not by human power or skill, but by the influence of God" (Mango 1972: 76). In his 562 A.D. poem for the inauguration of the Hagia Sophia, Paulus Silentiarius similarly mentions the "guiding advice of the immortal God" (De Stefani 2010: 301). In a simpler fashion the metaphor of the divine architect is used by the author of the anonymous ninth century A.D. Byzantine "Tale about the construction of Hagia Sophia": "The plan of the naos was shown to the emperor in a dream by an angel of the Lord" (Preger 1901: I, 83).

Thus according to our interpretation of the design principles of the Hagia Sophia meaning executing the concept of the building solely in rational numbers, while using principles of Classical *logistike* (applied mathematics) as well as Neo-Platonic philosophy one may find himself in this church very close to God's 'ratio' (his *Logos*), even before attending at the common and everyday rituals and prayers of the Christian service, through which men's approach to God is achieved in a more manifest way.[2]

Acknowledgment This research was sponsered by Zentrum für interdisziplinaere Technikforschung (ZIT) of the Darmstadt University of Technology. See preliminary reports: (Stichel and Svenshon 2002; Stichel 2003; Svenshon 2003).

Biography Rudolf H.W. Stichel (Dr. Phil) is *ausserplanmässiger Professor of classical archaeology at the Technische Universität Darmstadt. His research focuses especially on the art of Late Antiquity and the transition to the Middle Ages.*

Helge Svenshon (Dr.-Ing.) is *Privatdozent* of history and theory of architecture and classical archaeology at the Technische Universität Darmstadt. His main topic of research deals with the correlation of ancient practical mathematics and geodesy in design and planning of ancient architecture.

[2] For a new interpretation of central parts of the liturgy in Hagia Sophia, see (Schneider and Stichel 2003).

References

BERGGREN, L. 2002. Ancient and Medieval Approximations of Irrational Numbers. Pp. 31-44 in Y. Dold-Samplonius, ed. *From China to Paris: 2000 Years Transmission of Mathematical Ideas*. Stuttgart: Steiner.

BOISSANADE, J. F. 1832. *Anecdota Graeca*. Paris.

BRUINS, E. M. 1964. *Codex Constantinopolitanus palatii veteris no. 1*. Leiden: Brill.

BRUINS, E. M. & RUTTEN, M. 1961. *Textes Mathématiques de Suse*. Paris: P. Geuthner.

CANTOR, M. 1875. *Die römischen Agrimensoren und ihre Stellung in der Geschichte der Feldmesskunst*. Leipzig: Druck und Verlag von B. G. Teubner.

CANTOR, M. 1907. *Vorlesungen über die Geschichte der Mathematik*. Vol. I. Leipzig: Druck und Verlag von B. G. Teubner.

CRAWFORD, F. M., ed. 1959. *Plato. 'Timaeus'*. New York: Macmillan Publishing Co.

DE FALCO, V., ed. 1922. (Rev. U. Klien. 1975). *Iamblichi Theologumena arithmeticae*. Stuttgart: Druck und Verlag von B. G. Teubner, p. 79.

DE STEFANI, C., ed. 2010. *Paulus Silentiarius. Descriptio Sanctae Sophiae*. Berlin and New York: Walter de Gruyter, p. 301.

DIELS, H. & KRANZ, W. 1951/1952. *Die Fragmente der Vorsokratiker, griech. deutsch.* Dublin-Zurich: Weidmann.

D'OOGE, M. L., ROBBINS, F. E. & KARPINSKI, L. C., eds. 1926. *Nicomachus Gerasius. Introductio mathematica*. New York: Macmillan.

FOWLER, D. H. 1990. *The Mathematics of Plato's Academy. A New Reconstruction*. 2ed. New York: Springer-Verlag.

FRIEDLEIN, G., ed. 1867. *Boetii De institutione arithmetica libri duo*. Leipzig: Druck und Verlag von B. G. Teubner.

HELLER, S. 1956. Ein Beitrag zur Theodoros-Stelle in Platons Dialog "Theaetet". *Centaurus*, **5**: pp. 1-58. New York: John Wiley & Sons Ltd.

——. 1965. Die Entdeckung der stetigen Teilung. Pp. 319-354 in O. Becker, ed. *Zur Geschichte der griechischen Mathematik*. Darmstadt: Wissenschaftliche Buchgesellschaft.

HILLER, E., ed. 2009. *Theonis Smyrnaei Philosophi Platonici: Expositio Rerum Mathematicarum Ad Legendum Platonem Utilium*. Whitefish, MT: Kessinger Publishing.

HOCHE, R. G., ed. 1866. *Nicomachus Gerasius. Pythagorei Introduetionis Arithmeticae*. Leipzig: Druck und Verlag von B. G. Teubner.

HØRUP, J. 2002. *Lengths, Widths, Surfaces. A portrait of Old Babylonian Algebra and Its Kin*. New York: John Wiley & Sons Ltd.

KÄHLER, H. 1967. *Die Hagia Sophia*. Berlin: G. Mann.

KROLL, W. ed. 1901. *Proclus. In Platonis rem publicam commentarii*. Leipzig: Druck und Verlag von B. G. Teubner.

MAINSTONE, R. J. 1988. *Hagia Sophia: Architecture, Structure and Liturgy of Justinian's Great Church*. New York: Thames and Hudson.

MANGO, C. 1972. *The art of the byzantine empire 312-1453*. Englewood Cliffs: Prentice-Hall.

MEISSNER, B. 1999. *Die technologische Fachliteratur der Antike*. Berlin: Akademie Verlag.

NEUGEBAUER, O. 1935. *Mathematische Keilschrift-Texte*. Berlin: Springer.

NEUGEBAUER, O. & SACHS, A. 1945. *Mathematical Cuneiform Texts*. New Haven: American Oriental Society.

O'DONNELL, J. J., ed. 1984. (2ª ed.: 1990). *Boethius, Consolatio Philosophiae*. Pennsylvania: Bryn Mawr College.

PISTELLI, E., ed. 1894. (Rev. U. Klien. 1975). *Iamblichi In Nicomachi arithmeticam introductionem liber*. Leipzig: Druck und Verlag von B. G. Teubner.

PREGER, T. 1901. *Scriptores originum Constantinopolitanarum*. Leipzig: Druck und Verlag von B. G. Teubner.

PROCOPIUS, *Aedificia*. Ed. 1940. Cambridge: Loeb Classical Library.

Restle, M. 1979. *Studien zur frühbyzantinischen Architektur Kappadokiens*. Wien: Österreichischen Akademie der Wissenschaften, Philosophisch-historische.

Robson, E. 2001. Neither Sherlock Holmes nor Babylon: A Reassessment of Plimpton 322. *Historia Mathematica: International Journal of History of Mathematics*, **28**, 3 (August): pp. 167-206. London: Elsevier.

Schneider, W. C. & Stichel, R. H. W. 2003. Der 'Cherubinische Einzug' in der Hagia Sophia Justinians: 'Aufführung' und 'Ereignis'. Pp. 377-394 in E. Fischer-Lichte, ed. *Performativität und Ereignis. Theatralität IV*. Tübingen: Francke.

Schöne, H. 1903. *Heronis Alexandrini Opera quae supersunt omnia*. Leipzig : Druck und Verlag von B. G. Teubner.

Smily, J. G. 1944. Square Roots in Heron of Alexandria. *Hermathena*, **63**: pp. 18-26.

Stichel, R. H. W. & Svenshon, H. 2002. Das unsichtbare Oktagramm und die Kuppel an der 'goldenen Kette'. Zum Grundrissentwurf der Hagia Sophia in Konstantinopel und zur Deutung ihrer Architekturform. *Bericht über die 42. Tagung für Ausgrabungswissenschaft und Bauforschung* **42**: 187-205.

Stichel, R. H. W. 2003. Die Kuppel an der 'goldenen Kette': Zur Interpretation der Hagia Sophia in Konstantinopel. Pp. 244-251 in *Almanach Architektur 1998-2002: Lehre und Forschung an der Technischen Universität Darmstadt*. Tübingen: Ernst Wasmuth.

Sullivan, D. F. 2000. *Siegecraft. Two Tenth-Century Instructional Manuals by "Heron of Byzantium"*. Washington, D.C.: Dumbarton Oaks Research Library and Collection.

Svenshon, H. 2003. Das unsichtbare Oktagramm: Überlegungen zum Grundrissentwurf der Hagia Sophia in Konstantinopel. Pp. 234-243 in *Almanach Architektur 1998-2002: Lehre und Forschung an der Technischen Universität Darmstadt*. Tübingen: Ernst Wasmuth.

Tarán, L. ed. 1969. *Asclepius Trallianus. In Nicomachi arithmeticam introductionem scholia*. Philadelphia: University of Pennsylvania, p. 30.

Teske, D. 1997. *Gregor von Nyssa: Briefe*. Stuttgart: Hiersemann.

Thomas, I. 1980. *Selections Illustrating the History of Greek Mathematics, with an English Translation*. London: Harvard University Press.

Part III
Theories of Measurement and Structure

Chapter 17
Measure, Metre, Irony: Reuniting Pure Mathematics with Architecture

Robert Tavernor

Measure:	*mens* (L - mind), *mensurare* = measuring/measure
Metre:	*metron* (Gk), *metrum* (L - measuring rod), *mètre* (Fr) = metre
Irony:	*eironeia* (Gk - simulated ignorance), *eiron* - dissembler and simulator of power = irony

Rulers and Ruled

No civilisation has existed without measures, and each has described measures in a manner specific to its needs. To exist at all, measures must be practical and useful, and most have their origins in everyday experience. At some stage in the development of a civilised society measures will be refined, standardised and regulated, and represented physically. To endure and be accepted by hundreds, thousands, even millions of people—across great civilisations and around the globe—measures must reflect and extend the authority of leaders. Measure is therefore a statement and record of the changing balance of power and independence. It is an expression of culture.

Consequently, measures are also symbolic. Throughout history measures have embodied (literally, as we shall see) and demonstrated the potency of their creators. The Egyptians attributed the creation of measure to Thoth, the Greeks to Hermes,

First published as: Robert Tavernor, "Measure, Metre, Irony: Reuniting Pure Mathematics with Architecture" Pp. 47–61 in *Nexus IV: Architecture and Mathematics,* Kim Williams and Jose Francisco Rodrigues, eds. Fucecchio (Florence): Kim Williams Books, 2002.

R. Tavernor (✉)
Tavernor Consultancy, 24 Denbigh Street, Pimlico, London SW1V 2ER, UK
e-mail: robert@tavernorconsultancy.co.uk

K. Williams and M.J. Ostwald (eds.), *Architecture and Mathematics from Antiquity to the Future*, DOI 10.1007/978-3-319-00137-1_17,

and Judaic-Christian literature to Cain—the first offspring of Adam and Eve—and thus to the beginnings of civilisation. In ancient times, leaders of communities had the status of heroes and were elevated above the ordinary, being regarded as more than mere mortals. Kings and queens were presented as divine beings in communion with the heavens, and as channels of heavenly power and authority they bore gifts from the gods with which to benefit their subjects.

Heroes, kings and queens—as earthly representatives and even representations of gods—were correspondingly idealised and idolised. Their bodies were captured in paintings and statues, works of art that sought to be as perfect and beautiful as their subject. Their idealised forms, or the staffs and rods that defined their status, provided the earliest recorded linear standards: weight and capacity were derived from these principal lengths (Chisholm 1997).

Fundamental Measures

Although nomenclature varies from one ancient kingdom to another, the articulated parts of the body are usually used to define small measures and are identified in relation to the totality of the whole body. Thus, the digit (the breadth of the middle part of the first joint of the forefinger) represented 1 part, the palm (handbreadth) 4 parts, the span 12, the foot 16, the cubit 24, the double pace 80, and the fathom (the distance between the tips of the fingers of arms outstretched) 96 parts (Chisholm 1877, p. 27). Sets of linear relations of this kind were usually calibrated on rods and defined as a standard, from which replicas were made for general usage. As kings were regarded in ancient times as symbols of sacred and earthly authority, the original standards were safely stored in the treasuries of temples, and replicas of them displayed in public places.

When the notion of the perfect body was not the basis of a nation's linear measures, easily tradable and taxable items in the kingdom—particularly cereal grain—were used as standards. Invariably, however, these related back to the proportions of a perfect body. In Roman times the *uncia* or 'inch' was introduced as one-twelfth part of the foot measure. During the reign of King Edward II of England in the fourteenth century, an inch was defined as 'three barley corns, dry and round' (Cox 1957, pp. 23–4). Alternatives to a body-centred system were rare. For although the random selection of any familiar and conveniently sized items and actions may be related to one another to create a system or canon, they will not necessarily make for useful or memorable comparison. There can be no doubt that Western culture would have difficulty comprehending the irrationality of the ancient Indian scale of measures in which the *yôjana* (a day's march for an army) variously equals 16 or 30 or 40 *li*, and is also equal to eight *krôsas* (keu-lu-she): a *krôsa* is divided into 500 bows (*dhanus*): a bow is divided into four cubits (*hastas*): a cubit is divided into 24 fingers (*angulis*): a finger is divided into 7 barleycorns (*javas*): and so on to a louse (*yûka*), a nit (*likshâ*), a dust grain, a cow's hair, a sheep's hair, a hare's down . . . and so on for seven divisions, till we come to an excessively small grain of dust (*anu*): this cannot be divided further without arriving at nothingness . . . (Si-yu-ki 1885: I, p. 70, quoted in Cox 1957, pp. 23–24).

This is why measures are usually rationalised in relation to a single coherent form—and nothing is more readily accessible in everyday experience than the human body and its constituent parts.

It has been generally accepted in Western societies since Greek antiquity that a natural quality cannot be understood until it has been measured, or can be compared with something that is measurable. The ancient Greeks also realised that qualities could be described through a medium other than words, that is, through number. Their numbers were more than quantities, for they represented qualities too. Pythagoras defined the extraordinary properties of certain numbers, such as 6 and 10. He considered these integers to be perfect numbers, because they can be regarded as the sum of their parts: 6 is the sum of 1 + 2 + 3; and 10 the sum of 1 + 2 + 3 + 4. Plato described the perfection of the natural harmony that existed in the world and universe in the *Timaeus*. He adopted Pythagoras's perfect numbers, and the mathematical canon of united numbers and qualities he describes is consequently known through a resolution of their separate conclusions, as the Pythagoreo-Platonic system. Polykleitos of Sikyon, a sculptor working in the fifth century B.C. in ancient Greece, gave form to this system of numerical perfection. Famously, he created a perfect sculpture of a man in that its form visibly expressed a total body in which its parts had a harmonious correspondence with the whole (Fig. 17.1). This visual perfection could also be recorded through relationships of ideal numbers, a system that subsequently became known as 'the canon' of perfect proportions.[1]

Polykleitos's source for this canon probably had its origins in Egyptian antiquity. Whatever its origins, it was regarded by artists and architects as a symbol of natural physical and heavenly perfection—and so it remained across two millennia, through the civilisations of ancient Greece and Rome, and down to the European Renaissance. Indeed, it continues to provide a fundamental point of reference for notions of artistic proportion (and distortion) in modern art and design.

In Greek culture, philosophy, mathematics and art achieved a union that underlay the system of weights and measures used in the ancient world. Marcus Pollio Vitruvius, the Roman architect working and writing in the first century B.C. absorbed this tradition, and stated—what was probably a commonly held belief—that the finest buildings of antiquity reflected in their form the human proportions of the Greek canon (Vitruvius 2009: III.1). Since the numbers of these proportions were derived from Pythagoras and Plato's numerical definition of the universe, Vitruvius was aware that the measuring units he used to design

[1] See, for example, Stewart (1978, pp. 122–31) and Rykwert (1996, pp. 104–112).

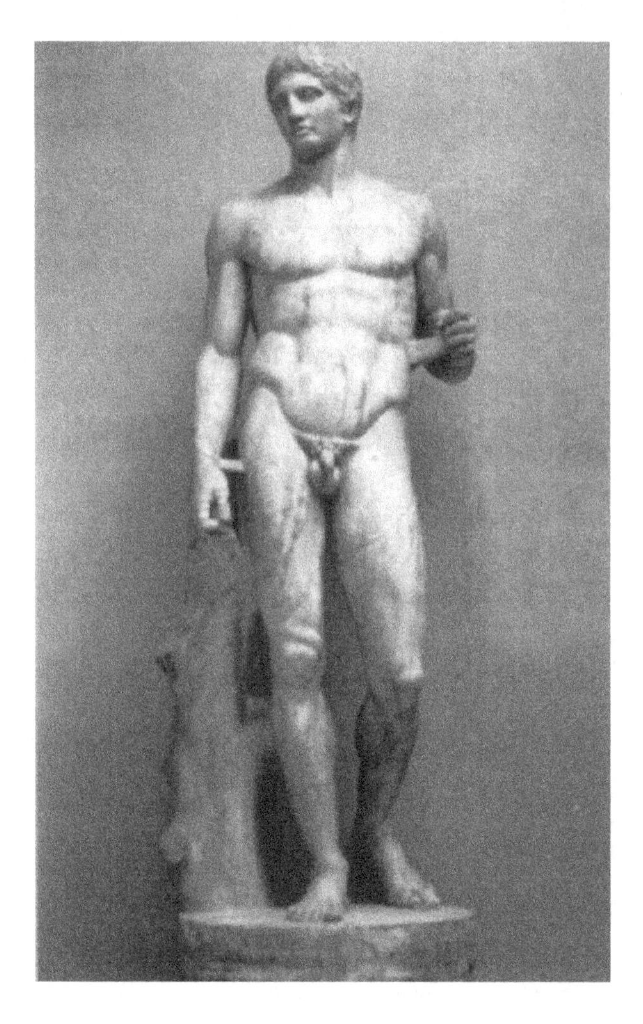

Fig. 17.1 *Doryphorus*. Florence, Uffizi. One of many Roman stone copies of Polykleitos's bronze statue of the spear-bearer Doryphoros, which is believed to embody his canon of bodily proportions. Photo: Courtesy Joseph Rykwert

buildings—the finger, palm, foot and cubit—and the perfect number relations between them, are a combination derived from the measures of the universe and of the idealised body of man. Consequently, body, architecture and the natural world were in perfect harmony, and the body of man was regarded as a symbolic microcosm of the harmonious universe.

Subsequently, Italian Renaissance artists and architects represented the idealised Vitruvian man graphically as *Homo quadratus*, a naked figure with outstretched limbs, bounded by a circle and square (Fig. 17.2). They also conflated this—essentially pagan—figure with the symbol of Christianity, the crucified body of Christ (Fig. 17.3) (Dodds and Tavernor 2002).

This association of the quality and quantity of measure with a sacred body contrasts starkly with the bodily detachment—intellectually and physically—of the modern era towards measure. The body is no longer used to define our official

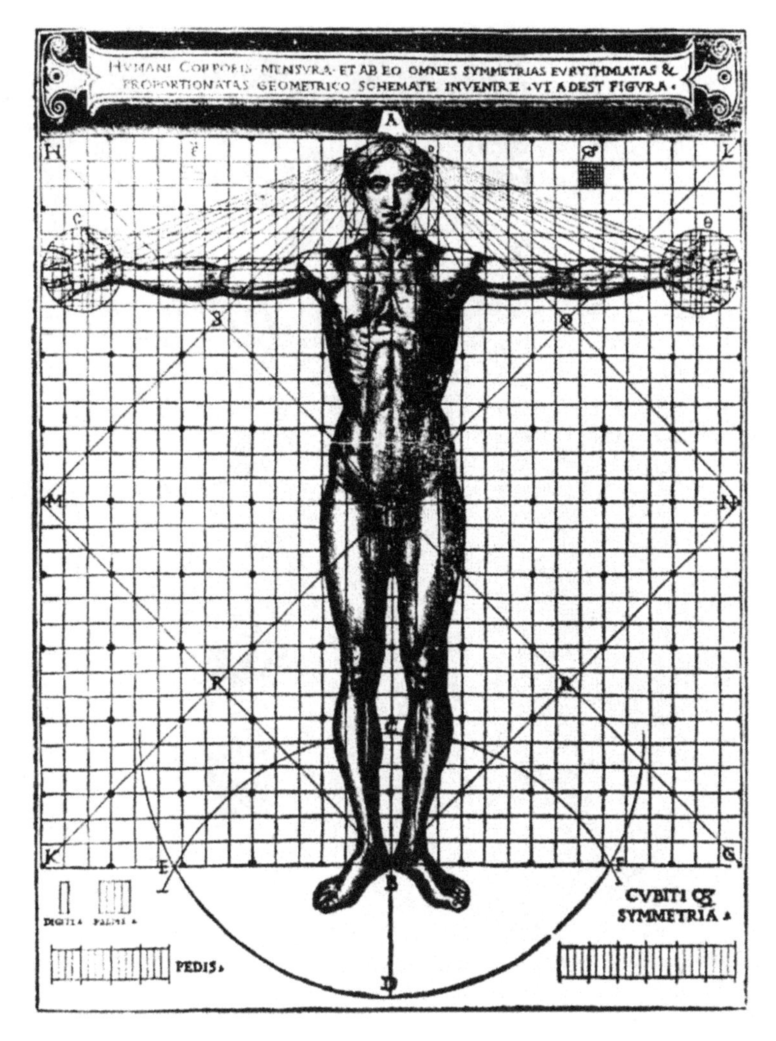

Fig. 17.2 *Homo quadratus*. Woodcut from Cesare Cesariano's *Vitruvius* (1521: Liber tertius, XLIX)

measuring standards, as modern—post-Enlightenment—science cannot recognise the relevance of bodily perfection to a universal system of measures.

The ancient union of mathematics and words is at the root of this modern dilemma. By using numbers and symbols as the principal language to relate abstract and concrete ideas, the sciences during the last two-and-a-half millennia have become increasingly mathematical and reductive: that is, differences in quality have been reduced to unitary differences of quantity. As qualities have been turned into abstract scientific quantities, so everything can be dissected into ever-smaller units of measurement, that is, into measures incomprehensible to and

Fig. 17.3 *Christ Crucified.* Filippo Brunelleschi. Wood and painted sculpture, early fifteenth century. Florence, Santa Maria Novella. Photo: Author

remote from everyday human experience: a level of absurdity that parallels the ancient Indian scale referred to above.

Science and the Metric

The seeds of this situation were unwittingly sown by Sir Isaac Newton, the father of seventeenth-century Enlightenment science. Newton still believed in a universe of qualities where God permeated everything, from earth to the limitless boundaries of space. However, being everywhere, Newton's God was necessarily incorporeal—and lacked the identifiable body granted to mortals (Newton 1704; Voegelin 1948). Newton's friend, the philosopher John Locke, extended this reasoning to society and politics. Locke rejected a traditional belief in the divine origin for human government and argued instead that sovereignty resides not in any one individual, but in all the people.[2] In so doing, he provided the intellectual basis for social and political revolution in France in 1792, which succeeded in uprooting the French monarchy, and destroying the symbolic validation of political power through the King's sacred body.

The execution of King Louis XVI the following year was both sacrilegious (as regicide always is) and, through the implementation of the recent invention of the guillotine, intentionally impersonal. An elected but undemocratic body of the nation's representatives—citizens—were to rule in place of the divine authority of a solitary monarch. However, the committees they constituted failed to unify the nation until a charismatic leader of men—Napoleon Bonaparte—filled the void in individual leadership Revolution had created. Napoleon was to become the nation's emperor, its surrogate king (Outram 1989). His was a modern reflection of the king's majesty and dominion, though (officially, at least) not his divinity.

Irrational notions about the divinity of kingship have been mostly expunged from modern thinking, although democracy (another ancient Greek concept), has flourished in the advanced societies of the modern era. Today, we place our faith in elected leaders—rarely superior models of humanity, moral or physical—and an advanced technological society; a value system shaped by (what we assume to be) the dispassionate objectivity of modern science. Rationality is always held higher than subjectivity in this system, and rational science is preferred to the subjectivity of art—which is consequently poorly understood and mistrusted.

This modern imbalance is very evident in the changed response to the representation of human proportion in art and architecture (Dodds and Tavernor 2002). While for thousands of years the body was seen to reflect and embody universal harmony and was consequently related qualitatively to a prevailing artistic canon, since the Enlightenment, scientists—in tandem with philosophers—have scrutinised the human body with detachment. It is now something to be dissected, analysed and quantified in ever-smaller microscopic and atomic detail as is the metric system, a product of revolution and modern scientific thinking.

The measurements of the internationally controlled metric system are precisely calibrated abstract quantities, verified scientifically. They make no reference to everyday experience or to art and symbolism, and have no obvious relation or relevance to human form, ideal or otherwise. Indeed, the metric system is deliberately anti-body. It developed from Enlightenment scientific concerns for precision and international uniformity, and the demands of ordinary people—originally mainly in France—for an equitable system that would provide uniformity across the civilised world.[3] In the search for 'rational' and universal measures appropriate for all nations on earth, the metre rod, which is at the root of

[2] This is encapsulated in two books by John Locke published in 1690: *An Essay Concerning Human Understanding* and *Two Treatises of Government*.

[3] For the origins of the metric system see in particular Bigourdan (1901) and De La Condamine (1747).

Fig. 17.4 Commemorative medal of the metric system, designed in 1799 but not cast until 1840. Pavillon de Breteuil, Sèvres, Paris. Image: Gonon (1840)

the metric system, was conceived as a fractional representation of the physical dimensions of the earth in relation to the mechanical laws that were understood to be controlling the forces of nature and the universe (Fig. 17.4).

A rational scientific conception was never delivered. The earth is not a pure sphere and the metre rod is not and could never be a precise fraction of the earth's circumference as French scientists had intended.[4] Nor have metric measures proved to be finite, or any more ideal than the standards derived from the idealised human form. Two hundred years after it first came into being the metric system has been redefined several times, until a new definition was conjured up by scientists from the insubstantial elements of the universe itself—of light and gas, using technical apparatus of their own contrivance.[5] The metre has become a measure without relation to corporeal form or even common human experience; it is an abstract scientific measure without tangible value.

[4] This was widely appreciated in France and abroad even while the metric system was being defined. See for example the report by the United States Secretary of State Thomas Jefferson to the US Congress in 1790 in (Peterson 1984, pp. 393–396).

[5] In 1953 a ten-nation advisory committee meeting at Sèvres recommended the abandonment of the physical metric prototype in favour of its definition through wavelengths of light. Finally, on 14 October 1960, it was agreed to return to a truly 'natural' and scientifically verifiable definition for the metre rod derived from the radiation of the orange-red light emitted by the radioactive krypton-86 atom. Correspondingly, it was agreed in 1960 that the metre should equal 1,650,763.73 wavelengths in vacuum of the radiation corresponding to the transition between the levels $2p^{10}$ and $5d^{5}$ of the krypton-86 atom. Since 1983 it has been defined, more simply (though no more comprehensibly), as the distance light travels in vacuum in 1/299,792,458 (that is, in approximately 0.0000003) of a second. See Danloux-Dumesnils (1969, pp. 36–42).

The Metre Reconsidered

I have introduced measure here as an essential part of the history of ideas, because it is more usually presented one-sidedly, as a catalogue of quantities of quasi-scientific status. Indeed, metrology is usually defined as a 'science or system of weights and measures'.[6] Numbers are calibrated with decimal point precision in comparative tables and accepted as verifiable truths around the globe. Neither scientists nor historians have considered measure as an *art*: as the outcome of social and political conditions, or as a potent instrument in the hands of creative artists, painters, sculptors and architects, those who provide the tangible imagery of a culture. Yet, a nation's measuring rod is the most succinct and precise statement of the dominating forces within a civilised society, such that the history of measure is always a history of ideas and of creativity, which has been promoted for and by the powerful.

Although the need for measures has remained constant for five millennia it is evident that the idea and reality of what measure represents has been transformed. Measure, I argue, needs to be recognised as more than an abstract calibrated length of inert material: it is a deliberate consequence of human thought—in Latin, *mens* (the root of 'measure')—and the dissembler—in Greek, *eiron* (the root of 'irony')—and simulator of power. Indeed, the metre rod might be better understood as the measure of all irony.

My interest in dimensional measure here is in its potential to combine the arts and sciences. Measure is essential to the making, understanding, use and enjoyment of a man-made object in a particular cultural setting. In particular, it is appropriate to regard measure through the medium of architecture, as in this discipline art, science and culture are entwined, and the human form is necessarily respected and celebrated. Since antiquity and until the late eighteenth century, idealised notions of the human body dominated measure as they did art and architecture, until this tradition was shattered—probably irrevocably—during the French Revolution, with the decapitation of King Louis XVI, the separation of Church from State, and the attempt to rescind all measures associated with the *ancien régime*. In place of traditional, anthropomorphically-related systems, the Revolutionaries sanctioned and promoted the scientifically-inspired metric system 'for the benefit of all mankind'. But in the process of its refinement and adoption internationally, the metre became disembodied, non-figurative and abstract—terms that, perhaps more than coincidentally, are also applied to the art and architecture of the twentieth century.

---

[6] See for example (Chisholm 1877) and (Klein 1988). Problems associated with the assumption that mensuration is a science is well discussed in (Fernie 1978).

Humanising the Earth: Le Corbusier's Modulor

During the twentieth century, some major artists and theorists questioned the absurdity of the scientific search for precision through abstraction, of separating body from measure, and challenged whether the metric system is appropriate for the everyday needs of the greater part of society. I will refer here to the example of the Swiss architect Le Corbusier (born Charles Édouard Jeanneret-Gris) who described the metre as "nothing but a length of metal at the bottom of a well at the Pavillon du Breteuil" (Le Corbusier 1954, p. 57), the headquarters of the International Bureau of Weights and Measures laboratory at Sèvres to the south of Paris. Le Corbusier designed a new measure called *le Modulor* (the Modulor) with which architects and engineers might humanise the metric system, by combining it with classical geometry and modern anthropometrics. The Modulor was developed during (and despite) the Nazi occupation of Paris during World War II, in his studio in Paris at 35 rue de Sèvres, en route to the International Bureau of Weights and Measures laboratory at Sèvres. Post-war, as his fame and opportunities increased, Le Corbusier applied the principles of the Modulor to the design of buildings that had a major impact on the development of modern architecture.

Le Corbusier sought a universal measure for architecture that would be applicable in any nation. The metre rod did not satisfy his objectives, for it is "a mere number without concrete being: centimetre, decimetre, metre are only the designation of the decimal system", whereas he would design the Modulor with numbers that "are *measures*" (Le Corbusier 1954: 60): by relating meaningfully to human physicality and culture. Le Corbusier assimilated recent and ancient aesthetic theories about natural beauty in art and architecture, and the Modulor is a measuring scale derived from ideal notions about the measurements of the human body.

An early encounter with Schuré's *Les Grands Initiés* (Paris, 1908) is thought to have predisposed Le Corbusier to believe that Pythagoras, an 'initiate' of universal natural order, was uniquely relevant for his own ambitions for architecture in the twentieth century. He considered Pythagoras to be a pre-eminent philosopher-mathematician, whose observations of the natural world enabled him to reveal the universal truths underpinning harmony and proportion (Benton 1987, p. 241). He was also familiar with the compositional studies of Renaissance art and architecture made by three nineteenth-century art historians, Adolf Zeising, Heinrich Wölfflin and August Thiersch, who were pioneers of art history as an academic discipline and who reached influential conclusions for modern designers on the fundamental rules of classical beauty. Zeising (1854) attempted to prove that the Golden Section is the key to all morphology, both in nature and art. He referred in particular to the thirteenth-century Italian mathematician, Leonardo of Pisa, better known as Fibonacci. As is well known, the Fibonacci number series tends towards the perfect ratio of 1:1.618, which was called the 'Divine' proportion by Luca Pacioli in the early sixteenth century and subsequently became known as the Golden Section. Wölfflin (1889) and Thiersch (1893) proposed that successful works of art and architecture embody fundamental geometries that recur throughout their composition. Wölfflin further argued that rectangles of similar proportions could be used to demonstrate the perfect compositional qualities of classically designed façades.

During his Purist period from 1918 to 1929, Le Corbusier produced paintings composed of equilateral triangles and the Golden Section. Similarly, as an architect he used *tracés régulateurs* (regulating lines) to study past buildings and to 'purify' the elevations of his own designs for buildings. Le Corbusier emphasised in his early designs 'the placing of the right angle' and the role of rectangles proportioned by the Golden Section, which he illustrated in a chapter on 'Regulating Lines' in *Vers une Architecture* (*Towards a new architecture* (1923)), arguably the most influential architectural treatise of the twentieth century. His publication, *Le Modulor* (1948/9), describes the Golden Section series—the *séries d'Or* (hence '*Modul-Or*' or Golden module)[7]—arranged as two related scales colour coded red and blue that relate to the body, head and outstretched arm of a man.[8] He proposed that measures or proportions of the Modulor man be taken from either coloured scale, separately or together.

Initially, Le Corbusier determined the overall length of the Modulor according to the height of "the Frenchman", 1.75 m tall. He extended this to an upper dimension of 2.164 m, the height of the Frenchman's raised hand, a dimension that was also arrived at by doubling the height of the Frenchman's solar plexus, or 'mid-point', above the ground. Unfortunately, the Frenchman's height of 1.75 m led to awkward subdivisions of British Imperial feet and inches and so proved difficult to use in the dominant English-speaking Anglo-Saxon nations, Britain and the USA, which had refused to embrace the metric system. Le Corbusier's assistant Py resolved this dilemma. (How aptly named he was for research into number and proportion, though one wonders whether Py existed beyond Le Corbusier's fertile imagination!) Unshackled by the national chauvinism of his co-researchers, Py made a seemingly random and somewhat audacious observation: "Have you never noticed that in English detective novels, the good-looking men, such as the policemen, are always six feet tall?" Immediately, as Le Corbusier relates the story, the Modulor was adjusted in length to 6 ft, or 1.83 m and, almost miraculously writes Le Corbusier, "the gradations of a new Modulor ... translated themselves before our eyes into round figures in feet and inches!" (Le Corbusier 1954: 56).[9] Le Corbusier made the blue scale twice the size of the red. The red scale descends according to the Fibonacci series from 6 ft, or 72 in. to

[7] See the comments of a correspondent in Le Corbusier (1955, pp. 91–2).

[8] See the sketch *Modulor man demonstrating the red and blue scales* that Le Corbusier made while aboard the ship *Vernon S. Hood* in January 1946 while en route to New York (Le Corbusier 1954: 51, Fig. 18).

[9] For a more complete account of Le Corbusier's development of the Modulor and the contribution of Hanning, among others, see (Matteoni 1894).

4 in. the blue from 144 in to 8 in 144, the multiple of 12 × 12 in or 2 × 6 ft, also appears in the Fibonacci series (Le Corbusier 1954: 82).

It is probably no coincidence that Le Corbusier's choice of primary colours for the Modulor scales relate to the scientific experiments of an American physicist, A. A. Michelson, who had evaluated the primary colours of light as a means of precisely formulating the length of the metre rod during the late nineteenth century. Michelson suggested that the light from the metal cadmium would provide a suitable alternative standard to a physical measuring rod, such that the metre length could be defined as equivalent to a specific number of wavelengths of the primary coloured light of the cadmium spectrum—red, green and blue (Michelson 1894). Using light to define the metre found an increasing number of supporters in the global scientific community post-World War II, and on 14 October 1960 international agreement was eventually reached to set the metre against the radiation of the orange-red light emitted by the radioactive krypton-86 atom (Danloux-Dumesnils 1969, pp. 36–42) Scientists regarded this as a return to a natural and verifiable definition for the metric system—although, compared to Le Corbusier's Modulor, their notion of natural measure is obscure to all but the high priests of science.

Le Corbusier enjoyed myth making, although the evolution of the Modulor was less magical than he would have us believe. He undoubtedly initiated the search for a more meaningful human measure, but an Englishman, Gerald Hanning, completed the groundwork that defined his revolutionary scale. From the outset, according to surviving evidence, Hanning worked up a set of dimensions based on the height of man in Imperial (British) inches. On receipt of this number sequence in the spring of 1944, Le Corbusier recommended Hanning read a recently published book by Elisa Maillard on the Golden Number sequence, *Du nombre d'Or* (Paris, 1943). Hanning then subdivided these dimensions using the ratio of the Golden Section to create the Modulor. It would appear that although Hanning had begun with subdivisions of the Modulor in inches, Le Corbusier was persisting with the dimensions of a metric "standard" man of 1.75 m up to 1950 (Benton 1987: 245). By the end of 1947 the Modulor was sufficiently resolved for Le Corbusier to send a manuscript version of it to Prince Matila Ghyka in London for comment. Ghyka was the author of *Esthétique des proportions dans la nature et dans les arts* (Paris, 1927), and *Le Nombre d'Or* (Paris, 1931) and was well placed to offer Le Corbusier an authoritative opinion on its relation to the Golden Section. He was sufficiently enthusiastic about the Modulor to write an explanatory and supporting account of it in the *Architectural Review*, published in 1948 (Ghyka 1948).

Le Corbusier first applied the Modulor to his innovative and influential housing concept, the *Unité d'habitation* built Marseilles between 1946 and 1952 (Fig. 17.5).

The *Unité* has an overall form 140 m long, 24 m wide and 56 m high, and the Modulor permeates every part of this complex building (Le Corbusier 1954: pp. 132–153). Yet Le Corbusier states that only 15 of its scalar measurements were required, which he enshrined in a block or "stele of measures" at the building's base (Le Corbusier 1954, p. 140). As Le Corbusier concluded about this innovative structure: "We may safely say that such exactitude, such rigour of

Fig. 17.5 Unité d'Habitation, Marseilles. Le Corbusier. Roofscape. Photo courtesy of Tim Benton

mathematics and harmony have never before been applied to the simplest accessory of daily life: the dwelling." (Le Corbusier 1954, p. 136).

It is perhaps no coincidence that this building was the first to be designed with the Modulor—the metre modified. The name Le Corbusier chose for this building, intended for communal living anywhere on the globe, the *Unité d'habitation*, can be read as a subtle inversion of the *Système International d'Unités*, the united international system that regulated the metric weights and measures, and which Le Corbusier had been so keen to subvert. He intended that the inhabitants of his *Unité* would dwell in harmony with nature in a vessel that celebrated the form, senses and intellects of humanity. It was a valiant attempt to reunite body and architecture—to put people before krypton atoms and the speed of light, to save mankind from the abstractions of pure science.

Le Corbusier's Modulor was truly natural. It was born from the union of the world's two major measures, the metric and imperial systems, comprehended through the human body. He succeeded in reuniting tradition and modernity, and in making measure useful and meaningful again. Le Corbusier had the qualities of a universal man in an era in which the tendency for specialisation and the separation of the arts and sciences undermined such a concept. It is ironic then that the ultimate sanctioning of the Modulor came from Einstein in 1946 as the Modulor and *Unité* were being conceived. As Le Corbusier proudly—almost breathlessly—boasts:

> I had the pleasure of discussing the 'Modulor' at some length with Professor Albert Einstein at Princeton. . . . In a letter written to me the same evening, Einstein had the kindness to say this of the 'Modulor': "It is a scale of proportions which makes the bad difficult and the good easy." There are some who think this judgement is unscientific. For my part, I think it is extraordinarily clear-sighted. It is a gesture of friendship made by a great scientist towards us who are not scientists but soldiers on the field of battle (Le Corbusier 1954, p. 58).

The twentieth century's greatest scientist and architect reached a natural accord. Such unions will need to be more commonplace if the human body is to be properly honoured in the architecture of the future.

Biography Robert Tavernor is Emeritus Professor of Architecture and Urban Design at the London School of Economics and Political Science (LSE), and founding director of the Tavernor Consultancy in London (http://www.tavernorconsultancy.co.uk), which has advised on many prominent urban design projects in the city. He is an internationally-renowned architectural historian and urbanist, who has published widely on architecture and urban design, including the impact of tall buildings on historic cities. He has a long and distinguished academic career, including working with co-authors on a series of high quality and much acclaimed modern translations of Vitruvius, Alberti and Palladio, and about whom he has also written. He is also co-editor with George Dodds of *Body and Building: Essays on the changing relation of Body to Architecture* (MIT Press, 2002 and 2005) and is the author of *Smoot's Ear: the Measure of Humanity* (Yale University Press, 2007) which develops the theme of this essay.

References

Benton, T. 1987. The Sacred and the Search for Myths. *Le Corbusier: Architect of The Century*, M. Raeburn & V. Wilson, eds. (Exhibition Catalogue) Manchester: Arts Council of Great Britain.

Bigourdan, G. 1901. *Le Système métrique avec des poids et mesures. Son éstablissement et sa propogation graduelle, avec l'histoire des opérations qui ont servi a determiner le mètre et le kilogramme, par G. Bigourdan.* Paris: Gauthier-Villars.

Cesariano, C. 1521. *Di Lucio Vitruvio Pollione de architectura libri dece. . . .* Como: G. da Ponte.

Chisholm, H. W. 1877. *On the Science of Weighing and Measuring and Standards of Measure and Weight.* London: Macmillan and co.

CHISHOLM, L. J., ed. 1997. Measurement Systems. *The New Encyclopaedia Britannica, 23*: pp. 693-7.

COX, E. F. 1957. A History of the Metric System of Weights and Measures, with emphasis on campaigns for its adoption in Great Britain, and in the United States prior to 1914. Unpublished Ph.D. thesis, Indiana University. University Microfilms, Ann Arbor, Michigan, no.: 22681.

DANLOUX-DUMESNILS, M. 1969. *The Metric System. A Critical Study of its Principles and Practice*. Trans. A. P. Garrett & J. S. Rowlinson. London: Athlone Press.

DE LA CONDAMINE, C. M. 1747. Nouveau projet d'une mesure invariable propre à servir de mesure commune à toutes des Nations. *Mémoires de l'Académie Royale des Sciences*. Paris : Académie Royale des Sciences.

DODDS, G. and TAVERNOR, R., eds. 2002. *Body and Building: On the Changing relation of Body and Architecture*. Cambridge, MA and London: MIT Press.

FERNIE, E. 1978. "Historical Metrology and Architectural History". *Art History*, **1**, 4: pp. 383-399.

GONON, P.M. 1840. *Médaille commémorative de l'établissement du système métrique et de son usage exclusif*. Lyon: M. Fontaine.

GHYKA, M. 1948. Le Corbusier's Modulor and the Concept of the Golden Mean. *The Architectural Review*, **103**, 614 (February): pp. 39-42.

KLEIN, H. A. 1988. *The Science of Measurement: A Historical Survey*. New York: Dover Publications.

LE CORBUSIER. 1923. *Vers une architecture*. Paris: G.Crès.

——. 1954. *The Modulor: A Harmonious Measure to the Human Scale Universally applicable to Architecture and Mechanics*. London: Faber and Faber.

——. 1955. *Modulor 2: Let the user speak next*. London: Faber and Faber.

MATTEONI, D. 1894. La ricerca di un'idea di proporzione: il Modulor. *Parametro*, 1980: 12-37. Faenza: Faenza Ed.

MICHELSON, A. A. 1894. *Valeur du Mètre*. Paris: Gauthier-Villars.

NEWTON, I. 1704. *Opticks or a treatise of the reflections, refractions, inflections and colours of light. Quaestio XVIII*. London: Royal Society.

OUTRAM, D. 1989. *The Body and the French Revolution. Sex, Class and Political Culture*. New Haven & London: Yale University Press.

PETERSON, M. D., ed. 1984. *Thomas Jefferson: Writings*. New York: The Library of America.

RYKWERT, J. 1996. *The Dancing Column. On Order in Architecture*. Cambridge (MA)-London: MIT Press.

SI-YU-KI, H. T. 1885. *Buddhist Records of the Western World*. Trans. S. Beal. Boston: Boston: J.R. Osgood.

STEWART, A. F. 1978. The Canon of Polykleitos. *Journal of Hellenic Studies*, **XCVIII**: pp. 122-31. Cambridge: Cambridge Journals Online.

THIERSCH, A. 1893. Die Proportionen in der Architektur. *Handbuch der Architektur*, J. Durm, ed. **4**, 1: pp. 38-87. Darmstadt: Stuttgart.

VITRUVIUS. 2009. *On Architecture*. Richard Schofield, trans. London: Penguin Classics.

VOEGELIN, E. 1948. The Origins of Scientism. *Social Research*, **XV**, 4 (December): pp. 462-494.

WÖLFFLIN, H. 1889. *Zur Lehre von den Proportionen*. Basel: Benno Schwabe & Co.

ZEISING, A. 1854. *Neue Lehre von den Proportionen des Menschlichen Körpers*. Leipzig: Duncker & Humblot.

Chapter 18
Façade Measurement by Trigonometry

Paul A. Calter

Introduction

We are all familiar with the trigonometry textbook problem, *"The angle of elevation to the top of a building from a point 200 ft from … Find the height of the building,"* and such methods are hardly new (Fig. 18.1). Here we describe a trigonometric method that not only measures heights of points on a building, but widths and *depths* of those points. It will give the height, horizontal position, and depth, (x, y, and z coordinates) of each selected point. To get the depth dimension, the procedure requires two theodolite setup positions, with a set of readings taken from each location. This second setup also provides a second set of numbers with which to check the first. This method will work with walls that are leaning out of plumb, have offsets, are curved, or have projecting elements, like sills or cornices.

This procedure does *not* require the theodolite to be at the same height at each position, thus is suitable for sighting from *sloping ground*. Further, it is not required that the theodolite positions be at the same distance from the wall. The procedure will give *two values* for each dimension (six figures for each point). The pairs of x coordinates and of z coordinates are not independent, and serve only as a check on the calculation. The two y coordinates are independent and can be averaged to give a final value.

This method was developed for the purpose of measuring Medieval and Renaissance structures in Italy, for research in the history of architecture. To measure a building, a historian is most likely to use a tape measure from scaffolding set up for that purpose, a direct but costly and laborious method. Dimensioned drawings are also made by stereophotogrammatry, such as those for

First published as: Paul Calter, "Façade Measurement by Trigonometry", pp. 27–35 in *Nexus I: Architecture and Mathematics*, ed. Kim Williams, Fucecchio (Florence): Edizioni dell'Erba, 1996.

P.A. Calter (✉)
Vermont Technical College, 108 Bluebird Lane, Randolph Center, VT 05061, USA
e-mail: pcalter@sover.net

K. Williams and M.J. Ostwald (eds.), *Architecture and Mathematics from Antiquity to the Future*, DOI 10.1007/978-3-319-00137-1_18,

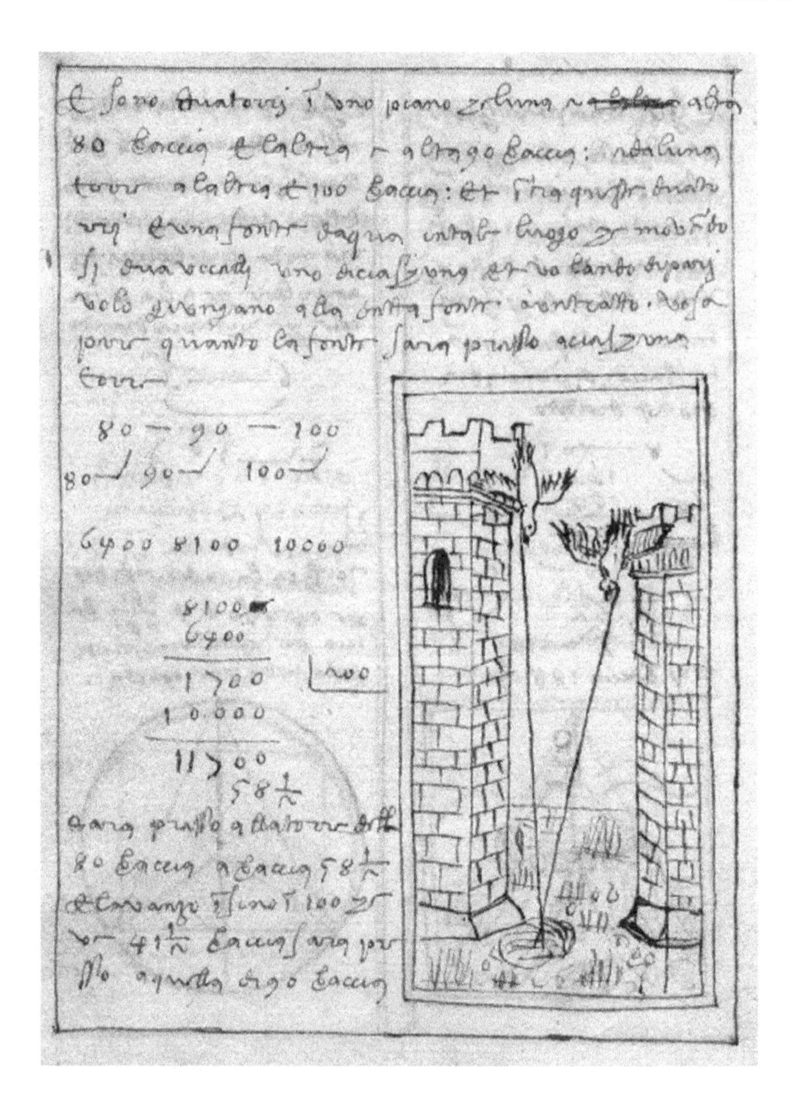

Fig. 18.1 Fifteenth-century survey exercise. Image: Calandri (1491: 100v)

Independence Hall in Philadelphia (U.S. National Park Service 1992: 33). This chapter describes an alternative, suggested to me by architect Kim Williams. A literature search revealed few references to a trigonometric method. Martin Kemp, talking about Filippo Brunelleschi, says "On his first visit to Rome, as described in his biography, he made measured drawings of Roman buildings, using his understanding of standard surveying techniques 'to plot the elevations', using measurements 'from base to base' and simple calculations based on triangulation. The basis for such procedures would have been the 'abacus mathematics' he learnt as a boy" (Kemp 1990: 11). His source for this information is Antonio Manetti's *Life of Brunelleschi* (Manetti 1970: 152–153). A search of Manetti's biography found reference to a visit to Rome, but no mention of his use of trigonometry to measure façades. In fact, there is some doubt expressed by the editor, Howard Saalman, that Brunelleschi ever went to Rome, and that this passage was added to enhance the stature of Manetti's subject.

The Method

1. Study the façade. Take photos. Measure by manual taping whatever can be easily reached. Make a preliminary drawing. Choose and number the target points. Place adhesive targets on the wall, where possible.
2. Select or lay out a base line. The intersection of the façade and pavement makes a good base line, if it is straight and horizontal. Use a stretched cord if no suitable physical base line is available, as shown in Fig. 18.2. The figure shows what is possibly the most difficult measuring situation, a curved building on sloping ground. Mark two theodolite setup points on the ground or pavement, which can be at different heights and at different distances from the base line. Record their horizontal distance c apart and their horizontal distances d_A and d_B to the base line.
3. Set up and level the theodolite at location A. With the telescope horizontal, sight and mark a point T on the wall.
4. Set a plumb line over the other theodolite location. Sight the plumb line with the theodolite and set the horizontal scale to zero.
5. Sight each target. For each, record the horizontal angle α and the vertical angle θ, Fig. 18.3.
6. After each target has been sighted, move the theodolite to the second location. With the telescope horizontal, sight a point on the wall vertically in line with point T, found in step 3. Measure the vertical distance Δ from that point to T.
7. Repeat step 5, recording the horizontal angle β and the vertical angle φ for each target point.
8. Enter all measurements into the computer spreadsheet and print out the x, y, and z coordinate of each target point.
9. Make a final dimensioned drawing by hand or by use of a CADD program.

Derivation of Façade Equations

The equations the spreadsheet uses to reduce the data are easily derived. Starting with the original taped measurements, Fig. 18.2,

$$c = \text{Horizontal distance between theodolite locations.}$$

d_A and d_B are the horizontal perpendicular distances from base line to theodolite locations.

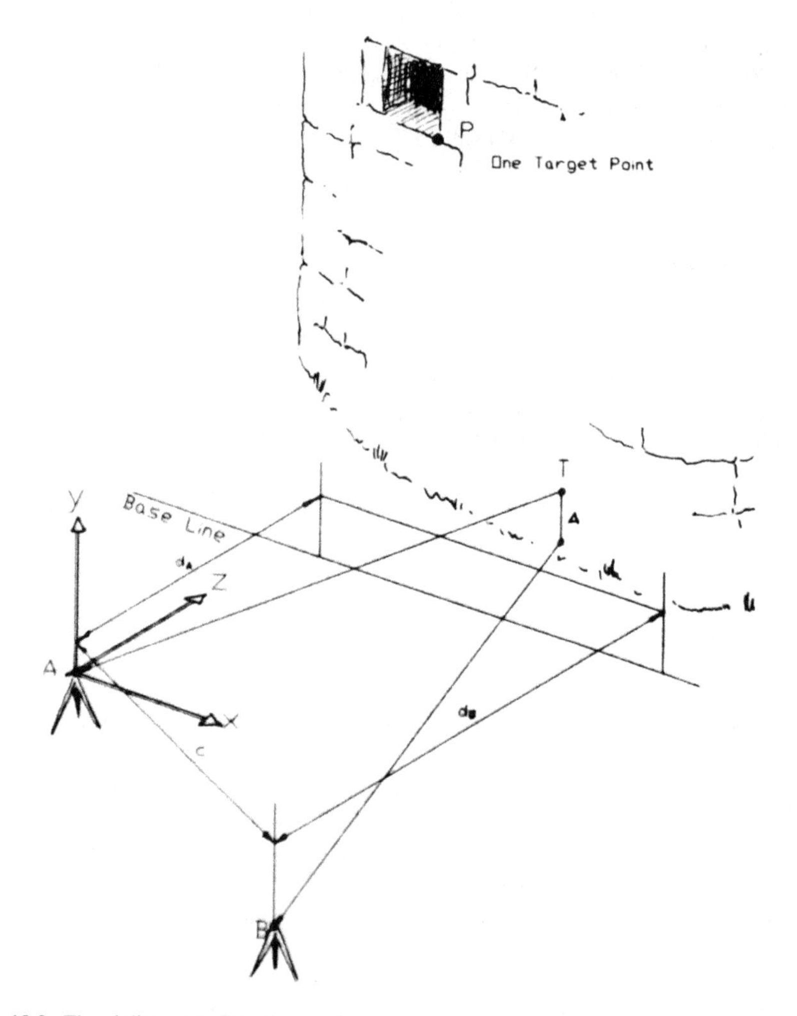

Fig. 18.2 Theodolite setup. Drawing: author

$$\Delta = \text{Vertical offset between theodolite tubes}$$

From these we get (Fig. 18.4),

$$\delta = \text{Horizontal offset} = d_B - d_A$$
$$\varepsilon = \text{Angular offset} = \arctan(\delta/c)$$
$$\text{L} = \text{distance AA}'\text{between A\&B parallel to baseline} = \sqrt{c^2 - \delta^2}.$$

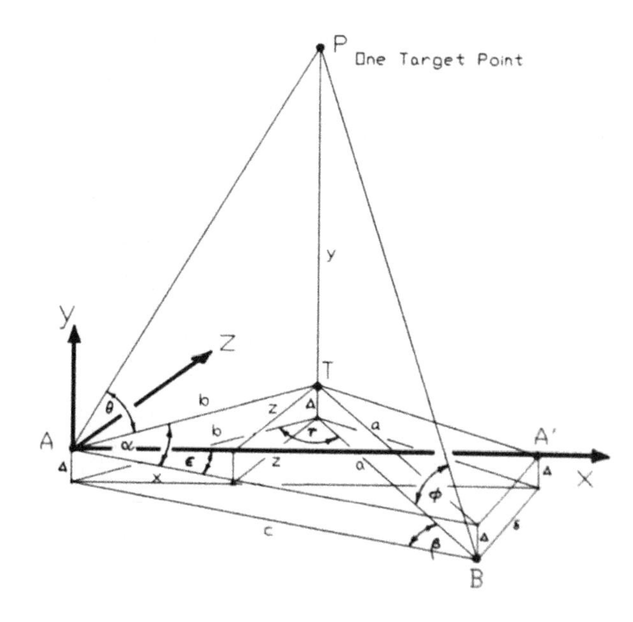

Fig. 18.3 Oblique view. Drawing: author

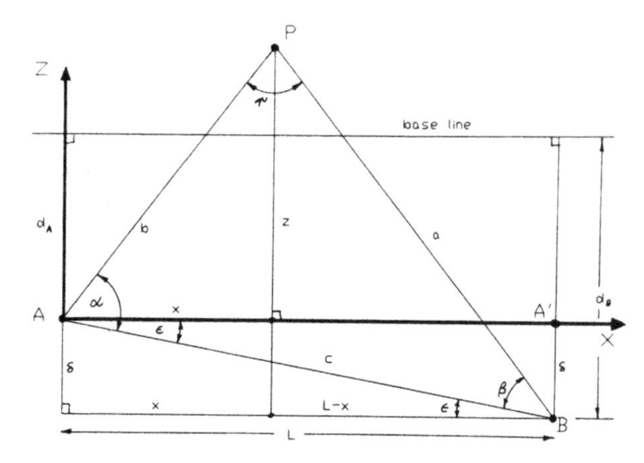

Fig. 18.4 Plan view. Drawing: author

For each target P (Fig. 18.3) we have,

$$a = \text{horizontal angle at A, from B to target}$$
$$\theta = \text{vertical angle at A, from horizontal to target}$$
$$\beta = \text{horizontal angle at B, from A to target}$$
$$\varphi = \text{vertical angle at B, from horizontal to target.}$$

Our coordinate axes will be as shown in Figs. 18.2 and 18.3, with the origin at A, with the x axis parallel to the baseline and directed to the right, the y axis vertical and directed upwards, and the z axis perpendicular to the x and y axes, and directed towards the building. A simple translation of axes will later place the origin at any selected point, such as a corner of the building.

We now calculate the x coordinate of point P. Referring to Fig. 18.3,

$$\gamma = 180 - \alpha - \beta$$

$$a = c\frac{\sin\alpha}{\sin\gamma}$$

$$b = c\frac{\sin\beta}{\sin\gamma}$$

From the plan view (Fig. 18.4) we see that

$$\cos(\alpha - \varepsilon) = \frac{x}{b}$$

$$x = b\cos(\alpha - \varepsilon)$$

From position B:

$$\cos(\beta + \varepsilon) = \frac{L - x}{a}$$

$$x = L - a\cos(\beta + \varepsilon)$$

Note that the x coordinate obtained from position B is not independent of that obtained from A, but is useful for checking the computation. Next we find the y coordinate of point P. From position A:

$$\tan\theta = \frac{y}{b}$$

$$y = b\tan\theta$$

From position B:

$$\tan\phi = \frac{y + \Delta}{a}$$

$$y = a\tan\phi - \Delta$$

Here, the values of y found from each setup position *are* independent. Next we find the z coordinate of point P. From position A,

$$\sin(\alpha - \varepsilon) = \frac{z}{b}$$

From which,

$$z = b \sin(\alpha - \varepsilon)$$

From position B,

$$\sin(\beta - \varepsilon) = \frac{z + \delta}{a}$$

From which

$$z = a \sin(\beta + \varepsilon) - \delta$$

As with the calculation for x, the two values of z are not independent.

Field Test at VTC

The method was tested by taking measurements of the front façade of Green Academic Centre (Fig. 18.5) at Vermont Technical College.[1] The figure shows eleven target points, all visible from both theodolite locations. These were sighted using a Wild T2 theodolite, capable of a precision of about 0.2 s of arc. The baselines were taped three times using a standard surveyor's tape graduated in millimetres, and the readings averaged. The data was reduced using Lotus 123 spreadsheet. Some of the distances measured by theodolite were also taped, for comparison.

The following table gives the three coordinates of each target point, in metres.

Point	x	y		z
1	−3.231	7.310	7.337	15.091
2	−3.230	6.451	6.443	15.091
4	1.381	7.363	7.351	13.271
5	1.381	(6.423)	(6.153)	13.212
7	6.779	7.348	7.342	13.254
8	6.776	6.445	6.444	13.263
9	6.770	−0.275	−0.274	13.280
10	16.208	7.335	7.350	15.024
11	16.208	6.450	6.457	15.024
13	11.319	2.945	2.946	14.579
14	14.913	(3.042)	(2.942)	14.508

[1] The measurements were made with the able assistance of Douglas Pennington of Vermont Technical College.

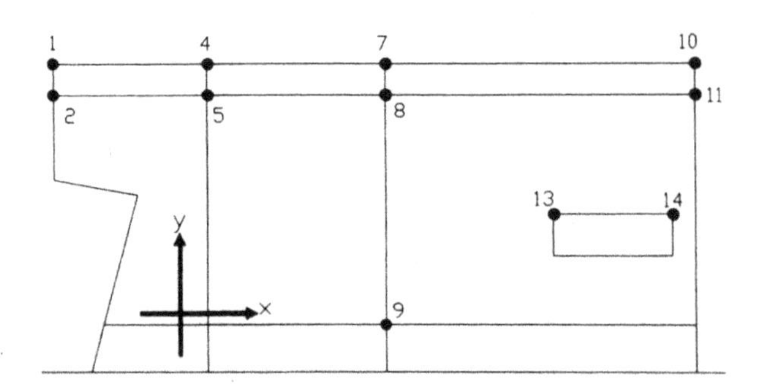

Fig. 18.5 Green Academic Centre. Drawing: author

The figures shown in parentheses for points 5 and 14 are considered measurement errors, and were discarded. Note that a pair of values is given for each y coordinate, corresponding to the two equations used for their calculation. These, of course, should be identical for each target point, and their difference gives us some measure of the precision of the method. For point 1, for example, the two values of 7.310 m and 7.337 m have an average of 7.324 m. Each differs from this by 0.0135 m, or 0.18 %. For the other points, the deviation from the average is also less than 0.2 %.

Next let us compare points that are expected to be at the same height on the building, or at the same depth, or on the same vertical. For example, points 1, 4, 7, and 10 are at the top of the building, and should all have the same *y* coordinate. The average y for these four points is 7.342 m, with a maximum deviation from this value of 0.032 m. These deviations may represent inaccuracies in the measurements, or may represent actual differences in height of these points. Note that points 1 and 10 are at opposite corners of the building, nearly 20 m apart, and that points 4 and 7 are on the offset portion of the façade. Other comparisons are given in the following table.

Target points	Average distance (m)	Max. deviation from average (m)
Horizontal 1, 2	3.2305	0.0005 (0.015 %)
Horizontal 4, 5	1.381	0
Horizontal 7, 8, 9	6.775	0.005 (0.07 %)
Horizontal 1 0, 1 1	16.208	0
Vertical 1, 4, 7, 10	7.342	0.0032 (0.44 %)
Vertical 2, 5, 8, 11	6.448	0.009 (0.14 %)
Depth 1, 2, 10, 11	15.058	0.0034 (0.23 %)
Depth 4, 7, 8, 9	13.269	0.0013 (0.10 %)

On the basis of this one test, it would appear that, with moderate care, accuracies within 0.5 % are easily obtained. There is no theoretical limit to the accuracy of the method. If better accuracy is needed, then repeated measurements can be made from the two theodolite setup points, or better, another set of readings taken from a third setup point.

Biography Paul A. Calter is Professor Emeritus of Mathematics at Vermont Technical College and was a Visiting Scholar at Dartmouth. A sculptor, he has interests in both the fields of mathematics and art. He received his B.S. from Cooper Union and his M.S. from Columbia University, both in engineering, and his Masters of Fine Arts Degree at Vermont College of Norwich University. Calter taught mathematics for over 25 years and is the author of eleven mathematics textbooks, numerous articles, and a mystery novel. He has participated in dozens of art shows, and has permanent outdoor sculptures at a number of locations in Vermont. For the "Mathematics Across The Curriculum program", Calter developed the course "Geometry in Art & Architecture" and has taught it at Dartmouth and Vermont Technical College, as well as giving workshops and lectures on the subject. His textbook *Squaring the Circle: Geometry in Art and Architecture* (Key Curriculum Press, 2008) explores the relationship between architecture, art, and geometry.

References

CALANDRI, Filippo. 1491. *De aritmetica*. Florence: Lorenzo da Morgiani and Giovanni Thedesco da Maganza.

KEMP, Martin. 1990. *The Science of Art*. New Haven: Yale University Press.

MANETTI, Antonio. 1970. *The Life of Brunelleschi*. H. Saalman, ed., C. Engass, trans. London and Pennsylvania: Penn. State Press.

U.S. NATIONAL PARK SERVICE. 1992. HABS/HAER Annual Report 1992.

Chapter 19
Ancient Architecture and Mathematics: Methodology and the Doric Temple

Mark Wilson Jones

Introduction

For researchers into mathematical aspects of architectural design a multi-faceted subject often referred to by the umbrella-term 'proportion' it can be disconcerting how little weight is given to this field by some authorities on general architectural history. Sir John Summerson, for example, the author of the evergreen *Classical Language of Architecture*, impatiently dismissed most past thinking about proportion as "a vast amount of pretentious nonsense" (Summerson 1980: 8).

Since the time of Summerson's comment (which was first published in 1963) proportional studies have made significant advances thanks to a higher standard of rigour, as indeed is now expected by the editors of academic journals, the *Nexus Network Journal* among them. Scholars who specialize in other areas do on occasions profitably tackle mathematical aspects of design a case in point being Janet DeLaine's inclusion of an instructive analysis of the layout of the Baths of Caracalla in a monograph primarily concerned with its construction (DeLaine 1997) yet there still remains the diffuse, if not clearly articulated, view that proportional studies can be left to one side by mainstream scholarship without too much loss.

Part of the reason for this state of affairs lies in the continuing perception that when it comes to proportion it is possible to prove anything.[1] Dissatisfaction also arises when proportion is treated as a subject unto itself; too often it is not sufficiently clear how the mathematical attributes claimed for a particular

First published as: Mark Wilson Jones, "Ancient Architecture and Mathematics: Methodology and the Doric Temple", pp. 149–170 in *Nexus VI: Architecture and Mathematics*, Sylvie Duvernoy and Orietta Pedemonte, eds. Torino: Kim Williams Books, 2006.

[1] Dinsmoor (1923, 1975: 161, n.1). For further scepticism see (Wilson Jones 2000a: 4–6).

M. Wilson Jones (✉)
Department of Architecture and Civil Engineering, University of Bath, Bath BA2 7AY, UK
e-mail: m.w.jones@bath.ac.uk

K. Williams and M.J. Ostwald (eds.), *Architecture and Mathematics from Antiquity to the Future*, DOI 10.1007/978-3-319-00137-1_19,

building engage with the design process that created it. This article proposes a methodology based on a series of criteria or tests which proportional studies should address in order to counter these problems, and in the general interest of plausibility.[2] Most of the material discussed is culled from antiquity and my own research in particular, but it is hoped that this methodology will have wider relevance. It will then be applied to a modular reading of Doric temples of the Greek classical period, while preparing the ground for a forthcoming fresh interpretation of the 'poor old Parthenon'.

The Criteria

The proposed criteria may be summarized under the following headings:

I. Degree of 'fit'
II. Corroboration in the form of texts
III. Corroboration in the form of non-textual sources, e.g., drawings
IV. Comparability
V. Mensuration
VI. 'Pay-off' in terms of design
VII. 'Pay-off' in terms of practicality
VIII. Corroboration at the level of detail.

The greater the number of criteria that are satisfied by a particular proposal, the greater the weight it carries. In practice, however, it will be rare for all of them to be addressed simultaneously; there might in many instances, for example, be no textual evidence to draw on.

I. Degree of 'fit' This is the most obvious test, yet it is one of the most problematic. Proportional studies often focus on single buildings or on a relatively small body of comparative material; in such circumstances statistical methods utilized in fields such as medical research are of limited applicability. Statistical tools are being developed for the metrical investigation of archaeological and architectural material (Pakkanen 2002; Baxter 2003: 231–233), but it is important to recognize that they do not provide absolute objective evidence, since they necessarily involve human discretion in establishing parameters, in selecting data sets, and in setting thresholds of significance. The match between a hypothetical project and its execution is inevitably subject to opinion, given uncertainty surrounding a number of issues: constructional tolerances; the rounding off or not of ideally intended dimensions; the precision of modern surveys; the difficulty of knowing whether highlighted limits and relationships were actually significant to the original designer. It is not my aim to pursue such

[2] Some of the aspects that concern this article were treated in (Wilson Jones 2000a: 4–11), but in a less methodical fashion.

questions here, beyond a plea for as much as possible of the information that a reader might need for evaluation, including hypothetical measurements/ratios and actual measurements/ratios, as well as discrepancies between them, along with clarification of any reliance on reconstruction and/or interpolation (Wilson Jones 2000a, b: 71–74).[3] The degree of discrepancy that is considered acceptable is again a matter of interpretation. Absolute differences seem more pertinent over small distances, percentage differences over large ones; anything over half a percentage point for medium to large distances is a poor match or a very poor one depending on circumstances.[4] (We might expect greater accuracy in relation to a simple rectangular plan than a complex affair involving polygonal geometry and/or curvature.) Scale drawings can never be more than indicative.

II. Corroboration in the form of texts Textual evidence is potentially decisive, even if its interpretation is rarely straightforward. A notable resource is Andrea Palladio's *Quattro Libri*; yet the published and built projects notoriously diverge.[5] In the ancient context the fragmentary nature of the evidence makes such exercises more vexatious, as illustrated by the difficulty of reconciling the reconstruction of the Mausoleum of Halicarnassus with Pliny the Elder's account of it.

Vitruvius bequeaths us the general principles of ancient architectural design and much more besides, but for one reason or another he may not always be our best guide as regards specifics (Wilson Jones 2000a, b: 34–38, with further bibliography). Suffice it to note the case of the Corinthian capital, and his omission of the proportional relationship that dominated its design in antiquity: the equality of the height of the capital to its cross-sectional width (Wilson Jones 2000a, b: 145–151, 154–155) (Figs. 19.1 and 19.2). This particular proportion may not be Vitruvian [*De Architectura* IV.1.11–12], but otherwise *the kind of relationships* he mentions, and *the kind of ratios* he recommends (1:1, 2:1, √2, 1/9, 6/7 and so on) do indeed strike a convincing tone. Inferences may also be made about general practice on the basis of documentation relating to single projects, for example the specification for the Arsenal at Piraeus or the correspondence relating to the construction of Milan Cathedral.

III. Corroboration in the form of non-textual sources Architects' drawings can yield significant insights into design intentions. Elaborate as they are, with successive re-workings and *pentimenti*, Borromini's plans for S. Carlo alle Quattro Fontane reveal the simple underlying geometry based on paired

[3] For further observations see Coulton (1975).

[4] In appraising measurements relating to the Corinthian order (Wilson Jones 2000a: 221–225), I proceeded on the premise that a discrepancy of less than 3 cm represented an excellent match, that one between 3 and 6 cm indicated a possible match, and that anything greater indicated a lack of match. For a notional column height of 9 m, 3 cm and 6 cm represent thresholds of 1/3 and 2/3 % respectively.

[5] See the discussion in Mitroviæ (2004).

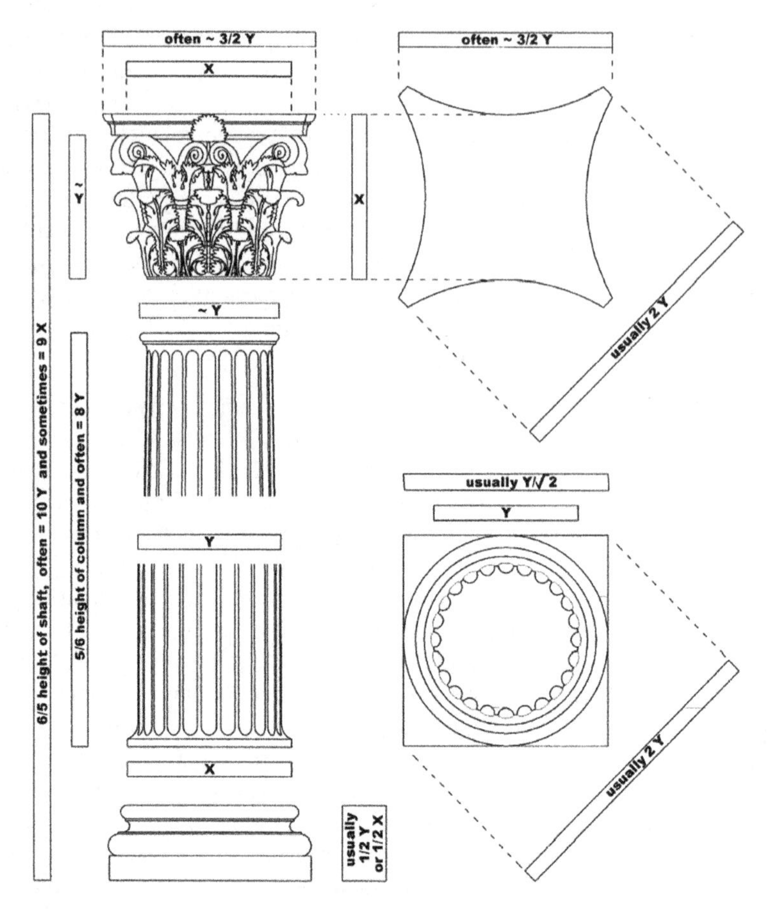

Fig. 19.1 Design framework for orthodox Corinthian columns in the Roman imperial period. Solutions with different characteristics can be achieved using a common set of proportional principles that operate in such a way as to allow flexibility. The key dimensions are the total height of the column, the height of the shaft, the lower diameter of the shaft (Y) and the diameter of the flare of the shaft where it meets the base (X). The height of the capital and its cross-sectional with also equal X. The most common relationships apart from 1:1 are 5:6, 2:3, 1:2, 1:8 and 1:10. The ratio between X and Y is not fixed, but is most commonly 7:6, 10:9 and 11:10 (drawing by author). In orthodox imperial Corinthian columns the shaft height is five-sixths that of the column, this being the key to a series of proportional schemes. Meanwhile the diameter of the shaft is often either one-tenth of the total height or one-eighth that of the shaft (especially if monolithic). In addition the height of the base may be 1/2 X, or 1/2 Y. X must vary with respect to Y so as to suit the overall column proportions. Drawing: author

equilateral triangles. Outside of Egypt, nothing quite comparable is known from antiquity, but there does survive a corpus of full-size templates on stone surfaces. One inscribed on the wall of the Hellenistic temple of Apollo at Didyma reveals a

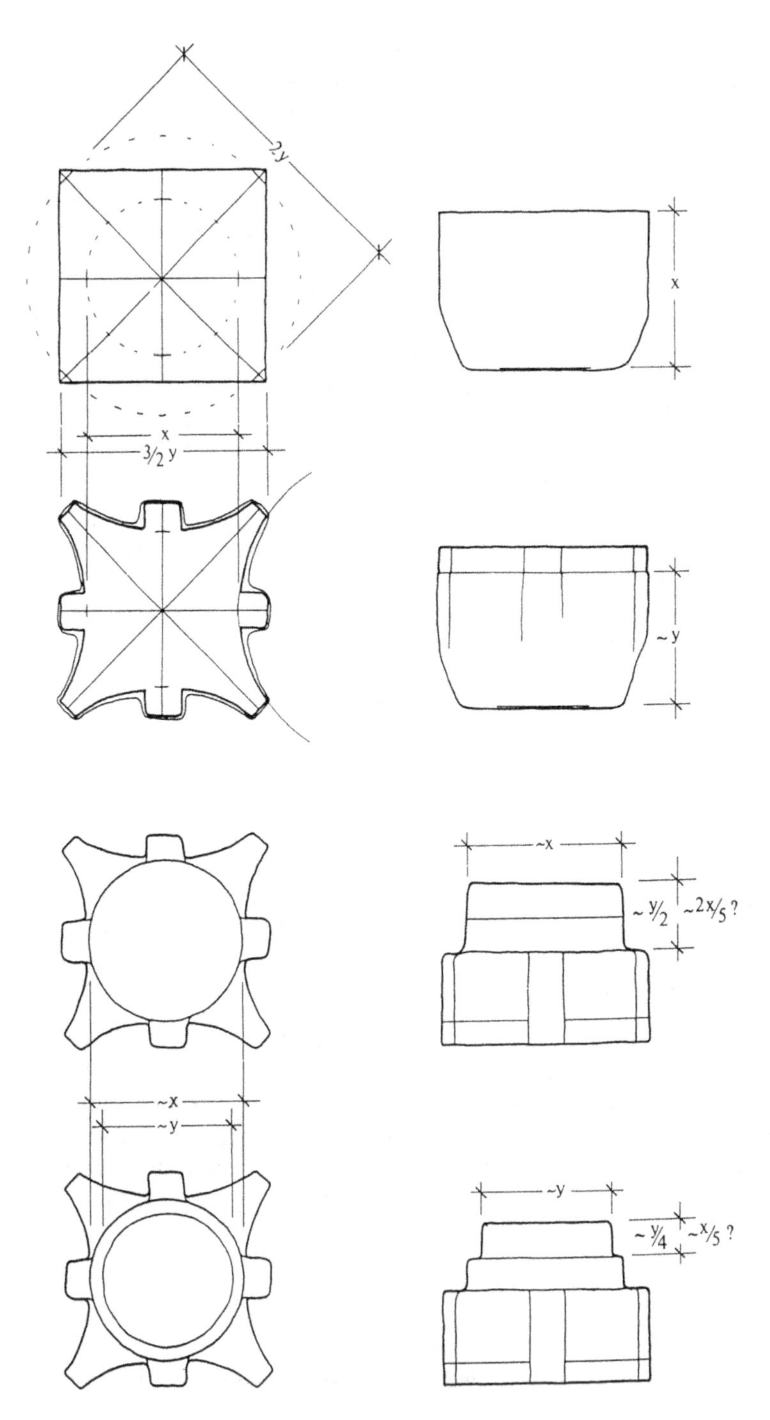

Fig. 19.2 Principal work-stages involved in the production of a typical Roman capital of the imperial period. Drawing: author

method of remarkable intelligence that was used to define the entasis of the peristyle columns.[6] Inscriptions too can be pertinent, as in the case of the measurements sometimes inscribed on the bottom of imperial Roman monolithic column shafts, discussed below.

IV. Comparability While analyses of unusual or 'one-off' buildings necessarily stand on their own, the relative conformity of many ancient building types facilitates comparison. In more recent periods, when it is far more common to know the identity of a building's designer, comparisons might be made across the oeuvre of a single architect. Comparative analysis is a powerful tool, one which counteracts the all-too-human tendency to select material which fits one's argument, while ensuring that the results pertaining to a single case are neither fortuitous nor misleading (Coulton 1974: 61). Jim Coulton has produced some compelling results on the basis of the comparative study of Greek buildings and architectural elements (Coulton 1974, 1979, 1989), and I have made this a cornerstone of my own research, for example on the proportions of the Roman Corinthian order. Bearing in mind that there existed a substantial minority of variations and exceptions, Fig. 19.1 summarizes the orthodox pattern of Corinthian design in the imperial period. It is also instructive to compare the remains of executed buildings with partly-finished components known from quarries, stockpiles and shipwrecks (see Fig. 19.2).

Comparative analysis served to confirm the findings of Jean-Claude Golvin's study, namely that the majority of monumental amphitheatres were laid out with variations on two oval geometries, one based on the 3:4:5 triangle, the other on the √3 or bisected equilateral triangle (Golvin 1988; Wilson Jones 2000a, b: 60–61, 88–89) (Fig. 19.3). When considering reasonably substantial sample groups, it is neither necessary nor desirable to try and resolve every single member. Exceptions and compromises are only to be expected of any grouping of the products of human creativity, and if half to two-thirds can be seen to conform to a pattern, that is quite sufficient to demonstrate that a certain procedure existed and was reasonably if not universally popular.

V. Mensuration Antiquarians as long ago as the Renaissance appreciated the desirability of expressing measurements in the units originally employed (Günther 1998). The stability of Roman metrological systems allows foot values to be presumed with relative confidence; it thus emerges from both surveys and inscriptions that column shafts gravitate towards increments of 4 ft or, more commonly, 5 ft (15, 20, 25, 30 and 40 ft being the most popular lengths). At the

[6] Haselberger (1980, 1997). On Graeco-Roman architectural drawings in general see (Haselberger 1997) and (Wilson Jones 2000a: 50–58, with further bibliography). Surviving scale drawings on stone tend to be plans of record, associated with maps or testaments, but one on the base of the Pergamonese statue group of the Dying Gaul does seem to have had a design purpose.

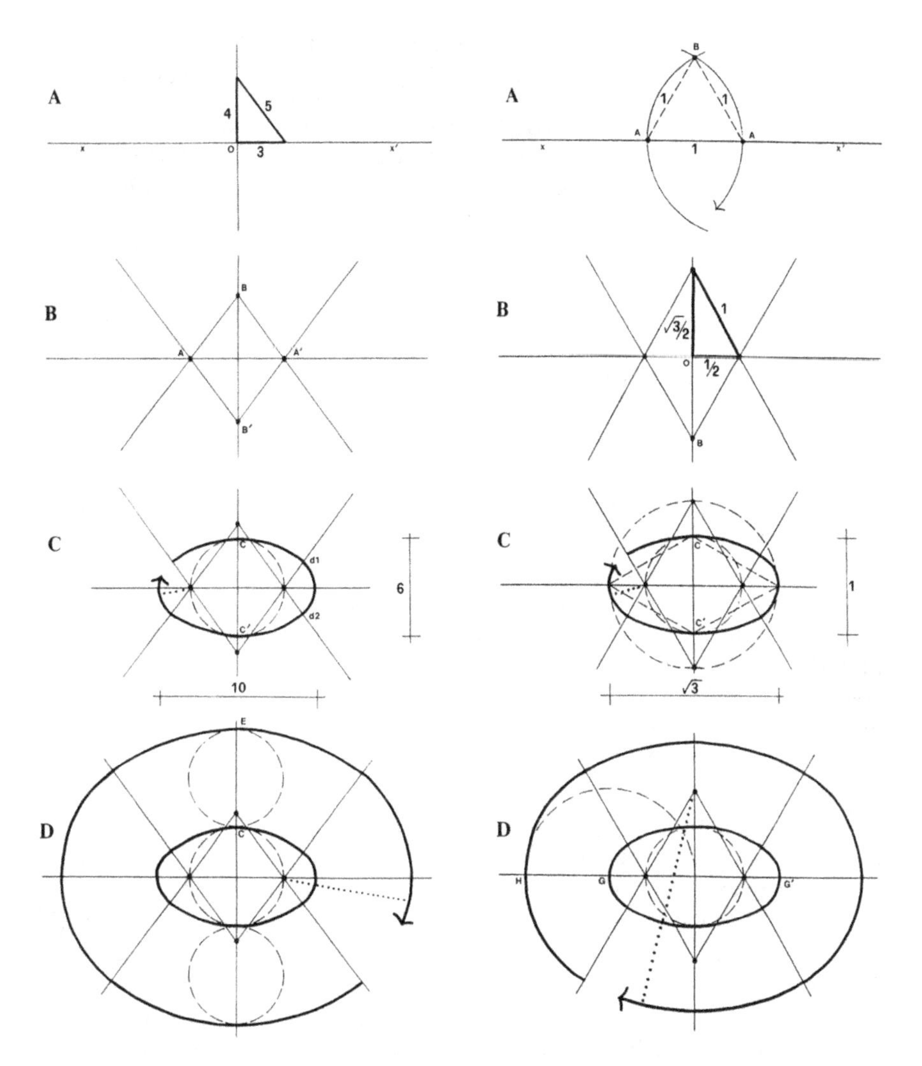

Fig. 19.3 Methods for laying out monumental civic amphitheatres using the 'Pythagorean-triangle-and-inscribed-circle' scheme (*left*), and the 'Equilateral-triangle-and-inscribed-circle' scheme (*right*). Drawing: author

same time the heights of complete Corinthian columns often match multiples of 6 ft (see Fig. 19.1), so it is easy to deduce the existence of a rule that fixed the height of the shaft as 5/6 that of the whole (base, shaft and capital combined).[7] The dimensions of monumental amphitheatres corroborate the focal triangles just

[7] (Wilson Jones 2000a: 147–151; Appendix B). Correlation with measured diameters also shows that shafts often had (or were supposed to have had) a slenderness ratio of 8 (Wilson Jones 2000a: 155).

mentioned; the 3:4:5 triangle used for the amphitheatre at Capua, for example, measures 75:100:125 ft (Wilson Jones 1993: 416–417, 420, 433, fold-out, B).[8]

VI. 'Pay-off' in terms of design The motivation for instilling architectural creations with mathematical harmony is abundantly clear over a wide sweep of history. In line with the Greek principle of *symmetria* (mathematical harmony), number, measure and proportion were thought to confer beauty, and to be a proper preoccupation for artists and builders alike.[9] There exists a wealth of sources to this effect, most obviously Vitruvius's treatise and those of his Renaissance followers. Eminent modern scholars have written handsomely on this intellectual underpinning, which can be taken for granted for the purposes of this article.

The most convincing interpretations to my mind are those that go beyond this general motivation, identifying benefits or 'pay-offs' of an architectural kind, ones that might arise out of a marriage between mathematics and the character of a specific design. For example, Bramante's choice of 1:1, 2:1 and √2:1 ratios for the Tempietto (Fig. 19.4) makes perfect sense in the context of a centralized plan based on concentric circles and cardinal axes, since these ratios fit an *ad quadratum* sequence habitually used to resolve this kind of composition (Wilson Jones 1990). Not only do the measures employed underwrite such a scheme (witness the interior radius of 10 *palmi*), but so does the type of floor Bramante chose. This is an intricate pattern of small coloured tesserae and white borders in the medieval Cosmatesque tradition that underwent something of a revival in the fifteenth century. Given that the Tempietto otherwise promoted the latest architectural fashion this was arguably an out-moded choice, one which Bramante presumably favoured because its *ad quadratura* pedigree so suited his groundplan.

Arithmetic, however, was generally the 'default' mode of design in the Roman period, and this holds true both for ancient Greece and the early modern period. Geometrical ratios, primarily √2 and √3, come to the fore in a substantial minority of cases, for solving what Vitruvius called "difficult questions of *symmetria*" (*De Architectura* I.1.4), or in more general terms *when there was some benefit associated with their use*, as at the Tempietto. Hence geometry was used for setting out theatres, due to their radial layout, and amphitheatres, due to their oval (or sometimes elliptical) plan. Hadrian's Villa near Tivoli constituted a veritable laboratory for testing innovative spatial effects based on the alternation of curve and counter curve. Planning therefore involved the extensive manipulation of compasses, just as it did in the case of the Annexe at Baiae discussed below (Jacobson 1986; MacDonald 1993; Wilson Jones 2000a, b: 93–94).

---

[8] Wilson Jones (1993, 416–417). Meanwhile, understanding that the Punic cubit continued to be used in North Africa helps identify instances where the imperial norms underwent a degree of revision, with amphitheatre layouts conceived in Roman feet being adapted for a workforce accustomed to the local cubit.

[9] On the importance of symmetria see Gros (1989), (Wilson Jones 2000a: 40–43), with further bibliography.

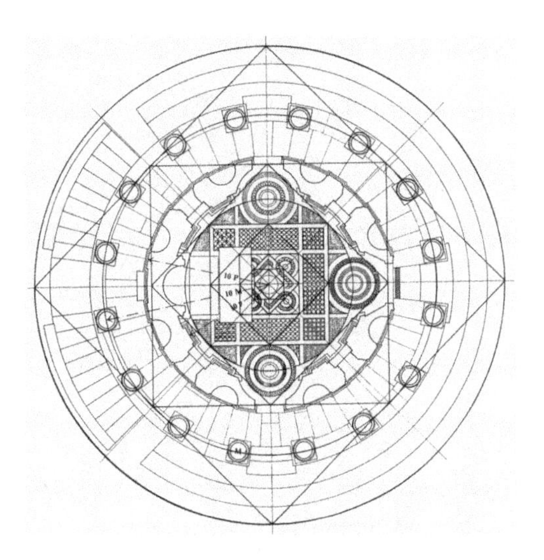

Fig. 19.4 The plan of the Tempietto, Rome (Bramante), with overlaid ad quadratum scheme. Drawing: author

VII. 'Pay-off' in terms of practicality It is surely no accident that the two most common geometries used for amphitheatres were based on the 3:4:5 and √3 triangles. Among the simplest mathematically defined triangles, both offer almost fool-proof ways of assuring that the axes of the oval met at a right-angle. The metrical standardization of Roman shafts provided other kinds of benefit. Having originated from quarries over a wide geographical spectrum, partly-finished monoliths of marble, granite and other fine stones circulated all around the Mediterranean.[10] Standardization served to streamline operations; architects, patrons and suppliers all must have become familiar with the ramifications of this practice, a point of indirect benefit from conception through to completion. The norms of design pertaining to the Corinthian capital brought comparable advantages (Fig. 19.2). Quarry workers roughed out capitals using coarse tools, the pick, point and mallet, while still working near the rock face (Wilson Jones 1991: 133 ff), a context in which simple proportional guidelines, chiefly 1:1, 3:2 and 1:2, simplified the fashioning of these relatively complex elements and reduced the risk of error (The same could hardly be said of the golden section, for example). It is true that Roman builders are renowned for their practical bent, yet few architects of any period would have conceived proportional schemes that were entirely divorced from practical considerations.

VIII. Corroboration at the level of detail The plausibility of a particular proposal can be significantly enhanced if it accounts for some feature or detail for which no other explanation comes readily to mind. The Roman Corinthian column is again instructive; the rule stipulating that 5/6 of its height be assigned to the shaft implied that a relatively tall base should accompany a squat capital, and vice versa, as

[10] For the marble trade in general see Ward-Perkins (1992), Fant (1993), (Wilson Jones 2000a: 152–165). See also various volumes in the ASMOSIA series of published conferences.

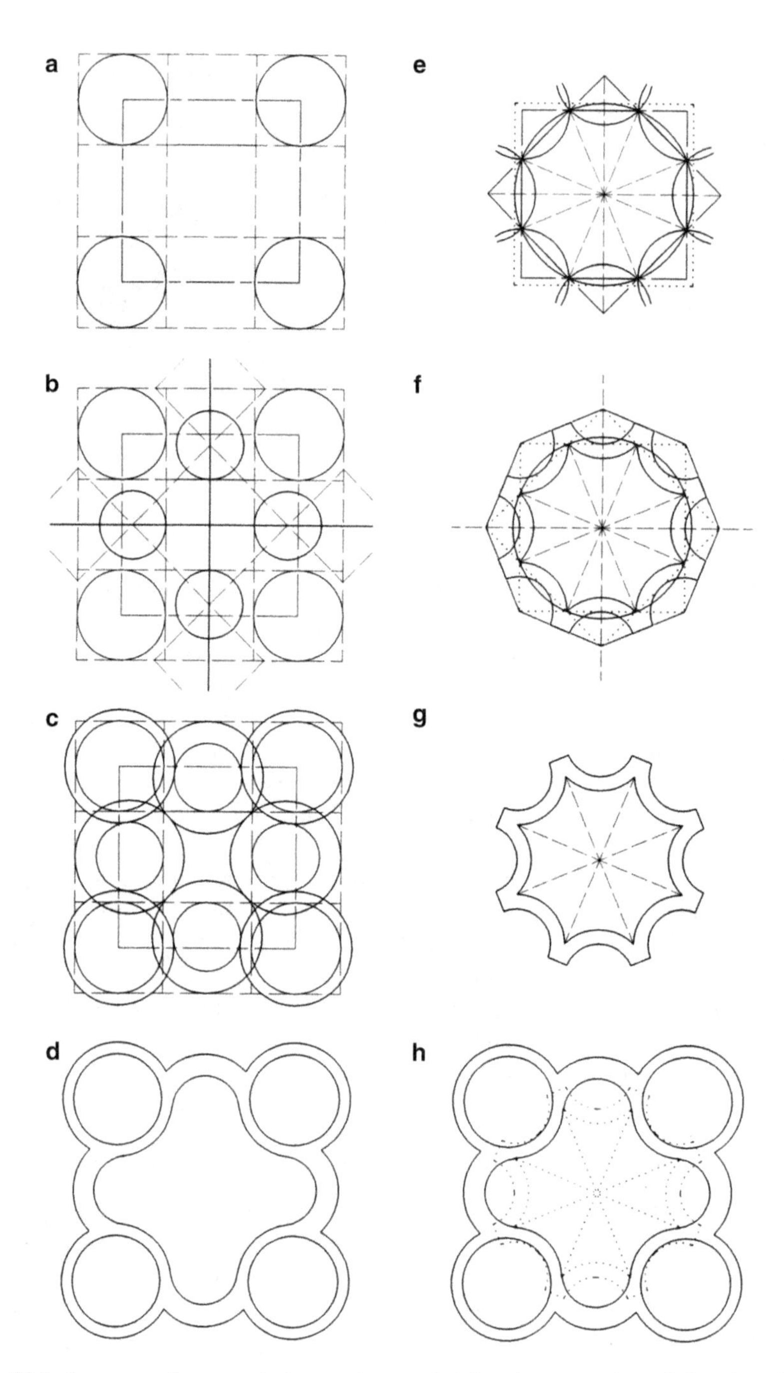

Fig. 19.5 Sequence of geometrical operations underlying the generation of the plan of the 'Annexe' to the so-called Temple of Venus at Baiae, near Naples. The *left side* relates to the layout of the ground floor, the *right side* relates to the layout of the upper floor, while h. shows the upper level (*dotted*) superimposed on the lower level. Drawing: author

indeed is frequently attested (Fig. 19.1). There is no other known precept, Vitruvian or otherwise, that would predict such a combination.

Turning to individual buildings, the tendency of Roman architects to manipulate wall thicknesses to suit their proportional schemes is illustrated at S. Costanza, where the desire to achieve internal and external diameters of 75 and 100 ft resulted in an unnecessarily thick outer wall one almost as wide as the ambulatory vault it supported. The 'Annexe' attached to the rotunda known as the Temple of Venus at Baiae offers another commentary on this approach (Fig. 19.5). It so happens that the geometries of the upper and lower levels did not coincide in some respects. A compromise was introduced in the interest of continuity of surface, resulting in different centres being used for the inner and outer faces of the small counter curves or lobes opening off of the central space. This explains one of the oddities of the Annexe plan curving stretches of wall that are not uniform in thickness, but that resemble instead portions of a crescent moon. The lobe walls are also curiously thick, given that they carry only tiny vaults. Again this was price to be paid for implementing geometries chosen on account of a concatenation of mathematical niceties and whole number dimensions (Jacobson and Wilson Jones 1999; Wilson Jones 2000a, b: 94–100).

Judging the Modular Hypothesis Against the Criteria

The same methodology can now be brought to bear on a single hypothesis, one that pertains to a historical period other than those so far considered. The next section reviews a comparative study of ten Doric temples from the classical period with hexastyle (six-column) fronts: the temples of Zeus at Olympia, of Hephaistos at Athens, of Apollo at Bassae, of Poseidon at Sounion, of Nemesis at Rhamnous, of the Athenians at Delos, of "Juno Lacinia", "Concord" and "the Dioscuri" at Agrigento, and the unfinished temple at Segesta (see Fig. 19.6 below). The focus is limited to the end façades of temples, since this aspect is likely to have governed to a significant extent that of the rest (Wilson Jones 2002: 682).

Scholarly investigations into the design of Doric temples have generated such a plethora of interpretations that some might judge the problem unsolvable. The elusive nature of Doric temple design may in part be attributed to the frequent lapses of regularity and symmetry which Vitruvius called "faults and incongruities" [*De Architectura* IV.3.1]. These flowed from the problem of configuring the peristyle and its frieze, and achieving axial coordination of triglyphs and columns at the same time as whole triglyphs at the corners.

While in the archaic period architects appear to have relied on rules of thumb in conjunction with a successive or stage-by-stage approach, by the second quarter of the fifth century they had acquired more control over the design process, becoming able to instil their projects with improved coherence, neater proportions, and greater regularity in column spacing. In agreement with previous studies, such as Dieter Mertens's examination of Sicilian temples (1984a, b), my study confirms a preference for accurate arithmetical ratios. A ratio of 2:3 between the widths of

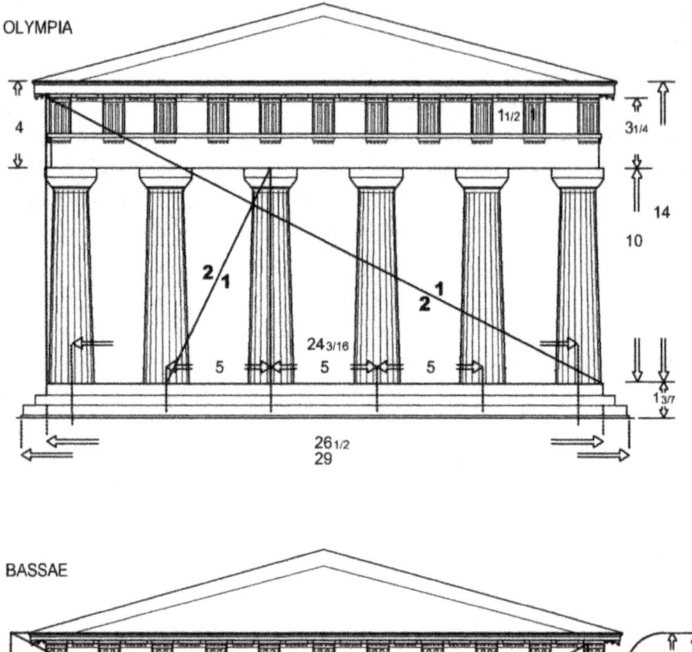

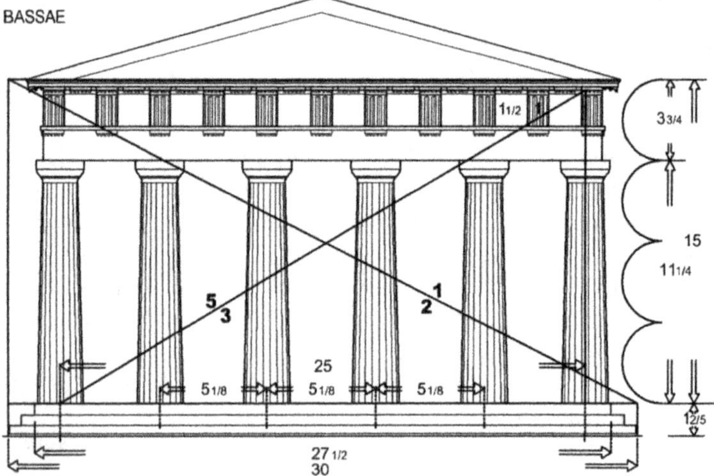

Fig. 19.6 (continued)

triglyphs and metopes became the norm, which makes column spacings equal to five triglyph widths. This was not just one consideration out of many, but rather the lynchpin of a modular design method based on the *nominal* triglyph width. It is important to note that the nominal, ideal, triglyph module is not the same as the actual triglyph width. The former tends to be a simple expression in feet and/or smaller metrical units (usually the *dactyl* or digit, 1/16th of a foot), while the latter reflected adjustments made in the course of detailed design and construction. Thus three of the sample group of ten temples (at Athens, at Sounion and the smallest of the three at Agrigento) share a common ideal module of 25 dactyls or ca. 511 mm, while in practice their average triglyph widths measure about 515, 511 and 510 mm.

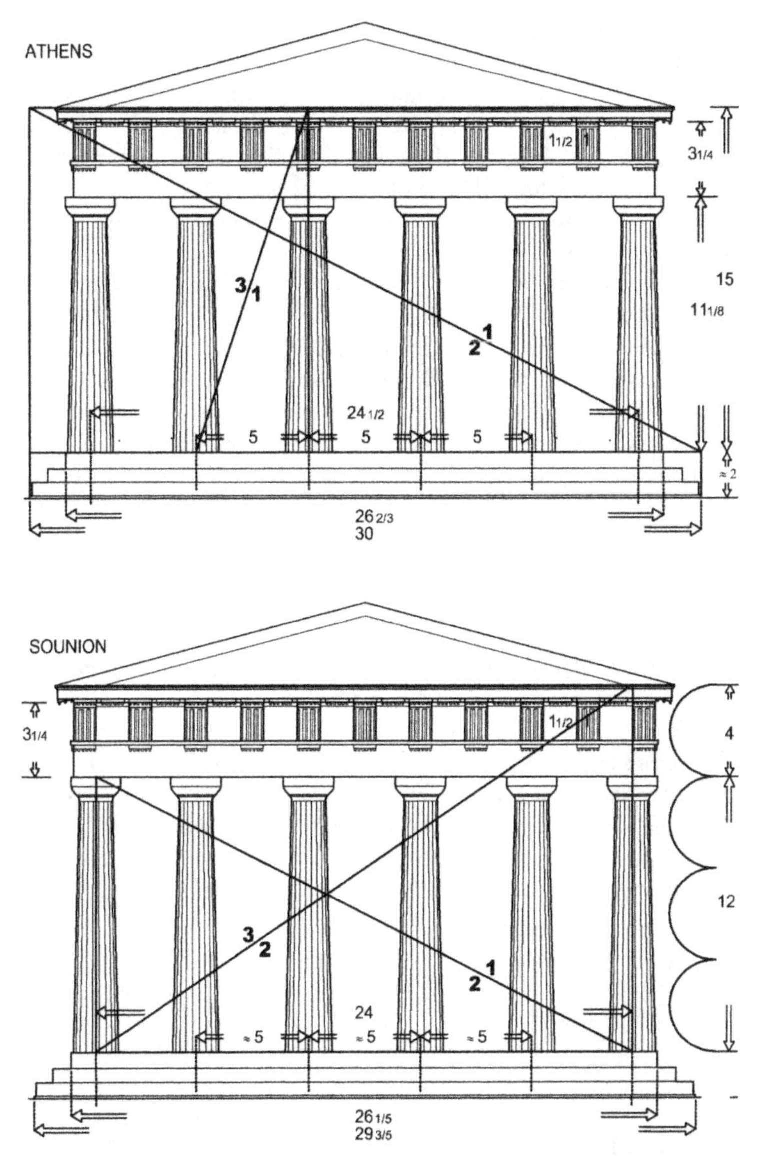

Fig. 19.6 Modular interpretations of the façades of four Doric temples of the classical period erected on the Greek mainland: the Temple of Zeus at Olympia, the Temple of Apollo at Bassae, the Hephaisteion in Athens, and the Temple of Poseidon at Sounion. Each façade is scaled to a common ideal nominal triglyph width or module of 1 unit (M). Drawing: author

Many of the most striking proportional relationships governing major horizontal and vertical limits of classical temple façades yield whole such modules, a case in point being the 2:1 relationship between the axial width of the peristyle and the column height at Sounion, or 24 M:12 M (see Fig. 19.6).

There is not space here to go fully into the details of the modular hypothesis. Instead its main features can be illustrated by the façades of four mainland temples shown in Fig. 19.6, as well as three tables relating to all ten temples studied. The first table sets out overall widths (Table 19.1), the second sets out salient limits of height (Table 19.2), while the third identifies key proportional relationships between measures of width and height (Table 19.3).

Contrary to what some see as a natural antagonism between modules and proportions, these tables demonstrate that the Greeks' handling of modular design was not a technique that operated in opposition to proportional methods. Modules and proportion went very much hand in hand, according to an approach which privileged what might be called 'modulated proportions'. Architects preferred modular values that returned simple proportions, and proportions that returned whole modules if not convenient fractions of them. It was not always possible to get both modules and proportions to work in tandem, but this was the aim. Similarly whole modules and very simple ratios would have been the ideal, but because of the complexities of design fractional values of modules (e.g., 26 ¾ or 27 ½) and ratios such as 12:7 and 20:9 were perfectly admissible. Designers evidently subscribed to different approaches; in establishing the width of temple fronts most seem to have privileged the axial width of the peristyle, but others privileged the stylobate width. Because of the values assigned to column diameters (as well as the existence of a projection or oversail between the face of the corner columns and the edge of the stylobate), it proved impossible to achieve whole modular values for both the axial width and the stylobate width.

It is important to understand that the principle of modulated proportions could apply not just to the overall design, but also to smaller scale composition. The 2:3 relationship between the widths of triglyphs and metopes set the tone for entablature design, which frequently yields a web of simple proportions. This made the bay width 5 M, while other salient proportions also coincide with simple modular expressions. Entablature schemes like those illustrated in Fig. 19.7 are proof that numerical harmony can be expressed equally well in terms of *both* proportions *and* triglyph modules; indeed, the modular approach probably rose to prominence in part because this was so.

In broad terms Doric buildings famously display an 'evolutionary' progression, with a long term trend towards lighter and more elegant proportions. In reality, as Coulton (1979) has pointed out, this was not a continuous gradual process, but rather one marked by relatively abrupt jumps or stages that might be likened to Stephen Jay Gould's concept of 'Punctuated Equilibrium'. When creating a new project architects evidently carefully appraised the composition of preceding buildings of note; in this process certain proportions might be repeated while others would be modified. We have been used to visualizing this process with reference to proportion; a previously employed ratio of 9:4 (2.250) might for example be refined to 20:9 (2.222) or 11:5 (2.200). But if anything such adjustment was easier in modular terms; starting perhaps from the 4 M height of the entablature of the Temple of Zeus at Olympia, and desiring a lighter effect, the architect of the Hephaisteion by this token opted for a reduced value of 3 7/8 M.

Table 19.1 Dimensions in plan of Doric temples expressed in terms of modules equivalent to the nominal width of triglyphs

Temple	Overall width in modules	Width of stylobate in modules	Axial width of peristyle in modules
WITH A THREE-STEP KREPIDOMA			
Olympia, Zeus	**29**?	26 1/2	24 1/8
Bassae, Apollo	**30**	27 1/2	**25**
Athens, Hephaisteion	**30**	26 3/4	24 1/2
Sounion, Poseiden	29 3/5	26 1/5	**24**
Rhamnous, Nemesis	**30**	26 1/4	**24**
Segesta, unfinished	**30**	26 1/2	**24**
WITH A FOUR-STEP KREPIDOMA			
Agrigento, Juno	**32**	27 1/2	**25**
Agrigento, "Concord"	**32**	27 1/2	**25**
Agrigento, Dioscuri	**32**?	**27**	24 2/5
Delos, Athenians' Apollo	30 1/4	26 1/2	**24**

Table 19.2 Dimensions in elevation of Doric temples expressed in terms of modules equivalent to the nominal width of triglyphs

Temple	Height of order inc. geison (M)	Height of order exc. geison (M)	Height of column (M)
MAINLAND & ISLANDS			
Olympia	**14**	13 1/4	**10**
Athens	**15**	14 3/8	11 1/8
Bassae	**15**	14 7/12	11 1/4
Sounion	**16**	15 1/4	**12**
Rhamnous	14 3/4	13 3/4	10 3/4
Delos	16 2/3	**16**	12 2/3
SICILY			
Agrigento, Juno	14 3/4	13 3/4	10 1/4
Agrigento, Concord	15 1/2	14 1/2	10 7/8
Agrigento, Dioscuri	**16**	**15**	11 2/5
Segesta	14 3/4	**14**	10 2/3

This modular hypothesis may now be subjected to the eight tests discussed earlier.

I. Degree of 'fit' The tolerances between predicted and actual values are small. Those for the whole number modular values listed in Table 19.1 are frequently less than 1 cm, and only on two occasions do they exceed 3 cm. A similar trend characterizes Table 19.2. A notional mean error lies between 1 and 1½ cm, that is to say less than 0.2 % for dimensions such as the heights of a column or an order,

Table 19.3 Dimensions in plan and elevation of Doric temples expressed along with the ratio between them

Different relationships occur as follows:			
The axial width of the peristyle could enter into a relationship with the total height of the order:			
Temple	*Axial width of peristyle in modules*	*Height of order inc. geison (M)*	*Ratio*
Bassae	**25**	**15**	5:3
Sounion	**24**	**16**	3:2
Or to the height of the order excluding the geison:			
Temple	*Axial width of peristyle in modules*	*Height of order inc. geison (M)*	*Ratio*
Delos	**24**	**16**	3:2
Segesta	**24**	**14**	12:7
Or to the height of the columns:			
Temple	*Axial width of peristyle in modules*	*Height of order inc. geison (M)*	*Ratio*
Bassae	**25**	11 1/4	20:9
Segesta	**24**	10 2/3	9:4
Sounion	**24**	**12**	**2:1**
Alternately the stylobate width could be related to the height of the order:			
Temple	*Axial width of peristyle in modules*	*Height of order inc. geison (M)*	*Ratio*
Olympia	26 1/2	13 1/4	**2:1**
Agrigento, "Juno-L."	27 1/2	13 3/4	**2:1**
Agrigento, Dioscuri	**27**	**15**	9:5

and less than 0.1 % (one in a thousand) for overall measurements of width.[11] As regards ratios between measures of width and height, these typically produce discrepancies between ½ and 2 cm (Wilson Jones 2001: 688, Table 3). As might be expected, tolerances tend to be smaller for well-preserved marble buildings, and larger for poorer preserved ones in humbler material.

II. Corroboration in the form of texts Here is essentially the same modular approach to design that Vitruvius was four centuries later to recommend for the Doric order, no doubt on the basis of Greek treatises. Meanwhile the preference for commensurable dimensions and proportions concurs broadly with the thrust of surviving specifications like that relating to the Arsenal at Piraeus (Jeppesen 1958). It is true that Vitruvius envisaged different proportions, more triglyphs per bay and concedes a half metope at the end of the frieze as opposed to the usual triglyph [*De Architectura* IV.3.3–8]. None the less, the core principle is manifestly

[11] Individual tolerances relating to the seven temples which display whole number modular values for the axial width are as follows: 6 mm (Bassae), 3 mm (Hephaisteion), 13 mm (Segesta), 10 mm (Juno), 24 mm (Concord), ca. 2 mm (Rhamnous) and 0 mm (Delos). Individual tolerances relating to the four temples which display whole number modular values for the height of the order including the geison are as follows: 5 mm (Olympia, as reconstructed), 10 mm (Hephaisteion), 32 mm (Bassae), 5 mm (Sounion), and for the three with whole number modular values for the height of the order excluding the geison: 50 mm (Delos), 1 mm (Agrigento, Dioscuri), 3 mm (Segesta). The large divergence at Delos suggests that either the supposed height of 15 modules was not in fact intended, or that a modification was introduced at a late stage of design.

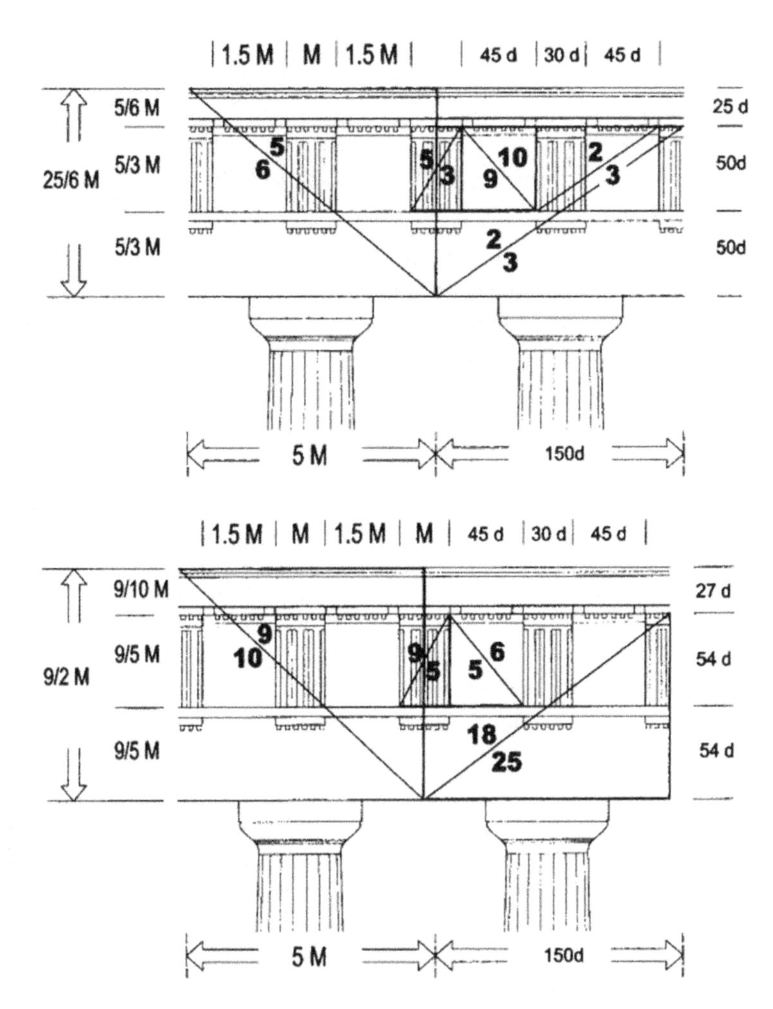

Fig. 19.7 The entablatures of the so-called Temples of Juno-Lacinia (*top*) and Concord at Agrigento (*bottom*) analyzed in terms of proportions, modules equivalent to the nominal ideal triglyph width (M), and dactyls (d). This module is in both cases equivalent to 30 dactyls or digits (1/16th parts) of the so-called Doric foot of ca. 328 mm. Drawing: author

the same: all dimensions should be convenient multiples or fractions of a module, and it is surely no chance coincidence that his module, ostensibly half a column diameter, equates to the triglyph width (Waddell 2002: 6; cfr. Falus 1979; Wesenberg 1994). That the module should derive from some physical member of the building such as a triglyph is underlined by the derivation of the word *embater*, which Vitruvius uses on occasions as an alternative for *modulus*.[12] Discounting his preferred wider central bay, his method predicts a stylobate width for a hexastyle

[12] It seems that in earlier usage *embater* denoted a physical element of a building, suggests as a step or plinth (or a triglyph?), see Coulton (1989: 86). Vitruvius may have learned from Hellenistic treatises of the modular/gridded plans of buildings such as the Temple of Athena Polias at Priene.

front equivalent to 27 triglyph modules, this being perfectly consistent with the values obtained in practice (i.e., a range of 26½–27½ M). In the past scholars have tended either to trace Vitruvius's account only as far back as the Hellenistic period, or doubt its legitimacy altogether. Yet it if the present hypothesis stands up to the other tests that follow, it would seem rather that he perpetuated a variation of fifth-century procedures.

III. Corroboration in the form of non-textual sources, e.g., drawings Evidence of this kind that may be invoked here concerns metrology. Archaeological artefacts demonstrate the existence of the foot units implicated in this interpretation: ones that are sometimes referred to as 'Doric', 'Attic' and 'Common' feet (this is modern terminology, subject to different preferences). Since mensuration is discussed below, suffice it here to list the salient pieces of evidence:

- the metrological relief from Salamis (which validates all three units) (Wilson Jones 2000b);
- the metrological relief now in Oxford (which validates the Attic foot) (Wesenberg 1976);
- a set of masons' marks on the unfinished temple at Segesta (which validates the Doric foot, being spaced at intervals of 10 such feet) (Mertens 1984a: 34–35, Taf. 33; cfr. Haselberger 1999: 53–54);
- the drawing already mentioned at Didyma (which validates digits of the Doric foot digits being 1/16th part of a foot, called *dactyls* in Greek);
- a builder's rule recently recovered off the coast of Israel (which validates the Doric foot) (Stieglitz 2006).

IV. Comparability A sample of ten relatively similar structures is sufficient for analysis to throw into relief both general patterns and special cases or exceptions. A case in point is the so-called Temple of Concord at Agrigento, the measurements of which suggest that the flank was laid out using a second module (one that approximated better to the actual triglyph width than to the nominal module used on the front) (Wilson Jones 2001: 708). A degree of special pleading is required too for the unfinished temple at Segesta on account of the presumed need for rounding off. Yet as mentioned earlier it would be unreasonable to expect conformity of every member of a given sample group; exceptions or compromises are bound to have occurred in a minority of cases.

V. Mensuration There used to be a rift between scholars of Greek architectural metrology, with one camp envisaging the use for construction of just three or four 'standard' units,[13] while the other, 'permissive', camp admitted a much greater number, subject to regional or local custom.[14] Although the debate will no doubt

[13] Dinsmoor (1961), Büsing (1982), Bankel (1983), Gruben (2001, 488, s.v. "Fuβ"); further bibliography in Wilson Jones (2000b).

[14] Jos De Waele has gone so far as to argue that, in theory at least, every Greek building could have been set out according to its own distinct foot unit, see (De Waele 1985, 1998); cf. (Ceretto Castigliano and Savio 1983; Höcker 1993).

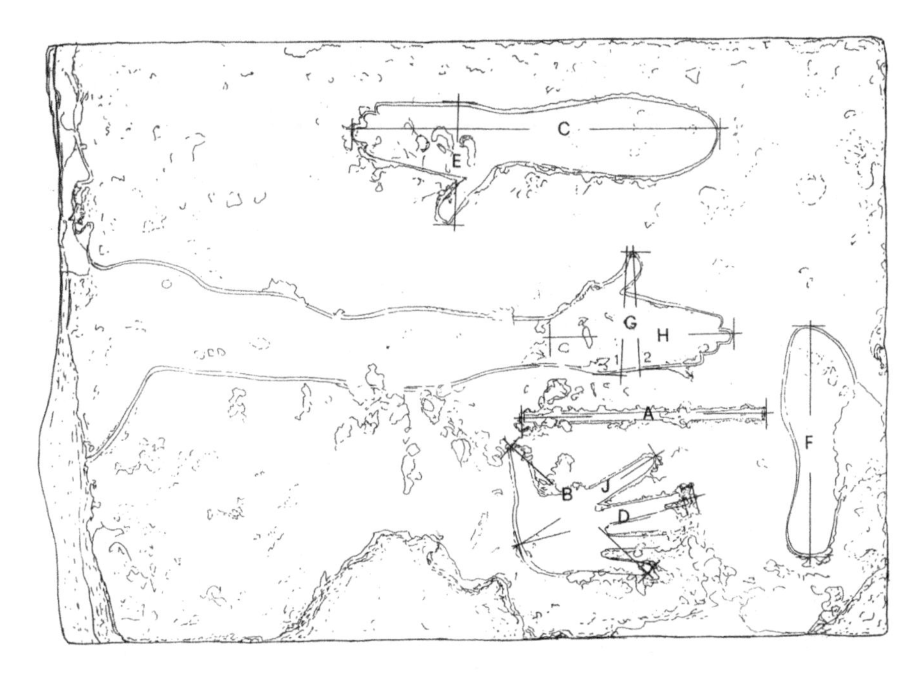

Fig. 19.8 Survey of the metrological relief from Salamis overlaid with letter codes showing the principal unit lengths. In particular the foot rule 'A' represents the so-called Doric foot of ca. 327–327.5 mm, the forearm 'C' represents a cubit (1½ ft) of the same unit, while the anthropomorphic foot print 'F' represents a foot of ca. 306–306.5 mm. Tracing by author and Manolis Korres, with annotations by author

persist, it is now substantially resolved by the above mentioned metrological relief from Salamis (Fig. 19.8). This clearly documents a 326–328 mm 'Doric' foot and a 306–308 mm 'Common' foot, while at the same time implicating the 'Attic' foot of 294–296 mm and the Samian cubit of ca. 523 (which itself equaled the Egyptian royal cubit) (Wilson Jones 2000b).[15]

The modular hypothesis goes hand in hand with this evidence in as much as most of the ten temples yield nominal triglyph modules equal to multiples of the feet just mentioned, and/or 5 *dactyl* increments (e.g., 15, 20, 25, 30 *dactyls*) (Wilson Jones 2001: 690 (including n.70) and Table 7).[16] Interestingly enough, this is strikingly reminiscent of the numerical pattern of standardized Roman monolithic shaft lengths in feet.

---

[15] See Wilson Jones (2000b). See Wesenberg (2001) for further discussion.

[16] Wilson Jones (2001, 690 (including n.70) and Table 7). Triglyph widths tend to cluster around 15d (3 × 5 d), 16d (1 ft), 18 d, 20 d (4 × 5 d), 24 d (1½ ft), 25 d (5 × 5 d), 30 d (6 × 5 d), 32 d (2 ft), and so on. There are occasions when integral foot values and multiples of 5 *dactyls* could coexist, as in the case of the temples of Juno-Lacinia and Concord at Agrigento, where the nominal triglyph module is equivalent to both 2 Common feet and 30 Doric *dactyls*.

Table 19.4

Temple	Overall width less peristyle width, in M	Metric value	Actual value	Difference in cm
Bassae	5	2.647	2.649	0.2
Athens	5 1/2	2.825	2.840	1.5
Segesta	6	5.254	5.237	1.7
Rhamnous	6	2.286	ca. 2.287	ca. 0.1
Agrigento, Juno	7	4.314	4.323	0.9
Agrigento, Concord	7	4.313	4.323	1.0

VI. 'Pay-off' in terms of design The Doric temple was arguably modular in conceptual terms, in as much as it might be interpreted as an assembly of repetitive elements (columns, triglyphs, metopes, roof tiles and so on) that were sometimes separately dedicated and financed by subscription (Fehr 1996). The triglyph and metope frieze was certainly vital to the look of the Doric temple. A building declares itself to be Doric by the presence of triglyphs, while one without them is much less so than one that lacks, say, mutules (as was quite common in Magna Graecia). If, as seems possible, they derived from tripods, ritual artefacts of great religious and social importance for the Greeks, triglyphs were also devices with a symbolic charge (Wilson Jones 2002). Furthermore, the triglyph frieze was critical to the formal resolution of temple designs, lying at the root of the infamous corner problem. Indeed modular design might be described as a strategy for overcoming this problem by retrieving mathematical harmony in spite of corner contraction.

VII. 'Pay-off' in terms of practicality The modular method helped in composing schemes that could be scaled to the constraints of the budget and the site. It enabled solutions to be transmitted easily from architect to architect, and so down the generations [see (Coulton 1983)]. Modular design facilitated the calculation of dimensions from the interrelationships between different members, and, if so desired, dimensioning in feet and digits could have been left until an advanced stage of design. As is shown by the entablatures illustrated from two of the temples at Agrigento, the choice of a triglyph module of 30 digits or *dactyls* allowed their desired modular and proportional characteristics to translate to a thoroughly simple specification in terms of whole *dactyls*. Gauging corner contraction was doubtless simplified by being couched in terms of the triglyph width for this was in any case inherent to the calculation.

VIII. Corroboration at the level of detail In theory, design processes based on arithmetical ratios could have produced similar results (provided these included a 2:3 triglyph:metope rhythm, and hence a column spacing of five triglyph widths). The setting out of the substructure was, however, an independent issue. There would seem to be no reason, deliberate intention apart, why the width of the krepidoma should so frequently match the whole number modules 30 or 32 M. The projection of the krepidoma with respect to the axial width of temples often matches whole numbers of modules, a related detail that acts as a small but significant proof of the modular hypothesis, since again there is no obvious alternative explanation for this pattern (Table 19.4).

Conclusion

In conclusion, the extent to which the modular hypothesis is able to stand up to the eight tests or criteria confers it a decent level of plausibility. We thus have the evidence, the motivation primarily a desire for universality and harmony in resolving the particular problems of Doric design, and a witness in the form of Vitruvius, our main direct authority on ancient architecture. The strength of this case invites the reappraisal of Greek design methods in the classical period, and the degree to which the modular interpretation complements, stand alongside or supersede them.

Where past analyses of temples have yielded foot units that convert to convenient fractions of triglyph modules (e.g., 5/6, 1/2, 3/8, 1/3), this to me suggests that their authors have unwittingly picked up echoes of modular procedures.[17] On the other hand it may be impossible to say whether the Athenians' temple on Delos can be better explained in terms of primarily proportional methods, as Mertens has proposed, or in terms of modules (Mertens 1984a: 220–227; 1984b; cfr. Bommelaer 1984, 1985). This is true in elevation, and in plan too (Fig. 19.9). Both methods would be capable of producing similar outcomes. In a recent study focusing on temple plans Gene Waddell has come to conclusions that in some respects parallel the present ones. The main difference lies in his belief that the triglyph module derived from the krepidoma, rather than, as I see it, the other way around (Waddell 2002). Returning to issues of detail, it is significant that while entablature proportions can vary quite considerably (e.g. triglypgh height to width; entablature height to bay width; cornice (or geison) height as a fraction of the total height), there is much greater constancy with regard to the relative widths of triglyphs and metopes. I contend that this so often matches a ratio of 2:3 precisely because it lends itself to an ensemble of modulated proportions.

Looking towards the future, perhaps the most interesting challenge will be to discern the nature of the interface a break or a merging? between the successive design procedures championed so cogently by Coulton [esp. (1985)], and the more unified approach implied by the concept of modulated proportions. Certainly, the orchestration of overall measures in plan and elevation would seem to imply that the outline design of classical temples was conceived in advance of construction, at

[17] I would argue that this applies to several of the proposals put forward by De Waele (1985, 1998), as well as those of Riemann (1951), Ceretto Castigliano and Savio (1983), Höcker (1993), De Zwarte (1996). For more detail see Wilson Jones (2001: 693–695).

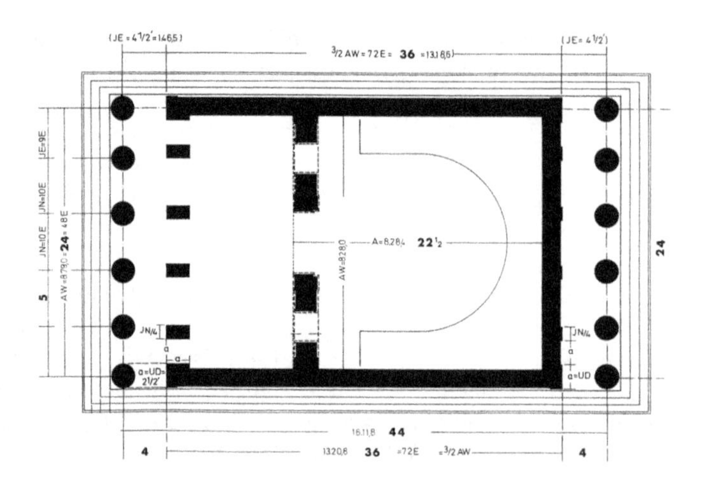

Fig. 19.9 Proportional and modular analysis of the plan of the Temple of the Athenians on Delos according to D. Mertens (1984a: Abb. 79). Mertens's module 'E' corresponds to half a triglyph module. Drawing: author

least in schematic form. Were successive techniques increasingly assimilated within the concept of modulated proportions from around the middle of the fifth century? But before advancing this discussion there is more to be done, and the modular hypothesis needs testing not just on façades but on plans as well (besides that of the temple at Delos). The debate also needs to be carried to that very influential structure, the Parthenon. Suffice it to note, as a kind of parting shot, that different scholars have detected a key role in its design for a module of 858 mm a measurement just fractionally larger than the actual triglyph width (Berger 1984a, b; Padovan 1999: 94–98; cfr. Sonntagbauer 1998). Despite the numerous features that set it apart from the Doric temples examined here, the Parthenon does yet partake of a common bond with them. Might the Parthenon also belong to the same family in terms of underlying design principles?

Biography Mark Wilson Jones is an architect and architectural historian whose research can cross over into the domain of archaeology. Having begun his training at the University of Cambridge, he went on to win the Rome Prize in Architecture at the British School at Rome. After several years in practice in London and Rome, he is Director of Postgraduate Research in the department of Architecture and Civil Engineering at the University of Bath, where he teaching history, theory and studio. Wilson Jones's interests revolve primarily around issues of design, especially as regards the ancient period. His book *Principles of Roman Architecture* (Yale University Press, 2000) was the first to be awarded both the Banister Fletcher Prize by the RIBA and the Alice Davis Hitchcock Medallion by the Society of Architectural Historians (UK). Another book with the same press, *Origins of Classical Architecture*, is to appear in 2014.

References

BANKEL, H. 1983. Zum Fußmaß attischer Bauten des 5.Jahrhunderts v.Chr. *Athenische Mitteilungen des Deutschen Archäologischen Instituts*, **98**: pp. 65–99. Darmstadt: Verlag Philipp von Zabern.

BAXTER, M. J. 2003. *Statistics in Archaeology*, London: Arnold.

BERGER, E. 1984. *Parthenon-Kongreß*. In E. Berger, ed. *Parthenon-Kongreß*. Mainz: Verlag Philipp von Zabern.

——. 1984b. Zum Maß und Proportionssystem des Parthenon Ein Nachwort zur Diskussion des Bauentwurfes. Pp. 119–174 in W. Hoepfner, ed. *Bauplanung und Bautheorie der Antike*. DAI Diskussionen zur archäologischen Bauforschung, **4**. Berlin: Deutschen Archäologischen Instituts.

BOMMELAER, J.-F. 1984. Temples doriques prostyles d'époque classique. Pp. 146–153 in W. Hoepfner, ed. *Bauplanung und Bautheorie der Antike*. DAI Diskussionen zur archäologischen Bauforschung, **4**. Berlin: Deutschen Archäologischen Instituts.

——. ed. 1985. *Le dessin d'architecture dans les sociétés antiques. Actes du Colloque de Strasbourg*. Leiden: Centre de Recherche sur le Proche Orient et la Grèce antiques.

BÜSING, H. 1982. Metrologische Beiträge. *Jahrbuch des Deutschen Archäologischen Instituts* **97**: pp. 1–45. Berlin: Deutschen Archäologischen Instituts.

CERETTO CASTIGLIANO, I. & SAVIO, C. 1983. Considerazioni sulla metrologia e sulla genesi concettuale del tempio di Giunone ad Agrigento. *Bollettino d'Arte*, series VI, **68**: pp. 35–48. Rome: Olschki.

COULTON, J. J. 1974. Towards Understanding Doric Design. The Stylobate and the Intercolumnations. *Annual of the British School at Athens*, **69**: pp. 61–86. Athens: British School.

——. 1975. Towards Understanding Greek Temple Design: General Considerations. *Annual of the British School at Athens*, **70**: pp. 59–99. Athens: British School.

——. 1979. Doric Capitals: A Proportional Analysis. *Annual of the British School at Athens*, **74**: pp. 81–153. Athens: British School.

——. 1983. Greek Architects and the Transmission of Design. Pp. 453–468 in *Architecture et socié té de l'archaïsme grec à la fin de la république romaine*. Actes du colloque international. Paris: Centre National de la recherche scientifique; Rome.: École française.

——. 1985. Incomplete preliminary planning in Greek architecture: some new evidence. Pp. 103–121 in J.-F. Bommelaer, ed. *Le dessin d'architecture dans les sociétés antiques. Actes du Colloque de Strasbourg*. Leiden: Centre de Recherche sur le Proche Orient et la Grèce antiques.

——. 1989. Modules and measurements in ancient design and modern scholarship, 'Munus non ingratum'. Pp. 85–89 in *Proceedings of the international symposium on Vitruvius' "De architectura" and the Hellenistic and Republican architecture*. Babesch (Bulletin Antieke Beschaving), Supp. **2**. Leiden: Stichting Bulletin antieke Beschaving.

DELAINE, J. 1997. *The Baths of Caracalla: A study in the design, construction, and economics of large-scale building projects in imperial Rome*. Journal of Roman Archaeology, Supp. **25**. Portsmouth, RI: JRA Editor.

DE WAELE, J. A. 1985. Le dessin d'architecture du temple grec au début de l'époque classique. Pp. 87–102 in J.-F. Bommelaer, ed. *Le dessin d'architecture dans les sociétés antiques. Actes du Colloque de Strasbourg*. Leiden: Centre de Recherche sur le Proche Orient et la Grèce antiques.

——. 1998. Der klassische Tempel in Athen: Hephaisteion und Poseidontempel. *Bulletin Antieke Beschaving*, **73**: p. 8394. Leiden: Stichting Bulletin antieke Beschaving.

DE ZWARTE, R. 1996. Der ursprüngliche Entwurf für das Hephaisteion in Athen: Eine modulare architektonische Komposition des 5. Jhs. v. Chr. *Bulletin Antieke Beschaving*, **71**: p. 95102. Leiden: Stichting Bulletin antieke Beschaving.

DINSMOOR, W. B. 1923. How the Parthenon Was Planned. *Architecture. The Professional Architectural Monthly Magazine*, **47**: part I, pp. 177–180; part II, pp. 241–244. Sydney: The Australian Institute of Architects.

——. 1961. The Basis of Greek Temple Design: Asia Minor, Greece, Italy. Pp. 255–368 in *VII Congresso Internazionale di Archeologia Classica*. Vol. I. Rome: "L'Erma" di Bretschneider.

——. 1975. *The Architecture of Ancient Greece. An Account of its Historic Development*. 3th ed.- New York: Cornell University Press.

FANT, J. C. 1993. Ideology, gift and trade: a distribution model for the Roman Imperial Marbles, The Inscribed economy. *Journal of Roman Archaeology*, Supp. **6**: pp. 145–170. Portsmouth, RI: J. H. Humphrey.

FALUS, R. 1979. Sur la théorie de module de Vitruve. *Acta Archaeologica*, **31**: pp. 248–270. Budapest: Academiae Scientiarum Hungaricae.

FEHR, B. 1996. The Greek temple in the early archaic period: meaning, use and social context. *Hephaistos,* **14** : pp. 165–191. Immenstadt: Verlag Hephaistos Publisher.

GOLVIN, J.–C. 1988. *L'amphithéâtre romain. Essai sur la théorisation de sa forme et de ses fonctions*. Paris: de Boccard.

GROS, P. 1989. Les fondements philosophiques de l'harmonie architecturale selon Vitruve. *Aesthetics Journal of the Faculty of Letters*, **14**: pp. 13–22. Tokyo: Tokyo University.

GRUBEN, G. 2001. *Der Tempel der Griechen*. 5th ed. Münich: Hirmer.

GÜNTHER, H. 1998. Die Rekonstrucktion des antiken Fussmasses in der Renaissance: Geschichte und Methode. Pp. 373–393 in D. Ahrens and R. Rottländer, eds. *Ordo et Mensura V. Internationaler interdisziplinärer Kongress für Metrologie* (1996). Münich: Scripta Mercaturae Verlag.

HASELBERGER, L. 1980. Werkzeichnungen am Jüngeren Didymeion. *Istanbuler Mitteilungen*, **30**: pp. 191–215. Istanbul: Deutschen Archäologischen Instituts.

——. 1997. Architectural likenesses: models and plans of architecture in classical antiquity. *Journal of Roman Archaeology*, **10**: pp. 77–94. Portsmouth, RI: JRA Editor.

——. ed. 1999. *Essence and Appearance. Refinements in Classical Architecture: Curvature*. Proceedings of the Second Williams Symposium on Classical Architecture. Philadelphia: University of Philadelphia.

HÖCKER, C. 1993. *Planung und Konzeption der klassischen Ringhallentempel von Agrigent. Überlegungen zur Rekonstruktion von Bauentwürfen des 5. Jhs. v. Chr*. Frankfurt: P. Lang.

JACOBSON, D. M. 1986. Hadrianic Architecture and Geometry. *American Journal of Archaeology*, **90**: pp. 69–85. Boston: Archaeological Institute of America.

JACOBSON, D. M. & WILSON JONES, M. 1999. An exercise in Hadrianic geometry: the design of the Annexe of the 'Temple of Venus' at Baiae. *Journal of Roman Archaeology*, **12**: pp. 57–71. Portsmouth, RI: JRA Editor.

JEPPESEN, K. 1958. *Paradeigmata. Three Mid-Fourth Century Main Works of Hellenic Architecture*. Aarhus: Aarhus University Press.

MACDONALD, W. L. 1993. Hadrian's Circles. Pp. 395–408 in R. T. Scott and A. R. Scott, eds. *Eius Virtutis Studiosi: Classical and Postclassical Studies in memory of Frank Edward Brown (1908–1988)*. Hanover and London: University Press of New England.

MERTENS, D. 1984. *Der Tempel von Segesta*. Mainz am Rhein: Verlag Philipp von Zabern.

——. 1984b. Zum klassischen Tempelentwurf. Pp. 137–145 in W. Hoepfner, ed. *Bauplanung und Bautheorie der Antike*, DAI Diskussionen zur archäologischen Bauforschung **4**. Berlin: Deutschen Archäologischen Instituts.

MITROVIÆ, B. 2004. *Learning from Palladio*. New York: Norton.

PADOVAN, R. 1999. *Proportion. Science, Philosophy, Architecture*. London: E & FN Spon Press.

PAKKANEN, J. 2002. Deriving ancient foot units from building dimensions: a statistical approach employing cosine quantogram analysis. Pp. 501–506 in G. Burenhult and J. Arvidsson, eds. *Proceedings of the 29th Conference Archaeological Informatics: Pushing The Envelope. CAA2001. Computer Applications and Quantitative Methods in Archaeology*. BAR International Series 1016. Oxford: Oxford Archaeopress.

RIEMANN, H. 1951. Hauptphasen in der Plangestaltung des dorischen Peripteraltempels. Pp. 295–308 in G. E. Mylonas, ed. *Studies presented to David M. Robinson*. St.Louis: Washington University.

SONNTAGBAUER, W. 1998. Zum Grundriss des Parthenon. *Jahreshefte des Österreichischen Archäologischen Institutes*, **67**: pp. 133–69. Wein: Österreichisches Archäologisches Institut.

STIEGLITZ, R. 2006. Classical Greek Measures and the Builder's Instruments from the Ma'agan Mikhael Shipwreck. *American Journal of Archaeology*, **110**.2 (April): pp. 195–204. Boston: Archaeological Institute of America.

SUMMERSON, J. 1980. *The Classical Language of Architecture*. (1st ed. 1963). London: Thames & Hudson Ltd.

WADDELL, G. 2002. The Principal Design Methods for Greek Doric Temples and Their Modification for the Parthenon. *Architectural History: Journal of the Society of Architectural Historians of Great Britain*, **45**: pp. 1–31. London: The *Society of Architectural Historians of Great Britain*.

WARD-PERKINS, J. B. 1992. *Marble in Antiquity. Collected Papers of J. B. Ward-Perkins*. In H. Dodge and B. Ward-Perkins, eds. *Archaeological Monographs of the British School at Rome*, **6**. Rome: British School.

WESENBERG, B. 1976. Zum metrologischen Relief in Oxford. *MarbWPr - Marburger Winckelmann-Programm* (1975/76): pp. 15–22. Marburg: Philipps-Universität.

——. 1994. Die Bedeutung des Modulus in der vitruvianischen Tempelarchitektur. Pp. 91–104 in *Le projet de Vitruve: objet, destinataires et réception du De Architectura. Actes du colloque international organisé par l'École française de Rome, l'Institut de recherche sur l'architecture antique du CNRS et la Scuola normale superiore de Pise*. Coll. École Française de Rome, **192**. Rome: Publications de l'École française de Rome.

——. 2001. Vitruv und Leonardo in Salamis. 'Vitruvs proportionsfigur' und die metrolosichen Reliefs. *Jahrbuch des Deutschen Archäologischen Instituts*, **115**: pp. 357–380. Berlin: Deutsches Archäologisches Institut.

WILSON JONES, M. 1990. The Tempietto and the roots of coincidence. *Architectural History. Journal of the Society of Architectural Historians of Great Britain*, **33**: pp. 1–28. London: The *Society of Architectural Historians of Great Britain*.

——. 1991. Designing the Roman Corinthian capital. *Papers of the British School at Rome*, **59**: pp. 89–150. Rome: British School.

——. 1993. Designing amphitheatres. *Römische Mitteilungen*, **100**: pp. 391–441. Berlin: Deutsches Archäologisches Institut.

——. 2000. *Principles of Roman Architecture*. New Haven: Yale University Press.

——. 2000b. Doric measure and Doric design, 1: The evidence of the relief from Salamis. *American Journal of Archaeology*, **104**: pp. 73–93. Boston: Archaeological Institute of America.

——. 2001. Doric measure and Doric design, 2: A Modular Re-reading of the Classical Temple. *American Journal of Archaeology*, **105**: pp. 675–713. Boston: Archaeological Institute of America.

——. 2002. Triglyphs, Tripods, and the Origin of the Doric order. *American Journal of Archaeology*, **106**: pp. 353–390. Boston: Archaeological Institute of America.

Chapter 20
Calculation of Arches and Domes in Fifteenth-Century Samarkand

Yvonne Dold-Samplonius

Introduction

Samarkand, with Bukhara the principal town of Transoxania, is first found in the accounts of Alexander the Great's campaigns in the east as *Maracanda*. Arab legend makes Alexander founder of the city. In ca. 900 AD the Samanid kingdom was founded, the beginning of a century of great prosperity for Transoxania, such as would only be seen 500 years later with Timur and his immediate successors. Although the capital was moved to Bukhara, Samarkand remained the premier centre of commerce and culture, especially in the popular estimation of the Muslim world. Among its native products, the paper of Samarkand, the manufacture of which had been introduced from China, was especially famous. After surrendering to Genghis Khan in 1220, the city was plundered and many of its inhabitants were deported. For the next 150 years it was but a shadow of its former self. The revival of the town's prosperity began when Timur Lang (1336–1405) became supreme in Transoxania after about 1369 and chose Samarkand as the capital of his ever-growing kingdom. It was in his reign that the art called "Timurid" had its origins. Timur enriched Samarkand with magnificent buildings and made it an international market surpassing Tabriz and Baghdad, at least during his lifetime; he transplanted thither the artists and craftsmen from the towns he conquered. The intellectual revival which characterized the fifteenth century is in part the work of the Timurid sovereigns and princes, many of whom were themselves poets, artists and scholars, and attracted to their courts men of genius. Among the former are Timur's son, Shah Rukh, who

First published as: Yvonne Dold-Samplonius, "Calculation of Arches and Domes in 15th Century Samarkand", pp. 45–55 in *Nexus III: Architecture and Mathematics,* ed. Kim Williams, Ospedaletto (Pisa): Pacini Editore, 2000.

Y. Dold-Samplonius (✉)
Ruprecht-Karls-Universität Heidelberg, Institute of Mathematics, Heidelberg
e-mail: dold@math.uni-heidelberg.de

K. Williams and M.J. Ostwald (eds.), *Architecture and Mathematics from Antiquity to the Future*, DOI 10.1007/978-3-319-00137-1_20,

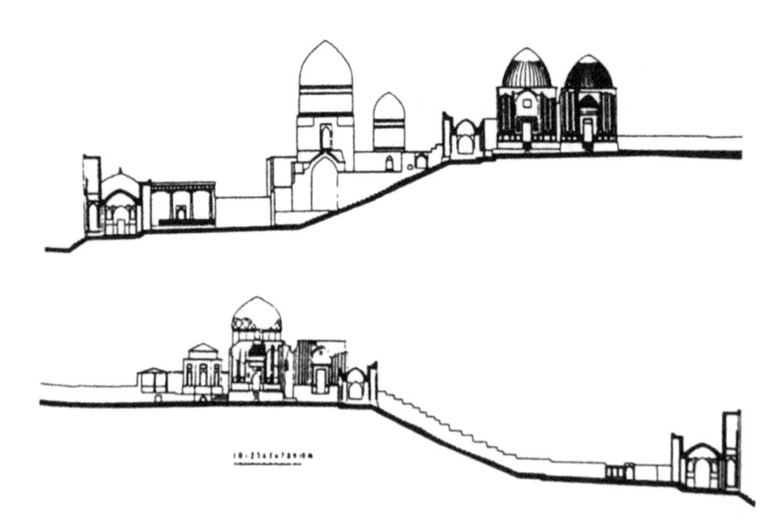

Fig. 20.1 Façades of the mausoleums (qubba) at both sides of the Shah-i Zindeh. Image: author, after Pougatchenkova (1981)

promoted historical studies, and his son, Ulugh Beg (1393–1449), astronomer, poet and theologian, who really made Samarkand what Timur had dreamt of: the centre of Muslim civilization. An artist, Ulugh Beg enriched Samarkand with superb buildings, such as the Timurid burial place, *Shah-iZindeh* (Fig. 20.1), a *madrasa* (high school) and others. A learned mathematician, he could solve the most difficult problems in geometry, but he was above all an astronomer. When Ulugh Beg decided to construct an observatory, he invited Ghiyāth al-Dīn Jamshīd Mas'ūd al-Kāshī to his court—sometime after 1416—as founding director, together with Mo'īn al-Dīn al-Kāshī. Director of the madrasa was Ulugh Beg's mathematics teacher, Qādi Zāde al-Rumi from Bursa (Turkey). The observatory, destroyed in the following century, was regarded in his day as one of the wonders of the world.

Ghiyāih al-Dīn Jamshīd Mas'ūd al-Kāshī ranks among the greatest mathematicians and astronomers in the Islamic world. He was a master computer of extraordinary ability, his wide application of iterative algorithms and his touch in laying out a computation so sure that he controlled the maximum error and maintained a running check at all stages; in short, his talent for optimizing a problem show him to be the first modern mathematician. Al-Kāshī died in June 1429 outside the Samarkand observatory, probably murdered on the command of Ulugh Beg. Two years earlier he had finished the *Key of Arithmetic*, one of his major works. The work is intended for everyday use; al-Kāshī remarks, *I redacted this book and collected in it all that is needed for him who calculates carefully, avoiding tedious length and annoying brevity*. By far the most extensive book is Book IV, *On Measurements*. Its last chapter, *Measuring Structures and Buildings*, is really written for practical purposes:

> The specialists merely spoke about this measuring for the arch and the vault and besides that it was not thought necessary. But I present it among the necessities together with the rest, because it is more often required in measuring buildings than in the rest.

Al-Kāshī uses geometry as a tool for his calculations, not for constructions. Besides arches, vaults and domes (qubba), al-Kāshī calculates here the surface area of a muqarnas (stalactite vault), that is, he establishes approximate values for such a surface. He is able to do so because, although a *muqarnas* is a complex architectural structure, it is based on relatively simple geometrical elements. For the calculation only elementary geometrical rules are used.

Calculation of Arches and Vaults

In this section the terms "arch" and "vault" are interchangeable. The difference between an arch and a vault is that the depth of an arch is not larger than its span, whereas in the case of the vault the depth exceeds the span. The depth of the arch is the distance between the front and back surfaces; that which is called depth in the arch is called length in the vault. Al-Kāshī remarks that,

> The predecessors determined those (i.e., arch and vault) as half a circular hollow cylinder, but we did not see something like it, neither in old nor in new buildings. We have mostly seen ones that are pointed in the middle, and in few cases they are smaller than half a hollow cylinder.

From the Byzantine Empire the Umayyads inherited a system of round arcading that, in the rarest of instances, showed a tendency towards becoming slightly pointed.[1] The innovation of the pointed, or ogival, arch came from the East. Under Umayyad rule the round arch persisted, but developed into the two-centred form showing an increasing tendency towards pointedness. A round arch is struck from a single centre. A pointed arch has more than one centre and can be thought of in its simplest form as being struck from two centres with overlapping arcs; these produce an increasingly pointed arch the further they are moved apart horizontally (Fig. 20.2). In the succeeding two centuries this trend was still apparent, but was complicated by the three- and four-centred arch. Based on this development it is to be expected that in early arithmetic books only hemispherical arches are treated.

First al-Kāshī explains extensively the different elements of an arch and how these are connected, or which part could disappear in a wall. He then gives five methods for drawing the façade of an arch.[2] The first two are three-centred arches. Figure 20.3 shows type 2, a three-centred arch, with point E as a double centre and the other two centres situated in the two lower points Z and H. When the two lower centres move, the arch will change its acuteness. Type 3 (not shown) deals with a

[1] This is essentially Creswell's theory, see Creswell (1960), Warren (1991: 59).

[2] All five constructions are performed on the video "Qubba for al-Kāshī", directed by Yvonne Dold-Samplonius (1995).

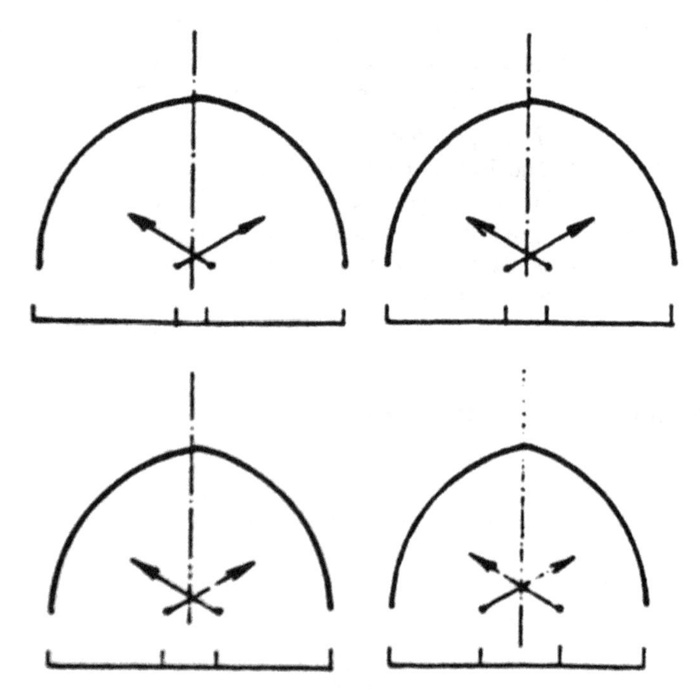

Fig. 20.2 Diagram showing pointed arches formed with constant radii on centres with successive separation of one-tenth, one-seventh, one-fifth, and one-third of a span. Image: author, after Warren (1991)

four-centred arch, which is similar to the three-centred arch except that the centre of the semicircle is split into two points displaced towards the extremes of the span. The greater the displacement, the shallower the profile. Type four (Fig. 20.2, lower right) and five are two-centred. As the second façade was the most common in al-Kāshī's time, he uses it to illustrate his calculation method. This façade is handy, according to al-Kāshī, when you need a span of 5–10, or up to 15, cubits.

Construction of the second façade (Fig. 20.3, taken from the oldest extant manuscript, Tehran, Malek Library 3180/1, with Roman letters added):

(1) Describe a semicircle on AD, the span of the arch;
(2) Extend AD in both ends by the thickness of the arch to the points 1 and M. E is the centre of the semicircle;
(3) Divide this semicircle in four equal parts through the points A, B, C, G, D;
(4) Extend BE and GE by EZ and EH, equal to AC, and by BK and GL, equal to DM, the thickness of the arch;
(5) Describe from the centre E the arcs ML and Kl, from the centre H the arc GT, and from centre Z arc BT;
(6) Connect HT and ZT and extend them by the thickness of the arch to the points O and S;

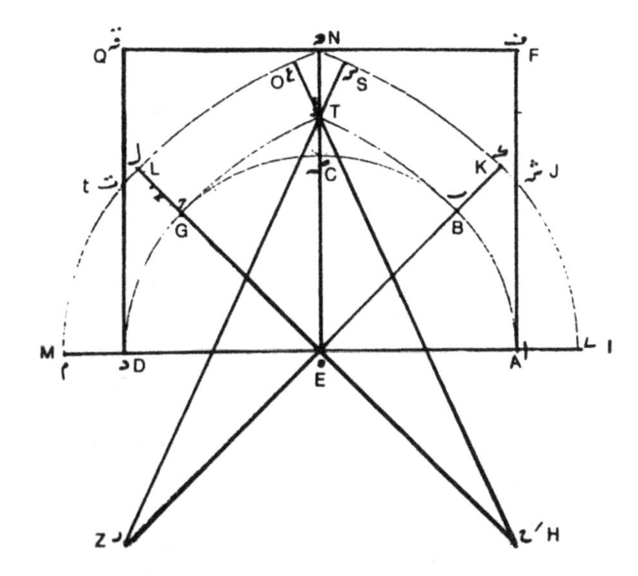

Fig. 20.3 Construction of the second façade. Image: author

(7) Describe arc LO from point H and arc KS from point Z;
(8) Erect the perpendiculars SN and ON on the lines TS and TO.

The sections AK, KT, TN, TL, and LD form together the façade of the arch. When we construct area AFQD with parallel sides and right angles, we obtain the spandrels of the arch.

After al-Kāshī has explained and carried out all five methods for constructing the façade of an arch and has completed the characterization of arch and vault, he continues with surveying them. He explains that he has already found out the relation between some measures of an arch and its span and between some of these and its thickness. He has laid these factors down in a table together with an explanation of the method. These quantities are also transformed into Indian numerals, which he has put down in the table as well. He also informs us about the particulars of finding these quantities. With this table the following parts of the arch (Fig. 20.3) can be found: the interior curve ABTGD, the inner height ET, the upper width TN, and the surface area of the arch as well as the surface area of the concavity, area ABTGDE.

With these values we can then calculate many different parts of the arch. To calculate the *volume* of the arch we proceed in the same way as for round arches: after the surface area of the arch has been found, by means of the table, we multiply this number with the depth of the arch and obtain its volume. Sometimes the arch disappears partly inside a wall and we want to know how much is visible and how large the *segments inside the wall* are, section tDM and section JAI: these segments are calculated by taking the difference of the circle segment MtE and the triangle tDE:

$$\frac{ED}{EM} = \frac{ED}{Et} = \cos \angle tEM, EM = MD + ED,$$
$$tM = \arccos \angle tEM \Rightarrow arctM, arctM \times ME = 2MtE,$$
$$\sin \angle tEM = \frac{tD}{tE} \Rightarrow \sin \angle tEM \times tExDE = 2\Delta tDE,$$
$$2MtE - 2\Delta tDE = 2tDM.$$

When we subtract this amount from the total surface area of the arch we obtain the surface area of the visible part of the arch.

It might be necessary to calculate the *spandrels*, section NQt and section NFJ: in this case we calculate the area AFQD and subtract from this amount the area of the visible part of the arch, calculated above, and the area of the opening of the arch, area ABTGDE, found by means of the table. This yields the surface area of the spandrels. When we multiply this amount with the depth of the spandrels, we obtain the volume of the two spandrels.

Al-Kāshī's book is for practical use, as explained above. Hence he rightly shows how to make life easier by working with rounded off values. More approximation is involved, as the types of arches are more varied than those five given by al-Kāshī. The method for calculating an arch is to select the type of arch nearest to it. Golombek and Wilber (1988: 153–157) have considered existing examples of Timurid arches in the order outlined by al-Kāshī. Examples have been recorded for all but the fifth method, which was, however, certainly common in small windows. In comparing the models described by al-Kāshī with actual examples of Timurid arches, we have to bear in mind that al-Kāshī's purpose was calculating volumes and surfaces, not constructing. This means that an elegant approximation, which leads us to an easy calculation, is the ultimate goal.

Bulatow (1978) has analysed arches from the twelfth to fifteenth centuries in Central Asia and suggests that some pointed arches were constructed as intersections of ellipses. Questioning the reason for using the ellipse, he notes that for spans exceeding 10 m these were easier to construct than four-centred arches. The architects were familiar with the stability of the ellipse as well, for its construction was known from Sassanian examples. According to his analysis, this kind of arch is found in some of the most important Timurid buildings of the period, as in the Gur-i Amir (Fig. 20.4) in Samarkand, and in the mausoleum of Timur Lang and Ulugh Beg, for instance, in the dome, interior niches, arches of the zone of transition, and entrance portal. The same arches have elsewhere been identified as three- and four-centred arches and can be considered as such for all practical purposes (see below).

The section on calculating arches ends in al-Kāshī's *Key of Arithmetic* with the following remark: *I talked a lot about the subject of this section, as this section is very important, and my predecessors did not treat it as they should.*

Fig. 20.4 Gur-I Amir, Mausoleum of Timur Lang and Ulugh Beg with a double-shelled dome. Image: author, after Pougatchenkova (1981)

Calculation of Domes

The Arabic word for dome or cupola is *qubba*, plural *qibāb* or *qubab*. By extension *qubba* also means a cupolaed structure or dome-shaped edifice, a domed shrine, a memorial shrine, or *kubba* (especially of a saint). In pre-Islamic times the *qubba* was a small domed leather tent carried by a camel, in which certain tribes kept sacred stones. Also, the dome located in front of the *mihrāb*, a recess in a mosque wall indicating the direction of prayer—as exemplified for instance in the Great Mosques of Damascus, Qayrawan, Cairo, and Cordoba—might have had a special meaning. From the late ninth and the tenth centuries A.D. the building of commemorative structures over certain burial places, especially those of Shī'ī saints, occurred. Throughout the entire Muslim world, all the special names for sepulchral buildings, which vary with country and language as well as with the person interred, come under the generic name of *qubba* (Fig. 20.1). There are basically two types of monuments: the circular, tower-like form, and the often more grandiose square or polygonal type. Both can be covered either by a circular dome or a conical or pyramidal roof. Its original, and later stereotyped, form is a square building covered by a dome. The oldest preserved example is the *Qubba of the Sāmānids* in Bukhara, constructed around 907 but certainly before 943. It consists of a square structure with a large central dome and four small corner ones set over a gallery. As early as the Seljuq period (eleventh century) the construction of domes with double shells was tried, which led to their successful development in Timurid times. The aim of a drum and a double-shelled dome is to give a towering effect to the exterior. A striking example is the *Gur-i Amir* in Samarkand (Fig. 20.4).

As long as domes consisted of cones or sphere-segments, their mensuration was automatically included when measuring solids, and the *qubba* did not have to be mentioned *per se*. At present it seems that, with the exception of al-Kāshī, only *qubbas* in the form of hollow hemispheres have been considered in arithmetic manuals. A hemispherical *qubba* is assumed to consist of the solid shell between two concentric, parallel hemispheres. In praxis, the inner and outer surfaces of the shell are never really parallel, because in the lower part, up to an angle of 61°, the pressure exerts a pulling force in the upper part.

When the inner and outer diameters of a hemisphere *qubba* are known, its volume and the inner and outer surface areas can be calculated as follows (Dold-Samplonius 1998). We know how to compute the surface area of a sphere with diameter equal to the outer diameter. Half of this amount is the outer surface area of the qubba. In the same way, the inner surface area of the *qubba* can be computed. To calculate the volume of the *qubba*, we compute the volumes of the outer and the inner sphere and take each time its half. The difference between these two amounts is the volume of the *qubba*.

The formulas for computing the area and the volume of a sphere are:

$$\mathrm{Area(sphere)} = (2\mathrm{r})^2 \times \pi$$

and

$$\mathrm{Volume(sphere)} = \mathrm{Area\ (sphere)} \times \mathrm{r}/3, 2\mathrm{r} = \mathrm{diameter}$$

Al-Kāshī does not carry out the calculation of the hemisphere *qubba*, but refers to his calculation of the sphere. There he uses, as expected, the right formulas for area and volume expressing л as the ratio between the circumference and the diameter of a circle. He distinguishes the following categories of *qubba*:

> They occur either in the form of a hollow hemisphere, or in the form of a segment of a hollow sphere, or in the form of a polygonal cone, or in the form, which arises by imagining the rotation of the façade of the arch, i.e. of an arch as mentioned in Section 1, around the line of its elevation, that is the line, which connects its upper limit with the middle of the line between its fundaments.

After remarking that the first three categories have already been dealt with earlier in the book, he indicates how to calculate the complicated type of *qubba*, i.e., the dome created by rotating an arch around its vertical axis. The method is illustrated in Fig. 20.5. The dome is divided in parallel slices by drawing circles from the axis on its surface. These circles have to be so close that the curves between two of them equal the corresponding chords. Seven or eight of these circles should normally suffice, according to al-Kāshī. In this way the dome is cut up in a cone and several frusta. We first measure all the circles on the surface of the dome. The next step is to measure the distance from the apex of the dome to the nearest

Fig. 20.5 Qubba, sliced in eight slices. Image: author

circle, i.e. the chord (Fig. 20.5: the segment c) equalling the curve on the circle. By multiplying half the circumference of the nearest circle by this amount we obtain the surface area of the cone. Thereupon we multiply half the sum of every two neighbouring circles by their distance to obtain the surface area of all frusta. The sum of these products yields the surface area of the qubba.

To obtain the volume of the *qubba*, which is a hollow solid, we first measure the volumes of the cone and the frusta, which fill the outer surface of the shell, and add these. From this sum we then subtract the sum of the volumes of the cone and the frusta filling the inside of the shell. The difference between these two sums is the volume of the *qubba*, as we have seen before in the case of the hemispherical *qubba*.

This general method is applied to a *qubba* based on the fourth type of arch, i.e. a two centred arch with its span divided by the two centres in three equal parts (Fig. 20.2, lower right). For practical application just the rules are enough. Hence, "to simplify the procedure", al-Kāshī gives only the calculation method but does not explain how he arrived at these results:

> To obtain the surface area of the interior of the dome we have to multiply the square of the diameter of the base of the hollow (= inner) dome by $1°46'32''$, if we compute sexagesimally, or by 1.775, computing in the decimal system. When we multiply the square of the diameter of the base of the (outer shell of the) dome by the same number, we obtain the exterior curved surface area of the dome, as the inner and outer surfaces are supposed to be parallel to each other. When we multiply the cube of the diameter of the base of the hollow dome as well as the cube of the diameter of the base of the dome by $0°18'23''$, in the sexagesimal system, or by 0.306, in the decimal system, and take the difference of these two products, we obtain the volume of the hollow qubba.

In both cases the results were checked by modern methods and we found that the factors are accurate and that the dome has been cut up in eight slices.

Could this factor be used for all kind of domes, with more or less deviation? Al-Kāshī makes no mention of the elliptical profile for either arches or domes. There are a number of domes for which the profile may be interpreted as the intersection of reflected elliptical curves. These include some of the most important buildings of the period, as the Gur-i Amir in Samarkand. Bulatow has demonstrated that the dome of the Gur-i Amir was probably designed using a pair of foci and string. However, looking at his analysis of the Gur-i Amir (Fig. 20.6) we

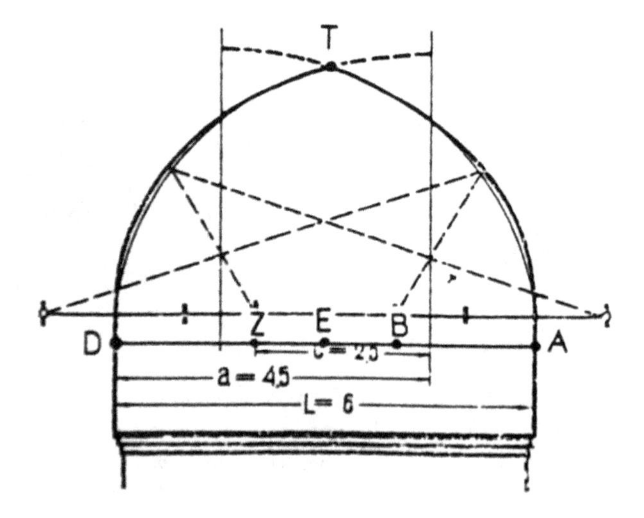

Fig. 20.6 Dome of the Gur-i Amir. Image: author, after the analysis of Bulatow (1978) with additions

see that this dome could also have been originated by the fourth method: with line AD as the span and the points B and Z dividing the span in three equal parts, we obtain the circle segments just inside the curve drawn by Bulatow. The difference between the two curves lies within the margin of error accepted by modern architects. It seems therefore that al- Kāshī's factors can also be used for calculating elliptical domes.

Conclusion

In medieval Italy it was common practice to pay artisans according to the surface area they had completed. Also in seventeenth-century Safavid Iran architects were paid a percentage on each building based on the cubit measure of the height and thickness of the walls:

> The Persians determine the price for masons on the basis of the height and thickness of walls, which they measure by the cubit, like cloth. The king imposes no tax on the sale of buildings, but the Master Architect, that is Chief of Masons, takes two percent of inheritance allotments and sales. This officer also has a right to five percent on all edifices commissioned by the king. These are appraised when they are completed and the Master Architect, who has directed the construction receives as his right and salary as much as five percent of the construction cost of each edifice (Necipoğlu 1995: 44, 159).

The same custom seems to have existed in the Arab world. It is also useful to know, more or less, how much material is needed like gold for gilding, bricks for construction or paint and such things. Payment per cubit was common in Ottoman architectural practice where a team of architects and surveyors had to make cost estimates of projected buildings and supply preliminary drawings for various options. In addition to facilitating estimates of wages and building materials before construction, al-Kāshī's formulas may also have been used in appraising the price of a building after its completion. His sophisticated formulas were, like the simple formulas found in the Arithmetic Books, useful for everyday life. This was al- Kāshī's objective for writing his *Key of Arithmetic*.

Biography Yvonne Dold-Samplonius studied mathematics and Arabic at the University of Amsterdam, specializing in the history of Islamic mathematics. She wrote her thesis with Prof. Bruins, Amsterdam and Prof. Juan Vernet, Barcelona. The academic year 1966/1967 she spent at Harvard studying under Prof. Murdoch. She has published about 40 papers on the history of mathematics. In recent years her interest has shifted to mathematics in Islamic architecture from a historic point of view. Under her supervision two videos concerning this subject, "Qubba for al-Kashi", on arches and vaults, and "Magic of Muqarnas" have been produced at the IWR (Interdisciplinary Center for Scientific Computing), Heidelberg, where she is an associated member. She is an effective member of the International Academy of History of Science.

References

BULATOW, M.S. 1978. *Geometric Harmony in the Architecture of Central Asia, 9th–15th century*. Moscow (in Russian).

CRESWELL, K.A.C. 1960. Architecture. Pp. 608–624. In P. J. Bearman ed. *Encyclopaedia of Islam*. 2nd edn vol. 1. Leiden: E. J. Brill.

DOLD-SAMPLONIUS, Yvonne. 1995. *Qubba for al-Kāshī*. Video directed by Yvonne Dold-Samplonius, distributed by American Mathematical Society.

——. 1998. Calculating Surface Areas and Volumes in Islamic Architecture. In Jan P. Hogendijk and A.I. Sabra eds. *Perspectives on Science in Medieval Islam*. Conference at Dibner Institute (1998). Cambridge, Massachusetts: MIT Press (in press).

GOLOMBEK, Lisa and Donald WILBER. 1988. *The Timurid Architecture of Iran and Turan*. 2 vols. Princeton, New Jersey: Princeton University Press.

NECIPOĞLU, Gülru. 1995. *The Topkapi Scroll. Geometry and Ornament in Islamic Architecture*. Santa Monica, California: Getty Centre for the History of Art and the Humanities.

POUGATCHENKOVA, Galina A. 1981. *Chefs-d'Oeuvre d'Architecture de l'Asie Centrale XIVe–XVe Siècle*. Paris: Presses de l'Unesco (in French).

WARREN, John. 1991. Creswell's Use of the Theory of Dating by the Acuteness of the Pointed Arches in Early Muslim Architecture. *Muqarnas* **8** (1991): 59–65.

Chapter 21
Curves of Clay: Mexican Brick Vaults and Domes

Alfonso Ramírez Ponce and Rafael Ramírez Melendez

Introduction

Since the beginning of time, man has had to confront the world around him in order to survive. To this end, he has had to create a vital second skin, thereby transcending his biological skin. This second skin has come to be termed Architecture.

The building of this second skin was begun with Man's dreams and out of the raw materials which nature provided for him. Buildings made of materials such as stone, wood, cane, clay, and brick are found in different regions of the world throughout Man's history. The specific techniques employed in these buildings form an integral part of our cultures and our building traditions.

The comprehensive knowledge of materials and their corresponding building techniques has become a vital necessity, given the backdrop of an ever increasing demand for living space, particularly housing. Moreover, it has become vitally important to rationalise the building process in order to achieve the lowest possible cost.

This chapter aims to describe, analyse, and formalise some of the fundamental properties of a popular construction technique for building brick vaults without any use of framework or any additional reinforcements whatsoever. The technique is of collective invention in Mexico dating back to the nineteenth century (Fig. 21.1a, b). Brick vaults, made only of pieces of clay and the intuition and skillful hands of

First published as: Alfonso Ramírez Ponce and Rafael Ramírez Melendez, "Curves of Clay: Mexican Brick Vaults and Domes", pp. 143–154 in *Nexus V: Architecture and Mathematics*, Kim Williams and Francisco Delgado Cepeda, eds. Fucecchio (Florence): Kim Williams Books, 2004.

A. Ramírez Ponce (✉)
E 21, M XII Educación, Coyoacán 04400, México, D.F.
e-mail: arponce50@hotmail.com

R. Ramírez Melendez
Pompeu Fabra University, Roc Boronat 38, 08018 Barcelona, Spain
e-mail: rafael.ramirez@upf.edu

K. Williams and M.J. Ostwald (eds.), *Architecture and Mathematics from Antiquity to the Future*, DOI 10.1007/978-3-319-00137-1_21,

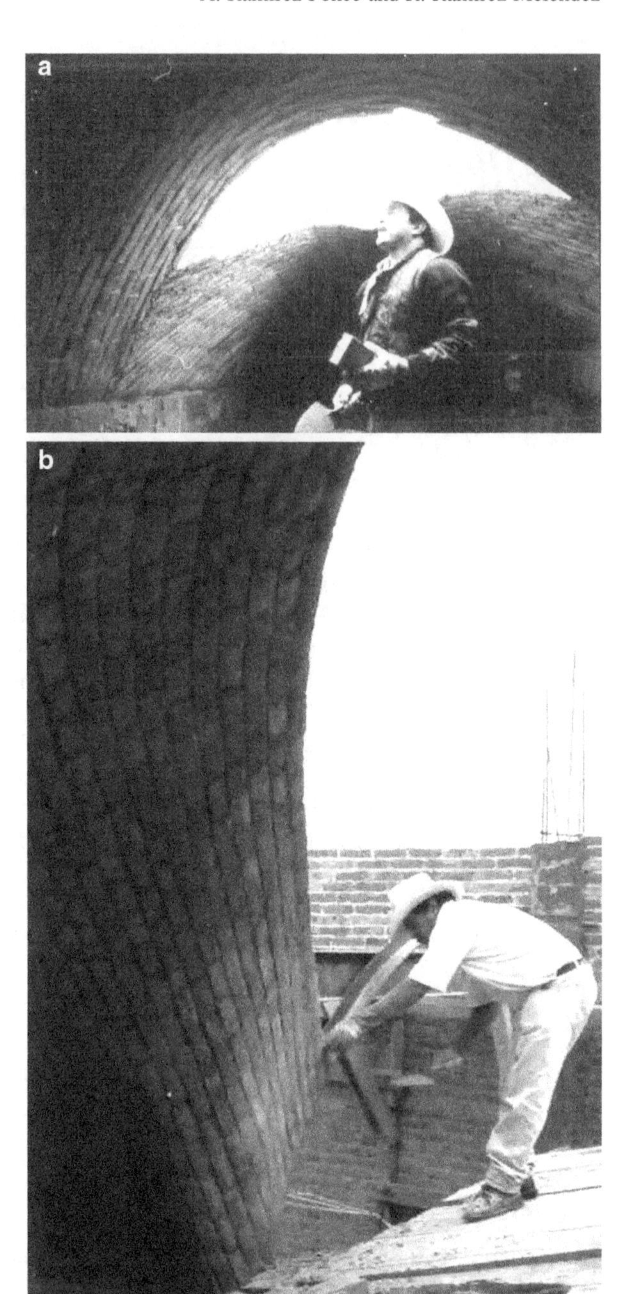

Fig. 21.1 (**a, b**) Constructing the Mexican vaults. Photo: authors

craftsmen, apart from offering an economical solution even to this day to the housing problem, possess both an architectonic and mathematical beauty.

The chapter is organized as follows: background; a description and examples of the technique we have termed "the leaning brick;" a mathematical formalization of the surface generated by this type of construction technique; and some conclusions and indications of areas of future research.

Background

Throughout history, there have been many different techniques for building covers with brick. According to their structural characteristics, the covers may be divided in two main groups: covers in which the brick works only as a skin, and covers in which the brick is both skin and part of the supporting structure. This second group may be further divided into "layered" covers and "leaning" covers. Our technique is inside this last subgroup.

Historically, the most ancient are the Nubic vaults of adobe, in southern Egypt which were built at least 3,300 years ago. One example of this which can still be seen today is in the Rameses funereal centre, in the Valley of the Kings, on the banks of the Nile opposite the city of Luxor. Later, around the third century, we have the vaults built in Persia. Later still, in the tenth century, are the incorrectly named "false Mayan arches", built with limestone in Yucatan (Fig. 21.2a, b).

The misnomer is due to the fact that these structures are not really arches; the stones in the Mayan structures do not transmit their loads from the top to the base, but are rather simply superimposed on the stones underneath with a small salient part (more or less a fifth) coming out. This results in a structure forming a steep "A" shape. Thus, the last stone placed on top of the structure which joins the two inclined planes is not a keystone but a simple lid. Generally, this kind of structure had a limited depth and was used to cover thresholds, doors or transition spaces. However, a structure with several metres of depth is structurally a vault and may cover an inhabitable space. Thus, the Mayan structures are really vaults initially formed by inclined flat surfaces as in the *cuadrángulo de las Monjas* (Quadrangle of the Nuns) in Uxmal, and later forming curved surfaces as in Labná. Finally, we have the Mexican vaults born in the region named *el Bajío* in central Mexico. These vaults and the Nubic vaults mentioned earlier are based on the same basic principle, bricks slightly tilted and leaning on one another. They are however quite different both in terms of the type of brick used—adobe bricks in Nubia and small bits of baked brick in Mexico—and also in the way the bricks lean on one another (Figs. 21.3 and 21.4).

In Nubic vaults the bricks lean against a wall which is higher than the lateral supporting walls. In the Mexican vaults the supports are the larger sides (in the case of rectangular vaults) or the corners—literally just the points—(for square-shaped vaults); this shall be shown in the corresponding diagrams. In Mexico, this technique dates from the second half of the nineteenth century and there are two

Fig. 21.2 (**a**, **b**) Examples of "false" Mayan arches. Photo: authors

Fig. 21.3 Ramesseum vaults. Photo: authors

Fig. 21.4 Nubic vault. Photo: authors

possible sites of origin: San Juan del Rio in the state of Queretaro and Lagos de Moreno in Jalisco. These structures have been constructed since the nineteenth century and are the object of our research.

The "Leaning Brick" Technique

The "leaning brick" is a popular construction technique which is at the same time millenary and modern. This technique is used for building roofs and covers with bricks without any framework or any kind of external support, making it a very economic way of covering space. Moreover it can be used between floors in a housing block or to cover an open area such as a terrace. It is a technique which can be learnt by professional builders as well as self-taught builders. It is an ingenious technique, as will be illustrated later, not invented by architects or engineers, but instead the fruit of common knowledge, all too often ignored or disregarded by professionals and academics. Hence the reason that it is not taught in colleges and is therefore in decline. The technique allows spaces of up to 10 m wide to be covered and is ideal for the vast majority of architectural spaces, especially living spaces in individual or collective housing, and spaces destined for educational or public service use.

The technique's low cost is based on three underlying conditions. The first, as mentioned earlier, is that no scaffolding or additional supports are required whilst the cover is being built. Secondly, low cost materials are used, such as the common handmade clay brick or, alternatively, wet clay brick commonly known as adobe, or an earth-cement brick in proportions of one to ten. Lastly, the labour-to-time ratio is highly efficient, on average just 2 h manpower are needed to complete one square metre of cover. Given our "build to finish" concept, that means that the vault's lower section would be completed. The technique does not even need additional iron or concrete reinforcements, just clay bricks and building expertise.

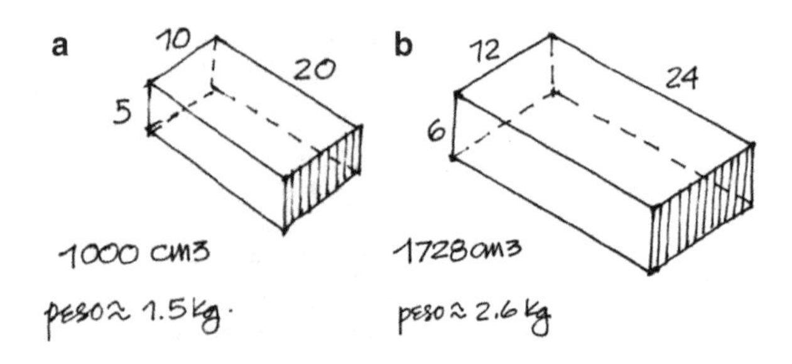

Fig. 21.5 (**a**) *left* the *cuña*; (**b**), *right* the standard Mexican brick. Drawing: authors

The brick used, called *cuña*, measures 5 × 10 × 20 cm (1,000 cm^3) (Fig. 21.5a). It has a resistance that fluctuates between 60 and 75 kg/cm^2 and an approximate weight of 1.5 Kg. In the case that a brick of these dimensions cannot be obtained or if it is too costly to manufacture it, the standard wall brick can also be used, either whole or cut in half; in Mexico, the standard wall brick's dimensions are 6 × 12 × 24 cm (Fig. 21.5b).

The mortar used is a mix of chalk, cement, and sand, similar to that used for walls. This low level of resistance means that it can be cut in half manually with the builder's trowel (an important requirement for the timely building of vaults). A skilled craftsman with the aid of a helper is able to achieve up to seven or eight square metres per day. In other words, each square metre of vault takes 2 h work, a figure which is three or four times lower than the man hours necessary to complete a concrete cover. It is important to emphasise this point as this kind of technique is often criticised for being too artesanal, overlooking the fact that to build a concrete cover requires three or four times more hours/man per square metre.

The Process

When applying this technique, generally the vaults are built to cover a flat area limited by a rectangular or square horizontal perimeter (Fig. 21.6). When the spaces to be covered do not take one of the aforementioned forms, then they are forcibly "regulated".

In other words, if the area to be covered is L-shaped, then the craftsman or architect requests for an intermediate beam to be placed, such that the L is now subdivided into two rectangular shapes. Twenty years ago when we began our experiments, we always based them on the conditions of the internal space, i.e., the space to be lived in, and as a consequence we changed the vault's shape and its horizontal perimeter line.

The perimeter of the spherical sections we build can be regular or irregular. Moreover the lines that make it up can be straight, curved or mixed lines and also horizontal or inclined. These lines are the surface's directives. Occasionally the directives are concrete sections and at other times they are commercial metal angles. On the other hand, the lines of brick which compose the surface have

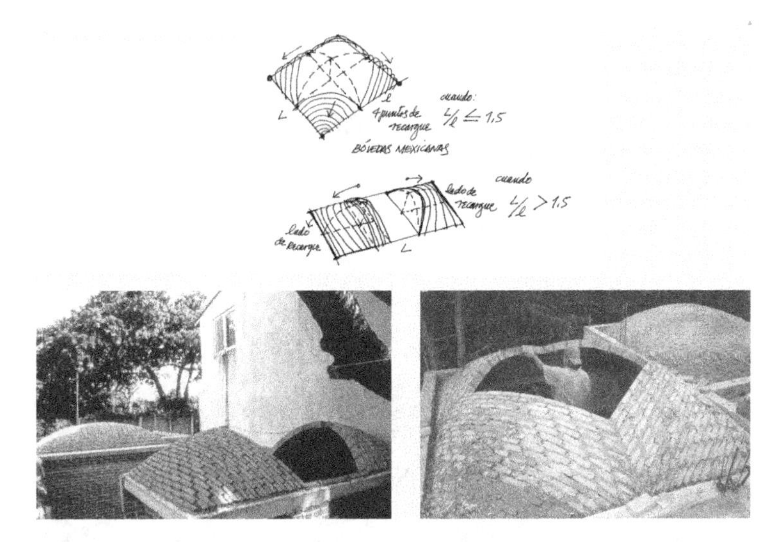

Fig. 21.6 Vaults built to cover a flat area limited by a rectangular or square perimeter. Photo and drawing: authors

Fig. 21.7 Photo: authors

different dimensions and as they move on the surface they become its generators (Figs. 21.7 and 21.8).

The mortar is placed in such a way that the inferior part of the bricks is enveloped in mortar whilst gaps are left in the upper part of the bricks. This is so that when the vault is covered from above the mortar penetrates into the brick joints. Only two people work on the vault, the *bovedero* (the vault builder) and his helper. The latter takes care of rendering and cleaning the interior of the vault as the work progresses.

This, it seems to us, is a very unique and intelligent construction technique. Rather than confronting and fighting gravity, it assumes immediate defeat. But it is thanks to the technique's surrender, and other assisting factors such as its light weight—that of a small brick—that it gains its stability and its vaulted form.

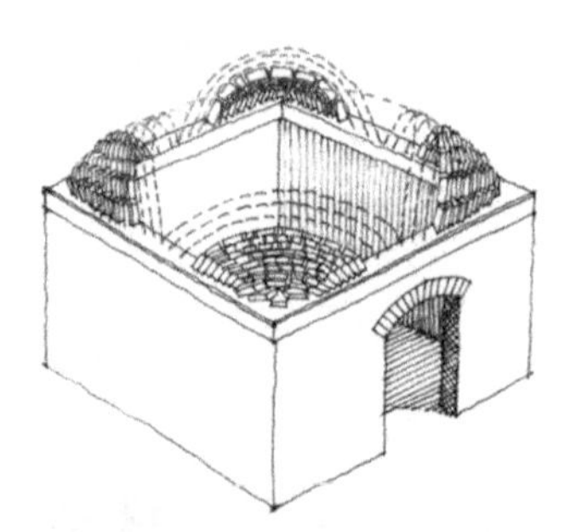

Fig. 21.8 Photo and drawing: authors

Within the process there are three key characteristics of the technique. Firstly, the bricks are placed one on top of the other in continual succession. Second, the bricks to be supported need to be small and light (completely the opposite of those large and heavy bricks whose purpose it is to support). With small reductions in its dimensions, the brick goes from 1,728 cm^3 to just 1,000 cm^3 and it weighs only 60 % of a wall brick. Third, the vault brick, unlike a wall brick, is used dry (wall bricks are wetted before being used) to increase the adhesion. The mortar is made up of cement, chalk, and sand in proportions of 1:1:8 or 1:1:10 (Figs. 21.9, 21.10, and 21.11).

Mathematical Analysis

Background Generally we think of a function as a correspondence which associates to each element of a set X, an element and only one element of another set Y. The set X is called the domain of the function. Typically, the domain of such functions is the set of points of the x-axis. These functions are generally called functions of one real variable. It is easy to extend this idea of a function to functions of two or more real variables: a function of two real variables is a function whose

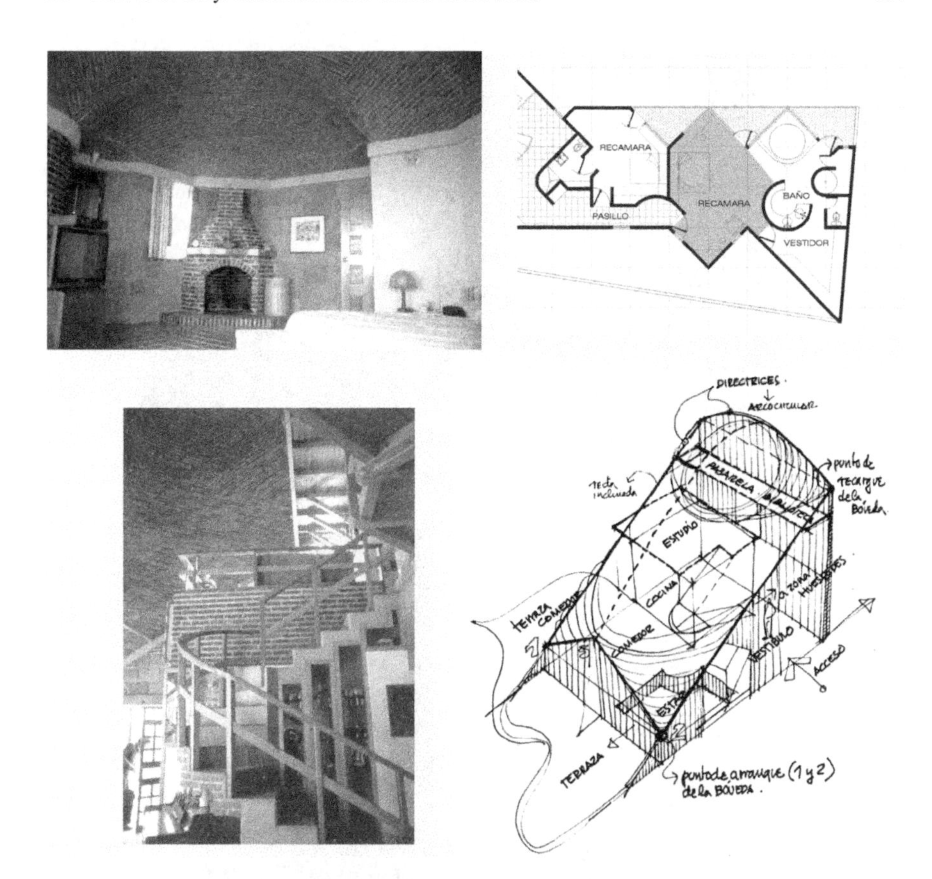

Fig 21.9 Examples of vaults. The covers can be regular surfaces, generally rectangles of different proportions and squares. However, any regular or irregular polygon can also be covered with the system by subdividing the space in small sections. Photo and drawing: authors

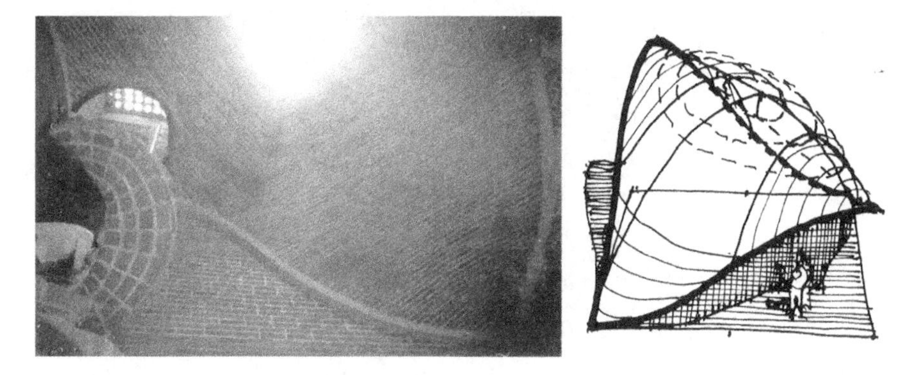

Fig. 21.10 A configuration of our invention: ellipsoidal paraboloids or Ellipars. Photo and drawing: authors

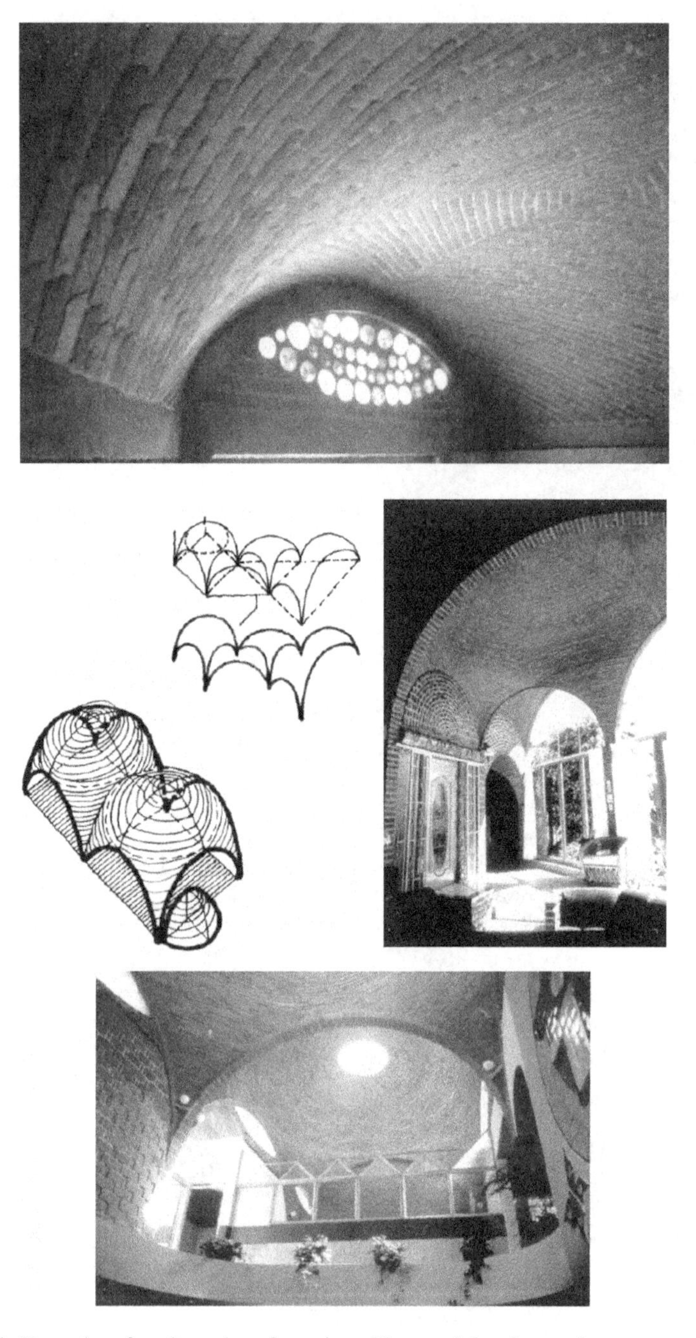

Fig. 21.11 Examples of vaults and configurations. Photo and drawing: authors

domain X is a set of points in the xy plane. If we call f such function, its value at point (x,y) is a real number denoted by $f(x,y)$. It is easy to imagine how such a function can represent a practical situation in architecture. For instance, consider a situation in which we are interested in determining the height of a ceiling of a room. We may represent the height of the ceiling at a certain point by $f(x,y)$ where the domain of the function is the set of all points (x,y) which correspond to points in the room's floor.

Surface integrals A surface integral can be imagined as the equivalent in two dimensions to the linear integral, a surface being the region of integration instead of a curve. There are three main representations of a surface. One is the implicit representation which considers a surface as a set of points (x,y,z) that satisfies an equation of the form $F(x,y,z) = 0$. Sometimes it is possible to isolate one of the variables in the equation, e.g., z, from the other two, e.g., x, y. When this is possible, we can obtain an explicit representation of the form $z = f(x, y)$. There exists a third surface representation method which is more useful for the study of surfaces. The parametric representation uses three equations to express x, y, and z as a function of two parameters u and v:

$$x = X(u, v), y = Y(u, v), z = Z(u, v)$$

Applying these equations, the surface defined by the points (x,y,z) is the image of a two-dimension connected set T defined by the points (u,v). In order to combine these three equations, a vector r is introduced which connects the origin (0,0,0) and a generic point (x,y,z) on the surface. This results in the so-called *vector equation* of the form:

$$\mathbf{r}(u, v) = X(u, v)\mathbf{i} + Y(u, v)\mathbf{j} + Z(u, v)\mathbf{k} \text{ where } (u, v) \in T$$

It turns out that the area of a (parametric) surface S, denoted by $a(S)$, is determined by the double integral

$$a(S) = \iint_T \left\| \frac{\partial \vec{r}}{\partial u} \times \frac{\partial \vec{r}}{\partial v} \right\| \, \text{du dv}$$

Thus, in order to calculate the area of S, it is necessary to firstly calculate the fundamental vector product $\partial r/\partial u \times \partial r/\partial v$ and then integrate its length in the region T. Sometimes $\partial r/\partial u \times \partial r/\partial v$ is expressed as

$$\partial r/\partial u \times \partial r/\partial v = \partial(Y, Z)/\partial(u, v)\mathbf{i} + \partial(Z, X)/\partial(u, v)\mathbf{j} + \partial(\mathbf{X}, \mathbf{Y})/\partial(u, v)\mathbf{k}$$

in which case we have

$$a(S) = \iint_T \sqrt{(\partial(Y,Z)/\partial(u,v))^2 + (\partial(Z,X)/\partial(u,v))^2 + (\partial(X,Y)/\partial(u,v))^2}\,\mathrm{du\,dv}$$

If the surface S is explicitly given by an equation of the form $z = f(x, y)$, x and y can be used as parameters and the fundamental vector product is

$$\|\partial r/\partial u \times \partial r/\partial v\| = \sqrt{1 + (\partial f/\partial x)^2 + (\partial f/\partial y)^2}\mathrm{dx\,dy}$$

and the integral for calculating the surface of the area takes the form

$$a(S) = \iint_T \sqrt{1 + (\partial f/\partial x)^2 + (\partial f/\partial y)^2}\,\mathrm{dx\,dy}$$

where the region T is the projection of S on the xy plane.

The Vault Surface as a Mathematical Function

Vault constructed on a square perimeter The first step towards a mathematical analysis of the vaults we described is to formalize their surface as a function of two real variables. Consider for simplicity, a vault of height 1 constructed on the perimeter of a square of 2×2 (the generalization to a height and rectangular perimeter of arbitrary dimensions is straightforward). The resulting vault would look like the one in Fig. 21.12.

For any point $p = (x, y)$ in the xy plane, in particular for any point inside the area delimited by the perimeter of the square of 2×2, we have $\tan\theta = y/x$ and thus $\theta = \arctan(y/x)$ (Fig. 21.13):

Also, if we denote $X = \sqrt{x^2 + y^2}$, then

$$X^2 + z^2 = D^2$$

where z represents the point in which the line perpendicular to the plane xy and passing by the point $p = (x, y)$, intersects the curve defined by the vault (Fig. 21.14).

That is, $z = f(x, y)$ where f is the function of two real variables that we are seeking. D is the distance from the point (x, y, z) to the origin. Thus z, representing the expression we are interested in, is defined by the equation

$$\mathrm{z} = \sqrt{\mathrm{D}^2 - \mathrm{X}^2}$$

D can be defined as a function of X and θ. Let $g(X, \theta)$ be such a function:

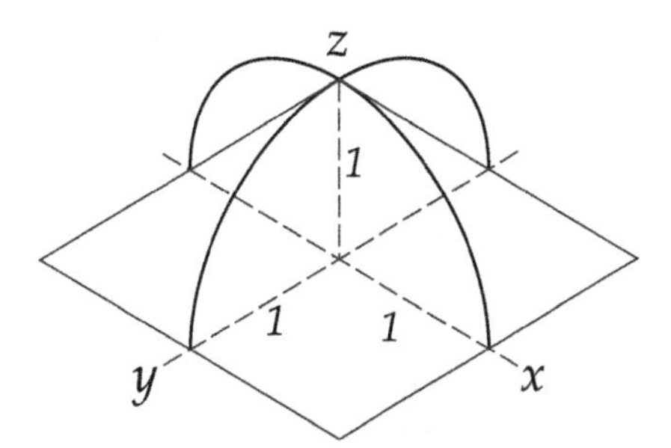

Fig. 21.12 Image: authors

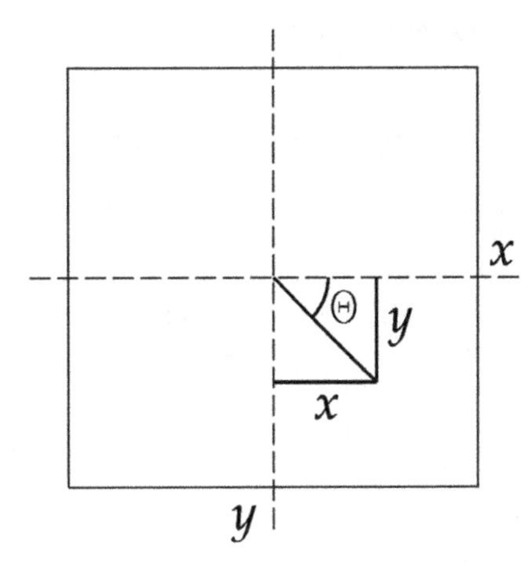

Fig. 21.13 Image: authors

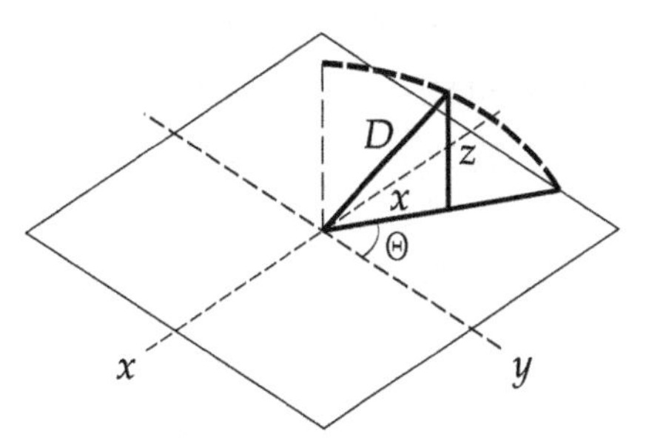

Fig. 21.14 Image: authors

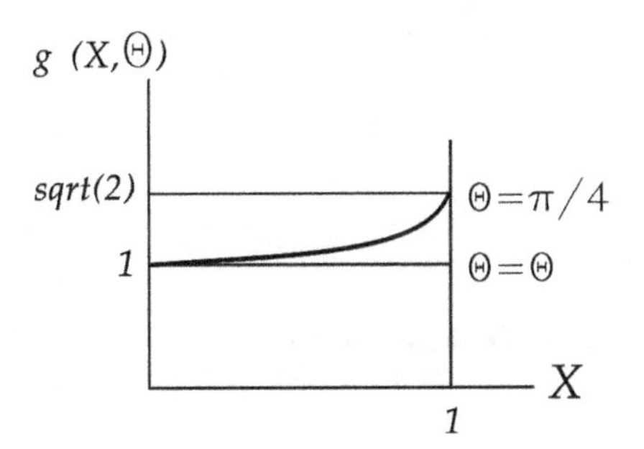

Fig. 21.15 Image: authors

$$g(X, \theta) = 1 + \left(\sqrt{2} - 1\right) X^2 \tan \theta.$$

Fig. 21.15 is a graphical representation of $g(X, \theta)$:

Thus, $\mathrm{z} = \sqrt{\left(\mathrm{D}^2 - \mathrm{X}^2\right)}$ becomes

$$\mathrm{z} = \sqrt{\left[\left(1 + \left(\sqrt{2} - 1\right)(\mathrm{x}^2 + \mathrm{y}^2)(\mathrm{y}/\mathrm{x})\right)^2 - (\mathrm{x}^2 + \mathrm{y}^2)\right]} \ \text{if}\, \mathrm{x} \neq 0$$

and given that $\tan \theta = y/x$ and $\mathrm{X} = \sqrt{\mathrm{x}^2 + \mathrm{y}^2}$,

$$\mathrm{z} = \mathrm{f}(\mathrm{x}, \mathrm{y}) = \sqrt{\left[\left(1 + \left(\sqrt{2 - 1}\right)(\mathrm{x}^2 + \mathrm{y}^2)(\mathrm{y}/\mathrm{x})\right)^2 - (\mathrm{x}^2 + \mathrm{y}^2)\right]} \ \text{if}\, \mathrm{x} \neq 0$$

$$\sqrt{(1 - y^2)} \ \text{if}\, x = 0$$

In order to calculate $a(S)$, the area of the surface defined by the vault, we need to evaluate the double integral

$$a(S) = \iint_T \sqrt{(\partial f / \partial x)^2 + (\partial f / \partial y)} \ \mathrm{dx}\, \mathrm{dy}$$

where *f* is defined as above and the region T is the projection of S on the *xy* plane, i.e., the square of 2 $\times$ 2 on which the vault is built.

Vault constructed as a section of a sphere Now consider again for simplicity, a spherical vault with a radius of 1 constructed on four arches (the generalization to a vault with arbitrary radius is straightforward). The resulting vault would look like the one in Fig. 21.16.

Being a spherical vault, here we already know the two-real-variables function which formalize its surface:

$$z = f(x, y) = \sqrt{1 - x^2 - y^2}$$

In order to calculate the exact vault surface we simply have to calculate the surface of the semi-sphere on the square region on top of which the vault is constructed (see Fig. 21.13). This results in

$$a(S) = \iint_T \sqrt{1 + (\partial f / \partial x)^2 + (\partial f / \partial y)^2} \ \mathrm{dx}\, \mathrm{dy}$$

where *f* is as above and the square region *T* is the projection of *S* on the *xy* plane.

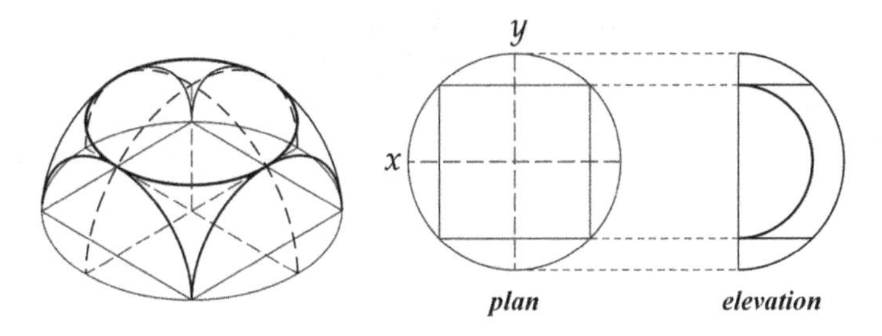

Fig. 21.16 Image: authors

Conclusions

We have described a popular construction technique for building brick vaults, without any use of scaffolding and or additional reinforcements. The technique, a collective invention in Mexico dating back to the nineteenth century, offers an economical solution even to this day to the housing problem. As a first step towards a mathematical analysis of the vaults we have formalized their surface as a function of two real variables. We have explored two cases: the surface of a vault constructed on a rectangular perimeter, and the surface of a vault constructed on four circular arcs resulting in a section of a sphere. We also have indicated how to calculate the area of the surface in both cases. In the near future, we plan to contrast our mathematical results with measurements in existent vaults and also provide a formalisation of the different patterns formed by the bricks on the vaults surface.

Biography Alfonso Ramírez Ponce is full Professor in the Faculty of Architecture of the National Autonomous University of Mexico (UNAM). He has a Masters and Ph.D. in Architecture and specializes in economical construction techniques, employing traditional materials such as the common clay brick. A hallmark feature of Alfonso Ramirez Ponce's designs is his passion for employing both traditional building materials and artisan craftsmanship in his life's works. His design portfolio includes works using traditional building materials such as brick and ferrocement, wood and adobe, reinforced ceramic, bamboo, earth-cement bricks, plaster domes, as well as recycled materials such as palm and coconut wood and tin. Alfonso has imparted invited specialist architectural courses in universities and institutes in most countries in Latin America as well as in Ethiopia, Singapore, Spain, Switzerland, United Kingdom, and United States of America.

Rafael Ramírez Melendez is currently Associate Professor in Computer Science at the Pompeu Fabra University, Barcelona, Spain. He obtained his B.Sc. in Mathematics form the National Autonomous University of Mexico, and his M.Sc. in Artificial Intelligence and Ph.D. in Computer Science from the University of Bristol, UK. For 5 years, he was Lecturer in the Department of Computer Science at the School of Computing of the National University of Singapore. His research interests include Artificial Intelligence, Machine Learning, Data Mining and their application to cognition and creative processes. He has published more than 100 research articles in peer-reviewed international Journals and Conferences. He currently acts as chair and program committee member for several artificial intelligence and machine learning related conferences, and as a reviewer for several international journals. He has given invited seminars across Europe, Asia and America.

Chapter 22
Mathematics and Structural Repair of Gothic Structures

Javier Barrallo and Santiago Sanchez-Beitia

Introduction

Showing a visitor a range of geometrical models, Spanish architect Antonio Gaudí once remarked, with excitement in his eyes: *Wouldn't it be beautiful to learn geometry in this way?* Without any doubt, mathematical education for architecture students would be more effective and pleasant if all the theoretical knowledge were explained with the help of real architectural examples. The relationship between classical architecture and mathematics is well known. Architecture, unlike other scientific disciplines, can be used as a never-ending source of numerical, algebraic, geometric, analytic and topologic problems, to name just a few fields of mathematics. A modern concept of architecture should necessarily include mathematics for its comprehension. Reciprocally, the teaching of mathematics in architecture should be based on the constructive event to be effective.

Interdisciplinary education provides a positive stimulus for both teachers and students, resulting in a much more persistent and interesting training. It is obvious that mathematical knowledge acquired inside an architectural environment is more likely to be applied by future architects after their university studies. As an example of this way of learning mathematics, in this chapter we will show some ideas and mathematical concepts related to one of the more complex branches of architecture: restoration, repair, and maintenance of Gothic buildings.

First published as: Javier Barrallo and Santiago Sanchez-Beita, "Mathematics and Structural Repair of Gothic Structures", pp. 21–30 in *Nexus V: Architecture and Mathematics,* Kim Williams and Francisco Delgado Cepeda, eds. Fucecchio (Florence): Kim Williams Books, 2004.

J. Barrallo (✉) • S. Sanchez-Beitia
School of Architecture, The University of the Basque Country (UPV-EHU), Plaza de Onati, 2, 20018 San Sebastian, Spain
e-mail: javier.barrallo@ehu.es; santiago.sanchez@ehu.es

K. Williams and M.J. Ostwald (eds.), *Architecture and Mathematics from Antiquity to the Future*, DOI 10.1007/978-3-319-00137-1_22,

Why Such Interest in the Gothic Style?

The constructive characteristics of the Gothic style are unique in the history of architecture. Gothic cathedrals pushed structure to the limit—soaring cross-vaulting, pointed arches, hollow walls and piers covered with tracery—and used the arch as an external brace—the flying buttress—to form one of the most beautiful stylistic elements of the Gothic style.

Gothic buildings changed the pier concept inherited from the Romanesque tradition, recovering the concepts of the pier as a skeleton; the pillar instead of the wall; the arch as vault centre instead of the arch as vault element; stress concentration points instead of stress lines. The whole system is a linear frame that supports forces in a delicate balance. This is the magnificence of the Gothic style (Fig. 22.1).

This fragile equilibrium implies a permanent stress on all the elements. Any failure provokes other structural failures in a chain reaction threatening the integrity of the building. Obviously the technology of the thirteenth to sixteenth centuries could not introduce numerical structural calculation because it simply didn't exist. The progress came by trial and error, empirical methods, and the results were transmitted by travelling master builders and stonemasons.

Why Mathematics?

A Gothic building is special in the sense that its construction usually extends over a long period of time. The building process was very slow due to technical and economic problems. Besides, initial design errors were detected and corrected during the construction, modifying the original design.

The construction of a Gothic cathedral usually began with the apse and proceeded towards the end of the nave and aisles. This method implies that the structural equilibrium of each part of the building has to be solved independently. Consequently, the whole structure suffers from changing its equilibrium by absorbing settling deformations due to its provisional situation. Furthermore, the technological progress during the long construction process causes heterogeneity between parts finished with older procedures and materials and parts executed at a later time. This is the reason why an accurate correlation between a mathematical model and the real building cannot actually be found. However, mathematical modelling is the only effective tool to estimate and understand the actual balance of a Gothic cathedral.

The study of Gothic buildings must take into account the following aspects:

- Heterogeneity of materials due to long construction periods;
- Technical alterations introduced by craftsmen;
- Wide variety of stone and mortars;
- Different stages of vertical and horizontal development;

Fig. 22.1 Main nave of the Church of Santa Maria la Real, Najera, Spain. A geometric analysis of the church shows the vaulting system to be completely distorted. Arches and vaults are deformed, the walls and columns are bent, and cracks appear in the aisles. Geometry is the first test to verify the structural state of a Gothic construction. Photos: authors

- Additions, substitutions, and partial collapses during the construction;
- Restorations and maintenance of the building.

These various aspects mean that, in many cases, it will be necessary to develop several models of the same building corresponding to different periods of time in order to determine the stress evolution of the building in a qualitative rather than a quantitative manner.

To find the effectiveness of a mathematical model, we must verify two criteria:

- Predicted deformations must agree qualitatively with building deformations.
- Deformations measured *in situ* in the building must be of the same order as those predicted by mathematical means.

Geometry of a Historic Building

Describing the geometry of a building is the first phase of historic restoration work. A precise geometric model must be simple conceptually, but substantially representative of the structural and constructive system of the building. In order to create an accurate survey, there exist two basic techniques for acquiring the geometrical data that defines a historic building: photogrammetry and topography (Fig. 22.2).

Photogrammetry is based on the principle of three-dimensional vision starting from two bidimensional images taken from points that are slightly separated. Just as we use our brains to comprehend spatial depth by means of two bidimensional images taken from our left and right eyes, a computer can reconstruct a three-dimensional environment from two photographs taken from separate points

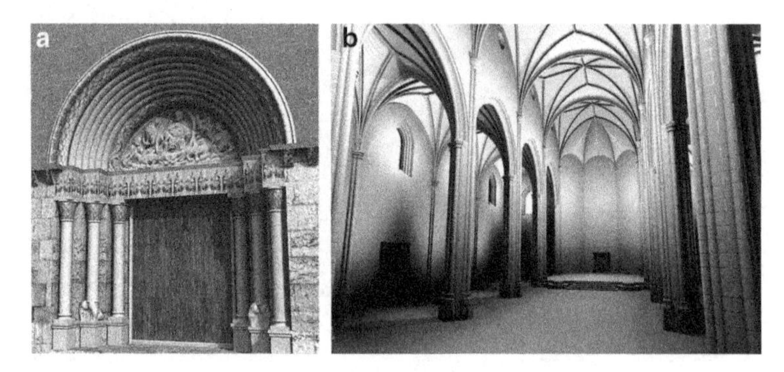

Fig. 22.2 Photogrammetry and topography are two techniques used to take precise coordinates from existing conditions. Photogrammetry is based on the measurement of an object in two photographs taken from separated points towards the same focus. Topography is based on the geometry of an object that looks smaller the further away it is. Photogrammetry is a technique usually used for detailed, ornamented, models while topography is generally used for larger, less detailed, structures. A Romanesque style portico (**a**, *left*) rendered by means of photogrammetry and the nave of a Gothic cathedral (**b**, *right*) measured using topography. Photo and rendering: authors

directed toward the same focus. A pair of photographs of this type is known as a stereoscopic image.

The geometric basis of photogrammetry is the following: an object placed near the line of infinity appears in the same position in the bidimensional images, whilst the relative positions of objects near the observer vary between the two images. By measuring the displacement of the object in both images, we are able to deduce the distance from the observer in the real world.

Topography is a well-known process for obtaing the spatial coordinates of any construction. The geometrical theory is extremely simple: an object looks bigger or smaller to us depending on its distance from the viewer. With an special device, the size of the object is measured and the distance of the point deduced. By tracing imaginary triangles from known coordinates, any visible point of the building can be measured with a high degree of precision.

Once we have a database with enough points from the building, we proceed to the elaboration of a three-dimensional, solid, model. Several shapes are used to create the final model: ellipsoids, cylinders, paraboloids, hyperboloids, and other shapes that fit the coordinates from the database until the geometry is successfully completed (Fig. 22.3).

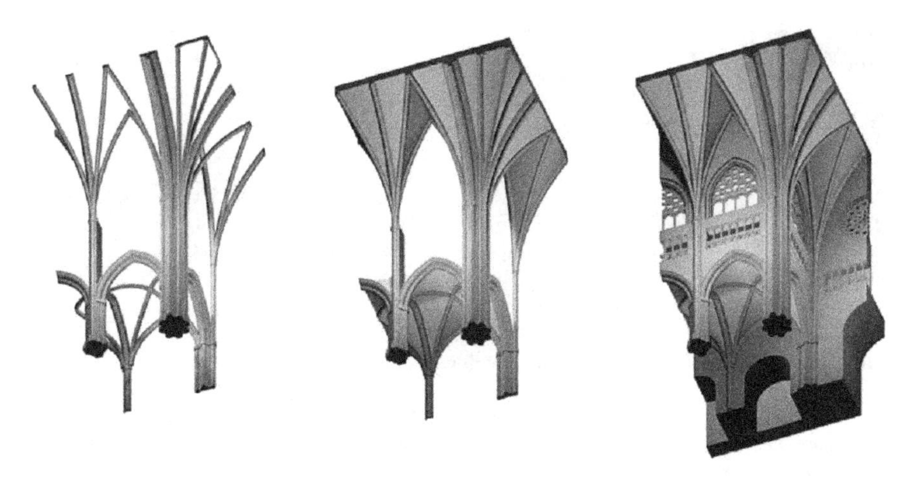

Fig. 22.3 Interior of the Bilbao Cathedral. The making of a computer model is similar to the physical creation of the cathedral. First, the columns and arches are constructed, followed by the vaults, piers and flying buttresses, and finally the walls, windows and ornaments. Rendering: authors

Experimental Measurements

Each part of a building has specific constructive and structural characteristics that must be represented in the model. Every element of the mathematical model needs to be associated with its mechanical properties.

The measurement of stresses and their directions is another important task (Figs. 22.4 and 22.5). Very high stresses might fracture the stone-mortar ensemble. Stresses whose directions do not follow the vertical structural elements usually transmit forces outwards that should be opposed by the external buttress system. If this is not the case, traction forces will appear in the structure, producing serious disorders in the building. Also, bending, cracks and deformation of structural elements might have occurred over time and should be measured for a representative period of time in order to achieve the best possible simulation of the building.

All these processes, known as monitoring, are completed by physical analyses of the soil and the materials used in the construction, mainly compression/deformation reports. The correlation of all the available reports, simulations, monitoring, weather conditions and soil prospecting may provide clues for detecting and solving the structural problems present in a building.

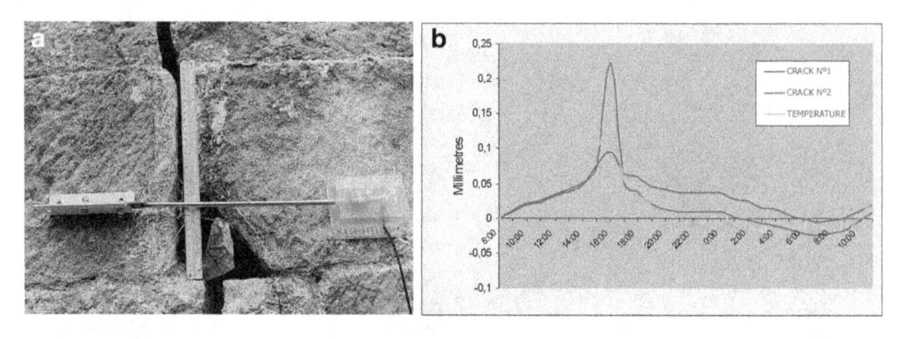

Fig. 22.4 An example of a device to monitor cracks in a building. (**a**, *left*) and the graph representing the movement of two cracks during a 24-h period; (**b**, *right*). The graph also includes a third measurement indicating the external temperature of the building. It is obvious that there is a clear correlation between the temperature and the movement of the cracks, which work as expansion joints. Photo and graph: authors

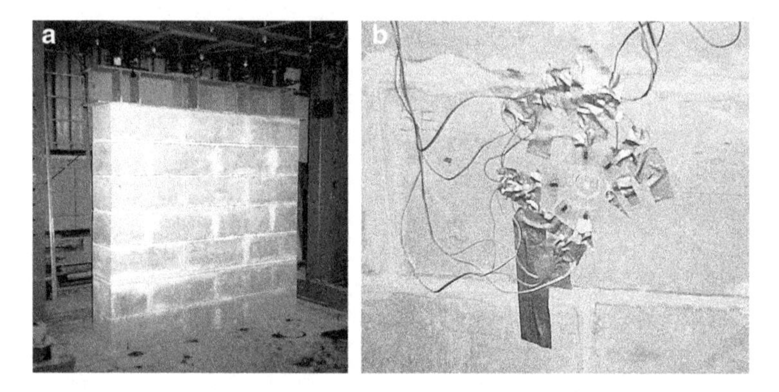

Fig. 22.5 An example of an experimental method to measure stresses, the Donostia Method, which was developed by the authors and has been widely used in several historic buildings: (**a**, *left*) a simulation of a masonry wall loaded in a laboratory; (**b**, *right*) the basic principle of the process: several electronic extensometric gauges placed around a hole drilled in the stone measure the deformation of the hole after the drilling. The dimension of the deformation, which of course is not perceptible by the naked eye, is transmitted from the gauges to an analog computer. Depending on the mechanical properties of the material, and the deformation measured, the stresses in all directions are calculated. Photo: authors

The Finite Elements Method

The Finite Elements Method is the methodology that nowadays gives the most satisfactory results in the analysis of historic architectural structures, especially Gothic cathedrals. The basis of this method consists in dividing a surface or volume into a reasonable number of small elements. The mechanical characteristics, as stress or displacement, of each and every element is calculated and transmitted to the neighbouring elements.

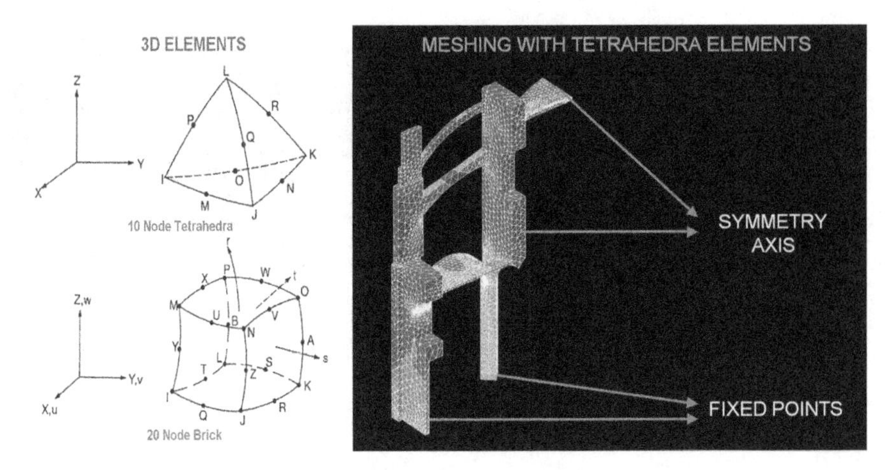

Fig. 22.6 Meshing a section of the external resistance system of a Gothic cathedral (flying buttresses, vaults, pier and column). The basic element selected for the mesh was the 10-node tetrahedra. Tetrahedra fit perfectly on complex geometries, such as the Gothic, with a reasonable number of nodes. Drawing and model: authors

The input for a Finite Elements Method analysis is a computer model representing the geometry of the building. This model is not a simple drawing: each geometric element must be perfectly defined and assembled with its neighbours. Also, it must include appropriate contour conditions and physical properties, such as the Young and Poisson Modulus and density, amongst others.

Once the geometric, mathematical, model is completed, it is meshed into small elements (Fig. 22.6). In space, we usually use tetrahedra (four nodes) or brick (eight nodes) elements, although these elements can also be implemented and extended to 10 and 20 nodes respectively by including an extra node to each aristae.

All this information is then translated to a system of equations in matrix form. The size of the matrix depends not only on the number of elements and nodes, but on the number of degrees of freedom. The degrees of freedom can be considered as the variables we want to solve for each node. A typical degree of freedom is six, which includes translation and rotation for each node along the three main axes X, Y and Z.

After the matrix containing the model is solved, stresses and displacements for each node can be easily estimated (Fig. 22.7). The results are usually represented graphically with colour maps that represent magnitudes. This technique is widely used, as it allows a quick visual understanding and interpretation of the model.

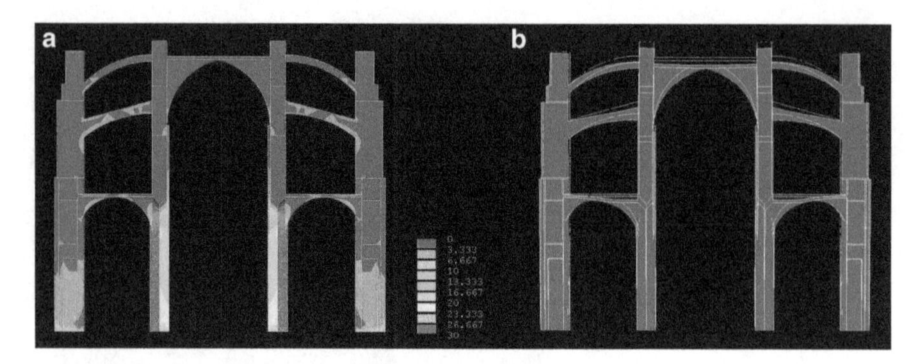

Fig. 22.7 Church of San Antonio Abad in Bilbao. (**a**, *left*) the equivalent stresses on a section of the church; (**b**, *right*) an estimation of the deformation suffered by the temple. Models: authors

Biography Javier Barrallo was born in Bilbao, Spain, in 1964. He is Professor of Mathematics at the School of Architecture of the University of the Basque Country in San Sebastian. He studied Computing Engineering at the University of Deusto, Bilbao, and obtained his Ph.D. degree with a work about fractal algorithms in 1992. Since then, he has published several books and works about computer art and design, organized several courses and conferences about Art and Mathematics (ISAMA, BRIDGES, Mathematics and Design). He has directed many exhibitions in galleries and museums all over the world under the main title "The Frontier between Art and Science". As a second investigation branch he has become a qualified expert in computer measurement and simulation of heritage buildings, especially Gothic structures. He has collaborated in the restoration projects of several well-known Spanish buildings.

Santiago Sanchez-Beitia is Professor at the School of Architecture of the University of the Basque Country in San Sebastian. Previously has been professor at the Pol. Univ. of Madrid (Spain) and invited professor at the Granada Univ. (Spain). He is the Director of both the Master Course of Restoration and Assessment of Historical Constructions and the Doctoral Programme in the same field at the University of the Basque Country. His Ph.D. degree was obtained with a work about prestressed steel for civil engineering in 1987. He has executed many structural analysis of historical constructions in Spain, Portugal, Belgium and Egypt. All the works have been published in peer review scientific journals and proceedings.

Chapter 23
Mathematics of Carpentry in Historic Japanese Architecture

Izumi Kuroishi

Introduction

Historically in Japan, carpentry had been recognized as a mystical profession. In a book entitled *Hidasho-monogatari* (飛騨匠物語), the master carpenter was so skilled that he could make his house float in the air (Masamochi and Hokusai 2002) (Fig. 23.1).

Such legends arose from the fact that carpenters kept their knowledge secret and transmitted their technology through apprenticeships. There are many old, mystical architectural treatises dating from the fifteenth century, but most of them contain lists of architectural styles and partial construction drawings, for which only limited literal explanations are provided. Besides these treatises, there exist drawings of large temples from the sixteenth century. However, among the drawings of whole architectural forms and individual parts, there is no coherent explanation. In constructing traditional Japanese buildings, carpenters retain their traditional ways of construction; once clients and a master carpenter decide the size and *Kiwari* (木割り) of the project, almost all the other design and structural systems are automatically fixed.

When the Western notion of architecture was introduced into Japan at the end of the nineteenth century, its idea of the mathematics as a fundamental knowledge of an architect became dominant, and the historical design and technology of the Japanese carpenter came to be regarded as too primitive. At the beginning of the nineteenth century, master carpenter and mathematician Heinouchi Masaomi wrote

First published as: Izumi Kuroishi, "Mathematics of Carpentry in Historic Japanese Architecture", pp. 117–129 in *Nexus V: Architecture and Mathematics*, Kim Williams and Francisco Delgado Cepeda, eds. Fucecchio (Florence): Kim Williams Books, 2004.

I. Kuroishi (✉)
School of Cultural and Creative Studies, Aoyama Gakuin University, 4-4-25, Shibuya, Shibuya-ku, Tokyo, Japan
e-mail: eaaa0925@nifty.com

K. Williams and M.J. Ostwald (eds.), *Architecture and Mathematics from Antiquity to the Future*, DOI 10.1007/978-3-319-00137-1_23,

Fig. 23.1 Drawings of the works by Hida carpenter by Hokusai from *Hidasho monogatari*

theories of *Kikujutu* (規矩術, Architectural Stereotomy), and used his knowledge of Japanese historical mathematics, *Wasan* (和算), to analyze the technology of carpentry (Sekino 1947; Hiroshi 1985). In Japanese architectural history, Heinouchi is recognized as a pioneer for introducing mathematics into Japanese traditional architecture, and thus modernizing it (Katsushige 1978; Norihito 2000). However, there were criticisms against Heinouchi's theories that they simplified the forms of historical architectures. Also, the nature of *Wasan* is very different from Western mathematics, and the relationship between *Wasan* and Japanese historical architecture has not been clarified yet. Thus, for the above cultural characteristics in both of the Japanese architectural production and of mathematics, I argue that the relationship between mathematics and the historical technology of Japanese carpenter must be reexamined in a different way. Such varieties and subtleties in the creation of architectural forms before the theorization by Heinouchi represent the cultural identity in the relationship between *Wasan* and Japanese architecture.

Italian mathematician Enrico Giusti writes:

> The objectives of mathematics are not the expression of the purified essence in the material impurity of the outside world independent from human beings but the formalization of the human being's practice. . .Mathematics always begins from the technological practice, and is not the daughter of the nature but of the art in its original meaning (Giusti 1999: 31–33; my trans.).

The nature of mathematics in architectural creation has varied widely even in Western architecture. In the present study, I am going to present a different perspective for the role of mathematics in Japanese historical architecture by focusing on the materials of carpenters' works. I will explain that, contrary to a kind of ideal mathematics that gives an order to the whole, there was another mathematics in Japanese carpenters' technologies, phenomenological mathematics, which emerged from the process of rationalizing practical activities and the public's habitual understanding of architectural form, the latter merging with the former to create the characteristic features of Japanese historical architecture.

History of Japanese Mathematics and the Role of Carpenter Technology

The Japanese learned about the square root and the theories of Pythagoras from Chinese mathematic textbooks in the sixth century, and practical technologies, such as calendar making, hydraulics, surveying, and financial administration, further accelerated its development. In the sixteenth century, *Tengenjutsu* (天元術, mechanical algebra) was introduced from China and theories of number sequences and approximate solutions developed, leading to the Japanese mathematics called *Wasan*. At the end of the nineteenth century, the Japanese government decided to replace *Wasan* with Western mathematics. One reason for this was that *Wasan* treated real, everyday, problems, and did not seek universal fundamental principles and logic; thus it was not deductive as is Western mathematics, but rather inductive. Since the seventeenth century geometric problem solving had been a popular, competitive, sport. Mathematic pictures known as *Sangaku* (算額,) hung in shrines (Fig. 23.2), popularizing the visual inductive method, as well as furthering the evolution of abstract algebraic theories. Also, Japanese calendar theory was based on that of China, so algebraic geometry developed instead of graphical geometry, so there was no concept of a coordinate system such as that of Descartes in Europe. *Wasan* solved solid geometry with algebraic geometry alone, not with analytic geometry, and did not encompass either the projection drawing method or the concept of the degree of angle. It used only the *Kokogen* method (勾股弦術, Pythagorean theory method) to solve geometrical problems.

The *Kokogen* method derived originally from *Kanejaku-san* (曲尺-算, *Kanejaku* mathematics), which derived from carpenters' techniques of *Kanejaku*, an L-shaped right-angle scale. Since the sixth century, Japanese people used *Kokogen* for the four operations of arithmetic, square root, cube root, and for drawing polygons (Katsushige 1985; Yoshio 1977). Descriptions of the *Kokogen* method appear in the *Gushikenki* (愚子見記), the oldest, well organized carpentry book, written in the seventeenth century (Katsushige 1978: 28–43; Kazuyoshi 1988). It is not an

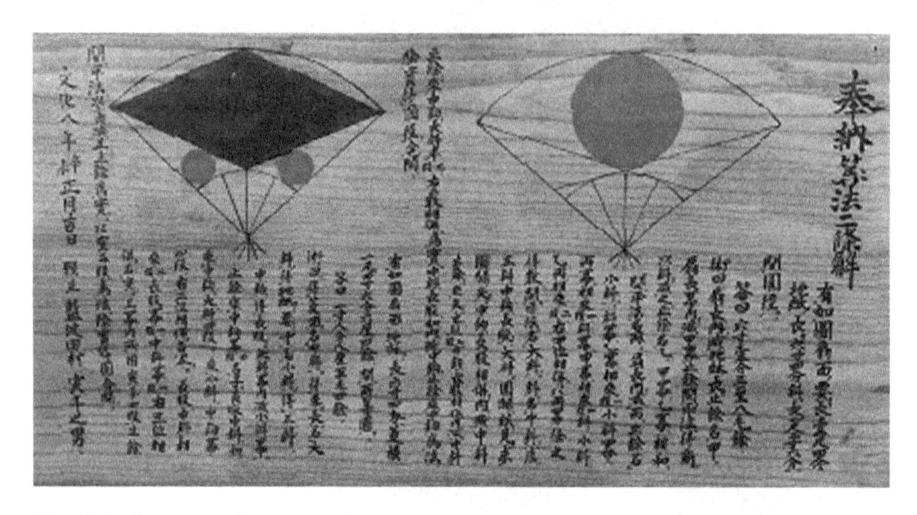

Fig. 23.2 Sangaku of Sakatare shrine in the city of Mita, 1811, renovated in 1967

accurate calculation but rather a kind of empirical method, which emerged from the act of measuring the drawn shapes, and was created by carpentry's practical technology (Fig. 23.3).

Development of Japanese Carpentry Technologies

Japanese carpentry technologies developed along with the elaboration of the techniques of *Kanejaku*. Especially after the medieval period the techniques of *Kanejaku* became more complicated in order to create detailed joints and ornamental parts, and the essence of carpentry technologies was known as *Sashigane* (差金術). Such emphasis on the construction technology in the creation of architecture arose for two reasons: the refinement of methods introduced by foreign technology; and the characteristic relationship between the proportional system and the construction technology.

Refinement of Method From antiquity in Japan, large wooden structures were built with simple construction methods. With the introduction from China of the wooden structure technologies of *Asuka*-temple (飛鳥寺) in 588, the Chinese Sui and Tang styles introduced with *Toshodai*-temple (唐招提寺) at the end of the seventh century, and the North Sung styles with Zen temples in *Kamakura* period, the ideas and technologies of Japanese architecture developed dramatically. Japanese carpenters' methods of learning and refinement of those Chinese architectural ideas and technologies influenced the nature of Japanese architecture.

In the construction of large temples, Chinese master carpenters or Japanese Buddhist monks educated in China directed Japanese carpenters not with literal texts but with substantial models. In the construction of *Aska*-temple in the sixth

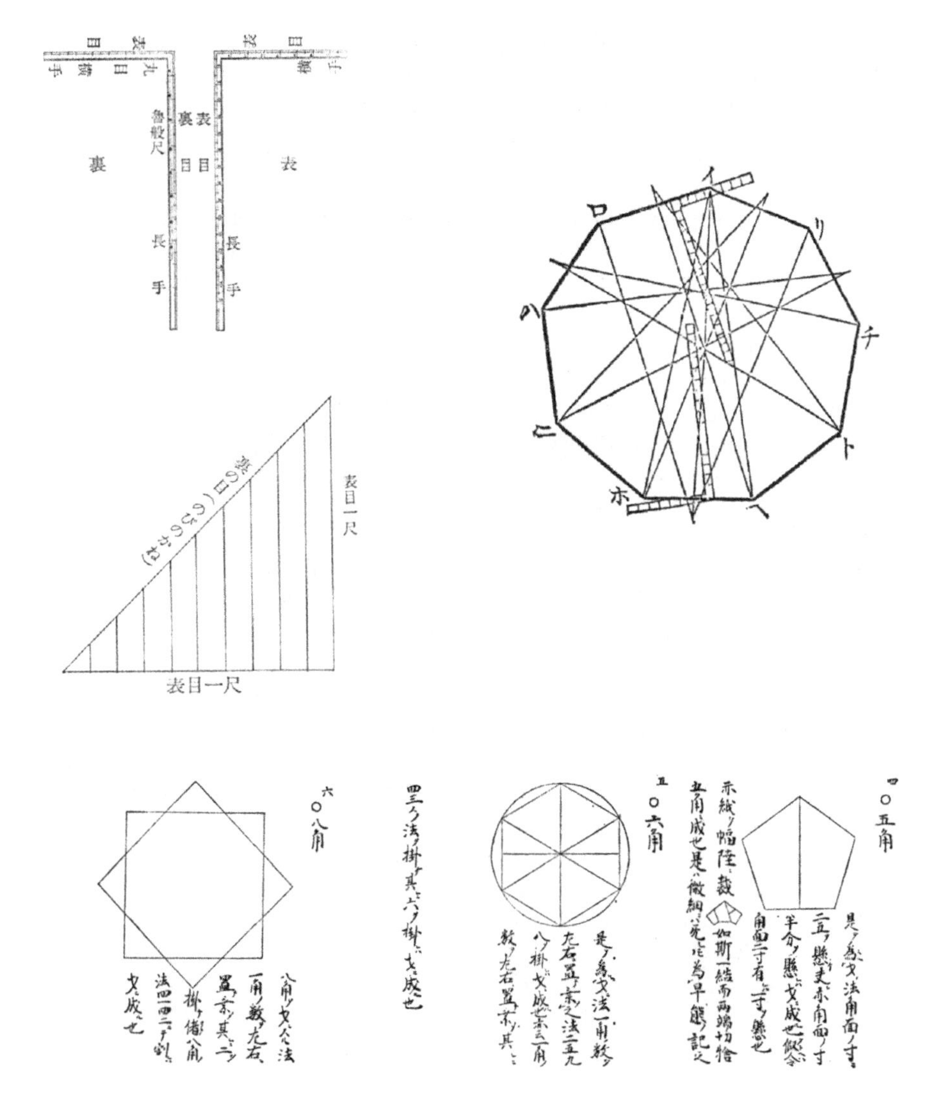

Fig. 23.3 The various measurements on *Kanejaku* and *Kanejaku-san* drawings from *Gushikenki* and *Sugakutaizen*

century, a 1:10 scale model was imported, enabling carpenters to conduct actual examinations of the design, structure, and details. Carpenters established their job hierarchies and the way of learning from models with visual and tactical methods.

Carpenters in ancient periods could neither read nor analyze their practice theoretically, but learned only through empirical practice and dialogue with those with greater experience (Naomi 1971: 156–217). Thus, the description of the ideas and technologies in texts, such as the technological descriptions and lists of the styles in the first organized architectural treatise *Shomei* (匠明) in 1608, became formalistic and dispersive. For example, *Shomei* has an analogical drawing of a human body indicating the relationship between the macrocosmic order with the arrangement of Chinese temple, but without an explanation.[1]

The Development of *Kiwari* (木割, proportion system) and *Sashigane* In buildings before the seventh century, all members had rounded shapes and pointed joints, enabling assembly with nails and a dynamic combination of parts. There were various joint systems, and not so much attention was paid to the accuracy of their placement (Hideo 1985). Meanwhile, Japanese carpenters created various ornamental forms, construction systems, and technologies by modifying those of China to suit the Japanese climate and ways of life. In the *Heian* period there appeared *Noyane* (野屋根), which had two layers of roofs extending around the peripheral space of the building in response to the frequent rainfalls. Because *Noyane* created the characteristic soft and delicate curves of the long, thin, extended, eaves, the arrangement of structural beams and forms of joints became more complicated, and more emphasis was placed on the construction technologies of those parts and joints. North Sung style's organized construction systems and ornamental details, as well as the accurate joint technology using many kinds of wood in *Shoin-zukuri* (書院造) from the *Muromachi* to the *Edo* period accelerated the development of highly elaborated construction technology for the edges of eaves and the joints.

In the history of the development of Japanese architecture between the *Heian* and *Edo* periods, there was another important development in the area of architectural proportion. Along with the introduction of the ideas and technologies of Chinese architecture in the sixth century, its system of structural proportions based on the size of the column also came into Japan. In the *Momoyama* period, there were three modules: intercolumniation, size of column, and distance between roof rafters; and almost all other parts' sizes and the distances between them were determined as integral multiples of these modules. This proportion system is called *Kiwari* (木割), and the styles of Japanese architecture are basically categorized according to *Kiwari* and roof designs (Fig. 23.4). However, because of the canonical nature of the technology and designs that had been introduced, *Kiwari* gradually became more unified and organized towards the *Edo* period; as Japanese architectural historian Nakagawa Takeshi noted, it became more focused on its numerical consistency independent from the total spatial organization (Takeshi 1985). The more the idea of style follows the rule of *Kiwari*, the less the liberal modification of the form was possible. On the other hand, carpenters demonstrated their desire for expression in the creation of the details by *Sashigane* independent from the total spatial organization.

As explained earlier, once the master carpenter or monk decided the size of the project and *Kiwari*, almost all other aspects of the design were automatically determined. Thus, *Kiwari* worked as ideal mathematics. Proportion, from its

[1] The original author in the Momoyama period is unknown. Heinouchi Masanobu published the first ed. in 1608; see Hirotaro and Yotaro (1971: 242).

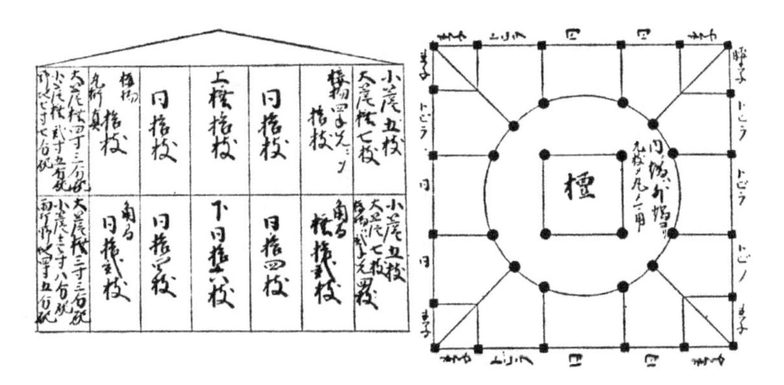

Fig. 23.4 Kiwari drawing of a two-story pagoda from *Shomei*

etymological meaning, is not merely formal but a flexible, organized, relationship of the parts into a whole, thus implying the adaptation to phenomenal necessity. However, as *Kiwari* became more numerically oriented in its canonization, the role of proportion was narrowed and fixed, and its role as ideal mathematics detached from phenomenal necessity in architecture.

The Idea of Geometry in *Sashigane-zu* (Drawing)

***Kanejaku* (曲尺)** According to legend, at the origin of *Kanejaku* are the instruments held by the god Fugi, depicted in cave drawings in Shangdon, China. Fugi holds a *Kanejaku* and his wife Joka holds a compass. *Kanejaku* has not changed so much. It has regular measure, *Omoteme* (表目), on the longer side and *Urame* on the shorter side, and on the back there are *Urame* (裏目), *Robang-jaku* (魯班尺), *Omoteme* and *Maru*-me (丸目). *Urame* is also called *Nobi-no-ku* (延びの矩) and shows the length of the diagonal of a square, which is √2 times the regular measure. *Urame* is used for the placement of the roof rafters, and for the cutting of a square timber from a round trunk. As Japanese architectural historian Muramatsu Teijiro notes, Japanese carpenters created this *Nobi-no-ku* in the medieval period to accommodate the development of the *Kokogen* method to create complicated roof shapes and joints (Teijiro 1973: 134).

The Method to Create *Sashigane-zu* and *Kataita* (型板, Template) Historical wooden architecture has assembled joints on columns to make long extended eaves, the corners of which curve upward, and the surface waviness of which forms various roof curves. In order to assemble the parts to form the eaves and roofs efficiently and accurately, carpenters draw *Sashigane-zu* on plates to create three types of *Kataita* at full scale: for the eave's front plate, which differs according to *Kiwari*; for the same type joints; and for each roof's rafters, which change along with the roof curves. Carpenters transmit their ways of drawing *Sashigane-zu* by

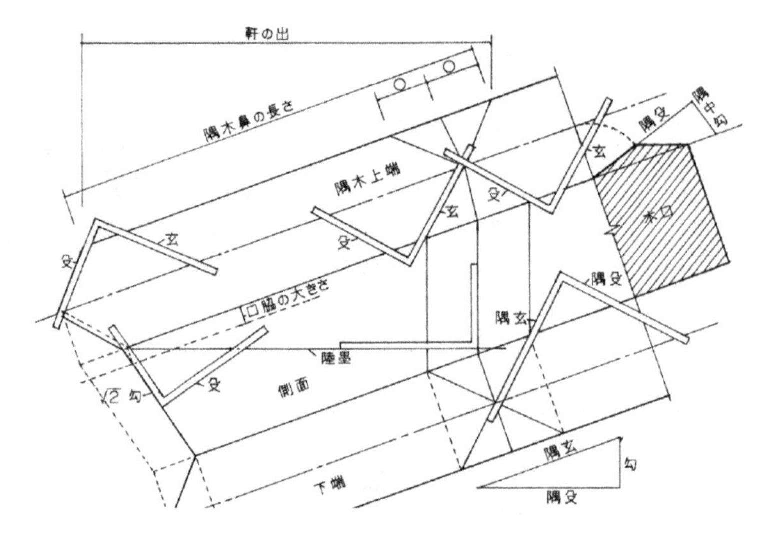

Fig. 23.5 An explanatory drawing of a *Sashigane-zu* of a roof rafter

recording them in a book called *Hinagata* (雛形, model), and according to *Hinagata*, carpenters draw the intersecting lines of each part on three surfaces continuously. Thus, they create a developmental elevation of the intersecting lines of the joint in order to see their three-dimensional relationships in a two-dimensional composition of four surfaces at a glance (Fig. 23.5).

To draw this developmental elevation, carpenters first set the horizontal line as the basis; however, they do not make it correspond with the total organization. Carpenters use *Kanejaku* not only for drawing straight lines but curved lines as well. The *Kanejaku* is tilted to plot each point of the curve at a small scale, then similar triangles are drawn continuously according to the curve, and the length of each part calculated with the Pythagorean theory. Then, the length of each part is multipled into a larger scale curve. In the process of drawing lines, *Kanejaku* technology uses a limited number of angles based on the simple combination of similar triangles.

Sashigane-zu is, thus, an approximate drawing, and the ways of making *Sashigane-zu* and *Kataita* differ for each carpenter, because they are based on the actual measurement. Also, when a full-scale curved form is drawn with *Kanejaku*, its height and length are usually modified slightly on site to make the curve appear smooth, and the arrangement for such an effect of optical illusion was kept as a trade secret in the carpenter's school.

***Sashigane-zu* as a Medium** The training of younger carpenters was based not on their logical comprehension of joint structure but on their habitual memorization of the method of drawing a *Sashigane-zu*. In that sense, again, *Sashigane-zu* was not a logical drawing but a medium enabling carpenters to connect their ideas to their practice by going through a serious of actions: drawing lines, checking by comparison with real objects, adjusting the lines of the drawing, measuring the parts of objects, calculating the length with the *Kokogen* method (Pythagorean theory), and formulating the lines and angles according to the result of this series of acts. Carpenters have to go back and forth between substantial object, algebraic mathematics, and geometric presentations to put their ideas into practice. In other words, their physiological and visual recognition are connected to the algebraic and geometrical logics by the approximate solution with *Kanejaku Kokogen* method. The lines of his *Sashigane-zu* are not the imaginative lines but the actual cutting lines directly connected to the act of construction. This role of *Sashigane-zu* as a medium is only possible because of its visual and empirical aspects, and projects a fundamental question on the meaning of the explanation of the invisible parts and systems in our architectural drawing.

Japanese carpenters have continued using this *Sashigane-zu*, but scholars regard it as too basic, primitive, and illogical to be theorized. However, *Sashigane-zu* is a simplified practice that can be adapted to the reality of the construction site. It utilizes only the limited number of angles and types of intersecting lines based on simple triangle assimilations, which carpenters can easily memorize. It also enables carpenters to modify the forms of joints according to the situation: the various nature of materials, site conditions, and complicated structural relationships between parts.

Innovations with *Sashigane*

In this section I am going to explain two examples that show the kind of innovative relationship created by *Sashigane* between architectural parts and the whole. One specific characteristic of historic Japanese architecture is that, in the formalization of *Kiwari*, carpenters came to focus on the elaboration of joints and ornament independent from the total coordination of architectural form.

***Kesen* (気仙) Architectures** In the historical architectures of *Kesen*, which is well known for having produced many talented carpenters who traveled for major projects all over Japan, there are typical examples of the extravagant elaboration of details over the total balance, such as the sculpted roof rafters of Fumon Temple. Too much emphasis on ornament sometimes transforms the structural order of Kesen architecture (Kenji 1978; Segawa 1998). In the *Saiko*-temple (西光寺) gate, built in 1567, the main central cross beam is unproportionally large and is set horizontally in order to strengthen the horizontal bearing force against frequent earthquakes in this region (Fig. 23.6).

There is no material explaining why the crossbeam is set horizontally, but from a historical comparative study with the other crossbeams, it probably started as vertical, then evolved into a square beam with sculptures projecting from the side walls, finally turning into a horizontal element. Thus, crossbeams began as an ornament and eventually changed the total structural system of the architecture. This is an example of the innovation of carpentry's ornamental technology moving

Fig. 23.6 Saiko-temple gate with a projecting horizontal crossbeam from *Catalog for Kesen daiku ten*

beyond the stylistic and structural canons in order to adapt to the local condition of architecture.

Kintaikyo (錦帯橋, Kintai Bridge) Another characteristic of historic Japanese architecture is that Japanese carpenters created new forms and structures by adapting Chinese models to the Japanese geographical conditions. Kintaikyo is a 35-m, five-arch bridge spanning the Nishiki river in the city of Iwakuni (岩国). It has been rebuilt at least nine times since 1639 due to its repeated destruction by the strong currents of the river (Izumi 1996). Iwakuni feudal load Kikkawa Hiromasa formed a group of engineers and scholars of different disciplines—mathematician, carpenter, masonry, metal workman, surveyor, earth worker—to create the bridge. (Due to the popularity of *Wasan* games, many mathematicians traveled all around Japan and worked for big projects at that period.) There is a small picture of a Chinese bridge of *Saiko-shi* (西湖誌) given by a Chinese monk Dokuryu to Kikkawa, but it does not indicate any technological methods for construction.

The form of the Kintaikyo bridge consists of four layers of eleven continuous cantilever beams supported by ten cross beams from both sides, which were connected by center cross beams (Fig. 23.7).

There are not sufficient materials to know how the construction of the bridges was carried out and how the details were modified from one to the next. However, according to a book on Kintaikyo's construction by historian Shinagawa Moto, there were small misalignments between each bridge's base and carpenters had to arrange the height and length of the curves and angles of beams slightly differently in each case so that the bridges would appear uniformly curved at the construction site (Moto 1984). The precise kind of wood used for each part of the bridge was decided depending on where the part was to be put, and full-scale *Sashigane-zu* were drawn. For the contraction and expansion of each piece, which also depended on what part of the curve the piece was put, carpenters modified their length and curves to 1/1,500 of the distance of beams.

In every renovation of the bridge, because of the lack of the *Hinagata* drawings of each beam, and because of the natural modification of the bridge curves due to

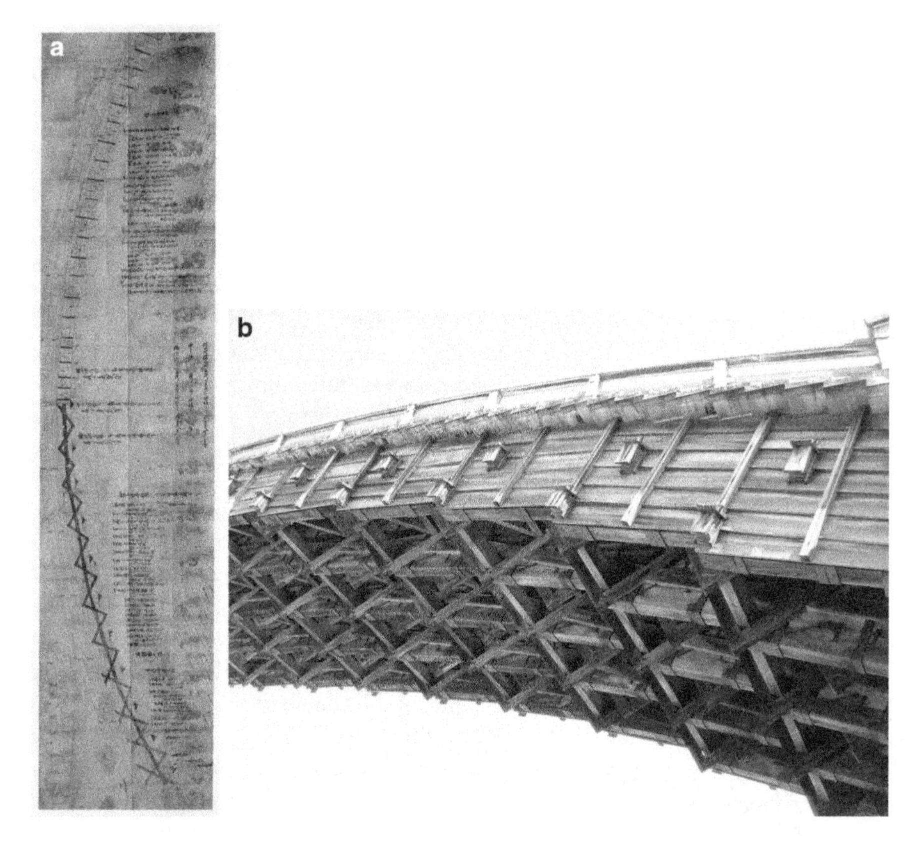

Fig. 23.7 Kintaikyo: Cantilevers and a construction drawing with records of modifications from *Catalog for Kintaikyo ten*

the climate and the weight of the passengers, carpenters had to re-measure every part and rearrange them to create new *Sashigane-zu*. The structural system and forms of the bridge were obviously much advanced compared with other constructions in Japan of the same period but, surprisingly, carpenters and other engineers were able to build it in a series of reconstructions with only the small picture to begin with.

Codification of *Sashigane-zu* with Mathematics

In the Edo period, because of the governing influences of the institutionally authorized carpenters to realize the increasing number of big scale projects, carpentry technologies were ruled more and more by the traditions and canons of each school. Heinouchi Masaomi, originally a mathematician and later the master carpenter of Tokugawa Shogun, published *Shokakujutu-shinsho* (諸家矩術新書) to analyze and organize *Sashigane* according to charts of stereographic projection (Masaomi 1848). He declared that without mathematics architecture cannot acquire a logical structure and that carpenters cannot use *Kanejaku* without learning mathematics. The word *Kikujutu* for his theories was derived from the etymology of Dutch surveying. Heinouchi intentionally used this word to emphasize the basis of his theory in Western mathematics, so avoiding empirical adaptations and ambiguous differences in *Sashigane-zu*, and reorganizing Japanese carpentry technology scientifically for the next generation.

Heinouchi introduced different types of drawings into *Sashigane-zu* through the method of stereographic projection: putting together plans, elevations and sections in one drawing; drawing coordination lines between the development drawings of each side of joint; organizing their horizontal lines; three dimensional perspectives of the joints; and three dimensional analysis drawings of structural frames (Fig. 23.8). He also calculated the angles and arrangement of the most complicated traditional fan frame roof rafter system. Thus he clarified the invisible parts of joints in *Sashigane-zu*, which used to be interpreted approximately by visual and empirical understanding, and calculated the angles by trigonometry. Therefore, he unified the logic of geometrical drawing with algebraic calculation and empirical application, and put together all the different ways of visualizing an architectural form.

However, Heinouchi still presupposed the usage of *Kanejaku* and set the *Kokogen* method as the premier principle of his deductive logical analysis of *Sashigane-zu*. He did not try to reframe the proportional system of *Kiwari* to establish the consistency with the geometrical orders in the construction of the parts by connecting the two mathematics—ideal and phenomenological. Thus he introduced ideas and representational methods of Western geometry, not as fundamentally different framework of architectural creation, but as another codification system of Japanese carpentry technology. Heinouchi presented accurate analysis of joint sections and calculations of the length and angles of their lines of intersection, however, this information were too complicated and difficult for the average carpenter, who was unable to utilize Heinouchi's theory in making *Kata-ita* and using *Kanejaku* at construction sites.

Conclusion

Japanese carpenters did not objectify and express their ideas and methods in writing, and their *Kataita* and *Kiwari* drawings were all executed on fragile wooden plates. Thus, research materials for the historical ideas and technologies of carpenters are rare. In this chapter, I have examined the limited historical materials along with the historical materials of *Wasan*, and tried to shed new light on the relationship between historic Japanese architecture and *Wasan*. The mathematical application of *Kanejaku* to the algebraic geometry of *Wasan* and the drawing method of *Sashigane-zu* both enabled the approximate unification

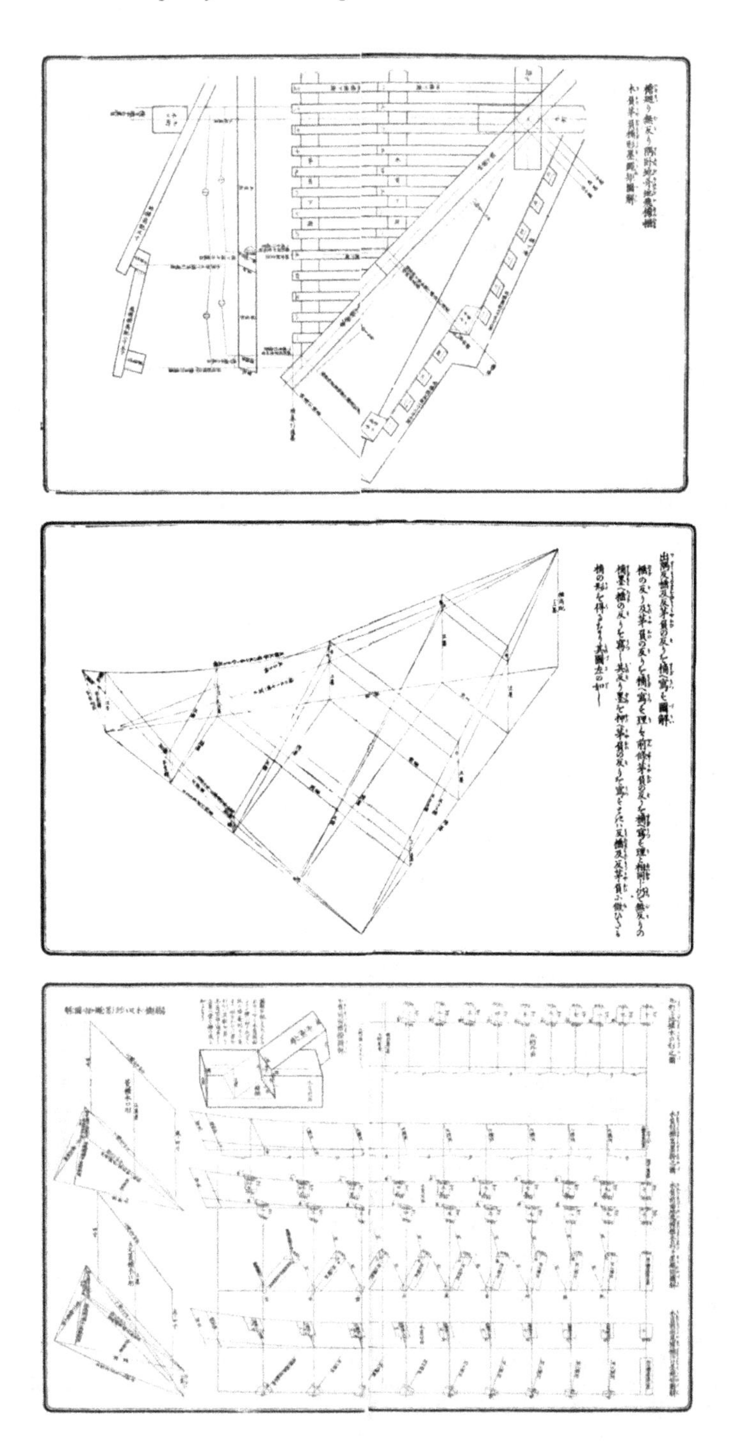

Fig. 23.8 Heinouchi Masaomi's Kikujutsu Drawings from *Shoka Kujutu Shinsho*

between the real, sensible, comprehension of architectural forms, their graphical representations, and the construction technologies. The adaptations by individual carpenters and the role of *Sashigane* as a medium helped to situate Japanese architecture in its local contexts. As I explained, such phenomenological mathematics, contrary to ideal mathematics, was present among the historic Japanese carpenter's ideas and technologies.

Actually, it is difficult to define whether ideal or phenomenological mathematics came first in the production of architecture. However, in the case of historic Japanese architecture, I have shown that the latter was incorporated with and lead the former to create the characteristic features of architecture. With that mathematics, the ideas and technologies of Japanese carpenters developed in a kind of closed system to create the ultimate elaboration of details of joints and ornaments; in turn, they did not create groundbreaking architectural forms and spaces as a whole. However, I believe that recognizing the dynamic contribution of phenomenological to ideal mathematics in architecture will present a new framework for understanding historic architectural forms.

Biography Izumi Kuroishi teaches architectural and urban theory in Japan. First she worked on the idea of architectural composition theories of French classicism and the design process theory of Modern architecture. After 8 years in design practice she defended her PhD thesis on the critical cultural approaches in Japanese modern architecture. She has written articles and presented papers on the modular systems of Le Corbusier and Ikebe Kiyoshi. Her interest in Japanese carpenters' technology is based on her research and practice of traditional wooden houses. She has also been working on the issues of phenomenology of urban spaces ethnography of interior space the garden suburb development and the design history of modern Japanese houses. She is the author of *Outside knowledge of architecture: Kon Wajiro ron* and her editorial work *Entwined perspectives in the colonized land of East Asia* is to be released at the end of 2013.

References

GIUSTI, E. 1999. *Ipotesi sulla natura degli oggetti matematici*. Torino: Bollati-Boringhieri. Japaneses ed. *Kazu wa dokokara kitanoka*, S. Ken, trans. Tokyo: Kyoritu Shuppan, 1999.

HIDEO, O. 1985. Kenchiku Kikujutsu. Pp. 302-317 in A. Yoshihiko et al., eds. *Koza Nihon Gijutu no Shakaishi 7 Kenchiku*. Tokyo: Nihon Hyoronsha.

HIROSHI, O. 1985. *Kinseiki Kokan Kikujutusho no sashizu ni mirareru zugakuteki seishituni kansuru kennkyu*. Master Diss. Tokyo: Kogyo University.

HIROTARO, O. and Yotaro, I. eds. 1971. *Shomei*. Tokyo: Kajima Shuppan.

IZUMI, M. 1996. *Ezu de miru iwakuni*. Iwakuni: Iwakuni Choko Kan.

KATSUSHIGE, K. 1978. Kiku to Kikujutsu. *Edo Kagaku Koten Sosho*, A. Kunio et al., eds., **16**: pp. 59-91. Tokyo: Kowa Shuppan.

———. 1985. Sashigane to Sashigane sanpo. *Sugakushi Kenkyu (Journal of History of Mathematics)* **107** (October 1985): pp. 28-43.

Kazuyoshi, F. 1988. Wasan sekisan sho shiryo toshiteno tokushitsu. Pp. 161-188 in N. Akira, ed. *Gushikenki no kenkyu*. Tokyo: Inoue Shoin.

Kenji, H. 1978. *Kesen daiku*. Kesennuma: NSK Shuppan.

Masamochi, I. and Hokusai, K. 2002. *Hidasho Monogatari*. Trans. S. Asahiko. Tokyo: Kokushokankokai (originally published in Edo period).

Masaomi, H. 1848. Shokakujutu-shinsho. *Edo Kagaku Koten Sosho*, 1978. A. Kunio et al., eds., **16**: pp. 160-279. Tokyo: Kowa Shuppan.

Moto, S. 1984. *Kintaikyo no hanashi*. Iwakuni: Yomei sha.

Naomi, O. 1971. *Bansho*. Tokyo: Hosei University Shuppan.

Norihito, N. 2000, *Bakumatsu Meijiki Kikujutu notenkaikatei no kenkyu*. E.D. Diss. Shinjuku: Waseda University.

Segawa, O. ed. 1998. *Kesendaiku Ten Catalog*. Iwate: Iwate kenritu Hakubutukan.

Sekino, M. 1947. Nihon daiku gijutu no hattatu shi. *Kenchiku zasshi*, **733** (August): pp. 17-20.

Takeshi, N. 1985. Kenchiku sekkei gijutu no hensen. Pp. 74-97 in A. Yoshihiko et al., eds. *Koza Nihon Gijutu no Shakaishi 7 Kenchiku*. Tokyo: Nihon Hyoronsha.

Teijiro, M. 1973. *Daiku dogu no rekishi*. Tokyo: Iwanami Shinsho.

Yoshio, M. 1977. Kanejaku Sanpo, *Sugakushi Kenkyu (Journal of History of Mathematics)*, **10** (November): pp. 3-8.

Chapter 24
On Some Geometrical and Architectural Ideas from African Art and Craft

Paulus Gerdes

Introduction

The rich cultural diversity of Africa is clearly visible in the wide range of house decorations, of architectural styles, and of settlement and enclosure shapes (Oliver 1971; Denyer 1978; Bourdier and Trinh 1985; Guidoni 1987; Eglash 1998). Unity within this diversity appears in the importance of the artistic and geometrical exploration of symmetrical forms and patterns. Shapes and decorations are not static; they may vary with the seasons, mark changes in the family composition, or be chosen for special ceremonies (Wenzel 1972; Courtney-Clarke 1986, 1990; Changuion, Matthews and Changuion 1989; Gerdes 1998d, 1996). Some traditional African architectural ideas may have been derived from or suggested by experience and knowledge in other cultural spheres, such as basketry (Gerdes 1990: 107–111, 1998c).

In this chapter, some examples of geometrical ideas in traditional African building will be presented, as well as some further suggestions for architectural shapes inspired by African art and craft.

First published as: Paulus Gerdes, "On Some Geometrical and Architectural Ideas from African Art and Craft", pp. 75–86 in *Nexus II: Architecture and Mathematics*, ed. Kim Williams, Fucecchio (Florence): Edizioni dell Erba, 1998.

Paulus Gerdes (1952–2014)

K. Williams and M.J. Ostwald (eds.), *Architecture and Mathematics from Antiquity to the Future*, DOI 10.1007/978-3-319-00137-1_24,

Rectangle Constructions

Most African peoples south of the Sahara traditionally build houses with circular or rectangular bases. Among the Mozambican peasantry, two methods for the construction of the rectangular bases are common. In both cases, the rectangular shape does not appear as the result of copying a rectangle or of starting with the construction of the right angles one by one (Gerdes 1998a: Chap. 12).

Figure 24.1 illustrates the first method. The house builders start by laying down on the floor two long bamboo sticks of equal length (a). Then these first two sticks are combined with two other sticks also of equal length, but normally shorter than the first ones (b). Now the sticks are moved to form the closure of a quadrilateral (c). The figure is further adjusted until the diagonals—measured with a rope—become equal (d). Then, from where the sticks are now lying on the floor, lines are drawn and the house builders can start. This construction method reflects the knowledge that when the diagonals of a parallelogram become congruent, the parallelogram becomes a rectangle. The rectangular shape appears as the result of a continuous transformation of a quadrilateral: a quadrilateral converges to a rectangle, the rectangle being the limit.

Figure 24.2 illustrates the second method. The house builders start with two ropes of equal length that are tied together at their midpoints (a). A bamboo stick, the length of which is equal to that of the desired width of the house, is laid down on the floor and, at its endpoints, pins are hit into the ground. An endpoint of each of the ropes is tied to one of the pins (b). Then the ropes are stretched and at the remaining two endpoints of the ropes, new pins are hit into the ground. These four pins determine the four vertices of the house to be built (c). This rectangle construction demonstrates the knowledge that as soon as two equally long curves that intersect in their midpoints become, in a continuous transformation, straight segments, they will constitute the diagonals of a rectangle. The defining elements of the rectangle are its width and diagonal length. With both construction methods, the four right angles of the rectangle appear exactly at the same time.

Changing Wall Decorations

Among the Ngongo, one of the ethnic-cultural groups of the Kuba Kingdom in central Congo/Zaire, the decoration of the walls of the houses and palaces with mat work was widespread. A collection of such architectural mats observed by the Hungarian ethnographer Torday was published in 1910. Figure 24.3 presents examples. The plane patterns have various symmetries. Horizontally one sees the sticks which are woven together by the vertical lianas. The use of architectural mats—Fig. 24.4 presents another example, this time a detail of a plaited mat that decorates part of the wall above the door of the house of a Bamileke chief in Cameroon—is one possible way to change decorations in agreement with the

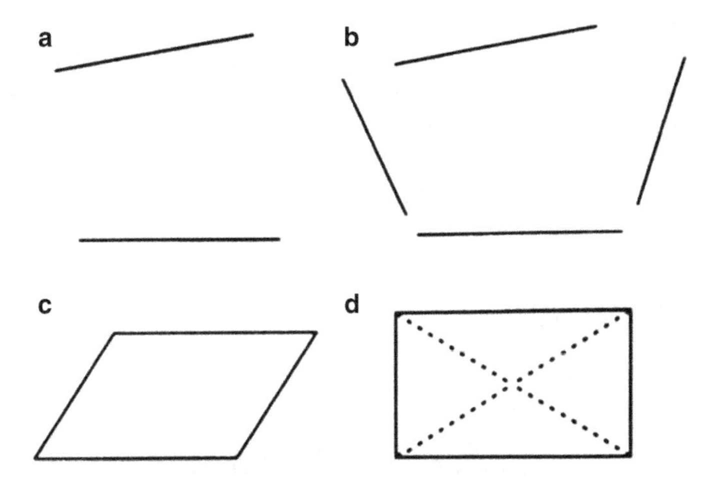

Fig. 24.1 Illustration of the first traditional method for construction of the *rectangular* base of a house in Mozambique. Drawing: author

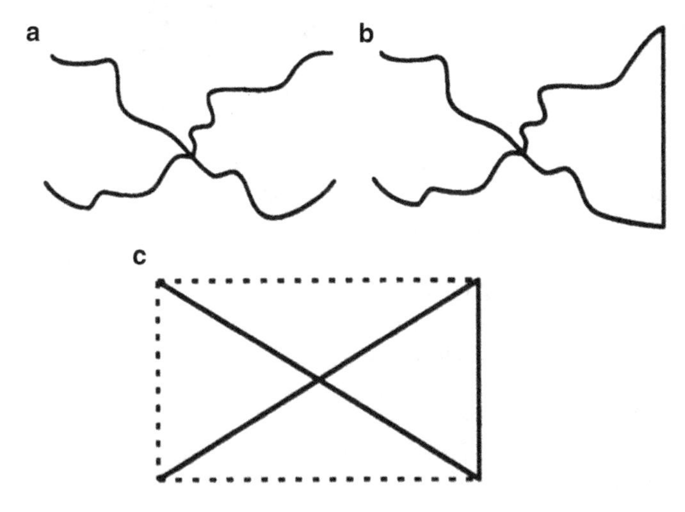

Fig. 24.2 Illustration of the second traditional method for construction of the *rectangular* base of a house in Mozambique. Drawing: author

season, ceremony, or life cycle. Another context in which this is possible is the *litema* tradition.

In Lesotho and neighbouring zones of South Africa, Sotho women developed a tradition of decorating the walls of their houses with designs. The walls are first neatly plastered with a mixture of mud and dung, and often coloured with natural dyes. While the last coat of plastered mud is still wet, the women engrave the walls, using their forefinger. Their art is seasonal: the sun dries it and cracks it, and the rain washes it away. An entire village is redecorated before special occasions such as

Fig. 24.3 Examples of architectural mats from the Ngongo (Congo/Zaire)

Fig. 24.4 Examples of an architectural mat from the Bamileke (Cameroon)

engagement parties, weddings, and important religious celebrations. The Sotho women call their geometric patterns *litema* (singular: *tema*). Symmetry is a basic feature of the *litema* patterns. They are normally built up from basic squares. The Sotho women lay out a network of squares and then they reproduce the basic design in each square (see the example in Fig. 24.5). Often the symmetries are two-colour symmetries: horizontal and vertical reflections about the sides of the squares reverse the colours (see the examples in Fig. 24.6) (Gerdes 1998a: Chaps. 1 and 6).

'Squaring the Circle': Suggestions from Basketry for Architecture

Transport and storage baskets with square bottoms and circular openings may be woven in such a way that the strands on the walls are either perpendicular to or make angles of 45° with the sides of the square. The Makonde basket weavers in northern Mozambique use the second method to make their *likalala* baskets. They start making the basket by plaiting (twill weave) a square mat with $6 + 2(4 \times 4)$ or 38 strands in both directions (Fig. 24.7a). At the midpoints A, B, C and D of the sides of this square the corners of the basket will rise after folding and interweaving the outstanding triangles 1, 2, 3, and 4 (Fig. 24.7b).

A diagonally woven basket like a *likalala* has the property that each horizontal intersection, including the bottom square and the top circle have the same

Fig. 24.5 Examples of wall decorations from Sotho. Drawing: author

Fig. 24.6 Examples of wall decorations from Sotho with two-colour symmetry. Drawing: author

circumference. The inverted, upside-down *likalala* may constitute an attractive architectural structural shape. The base circle converges to the square roof: although all horizontal intersections are smooth (admitting in all their points tangent lines), the limit curve, the square, is not.

Among the Bassari, who live in the Senegalese-Guinea border region, a very special 'basket' of the *likalala* type is woven. It is the extreme case: the lowest numbers of strands, 2 or 4 in each direction, are used. Figure 24.8a displays the 2×2 mini-basket. They call such an over-one-under-one plaited mini-basket *epog edepog*; it is used as a penis holder.

Sometimes Bassari weavers do not make a rim but finish the mini basket in the same way as it was started. They close it with a square top. The closed 'basket' is twisted: the top square is rotated around the basket's axis about an angle of 90° in relationship to the bottom square (Fig. 24.9). The height of the basket is one and a half times the side of the square. The small closed baskets are used to make a jingle collar, called *bamboyo*. A collar is composed of about 20 of such small closed 'baskets'. The sides of the squares are about 2–3 cm. The thread passes through the centres of the bottom and top squares and the little bell baskets contain small stones that ring when dancers wear the collar during the *omangare* feast. The bell basket may also constitute an interesting architectural shape. The base square converges smoothly to the middle circle and then returns to the square roof. All horizontal intersections have the same circumference.

It is possible to weave a closed twisted basket still smaller than a *bamboyo* bell. To avoid too much pressure on the strand, we may make it out of a strip of card

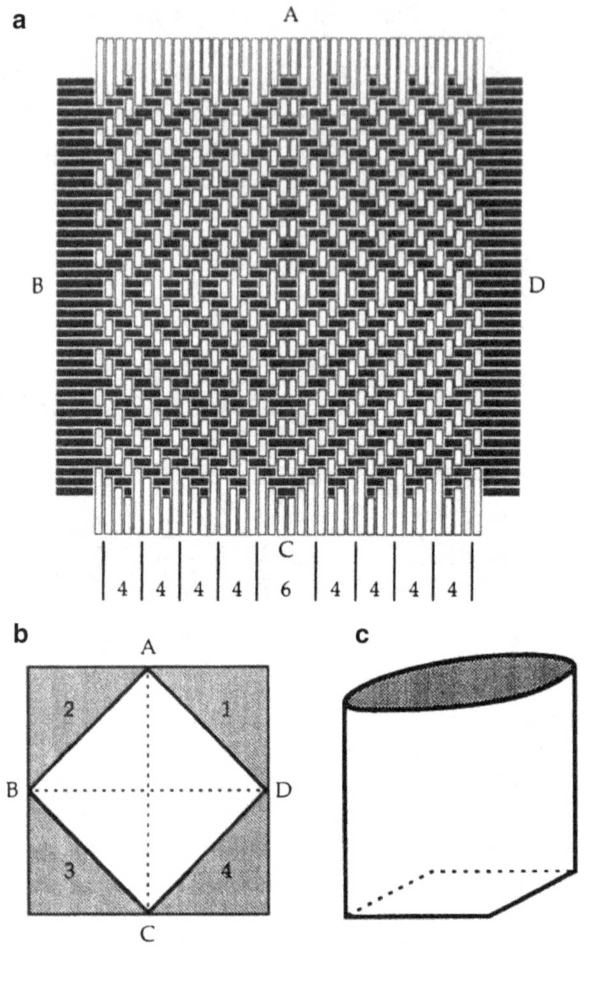

Fig. 24.7 (**a**) First *square* mat. (**b**) Outstanding *triangles* to be folded. (**c**) Shape of the *likalala* basket. Drawing: author

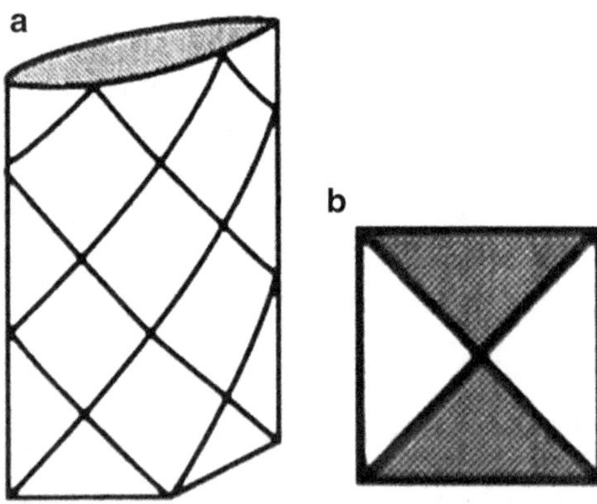

Fig. 24.8 (**a**) Shape of *epog edepog* mini-basket. (**b**) Bottom of *epog edepog* mini-basket. Drawing: author

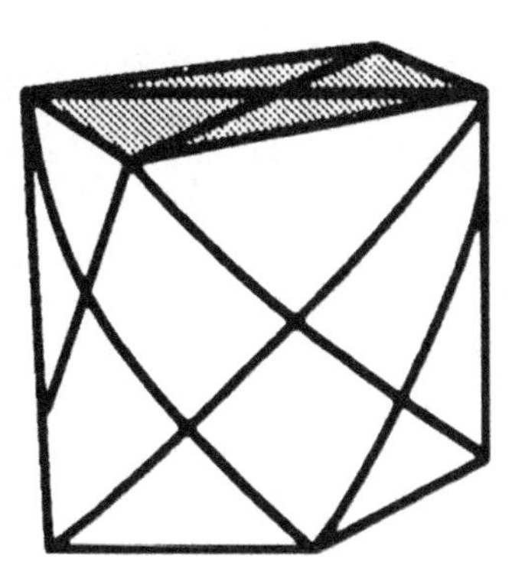

Fig. 24.9 Small twisted closed basket. Drawing: author

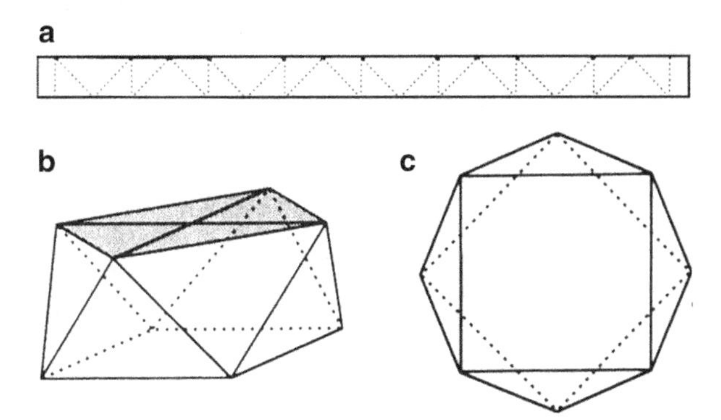

Fig. 24.10 (**a**) Strip with folds. (**b**) Twisted decahedral basket. (**c**) Decahedron seen from above. Drawing: author

board paper and fold it several times in such a way that the strip after unfolding presents the folding lines shown in Fig. 24.10a. Only one such a strip is sufficient to weave the twisted decahedral basket in Fig. 24.10b, which may inspire a beautiful architectural shape. Figure 24.10c shows the decahedron from above. All its horizontal intersections have the same circumference and each of the eight wall faces is a quarter of the base and top squares (Gerdes 1998a: Chap. 16).

Fractal Structures: Suggestions from Basketry for Architecture

In the North of Mozambique, in the South of Tanzania, in the Congo/Zaire region and in Senegal, a pyramidal basket is woven—called *eheleo* in the Makhuwa language in the North of Mozambique (Fig. 24.11). In Tanzania and Mozambique it is used as a funnel in the production of salt (Gerdes 1998a: Chap. 10). The *eheleo* has the form of a triangular pyramid: its base (or top in the case of the funnel) is an equilateral triangle and its other three faces are isosceles right triangles. Figure 24.12 presents a rotated *eheleo* pyramid as part of a cube. The shape of

Fig. 24.11 *Eheleo-funnel*. Drawing: *Marcelino Januário*

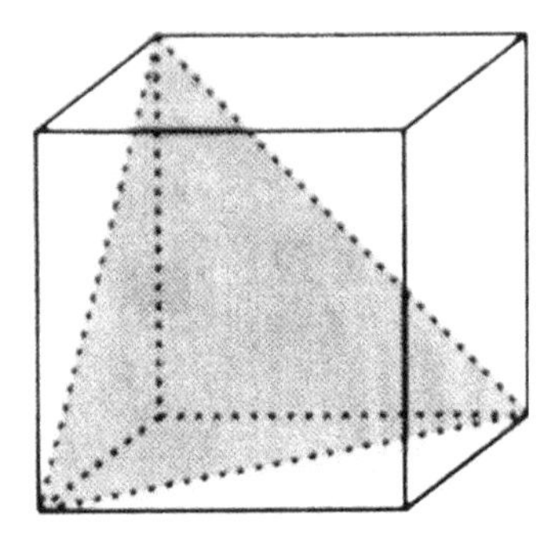

Fig. 24.12 *Eheleo-pyamid* inside a cube. Drawing: author

the *eheleo* is interesting for architectural exploration. Figure 24.13 shows the composition of a structure that explores the right angles of the eheleo, and the idea of similarity as current in African art. The height of each new pyramid that is added to the structure is a fixed proportion of the height of the last one (In Fig. 24.13 this proportion is 2/3). Progressing with this building up and introducing pyramids inside others (for instance, using glass or another transparent material) allows us to visualize a fractal architectural structure. Another way to produce a fractal architectural structure with eheleo pyramids is by joining differently sized eheleo pyramids placed on their equilateral-triangular base. Figure 24.14 displays the first four phases in building up this fractal structure (seen from above).

Fig. 24.13 Composition of an *eheleo* inspired shape. Drawing: author

Hexagonal Weaving: Suggestions from Basketry for Architecture

In various parts of Africa and also in several other parts of the world, artisans weave baskets with a pattern of regular hexagonal holes. The strands are woven one-over-one-under in three directions leading to a very stable fabric (Fig. 24.15). In Kenya, Madagascar and Mozambique, craftsmen use the hexagonal weaving technique to produce transportation baskets and fish traps. The technique is used for various other purposes: in Kenya for making cooking plates; among the Pygmies (Congo/Zaire) and in Cameroon for weaving carrying baskets; in northeastern Congo/Zaire among the Meje for covering pots; among the Mangbetu for weaving hats. Basket makers who use the hexagonal weaving technique also know that when they want to curve a woven surface more than what is done with a cylindrical surface (in other words, when they want a basket with corners), this is possible if they introduce one or more smaller holes, normally pentagonal in shape (Fig. 24.16). This idea may be explored in architecture in the style of Buckminister Fuller. Figure 24.17 shows a woven, semi-spherical dome structure with a central pentagonal hole on the top surrounded by a layer of pentagonal holes. Many other related dome structures may be conceived: a pentagonal hole on the top surrounded

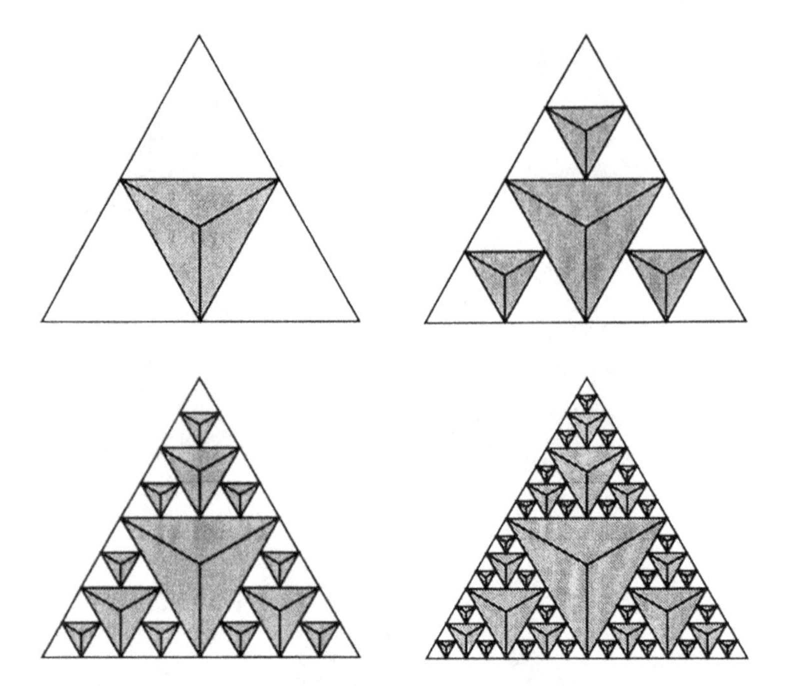

Fig. 24.14 Composition of another fractal architectural structure inspired by the *eheleo*. The first phases are illustrated. Drawing: author

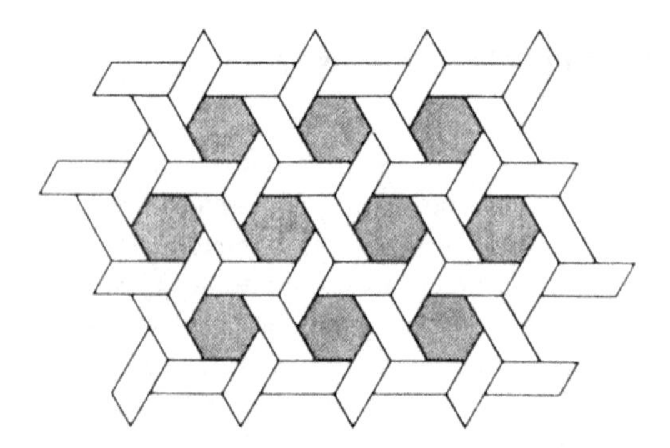

Fig. 24.15 *Hexagonal* weaving technique. Drawing: author

by a layer of hexagonal holes; a hexagonal hole on the top surrounded by a layer of alternately pentagonal and hexagonal holes; pentagonal hole on the top surrounded by one layer of hexagonal and one layer of pentagonal holes (Gerdes 1998a: Chap. 13, 1998b: 40–45).

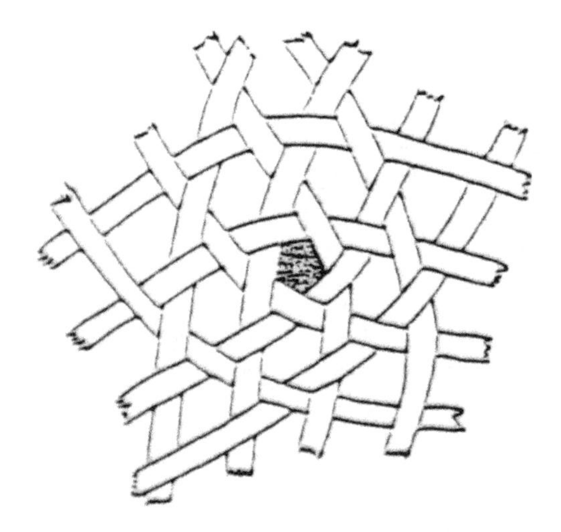

Fig. 24.16 *Pentagonal* hole in *hexagonal*

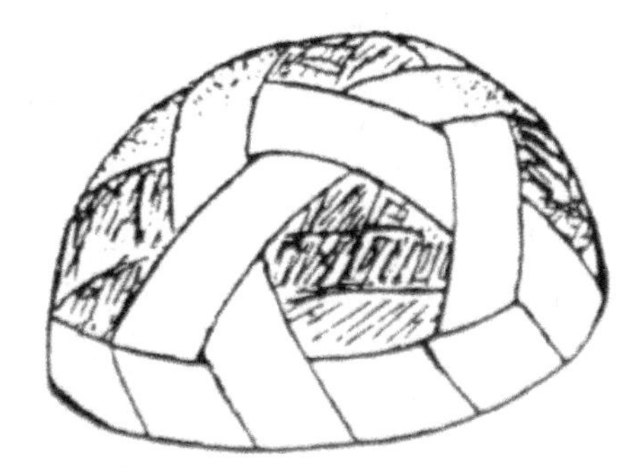

Fig. 24.17 Semi-spherical dome structure web with *pentagonal* holes only. Drawing: Marcos Cherinda

Acknowledgments The chapter was written during sabbatical leave at the University of Georgia, Athens. USA. The author thanks the Research Department of the Swedish International Development Agency for the financial support to MERP.

Biography Paulus Gerdes (1952–2014) was a Professor of Mathematics and Vice-President for Southern Africa, African Academy of Sciences. He is Chairman of the African Mathematical Union Commission on the History of Mathematics in Africa; President of the International Study group on Ethnomathematics; and Fellow of the International Academy for the History of Science. He is the former President (Rector) of the Universidade Pedagógica in Mozambique. He is the author of several books on geometry and culture, such as *Geometry from Africa, Otthava: Making Baskets and Doing Geometry in the Makhuwa Culture in the Northeast of Mozambique*, African Basketry: A Gallery of Twill-Plaited Designs and Patterns, *Sipatsi: Basketry and Geometry in the Tonga culture of Inhambane (Mozambique, Africa), Geometry and Basketry of the bora in the Peruvian Amazon* and *Timhlèlò, Interweaving Art and Mathematics: Colourful Circular Basket Trays from the South of Mozambique*.

References

Bourdier, Jean-Paul and Trinh MINH-HA. 1985. *African Spaces: Designs for Living in Upper Volta*. New York: Africana Publishing Company.

Changuion, P., T. Matthews and A. Changuion. 1989. *The African Mural*. Cape Town: Struik.

Courtney-Clarke, Margaret. 1986. *Ndebele: The Art of an African Tribe*. New York: Rizzoli.

——. 1990. *African Canvas*. New York: Rizzoli.

Denyer, Susan. 1978. *African Traditional Architecture*. London: Heinemann.

Eglash, Ron. 1998. *African Fractals* (in preparation).

Gerdes, Paulus. 1990. *Ethnogeometrie. Kulturanthropologische Beitrage zur Genese und Didaktik der Geometrie*. Bad Salzdetfurth, Germany: Verlag Franzbecker.

——. 1996. *Femmes et Géométrie en Afrique Australe*. Paris and Montreal: L'Harmattan.

——. 1998a. *Geometrical And Educational Explorations Inspired By African Cultural Activities*. Washington DC: MAA.

——. 1998b. Molecular Modeling Of Fullerenes With Hexastrips. *Chemical Intelligencer* (January 1998): 40–45.

——. 1998c. *On Culture and the Awakening of Geometrical Thinking*. Minneapolis: MEP Press.

——. 1998d. *Women, Art and Geometry in Southern Africa*. Trenton, NJ, USA & Asmara, Eritrea: Africa World Press.

Guidoni, Enrico. 1987. *Primitive Architecture*. New York: Electa/Rizzoli.

Oliver, Paul (ed.). 1971. *Shelter in Africa*. London: Barrie and Jenkins.

Wenzel, Marian. 1972. *House Decoration in Nubia*. Toronto: University of Toronto Press.

Chapter 25
Design, Construction, and Measurement in the Inka Empire

William D. Sapp

Introduction

When the Spaniards entered the Andes in AD 1532, the Inka empire was less than 100 years old. When Pachacuti Inka Yupanqui assumed control of the Inkas in AD 1438 they were only one of several regional polities competing for power in the Central Andes. Under the leadership of Pachacuti Inka Yupanqui, his son Topa Inka Yupanqui, and his grandson Huayna Capac, the empire exploded in size. Between AD 1450 and AD 1528 they conquered and incorporated more than 80 separate polities, including Chimor, until then the largest empire in the New World. By the time the Spaniards arrived, the Inka empire stretched over 4,200 km and 32° of latitude, roughly the same distance from Glasgow, Scotland to Cairo, Egypt. It reached from the modern border of Ecuador and Columbia to the Maule River in central Chile, and from the Pacific Ocean to the cloud forests of the Amazon Basin and the *pampas* of northwestern Argentina.

Architecture and Expansion

Despite the relatively short duration of the Inka empire, there was a massive effort on the part of the state to acculturate conquered polities through the imposition of a single language and state-sponsored religion, the organization of a complex

First published as: William D. Sapp, "Design, Construction, and Measurement in the Inka Empire", pp. 133–145 in *Nexus III: Architecture and Mathematics,* ed. Kim Williams, Ospedaletto (Pisa): Pacini Editore, 2000.

W.D. Sapp (✉)
San Bernardino National Forest, 602 S Tippecanoe Ave., San Bernardino, CA 92325, USA
e-mail: billsapp@fs.fed.us

K. Williams and M.J. Ostwald (eds.), *Architecture and Mathematics from Antiquity to the Future*, DOI 10.1007/978-3-319-00137-1_25,

administrative system integrating local elites, and the colonization of rebellious and under populated territories with loyal citizens.

Imperial expansion and acculturation was facilitated by a massive building program. The state constructed administrative centres, storehouse complexes, ritual centres, country estates for the emperors and their lineages, terracing and irrigation systems to bring previously unproductive land under cultivation, and a 40,000 km road system. As part of this effort, a number of skilled craft specialists, including masons and carpenters, were removed from their local ethnic groups, exempted from the general labour tax and worked exclusively for the state.

The program of state construction undertaken by the Inkas adhered to a rather strict set of architectural canons that used a repetitive set of architectural spaces, elements and ornamentation. These features were plazas, free standing rectangular rooms, and trapezoidal niches, windows, and doorways. So amazingly ubiquitous are these features and the degree of formal unity in Inka architecture that the German naturalist von Humbolt wrote:

> ... but what seems to me worthy of the highest interest, is the uniformity of construction that one perceives in all the Peruvian monuments. It is impossible to examine attentively a single edifice of Inca times, without recognizing the same type in all the other monuments that cover the back of the Andes ... One should think that a single architect has built this large number of monuments (Protzen 1993: 11).

Architectural Canons

The plaza was the central focus of public ritual in Inka cities, administrative centres, and country estates. Many public ceremonies were held in the plazas of Cuzco, the Inka capital. Most corresponded to important dates in the agricultural cycle, including the two solstices. There were also ceremonies associated with the death and coronation of emperors. Many ceremonies lasted for several days and all were accompanied by sacrifices, recitations, dancing, and the consumption of large quantities of alcohol by all participants (Rowe 1946: 308–312). Plazas were usually constructed near the centre of a site, often astride the main road. In Cuzco, the main plaza stood at the intersection of the roads that led to the four quarters of the empire. Plazas were typically surrounded by enclosures and freestanding rectangular buildings.

The basic architectural element was the freestanding rectangular room. Rooms were often grouped symmetrically inside of a walled compound or *cancha* (Fig. 25.1). Rectangular rooms were used for virtually all types of activities, including habitation, workshops, palaces, temples, and administrative facilities. Larger rooms, called *kallanka*, were used as barracks and to house public rituals when inclement weather prevented ceremonies from being held outdoors. Freestanding rooms and *canchas* occur at all socio-economic levels of society, from farmsteads constructed of sod to the palaces of the emperors constructed of the most finely dressed stones. Around the two central plazas in Cuzco large *kallankas*

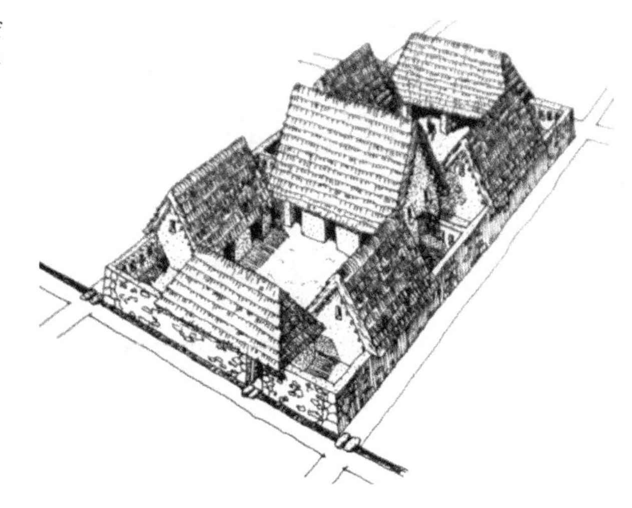

Fig. 25.1 Isometric view of a double *cancha* at the royal estate of Ollantaytambo. Image: author, after Lee (1996)

and *canchas* served as palaces, temples, and residences for nobility. Rooms were constructed with gable, hip, or shed roofs of thatch (Fig. 25.2).

Rectangular rooms and *cancha* enclosure walls were decorated with a set of architectural characteristics that are repeated in virtually all bureaucratic and elite constructions: symmetrically placed trapezoidal niches, windows and doorways. Most niches are window-sized, although door-sized niches also occur. Double jams were used on some doors to indicate high status. Doors were normally centred on one of the long walls of a room. Small rooms had one door, while larger rooms had more. Two *kallanka* that face the plaza at the provincial centre of Huánuco Pampa are about 75 m long. One *kallanka* has four doors evenly spaced on a long wall, while the other has nine doors.

Despite the fact that there were only a few architectural features utilized in a limited set of combinations, considerable variation is demonstrated by construction style and the design of individual buildings.

Construction and Design

The quality of Inka stonework reflects a combination of style and social status (Niles 1999: 229). The finest stonework, associated with the highest level of social status, is found in Cuzco. Much of the city was destroyed during the siege of Manco Inca in AD 1536–37. It was further destroyed by the Spaniards, who tore down most of the remainder in order to rebuild the city to reflect their own ideas of urban design. However, because Inka Cuzco was laid out on a grid pattern generally maintained by the Spaniards, many *cancha* enclosure walls have been preserved on the exterior of current buildings (Fig. 25.3). These walls provide examples of the finest stonework found in the empire. Walls of this style are constructed of solid,

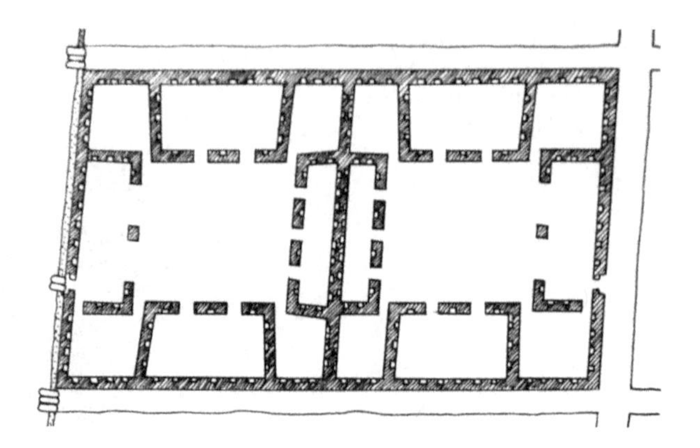

Fig. 25.2 Plan view of the same *cancha*. Image: author, after Protzen (1993)

dressed stone. Individual blocks are typically large and dressed on all sides. They represent a greater expenditure of labour than other styles of stone construction. Cuzco-style masonry occurs in two general sub-groups: irregular polyhedral blocks and coursed masonry of rectangular blocks. Walls were typically battered and joints were rusticated. The size of stones tended to decrease as the height of walls increased (Figs. 25.3 and 25.4). Cuzco-style masonry was used for some of the *canchas* and rooms at a number of country palaces, such as Machu Picchu and Ollantaytambo, and administrative centres, such as Huánuco Pampa.

Unworked fieldstone and mud mortar construction is associated with the lower levels of social status in the Inka empire. Around Cuzco this style of building was used for support facilities, such as storehouses and small rooms at royal estates. At provincial centres it was used for storehouses and lower status residential constructions.

In between high status Cuzco-style masonry and lower status fieldstone and mud mortar construction is Intermediate-style masonry, which consists of dressed, or partially dressed stones (Niles 1987: 211–212). The stones are polyhedral (Fig. 25.4), fitted (Fig. 25.5), or coursed (Figs. 25.6, 25.7, 25.8 and 25.9). Walls are generally constructed with two separate faces of stone and a core of rubble and dirt. Intermediate-style masonry is typically associated with royal estates and provincial administrative centres. It is the most common type of high status Inka construction outside of urban Cuzco.

All three styles of masonry exhibit the basic architectural features of Inka design. While it is clear from the symmetry demonstrated by Inka buildings that specific measurements and design criteria were employed, we do not know exactly how these designs were recorded or transmitted. Garcilasco de la Vega wrote that the Inkas built models of proposed constructions, but the existing models of *canchas* are rather crude, and they are not sufficiently detailed to transmit necessary design data for construction purposes (de la Vega (1609) 1961: 43–44). Juan de Betanzos wrote that the Inkas used cords to lay out proposed buildings, but he tells us nothing about what units of measurement were used (de Betanzos (1557) 1996: 45).

Fig. 25.3 Cuzco. Coursed, Cuzco-style masonry incorporated into a modern building. Note distinct batter, minimal rustication, and decreasing size of courses. Photo: author

Fig. 25.4 Lower retaining wall of Tarahuasi at Limatambo. Polyhedral, intermediate-style masonry showing significant rustication. Note decreasing stone size with increasing wall height. Photo: author

The design and construction of Huayna Capac's royal estate at Yucay was the responsibility of Sinchi Roca, the emperor's half brother (Niles 1999: 123). Corvée labour used to construct the royal estate was provided by a general labour tax, the primary source of imperial revenue. Under this system communities provided workers for state construction projects. On projects requiring expert knowledge,

Fig. 25.5 Machu Picchu, Room 1. Double masma constructed using intermediate-style masonry of roughly fitted stones. The ratio of niche to space width is approximately 1:3. Photo: author

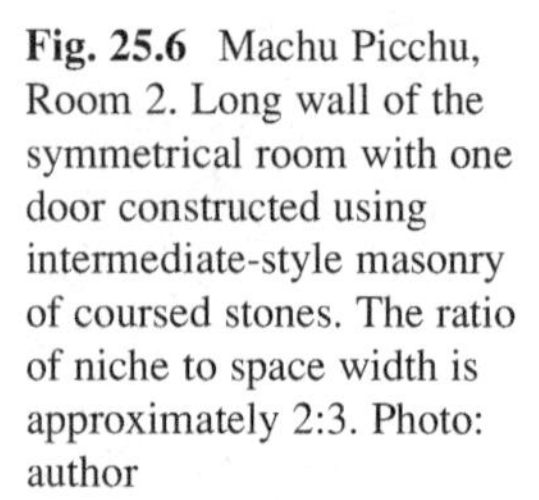

Fig. 25.6 Machu Picchu, Room 2. Long wall of the symmetrical room with one door constructed using intermediate-style masonry of coursed stones. The ratio of niche to space width is approximately 2:3. Photo: author

Fig. 25.7 Machu Picchu, Room 2. Short wall of the symmetrical room with one door constructed using intermediate-style masonry of coursed stones. The ratio of window to space width is approximately 2:4. Photo: author

Fig. 25.8 Machu Picchu, Room 3. Long wall of the symmetrical room with two doors constructed using intermediate-style masonry of coursed stones. The ratio of niche to space width is approximately 2:2. Photo: author

Fig. 25.9 Machu Picchu, Room 3. Short wall of the symmetrical room with two doors constructed using intermediate-style masonry of coursed stones. The ratio of niche to space is approximately 2:3. Photo: author

these workers laboured under the direction of trade-specific specialists. From this evidence it is possible to envision a system where a master architect oversaw the design and building of state structures, while quarrying, dressing, and construction were carried out by masses of corvée labour working under the direction of master craftsmen in the appropriate trade.

What Units of Measurement Were Used by the Inka?

Despite extensive studies of Inka architecture, we still don't know what measurements and mathematical relationships were used in the design of buildings and the placement of niches, windows and doors. Building layouts reflect Inka architectural canons, but substantial variations occur both within and between sites. Differences in niche width and spacing suggest that the onsite team of experts were given substantial leeway in design and construction, maintaining aesthetic values while fitting buildings to the local landscape.

We know that the Inka knew of the plumb bob, and possibly the sliding scale (Rowe 1946: 225). We also know that there are *Quechua* (Inka) words for anthropomorphic measurements including the half arm-span or yard (*sikya*), the forearm or cubit (*cuchuch*), the foot (*chaqui*), and the hand-span (*capa*). The possible use of anthropomorphic measurements in building design led Vincent Lee to posit that architectural measurements could be expressed as ratios of small whole numbers (Lee 1996: 10–13). According to Lee, these ratios represented the widths of niches, doors, windows, and the spaces in between them. The use of ratios based on anthropomorphic units to describe and design buildings would have allowed the use of *quipu* to transmit measurements. The Inka had no written language, but used *quipu*, mnemonic devices, to record accounting, tax, and census data. *Quipu* were sets of coloured strings in which different types of knots were tied. String colours indicated categories and knots indicated numbers of units. *Quipu* are ideally suited to carry the kind of information described by Lee.

According to Lee's hypothesis and our knowledge of Inka construction technology, an architect in the field could stretch a cord across the span to be occupied by a wall and drop plumb bobs to mark the placement of niches, windows, and doors. To establish a base unit to apply to the ratios, the appropriate anthropomorphic measurement would be taken from an onsite architect or craftsman using a sliding scale. This would account for variation between sites. Using an average height of 157 cm for a pre-Hispanic Andean native, anthropomorphic measurements are fairly simple to calculate (half arm-span: 81 cm; forearm: 40 cm; foot: 27 cm; hand-span: 20 cm) (Lee 1996: Fig. 16). Lee measured the placement of niches in walls at some 50 sites representing a cross section of construction styles. He found that he could express the relationship of niches and the spaces in between them as ratios of small whole numbers using anthropomorphic measurements as a basic unit. Using the largest unit of anthropomorphic measurement possible while still retaining ratios based on small whole numbers, it is possible to test the hypothesis on other existing Inka walls. Using Lee's methodology, we will examine the findings from three rooms at Machu Picchu.

Machu Picchu

Machu Picchu is one of the most recognized archaeological sites in the world. It is believed to have been a country estate constructed during the rule of Pachacuti Inca Yupanqui. This would establish its construction sometime between AD 1450 and AD 1473. It is located above the Urubamba River at the upper limits of the two canopy rain forest found on the western edge of the Amazon Basin. Most of the buildings at Machu Picchu are constructed of Intermediate-style masonry, although some are constructed of Cuzco-style masonry. The rooms reviewed here are constructed in the Intermediate-style. They include one half of a double *masma* of fitted stones and two rectangular buildings of coursed stones. The *masma* is a form of rectangular room in which one long side is left completely open (Gasparini and Margolies 1980: 165–169). A double room is formed when a rectangular building with a gable roof contains a dividing wall that runs between the short walls at the gable peak. The double *masma* at Machu Picchu has a gable roof, a dividing wall, and the two rooms thus formed each have one open side (Fig. 25.5).

In these rooms we will look only at the placement of niches and windows and not at the placement of doors. The measurements presented here were calculated at the bottom of the niches. Measurements were taken at the middle and top of niches as well, but the results could not adequately be described using anthropomorphic units.

Room 1: Double Masma

The double *masma* was reconstructed sometime after 1951. A photograph from that year indicates that the niches on the short walls were reconstructed in their original location. It is assumed that the long wall was similarly reconstructed to reflect the original placement of architectural ornamentation.

There are three windows on the long wall and two niches on the short walls of the double *masma*. The windows and niches are all approximately the same size. They are 40–44 cm wide at the bottom, 35–37 cm wide at the top, and 78–80 cm high. They are approximately 110 cm above ground level. On the long wall, the spaces between the niches, and between the niches and the ends of the wall range from 135 to 150 cm wide. Using a forearm as the unit of measurement, the long wall exhibits a ratio of niches to spaces of approximately 1:3. The spaces between the two niches on the short walls are approximately the same size as the spaces on the long wall. The spaces on either end of the short walls are smaller, ranging from 45 to 97 cm in width.

Room 2: Symmetrical Room with One Door

This room is situated immediately to the east of the double *masma*. There is a single entry flanked by two niches on one of the long walls. The other long wall contains four niches (Fig. 25.6). The short walls each have two windows (Fig. 25.7). The niches and windows are 56–60 cm wide at the bottom, 45 cm wide at the top, and 102 cm high. The bottoms of the niches are approximately 108 cm above ground level.

The spaces between the niches on the long wall range from 90 to 98 cm wide. Using an anthropomorphic foot as the unit of measurement, the wall demonstrates a ratio of niche to space width of approximately 2:3. The spaces at either end of the wall are 60 cm wide, smaller than the spaces between the niches. The spaces between the windows on the short walls are approximately 128 cm wide. The spaces on either end of the short walls are the same width as the spaces between the windows. The ratio of the widths of windows to spaces on the short walls is approximately 2:4. While visually the ratios appear to apply on the long wall and short walls, exact measurements demonstrate that the spaces are slightly wider than these ratios indicate.

Room 3: Symmetrical Room with Two Doors

This room is located immediately to the north of the double *masma*. One of the long walls has two doorways and three niches. The other long wall contains six niches and a window (Fig. 25.8). The short walls have four niches and a window (Fig. 25.9). All of the niches and windows are 55–58 cm. wide at the bottom. 45 cm wide at the top, and 93 cm high. The bottoms of the niches are approximately 135 cm above ground level.

While the niches in this room are nearly identical to the niches in Room 2, the spaces between them are not. On the long wall with seven niches and windows, the spaces between the niches are approximately 65 cm wide, while the spaces at either end of the wall are 25 and 31 cm wide. The ratio of niche to space width on the long wall is close to 1:1, but it is not exact. The difference in niche width to space width is not great, but the spaces are consistently 8–10 cm, wider than the niches they separate.

The spaces on the short walls are approximately 90 cm wide, while the spaces at either end of the walls range from 32 to 37 cm wide. Using one foot as the unit of measurement we can generate a ratio of niche to space width of 2:3. As on the long wall, the ratios appear to apply visually, but exact measurements demonstrate that the spaces are slightly wider than even multiples would indicate.

Discussion

Room 1 differs from Rooms 2 and 3 in both the type of construction (fitted versus coursed masonry) and the size of the niches and windows. The width of the niches and windows in Room 1 averages 42 cm, which corresponds most closely to a unit based on one forearm. The niches and windows in Rooms 2 and 3 are wider, ranging from 52 to 60 cm, or approximately two anthropomorphic feet.

The spaces between the windows and niches in Room 1 exhibit greater variation than is found in Rooms 2 and 3, ranging from 135 to 150 cm in width. On the long wall in Room 2 the spaces between the niches range from 95 to 98 cm. On the short walls of Room 2 the spaces range from 127 to 128 cm. On the long wall of Room 3 the spaces range from 65 to 68 cm. On the short walls of Room 3 the spaces range from 89 to 90 cm. On the long wall of Room 1 and the short walls of Room 2, the spaces at the ends of the walls are the same width as the spaces between the niches. However, on the long wall of Room 2 and both walls of Room 3 the spaces at the ends of the walls are substantially narrower than the spaces between the niches and windows. It would appear that it was more important to maintain a prescribed distance between niches than it was to maintain the same relationship at the ends of walls. The width of spaces at the ends of the walls appears to be a function of overall wall length rather than contingent on niche width.

Rooms 2 and 3 are virtually identical in terms of construction style and in the size of niches and windows. The stones in both of these rooms are more finely dressed than in Room 1. It is possible that these rooms were constructed under the direction of the same individual using the same base measurement to establish niche width. In both rooms the niches are spaced closer together on long walls than on short walls. When the average space width on the long wall is compared to the average space width on the short wall the results are surprisingly similar. In Room 2 the spaces on the short walls are 1.36 $\times$ the width of the spaces on the long wall. In Room 3 the spaces on the short walls are 1.35 $\times$ the width of the spaces on the long wall. Because we allowed for a degree of error, the ratios do not reflect the exactitude of this relationship. The ratio of space width in Room 2 is 2:3, while in Room 3 it is 3:4. Lee's data is limited to single walls at most sites, so direct comparisons are not possible. However, it further suggests that the width of spaces between niches represented an important variable, while the spaces at the ends of walls were less important.

The distances used to determine the ratios generated for all of the walls are approximate. In fact, the spaces between niches in all three rooms are consistently larger than even multiples of the unit of measurement would suggest. On the long wall of Room 2 the difference averages 11 cm per space and on short walls the difference averages 14 cm. On the long wall of Room 3 the difference averages 9 cm per space and on the short walls the difference averages 5 cm. It is impossible to tell from the present sample whether these differences were deliberate, or simply the result of translating design into actual construction. Lee published ratios rather than actual measurements for the walls in his sample, so a comparison of results cannot be made.

Conclusion

Lee offers an interesting hypothesis for designs based on ratios of small whole numbers and an anthropomorphic-based system of measurements. The evidence from Machu Picchu appears to support his hypothesis. The width of niches and spaces in each of the rooms can be expressed as ratios of anthropomorphic measurements. While the spaces are consistently larger than the ratios would suggest, the differences do not appear to be significant. Perhaps they are a function of construction procedure or overall wall length rather than of actual design criteria.

A comparison of space width between niches on the long and short walls of Rooms 2 and 3 suggests a relationship between space width and wall length that Lee did not consider. Testing whether this relationship is an Inka architectural canon, or merely the preference of the designer of these two rooms, requires that the long and short walls of a large number of rooms from various sites be measured, analysed, and compared.

Biography William D. (Bill) Sapp received his PhD in anthropology at UCLA where he studied with the world's leading Moche scholar, Christopher B. Donnan. Bill has focused on the Middle Horizon (AD 600–900), Late Intermediate Period (AD 900–1460) and Late Horizon (AD 1460–1532) prehispanic cultures of the Peruvian north coast: Lambayeque, Chimú, and Inka. His excavations at Algarrobal de Moro and Farfán (both directed by Carol Mackey), and Cabur capitalized on his expertise in monumental architecture, elite burials, and high status ceramics. He currently serves as a director of Conservation Volunteers International Program, a non-profit corporation that organizes volunteer work parties to help maintain ruins and trails in the Machu Picchu Historical Sanctuary, a 35,000 ha park that includes the citadel of Machu Picchu, scores of other magnificent Inka archaeological sites, and many kilometers of Inka roads and trails.

References

De Betanzos, Juan. 1996. *Narrative of the Incas* (1557). Roland Hamilton and Dana Buchanan, trans. Austin: University of Texas Press.

De La Vega, Garcilasco. 1961. *The Royal Commentaries of the Incas* [1609]. Maria Jolas, trans. New York: The Orion Press.

Gasparini, Graziano and Luise Margolies. 1980. *Inca Architecture*. Patricia J. Lyon, trans. Bloomington: Indiana University Press.

Lee, Vincent. 1996. *Design by Numbers: Architecture and Order Among the Incas*. Wilson, Wyoming: self-published.

Niles, Susan. 1987. *Callachaca*. Iowa City: University of Iowa Press.

——. 1999. *The Shape of Inca History*. Iowa City: University of Iowa Press.

Protzen, Jean-Pierre. 1993. *Inca Architecture and Construction at Ollantaytambo*. New York: Oxford University Press.

Rowe, John H. 1946. The Inca Culture at the Time of the Spanish Conquest. Pp. 183–330 in Julian Steward, ed. *Handbook of South American Indians*. Vol. 2. Bureau of American Ethnology Bulletin 143. Washington, D.C: United States Government Printing Office.

Part IV
From 1100 A.D.–1400 A.D.

Chapter 26
Vastu Geometry: Beyond Building Codes

Vini Nathan

Introduction

During the medieval ages in India, no single dynastic power served as the undisputed dispenser of cultural and artistic ideas. However, despite their regional flourishes, Hindu temple designs displayed a remarkable unity of aesthetic purposes. This unified philosophy was codified into a system of rules or canons (a compendium of architectural guidelines) called the *Vastushastras*. These canons were the purview of the priestly class (Michell 1988), were intentionally made very complex so that they were incomprehensible to even skilled building craftspeople (Grover 1980) and were seldom challenged (C.H.G. Rao 1995; S.K.R. Rao 1995).

Of all the canons and rules in the *Vastushastras*, the one that found the most favor with building designers from ancient times to the present day is the *Vastu purusha mandala*. Michell describes the Vastu purusha mandala as "a collection of rules which attempt to facilitate the translation of theological concepts into architectural form" (Michell 1989: 49). This law of proportions and rhythmic ordering of elements not only found full expression in temples, but extended to residential and urban planning as well.

This chapter argues that the influence of the Vastu purusha mandala extended beyond building activity to encompass the cultural milieu as well. The first section discusses the principles underlying the Vastu purusha mandala. The application of the Vastu purusha mandala in residential design and city planning is discussed in

First published as: Vini Nathan, "Vastu Purusha Mandala", pp. 151–163 in *Nexus IV: Architecture and Mathematics,* Kim Williams and Jose Francisco Rodrigues, eds. Fucecchio (Florence): Kim Williams Books, 2002.

V. Nathan (✉)
College of Architecture, Design and Construction, Auburn University, 202 Dudley Commons, Auburn, AL 36849, USA
e-mail: vininathan@auburn.edu

K. Williams and M.J. Ostwald (eds.), *Architecture and Mathematics from Antiquity to the Future*, DOI 10.1007/978-3-319-00137-1_26,

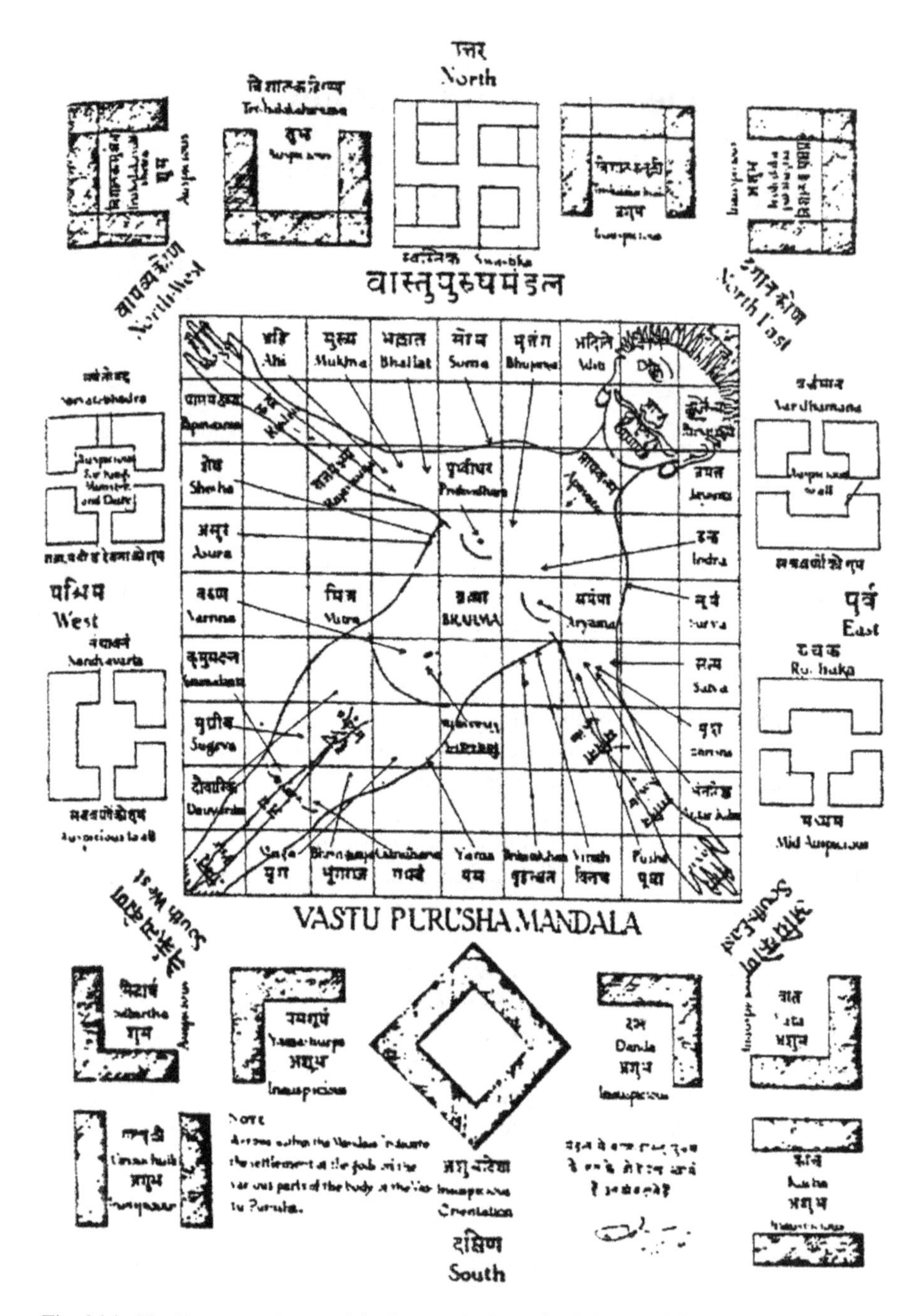

Fig. 26.1 The Vastu purusha mandala. Image: Author, after Dhama (1962)

the second section. The implications of the mandalas on the social milieu are also identified. Finally, the current status of the mandala in contemporary Indian architecture and urban design are identified. (Throughout this chapter, the terms vastu and mandala refer to the Vastu purusha mandala.)

The Vastu Purusha Mandala

The Vastu purusha mandala is a compendium of laws governing the cosmos that corresponds to the built environment on earth. This master grid for design comprises a square with the symbol of a cosmic man who is pressed down on each of its subdivisions by various divinities (Chakrabarti 1999; Crouch and Johnson 2001) (Fig. 26.1).

Vastu refers to the site and to the buildings, *purusha* denotes man, and *mandala* represents a closed polygon (Janardana 1995). The six factors underlying these building canons include cosmic influences, solar energy, geo-magnetic fields, geology of the crust, hydrology and eco-systems, and socio-cultural beliefs. Using the contextual information as a point of departure, detailed directives are provided regarding site selection; where and when to commence excavation; location and extent of open spaces; location of the water source; laying of the foundation stone; orientation of the entrance door; number and placement of windows; direction and location of stairs; suitable location of resting areas; location of bathrooms; and types of vegetation surrounding the house (Deshpande 1995).

Though the mandala can be derived from any closed polygon, the favored shape is a square since it is considered to be the most basic, rational and elementary of all geometric forms. Unlike the circle that represents mobility, the square symbolizes stability. Stability was particularly significant since the buildings that were the bedrock of the mandalas, temples, were meant to be the permanent abodes for elusive gods. In addition, for some unexplained reason, only the square form was believed to house the movement of the Sun and the Moon (Janardana 1995).

The mandala can be generated in 32 ways. The most basic consists of one square; all the others result from the division of this square into 4, 9, 16, 25, 49, 64, 81 and so on, up to 1,024 smaller squares. Individual gods occupy the quarters or small squares, called *padas*. The central square is assigned to the god Brahma (the Creator). The squares of the mandala are proportioned to reflect the "perfection of the universe, and architects strove to achieve a similar effect by controlling the dimensions of the building" (Crouch and Johnson 2001: 41). Figure 26.2 shows the *sakala*, *pechaka* and *pitha* mandalas which represent the single, four, and nine grid mandalas respectively.

Of the numerous forms of the mandala, two are of particular importance: the *manduka* mandala (Fig. 26.3), consisting of 64 *padas*, and the *paramasayika* mandala, consisting of 81 *padas*. The manduka mandala is generated by dividing the sides of the squares into even numbers of parts. Its central axes are oriented towards the cardinal points. The paramasayika mandala (Fig. 26.4) is generated by

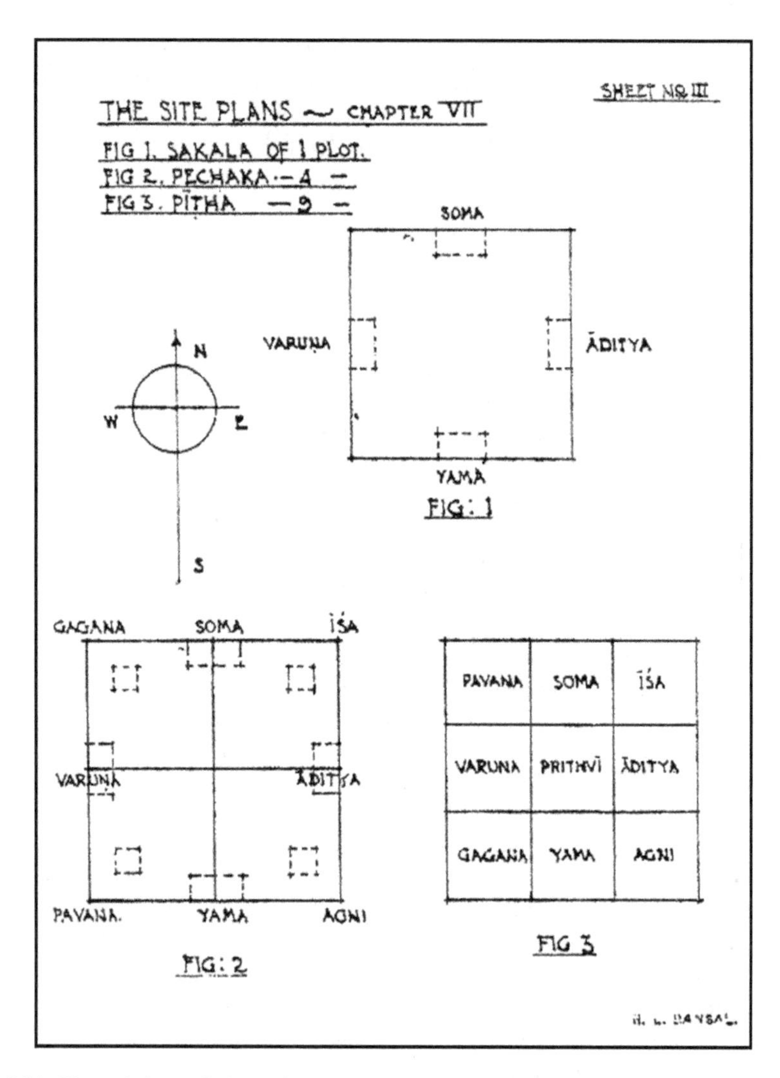

Fig. 26.2 The *sakala, pechaka and pitha* mandalas, representing respectively the single-, four- and nine-grid mandalas. After Acharya (1980)

dividing the sides into uneven numbers of parts. In both kinds of mandala, the central god, Brahma, is surrounded by 44 Vedic gods. Important gods occupy the innermost ring and gods of lower rank in the celestial hierarchy occupy the outer rings.

The size of the mandala is considered immaterial as far as its magical potency is concerned. In a plan for a large area the mandala regulates the disposition of the various buildings, and in the plan of a single building it defines the proportion of different architectural elements. Chakrabarti (1999) argues that since the mandalas could be derived in different ways, its versatility accommodated regional variations

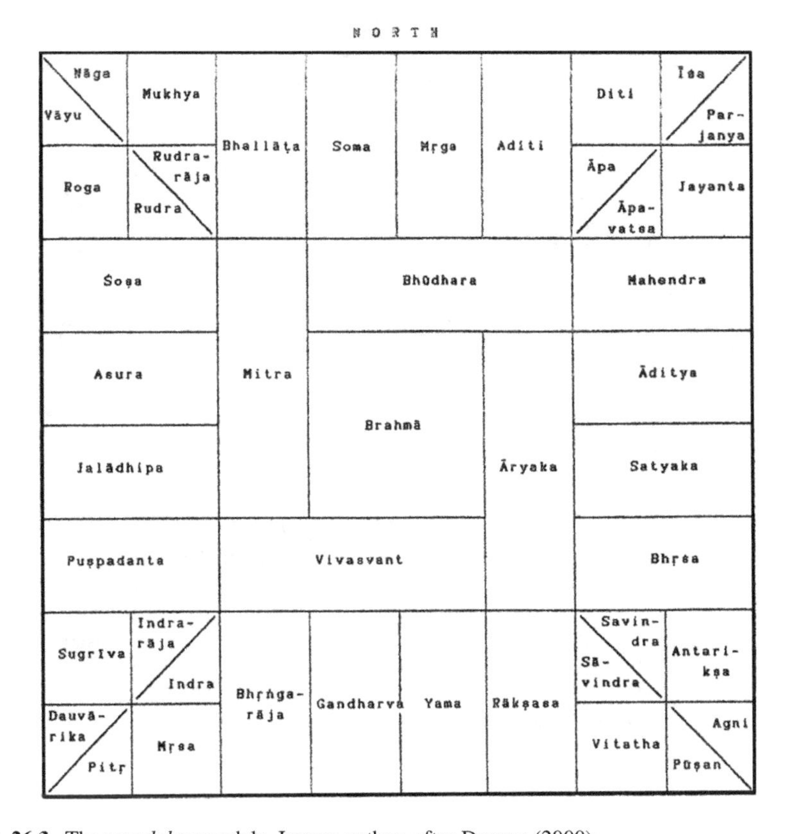

Fig. 26.3 The *manduka* mandala. Image: author, after Dagens (2000)

in climate, topography, availability of building materials and the prevailing socio-cultural milieu. For example, Fig. 26.5 shows how the basic 9×9 grid mandala could be applied to accommodate two different regions.

In calculating the proportions for a building form or town plan, the ceremonial priest was not concerned with the final result, but with the remainder, which had to conform to the rules of proportion. The interpreters of the mandala regarded the remainder as the most important result of the whole arithmetical operation. Thus in Indian architecture the doctrine of proportion was, strictly speaking, a "doctrine of the remainder" (Volwahsen 1994).

Eight basic equations, each using the remainder from the preceding equation were used for calculations. It was difficult to coordinate all eight requirements. The nature and degree of the compromises that were inevitable remain unexplained. The need to keep all mandalas deliberately complex, confusing and open to interpretation was motivated by the argument that only the learned, priestly class could properly cull the magic force from the symbols.

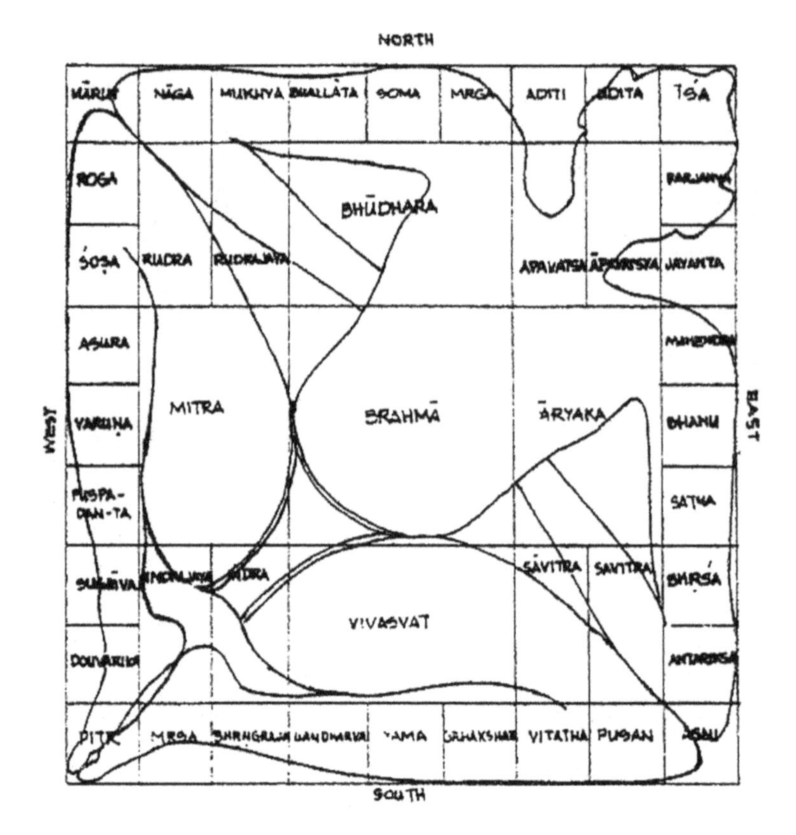

Fig. 26.4 The *paramasayika* mandala. Image: author, after Acharya (1980)

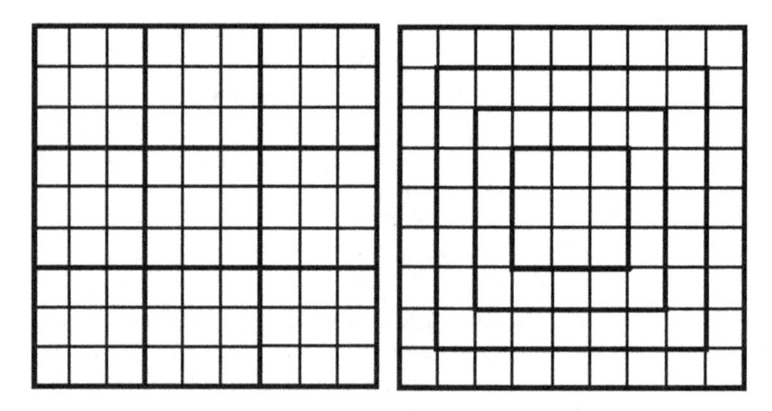

Fig. 26.5 The basic 9 × 9 grid mandala can be applied to accommodate two different regions. After Chakrabarti (1999)

The Mandala in Town or City Planning

Due to the scale and magnitude involved in town planning, the mandala was used in a freer fashion, and often societal and cultural systems such as caste and later class superceded the canons. In planning a town or village, the appropriate vastu purusha mandala had to be ascertained. Of the 32 possibilities the priest-astrologer chose a mandala that was most auspicious and at the same time had as many *padas* as there were to be residential quarters. Once again, mandalas with 64 and 81 *padas* were held in particularly high regard.

As far as the natural features of its setting permitted, the town or city was supposed to be an exact rectangle, if not actually a square in outline. The town wall was erected along the outer order of the mandala. Streets were aligned from north to south and from east to west along the lines demarcating the *padas* from one another. One dwelling block was exactly coextensive with one pada. The mandala also provided detailed instructions about the network of streets. The widths of the street followed a strict hierarchical order; the widest streets were the ceremonial way, and the streets narrowed as one moved from the residential quarters of the higher to the lower castes. Only the main thoroughfares that gave the city its basic aspect had to conform to the divisions specified in the Vastu purusha mandala. The dwelling blocks produced by the division of the *padas* could be subdivided by alleys and foot paths in any sort of pattern.

Cruikshank's analysis of the planning principles of the city of Jaipur (Cruikshank 1987) suggests that the architect, Vidhyadhar, who was also a mathematician and an astrologer, used the nine-square mandala to guide his design decisions. However, the mandala had to be revised to accommodate the unique site conditions of the city. Though the appropriateness of superimposing mandala geometry in a post-mortem analysis of the city is questionable (Chakrabarti 1999), during the planning of a satellite town to Jaipur, the architect B. V. Doshi combined mandala planning principles with modern architectural building elements. Fig. 26.6 shows the plan of Jaipur with the superimposed nine-square mandala. The concepts underlying the plan of the satellite city (Vidhyadhar Nagar) is shown in Fig. 26.7.

In some instances, the guidelines in the mandala contradicted each other. For example, a particular canon approved a square form for a city for all castes, whereas another canon in a subsequent section of the mandala restricted the use of such a form for a city that will be inhabited by Brahmins only. There was no explanation provided for this apparent contradiction. In addition, the amount of detail provided was proportional to the hierarchy of the caste discussed. As a result, the dwelling prescriptions for the Brahmins (priests, highest caste) were the most profuse, and those related to the Sudras (servants, lowest caste) were very sketchy and open ended.

Until the Mogul period most Indian cities were built of unbaked clay, wood and other perishable materials. Stone, the most durable material, was reserved for temples and other significant buildings. Thus only fragmented remains of urban

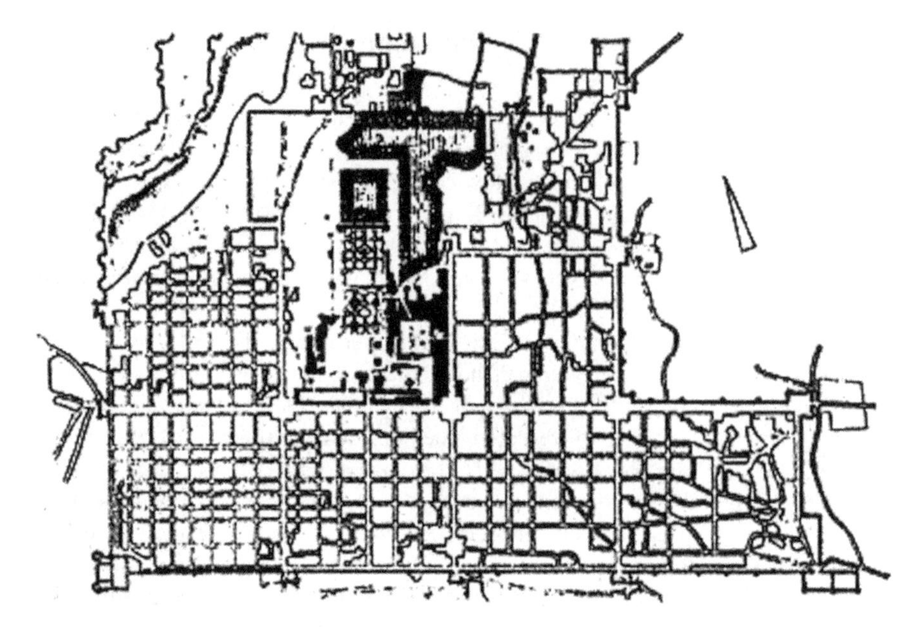

Fig. 26.6 The plan of Jaipur with the superimposed nine-square mandala. Image: author, after Volwahsen (1994)

complexes or townships have survived and it is not easy to gauge how strictly the rules of the mandala were followed.

The Mandala in Residential Design

The influence of kinship and caste as significant prime social influences could be seen in residential building activity. Once the site had been selected based on its orientation and soil characteristics, the actual building form and design were determined. The soil was examined meticulously; its color, small, feel and taste were taken into consideration. The color of the soil indicated for which caste the site was particularly well suited. Four colors were distinguished: white soil was for Brahmins (priests); red for Ksatriyas (warriors); yellow for Vaisyas (merchants); black for Sudras (servants). In addition, the taste of the soil was linked to caste system: sweet for Brahmins, astringent for Ksatriyas, pungent for Vaisyas and bitter for Sudras. The next step involved selection of appropriate building materials. The canons prescribe that stone or wood was worthy of gods, Brahmins, and kings, but unsuited to Vaisyas and Sudras.

The most popular ground plan was the *catushala*, in which the inner court was enclosed on all four sides by buildings. All the rooms open onto the court, and corridors afford communication between them. With this type of house, several storeys could be built within a closely-confined space. The outer walls facing the

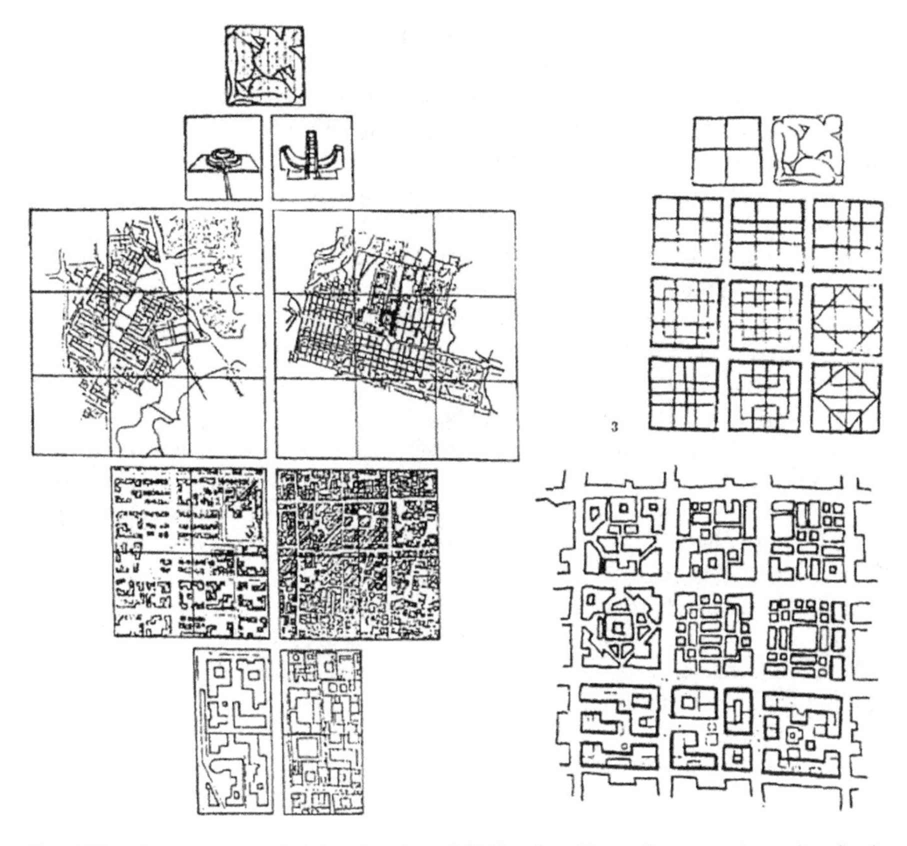

Fig. 26.7 The concepts underlying the plan of Vidhyadhar Nagar. Image: author, after Curtis (1985)

street did not have any windows, for reasons of security. With this symmetrical ground plan the location of the entrance to the house was not problematic, even though it was oriented in a direction corresponding to the house owner's caste.

In addition to the popular catushala type of house, a prospective owner could choose between the following types: the *trishala*, with three wings arranged in an U pattern; a *dvishala* with two wings set at right angles; an *ekshala*, a single rectangular building. Certain forms were taken for granted and strongly resisted change. Consequently some of these forms persisted for long periods of time.

If the builder/owner was not satisfied with any of these basic types of houses, or if the front and rear door could not be located in the proper place, then one could refer to the manual in which all the permissible combinations were listed. Each of the 14,000 possibilities had a name of its own. When a particular alternative was chosen, the builder had to ensure that no deviations occurred from the closely-defined model form. There was little or no premium on originality. The canonical rules left no room for any later modification, particularly for expansion since, as Rao points out, "a house when completed and constructed is an organic whole and must not be mutilated on any account" (C.H.G. Rao 1995: 25).

The main entrance of a dwelling should never be sited axially; the orientation of the entrance depended on the caste of the occupant. Wherever a house was located in the city district, the entrance door had to be on the south side if the resident was a Brahmin, on the west for Ksatriyas, on the north for Vaisya households and on the east for Sudras.

The square of the mandala was the basis for the residence of a Brahmin. It was considered the perfect form, appropriate only for those close to the gods. For this reason the ground plan of a Brahmin house should not deviate from the square by more than one-tenth the length of a side. In clearly defined deviations from this absolute standard, the proportions of the sides in a Ksatriya house should be 1:1 1/8; for a Vaisya house, 1:1 1/6; for a Sudra house, 1:1 1/4.

The actual building form was again dictated by different rationales for different aspects of the house. Typically, the house is built around a central open space, ruled by Brahma (the Creator), as depicted in the mandala. Each side of the square forms a wing of the house (Fig. 26.8). Depending on the scale of the house, additional grids are added to make it a seven-wing house (Fig. 26.9) or a ten-wing house (Fig. 26.10) (Chakrabarti 1999).

The caste stratification of Hindu society was reflected not only in the ground plan of a house but also in its vertical section. The rules pertaining to the number of storeys permitted in houses made it easy to determine the caste to which the owner belonged (Table 26.1).

The mandalas also provided detailed instructions regarding the placement of doors, windows, furniture and other household items. It is uncertain to what degree to these rules were followed, especially with movable furniture and goods.

On examining the mandala for residential building one finds that most of the guidelines are related to houses for the more affluent, higher castes. In addition, the mandalas prescribed generous dimensions for the higher castes and smaller building sizes for the lower castes. It appears that the assumption is that the higher castes are by default the more affluent and the lower caste, poorer. Since only the Brahmins, the priestly higher caste, were allowed to interpret the mandalas and allow "deviations" where appropriate, what would be the residential options for a lower caste affluent householder?

The mandalas prescribed large clusters of semi-independent rooms, thereby implicitly endorsing the joint or extended family system. This is ironic since records indicate that the mandalas were composed when the joint family system was pervasive only in the lowest castes.

The mandala does not address issues related to mobility between the different castes. For example, it appears that a house built by a Brahmin would have to be sold or rented only to another Brahmin. One assumes that vertical mobility was possible in one direction (downward); therefore, Brahmins are allowed to buy or rent houses built for Ksatriyas or Vaisyas, since the more modest Ksatriya or Vaisya house will still conform to the maximum dimensions that are allowed by the mandalas for the higher caste Brahmins.

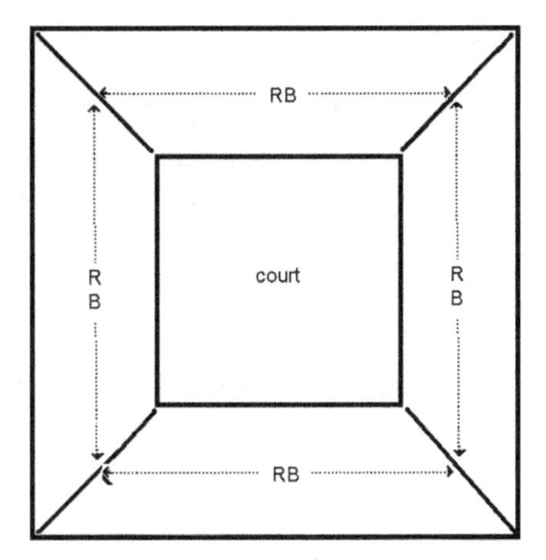

Fig. 26.8 A typical house built around a central open space. "RB" indicates roof beams. Image: author, after Dagens (2000)

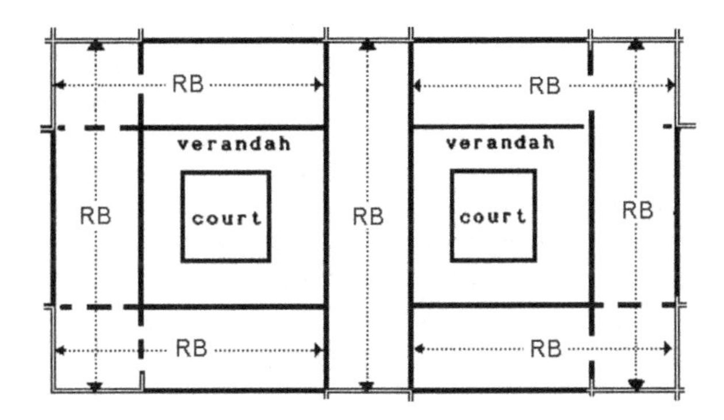

Fig. 26.9 Additional grids are added to the basic house to make the seven-wing house. Image: author, after Dagens (2000)

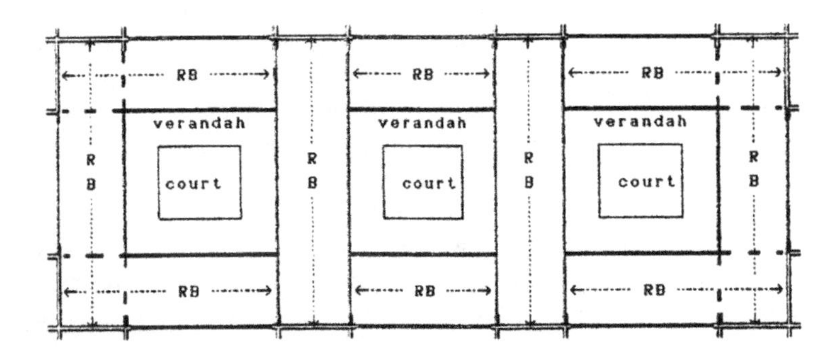

Fig. 26.10 The ten-wing house. Image: author, after Dagens (2000)

Table 26.1 Maximum building height allowed for different castes

Caste	Maximum building height allowed for residence
King (royal)	Ground Floor + 7½ floors
Brahmin (priest)	Ground Floor + 6½ floors
Ksatriyas (warrior)	Ground Floor + 5½ floors
Vaisya (merchant)	Ground Floor + 4 floors
Sudra (servant)	Ground Floor + 2½ floors

Evolution and Contemporary Applications of Vastu

The mandala was not only the precursor to the modern day grid but was also a grid in its own right. Some scholars maintain that the canonical orders of the mandala served as the precursor of Leonardo da Vinci's system of proportions demonstrating how the human figure could be contained in a circle (Grover 1980). Parallels between the grid in modern architecture and the mandala in early Hindu architecture are also common. It has been argued that Le Corbusier articulated his system of the Modular based on a rationale comparable to that of the mandala. In a similar vein, the success of Le Corbusier's design of the city of Chandigarh has been attributed to its adherence to vastu principles. For example, vastu "specialist" Harish Saini (1996) analyzed Chandigarh (Fig. 26.11).

He was quoted in the popular press as saying that Chandigarh 'stands in contrast to the other cities in terms of its "orderliness" and "disciplined development." Further to it, its positive response to the Vastu Shastra, i.e., proper archaeo-astronomical placement, makes it free from all misfortunes and as such the city is bound to flourish' (Saini 1996).

With the advent of Modernism in Indian architecture, the three major qualities of traditional Indian architecture, decoration, plasticity and good craftsmanship, were soon lost. The tacit assumption was that these three qualities would be a natural outcome of establishing a unique identity to Indian architecture. Interestingly, the revival of interest in Indian building traditions have largely ignored these qualities and has instead focused on the orientation and organization of space using ancient sacred texts as references. Chakrabarti identifies attempts by noted Indian architects to mesh the issue of identity and distinctiveness amidst the fervor and fascination with new technology and architectural expressions (Chakrabarti 1999: 28). One example includes emphatic references and esoteric vocabulary borrowed from ancient building treatise, the Vastu purusha mandala being the most popular.

Increasingly, the works of prominent architects such as Charles Correa and Balakrishna Doshi have revived interest in the mandala in contemporary India. The implicit rationale is that associating design thinking with long standing canons such as the mandala impart design with a philosophical, vernacular edge. The degree to which interest in the mandalas go beyond being a passing fad remains to be seen.

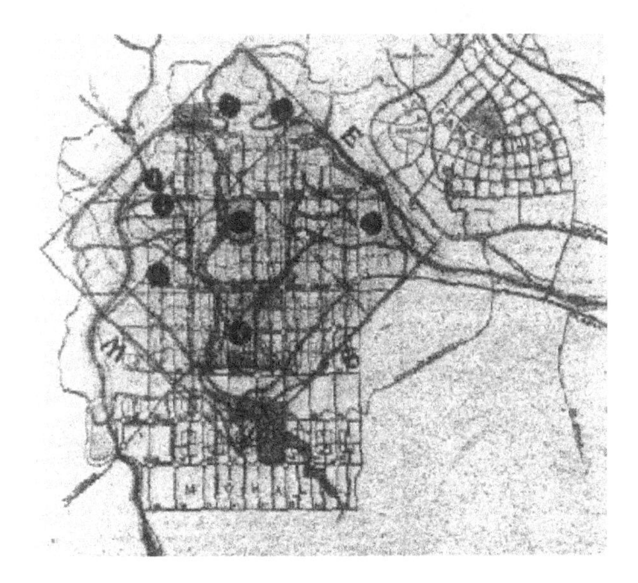

Fig. 26.11 Harish Saini's analysis of Chandigarh. Image: author, after Saini (1996)

Conclusion

The Vastu purusha mandala is an exhaustive, complex and confusing system of building rules or canons that are based on a curious mix of sacred and pragmatic considerations. The geometric principles of the mandalas were such that "each part of the design was calculated to conform mathematically to a specific proportional system based on an established unit" (Crouch and Johnson 2001: 41).

Despite their genesis from sacred buildings (Hindu temples), the mandalas provided a plethora of laws of proportions and ordering that were applied to towns and secular buildings. The mandalas were created during medieval times and their directives for residential designs reflect the socio-cultural milieu at that time. Though they were seldom challenged, they seem to have evolved through revisions and deviations made to accommodate specific circumstances. Only Brahmins were allowed access to the information and the authority to interpret these guidelines and therefore, the degree to which their decisions were skewed by caste divisions (and discrimination) is hard to determine.

In recent times, the mandala has resurfaced in Indian architecture as the primary means for imparting a unique Indian sensibility to otherwise modern design. In contemporary India, the mandala, divorced of its overly religious or superstitious overtones, is used as a geometric organizing tool or grid in the design of civic buildings. Interestingly, it is in the residential sector that the sacred rituals and other protocols in addition to the geometric principles associated with the mandala seems to find full expression.

Biography Vini Nathan is Dean and McWhorter Endowed Chair in the College of Architecture, Design and Construction at Auburn University. An architect-interior designer, she was previously Dean of and Professor in the School of Architecture at Philadelphia University. She has held faculty positions at Michigan State University and New York Institute of Technology. Through Walker Group/CNI New York, Vini collaborated with Cesar Pelli Associates on the design of the Kuala Lumpur City Center (Petronas Towers) in Malaysia. Her areas of scholarly interest include leadership through design, disruptive entrepreneurship and chaos theory. She earned graduate degrees in design and architecture from Virginia Tech and the University of Michigan.

References

ACHARYA, P.K. 1980. *Architecture of Manasara.* Manasara series IV, II ed. New Delhi: Oriental Books Reprint Corp.

CHAKRABARTI, V. 1999. *Indian architectural theory.* New Delhi: Oxford University Press.

CROUCH, D. P. and JOHNSON, J. G. 2001. *Traditions in architecture.* New York: Oxford University Press.

CRUIKSHANK, D. 1987. Variations and traditions. *The Architectural Review*, **182**, 1086 (August): pp. 51-58.

CURTIS, W. J. R. 1985. *Balakrishna Doshi: An architecture for India.* Paris: Electra Moniteur.

DAGENS, B. 2000. *Mayamatam: Treatise of housing, architecture and iconography.* Vol. II. New Delhi: Indira Gandhi Center for the Arts.

DESHPANDE, S. A. 1995. Vastu: Yesterday and day before. In G. D. Vasudev, ed. *Vastu, astrology and architecture.* Delhi: Motilal Banarsidass Publishers.

DHAMA, B.L. 1962. *Domestic Architecture.* Jaipur: Ajanta Printers.

GROVER, S. 1980. *The architecture of India: Buddhist and Hindu.* New Delhi: Vikas Publishing House.

JANARDANA, M. V. 1995. References to vastu in the scriptures. In G. D. Vasudev, ed. *Vastu, astrology and architecture.* Delhi: Motilal Banarsidass Publishers.

MICHELL, G. 1988. *The Hindu Temple: An Introduction to its Meaning and Forms.* Chicago: University of Chicago Press.

———. 1989. *The Penguin Guide to the Monuments of India:* Vol. I: *Buddhist, Jain and Hindu.* London: Viking.

RAO, C. H. G. 1995. General vastu guidelines. In G. D. Vasudev, ed. *Vastu, astrology and architecture.* Delhi: Motilal Banarsidass Publishers.

RAO, S. K. R. 1995. Vastu vidya. In G. D. Vasudev, ed. *Vastu, astrology and architecture.* Delhi: Motilal Banarsidass Publishers.

SAINI, H. 1996. Vaastu ordains a full flowering for Chandigarh. *The Tribune: Saturday Plus* (newspaper), 24 February 1996.

VOLWAHSEN, A. 1994. Living Architecture: India. *Architecture of the World.* Vol. VII. Germany: Henri Stierlin Ed.

Chapter 27
Algorithmic Architecture in Twelfth-Century China: The *Yingzao Fashi*

Andrew I-kang Li

The *Yingzao fashi* as Algorithmic Architecture

The *Yingzao fashi* (Building standards) was written by Li Jie (d. 1110), court architect during the late Northern Song dynasty (960–1127), and published in 1103. Li evidently meant to educate government officials who commissioned buildings and to set standards for the builders who built them. He set out rules for designing foundations, masonry buildings, wood-frame buildings (*da muzuo,* or structural carpentry), finish carpentry (*xiao muzuo*), and painted decoration. He also defined terms and provided methods for estimating materials and labor. The book includes numerous drawings, but these reflect a much later style—probably Ming (1368–1644) or Qing (1644–1911)—and so can be used as references for the Song only with caution.

In the classical Chinese literature, the *Yingzao fashi* is one of only two surviving books that deal with architecture. The other is the *Gongcheng zuofa zeli* (*Structural regulations*), published in 1733. These two books are important simply by existing, since architecture—or, perhaps more properly, building—was not an appropriate subject for literati. However, they are interesting on their own account, because they document what had developed as, and probably still was, an oral tradition of structural carpentry. In the case of the *Yingzao fashi,* that tradition used a few rules to create many designs. We will examine this approach in more detail, but for the moment let us just call it *rule-based.*

First published as: Andrew I-kang Li, "Algorithmic Architecture in Twelfth-Century China: *The Yingzao Fashi*". Pp. 141–150 in *Nexus IV: Architecture and Mathematics*, Kim Williams and Jose Francisco Rodrigues, eds. Fucecchio (Florence): Kim Williams Books, 2002.

A.I. Li (✉)
Initia Senju Akebonocho 1313, Senju Akebonocho 40-1Adachi-ku, Tokyo 120-0023, Japan
e-mail: i@andrew.li

K. Williams and M.J. Ostwald (eds.), *Architecture and Mathematics from Antiquity to the Future*, DOI 10.1007/978-3-319-00137-1_27,

As an example of this approach, consider the curved roof section, so often identified as the characteristic feature of Chinese architecture. Li Jie does not list legal roof sections for the builder to choose from. Rather, he spells out in a two-rule procedure called *juzhe* how to create the roof section for a building of any given depth. We will see this procedure in detail later. For now, the important point is that, given these two rules and a building of any legal depth, we can always find the correct roof section.

Another example of this approach is the modular unit *fen*. The *fen* can have eight different values, from 9.6 to 19.2 mm, depending on the grade or rank (*deng*) of the building. So, for example, a (modular) dimension of 10 *fen* can have eight possible (absolute) values, ranging from 96 mm. at the eighth grade to 192 mm. at the first grade. Li Jie stresses that the *fen* is fundamental and usually uses it when specifying dimensions. The user chooses the appropriate scale or rule, reads off the dimension in *fen*, and obtains the correct length. Again, few rules, many designs.

Liang Sicheng (1901–1972), who pioneered the study of the *Yingzao fashi*, perceived the significance of this approach and called the manual a "grammar book of Chinese architecture" (Sicheng 1984a). I go one step further and formalize Li's rule-based approach. This allows us to see the *Yingzao fashi* as algorithmic, gives a graphic version of the text, and provides other benefits, as we will see.

Formalizing the *Yingzao fashi*

We begin our formalization with a definition: a style is a set or language of designs perceived to be similar (Stiny and Mitchell 1978). There are two basic ways of defining a language of designs: by listing the member designs (enumeratively) and by showing how to create them (generatively).

The important difference between the two is that the second, by showing how to create the designs, helps explain why they look similar and thus why we perceive a style. This was Li Jie's approach. We can say that the *Yingzao fashi* is a generative definition of the official Song architectural style.

To formalize the text, we "translate" it into a formal language (Here, distinguish *formal language* and *language of designs*). The language we use is shape grammar, which is not only formal but also graphic: it manipulate shapes, like plans, sections, and elevations. This contrasts with most other formal languages, which manipulate symbols, like letters and numbers. Thus our grammar will appeal to designers.

How does shape grammar work? Here is an extremely brief and informal introduction. A shape grammar consists of an initial shape and replacement rules. An initial shape is often a point in the working plane or space. A replacement rule consists of two shapes—one on the left, one on the right—with an arrow in the middle.

To create a design, compare the left side of a rule to the current shape; if you are just beginning the process, this is the initial shape. If there is a match, subtract (that is, erase) the left-side shape from the current shape and add (draw) the right-side shape. This yields a new current shape. Continue until finished. There are precise definitions for *shape, compare, match, subtract* and *add,* but an intuitive interpretation of these terms will suffice for a general appreciation of the grammar.

I have written a grammar of the *Yingzao fashi* that generates designs, each consisting of plan diagram, section diagram, plan section, roof section, elevation and text descriptions. In this chapter I show only the part which creates roof sections.[1]

A Grammar of Roof Sections

When considering the roof section in a Chinese building, it is important to remember that the purlins (*tuan*) support the rafters (*chuan*). The rafters span from purlin to purlin, forming the curved section. This is the opposite of the western practice, where the rafters support the purlins, and span from ridge to eaves in a straight line, forming a triangular section.

Li Jie's procedure for creating the roof section requires that we know the depth vy of the building, where v is the number of rafters and y is the horizontally projected length of each rafter. We then calculate the height of each purlin, with the eaves purlin taken as zero. There are two steps: *ju,* 'raise,' and *zhe,* 'lower.'

First, find the height $h_0 = vy/4$ of (i.e., raise) the ridge purlin; call it the roof height. Draw a line connecting the ridge purlin and the eaves purlin; call it the working roof line.

Next, find the height of the first purlin below the ridge purlin. We already know its horizontal location: it is offset by y from the ridge purlin. Find the intersection of the working roof line and the vertical line at a distance of y. Lower this point a distance of $h_0/10$; the resulting point is the elevation of the purlin. From this point to the eaves purlin, draw a new working roof line. Repeat with the remaining purlins, each time halving the lowering: $h_0/20$, $h_0/40$, etc. For any set of legal starting conditions, there is exactly one legal roof section.[2]

Now let's translate this procedure into a shape grammar (Fig. 27.1).

The grammar consists of an initial shape, an initial description, and four rules. The initial shape consists of the point (b, c), indicated by a cross and the state label "E" that indicates that this is stage E, which deals with the roof section. Stage A creates the plan diagram, stage B the section diagram, and so on. Rule E1 corresponds to "raise," and rule E2 to "lower." Rules E3 and E4 perform housekeeping functions like erasing construction lines. The grammar generates, not only the section, but also a description l comprising the height differentials between purlins.

---

[1] For the complete version, see Li (2001). *A Shape Grammar for Teaching the Architectural Style of the 'Yingzao fashi'*. Ph.D. Diss.: Massachusetts Institute of Technology.

[2] This account of *juzhe* is slightly simplified, but it serves our purpose.

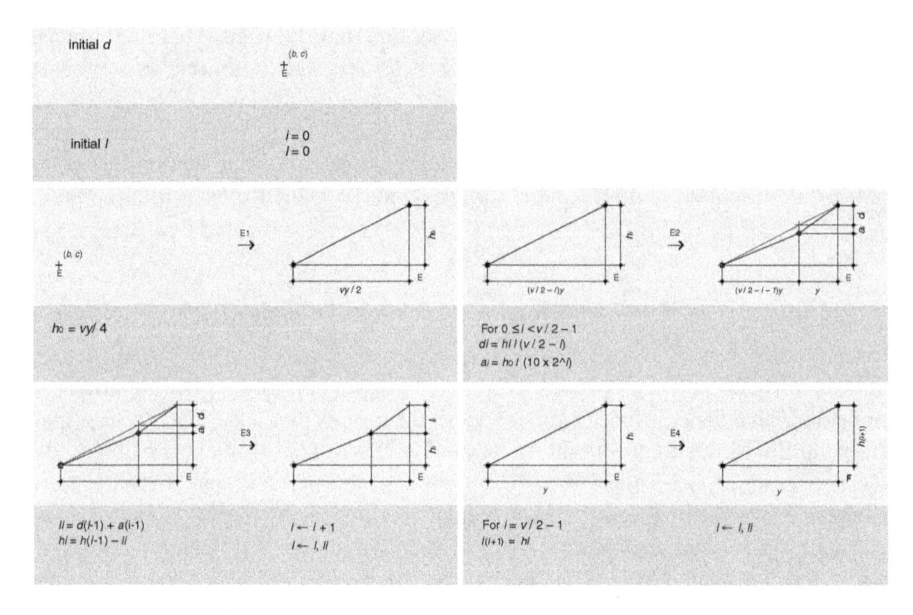

Fig. 27.1 A grammar of roof sections. Image: author

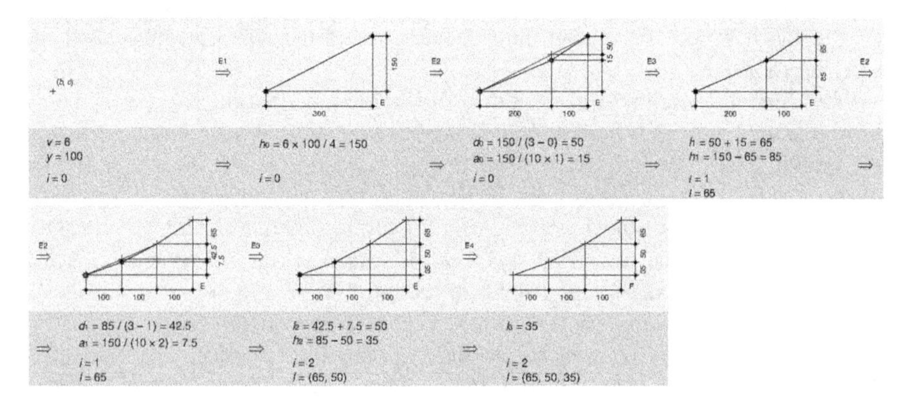

Fig. 27.2 Derivation of the roof section of a building six rafters deep with rafters 100 *fen* long (i.e., $v = 6$, $y = 100$). Image: author

To show how the grammar works, let's use it to create the roof section of a building 6 rafters deep with rafters 100 *fen* long; that is, $v = 6$ and $y = 100$ *fen* (Fig. 27.2).

We apply rule E1 to raise the ridge purlin to a height of 150 *fen* above the eaves purlin. We apply rule E2 to find the elevation of the next purlin down. The intersection at the working roof line is $150/3 = 50$ *fen* below the ridge purlin. Rule E3 increments the counter i and calculates the new working height (150–$65 = 85$ *fen*) and erases the construction lines. The description is $l = 65$. We still have another purlin to locate, so we apply rule E2 again. The third purlin is lowered 42.5 + 7.5 = 50 *fen*. Rule E3 establishes the new working roof line, erases the construction lines, increments i, and updates the description $l = (65, 50)$. Now we have finished with all the purlins, so we apply rule E4, which removes the labels, updates the description $l = (65, 50, 35)$, and changes the state label from "E" to "F." The design is ready for the next stage of generation.

Formalizing the Human Role

We have now seen how the explanation of *juzhe* that Li Jie wrote in words can be expressed formally and graphically as a shape grammar. We have also seen that, given any appropriate starting conditions, we can always create a roof section. Different starting conditions lead to different sections, but there is always a section at the end. This is because the generative definition is complete. But in practice, the information is not always complete. When there are gaps, then the design can be completed only if the missing information is supplied. Where does that information come from? Is it reliable?

The answer to both questions involves us, the users. We have three roles in the generative definition of a style. This is easy to explain in formal terms. First, we perceive the initial similarity. Second, we propose the hypothetical definition. And third, we evaluate whether new designs created by the definition belong to the language.

As an example, take the sections of a building type called a *ting* hall (Fig. 27.3). The structural frame of a *ting* hall is composed of repeated transverse frames (*liangjia*) perpendicular to the front elevation. Each of these transverse frames is composed in turn of columns (*zhu*) and transverse beams (*fu*). The *Yingzao fashi* shows 18 transverse frames drawn in section.

Our first act is to accept this corpus of 18 sections as being similar. The question immediately comes to mind: what is the relation between this corpus and the language of sections? By accepting the 18 sections as similar, we have also assumed that they are legal (i.e., in the language). This implies that the language contains all 18 sections and possibly more; that is, the corpus is a subset of the language.

Our second act is to propose a grammar of the language (Li 2001). It generates, among others, the following five sections of six-rafter buildings (Fig. 27.4). That is, we formulate a hypothesis that makes five predictions.

Our third act is to evaluate these sections. Do they belong to the language? Are the predictions true? The first section is not exactly like any of the 18, but it is not obviously illegal either; it is probably legal. The second is in the corpus; definitely legal. The third is not very different from any of the 18; probably legal. The fourth has no spaces deeper than one rafter, which makes the building difficult to use; it is almost definitely illegal. The fifth has a clear span, which is seen in the smaller section (four rafters) but not in the larger sections (eight and ten rafters); maybe legal, maybe not.

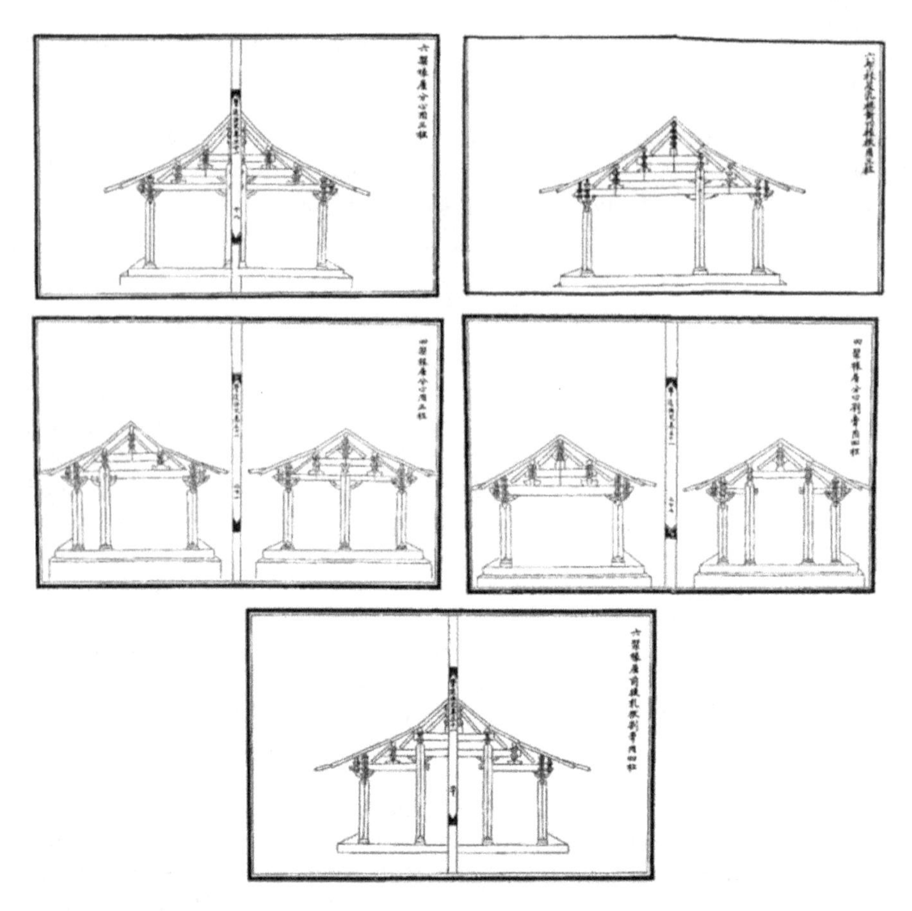

Fig. 27.3 The *Yingzao fashi* contains 18 sections of *ting* halls from four to ten rafters deep. Five are of 10-rafter halls, six are of 8-rafter halls, three are of 6-rafter halls, and four are of 4-rafter halls. The seven sections of 4- and 6-rafter halls are shown here. Image: author, after Sicheng (1983a: 319–321)

If we accept as legal all sections but the fourth, then we can revise our grammar so that it no longer creates that section. One way is to allow one-rafter bays to be created only once, at the exterior of the building. With the revised grammar, we can generate more designs. If the designs seem dissimilar from those known to be in the style, we revise the grammar again. In this way, we refine our hypothesis until it defines the style as best we understand it.

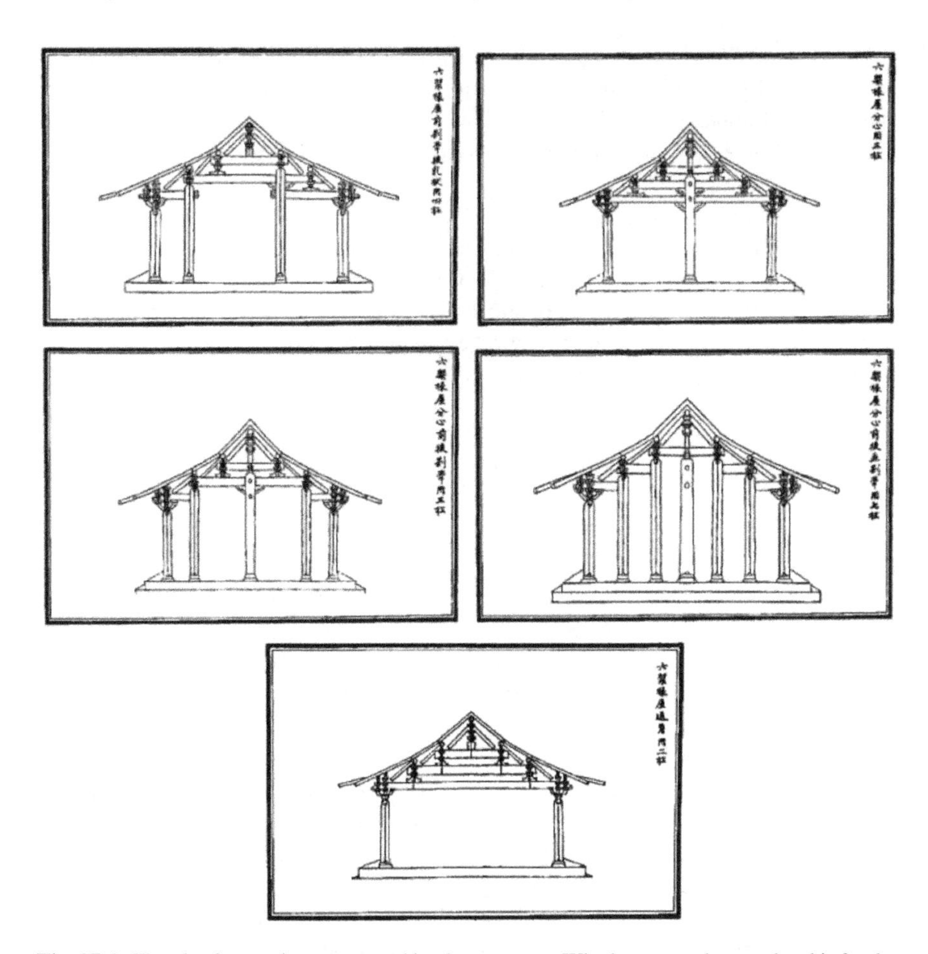

Fig. 27.4 Five 6-rafter sections generated by the grammar. Whether or not they are legal is for the user to determine. Image: author

Formalizing the Student's Role

This suggests another role for our grammar: to provide a useful experience for students learning about the style of the *Yingzao fashi*. I believe that the most useful such experience is to participate in composing and testing the hypothetical definition. Thus, our grammar generates all and not only (in other words, more than) the designs that are likely to be in the language. As the grammar reflects the imperfections of the text, so do its products, which the student can evaluate.

This differs from the usual analytical approach, in which the author is also the judge, because he is aiming for an authoritative definition. We might call this the expert approach. The advantage of our approach is that the student, not the teacher, aims for the authoritative definition. We give her no more information than there already is, so she must confront the gap between what she knows and what she needs to know. What information is missing, and why? Was it knowledge common to Li Jie and his readers, but now lost to us? Was it overlooked by Li? Was it specialized builder's knowledge that, whether by design or by ignorance, Li omitted? What assumptions are needed, and are they justified? We might call this the naïve approach.

In our example of the sections, the student need not agree with my evaluations. She may, for instance, consider that six-rafter-long beams are impractical, making a clear span unlikely in a six-rafter-deep building (Fig. 27.4, fifth section). She could then modify the grammar to limit the length of clear spans to four rafters. The important thing is that she can consider the question because it has been made clear, indeed almost inescapable. The lesson here is that style is not "out there"; it is a human construct.

Conclusion

We have seen that the *Yingzao fashi* as a definition of style is primarily generative, and have used shape grammar to characterize that definition formally and graphically. This has clarified, not only where the gaps are in the definition, but also how we users are responsible for filling those gaps. This in turn has suggested an explicitly experimental approach to teaching the style.

From here it is easy to imagine automating a grammar to emphasize the user's interaction with it: what she decides and when she decides it. In this case, we need not actually implement the shape grammar mechanism; we can merely simulate it. This allows the user to concentrate on the overall structure and logic of the grammar as a characterization of the style.

To test the feasibility of this approach, I have used Macromedia Flash to begin a prototype simulation of the section grammar, generating—in real time—a large number of designs. Freed of the distraction of executing the grammar manually, the user can consider issues of more direct interest: what her choices are at any stage, how those choices affect the design, which designs are in the language and which are not. One drawback is that the simulation cannot be modified by the user; it generates this one language of designs and no other. There is no immediate solution to this, but the benefit is clear: it shows how designers can use grammars to think about design more practically.

Other possible future work is a comprehensive comparative study of Chinese wood-frame architecture. The *Yingzao fashi* prescribes a style that evolved until just after the beginning of the Ming. At that point, there was a great stylistic break, after which the style changed markedly and virtually ceased to evolve (Sicheng 1984b: 103). Coincidentally, for this period we have the *Gongcheng zuofa zeli* of 1733 (already mentioned). This sets up a series of comparisons that can be done with shape grammar. For instance, now that we have a grammar that generates buildings in the style of the *Yingzao fashi,* we can formalize the relation between the manual and the extant pre-Ming buildings: how does the grammar have to be modified to produce those buildings? Then, since the extant buildings change through time, we can see how the grammars evolve, as Terry Weissman Knight does (1994). Similarly, we can construct grammars of the style of the *Gongcheng zuofa zeli* and of that of Ming–Qing buildings. We can compare them with each other and with their pre-Ming counterparts.

Thus we can do a shape-grammatical study of Chinese wood-frame architecture from the eighth to the twentieth century; if we consider indirect evidence, we can begin even earlier. This would be a complete formal statement of a long tradition, and an appropriate extension of the studies, begun by Liang Sicheng, of this "grammar book" of Chinese architecture.

Biography Andrew I-kang Li is an independent researcher in computational design based in Tokyo. His work ranges from a computational analysis of the twelfth-century Chinese building manual *Yingzao fashi* to a software application for creating and editing shape grammars. He taught at the School of Architecture, The Chinese University of Hong Kong, from its founding in 1991 until 2010, and has been president of the Association of Computer-Aided Architectural Design Research in Asia (CAADRIA). He has also taught at Korea Advanced Institute of Science and Technology (KAIST) and Tunghai University, Taiwan. He was born in Montréal, Canada, and has an LMus in piano performance (McGill University, Canada), an AB in Chinese (Harvard University, USA), an MArch (Harvard), and a Ph.D. in computational design (Massachusetts Institute of Technology, USA). He studied Chinese architectural history at Nanjing Institute of Technology (now Southeast University) as a China/Canada government exchange scholar.

References

Li, A. I. 2001. A Shape Grammar for Teaching the Architectural Style of the '*Yingzao fashi*', Ph.D. Diss.: Massachusetts Institute of Technology.

Sicheng, L. 1983. *Yingzao fashi zhushi* (The annotated *Yingzao fashi*). Beijing: Zhongguo Jianzhu Gongye.

———. 1984a. Zhongguo jianzhu zhi liangbu 'wenfa keben' (The two 'grammar books' of Chinese architecture). *Liang Sicheng wenji (The collected works of Liang Sicheng)*. Vol. II. Beijing: Zhongguo Jianzhu Gongye, pp. 357–363.

———. 1984b. *A Pictorial History of Chinese Architecture: A Study of the Development of Its Structural System and the Evolution of Its Types*, Wilma Fairbank, ed. Cambridge, MA: MIT Press.

Stiny G. and Mitchell W. J. 1978. The Palladian Grammar. *Environment and Planning B: Planning & Design*, **5**: pp. 5–18.

Weissman Knight, T. 1994. *Transformations in Design: A Formal Approach to Stylistic Change and Innovation in the Visual Arts*. Cambridge: Cambridge University Press.

Chapter 28
The Celestial Key: Heaven Projected on Earth

Niels Bandholm

Major Geographical and Historical Setting[1]

The island of Bornholm, with an area of 587.5 km^2, is situated 40 km southeast of Sweden in the Baltic Sea but is territorially a part of Denmark (Fig. 28.1). It harbours a rich field of archaeological remains. Excavations reveal that it was settled before the Stone Age, around 3600 BC, when many dolmens, passage graves and some woodhenges were constructed. From the Bronze Age (1700–500 BC) there is rock art on flat, glacier-scoured rock surfaces, burial mounds, cairns and monoliths.

From the Iron Age (500 BC–800 AD) there are remains of stone ship settings, stone burial mounds, stone circles and 250 monoliths, plus numerous finds of glass, jewellery, weapons, coins and thousands of small Gold-Figure Foils.

The first written record of the island dates from the Viking period, when Wulfstan (890 AD) relates that "Burgenda land (Bornholm) is independent and has its own King." Gamleborg in Almindingen was built as the main fortification in this period.

Bornholm came under Danish rule at the time of Harold Bluetooth (911–986), the first Danish king to convert to Christianity. During the transition to Christianity, between 1050 and 1150 AD, around 40 runic stones were erected.

First published as: Niels Bandholm, "The Celestial Key: Heaven Projected on Earth". Pp. 95–116 in *Nexus VII: Architecture and Mathematics,* Kim Williams, ed. Turin: Kim Williams Books, 2008.

[1] Nielsen (2006). The second edition (1994) contains an English summary; see also http://www.bornholm.info/historie and http://www.sacredsites.com/europe/denmark/bornholm.htm

N. Bandholm (✉)
Klostervej 30, Kloster, 6950 Ringkoebing, Denmark
e-mail: niels.bandholm@gmail.com

K. Williams and M.J. Ostwald (eds.), *Architecture and Mathematics from Antiquity to the Future*, DOI 10.1007/978-3-319-00137-1_28,

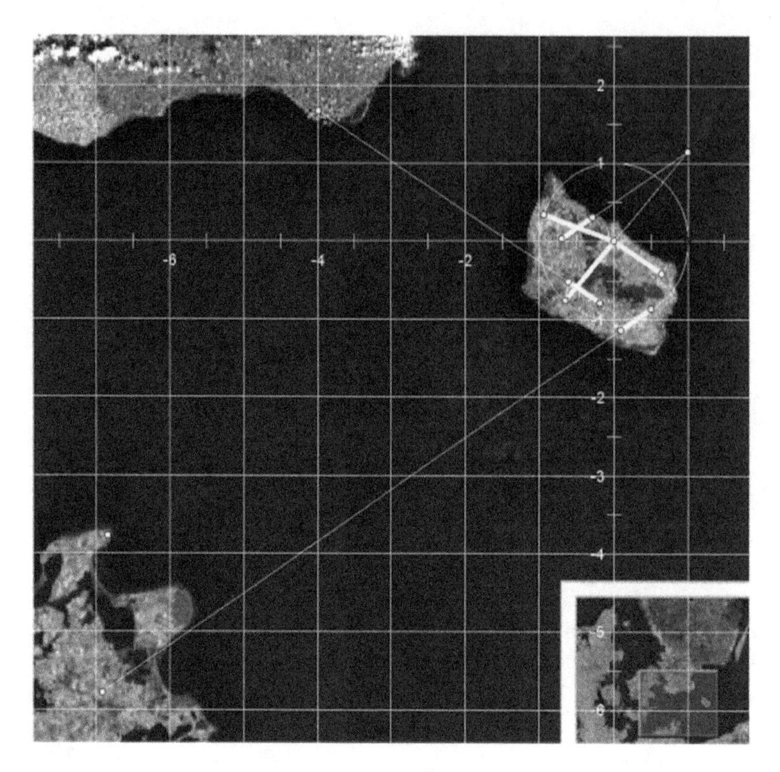

Fig. 28.1 Bornholm with alignments in Baltic Sea. Image: author

The village of Aakirkeby plus three of the four districts of the island were given to the Archbishop in Lund as a gift of absolution from the Danish King in 1149.[2] Archbishop Eskil's sovereignty over the major part of Bornholm lasted until he resigned in 1177. The medieval stone churches on Bornholm are generally believed to have been planned and initiated around the time when he was Archbishop (Fig. 28.2).

Cultural Succession and Integration

Since the Bronze Age, Bornholm has been a stepping stone in the Baltic Sea for cultural influences streaming north/south and east/west—a melting pot for cultural succession and integration.

[2] Svend II Grathe (1146–1157). In 1103/1104 Lund became Episcopal See of the North (Denmark, Sweden, Norway, Iceland, Greenland, the Faroe Islands, the Orkneys and Shetland, the Hebrides and the Island of Man). The first Archbishop in Lund was Asger (1103–1137) followed by Eskild (1137–1177) and Absalon (1178–1201).

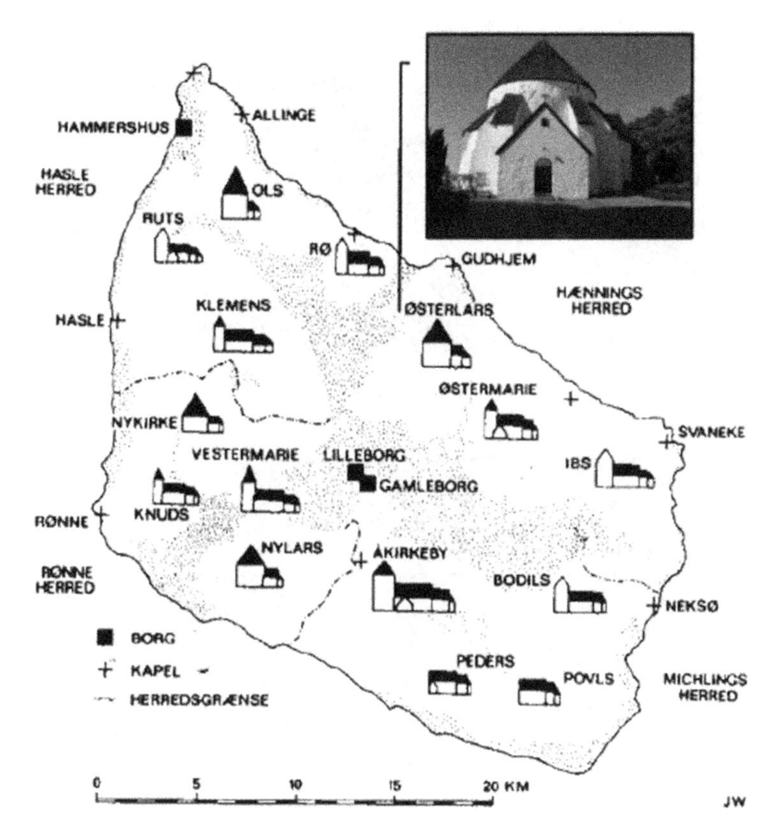

Fig. 28.2 Bornholm with churches. Image: Wienberg (1986), reproduced by permission of Jes Wienberg

Before the reign of the Archbishop of Lund, the island is believed to have been Christian in the forms of Aryanism, Iroceltic Christianity and Orthodox Christianity (Lidegaard 1999, 2004: 121–186). Christian influence could have begun as early as the era of the Great Migration in the sixth century.

Beliefs from earlier death and fertility cults were incorporated into the new faith. There are many examples of cultural integration, such as a phallic stone at Bodilsker church (Fig. 28.3) and capitals showing female genitals at Dalby (S. Sweden) (Fig. 28.4), besides the aforementioned Christianised runic stones, in or near most churches. Genitalia were common ornaments in early Roman church buildings (Geese 2004: 334–345). Placement near former holy springs (Rø church and AA church; see Table 28.1) and wells in the crypts at Lund and Dalby, show connection to the fertility cult and pagan holy places (Andrén et al. 2002: 299–332).

In general it is common knowledge that many Christian customs and symbols, including the cross, are legacies of earlier belief systems. Graveyards, crypts and the holy remains of saints show connections to older death cults.

Fig. 28.3 Phallus at Bodisker Church (Bo). Photos: author

Fig. 28.4 Genitals and well in Dalby Crypt near Lund. Photos: author

The Celestial Key, as presented here (Fig. 28.5), could be the result of a similar cultural integration of former astronomical practices into the Christian adaptation of Muslim astronomy. This is further elucidated in the discussion, but first a presentation of the geometrical riddles, which call for an explanation.

Distances (km)

The tables in Fig. 28.5 show the distances between churches as fractions of the ØL–Chr distance or of the NL–ØL distance on the central NL–ØL–Chr line (Haagensen 2006: 125).[3] The upper table contains values corresponding to ØL–Chr, drawn on the map as thin double lines. The lower table contains values

[3] Haagensen (2007: 227–229) introduced the point Chr as $\sqrt{7}/\sqrt{3}$ times the vector NL–ØL from ØL along the geodetic. It is about 61 m SSE of ST. Thus the distance ØL–Chr is constructed. It is used for comparison with former publications and to illustrate the meridian-convergence. (The distance ØL–ST is 21.930 km.)

Table 28.1 Signs and saints connected to churches with coordinates in unity of NL-ØL = 14336 m. The bold coordinates are very close to whole numbers, but much more remarkable are the distance ratios as shown in Fig. 28.5

Sign	Church	Patron Saint	Comment	Coordinates[a]
ØL	Østerlars	St. Laurentius	Round church oldest: 1150	0.0000; 0.0000
NL	Nylars	St. Nicholas of Myra	Round church	−0.6530; −0.7574
Ny	Nyker	All Saints.	Round church	−0.8558; **−0.2486**
Ol	Olsker	St. Olaf	Most recent Round church	−0.7149; **0.5024**
AA	Aaker	St. John the Baptist,	Largest, renovated in 1874	−0.1891; −0.7830
Ru	Rutsker	St. Michael	New tower 1886	−0.9389; 0.3414
Rø	Rø	St. Andreas	Rebuilt (moved?) 1888	−0.2909; 0.3042
Kl	Klemsker	St. Clement	Rebuilt (moved?) 1882	−0.7064; 0.0289
ØM	Øster Marie	Blessed Virgin Mary	Rebuilt, original ruins remain	0.2409; −0.2579
Kn	Knutsker	St. Canute Lavard	New tower 1879	−0.9328; **−0.5000**
VM	Vester Marie	Blessed Virgin Mary	Rebuilt moved 34 m in 1885 (Haagensen 2007: 164–172)	−0.6085; −0.5110
Ib	Ibsker	St. Jacob	Mostly undamaged	0.6354; −0.4219
Bo	Bodilsker	St. Botolph	Renovated	0.4937; −0.8512
Pe	Pedersker	St. Peter	Tower from 1500 cent.	0.0651; −1.1256
Po	Poulsker	St. Paul	Most recent 1250	0.3532; −1.1557
ST	Store Tårn	Island "Earth goddess"?	Big Tower, Christiansø built 1684	**0.9961**; 1.1609
Chr	Calculated point on Christiansø 61 m SSE of Big Tower			**0.9975**; 1.1569
Ales	Stone Circle	Bronze Age?	Ship setting in Sweden	**−4.0060**; 1.6677

[a]The coordinates are based dimensions from a 1985 survey found (Haagensen 1993: 174–175). They are transformed to spherical coordinates of the geoids of 1950 and then projected to a plane touching the sphere in ØL through a stereographic projection. This plane projection is shown in Fig. 28.5

corresponding to NL–ØL, drawn as thick single lines. "Measured" distances are shown in bold type.[4] These represent quarter, third, half or whole measures of the two unit distances ØL–Chr and NL–ØL. Clearly there is a striking pattern to be seen, with surprisingly little deviation. The ratio of the two unit distances is very close to $\sqrt{7}/\sqrt{3}$. This ratio will be derived below when the Celestial Key is presented.

[4] Distances are calculated from coordinates in the geoids 1950 as above but *along* geoids using the program KmsTrans from Danish State survey Kort og Matrikelstyrelsen.

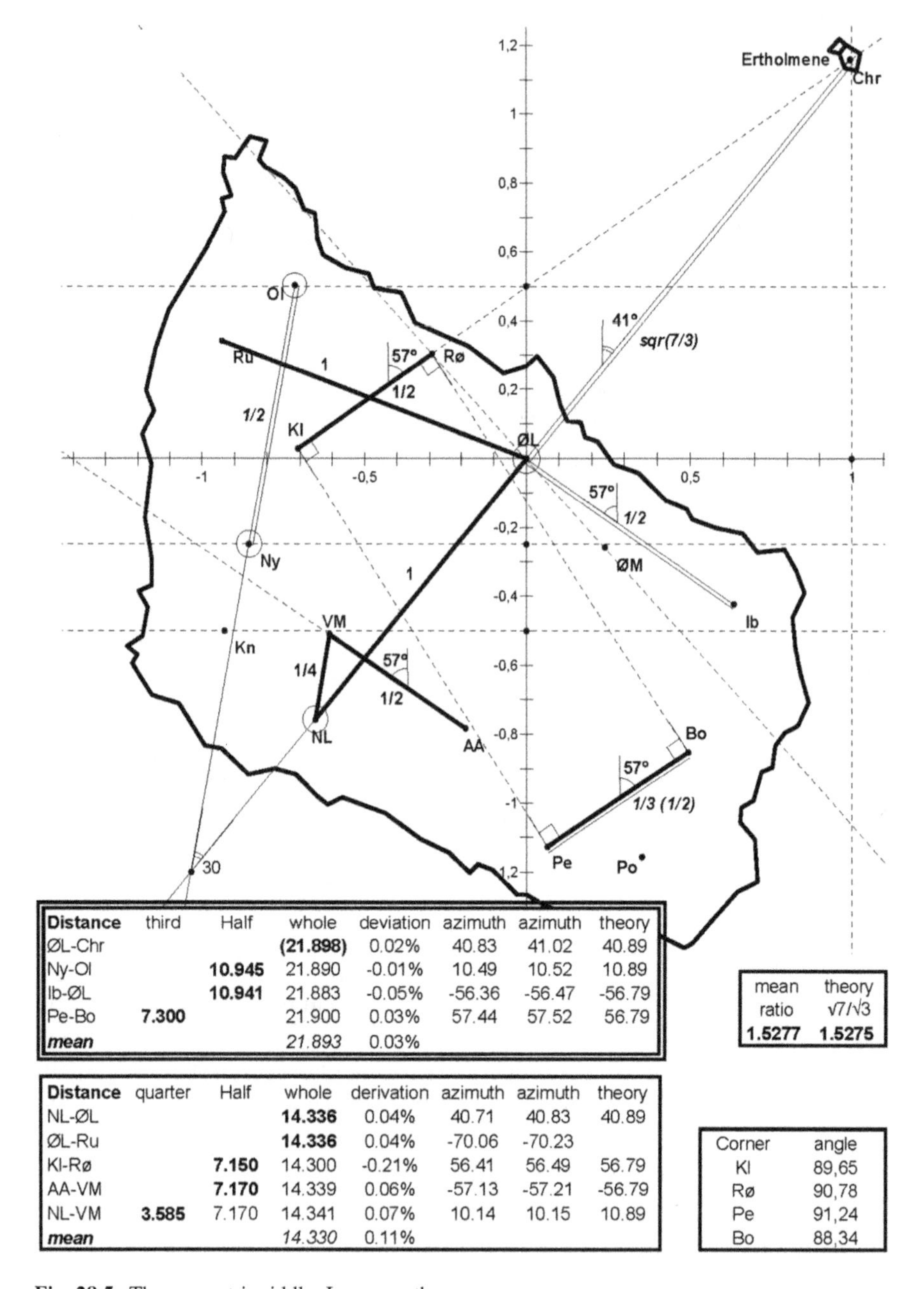

Distance	third	Half	whole	deviation	azimuth	azimuth	theory
ØL-Chr			**(21.898)**	0.02%	40.83	41.02	40.89
Ny-Ol		**10.945**	21.890	-0.01%	10.49	10.52	10.89
Ib-ØL		**10.941**	21.883	-0.05%	-56.36	-56.47	-56.79
Pe-Bo	**7.300**		21.900	0.03%	57.44	57.52	56.79
mean			*21.893*	0.03%			

mean ratio	theory √7/√3
1.5277	**1.5275**

Distance	quarter	Half	whole	derivation	azimuth	azimuth	theory
NL-ØL			**14.336**	0.04%	40.71	40.83	40.89
ØL-Ru			**14.336**	0.04%	-70.06	-70.23	
Kl-Rø		**7.150**	14.300	-0.21%	56.41	56.49	56.79
AA-VM		**7.170**	14.339	0.06%	-57.13	-57.21	-56.79
NL-VM	**3.585**	7.170	14.341	0.07%	10.14	10.15	10.89
mean			*14.330*	0.11%			

Corner	angle
Kl	89,65
Rø	90,78
Pe	91,24
Bo	88,34

Fig. 28.5 The geometric riddle. Image: author

Angles

Four of these distances have azimuths[5] near ±56.79° with a mean deviation of about 0.5° [The azimuth in each end of a vector distance is slightly different because crossing meridians converge on the North Pole, e.g., NL(40.71) → ØL (40.83°) → Chr(41.02°) → (compare with table in Fig. 28.5)].

Two of these (Kl–Rø and Pe–Bo) are nearly parallel sides in a "rectangle" Kl–Rø–Bo–Pe with corner angles close to 90°. Consequently the azimuths of the other sides are close to the complementary angle (56.79–90° = −33.21°). Both these angles will be derived below when the Celestial Key is presented.

Alignments (Fig. 28.1)

The projections of the lines Kl → Rø and NL → ØL meet on the islet Christiansø, about 21.9 km from ØL. Christiansø is a part of "Ertholmene", probably named after the pagan Earth Goddess, Nerthus.

The line AA → VM projects about 62.2 km to the largest stone ship setting in Scandinavia: Ales Stena in southern Sweden, which is itself aligned with the rising midwinter sun. In clear weather it is possible to see Bornholm from Ales with the naked eye.

Although it is not possible to see the island Rügen from Bornholm, more than 130 km from ØL, it is remarkable that Bo → Pe and ØL → VM both point to Marienkirche (founded 1185) and ØL–Kn point to the heathen cult place Arkona (conquered in 1136 and destroyed in 1168) (see Fig. 28.1).

Many churches on Bornholm are approximately aligned with sacred springs: "Holy-Spring" with Pe–Ibs and Po–Bo, and both "Solomon's Spring" and "Josephs Spring" with the three churches Rø–ØL–ØM.

Coordinates and Deviation (Table 28.1)

In the coordinate system, with distance ØL–NL = 14.336 km as unity, some places have striking coordinate values (x,y). Chr $x \approx 1$, Ales Stena $x \approx 4$, Ol $y \approx 0.5$, Ny $y \approx -0.25$, Kn $y \approx -0.5$. Turning the coordinate system +0.15° (left) makes Chr $x = 1.000$ and turning −0.15° makes Ol $y \approx 0.500$, Ny $y \approx -0.250$. A deviation in

[5] Azimuth is the angle from the North measured positive clockwise. They are in the interval −90° to 90°, as church distances are vectors oriented from south towards north. Negative azimuth is chosen to make symmetry apparent.

angles of about 0.3° must be expected, equal the difference between azimuths in each end of the line Nl → ØL → Chr (to Rügen about the double deviation).[6] Apparently there is greater uncertainty in angles than in distances. In the first publication of these results (Bandholm 2007), the calculations of distances were based on a spherical earth and were less accurate than Fig. 28.5, based on the geoids.

Previous Research

This study is inspired by Erling Haagensen's work, and owes much to his discoveries. He is the first person to have done serious research into the geometry behind the geographic placement of the churches on Bornholm as described in five books (Haagensen 1993, 2003, 2007; Haagensen and Lincoln 2000). He gives a Vesica Piscis construction of the azimuth of 40.89° for the axis NL–ØL–Chr, and he derives the ratio $\sqrt{7}/\sqrt{3} = 1.5275\ldots$, although he does not clarify why this ratio was chosen. Haagensen's work has been criticised by several historians.[7]

Haagensen does not explain the azimuth around ±57° found for four distances. Nor does he connect the layout of churches to the stereographic projection used in the astrolabe.

The Astrolabe and the Stereographical Projection [8]

Important new astronomical knowledge and technology were transferred from the Moslem to the Christian world, at the time when the Nordic countries were Christianised. An example is the astrolabe—a technical wonder of the period. Its practical application as an analogue computer for calculating time and direction must have been at least as great a marvel then as the digital computer is today. It was later the model for making the first church clocks (North 1974).

---

[6] The azimuth of ØL → ST(40.88°) → is closer to theory (40.89°) compared with ØL → Chr (41.02°) → .

[7] Most distinctly by professor Jes Wienberg (2002a: 175–188). See also (Wienberg 2002b). Weinberg could not see much historical evidence to support Haagensen's hypothesis that the round churches were constructed by the religious brotherhood of the Knights Templar, inspired by M. Wivel (1989). Haagensen's scientific reputation may have been compromised by his books having been the inspiration for several popular films on the Templars' hidden treasure and by his having co-written the popular book *The Templar's Secret Island* together with Henry Lincoln (Haagensen and Lincoln 2000).

[8] For conform properties of the stereographic projection, see M. Jaff, "From the Vault of the Heavens". Pp. 49–63 in *Nexus Network Journal*, **5**, 2003.

I propose the geometry behind the construction of the astrolabe as the template and inspiration for the geometrical layout of the churches on Bornholm.

The planispheric astrolabe is based on a stereographical projection of the heavenly sphere on the equatorial plane (Fig. 28.6). Each point on the sphere (e.g. a star) is connected with a line to the South Pole. Where the line intersects the plane of the equator, the point is represented with a star pointer on the so-called *rete*, a fretted network free to turn around the centre—the North Pole—over what looks like a spider's web called the *climate*. This engraved network is a projection of the sky's spherical coordinates (azimuth and latitude). Curved azimuth lines radiate from the climate's web representing the zenith, the point directly above the observer. Around the zenith are eccentric circles representing equal latitudes on the sky. The climate's coordinate web changes with the geographical latitude of the observer and must be replaced as the observer moves along the north–south dimension (as he "changes climate"). All climates have the same three concentric rings, representing the Tropic of Cancer, the Equator of the Sky, and the Tropic of Capricorn. The Tropic of Capricorn is usually the outer limit of the astrolabe. The three rings should have been on the rete as they turn with the stars, but centred as they are on the North Pole, they do not change in turning and can be inscribed identically on all climates independent of geographical latitude. On the rete is shown an eccentric circle touching both tropics. This is the sun's yearly path (the ecliptic) through the zodiac.

The astrolabe has a sighting instrument underneath it to enable measurement of the altitude of the sun or, at night, a star. The rete is turned until the heavenly object has the right altitude on the climate's "spider web". The user can then read the time of day or night and the azimuth of the object, and orient the astrolabe towards it and use it as a compass.

Transmission of Astrolabe Theory to the Nordic Countries

The theory of the stereographic projection can be traced back to the Greek Hipparchus (180 BC) and is described by Ptolemy (about 100–160 AD), who may have used an astrolabe. Later in the Hellenistic period it spread from Alexandria to both the East and the West. It was copied and developed by the Arabs, and the Western type of astrolabe is derived from the Moorish type found in Spain. Towards the end of the tenth century, knowledge of the astrolabe began to enter into the Latin West. It was introduced into the schools of Lorraine, for example, at Liége (Welborn 1931). From this area the English, and later the Danish king Canute the Great (1018–1035), engaged many educated men to become bishops in England and the Nordic countries. This could have provided an early acquaintance with the astrolabe in these learned circles. In 1092 the English Prior, Walcher of Malvern, used an astrolabe and a lunar eclipse to make a lunation table. Better known is the influence of Gerbert of Aurillac (930–1003), who was the only mathematically learned Pope (Sylvester II 999–1003). In his youth he went to Moorish Spain to

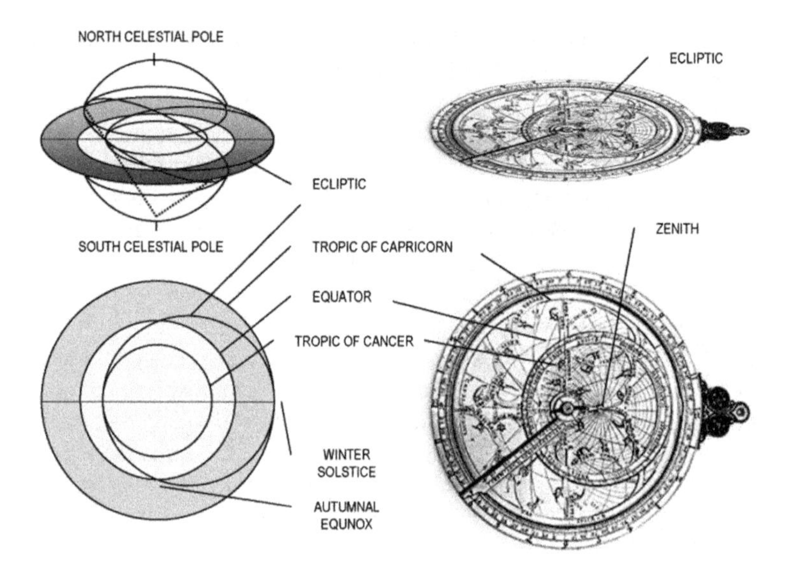

Fig. 28.6 The stereographic projection and the astrolabe. Image: author

learn mathematics and astronomy. Found among his works is the first Latin work on the astrolabe, *Liber de astrolabio* (Pedersen 1996: 225–227). The construction and use of the astrolabe was described by the monk Hermannus Contractus of Reichenau (1013–1054). After that it was possible to construct an astrolabe with a northern climate as Walcher might have done. Gerbert was a teacher in Reims and one of the first to teach Arabic science in the Latin world. Later, during the twelfth century, many Arabic astronomical tables and theories were translated and taught in Europe.

A third, but less well documented way for the astrolabe to reach Scandinavia could have been its use as a navigational instrument, e.g., during the Norman reign in Sicily.

Thus there were ample possibilities for the planners of Bornholm's churches to learn of the astrolabe and to apply it to the geometrical layout. A key figure could well have been the Archbishop Eskil (1100–1178), who in his early youth went to the Latin school in Hildesheim (1112–1130). Later as Archbishop (1137–1177) he travelled extensively and had close contacts with Bernard of Clairveaux (1090–1153) (canonised in 1174). Thus he had many possibilities for meeting teachers of organisation, mathematics and astronomy. Perhaps, then, it was in some of these circles that what follows might have been developed.

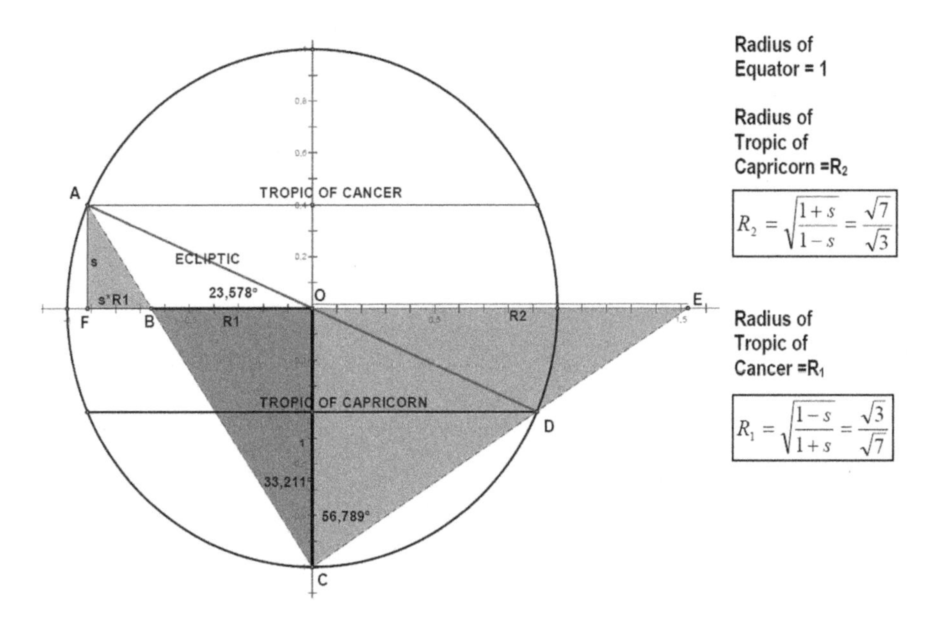

Fig. 28.7 Orthographic projection. Image: author

The Celestial Key

The term is a combined expression for two views of the stereographic projection:

(1) The orthographic projection on a plane through the poles and solstices (see Fig. 28.7 and the centre of Fig. 28.9 in half size);
(2) The plane projection of the sphere on the equatorial plane (see Figs. 28.9 and 28.10), and the special tri-Vesica Piscis construction of the equator and the tropics in the equatorial plane (see Fig. 28.8).

Ptolemy used the term *analemma* for a combination of these.

The Orthographic Projection

A plane cut through the poles and solstices in the stereographic projection (Fig. 28.6), now called an orthographic projection (Fig. 28.7), reveals several striking coincidences with the geometric riddle in Fig. 28.5.

In the general calculation shown below, the sine to the ecliptics obliquity is called s. By use of similar triangles it is easy to calculate the ratio of the tropics if the angle of the ecliptic is known.

By choosing the value s = 0.4, for example, the obliquity of the ecliptic is Arcsin(0.4) = 23.578...°, the ratio (R_2) between Equator radius and radius of the

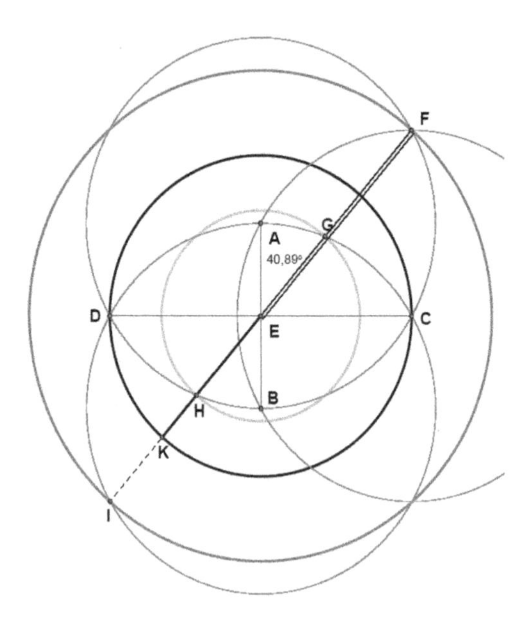

Fig. 28.8 Tri-Vesica construction of equator and tropics in medieval astrolabe. Image: author

Tropic of Capricorn is exactly √7/√3 in the planispherical projection and, furthermore, the projection angle is Arctan(√7/√3) = 56.789...° .

Proof The circle has a radius OC = 1, AF = s is sinus to angle AOF, which is the obliquity of the Ecliptic and FO is cosine AOF = $\sqrt{(1-s^2)}$. BO is the radius R_1 of the Tropic of Cancer and OE is radius R_2 of the Tropic of Capricorn.

The triangles ABF and CBO are similar, hence FB = AF · BO/OC = s · R_1 and FO = $\sqrt{(1-s^2)}$ which is also FO = FB + BO = $R_1 \cdot (1 + s)$

Hence the radius of projection of the tropic of Cancer is $R_1 = \sqrt{((1-s^2))/(1 + s)} = \sqrt{((1-s)/(1 + s))}$

The triangles CBO and ECO are also similar, so R_2 and R_1 are reciprocal $R_2 = 1/R_1 = \sqrt{((1 + s)/(1-s))}$

The obliquity of the ecliptic has decreased since antiquity.[9] By using Newcomb's formula for the change of obliquity, it is possible to identify 918 AD as the year with the value 23.578...° (= 23°34.69′ decimal minutes). Has any Arabic astronomer used this value for the obliquity? Not exactly, but almost. The value 23°35′ has been used by Habash (mid-ninth century), Al-Battani (850–929), Ibn Yunus (940–1009), Al-Biruni (973–1048), and Al-Tusi (1201–1272) in 1250 (King 1999: 230, 355).

So how can one be sure that the exact value of Arcsin(0.4) was intended? This follows from a very special tri-Vesica Piscis construction for the value of √7/√3 (and √3/√7). It could well have been used by the makers of astrolabes.

[9] Newcomb's formula: $e = 23°.452294–0°.0130125 \cdot T–0°.00000164 \cdot T^2 + 0°.000000503 \cdot T^3$, T in centuries before 1900 AD (see graph Bandholm 2007).

The Tri-Vesica Piscis Construction

The points K, E and F correspond to the churches NL, ØL and (Chr), and the axis has azimuth $\text{Arcsin}(\sqrt{3}/\sqrt{7}) = 40.893\ldots°$ Compare with Fig. 28.5.

Construction On a vertical line two points A and B are marked. Two circles with the radius AB and centres in A and B are drawn, and then a third circle with the same radius and its centre in their crossing point C. The horizontal line CD is perpendicular to AB and crosses in the central point E.

A line from E to the crossing point F is drawn and it crosses the circle at G.

Now if EC is equal to 1 (the radius of equator), then EF is $\sqrt{7}/\sqrt{3}$ (radius of tropic of Capricorn), EG is $\sqrt{3}/\sqrt{7}$ (radius of tropic of Cancer) and FH is the diameter of the ecliptic.

Proof (See Fig. 28.8) Triangle AEC (not drawn) is right-angled in E and AE is half AC. Using Pythagoras gives $EC^2 + AE^2 = AC^2$ substituting gives $1^2 + ½\ AC^2 = AC^2$ and AC becomes $AC = 1/(1-½^2) = 2/\sqrt{3}$. But AC = CF in right-angled triangle ECF (not drawn). Using Pythagoras in triangle ECF gives $EF^2 = EC^2 + CF^2 = 1^2 + (2/\sqrt{3})^2 = 7/3$ hence $\mathbf{EF} = \sqrt{\mathbf{7}}/\sqrt{\mathbf{3}}$.

Triangle CGE (not drawn) is right-angled in G and similar to triangle ECF. CGE is a right angle because angle CGI spans the diameter CI (not drawn). Substituting values in the proportion EG/EC = EC/EF gives $EG/1 = 1/(\sqrt{7}/\sqrt{3})$ hence $\mathbf{EG} = \sqrt{\mathbf{3}}/\sqrt{\mathbf{7}}$.

This construction, shown in Fig. 28.8, is more accurate and faster than Fig. 28.7 for making the template of an astrolabe.

I have not been able to trace this construction, nor the use of the exact value of Arcsin(0.4) for the obliquity.[10] If it can be done, a historical connection to the geometrical layout might be found.

Application of the Key

More support for the idea that the key (=orthographic and plane projection) was the inspiration for the layout appears if the trace of the sphere as NOPSTM is drawn at half-scale but still centred on ØL in Fig. 28.9.

The projection lines are drawn from the celestial North Pole N in (0; 0.5) grey lines. The line Kl–Rø not only has the right azimuth, (as Pe–Bo) but it is close to the projection line L–M–N (Rø 65 m and Kl 103 m distant) and, even more strikingly, this line goes nearly through the point F (Chr).[11]

---

[10] In private correspondence Dr. John D. North declared that he had not seen the construction before.

[11] This is a mathematical coincidence as the punctured line from N with angle $\text{Arctan}(\sqrt{3}/\sqrt{7}) = 56.789..°$ is cutting the vertical (x = 1) in $\sqrt{3}/\sqrt{7} + ½ = 1.15465\ldots$ whereas the tri-vesica construction above gives $2/\sqrt{3} = 1.15470$. In reality the difference is only 0.77 m at Chr.

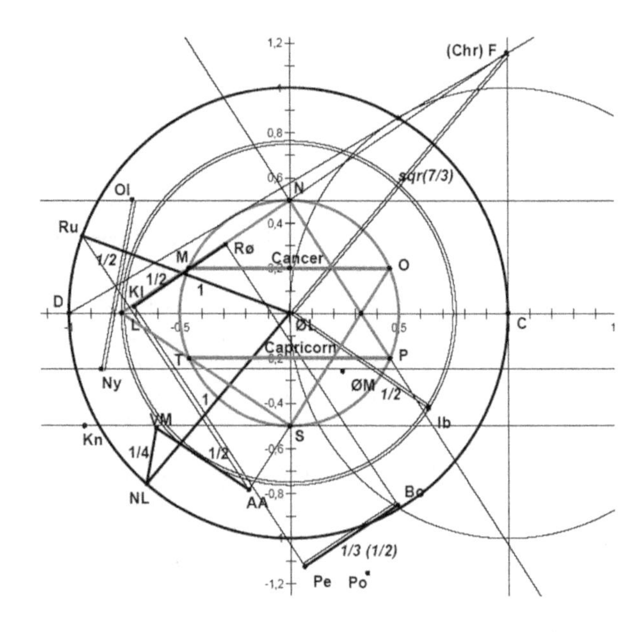

Fig. 28.9 Application of the celestial key. Image: author

The plane projection of Capricorn (double line) has radius ØL–Ib.

The projection line NP perpendicular to LN has azimuth $-33.21°$ and it happens to be close to azimuth for:

$$\text{Bo }(-34.15°) \rightarrow \text{Rø }(-34.29°) \rightarrow$$
$$\text{Pe }(-33.80°) \rightarrow \text{Kl }(-33.94°) \rightarrow$$
$$\text{AA }(-33.79°) \rightarrow \text{Ru }(-33.93°) \rightarrow$$

Furthermore the projection line OS through the South Pole S goes to AA (missing 47 m) and is perpendicular to AA–VM, which is parallel to SL as is Ib–ØL.

Plane Projection of the Ecliptic

It has not been possible to find the distance representing the radius of the Tropic of Cancer, nor the zenith or horizon in the plane projection, but the radius of the ecliptic (grey circle in Fig. 28.10) can be calculated by means of the two tropical radii as it touches both circles in the projection (thin and double line). For example, $R(\text{ecliptic}) = ½(R_1 + R_2) = ½\,(\sqrt{3}/\sqrt{7} + \sqrt{7}/\sqrt{3}) = 5/\sqrt{21}$. With the unity of ØL–NL of 14336 m it is 15641.8 m. This corresponds almost perfectly to the distance between Rø and AA (triple line) which is measured as 15638.7 m. The difference is

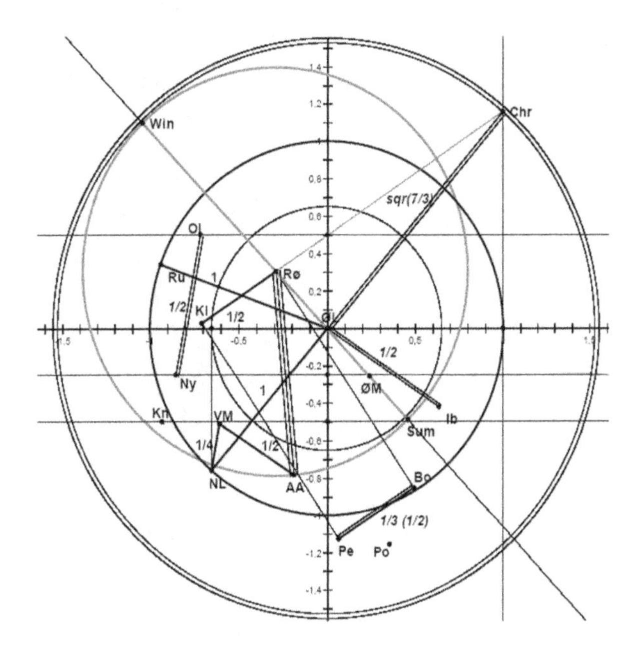

Fig. 28.10 The ecliptic. Image: author

only 3 m. The only church on the ecliptic is AA, the largest church on Bornholm which might represent the sun.[12]

Furthermore the ecliptic almost touches (230 m short) both tropics, as it should, in the two solstice points (Summer and Winter) and the line through these points is very close to the line Rø–ØL–ØM (ØM is 60 m distant).

Transformation of Chr and ØL to Ol and Ny

The placement of Ol and Ny does not come out directly of the key, but points to another way to construct $\sqrt{7}/\sqrt{3}$ as shown in Fig. 28.11 where a special transformation reallocates Chr to Ol and ØL to Ny—and explains why the azimuth of Ny → Ol should have the value 10.89° (= 40.89 − 30°).[13]

[12] The Sun's position at the ecliptic on an astrolabe represents a date around 24 April at 1:30 pm.

[13] Also described as a construction by Haagensen (2003) and nicely illustrated on http://www.new-science.co.uk

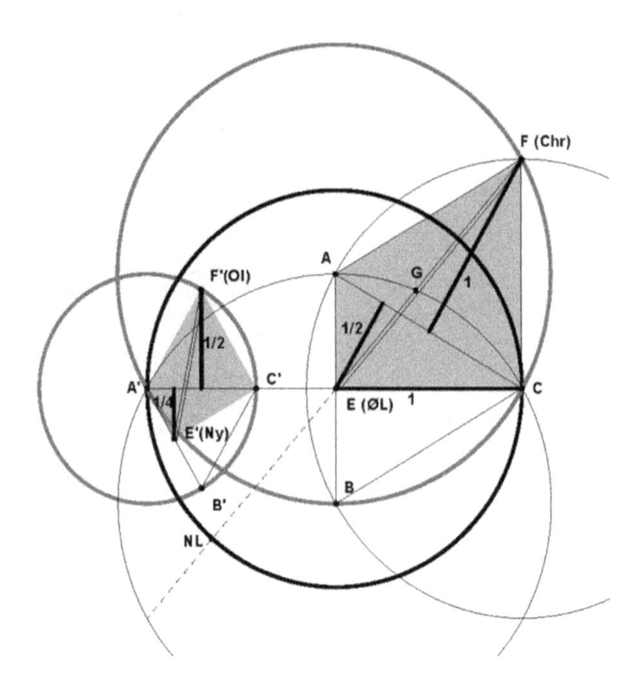

Fig. 28.11 Construction of the coordinates of Ol and Ny. Image: author

Description of Transformation in Fig. 28.11

EC = 1 by definition, hence isosceles triangle ACF (grey) and ABC (grey in one of its halves) both have vertical of 1. The diagonal EF is shown to have the length $\sqrt{7}/\sqrt{3}$. Now the big grey triangles are multiplied by ½ and turned 30° anticlockwise and displaced so point A is moved to A′. It is then obvious that the ordinates of F′ and E′ are 0.5000 and—0.2500 with a difference of ¾. The abscissas are less obvious, but the coordinates can be calculated to be:

F′: (−0.7113; 0.5000) and E′: (−0.8557;−0.2500) with
Ol: (−0.7149; 0.5024) and Ny: (−0.8558; −0.2486) for comparison.

This implies that Ol is about 61 m and Ny 20 m from the transformed placement. But the distance Ol–Ny is only about 5 m longer than the expected value of $\frac{1}{2}\cdot\sqrt{7}/\sqrt{3}$ over nearly 11 km.

The azimuth of Ny → Ol is seen to be 40.89 − 30° = 10.89° but can be derived directly by realizing that the difference in ordinates between churches was intended to be ¾ and the distance Ny–Ol is believed to be $\frac{1}{2}\sqrt{7}/\sqrt{3}$; the ratio is 0.982. By looking at Fig. 28.11 it is clear that the azimuth is Arccos(0.982…) =10.89°.

Deviation Between Theory and Facts

It is difficult to decide when a hypothesis deviates so much from fact that it has to be rejected. Is the difference a result of inaccuracy or methodological problems (plane $\rightarrow$ spherical representation)? Or is it simply wrong?

If the layout was intended, the differences between actual values (medieval position, e.g., VM is moved 34 m), and those predicted by the plane models (Figs. 28.8, 28.9, 28.10, and 28.11) require consideration. If angles from the **plane** model are transferred to the **spherical** earth and measured from the North (azimuth) a deviation of about 0.3° might be expected as mentioned.[14] Furthermore, some inaccuracy in practical layout must be allowed for, and this might accumulate in the measuring process to result in the observed difference in angles of about 0.5° seen in Fig. 28.5 (giving a relative error of 0.87 %). Distances seem to be 10 times more accurate.

A coarse estimate of the probability for a chance arrangement of 7 distance vectors transformed with multiple distance relationships and intended angles from the key is indeed very small.[15]

Some displacement may even have been intentional to take into account practical necessities or accommodate conflicting measures, e.g., alignment with springs, local geography or geometrical figures (Haagensen 2007).[16]

In the case of Pe–Bo, it is difficult to decide which distance was intended. It would be logical to assume it to have the same length as Kl–Rø so Kl–Rø–Bo–Pe could become a perfect rectangle, but the distance 7300 m is close to 1/3 of ØL–Chr (7299 m) compared with 1/2 of NL–ØL (7168 m). It seems that maybe Bo should be placed about 170 m further west, because the azimuth Pe(−33.80°) $\rightarrow$ Kl fits very closely to that of AA(−33.79°) $\rightarrow$ Ru, but deviates from Bo(−34.15°) $\rightarrow$ Rø. Furthermore the ratio between 1/2 (NL–ØL) and 1/3 (ØL–Chr) = 0.982...is the same as seen in the placement of Ny–Ol. Surely then the difference of 132 m was known. Archaeological explanations may be discovered.

Further study on the churches not mentioned—e.g., Kn nearly south of Ru with y = −0.5, Azimuth Po(−40.79°) $\rightarrow$ Ru, y(Po) = −1.1557 ≈ -y(Chr)—and

---

[14] The difference of 0.31° between azimuth in each end of NL $\rightarrow$ ØL $\rightarrow$ Chr. Line.

[15] From the coordinates in Table 28.1, the vectors NL $\rightarrow$ ØL, ØL $\rightarrow$ Chr, Kl $\rightarrow$ Rø, Pe $\rightarrow$ Bo, AA $\rightarrow$ VM, Ib $\rightarrow$ ØL, Ny $\rightarrow$ Ol and NL-VM are calculated and correspondingly multiplied with 1(def), $\sqrt{3}/\sqrt{7}$, 2, 2, 2, $2\sqrt{3}/\sqrt{7}$, $2\sqrt{3}/\sqrt{7}$ and 4, and turned 40.89°, 40.89°, 56.79°, 56.79°, −56.79°, −56.79°, 10.89° and 10.89°, in that order. These transformed vectors are nearly parallel and expand a very little space of uncertainty around the vector (0;1) (≈0.126 km^2). Random chance vectors are allowed length 0.25 to $\sqrt{7}/\sqrt{3}$ and azimuth from 0° to 90°. These chance vectors are each represented by 15 vectors by combinations of factors 1, $\sqrt{3}/\sqrt{7}$, $2\sqrt{3}/\sqrt{7}$, 2 and 4 - and turned angles of 10.89°, 40.89° and 56.79°. The chance vectors are taken from an area ≈ 151 km^2 (reduced by symmetry and geography). The probability that seven representatives all fall within the uncertainty area is less than $10^{-13} > (15 \cdot 0{,}126/151)^7$.

[16] Haagensen has assumed that it is the case for Ol to show a method to measure the radius of the earth. I do not agree.

studies of churches in southern Sweden might give additional evidence, or they might reveal that the correspondence with predictions from The Celestial Key is a lucky, though very improbably coincidence.

The mathematical probability of the geometrical layout of Haagensen has been disputed by J. Schmidt (2002: 189). But it is defended by Professor Niels Lind through genuine Bayesian statistical analysis (Lind 2002). His analysis has been criticised by J. Jerkert (2003). Despite this rather technical discussion, Jerkert points to the importance of the context in which the geometrical layout of the churches should be seen. Was Christiansø known for its religious significance at that time? Was deliberate alignment of churches relevant to the worldview of that era? Was the technique available in Bornholm at this time?

Discussion—Astroarchaeological Prelude

Before the new knowledge of the astrolabe, there was a long indigenous tradition of observing the directions of the rising or setting sun on important dates (McCluskey 1998). There is hardly any written documentation, so it must be inferred from archaeological finds and astronomical alignments of stones and rock carvings, but it does give further support to the Celestial Key.

John D. North describes how long barrows and cursus (one 10 km long) are oriented towards the rising and setting of stars as far back as 4000–3000 BC (North 1996: 138–188). The heliacal rising of stars (their appearance on the horizon before sunrise) is the episode used to fix the day of the year. But caused by the precession, it changed slowly (1° in 72 years around the ecliptic pole) and must be steadily corrected, so date fixing was supplemented by observing the periodic movement of the sunrise along the horizon from solstice to solstice.

The rising and setting of a star is symmetrical along the north south direction. Alignment to a star at a certain height over the horizon can be used with high precision to fix azimuth of the direction. If this method were used in the layout of churches it would explain the obvious north–south symmetry as well as the parallel directions. For the azimuth around $\pm 56.8°$ the star used might be identified, e.g., Sirius at an altitude of 2.36° looking south (1150 AD)—or Arcturus at an altitude of 7.2° looking north. It might even explain why these angles cluster in two groups.

The solstice alignment shows up at Newgrange (3500 BC) and Stonehenge (3000 BC). The Nebra sky disk (1600 BC) has been interpreted as showing the angle (82°) between the solstices at the latitude of the hill Mittelberg (51°), and from here alignment towards two distant mountain peaks by sunset at midsummer and on the midquarter days, around 1 May and 2 August.

At the latitude of Bornholm the angle between solstices has been exactly 90° (1600 BC in the south and 500 BC in the north).[17] Several rock art carvings on Bornholm show alignments to solstice and mid-quarter days (Fig. 28.12) (Jensen

[17] With the sun's center on the horizon, the latitude λ can be found from $\lambda = \text{Arccos}(\sqrt{2} \cdot \text{sine})$ with e from Newcomb's formula (note 11).

Fig. 28.12 Sun cross at Bornholm. Photo: author

2000–2001). Furthermore Haagensen has identified upper windows in the ØL church to be in alignment with the sunrises of the winter and summer solstices with azimuth of 43.1° and 132.73° (Haagensen 1993: 87, 90; 2003: 76–78, 82–83; 2007: 112–129).

The digital calendar took over when important pagan feast days were substituted by Christian holy days (McCluskey 1989, 1998: 60–76).

In Table 28.2 there are eight feast/holy days: the four dates of solstice and equinox and, between these, the mid-quarter days, which were well-known festivals in the Celtic calendar. This division is traceable to the Stone Age, inferred by the astronomical alignment of graves and stone settings.[18]

The azimuth at solstice, equinox and mid-quarter days depends on the latitude and height above sea level. It is also influenced by refraction and, therefore, factors such as air temperatures. Investigations carried out by Jens Lindhard of hundreds of alignments between neighbouring churches and barrows in the northwestern part of Denmark emphasize angles of 41°, 86° and 126°. These are interpreted as Midsummer Solstice, Equinox and Winter Solstice respectively, at the latitude around (57°) (Linhard 2007).

The azimuths of 43.2° and 58.4° are noticeable as being close to azimuths of 40.9° and 56.8° found for 6 pairs of lines connecting the churches at Bornholm. These azimuths could have been a near continuation of an older Norse practice for fixing a religious date. Using the Key, angles were now linked to the sky and the tri-vesica construction of the astrolabe, and were thereby independent of the place of observation. It could be a symbol of the unchanging heavenly world with great persuasive power for heathen priests converting to Christianity, even more so if these alignments included pagan sites such as Ertholmene and Ales Stena.

[18] The eight-part year was first proposed by Lockyer (1906: 30), discussed in great detail by Thom (1967: 107–117, supported by Heggie (1981: 219), discussed in North (1996: 301, 509, 540) and criticized by Ruggles (1999: 47–67, 88, 142–3).

Table 28.2 Substituting of Celtic/Norse festivals with Christian feasts

Approximate date (Julian calendar)[a]	Astronomical	Celtic/Norse	Christian	Azimuth[b] ØL year 1150
21–22. December	Midwinter solstice	(Sol Invictus)/ Yule	Christmas	132.1°
1. February	Mid-quarter day	Imbolc/diesting	Candle mass, St.Brigit	117.7°
21–22. March	Spring equinox	Goddess Eostra?	Marys Annunciation, moon → Easter	88.3°
1. May	Mid-quarter day	Beltane/Mayday	Voldermas, May Queen, Cross mass	58.4°
21–22. June	Midsummer solstice	/midsummer festivities	Saint John the Baptist.	43.2°
1. August	Mid-quarter day	Lugnasa/hay harvest	Lammas, St. Oswald, St. Justus	58.4°
23. September	Autumn equinox	/Thanksgiving	Conception of St. John the Baptist.	88.3°
1. November	Mid-quarter day	Samhain/butcher feast	Halloween, All Saints, All Souls	117.7°

[a]The date in the Gregorian calendar for the astronomical mid-quarter days is from 2 to 8 days later. See McCluskey (1989). The mid-quarter days are calculated as geometrical midpoint on ecliptic. e.g., $\sin\delta = \pm\text{sine}/\sqrt{2}$; see note below

[b]The angle from north to sunrise with the sun's upper limb over the sea horizon ($-16'$) and corrected for refraction ($-35'$) and height above sea level. Azimuth $= \text{Arccos}((\sin\delta - \sin\lambda \cdot \sin\theta)/\cos\lambda \cdot \cos\theta)$ where declination: $\delta = \pm e$ (solstices, e from note 9), 0 (equinox) and for δ at mid-quarter days (see note above); λ is geographical latitude and θ is the angle from the horizon (negative under) $= -0.0321 \cdot \sqrt{(\text{height in m})} - 35/60 - 16/60$ in degree, see North (1996: 553–563) and Ruggles (1999: 22)

Discussion: Symbolic Use of the Celestial Key

In the early Middle Ages many phenomena were interpreted symbolically as signs revealing higher truth (Pedersen 1996: 194–197). That the special value of the obliquity Arcsin(0,4) will generate the exact ratio of $\sqrt{7}/\sqrt{3}$, could be a sign for the shrewd cleric, who could also see the Tri-Vesica Piscis construction. This is even partly depicted in ØL as shown in Fig. 28.13. Besides, 7 and 3 are holy numbers.

If the Celestial Key was used in the arrangement of churches would it not also have been used in the design of churches? A striking illustration is the plan of the Cathedral of Stavanger (1130) (see Fig. 28.14).[19] The length of the nave is $\sqrt{7}/\sqrt{3}$ times the width, as shown with the overlay of the plane projection. Furthermore the orthographic projection of the Tropics is aligned with the columns, which are crossed by the Equator, and the projection of Tropic of Capricorn. The light in

[19] Measured by Gerhard Fischer in 1939–1940.

Fig. 28.13 Christ as Majestas Domini in a (Tri) Vesica Piscis in Østerlars. Notice the sun, moon and rainbows. Photo: S.Plum/ Majpress firstmay@image.dk

the church comes in through the clerestory to descend between the tropics, symbolizing the suns place projected from heaven's sphere in the church (Nielsen 2001: 28).

A similar plan can be drawn of the Cathedral of Lund (c.1110 AD). But here some holy numbers come into play as close rational approximations to $\sqrt{7}/\sqrt{3} \approx 55/36 \approx 84/55$. The inner width of the nave is about 55 "natural cubits" (cu) and produces radii of tropics of 36 cu and 84 cu.[20] Mogens Koch has shown that 12 cu is one of the modules in the church and width between columns is 24 cu. (Koch 1993: 96–100).

[20] Mogensen measured the width of the nave to 54.86 "natural cubits", see Mogensen (2003: 49). The "natural cubit" = 0.4666 m was discovered by Mogens Koch on the capital in the crypt.

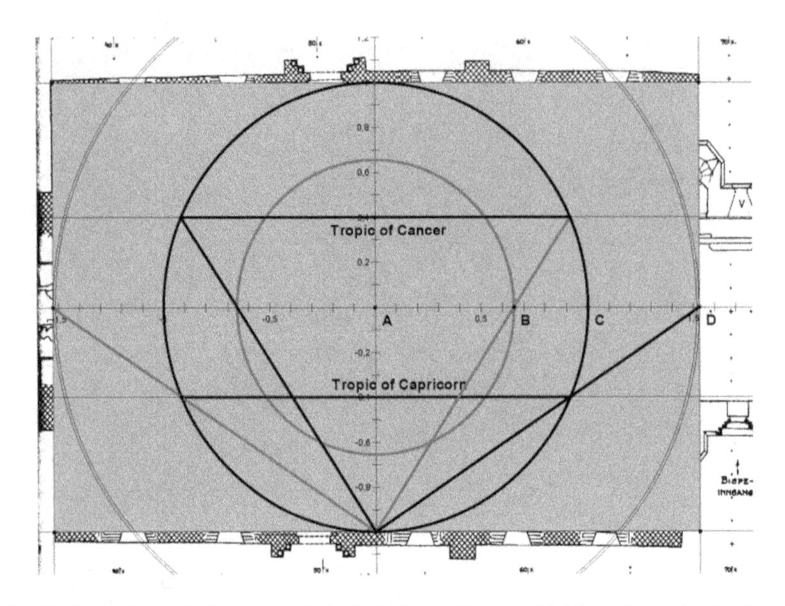

Fig. 28.14 Plan of nave in Stavanger Cathedral. Image: Fischer (1964), with overlay by the author

The "natural cubit" is 1½ feet. This in itself is another, but not as good, approximation for $\sqrt{7}/\sqrt{3} \approx 1.5$.

Surely the proportions of churches and their symbolic meaning contain many mysteries that have yet to be discovered. It is possible that the Celestial Key could unlock some.

Conclusion

The Celestial Key combines several striking astronomical, mathematical and geographical coincidences.

The obliquity of the ecliptic was 23°35′ as measured by the Arabs in the eleventh century. Assuming an obliquity of Arcsin(0,4) it generates the exact ratio of $\sqrt{7}/\sqrt{3}$ and the angle 56.8° in the stereographic projection, and leads to a "Tri-Vesica Piscis" astrolabe construction with the angles of 40.9°. This angle is close to the azimuth of the rising midsummer sun on Bornholm and the azimuth of 56.8° is close to the rising sun at important mid-quarter days in the Celtic calendar (1 May and 1 August). Both directions have also been found in rock art from the Bronze Age (1000–500 BC), and the symbolic value was reinforced by pointing out pre-Christian holy places, "Ertholmene" and "Ales Stenar".

The stereographic projection of the heavenly sphere is a remarkable metaphor for the connection between heaven and earth, so it may perhaps be found in the measure and proportions of Romanesque church buildings.

The Celestial Key could be the result of thousands of years of semiotic evolution reinforced in a network of symbolic interpretations—many now extinct—but then known at that time to the myth-creating minds of the Middle Ages.

Biography Niels Bandholm earned his Master of Science in Chemistry and Physics from Aarhus University, Denmark in 1971, with a M.Sc. thesis in the History of Science in 1972. He is the author of books on chemistry, computer science and energy storage for the Danish upper secondary schools, where he was employed as a teacher and a senior master in physics, chemistry and astronomy from 1966 until his retirement in 2005. In addition, he has organized many conferences on interdisciplinary didactics and teaching. The present study is an example of an interdisciplinary project and was kindly hosted by the History of Science Department (now Department of Science Studies) at Aarhus University 2006–2007. He has been greatly interested in environmental issues since 1987, initiating organisations, newsletters and eco-villages. He is now building his own dome house in a sustainable society with 300 persons 15 km north of Aarhus.

References

Andrén, A., C. Raudvere and K. Jennbert (eds). 2002. *Plats Och Praxis: Studier av Nordisk Förkristen Ritual*. Lund: Nordic Academic Press.

Bandholm, N. 2007. Bornholms Gåde. *LMFK-Bladet* **2** (March). See http://www.lmfk.dk/artikler/data/artikler/ 0702/0702_30.pdf

Fischer, G. 1964. *Domkirken i Stavanger*. Oslo: Dreyer.

Geese, U. 2004. *Die Kunst der Romantik*. Körnemann. (English ed. Ulrike Laule and Uwe Geese, *Romanesque*, Berlin: Feierabend Verlag, 2003 .

Haagensen, E.1993. *Bornholms Mysterium*. Lynge: Bogan.

———. 2003. *Bornholms Rundkirker*. Lynge: Bogan.

———. 2006. *Tempelherrenes Skat*. Revised 2nd edition. (with English summary). Lynge: Skive.

———. 2007. *Anklaget for Tavshed*. Hellerup.

———. and H. Lincoln. 2000. *The Templars' Secret Island*. Moreton-in-Marsh: Windrush Press.

Heggie, D. 1981. *Megalithic Science*. London: Thames and Hudson.

Jaff, M. 2003. From the Vault of the Heavens. *Nexus Network Journal* **5**, 1: 49–63.

Jensen, M. 2000–2001. Nyt om Helleristninger på Bornholm. *Bornholms Museums Årsskrift*. Rønne: Bornholms Historiske Samfund.

Jerkert, J. 2003. *Bayesian Confusion*. Stockholm: Fri Tanke. Translated and revised from *META*, **4**: 49–55.

King, D. A. 1999. *World-maps for Finding the Direction and Distance of Mecca*. Leiden: Brill.

Knudsen, A. V. 1997. *Bornholms Gamle Kirker*. Rønne: Bornholms Historiske Samfund.

Koch, M. 1993. *Geometri og Bygningskunst-Tekster af Mogens Koch*. Copenhagen: Christian Ejlers Forlag. (From *Arkitekten* 1963).

Lidegaard, M. 1999. *Da Danerne Blev Kristne*. Copenhagen: Nyt Nordisk Forlag Arnold Busck.

———. 2004. *Hvad Troede de På?* Copenhagen: Gyldendal.

LIND, N. 2002. Bayesian Analysis of the Location of Bornholm's Medieval Churches. *META*, **2**: 49–55.

LINHARD, J. 2007. *Kirkerne Ligger På Rette Linier*. Unpublished presentation at the Steno Institute.

LOCKYER, J. N. 1906. *Stonehenge and Other British Stone Monuments Astronomically Considered*. London: Macmillan.

MCCLUSKEY, S. C. 1989. The Mid-quarter Days and the Historical Survival of British Folk Astronomy. *Archeoastronomy*, (Journal for the History of Astronomy Supplement) **13**, 20: 1–19.

———. 1998. *Astronomies and Cultures in Early Medieval Europe*. Cambridge: Cambridge University Press.

MOGENSEN, L. 2003. *Kosmisk By*. Lund.

NIELSEN, D. 2001. The Squaring of the Circle in two Early Norwegian Cathedrals? *Nexus Network Journal* **3**, 1: 27–42.

NIELSEN, F. O. 2006. *Fortidsminder på Bornholm*. Third edn. Bornholms: Bornholms Regionskommune.

NORTH, J. D. 1974. The Astrolabe. *Scientific American* **230** (January): 96–106.

———. 1996. *Stonehenge: Neolithic Man and the Cosmos*. London: Harper Collins.

PEDERSEN, O. 1996. *Naturerkendelse og Theologi*. Århus: Århus Universitet.

RUGGLES, C. 1999. *Astronomy in Prehistoric Britain and Ireland*. New Haven: Yale University Press.

SCHMIDT, J. 2002. Bornholms Geometry og Haagensens Matematik. *Bornholmske Samlinger* **III**, 16: 189–190.

THOM, A. 1967. *Megalithic Sites in Britain*. Oxford: Oxford University Press.

WELBORN, M. C. 1931. Lotharingia as a Centre of Arabic and Scientific Influence in the Eleventh Century. *Isis* **16**, 2: 188–199.

WIENBERG, J. 1986. Bornholms Kirker i den Ældre Middelalder. *Hikuin* **12**: 45–66. Højbjerg.

———. 2002a. Mellem Viden og Vrøvl. *Bornholmske Samlinger* **III**, 16: 175–188.

———. 2002b. Middelalder uden Mystic. *Bornholmske Samlinger* **III**, 16: 203–206.

WIVEL, M. 1989. Bornholms Runde Kirker og Tempelridderne. *Bornholmske Samlinger* **III**, 3: 49–63.

Chapter 29
Friedrich II and the Love of Geometry

Heinz Götze

Introduction

The Castel del Monte was built in the northern part of Apulia by the Holy Roman Emperor Friedrich II of Hohenstaufen in the last decade of his life.[1] Even today it remains an object of wonder (Fig. 29.1). Standing on a conical hill in the flat table-like countryside called the Murge that slowly falls toward the sea, the castle is visible from afar, golden in the bright sunshine. Its unique form—eight-sided with octagonal towers at each corner—is sharply defined by shadows. The stark, sharply delineated walls emphasize its stereometric character. For Castel del Monte, there is no need to invent or to guess at the geometric relationships. They simply exist in the building, which cries out for a mathematical analysis to help evaluate it as an architectural object in the same way that historical, chronological, art-historical, and architectural-historical analyses do.

One must, of course, ignore small deviations of the castle's actual measurements from the obvious design objectives. The architect here built in certain small

First published as: Heinz Götze, "Friedrich II and the Love of Geometry", pp. 67–79 in *Nexus I: Architecture and Mathematics*, ed. Kim Williams, Fucecchio (Florence): Edizioni dell'Erba, 1996. This chapter originally appeared in *The Mathematical Intelligencer*, vol. 17, no. 4 (1995) pp. 48–57. This was a somewhat expanded version of an article in *Architektur Aktuell* 169/170 (1994) pp. 88–95. English translation by L.L. Schumaker. Computer-generated Fig. 29.1b was prepared by Susanne Krömker of the Interdisciplinary Center for Scientific Computation of the University of Heidelberg. Reprinted by permission.

Heinz Götze (1912–2001).

[1] This discussion is based on work presented in Heinz Götze, *Castel del Monte, Gestalt und Symbol der Architektur Friedrichs I*, 1st edn (Munich, 1984) and 3rd edn (1991) pp. 9–12 and 84 ff. See also Heinz Götze, "Die Baugeometrie von Castel del Monte," in *Sitzungsberichte del Heidelberger Akademie der Wissenschaften, Phil.-hist. Klasse*, Jahrg. 1991, Bericht 4.

K. Williams and M.J. Ostwald (eds.), *Architecture and Mathematics from Antiquity to the Future*, DOI 10.1007/978-3-319-00137-1_29,

Fig. 29.1 The Castel del Monte. Photo: author

deviations; for example, to emphasize the east wing of the building. This did not lead to a devaluation of the basic concept—quite the contrary. The deviations clearly are based on a well-defined overall plan.

A Geometric Analysis and Definition

The two-dimensional layout of Castel del Monte can be identified as a symmetry group with 16 elements: eight reflection planes and eight rotation planes. It is an automorphic group. These symmetry relations reveal themselves in all tour towers (Fig. 29.2). The multiplicity of symmetries is expanded by homotheties, that is, "similarities" among the large octagon of the main building, the octagon of the inner courtyard, and the eight octagonal towers placed in the same system of axes.

The ratios of the size of the three octagons "similar" to each other is given by $4/h$ (I):$2h$(II):$2(2\sqrt{2}-1)h$(III), where h is the mesh size of the basic grid (Fig. 29.3). The ratios of the sides of the three octagons is $2a$:a:c($\sqrt{2}-1$), where a is the side length of the courtyard octagon. The symmetry group involved here is a planar group of type D16, in the notation of J. M. Montesinos-Amilibia (1987).

Simple reflection symmetry, which is often encountered in nature, has long played an essential aesthetic function in architecture. There are other mathematical relationships that have equally strong aesthetic effects, and which also appear in nature—for example, the "golden ratio" which is connected with the pentagon. The architect of the Castel del Monte was clearly aware of the aesthetic importance of symmetries. He used them to achieve the impressive appearance of the castle, which still affects us today.

The planimetric aerial photo (Fig. 29.4) shows that the tangents of the octagon forming the inner courtyard intersect at the centres of the octagonal corner towers:

Fig. 29.2 The symmetry group that characterises Castel del Monte. Image: author

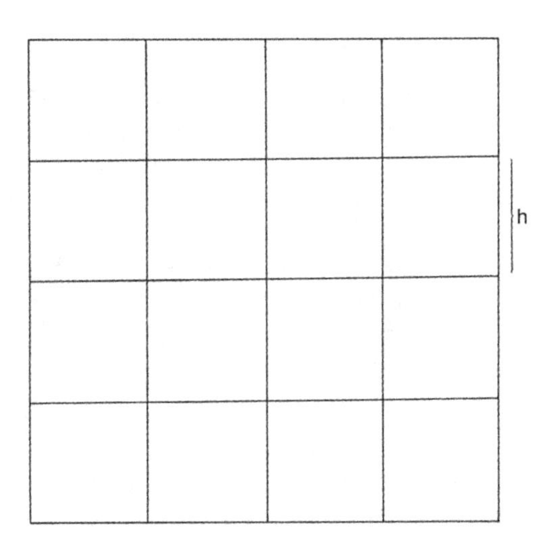

Fig. 29.3 The basic grid. Image: author

they form an eight-pointed star, the tips of which lie at the centres of the towers. This provides a geometric relationship between the inner courtyard and the corner towers, established by the similarity relationships discussed above.

This relationship was first discussed in my book on Castel del Monte (Götze 1984), where I also gave a geometric construction for the layout (Figs. 29.5, 29.6, and 29.7). I am indebted to three very famous mathematicians who have dealt with this geometric configuration and my construction: F.L. Bauer of Munich, Marcel Erné of Hannover, and Max Koecher of Münster. It was Max Koecher who first thought about the strong aesthetic effect of the multiple symmetries in the geometric plan, and defined it as a geometric configuration with its own intrinsic

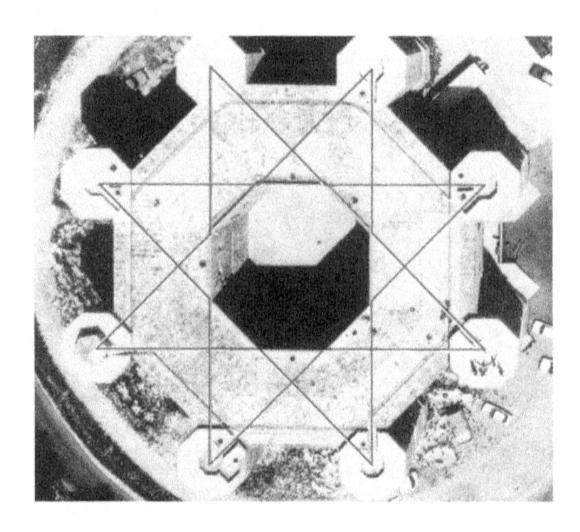

Fig. 29.4 The aerial view of the Castel del Monte with the geometry overlaid. Image: author

aesthetic (Koecher 1991: 221–233). He described it as a configuration of octahedra, generalized from regular polygons, built up according to the following principles:

1. The configuration comprises the centre and the vertices of a central octagon O, along with the centre points and vertices of eight smaller translated copies O_V, of the central octagon, all equal in size.
2. The centres M_V, of O_V, lie on the rays passing through the centre M of O and the vertices of O.
3. All M_V, are at the same distance from M.
4. As many as possible of the vertices and centres of O and of O_V, are collinear; that is, they lie on common straight lines. This last condition is essential for the aesthetic configuration. Koecher then discussed the many different possible collinear arrangements.

In Fig. 29.8, two such "collineations" (i.e., points of the basic octagonal construction which lie on a single line) are marked: the line connecting the centre of the octagons M_aM_v (or equivalently, the line through c and b), and the tangent lines to the two exterior octagons.

An alternative to the construction of the layout shown in Figs. 29.5, 29.6, and 29.7 (which require both compass and ruler) is due to Marcel Erné of the Technical University of Hannover; it does not require a compass (Fig. 29.9).

After drawing a right angle, as in the first construction, we divide a large square into 16 subsquares with side lengths h (as in Fig. 29.3 above). The quantity h can be considered as a modulus, the size of which can be computed from the measurements of the castle. In the next step a second, equally large square is created and rotated by 45° so that the vertices of the new square lie on the extensions of the main axes of the original square. We have here just the classical approach to constructing an octagon by rotating a square. One of the advantages of this second construction is that one can readily compute the lengths of all line segments.

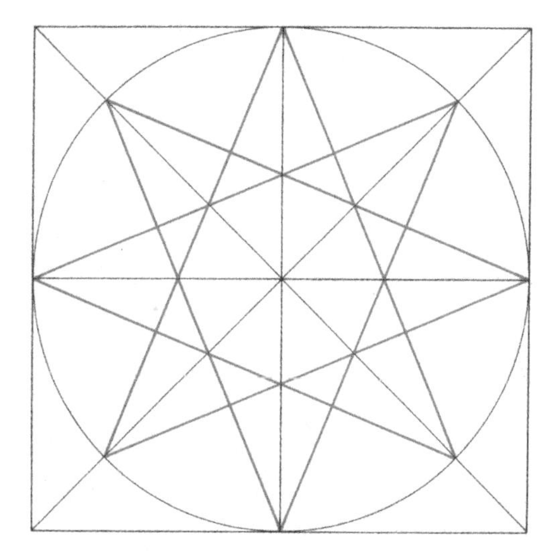

Fig. 29.5 Step 1 in the geometric construction of the Castel del Monte. Image: author

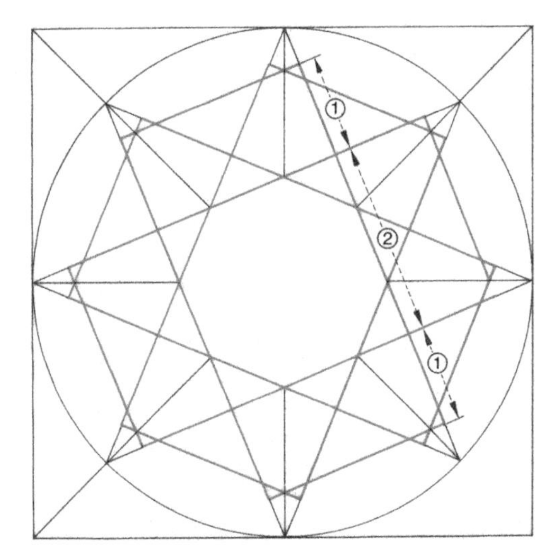

Fig. 29.6 Step 2 in the geometric construction of the Castel del Monte. Image: author

Relationships

The sides of the octagon forming the interior walls of the outside of the castle are determined by the intersecting grids and are twice as long as the sides of the octagon forming the interior courtyard. The width of the eight outside towers is equal to the length of a courtyard side a.

Instead of the quantity a, we can also use the mesh size h as the basic unit. The quantity a can be found from the actual building by measuring the width of a tower or the length of a side of the interior courtyard. It is not so clear how to measure the quantity h, although it is of great importance as the basic unit used in constructing

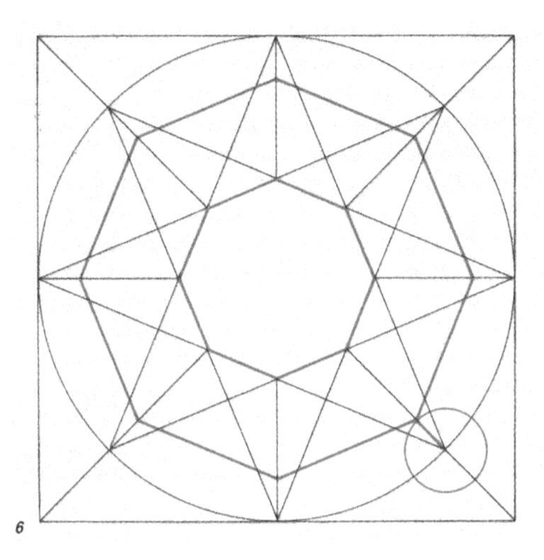

Fig. 29.7 Step 3 in the geometric construction of the Castel del Monte. Image: author

Fig. 29.8 Two collinear arrangements. Image: author

the rectangular grids. The grid plays an essential role in the development of the layout, and presumably also played a key role in the translation of the drawing to the actual physical construction of the building. For this kind of complicated geometric configuration, it is impossible to build without marking it full-scale on the construction site. It is also clear that this could not have been done "by hand," but must have been accomplished with mechanical measurement tools which were readily available. Using h as the basic unit, we have the following (refer to Fig. 29.3):

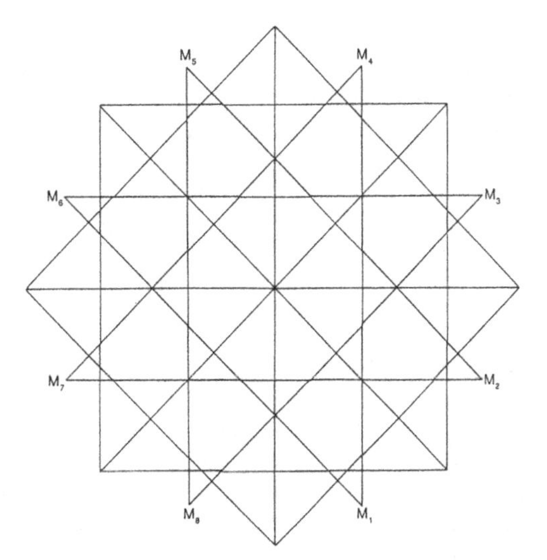

Fig. 29.9 The alternate construction for the layout. Image: author

1. The distance between the outside of the courtyard walls and the inside of the exterior walls is h.
2. The diameter of the interior courtyard from the centre of one wall to the centre of the opposite wall is 2 h.
3. The distance between the centres of the towers is also 2 h.
4. The lengths of the sides of the towers is $b = \frac{a^2 h}{2}$, and the thickness of the outer walls is $c = \frac{a^2}{(2\sqrt{2})h}$.
5. The overall width of the castle is $(4\sqrt{2})h$. This mathematical prescription corresponds to the geometrical configuration.

The appearance of √2 is noteworthy, albeit not surprising since we are dealing with the construction of squares and octagons.

The Russian architectural historian M.S. Bulatov pointed to a high multiplicity of symmetries as well as the repeated use of the quantity √2 as characteristics of Islamic architecture (Bulatov 1988: 98–104). Bulatov based his interpretation on his own careful studies and surveys of central Asian architecture as well as on the writings of Arabic scholars.

As already suggested, the practical marking of the floor plan on the construction site probably began with the two gridded squares that determine the outer octagon. This was technically possible using the grid size which provided the basis for all other measurements. The essential role of the initial square, and with it the outside octagon, is also suggested by the exact equality of the distances between opposing walls of the outside octagon—it is 36 m, which is approximately 120 Roman feet.

Castel del Monte was erected by Cistercian masons. Friedrich II felt a close affinity to this order, which played a major role in the architecture of the high

Middle Ages. For comparisons, note that the nave of the abbey church of Eberbach in the Rheingau region of Germany, also erected by Cistercians, has a length of approximately 71 m, that is, around 240 Roman feet—exactly twice the distance between the insides of opposite walls of Castel del Monte. Furthermore, the central room of the Basilica of Fanum of Vitruvius is 120 Roman feet long and 60 Roman feet wide.

In all three buildings, the lengths are multiples of 60 Roman feet, and thus correspond to numbers in the hexadecimal system, known since the time of Babylonian astronomers, and which remained in use in Europe until the tables of Regiomontanus appeared (1436–1476). In view of this, it seems quite reasonable to consider the hexadecimal system of the outside octagon as the practical starting point for the construction of the layout. This leads us to wonder where this very unique idea for designing a European building at this time in history might have originated. The layout of other castles built around the same time (for example, the Wartburg in central Germany, the Marksburg on the Rhein, and the Chateau Chillon near Montreaux, to mention only three arbitrary examples from different regions) are all very far from the kind of geometrical configuration which Castel del Monte exhibits. The keeps of castles in southern England and donjons in northern France are two other examples lacking such a geometrical structure.

The practical aspects of building, the technique of ribbed arches, and the design of capitals were all part of the work of the Cistercian masons, who were dedicated to the Gothic style of Middle Europe. This says nothing about the design of Castel del Monte itself. Even the book of the contemporaneous architect Villard d'Honnecourt contains nothing remotely similar. There are no written records concerning the history of the design of the castle, and we do not know who the architects were. But we need not depend on conjecture alone to learn something about the creation of this remarkable building, given its well-defined geometric configuration and its inner aesthetic.

Searching for related structures, we find in the *Carta Pisana*, a navigational chart drawn at the end of the thirteenth century, an interesting depiction of an octagonal compass that exactly matches the shape of the layout of the Castel del Monte (Fig. 29.10).

The navigational charts of the Mediterranean are primarily of Arabic origin (including the Arabic Caliphate of Córdoba). Two other examples include the Maghrebian navigational chart of the western part of the Mediterranean (Fig. 29.11) produced in the first half of the fourteenth century, and a drawing of two symmetric wind roses that does not include any underlying geographical information (Fig. 29.12).

On the other hand, both of the wind roses in Fig. 29.12 clearly show an underlying square grid and the extended grid lines of the crossed squares, the intersections of which determine the vertices of the eight-pointed star.

Fuat Sezgin has carefully studied navigational charts (Sezgin 2000). He cites the historian Ibn Fadlallah al-'Umari (d. 1349) as saying that navigational charts always include wind roses. In addition, as part of work done for the seafarer Abu Mahammad 'Abdalla B. Abi Nu' Aim al-Ansari al-Qurtubi of Córdoba, Ibn

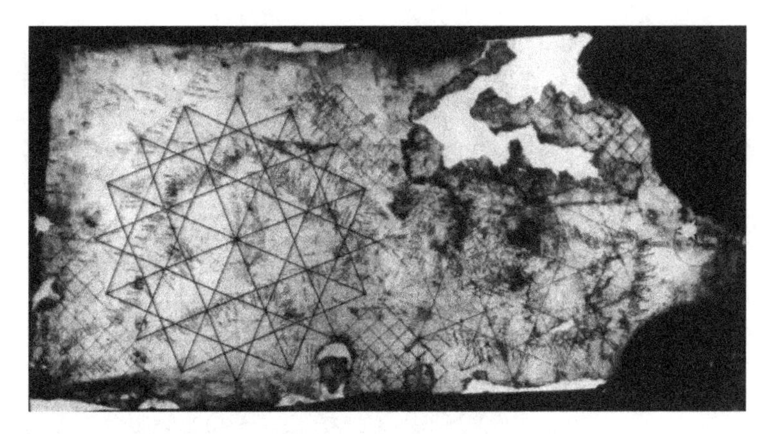

Fig. 29.10 Geometrical construction of the Castel del Monte overlaid on a reproduction of the *Carta pisana* (Bibliotheque Nationale de France, Dep. des Cartes et Plans, Res. Ge. B1118). Image: author

Fig. 29.11 Geometrical construction of the Castel del Monte overlaid on a reproduction of a Maghrebian navigational chart. Image: author

Fig. 29.12 Geometrical construction of the Castel del Monte overlaid on a reproduction of two wind roses. Image: author

Fadlallah al-'Umari also reported that only the four principal directions and the four directions halfway between have Arabic names, even when the wind rose shows more than eight directions (up to 32). Clearly, wind roses in the form of eight-pointed stars were developed by Arabic-Spanish sailors. The use of this style of wind rose for the design of the layout of Castel del Monte provides a clear indication of its origin, and possibly also for the meaning of the building itself. Eight-sided stars are older than the navigational charts. They can be found already as projections of the ribs of the cupola in front of the Mihrab in the Umaiyaden Mosque in Córdoba (AD 961–966). This mosque also contains the first stage of the pattern in the form of crossed squares, which can also be seen in the cupola of the Umaiyadi Alferia Palace (second half of the eleventh century) in Zaragoza.

The appearance of eight-sided stars is not confined to the Arabic region of the Mediterranean—they also appear as far away as Persia, India, and central Asia.

As the examples show, the eight-pointed star constructed from crossed squares is a widely used motif in the Muslim world, appearing in many contexts. Its use in the

Fig. 29.13 Geometrical construction of the Castel del Monte overlaid on a photograph of a mosaic from the Alhambra. Image: author

cupolas of religious buildings as well as in the form of wind roses suggests a connection with concepts of the heavens. For the discussion here, it suffices to recognize the underlying geometric configuration, exalted by the places in which it was used, and characterized by its multitude of symmetries.

The navigational charts, along with a mosaic of similar form in the Alhambra (Fig. 29.13) exhibit an additional step in the development of the eight-pointed star figure: the intersection points of the tangents, that is the vertices of the star, are distinguished by bundles of rays.

Here a new idea of the architect of Castel del Monte comes into play: he emphasizes the special position of the tips of the stars by repeating the eight-sided pattern in smaller, similar form. These corner towers are not only connected by a line from their centres M_v to the centre M of the central octagon, but they also form a complete geometrical system together with the octagons of the outside wall and the inside courtyard (see Fig. 29.2 above).

With this novel idea, the architect increased the symmetries by an order of magnitude, and thus the aesthetic effect of the entire building. Since simple reflection-symmetry is already regarded as a harmonizing element in architecture, it is no surprise what an immense effect this additional symmetry has.

Suppose we carry the idea of the architects of Castel del Monte one step further and construct new eight-pointed stars formed from the tangents to the sides of the towers (Fig. 29.14). We immediately recognize geometric relationships between these outer stars as manifested in additional collineations.

This is not to say that the architect of Castel del Monte took this additional step. It is not needed to establish the geometric relationships found in the castle, as these

Fig. 29.14 A further development of the Castel del Monte geometry. Image: author

are based on the geometric system that includes the eight exterior octagons as well as the two basic octagons. The figure of the eight exterior stars corresponds to patterns found in Indo-Arabic constructive geometry, as can be seen in the mosaic in the centre of the mausoleum of Humayun in Delhi (1565), in which the stars are aligned along lines and touch each other only at one tip (Fig. 29.15).

The close geometric connections between the eight "satellite" stars, as shown by the touching of the points of neighbouring stars, provide additional evidence that the size of the towers was not chosen arbitrarily, but follows the geometrical system. These connections further support the completeness of the geometric design of Castel del Monte as an example of a configuration with an inner aesthetic.

The repetition of the basic eight-pointed star can be continued and, as Max Koecher observed, results in a fractal with infinite iteration possibilities (Fig. 29.16).

An indirect proof of the geometric rules underlying the design of the castle is the fact that a computer-graphics model of the castle requires nothing more than these rules to effect complete reconstruction, as the Heidelberg Centre for Scientific Computing has shown.

The determination and analysis of the origins of the basic geometric form underlying the design of Castel del Monte establishes a connection between it and Indo-Arabic geometry. Such a complex geometric analysis, however, is unusual even for this area. It is thus natural to ask if the emperor himself

Fig. 29.15 Eight-pointed stars in a mosaic from the mausoleum of Humayun in Delhi. Photo: author

Fig. 29.16 Computer rendering of the fractal iteration of the Castel del Monte geometry. Image: author

provided the inspiration for the design of this, the crown of his castles. His interest in mathematics and architecture is well known. The collection of permanent scholars at his court included Arabic mathematicians such as Theodorus of Antioch, who carried out mathematical correspondence with Leonardo of Pisa.

This was the period in which Leonardo of Pisa collected Indo-Arabic mathematical results and disseminated them throughout Europe. The accomplishments of Greek mathematics had been preserved and extended by

Arabic scholars. Leonardo was in close contact with Friedrich II and his court. It may be assumed that Friedrich II, given his interest in mathematics, took an active part in the design of this castle, which could symbolize the principles of his empire.

Thus, Castel del Monte, with its extraordinary aesthetic radiance, is not only an art and architectural monument without equal, but also a scientific and cultural one. It stands at the crossroads of the Arabic-geometric and Middle-European-Gothic worlds, and represents the ruling spirit of one of the most important emperors of the middle ages.

Biography Heinz Götze (1912–2001) was a partner, longtime CEO and member of the board of the Springer publishing group. He was a promoter of science and research in the field of humanities and cultural sciences, especially archeology, classical philology and art history in Europe and Asia, and a collector of East Asian calligraphy.

References

BULATOV, M.S. 1988. *Geometrische Harmonisierung in der Architektur Zentralasiens im 9–15 Jahrhundert* (in Russian) 2nd edn. Moscow: Nauka.

GÖTZE, Heinz. 1984. *Castel del Monte, Gestalt und Symbol der Architektur Friedrichs I,* 1st edn. Munich: Prestel-Verlag.

———. 1991. Die Baugeometrie von Castel del Monte. In *Sitzungsberichte del Heidelberger Akademie der Wissenschaften, Phil.-hist. Klasse.* (1991) **4**.

KOECHER, Max. 1991. Castel del Monte und das Oktogon. In *Miscellanea Mathematica.* Heidelberg: Springer-Verlag.

MONTESINOS-AMILIBIA, J.M. 1987. *Classical Tesselations and Three-Manifolds.* Heidelberg: Springer-Verlag.

SEZGIN, Fuat. 2000. Mathematische Geographie und Kartographie im Islam und ihr Fortleben im Abendland. Historische Darstellungen 1 & 2. *Geschichte des Arabischen Schrifttums*, vols. 10 and 11. Leiden: E.J. Brill.

Chapter 30
Metrology and Proportion in the Ecclesiastical Architecture of Medieval Ireland

Avril Behan and Rachel Moss

Introduction

The 1140s marked a turning point in Irish monastic architecture. Up to the twelfth century Irish monasteries had typically comprised an apparently random collection of small buildings, the churches small in scale and simple in planning. The introduction of European monastic orders, in particular the Cistercians, was to lead to a revolution in both the layout and the aesthetic of monastic architecture, a topic which has received much attention from architectural historians over the years. However, the technologies required to achieve this revolution—in particular, the proportional systems and metrology used—have come under less scrutiny. While a small number of scholars have acknowledged a consciousness of the use of proportional systems, few have explored in any depth how the adoption of particular systems may have affected the overall design of buildings, in particular their detailing; what they tell us about the origins and training of the craftsmen who were using them; and what a study of the development of such systems can add to the poorly documented building history of Ireland.

First published as: Avril Behan and Rachel Moss, "Metrology and Proportion in the Ecclesiastical Architecture of Medieval Ireland", pp. 171–183 in *Nexus VII: Architecture and Mathematics*, Kim Williams, ed. Turin: Kim Williams Books, 2008.

A. Behan (✉)
Dublin Institute of Technology, College of Engineering & Built Environment, Bolton Street, Room 33701, Dublin, Ireland
e-mail: avril.behan@dit.ie

R. Moss
Department of the History of Art and Architecture, Trinity College, Dublin, Ireland
e-mail: rmoss@tcd.ie

K. Williams and M.J. Ostwald (eds.), *Architecture and Mathematics from Antiquity to the Future*, DOI 10.1007/978-3-319-00137-1_30,

Parameters of the Study

Metrology and systems of proportion have only been touched on in literature dealing with Irish architectural history. A number of commentaries on a tenth- to twelfth-century Irish law tract, which deals with the costing of ecclesiastical buildings, including round towers, conclude that the standard proportionate system for early single cell churches was 1.5:1.[1] The foot or *traig* was the unit of measurement used, but as yet the exact value of this is unclear. Stalley examined the proportions and systems of measurement of round towers, concluding that many towers appear to have adhered to a 1:2 ratio of circumference to height, and, certainly in the case of Glendalough tower, the English foot (0.3048 m), which may have been equivalent to a *traig*, was the unit of measurement used (Stalley 2001). Almost without exception the study of proportionate systems in later medieval Irish architecture has been limited to an examination of the use of $\sqrt{2}$ and the golden section in the laying out of monasteries and parish churches from the twelfth to the fifteenth centuries.[2] While there is a general consensus that both methods were engaged, there has been little attempt to expand this line of enquiry into the use of similar systems in the design of elevations and architectural detailing, or to look at the units of measurement used. As has been demonstrated by a number of studies from continental Europe (Paul 2002; Davis 2002; James 1973), this methodology can prove particularly successful in the study of window tracery. Tracery, having both structural and artistic functions, is an indicator of the abilities of the mason in two important elements of the craft: design and stereotomy (Curl 1992). In addition, in an Irish context, the sponsorship of windows is one of the most frequently documented activities relating to building history, allowing firm conclusions to be drawn regarding the context in which such designs were created.[3]

This study will focus on the tracery of a group of buildings with similar ‘looped’ tracery. Figure 30.1 shows the locations of the selected sites overlaid on the medieval kingdom boundaries c.1534 suggested by K.W. Nicholls (1976).

The occurrence of this particular form of tracery is relatively widespread, both regionally and temporally. For the purposes of this study two clusters located in regions under different political control during the later middle ages, one Gaelic and one Anglo-Norman, have been selected (see Fig. 30.2 and Table 30.1). Although difficult to date with any precision, buildings that range in date from the late thirteenth to the sixteenth centuries have been included in the study in order to

[1] The original manuscript text of the law is in Trinity College Dublin MS H.3.17. The most comprehensive of the texts is in Long (1996: 141–164).

[2] For Cistercian and Franciscan planning see Stalley (1987, 1990) For proportional systems in medieval parish churches see O’Neill (2002). For medieval friaries in Connaught see Mannion (1997).

[3] For example references to several schemes of refenestration are mentioned in the medieval Register of Athenry Friary; see Coleman (1912). For other references, see Moss (2006).

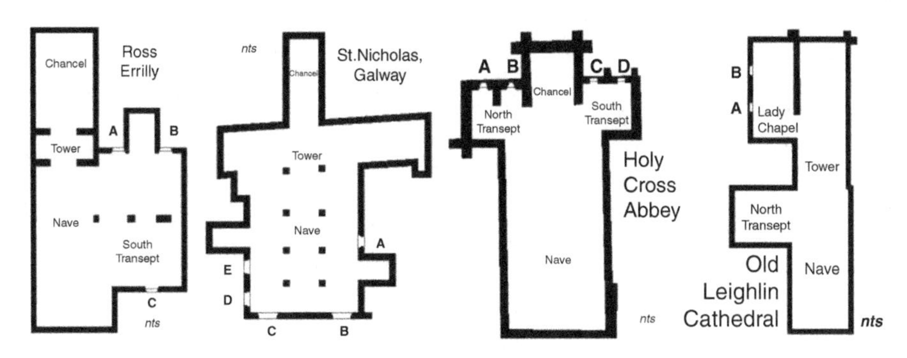

Fig. 30.1 Site locations shown against medieval kingdom boundaries c.1534. Image: authors, after Nichols (1976)

establish whether any degree of continuity etc. can be detected. The buildings chosen also vary in the type of establishment, with the sample covering each of the main orders (Augustinian, Cistercian, Dominican and Franciscan) as well as a collegiate church and a cathedral.

Field Data Collection

Since this study is empirical rather than stylistic, the primary requirement is the collection of detailed measurements of looped tracery at the selected sites. For this particular evaluation, measurements in all three dimensions (plan and elevation) are required, an exercise best achieved through the generation of 3D models of the tracery (Fig. 30.3). Although a number of methods exist for the production of such models,[4] stereo photogrammetry[5] was chosen. This method results in the creation of true-to-scale 3D models, created using a small number of reference (control) measurements and a pair of photographs, which also have a wide range of interpretative uses.[6] This technique also has the advantage of using relatively

[4] Other options include terrestrial laser scanning and discrete point/line measurement using a reflectorless total station. The total station option was rejected because the required field time was prohibitive for the number of sites being visited for the ongoing project. Terrestrial laser scanning was not used due to the unavailability of equipment, because there would be no gain in accuracy, and because significantly more field time would be required without a commensurate reduction in processing time.

[5] Photogrammetry is the science of generating measurements from imagery. Stereo photogrammetry uses two photographs captured and viewed in a simulation of the way human eyes achieve depth perception from offset images.

[6] This is to be compared with the results of terrestrial laser scanning, which although usually accompanied by supporting photographs, requires a detailed understanding of the handling of point clouds (set of 3D points) to ensure the best results.

Fig. 30.2 Ground plans of Ross Errilly, St. Nicholas', Galway, Holy Cross and Old Leighlin showing window locations (not to scale). Image: authors

inexpensive field equipment[7]: for this study photographs were taken using a Nikon D70 with 18–70 mm Nikkor lens, while a Leica TPS 1205 reflectorless total station[8] was used to collect the control (scale and orientation) information.

The field activities required for each window of interest were as follows:

- A pair of photographs of the window was acquired. The required conditions for the photo pair were as follows:
  - the plane of the camera sensor (the camera back) was aligned approximately parallel to the main plane of the window;
  - the two photographs were taken such that they overlapped by between 70 and 80 %;

[7] Suitable digital cameras cost between €500 and €1,000; reflectorless total stations of sufficient accuracy cost about €12,000. This is still inexpensive when compared to a terrestrial laser scanner price of more than €80,000.

[8] The reflectorless total station generates a 3D coordinate for any point, identified by the operator with the crosshairs of a telescope, using horizontal and vertical angle measurements and a distance measured using a time-of-flight laser. The calculation is based on trigonometric formulae and is a standard surveying technique.

Table 30.1 Sample of medieval sites containing looped tracery

Site name	Medieval kingdom	Window location	Window orientation	Modern county
Meelick Franciscan Friary	Connaught	Chancel	East	Galway
Ross Erilly Franciscan Friary	Connaught	South Transept	East (A & B)	Galway
		South Transept	West (C)	
St. Nicholas' Collegiate Church, Galway	Connaught	Nave	South (A)	Galway
		Nave	North (D & E)	
		Nave	West (B & C)	
Fethard Augustinian Abbey	Ormond	South Transept	East	Tipperary
Holycross Cistercian Abbey	Ormond	North Transept	East (A & B)	Tipperary
		South Transept	East (C & D)	
St. Laserian's Cathedral, Old Leighlin	Ormond	North Chapel	North (A & B)	Carlow
St. Dominic's Dominican Friary, Cashel	Ormond	South Transept	South	Tipperary

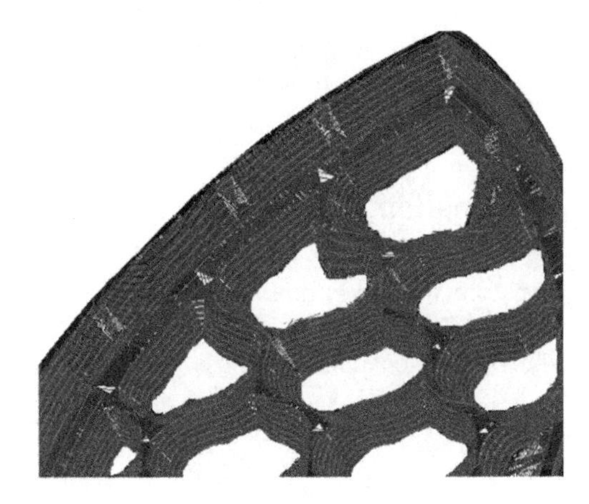

Fig. 30.3 3D model of tracery with overlaid contours. Rendering: authors

– auto-focus was switched off and focus was set to infinity, and
– a light-meter was used to ensure sufficient radiometric quality of the images.

- The relative positions of a minimum of three control points were measured in three dimensions—here the reflectorless total station was used. The points were clearly identifiable in the photographs and could be measured unambiguously using the total station. In this study, typically between 6 and 12 control points were measured to ensure redundancy.[9]

[9] This level of redundancy was required mainly because the chosen control points were naturally occurring (e.g., sharp corners on stonework or patterns caused by lichens) or pre-existing features (e.g., screws holding protective grilles or metal bars used to prevent unauthorised entry to sites). To generate the highest accuracy photogrammetric products it is advisable to use man-made targets (typically plastic cards or reflective stickers) but these could not be used in this survey

Processing

To generate the 3D model from the stereo imagery the processing package Leica Photogrammetric System (LPS) was used in combination with Autodesk Civil 3D 2007/2008.[10] The processing steps involved were as follows:

- The control was checked using Civil 3D to ensure that the *x-y* plane of the coordinate system was parallel to the plane of the camera sensor (this was a requirement of the LPS software);
- Orientation was established by measuring the exact relative geometries of the images at the time of capture and defining the positions of the control points on both photographs to assign a scale to the stereo model in three dimensions;
- A 3D digital model of the tracery was generated using LPS's Automatic Terrain Extraction method, which uses image matching techniques[11] to define 3D coordinates for a grid of points laid across the model.
- The quality of the 3D model was improved by removing erroneous points and adding breaklines. In LPS an operator, viewing in stereo, can define points or lines in 3D, ensuring that major features (such as significant changes of direction in the moulding profiles) are accurately included.[12]

Information Extraction

For each window a number of key elements was extracted from the 3D model. Table 30.2 lists the nine key dimensions extracted for each window, while Fig. 30.4 shows the locations of those dimensions. Table 30.3 lists the nine derived

because of the delicate nature of some of the sites (and the potential damage that the targets might cause) and the inaccessibility of the features (lifting or hoisting equipment could not have been used in many of the locations because of issues of topography and the position of the features in very close proximity to modern graves). The extra points enabled detailed accuracy checking after the modelling procedure.

[10] This is a Computer Aided Drafting package with a number of enhancements for the better handling of survey generated data and the manipulation and visualisation of three-dimensional models.

[11] Image matching involves automatically checking the levels of similarity between pixels in the overlapping images to find the best correspondence. Once identikit pixels have been found, a space intersection can be carried out using the orientation information previously calculated from the control information to generate a 3D coordinate for the matched point.

[12] While image-matching techniques are relatively robust the LPS software was primarily designed for aerial photogrammetric work and, thus, needs operator input to ensure the highest quality of the resultant 3D model.

Table 30.2 Details of extracted dimensions

Item of interest	Quantity		
Full window	(1) Width	(2) Overall height	(3) Height to springing of the arch
Light	(4) Width	(5) Overall height	(6) Height to springing of the arch
Arch	(7) Span	(8) Height	
Mullion	(9) Width		

For the measurement of light widths, where possible, an average was taken between the width at the base of the light and at the spring of the arch

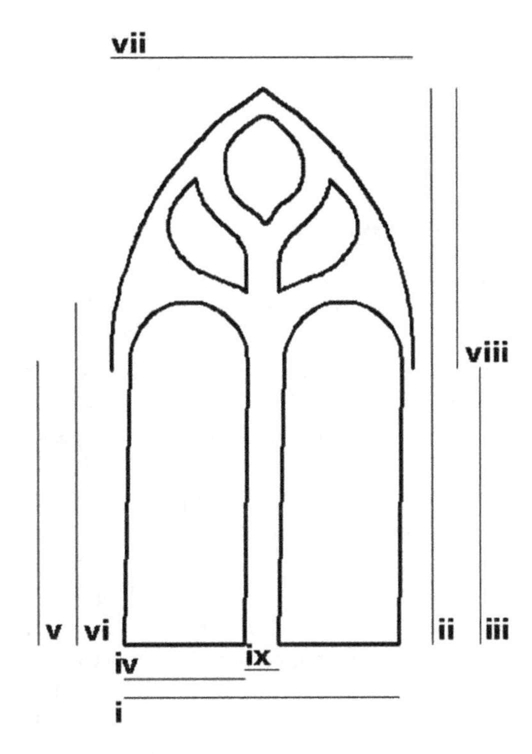

Fig. 30.4 Diagram showing the locations of extracted dimensions in Table 30.2. Image: authors

Table 30.3 Details of proportions studied

Item of interest	Proportion		
Full window	Tracery field height to light height (at springing point)	Tracery field height to light height (at arch peak)	Overall width to overall height
Light	Light width to light height (at springing point)	Light width to light height (at arch peak)	Window width to light height
Mullion	Mullion width to overall window width	Normalised mullion width to overall window width (based on number of mullions)	

proportions (since the intention of the mason with regards to important dimensions and proportions is not known, a number of variations have been examined, e.g., height to springing of the arch or to its peak). A list of all extracted dimensions and proportions is available from the authors.

Analysis

Seven different sites with a collective total of seventeen looped-tracery windows were surveyed. Nine sets of proportions and nine measurements were recorded for each window.

Proportional Analysis

Following the model of previous studies, proportional analysis was carried out initially through a search for proportions known to have been used in Irish medieval architecture, such as the Golden Section or 1:√2 relationships. In Britain and continental Europe authors have found evidence for the use of both of these relationships as well as 1:√3 and less geometrically-based proportions such as 1:2, 1:3 and 1:4. Reuse of the same measurement, i.e., a 1:1 relationship, has also been examined. Each potential proportional relationship within the sample of 17 windows was examined in normal and inverted forms producing a total of 306 proportions.

A search for each possible relationship was made within the 306 proportions extracted. Since the measurements can vary from their true value due to variables such as the photograph orientation process, human error in the measurement phase, and weathering of the stone, a range of values distributed about each ideal proportion were examined. ±5 % was added to each ideal proportion to generate a range of values that mirrors statistical norms of 95 % confidence intervals.

Golden Section, 1:√2 and 1:√3

Limited evidence for the occurrence of the Golden Section, 1:√2 and 1:√3 proportionate systems was detected in the analysis of tracery from the study sample. The nature and distribution of the elements where such systems were detected suggest a random rather than deliberate use in the design process.

Fig. 30.5 St. Nicholas' Collegiate Church, North Windows E (*left*) and D (*right*). Photos: authors

1:2	0.5	(range 0.525–0.475)
1:3	0.333	(range 0.350–0.316)
1:1		(range 1.050–0.950)

The relationship of 1:2 occurs eight times overall; in three cases in the category "Overall window width to overall window height" and in the other three as "Window width to light height (at arch peak)". In each of these six cases another proportion or regular relationship is also evident in the data. For the two north nave windows of St. Nicholas' (Fig. 30.5) the 1:2 relationship of window width to light height is accompanied by a 1:3 relationship between the window's width and its height. The other proportions for these two windows also display similarities when compared. Interestingly the tracery in these two windows is quite different in character, although both fall within the category of the looped style.

At Fethard Augustinian the 1:3 relationship is also demonstrated for light width to height in combination with the 1:2 ratio for overall window width to height in the east window of the south transept. At Old Leighlin Cathedral the 1:2 ratio occurs twice (window width to light height and tracery height to light height) in the eastern north-facing window of the Lady Chapel (B in Fig. 30.2) with the 1:3 ratio evident in the proportion of overall window width to height. At St. Dominic's, Cashel, the 1:2 relationship of window width to height is accompanied by a 1:1 ratio between window width and light height in the nave south window. The fact that both proportions are width to height could point to a deliberate plan by the mason.

Fig. 30.6 Beautifully proportioned north transept (north eastern) window at Holycross. Photo: authors

Perhaps of most interest is the occurrence of four identifiable proportions in the northeastern window of the north transept at Holycross (A in Fig. 30.2). In this one window there are two occurrences of 1:1 ratios and one each of 1:2 and 1:3. The result, not unsurprisingly, is an aesthetically pleasing window of beautiful proportions (Fig. 30.6).

At the same site, the two east windows in the south transept also utilise the 1:1 proportion and are very close to using the 1:2. As at St. Nicholas' the tracery designs of the two windows are quite different but these similarities in measurements and proportions hint that the work had the same basis.

Holycross underwent a major programme of renovations during the fifteenth century which, although not documented, can be closely linked to members of the powerful Butler family through the incorporation of heraldry in the fabric of the church. A recent study of moulding profiles in the abbey church has led Danielle O'Donovan (2007) to suggest that Holycross provided a major hub for masons brought into the area by the Butlers, whose work was subsequently emulated throughout the territory. The relatively rare occurrence of such a perfect set of proportions in the window help to reinforce this argument, suggesting perhaps the work of a craftsman trained in the basic principles of design.

1:4 0.250 (range 0.263–0.238)

The final standard ratio worth mentioning occurs in four windows at three sites and in all cases represents the relationship between light width and light height. All four windows have two lights, but the examples at Ross Errilly and Old Leighlin are very simple in tracery style, while that at Holy Cross is more complex.

Metrological Investigation

Eric Fernie, James Addiss and others have highlighted the many potential pitfalls that exist in establishing the units of measurement used in a medieval building, suggesting that "one can get any foot from any building" [Raper 1760, in Fernie (2002)]. This study has been careful to follow Addiss's recommendation of using "explicit and comprehensive" measurement as a means of increasing the probability that the conclusions drawn will be correct (Addiss 2002).

In order to conduct an objective metrological investigation, an adaptation of F. Bettess's methodology was used (Bettess 1991). This method is based on the principle of 'least squares' and offers significant flexibility by supporting full units and their fractions. Each measurement taken from the sample window is divided, in turn, by a range of potential units resulting in an integer value plus a remainder ($A = I + r$). Since it is known that medieval masons used full units and halves, thirds and quarters thereof, the remainder is evaluated for similarity to each of these for each candidate unit (i.e., r is compared with I/2, I/3, I/4, 2I/3 and 3I/4). Based on an analysis of known medieval foot units, a range of sample units from 0.249 m to 0.325 m, were selected for the study (Zupko 1978; Bettess 1991; Stalley 2001; Addiss 2002).

The analysis steps were as follows:

1. Each window measurement (Table 30.2) was divided by the sample unit;
2. The difference was calculated between the remainder and the unit, and between the remainder and each of the standard fractions of the unit (half, quarter, third, two-thirds, three-quarters);
3. The minimum difference calculated in step 2 is extracted—this is taken to denote the most probable unit plus fraction combination (the fraction can, of course, equal zero indicating that an integer number of units was used);
4. The variances of the minimum differences chosen in step 3 were calculated for:
    - Each site
    - Each region
    - The full data set;
5. The variances in each category were compared and the minimum value extracted, resulting in Table 30.4 which lists the most probable metric unit value for each site, each region and for the full dataset;
6. The probable units were compared with proven units of measurement.

Table 30.4 Most probable units from analysis related to known units (all unit and difference values are quoted in metres)

Probable unit	Location	Closest known unit	Difference	Sample size (number of windows and measurements)
0.309	St. Dominic's, Dominican Friary, Cashel, Tipperary	0.3048 Standard English Foot	0.0042	1 & 9
0.319	Fethard Augustinian Abbey, Tipperary	0.3167 English Medieval Foot	0.0023	1 & 9
0.320	St. Laserian's Cathedral, Old Leighlin	0.3167 English Medieval Foot	0.0033	2 & 18
0.285	Holy Cross Cistercian Abbey, Tipperary	0.2800 Anglo-Saxon Foot	0.0050	4 & 36
0.284	Kingdom of Ormond	0.2800 Anglo-Saxon Foot	0.0040	8 & 72
0.295	Meelick Franciscan Friary, Galway	0.3048 Standard English Foot	0.0098	1 & 9
0.299	Ross Errilly Franciscan Friary, Galway	0.3048 Standard English Foot	0.0058	3 & 27
0.317	St. Nicholas' Collegiate Church, Galway	0.317 English Medieval Foot	0.0000	5 & 45
0.269	Kingdom of Connaught	0.2800 Anglo-Saxon Foot	0.0110	9 & 81
10.275	Full Sample	0.2800 Anglo-Saxon Foot	0.0050	17 & 153

While it is possible that measurements made by medieval masons may have been based on a small unit such as a palm or a span, this investigation focussed on a limited selection of 'foot' values that are known to have been used in the period. The options chosen are: the standard English Foot (0.3048 m), which was shown by Roger Stalley to have been in use in early medieval times; the English Medieval Foot (0.3167 m), used for building works in England and believed to be derived from the Greek Common Foot; and the Anglo-Saxon Foot (0.2800 m), reported by Bettess in his studies at Jarrow and Yeaverling (1991).[13]

[13] For a list of comparative linear measures, see Zupko (1978) and Strayer (1989: 580–596). For a similar list including the English Medieval Foot (based on the Greek Common Foot and used for buildings) see Skinner (1967).

Units of Measurement

As with the proportional systems examined above, the random nature of measurements close to the Standard English and Anglo-Saxon foot values suggests that they were not used in the building sample chosen here.

The most compelling evidence is for use of the English medieval foot of 0.317 m. It appears as the most viable candidate at three sites, Old Leighlin and Fethard Augustinian church in Ormond, and St Nicholas' Collegiate Church in Gaelic Connaught, where the match is very good.[14]

Little is known of the history of the two Ormond sites; the construction of the Lady chapel in which the Old Leighlin windows are found is usually associated with the episcopate of Matthew Sanders, the Drogheda-born bishop between 1527 and 1549, who is also credited with the 'erection and glazing' of the south window in the church (Ware 1739–1746: I, 461). Of the construction of the south transept at Fethard, nothing is known. In Galway, however, we are on safer ground. A manuscript preserved in Trinity College entitled "Account of the town of Galway" records that in the year 1538 during his mayoralty of Galway, "John French alias Shane Itallen, soe called on account of the abundance of salt that he brought into the country, built the north side of the church" (p. 10). Isolated among other Connaught examples, it is tempting to see the effect of this influential and well-travelled patron at work in the design of the windows, possibly introducing professionally-trained masons into the area to conduct this work.

Conclusion

The sample of just 17 windows examined from only seven sites is, of course, small, and results gleaned from this survey cannot be seen as conclusive. However, preliminary findings suggest that the areas of medieval metrology and proportionate systems in design do have the potential to provide empirical evidence for the work of professionally-trained masons in Ireland, and to distinguish them from craftsmen who had the ability to copy architectural form, but without understanding the underlying principals of design.

Biography Avril Behan is a lecturer in Geomatics (specifically remote sensing, photogrammetry, CAD, BIM, and land surveying) at the Department of Spatial Information Sciences, Dublin Institute of Technology. This work was carried out as part of her Ph.D. on Metrology and proportion in the window tracery of medieval

[14] <3, <2 and <1 mm respectively: It is acknowledged that this level of accuracy is not possible from the original measurement method but the results are analysed here *relative* to the other measurements in the group, rather than in their absolute form.

Ireland: An empirical study of Ormond and Connaught, obtained from the Department of History of Art and Architecture, University of Dublin, Trinity College. Her other research interests include heritage applications of terrestrial laser scanning and photogrammetry, CAD and visualisation, satellite remote sensing, airborne laser scanning, and the usage of Web 2.0 applications for higher education. She has presented at conferences such as CIPA/VAST 2006 on "Close-Range Photogrammetric Measurement and 3D Modelling for Irish Medieval Architectural Studies" and ISPRS Congresses in 2008 and 2000 on remote sensing education and on accuracy checking of Laser Altimeter Data, respectively.

Rachel Moss is an assistant professor at the department of History of Art and Architecture Trinity College, Dublin. A specialist in the field of medieval architecture and sculpture, research projects with which she has been involved include the electronic capture and archiving of medieval architectural details with the department of Computer Science at Trinity College Dublin. She was also involved, in a consultative role, with the establishment of a state-sponsored national database of movable Irish field antiquities. She has published numerous articles on medieval art and architecture and edited/co-edited, *Art and Devotion in Late Medieval Ireland* (Dublin, 2006), *Making and Meaning in Insular Art* (Dublin, 2007) and *Art and Architecture of Ireland, Volume 1: The Medieval Period* (New Haven and London, forthcoming 2014).

References

Addiss, J. 2002. Measure and Proportion in Romanesque Architecture. Pp. 57-82 in N. Wu, ed. *Ad Quadratum: the Practical Application of Geometry in Medieval Architecture*. Aldershot: Ashgate.

Bettess, F. 1991. The Anglo-Saxon Foot: A Computerised Assessment. *Medieval Archaeology* **35**: 44-50.

Coleman, A., ed. 1912. Regestum Monasterii Fratum Praedicatorum de Athenry. *Archivium Hibernicum* **1**: 201-221. Dublin: Catholic Historical Society of Ireland.

Curl, J. S. 1992. The Art of Cutting and Dressing of Stones. p. 297 in *Encyclopaedia of Architectural Terms*. Dorset: Donhead.

Davis, M. T. 2002. On the Drawing Board: Plans of the Clermont Cathedral Terrace. Pp. 183-204 in N. Wu, ed. *Ad Quadratum: the Practical Application of Geometry in Medieval Architecture*. Aldershot: Ashgate.

Fernie, E. 2002. Introduction. Pp. 1-9 in N. Wu, ed. *Ad Quadratum: the Practical Application of Geometry in Medieval Architecture*. Aldershot: Ashgate.

James, J. 1973. Medieval Geometry: the Western Rose of Chartres Cathedral. *Architectural Association Quarterly* **5**, 2: 4-10.

Long, W.H. 1996. *Glendalough, Co. Wicklow: An Interdisciplinary Study*. Ph.D. thesis, Trinity College Dublin.

Mannion, S. 1997. *A Study of the Physical Remains of the Medieval Friaries of Connacht*. Ph.D. thesis, Queen's University, Belfast.

Moss, R. 2006. Permanent Expressions of Piety: the Secular and the Sacred in Later Medieval Stone Sculpture. Pp. 72–97 in R. Moss, C. O'Clabaigh and S. Ryan, eds. *Art and Devotion in Late Medieval Ireland*. Dublin: Four Courts Press.

NICHOLLS, K.W. 1976. *Lordships of Ireland c.1534*. Oxford: Oxford University Press.

O'DONOVAN, D. 2007. *Building the Butler Lordship 1405 –c.1552*. Ph.D. thesis, Trinity College Dublin.

O'NEILL, M. 2002. The Medieval Parish Churches in County Meath. *Journal of the Royal Society of Antiquaries of Ireland* **132**: 1-56.

PAUL, V. 2002. Geometry Studies: The Blind Tracery in the Western Chapels of Narbonne Cathedral. Pp. 205-216 in N. Wu, ed. *Ad Quadratum: the Practical Application of Geometry in Medieval Architecture*. Aldershot: Ashgate.

SKINNER, F. G. 1967. *Weights and Measures: their Ancient Origins*. London: Her Majesty's Stationery Office.

STALLEY, R. 1987. *The Cistercian Monasteries of Ireland*. London and New Haven: Yale University Press.

———. 1990. Gaelic Friars and Gothic Design. Pp. 191-202 in P. Crossley and E. Fernie, eds. *Medieval Architecture and its Intellectual Context*. London: Hambledon Continuum.

———. 2001. Sex, Symbol and Myth: Some Observations on Irish Round Towers. Pp. 27-48 in C. Hourihane, ed. *From Ireland Coming: Irish Art from the Early Christian to the Late Gothic Period and its European Context*, Princeton: Princeton University Press.

STRAYER, J.R., ed. 1989. *Dictionary of the Middle Ages*. 13 Vols. New York: Charles Scribner's Sons.

WARE, J. (Sir). 1739-1746. *The Works of Sir James Ware concerning Ireland Revised and Improved*. 3 vols. Dublin.

ZUPKO, R. E. 1978. *British Weights and Measures: A History from Antiquity to the Seventeenth Century*. Madison, Wisconsin: University of Wisconsin Press.

Chapter 31
The Cloisters of Hauterive

Benno Artmann

Introduction

One of the most typical elements of Gothic architecture is the tracery found in windows, on walls, and in many other places in Gothic churches. What is mathematical about it? Tracery is exclusively constructed from circular arcs (and straight line segments)! It is the most mathematical kind of art known to me. In many of the thousands of Gothic churches and other buildings of that time surviving in Europe one can find nice examples, take photos, and analyse them geometrically at home. In what follows I will first give a short introduction to tracery (German: *Maßwerk,* French: *réseau)* and then direct your attention to one mathematically outstanding example.

General Remarks About Tracery

The Gothic style originates from France, more precisely from the parts of France close to Paris, from about 1150. Tracery, however, first appears in the 1210s in Reims, so that for instance you will find no tracery windows in the older parts of the cathedral of Chartres. Gunther Binding (1989) defines three principal periods of stylistic development: (1) 1270 High Gothic, (2) 1270–1360/1380 Radiant and (3) 1350–1520 Flamboyant. Obviously, the years are to be taken approximately; in various parts of Europe the architecture of the same year may be very different.

First published as: Benno Artmann, "The Cloisters of Hauterive", pp. 15–25 in *Nexus I: Architecture and Mathematics,* ed. Kim Williams, Fucecchio (Florence): Edizioni dell'Erba, 1996. This chapter originally appeared in *The Mathematical Intelligencer,* vol. 13, no. 2 (1991), p. 44–49. Reprinted by permission.
Benno Artmann (1933–2010).

K. Williams and M.J. Ostwald (eds.), *Architecture and Mathematics from Antiquity to the Future,* DOI 10.1007/978-3-319-00137-1_31,

The first tracery windows were built by the architect Jean d'Orbais in the cathedral of Reims during the years 1211–1221 (Fig. 31.1). Their construction is based on the equilateral triangle as shown in Fig. 31.2. An iteration and variation of the first construction can be seen in the north window of the church of Haina, a former Cistercian monastery, located about 30 km north of Marburg, Germany (Fig. 31.3). The two smaller parts of this window are clearly repetitions of the first Reims tracery. The upper part can be found in Reims as well, and the whole composition is found in St. Denis (Binding 1989: 47, 51). From the large upper circle we are able to learn how architects free themselves from the overly strict rules of geometry in favour of aesthetic considerations. How is that part of the window constructed?

Start with the two small pointed arcs, which are constructed from equilateral triangles (Fig. 31.4). The points of abutment of these arcs are A, B, C. Now the architect selects the point M, or, equivalently, the radius r of the great circle. The size of this circle is determined by the architect's artistic judgment. The distance AM will then be a + r, because the arc and circle are in contact. For the completion of the window the architect needs the points X, Y of abutment for the great arc such that XYZ is an equilateral triangle and the circle about M is touched by the arcs. The distance from Y to the point of contact T must be 2a, hence the distance from M to X (or Y) should be 2a–r, and x, y can be found. Observe that, in the real window, the architect marks A, B, and C but conceals X and Y.

The most important mathematical tools in this—and all other constructions of traceries—are circles in contact and the division of a circle into equal parts. Sometimes a little more has to be known. Readers may amuse themselves by constructing a so-called 8-foil as in Fig. 31.5 or by finding M and r in Fig. 31.6, where ABC is again an equilateral triangle. Thereafter, the design of traceries became rapidly more complicated. Figure 31.7 shows an example from Strasbourg about 1285. It is basically an iteration of equilateral triangles.

Some 80 years after the north window, the great west window of Haina was built, about 1330 (Fig. 31.8). Observe the wavy pentagram consisting of curved equilateral triangles and the division of the great circle into 15 parts. One final example shows the different stylistic means of the late Gothic times in Germany (Fig. 31.9).

The Role of Geometry in Gothic Architecture

In their fundamental work on early Gothic architecture, Kimpel and Suckale say

> tracery is the specially favored medium of the Gothic 'love of geometry.' One of the main pursuits of architects up to the sixteenth century was to decorate a building with a profusion of variants and to invent new ones. Tracery is that part of the Gothic style that is most distant from the anthropomorphic architecture of antiquity. To put it positively: It is one of

Fig. 31.1 Reims 1211–1221. Image: Binding (1989: 46)

> the few creations of ornaments in Europe that doesn't owe anything to antiquity. (Kimpel and Suckale 1985: 26)

The mastery of geometry raised Gothic architects above the artisans and made them—in medieval terms—scientists. In fact the stonemasons considered geometry an essential part of their trade. Figure 31.10 shows a late medieval woodcut associating geometry with stonemasonry.

We have seen a little of the freedom and the constraints of the geometric methods for the construction of traceries in the example from Haina (north window). More elaborate designs follow the same rules: every detail has to be geometrically constructed, except for the smallest ornamental pieces or things such as the profiles of the mullions. In modern terms we could clearly speak of geometrical or "concrete" art long before the "Neo-Geo" of the twentieth century.

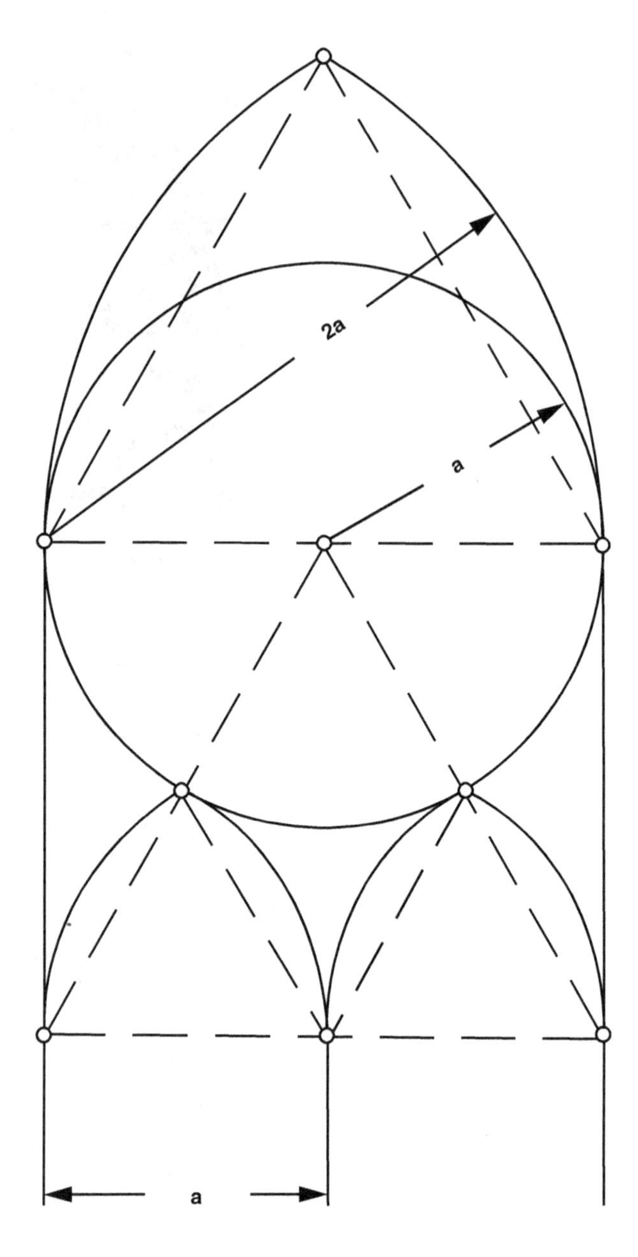

Fig. 31.2 Construction based on the equilateral *triangle*. Image: author

Once one has seen a specimen of tracery, it is in most cases relatively easy to reconstruct it and hence "understand" such a window to a much higher degree than other works of art. But always remember that geometry is the servant and not the master of the architect. The geometric methods were easily understood and adopted

Fig. 31.3 Haina, north window, dating from 1250. Image: Ungewitter (1890–1892: Tafel CXXIV)

by the nineteenth century architects who created so many "neo- Gothic" churches in Europe and America.[1]

However elaborate the construction of a tracery window might be, it was not an exercise in mathematical geometry as we know it. Just as little was it geometry in the sense of the contemporary mathematicians such as Leonardo Fibonacci (Kimpel and Suckale 1985: 45). Euclid was known in Latin translation since about 1150, but long before that people learned geometry from the so-called pseudo-Boethius. which in its essential parts is a boiled-down version of the first books of Euclid without proofs.[2] The first proposition in Euclid's first book is the construction of an

[1] Nineteenth-century architects wanting to build Gothic churches could find complete instructions in G. Ungewitter, *Lehrbuch der Gotischen Konstruktionen*, 2 vols. Leipzig, 1890–1892.

[2] For an edition of the principal medieval-textbook on geometry before Euclid was translated, see Menso Folkerts, "Boethius," *Geometric 11*, Wiesbaden: F. Steiner (1970).

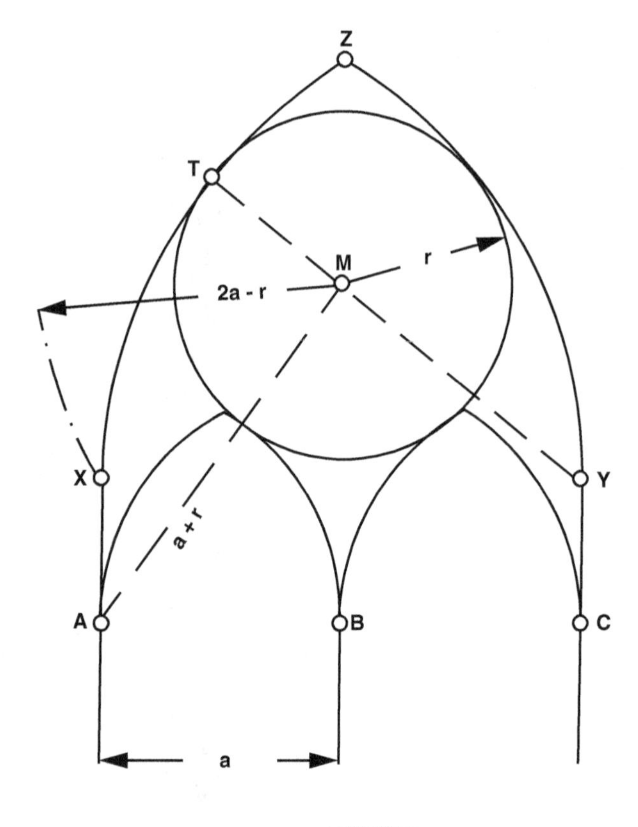

Fig. 31.4 Construction of window in Fig. 31.3. Image: author

Fig. 31.5 An eight-foil drawing. Image: author

equilateral triangle using the arcs that we see so frequently in Gothic traceries. The use of circles in contact and subdivisions of a circle into equal parts was the principal method of construction. The traceries became more and more

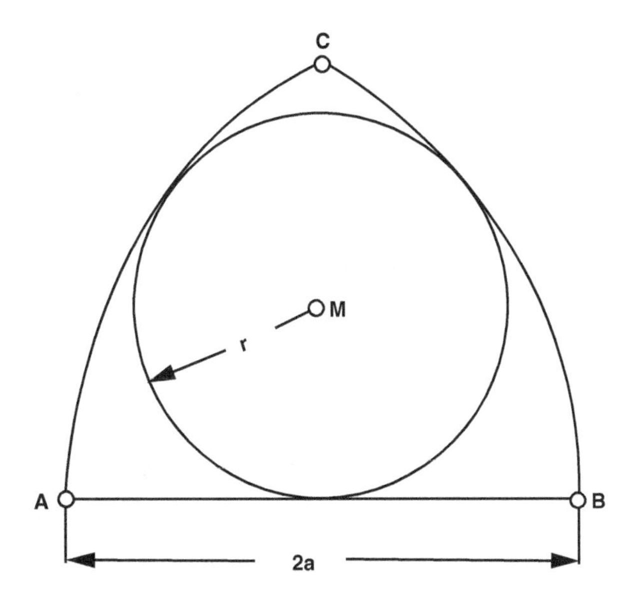

Fig. 31.6 Find M and r. Image: author

Fig. 31.7 Strasbourg, north window of the cathedral from 1285. Image: Binding (1989: 426)

complicated, but geometric methods remained the same. Essentially the same phenomenon has been observed by Jens Hoyrup in Babylonian mathematics: a fixed method (in that case a way to solve quadratic problems) remains in constant

Fig. 31.8 Haina, west window dating from 1330. Image: Ungewitter (1890–1892: Tafel CXXV)

Fig. 31.9 Esslingen, south window dating from 1410. Image: Binding (1989: 331)

use, and its school of practitioners adds more and more complicated examples. Hoyrup calls this "subscientific mathematics"; it is the trademark of Babylonian schools of scribes as well as of medieval guilds of masons (Hoyrup 1990: 63–87).

I believe the concept of "subscientific mathematics" is most suitable to describe what we see in Gothic architecture. A fixed supply of elementary geometry was used for traceries and for other problems in the plan of a Gothic building as well. It

Fig. 31.10 Geometry and stonemasonry (Nümberg 1493). Image: Binding (1993: 35)

is of no use to speculate further about constructions according to the golden section and the like. K. Hecht has rejected dozens of theories about "secret methods" of medieval architects by confronting them with the actual buildings (Hecht 1979). (If you measure enough distances and allow generously for tolerances, you will find the golden section in any old building.) We have some Gothic design booklets from about 1500, edited by Shelby (1977). A careful analysis by Hecht confirms what was said above: they are as "subscientific" as everything else (Hecht 1979: 171–201).

The Cloisters of Hauterive

The tracery windows in the cloisters of Hauterive are quite different from the traceries we have discussed so far. They are not only constructed by geometric methods, but in fact they show geometry itself.

The Cistercian monastery of Hauterive is situated on the banks of the river Saane (Saarine) some ten kilometres south of Fribourg in Switzerland. It is still inhabited by Cistercian monks. The history of its buildings is presented in detail by Catherine Waeber-Antiglio (1976). The cloisters and the choir of the church were built in the years 1320–1328. Some reconstruction took place around 1910, but what you see in Hauterive are essentially the original medieval windows (Figs. 31.11 and 31.12). The west, north, and east parts of the cloisters are in their original condition. The south part was taken away in the eighteenth century and two of its windows have been placed on the west and east side (numbers I and XIX). On the ground plan (Fig. 31.13) you can see the foundations of the old south wing.

In each bay of the cloisters (except for the ones in the corners) we find the same composition of windows: three small ones, separated by double columns and round arches, looking almost Romanesque, on the lower level and above them either pointed or round tracery windows. The pointed windows have elaborate and delicate traceries, but they are not what we are looking for. The round ones have the geometrical motifs. Before giving a mathematical analysis, let us look at the cloisters as a whole.

In the north wing we see only round arches; the east and west wings have alternating pointed and round arches. Windows I and XIX come from the old south wing (Waeber-Antiglio 1976: 136). The great window in the choir of the church was built in the same period and clearly shows the same style as the cloisters. Waeber-Antiglio gives a detailed stylistic and historical analysis of the cloisters and the choir of Hauterive, placing them firmly in context with their neighbours in space and time and especially with the building tradition of Cistercian monasteries. She points out the cloisters' specific originality: by using a very conservative (for the time) design of the triplets below and placing above them the most modern traceries, the cloisters of Hauterive are uniquely distinguished (Waeber-Antiglio 1976: 123–178). Spahr and Cist (1984: 23) say that the cloisters are among the finest preserved north of the Alps.

Neither Waeber-Antiglio nor Spahr mention the mathematical significance of the designs of the windows, which we now discuss. First, observe that the architect stresses interest in the circle by selecting round arches. He approaches this topic systematically by subdividing the circle in three (window II), four (IX), five (VI), six (XI) equal parts. With little ornamental subdesigns the architect tries to alleviate the dry geometric diagrams, sometimes successfully, as in IX, sometimes less

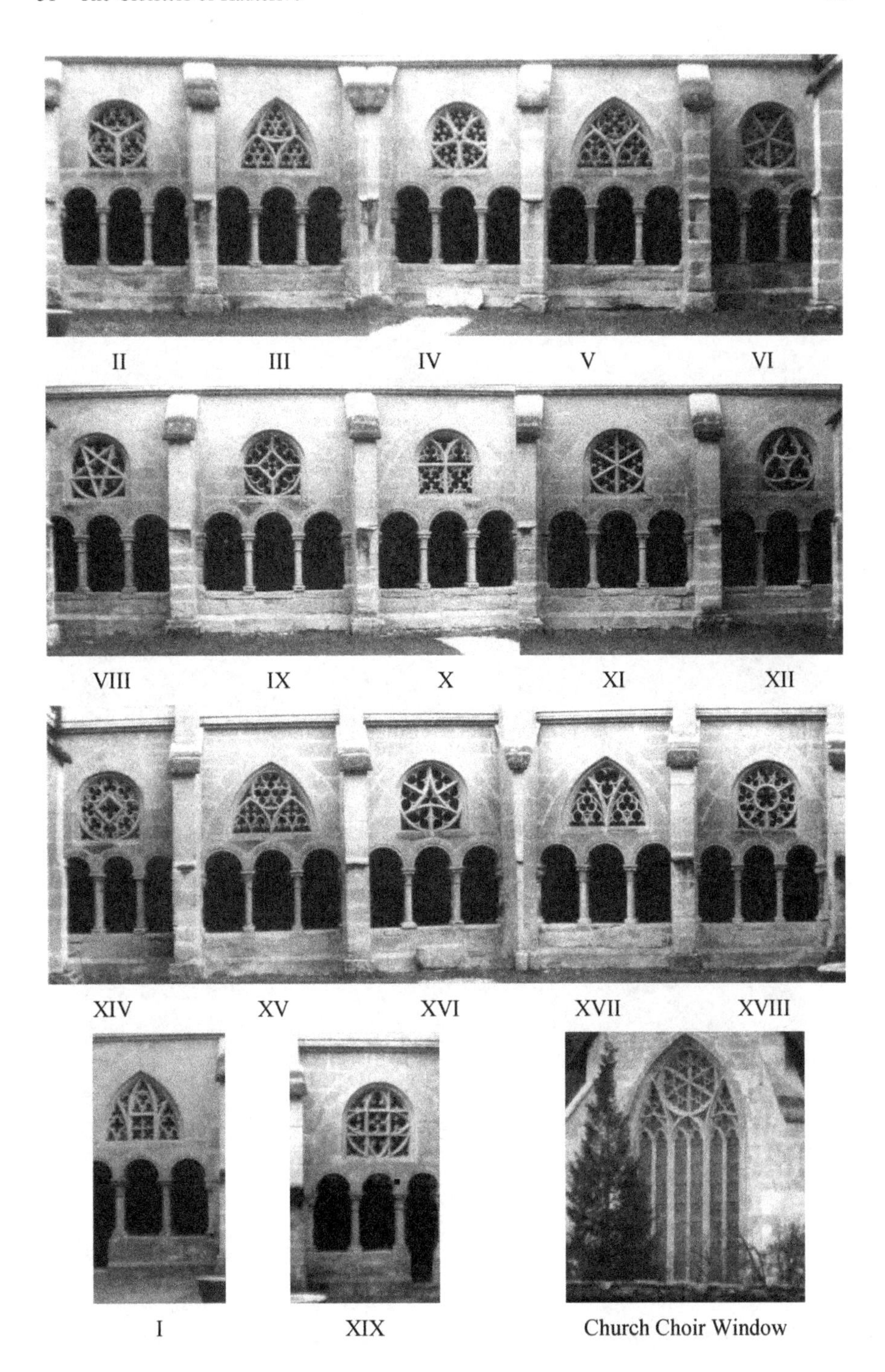

Fig. 31.11 The bays of the cloisters of Hauterive as they go around in numerical order. Photo: author

Fig. 31.12 The windows of cloisters of Hauterive arranged according to their geometrical construction. Photo: author

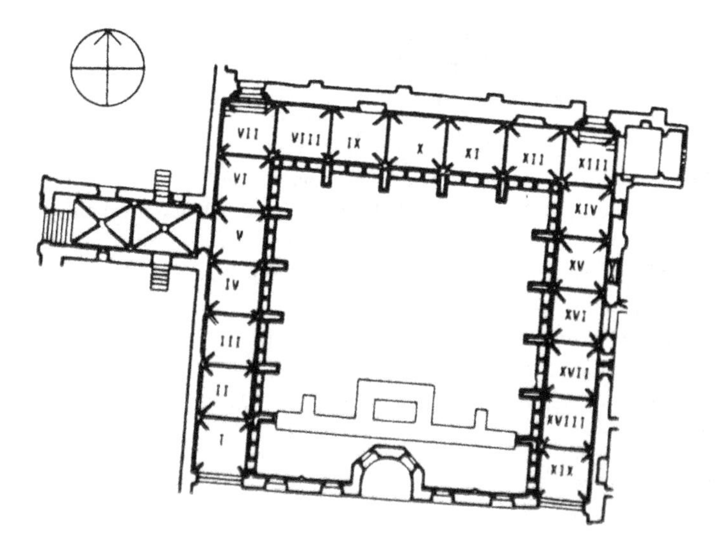

Fig. 31.13 Ground plan of the cloisters of Hauterive. Drawing: author, after Waeber-Antiglio (1976: 132)

convincingly as in XI. Fivefold symmetry is repeated by the pentagram in VIII and again in the beautiful and ingeniously constructed rose in window IV.

The subject of triangles is taken up and iterated in window XII. Observe that II has the sequence 3–9–27 built into it. Three- and six-fold symmetries are combined in window XVI, which has a very simple construction once you view the circle as the incircle of an equilateral triangle. Four and eight are combined in XIV.

Window X tries to have it its own way: in spite of the supplied round arch, we see no circle but two squares with conventional pointed arcs above them. A second view reveals the role of this window: its bilateral symmetry stresses its position in the middle of the cloisters right across from the (former) fountain chapel. It is quite understandable in the overall composition of the cloisters.

So far I have described what we see. Is there an interpretation that gives a general and coherent explanation of the designs? I offer this one: the architect was interested in theoretical geometry, especially in Euclid's theory of the subdivision of a circle as presented in the *Elements,* Book IV. In that Book Euclid treats, in a rather Bourbakist fashion, the construction of the regular n-gons, more specifically the inscription of regular 3-, 4-, 5-, 6-, 15-gons into given circles. I understand the cloisters of Hauterive as a commentary to Euclid Book IV, carved out of stone.

Can I maintain this interpretation in the case of window XVIII with its implicit 9-gon? First observe the five-foil in the centre and the three bars of an equilateral triangle in the surrounding annulus. The combination of these two designs would result in the regular 15-gon as constructed by Euclid. By subdividing each of the three parts of the annulus again into three equal parts the architect goes beyond Euclid: the construction of the regular 9-gon by ruler and compass is impossible. I believe that in this particular case aesthetic considerations overwhelmed mathematics: 15 little ornamental triangles would have required a very narrow annulus and a much greater five-foil with a rather empty centre. As it is, this window is one of the most beautiful of the cloisters. Go to Hauterive and judge for yourself: is it about mathematics or not? In any case you will enjoy it.

Biography Benno Artmann (1933–2001) graduated from the University of Tübingen (including with Günter Pickert, Helmut Wielandt, Hellmuth Kneser, Max Muller) and the University of Giessen, where he received his doctorate in 1965 with Günter Pickert (automorphisms and coordinates with flat associations). From 1974 to 1998 he was a professor at the Technical University of Darmstadt. His main interests were mathematics education and geometry, especially the geometry of the associations and rings (after Johannes Hjelmslev, Hjelmslev planes, etc.). He was the author of *Euclid: The Creation of Mathematics* (Springer-Verlag, New York, 2000).

References

BINDING, Gunther. 1989. *Masswerk.* Darmstadt: Wiss. Buchgesellshaft.

———. 1993. *Zur Methode der Architekturbetrachtung Mittelalterlicher Kirchen.* Köln: Kunsthistorisches Institut der Universität zu Köln.

HECHT, Konrad. 1979. *Mass und Zahl in der Gotischen Baukunst.* Hildesheim: Olms.

HOYRUP, Jens. 1990. Sub-scientific Mathematics. In *History of Science* **28** (part 1) 79 (1990): 63–87.

KIMPEL, D. and R. SUCKALE. 1985. *Die Gotische Architektur in Frankreich 1130–1270.* Munchen: Hirmer.

FOLKERTS, Menso. 1970. *"Boethius" Geometric 11.* Wiesbaden: F. Steiner.

SHELBY, Lon R. (ed.). 1977. *Gothic Design Techniques: The fifteenth-century booklets of Mathes Roriczer and Hans Schmuttermayer.* Carbondale: Southern Illinois University Press.

SPAHR, Columban and O. CIST. 1984. *Hauterive.* Munchen and Zurich: Schnell and Steiner.

UNGEWITTER, G. 1890–1892. *Lehrbuch der Gotischen Konstruktionen,* 2 vols. Leipzig: T. O. Weigel.

WAEBER-ANTIGLIO, Catherine. 1976. *Hauterive: La Construction d'une Abbaye Cistericicnne au Moyen Age.* Fribourg: Ed. Universitaires.

Chapter 32
The Use of Cubic Equations in Islamic Art and Architecture

Alpay Özdural

Introduction

The predominance of geometry in the ornamental arts that adorn the buildings of the Islamic world, from Spain to Central Asia, has always been a fruitful field of research. Starting in Umayyad times in Damascus as hesitant experiments, simple geometric motifs had developed almost instantly into intriguing and awe-inspiring ornamental geometric compositions that reached continually new climaxes, with increased complexity or new horizons, in Baghdad, Isfahan, Cordoba, Alhambra, Tabriz, Samarqand, Delhi, Istanbul, and again Isfahan. It is generally viewed that those intricate patterns were conceived and produced by artisans who were not only masters in their crafts but also in geometry. To imagine all those medieval artisan/architects also as mathematicians well versed in Euclid, though an attractive thought, had always seemed rather implausible to me since they were known to be mostly illiterate. Lately I have been developing the ideas that most of the aesthetic, structural or spatial innovations that we observe in the major architectural centers of the Islamic world were mainly due to the active role of mathematicians at the conception stage, and that some of the great accomplishments of Islamic art and architecture can be explained as the products of the collaboration between mathematicians and artisans at special meetings.[1]

This sort of collaboration is best exemplified by a Persian work on ornamental geometry, *Fī tadākhul al-ashkāl al-mutashābiha aw al-mutawāfiqa* (*On interlocks*

First published as: Alpay Özdural, "The Use of Cubic Equations in Islamic Art and Architecture". Pp. 165–179 in *Nexus IV: Architecture and Mathematics*, Kim Williams and Jose Francisco Rodrigues, eds. Fucecchio (Florence): Kim Williams Books, 2002.
Alpay Özdural (1944–2003).

[1] The first part of the argument was first mentioned in (Holod 1988). For my publications on this point, see (Özdural 1995, 1996, 1998, 2000). Similar views are expressed in (Necipoğlu 1995: 167–175).

K. Williams and M.J. Ostwald (eds.), *Architecture and Mathematics from Antiquity to the Future*, DOI 10.1007/978-3-319-00137-1_32,

of similar or complementary figures; referred to hereafter as *Interlocks of Figures*).[2] It is an anonymous work, or rather a collection of 68 separate constructions, which appears to have been compiled from notes taken by a scribe at a series of meetings between mathematicians and artisans around the turn of the fourteenth century. This approximate date corresponds to the golden age of the Ilkhanid era, which began with the reign of Ghazan Khan (1295–1304). He and his vizier, Rashid al-Din (1247–1318), undertook huge construction campaigns around Tabriz and gathered a great number of scholars, scientists, and artisans there. Rashid al-Din specifically dedicated his suburb, the Quarter of Rashid, to the encouragement of the arts and sciences. *Interlocks of Figures* seems to be produced in the vigorous atmosphere created by this intensive architectural activity. The meetings were probably held at Tabriz under the sponsorship of either of the leaders, who perhaps demanded the application of the highest possible advancements in geometry to the ornamental arts.

The last point is my conjecture based on the fact that three of the constructions in *Interlocks of Figures* involve the solutions of cubic equations, one of the great achievements of Islamic mathematics up to that time. These three constructions were essentially verging procedures, that is to say, mechanical equivalents of the solutions by means of conic sections. Owen Jones remarks on the significance of the use of cubic equations in the ornamental arts:

> As with proportion, we think that those proportions will be the most beautiful which it will be most difficult for the eye to detect; so we think that those compositions of curves will be most agreeable, where the mechanical process of describing them shall be least apparent; and shall find it to be universally the case, that in the best periods of art all mouldings and ornaments were founded on curves of higher order, such as the conic sections; whilst, when art declined, circles and compass-work were much more dominant (Jones 1982: 69).

Verging Procedures

The first of these verging procedures surfaces in Construction 16 of *Interlocks of Figures* (Fig. 32.1):

> Triangle AK[B] is a right-angled triangle in which the ratio of the difference between the shortest side and the hypotenuse to the difference between the [former] difference and the shortest side [the rest of the sentence is missing in the text; add "is the same as the ratio of the shortest side to the intermediate side"].
>
> **Section**
>
> The procedure is this:
>
> By means of GD (A, B, D in the text; to make the procedure more understandable, it should read "mark a given length, GD, on an arbitrary line AB, and by means of GD") erect perpendicular GE [equal to GD]. Make point E the center and then with [compass opening] EG describe arc ZH in the direction of B. Bisect GD at point T (E in the text).

[2] The only copy of this manuscript is preserved in Ms. Persan 169 in the Bibliothèque Nationale, Paris, a compilation of twenty-five works on mathematical subjects, mainly practical geometry.

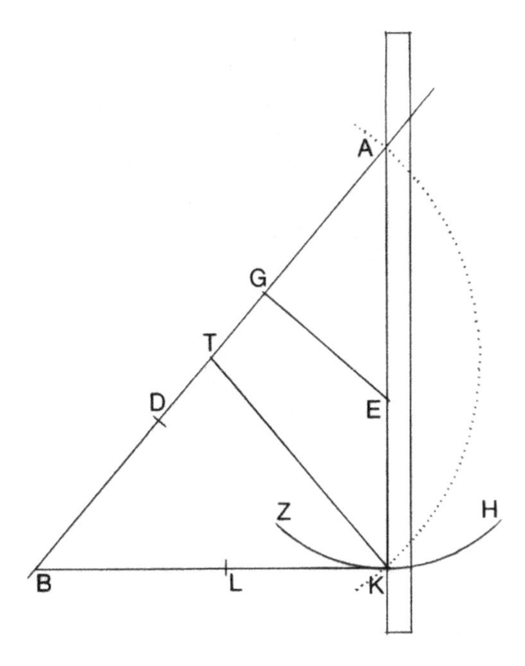

Fig. 32.1 Construction 16 of *Interlocks of Figures*. Image: author

> Put [one] arm of the compass fixed on point [T], and place the straightedge so that its edge always touches point E (more information is needed at this crucial juncture, add "so that at every instance it cuts simultaneously arc ZH at point K and the line at point A. With the other arm of the compass, compare the changing lengths of TK and TA"). Give motion to the straightedge until [the position is reached] that the lengths of TK and T become equal. Mark points K, A, and B [by making TB equal to TA]. When lines AK and KB are drawn, triangle AKB is the required right-angled one. [In this triangle,] AD is equal to BG and BK. When BL, which is equal to BD, is subtracted from BK, KL is equal to GD. God knows best [Bibliothèque Nationale, Paris, Ms. Persan 169, sec. 24, fol. 185r].[3]

The scribe's lack of familiarity with the verging procedure is apparent in the amount of the missing information. He had some acquaintance with ordinary geometrical methods, but when the construction involved advanced techniques, his knowledge proved insufficient to record the explained procedure accurately. Despite all the missing information in the text, we are able to restore it to a fully detailed verging construction. According to the text, the hypotenuse is AB, the intermediate side AK, and the shortest side BK. Also, BK = BG, AG = BD. Then, AB−BK = AG = BD and BG−AG = GD. According to the restored text, AG : GD = AK : KB. If perpendicular GE is drawn, by similar triangles, AG : GE = AK : KB. Then GD = GE. Since angles K and G are right angles, GB = BK, and BE is common, triangles EGB and EKB are congruent. Then GE = EK.

[3] In this translation from the original Persian manuscript, simple restorations are added in square brackets; more detailed ones are explained in parentheses.

The problem in the text is to construct a right-angled triangle ABK with these properties. In other words, to construct triangle ABK so that (AB−BK):(BK−[AB−BK]) = AK:KB. Since GD is assumed in the text as the given length, the problem is to determine points A, K and B. The solution is reached by means of a verging procedure, that is to say, constructing points A and K by way of iteration in such a way that AEK is straight and AT = TK. If it were only this text that accompanies the construction, it would seem merely a geometric problem. Its real objective is stated in the text added to the reverse side of the folio:

Section

In this knot pattern (ʿaqd) we need a right-angled triangle such that if [a length equal to the shortest side] is cut from the hypotenuse of the triangle towards the shorter side and a perpendicular is erected at the point of cutting, it cuts off the intermediate side at a point where [the distance] from it to the right angle is equal to the perpendicular itself.[4]

Obtaining a triangle of this sort is difficult. It falls outside the Elements of Euclid and concerns the science of conics (makhrūṭāt). If the perpendicular length is assumed, as in this example, the construction is achieved by means of a moving straightedge [Bibliothèque Nationale, Paris, Ms. Persan 169, sec. 24, fol. 185v].

The scribe appears more confident in explaining the properties achieved once the construction is completed than when he was explaining the details of the "moving geometry" (the term used by Muslim mathematicians for verging procedures). He also relates the explicit words of the author of the construction that it cannot be achieved by means of compass and straightedge, the tools of Euclidean geometry, because it concerns "the science of conics," i.e., cubic equations. Indeed it does. It is in fact the solution by means of moving geometry for the equation:[5]

$$x^3 + 2x^2 - 2x - 2 = 0 \quad (\text{if GD} = 1 \text{ and GA} = x).$$

Construction of the Knot Pattern (*ʿaqd*)

We understand that the purpose of the whole exercise was to create a special knot pattern (*ʿaqd*, as it was called in those days), from which an ornamental composition would be generated. The basic unit in this pattern is a right-angled triangle of which the hypotenuse is divided into three parts by a medial segment in such a way that if a perpendicular is erected from the end of the segment to the intermediate side, the shorter segment it cuts off from the intermediate side would be equal to the length of the perpendicular, which is also equal to the given medial

[4] In Arabic, *ʿaqd* literally means knot. Here it is used to mean the unit to be repeated to generate an interlocking ornamental composition. To convey both meanings, I translate it as "knot pattern."

[5] It is assumed in the text that the length of segment GD is 1 and GA is x. Then perpendicular GE = 1, EK = 1, BD = x, BK = BG = $1 + x$, EA = $\sqrt{(1 + x^2)}$. Since AG:GE = AK : KB, we have $x : 1 = [1 + \sqrt{(1 + x^2)}]:(1 + x)$. This equation can be reduced to $x^3 + 2x^2 - 2x - 2 = 0$. The equation has one positive root, $x = 1.1700865$ accurate in seven decimals. I also compute the angles: tan $\angle AEG = x$, so $\angle AEG = \angle B = 49.481553°$, $\angle A = 40.518447°$, $\angle AKG = \frac{1}{2}\angle AEG = 24.740777°$.

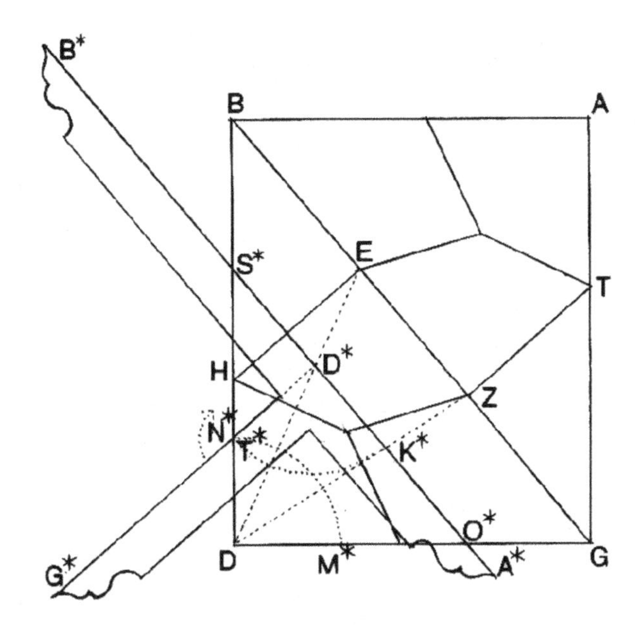

Fig. 32.2 Construction 37 of *Interlocks of Figures*. Image: author

segment (KE = EG = GD). In this seemingly complicated problem, there are two requirements to be met: (1) triangle AKB should be right-angled; (2) the three segments, DG, GE and EK should be equal to each other. Both of these requirements are met simultaneously by means of moving geometry: (1) since TK = TA = TB by the aid of the compass on point T, triangle AKB is half of a rectangle, hence right-angled; (2) since GE is taken equal to GE and arc ZH is drawn with radius EG from center E, EK is equal to EG wherever AEK cuts arc ZH.

One wonders what sort of an ornamental composition would require such delicate properties? It becomes apparent in Construction 37 of *Interlocks of Figures* (Fig. 32.2):

> The construction of this knot pattern is [performed] by a T-ruler (*gunyā mistar*). I say that in this knot pattern defined by the repeat unit (*khāna*, literally "house" or "home" in Persian) ABDG (ABD in the text), it is required that the "orange" ABZT be congruent to the "orange" (*turanj*) DGEH in such a way that BZ will be equal to GE and [a portion of] each will be common to both, thus BE will be equal to GZ. The other [requirement] is that since in the "orange" EGDH sides EG and GD is equal, EH and HD will also be equal. Necessarily, the angles at E and D will be equal and right [angles]. As this preliminary is now known, let us assume that side GD of the knot pattern of the repeat unit is known but the indefinitely extended side DB is unknown (that is, line DB is drawn but point B has not been defined on it).
>
> Then we take the ruler and from point D*, [which marks the intersection of perpendicular arms,] with an arbitrary compass opening mark lengths D*K* and D*T* equal to each other. Then go back to the repeat unit of the knot pattern, and, in the same manner that the lengths are marked on the ruler, mark points M* and N* on sides DG and DB. Take the ruler again and position the letter T* on the perpendicular arm at the letter N* so that both points are fixed on each other. Then give motion to the ruler pivoted on this point from left to right until lengths S*D* and K*O* on either side of the ruler become equal. Point T* should never be

> separated from point N*. [At this position] draw line S*O*. From point G, which is known, and parallel to S*O* draw line GB to define the rectangle. Divide line BG according to the proportion of S*D*, D*K*, and K*O* (it can be done for instance by drawing lines from point D through points D* and K*). On the ruler, which is parallel to GB, length S*D* is equal to K*O*, and K*D* is equal to D*T* and T*D. Thus, on line GB length BE will be equal to ZG, and ZE equal to EH and to HD. These constitute what is required. God knows best [Bibliothèque Nationale, Paris, Ms. Persan 169, sec. 24, fol. 190r].[6]

From the figure and explanation we understand that the gist of this and the previous construction was to create a special rectangular repeat unit so that when two congruent right-angled isosceles quadrangles (which are called "orange" in this example) facing opposite directions are contained in it, the segment that they share on the common diagonal is equal to their shorter sides. What I call "isosceles quadrangle", for want of a better term as it does not exist in modern mathematics, is a figure peculiar to the ornamental arts throughout the Islamic world, but known under different names such as "orange," "pine cone," "almond," "barleycorn" and "rhomboid." In this combined form of two isosceles triangles, with the aid of the axis through their vertices, it can be subdivided into multiple isosceles quadrangles. Depending on the angles of the initial isosceles quadrangle, ornamental configurations can be obtained from these subdivisions when the unit pattern is repeated. It was particularly on this point that the authors of these two constructions had concentrated their efforts. Whichever method of moving geometry is used, the pattern it yields generates by repetition a delicate composition (Fig. 32.3).

The second method too is the mechanical equivalent of a cubic equation, which is the reduced form of the same problem. In this case, since GD $= 1$ and EH $= x$, it corresponds to: $x^3 - 3x^2 - x + 1 = 0$ (Özdural 1996: 199). Here, the procedure of the moving geometry is given in full detail in the text. Apparently this time the scribe was able to record the explanation of the author correctly; but again shows his unfamiliarity with the subject by placing the T-ruler on the wrong side.[7]

In this second solution, the T-ruler has replaced the traditional straightedge in performing the moving geometry. With its permanently perpendicular arms it proves to be a practical and efficient tool in meeting the two requirements of the problem: it ensures that HE is always perpendicular to the diagonal while the movement of its long edge, on which the required proportion of the segments is marked, determines the position of the diagonal. Its useful peculiarities should have

---

[6] In the figure of the original manuscript, the points that belong to the T-ruler and those used to perform the moving geometry were distinguished by red ink. Some of these were identical to the ones that were used for the pattern itself; and no differentiation was made in the text. In order to avoid the confusion they create, the letters written in red ink are distinguished here by adding stars above, both in the figure and the text.

[7] In the original figure, the T-ruler is placed upon triangle GAB. The explanation in the text, however, makes sense only if the ruler is placed upon triangle GDB.

Fig. 32.3 The decorative scheme generated by construction 16 or construction 37. Image: author

attracted the attention of the participants of the discussions; a full page is allocated in *Interlocks of Figures* for its description and potential use (Fig. 32.4):

> The true nature of the proportion of this [preceding] knot pattern belongs to conics. This we can draw with the aid of an instrument called T-ruler. That is an instrument with which many knot patterns formed by conics can be drawn. In fact, this is the opinion of Katib[i]; whether it is true or not is not clear.
>
> Be that as it may, one produces the ruler in the same way as the alidade of an astrolabe (*'iḍada-i asṭurlab*). At the middle of it erect a perpendicular ruler similar to the "arrow" (sahm) of the alidade of the "boat astrolabe" (*asṭurlab-i zawraqī*, which was developed by al-Sijzi ca. 980). This is called the "mast of the bracket" (*saṭāra-i gunyā*).
>
> For example, the ruler ABGD consists of ruler AB and perpendicular [arm] GD. Should an inclination (inhiraf) be given to the edge AB of the ruler, like the inclination of the "tailored (*mujayyab*) alidade," the edge GD of the perpendicular [arm] would have the same declination. While [the declination] from the perpendicular line GD becomes distinctly apparent, point D on the edge of the alidade remains fixed. Angle GDA is found so perfectly perpendicular that with this ruler many amazing proportions (*nisbathā-i gharīb*) can be created [Bibliothèque Nationale, Paris, Ms. Persan 169, sec. 24, fol. 191v].

The T-ruler appears from the text as an instrument devised for executing patterns that involve conic sections. It is interesting that this instrument, which looks so familiar to us as it is very similar in principle to the T-square that is ordinarily used by the architects today, was newly invented and being introduced at that particular meeting. To describe this new instrument better, it was compared to the astrolabe, an instrument peculiar to astronomy. We thus understand that the latter, even a specialized version of it, the boat astrolabe, was known more commonly than the former in those days. Hence, we can say that this ordinary looking instrument, that facilitates drawing parallel and perpendicular lines, was not actually known in the Islamic world until the turn of the fourteenth century. With its introduction about

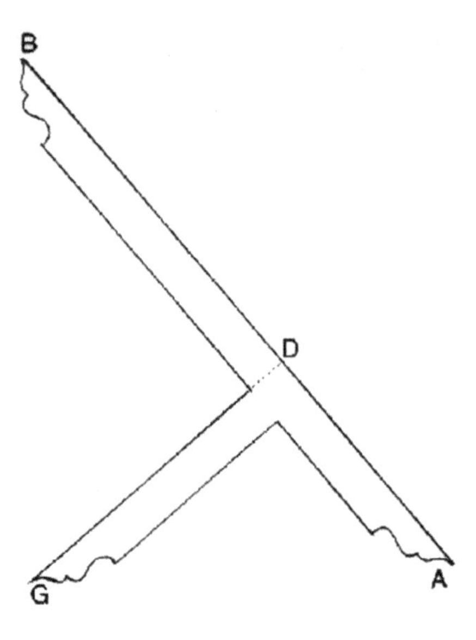

Fig. 32.4 The ABGD T-ruler described in Ms. Persan, fol. 191v. Image: author

1300, the ornamental arts and architecture appear to have gained a new impetus. So far the abundance of architectural drawings in the Islamic world after the fifteenth century was explained by the encouraging effect caused by the development of cheap paper industry during the Ilkhanid era (Necipoğlu 1995: 4–9); we can now add to it the factor of the invention of the T-square.

Al-Katibi was the name of the person who introduced the T-square at that particular meeting. The same name appears in the signature of the designer, Ali ibn Ahmad ibn Ali al-Husaini al-Katibi, of the luster tile mihrab from Imamzada Yahya at Varamin in Persia (Ritter et al. 1935: 67). Its date, 1305, suggests the attractive possibility that he and the inventor of the T-square were the same person. If this were the case, it means that a mathematician who participated in the discussions that produced *Interlocks of Figures* was also a practicing calligrapher; he thus personifies the intimate link between theory and praxis in those times.

Despite its generalized use in the future, when it was first introduced, the T-ruler was looked upon merely as a convenient tool to construct complex ornamental patterns. We see it in function again in another pattern that concern cubic equations, Construction 40 (Fig. 32.5):

> The proportion of this knot pattern is also [derived] from conics. It requires the construction of a right-angled triangle so that the altitude plus the shortest side is equal to the hypotenuse. Ibn [al-]Haytham composed a treatise on the construction of this triangle, and his construction is by means of conic sections, a hyperbola and a parabola (qaṭṭā‘a-ī makhrūṭāt zayid wa mukafi). However, the objective can be achieved here with the aid of this T-ruler. According to the aforementioned preliminary, the object of our knot pattern is those four figures: "pine cones" (sanaubarī) with two right angles surrounding a right-angled equilateral and equiangular quadrangle (i.e., the square). For example, the pine cone-like quadrangles AIHK, GHMN, DMLS, and BLKO surround square KHML.
>
> Now, as angle H of the square and both [angles] of the figure are right angles, necessarily lines KG (K in the text) and HD are straight. Thus triangle AKG is

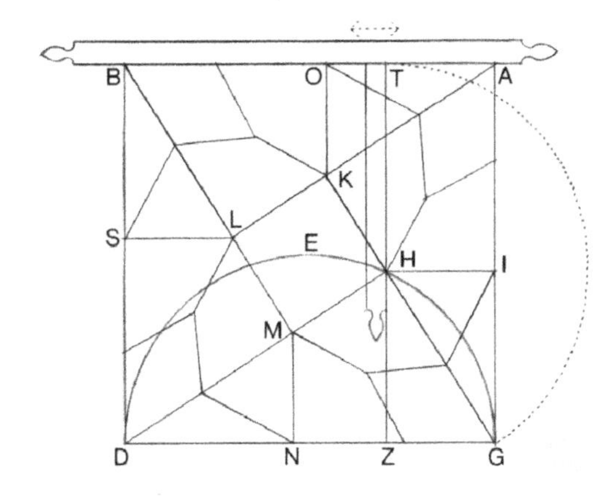

Fig. 32.5 Construction 40 of *Interlocks of Figures*. Image: author

right-angled and equal to triangle GHD. Since this triangle is right-angled, it is inscribed in a semicircle. Then point H has to be sought on arc GE (AE in the text). Subsequently, at every instance we have angle T of the ruler right-angled, its AB straight, and side AB of the [given] square and of the ruler are fixed on each other (the crucial information missing in the text can be deduced easily from the figure itself: "Give increments of sliding motion to the T-square so that it cuts the semicircle at changing positions of point H. At every instance put one arm of the compass on point H and with the other arm compare the changing lengths of HT and HG. When HT = HG, mark the point H as its required position"). God knows best [Bibliothèque Nationale, Paris, Ms. Persan 169, sec. 24, fol. 191r].

The Contributions of Omar Khayyam and al-Katibi

Although the scribe missed the crucial part of the procedure based on moving geometry, displaying again his incompetence with advanced geometrical techniques, the elegance of the restored construction indicates a high caliber mathematician behind it. When TH and HG be equal, the proof of the requirement is visible: TH + HZ = HG + HZ = GD, i.e., "the altitude plus the shortest side is equal to the hypotenuse." The scribe wrongly attributes the authorship of a treatise concerning this problem to Ibn al-Haytham. Among about 180 works of this prolific author, none answers the description. In an untitled treatise, however, Omar Khayyam describes precisely the problem concerning this special triangle, reduces its solution to a cubic equation, $x^3 - 20x^2 + 200x - 2000 = 0$, and offers two solutions by means of conic sections and one by approximation using astronomical tables (Amir-Moéz 1963). Probably the scribe was mistaken because "Khayyam" sounds similar to "Haytham," and the latter is widely known by his works on conic sections. It was evidently Omar Khayyam, also a prominent mathematician who is celebrated by his

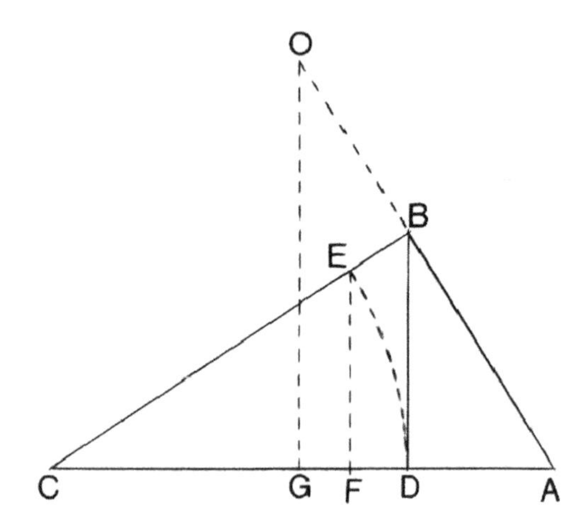

Fig. 32.6 Omar Khayyam's triangle. Image: author

works on cubic equations, who was the discoverer of this unique triangle (Fig. 32.6). When his equation is solved by modern means, angle BAC corresponds to 57.0648796°.[8]

The solution by means of moving geometry in *Interlocks of Figures* is in fact nothing but a direct translation of the problem into conic sections. The parabola is defined as the path (locus) of a point moving so that its distance from a fixed line (the directrix) is equal to its distance to a fixed point (the focus). What is achieved by the aid of T-ruler in Construction 40 answers precisely the parabola according to this definition (Fig. 32.7).

The distance of a point (H) from the directrix (side AB of the square) is set by the perpendicular arm of the T-ruler (TH), and the point's distance to the focus (G) is defined by the compass opening (HG), when the two distances be equal the point is located on the parabola. Since this particular position of point H is also located on the semicircle, the required solution thus becomes "the intersection of a parabola and a circle," as described by Omar Khayyam in his treatise.

We thus understand why al-Katibi claimed that with the T-ruler many knot patterns formed by conics could be drawn. In this example, a parabola can actually be drawn passing through points D and H by sliding the T-ruler and measuring the distances at regular intervals. The vertex of this parabola, point (O), is equidistant from the directrix and the focus. The focus and the vertex determine the axis of the parabola (GA), and the line through the focus parallel to the directrix is the *latus rectum* (GD). General equation of a parabola is $y^2 = 2px$. In our case, by assuming $GD = 1$, $GZ = y$, and $ZH = x$, it becomes $y^2 = 1 - 2x$. Its intersection with the circle, $x^2 + y^2 = y$, reduces the problem to the cubic equation $x^3 - 4x^2 + 6x - 2$.

This neat and simple solution suggests the authorship of a resourceful and talented mathematician, and the fact that it was based on the use of the T-ruler

[8] Assuming $AC = 1$, the following values are also computed for later use: $AB = 0.543689$, $BD = 0.4563109$, $CB = 0.83922867$, $CD = 0.7044022$, and $AD = 0.2955977$.

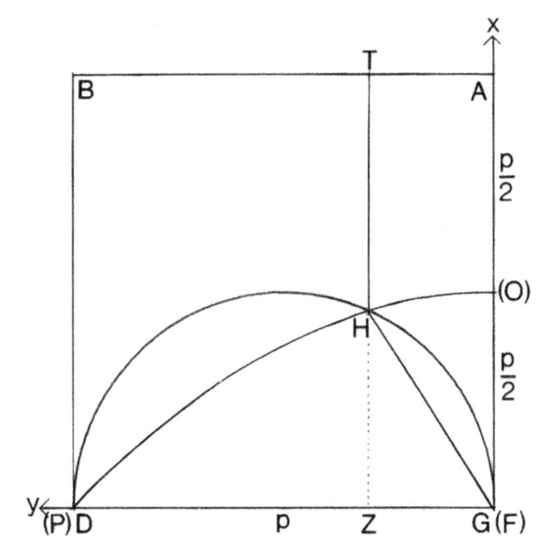

Fig. 32.7 The solution of construction 40 by means of conic sections. Image: author

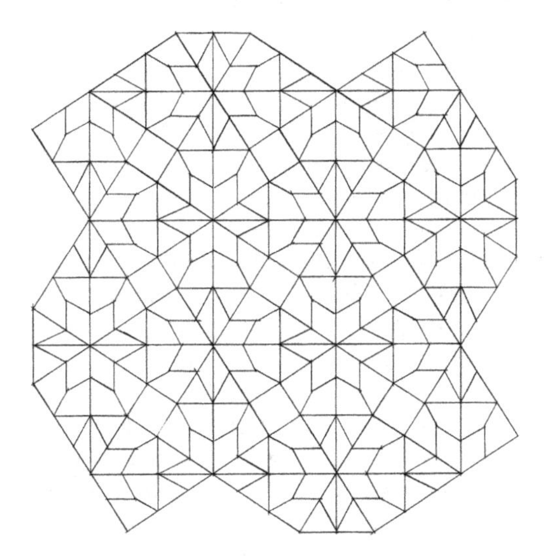

Fig. 32.8 The decorative scheme generated by construction 40. Image: author

points to its inventor, al-Katibi. If al-Katibi were both the author as well as the designer of the luster mihrab from Varamin, then the connection between mathematics and arts that he personally represents provides us a sharp insight into the milieu that created the high standard in the ornamental arts in those times. In any case, whoever the author of this solution based on moving geometry might be, he had the imagination to envisage a decorative composition as elegant as its solution (Fig. 32.8).

Omar Khayyam's Triangle and the North Dome Chamber

This decorative scheme is not the only outcome of Omar Khayyam's triangle. He says:

> A triangle with mentioned properties is very useful in problems similar to this one. This triangle has other properties. We shall mention some of them so that whoever studies this paper can benefit from it in similar problems (Amir-Moéz 1963: 326).

A more attentive study of this triangle reveals indeed some very interesting properties (Fig. 32.6): between the hypotenuse AC and the shortest side AB, CB becomes the geometric mean, CD the harmonic mean, and GO, the perpendicular on the midpoint of AC, the arithmetic mean.[9] Then, AC:GO::CD:AB, which Greeks called "the musical proportion" and judged most perfect. Simpler numerical versions of this proportion, such as 12:9::8:6, are known to have been used in Renaissance architecture. In the hands of Omar Khayyam this proportion attains a mathematical complexity with its irrational magnitudes obtained by means of cubic equations. The richly interrelated proportions embodied in Omar Khayyam's triangle present themselves provocatively as tools suitable for architectural application.

The North Dome Chamber in the Friday mosque of Isfahan has always impressed its visitors with its mature proportions. Eric Schroeder has judged it as marrying genius and tradition more elegantly than many other famous domed structures (Schroeder 1938–1939: III, 1005). Its foundation is dated 1088–1089. It was the time when Omar Khayyam was enjoying high prestige as the leading astronomer in Isfahan who had already published his major works on geometry and algebra, and whose reformed calendar was lately adopted by royal decree.

If Omar Khyyam's triangle is superimposed over a drawing of the North Dome so that the hypotenuse corresponds to the span, we perceive that the generating force behind that astoundingly powerful space is the musical proportion contained in this singularly unique triangle (Fig. 32.9 and Table 32.1). The geometric scheme is my own, but its very close agreement with actual dimensions gives it credibility and suggests strongly that Omar Khayyam, one of the greatest intellects the Islamic world had produced, was actually the designer of one of the greatest accomplishments of Islamic architecture.[10]

---

[9] The following values are computed in addition to the ones in n. 16:GO = 0.7718445, CF = 0.5911954 = CB^2.

[10] For my detailed argument on this point, see (Özdural 1998: 699–715). The photogrammetric drawing and the dimensions obtained from Rassad Survey Company, "Masjed-e Jame'Esfahan" (paper presented to the International Committee for Architectural Photogrammetry at the Symposium on the Photogrammetric Survey of Ancient Monuments, Athens, 1974, pl. 13), are published in ibid., Fig. 32.5.

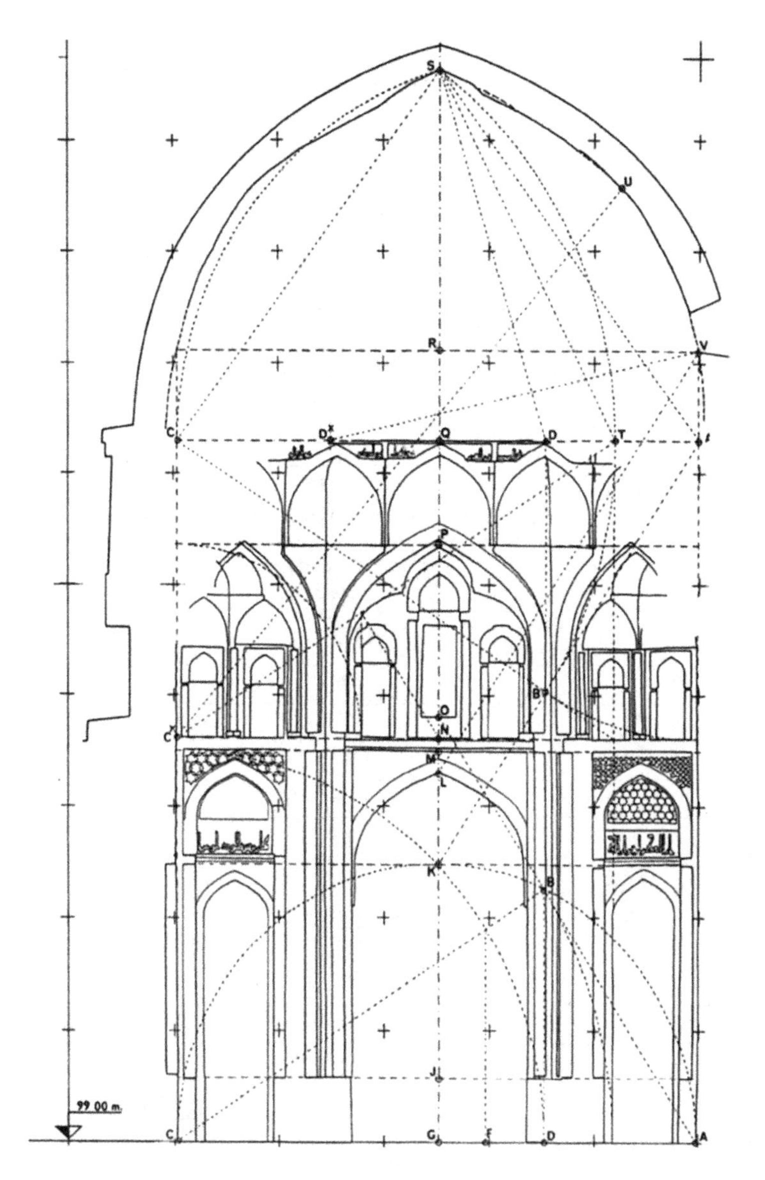

Fig. 32.9 The geometric scheme generated by Omar Khayyam's triangle. Image: author

Biography Alpay Özdural (1944–2003) was Associate Professor at Eastern Mediterranean University, Faculty of Architecture in North Cyprus, where he taught design and history of architecture courses. By profession he was an architect specialized in restoration and preservation of historic monuments and sites. His other fields of interest were architectural photogrammetry, history of mathematics, and historical metrology. The last 10 years of his life he concentrated his efforts on muqarnas, a type of three-dimensional geometric decoration peculiar

Table 32.1 Relationships between key points shown in Fig. 32.9

Dimensions of the North Dome Chamber	Components of Omar Khayyam's Triangle	Theoretical values	Actual Measurements	Error
AC	AC	9.900m	9.90m.	Nil
GM	CD	6.974	6.99	0.2%
NQ	AB	5.383	5.36	0.4%
GO	GD	7.641	7.63	0.1%
GM acts as the harmonic mean between AC and NQ. GO acts as the arithmetic mean between AC and NQ, thus: AC:GO::GM:NQ becomes the upward musical proportion				
NR	CD	6.794	7.00	0.4%
LR	GO	7.641	7.63	0.1%
KQ	GO	7.641	7.64	Nil
NR acts as the harmonic mean between AC and NQ. RL and KQ act as the arithmetic means between AC and NQ, thus: AC:LR::AC:KQ::NR:NQ becomes the downward musical proportion				
GK	½AC	4.950	4.95	Nil
JK	½GO	3.821	3.81	0.3%
KO	½AB	2.692	2.66	0.4%
OQ	½AC	4.950	4.96	0.2%
NP	½CD	3.487	3.47	0.5%
GO	AC+ ½AB	12.591	12.59	Nil
GN	½ (AC+BD)	7.209	7.23	0.3%
By the rule of halves: GK:KJ::OQ:JK::NP:KO becomes the musical proportion of halves. Since AC = AB + BC, AC acts as the arithmetic mean between GQ and GN				
QS	$\sqrt{AC\text{-}DC\text{-} \frac{1}{2}AC^2}$	6.674	6.68	0.1%
CS	BC	8.309	8.29	0.2%
JM	CF	5.853	5.85	Nil
PQ	AB - ½ CD	1.896	1.89	0.3%
QS and CS are components of the isosceles triangles CSA, TCS and CDS; CS:AC::BC:AC::BD:AB JM:GM::KL:PQ::PQ:KN::KN:KO::QR:PQ::NR:CS::CS:AC				

to Islamic architecture, and on the collaboration of mathematicians and artisans in the medieval Islamic world. He published articles in such journals as *Historia Mathematica, Technology and Culture, Muqarnas* and *Journal of the Society of Architectural Historians*.

References

AMIR-MOÉZ, A. R. 1963. A Paper of Omar Khayyam. *Scripta Mathematica*, **26**: pp. 323-337. New York: Yeshiva University.

HOLOD, R. 1988. Text, Plan, and Building: On the Transmission of Architectural Knowledge. Pp. 1-12 in M. Bentley Ševĉenko, ed. *Theories and Principles of Design in the Architecture of Islamic Societies*. Cambridge, Mass.: The Aga Khan Program for Islamic Architecture at Harvard Univ. and Massachusetts Institute of Technology.

JONES, O. 1982. *The Grammar of Ornament*: New York: Van Nostrand Reinhold.

NECIPOĞLU, G. 1995. *The Topkapı Scroll—Geometry and Ornament in Islamic Architecture*. Santa Monica: The Getty Center for the History of Art and Humanities.

ÖZDURAL, A. 1995. Omar Khayyam, Mathematicians, and *Conversazioni* with Artisans. *Journal of the of the Society of Architectural Historians*, **54**, 1 (March): pp. 54-71. Berkeley: University of California Press.

———. 1996. On Interlocking Similar or Corresponding Figures and Ornamental Patterns of Cubic Equations. *Muqarnas*, **13**: pp. 191-211. Cambridge, Mass.: The Aga Khan Program for Islamic Architecture at Harvard Univ. and Massachusetts Institute of Technology.

———. 1998. A Mathematical Sonata for Architecture: Omar Khayyam and the Friday Mosque of Isfahan. *Technology and Culture*, **39**, 4 (October): pp. 699-715. Chicago: University of Chicago Press.

———. 2000. Mathematics and Arts: Connections between Theory and Practice in the Medieval Islamic World. *Historia Mathematica*, **27**, 2 (May): pp. 171-201. Sheffield: Academic Press.

RITTER, H., RUSKA, J., SARRE, F. and WINDERLICH, R. 1935. *Orientalische Steinbucher und Persische Fayencetechnik*. Istanbul: Archäologischen Institutes des Deutschen Reiches.

SCHROEDER, E. 1938-1939. Seljuq Architecture. *A Survey of Persian Art*. A. Upham Pope & P. Ackerman, eds. Vol. III: p. 1005. London: Oxford University Press.

Chapter 33
Explicit and Implicit Geometric Orders in Mamluk Floors: Secrets of the Sultan Hassan Floor in Cairo

Gulzar Haider and Muhammad Moussa

Introduction

During two and a half centuries of Mamluk rule in Egypt (1250–1517 A.D.), the integrated complexes of mosque-academy-mausoleums and even attached dormitories and hospitals were refined into a building type that represented an essential and creative relationship among power, patronage, faith, and architecture. Added to and transformed beyond the Fatimid memory, Cairo became the "City of a Thousand Minarets." The Sultan Hassan Complex is considered to be a masterpiece of this period and universally accepted as one of the finest example of Islamic architecture.

The Mamluk *Madrassah* (religious academy) is key to such complexes and is characterized by central courtyard with four *iwans* (deep vaults). The *iwan* that marks the Mecca orientation is usually slightly larger than the other *iwans* and houses the *mihrab* (prayer niche) and the *mimbar* (stepped pulpit). The courtyard and the *iwans* are multi-use spaces for congregational prayer, small group prayers and academic gatherings around the teachers of various subjects. Beyond this central court of mosque-*madrassah* the complex comprises a substantial mausoleum of the patron, dormitories and other essential facilities.

First published as: Gulzar Haider and Muhammad Moussa, "Explicit and Implicit Geometric Orders in Mamluk Floors: Secrets of the Sultan Hassan Floor in Cairo", pp. 93–104 in *Nexus V: Architecture and Mathematics,* Kim Williams and Francisco Delgado Cepeda, eds. Fucecchio (Florence): Kim Williams Books, 2004.

G. Haider (✉)
School of Architecture, Beaconhouse National University, Lahore, Pakistan
e-mail: ghdesigngroup@gmail.com

M. Moussa
HHCP Architects, 120 N Orange Ave., Orlando, FL 32801, USA
e-mail: moussa.arch@gmail.com

K. Williams and M.J. Ostwald (eds.), *Architecture and Mathematics from Antiquity to the Future*, DOI 10.1007/978-3-319-00137-1_33,

The Sultan Hassan is the most magnificent in scale, quality of design, construction, and decorative detail among all such complexes in Cairo (Figs. 33.1 and 33.2). In addition to four schools of religious law with their respective dormitories, a large mosque-*iwan*, and the Sultan's tomb, it also includes a hospital, an orphanage and a water well. The main courtyard is a prominent open space in the heart of a dense and massive edifice; it acts as the essential center towards which all the significant spaces of the complex open. It is a quasi-cubic void, which measures 32 $\times$ 34.6 m in plan, and is about 40 m high, its top stellated edge framing the open sky. A domed octagonal water basin marks both the center of the courtyard and the *axis mundi* in this complex. The floor of the courtyard space is the geometrical marble pavement, the largest of its kind in Egypt (Fig. 33.3). Its texture and colors bring a sense of liveliness into the heart of the building, like a flowered valley at the foot of a stone canyon. This impression is accentuated by the geometry of the floor, which gives an order like that of a formal garden (Fig. 33.4).

This chapter is the first detailed measurement and morphological report as well as an interpretive analysis of this geometric-architectural treasure.

Documentation of the Courtyard Floor Pattern

The documentation process of the Sultan Hassan marble floor was initiated in December 1992 during a Carleton University studies abroad trip directed by Dr. Gulzar Haider. Between 1993 and 1996, scaled drawings of the floor pattern were manually attempted. The challenge of accurate manual drawings at convenient scale as well as dimensional discrepancies necessitated a second field trip in 1998. Back in Canada, it took Muhammad Moussa about 15 weeks to get the first AutoCAD drawing of the entire pattern. The experience of making this drawing in itself had started to reveal the "not so simple" nature of the floor and further research was planned, which evolved into a master's thesis. It was considered prudent at that time that a final on-site check be made of connectivities of the drawn pattern in comparison with the actual floor. This led to the final field trip in summer 1999. Muhammad Moussa defended his Master's of Architecture thesis reporting the analysis and interpretations of the Sultan Hassan Floor in 2001, Carleton University.

Construction of the Digital Floor Model

Line Drawing Models of Individual Patterns The analysis of the pavement aimed at the study of geometric elements and patterns and their organization into the whole. Questions and curiosities about symbolism and meaning consciously imparted by the makers of the floor were deliberately held back. The intent was to allow the architecture to take the lead in unfolding its scientific and/or cultural

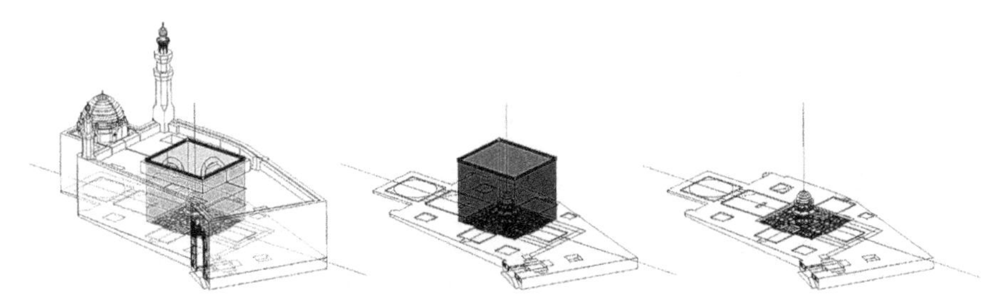

Fig. 33.1 Three-dimensional AutoCAD study model of Sultan Hassan complex. Image: authors

Fig. 33.2 Sultan Hassan complex in Cairo skyline. Photo: authors

significance, through investigating its mere physical presence as manifestation and choreography of geometric operation. The nature of the geometric operations in two dimensions makes it possible to carry out such a 'performance;' imagining the motion that an object must go through in order to coincide with its image. For instance, performing the operation of translation denotes a different movement than performing an operation of reflection. A significant distinction between these two operations is that translation is achieved through sliding in the two dimensional plane, while reflection requires a flip through three dimensional space. In fact, two dimensional symmetry operations are grouped according to as "proper" and "improper" operations, proper operations being traceable in the two dimensional plane (translation and rotation), while improper operations involve a path between the object and its image that is not traceable in two dimensions (reflection and glide reflection). The fact that the total sum of the two-dimensional symmetrical operations is only four provides another advantage to derive their different compositional possibilities. Purely as symmetry operations, their combinations have been explored, counted and classified as the known symmetry groups: seven linear (frieze) groups, and 17 periodic (planar).[1]

[1] The study employed the common classification of symmetry groups, also known as "Crystallographic groups", as outlined by Stevens (1980).

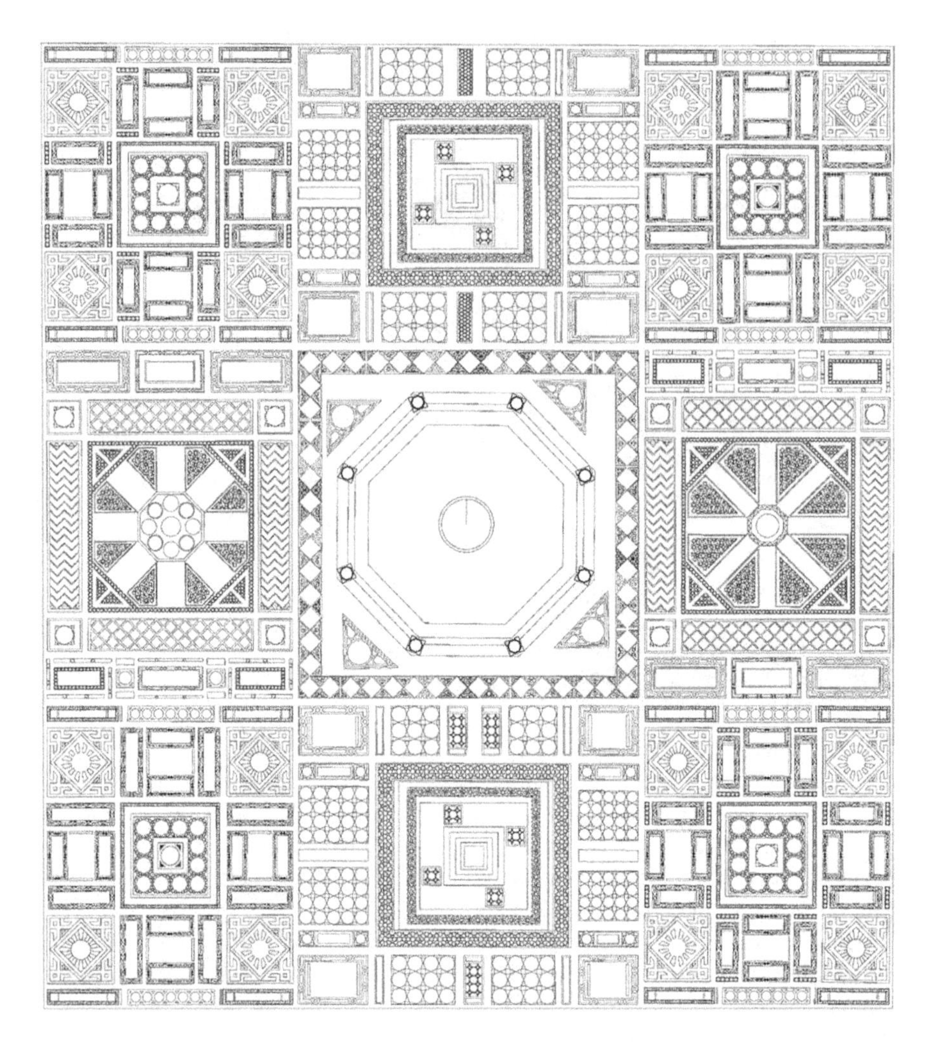

Fig. 33.3 Line drawing of the geometric marble patterned floor. Image: authors

Layered Model of Line Drawings and Color Patterns The floor as a two dimensional plane is divided into nine sectors labeled A1, A2, A3, B1, B2, B3, C1, C2, and C3 (Fig. 33.5). Numbers refer to the sectors arranged parallel and letters refer to the sectors arranged normal the Mecca orientation. Sector B2 is in the center of the floor, row A is closest to the *Mihrab* and row C is farthest. Constituent patterns in each sector were isolated and classified under symmetry groups. This stage revealed the commonalities and differences among various sectors as well as the unique character of each sector.

Initially, when the line drawing of the entire floor was assembled in a digital format, line merely denoted the color difference among neighboring marble pieces. As the on-site documentation photographs were carefully reviewed, color started to

Fig. 33.4 View into the courtyard from the roof. Photo: authors

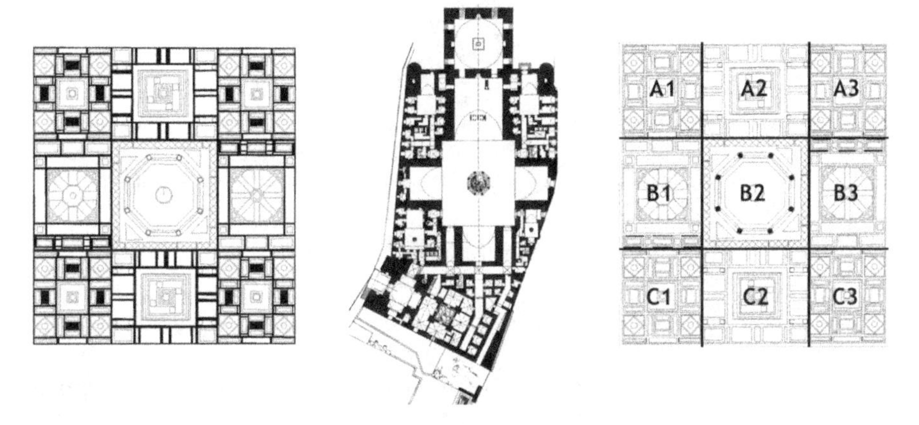

Fig. 33.5 Diagram of patterned areas, Sultan Hassan main floor plan, and division of the floor into nine sectors. Image: authors

emerge as a dimension far more significant than some random choice of natural tones. Overlooking the slight color variations among different cut pieces, the floor can be seen as a pattern of white and five other colors: sand yellow, sky blue, olive green, wine red, and black. While the black marble pieces are used just like other colors, the white operates as a background on three levels. First, the white is a general background between all the enclosed patterns within a sector; second, it is the filler between different patterns within a single patterned frame; third, it is the filler within the patterns themselves. Color information was added to the AutoCAD model of the floor by creating a separate digital layer for each color in each of the nine sectors. A layer indicating the pattern of white could be obtained by taking the entire model and rendering all the five colors as black (Fig. 33.6).

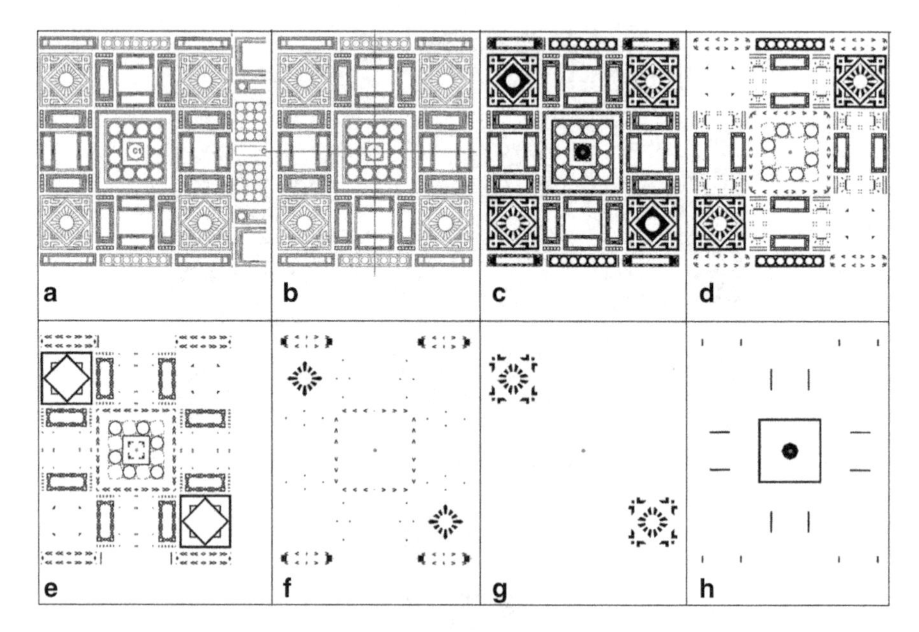

Fig. 33.6 Color analyses of the Sultan Hassan pavement. (**a**) detail of sector C1; (**b**) line drawing; (**c**) black and white representation of color; (**d**) Red color selection; (**e**) black color selection; (**f**) yellow color selection; (**g**) blue color selection; (**h**) green color selection. Image: authors

Analysis as Pursuit of a Typology of Patterns

The analytical process of line drawings and individual color configurations can be outlined as follows:

1. Line geometry within the individual patterned frames and their symmetry group.
2. Patterned area as a motif in itself.
3. Peeling of colored layers and the repeat of steps 1 and 2, for each patterned area.
4. Applying steps 1, 2 and 3 above to individual sector as well as the overall floor.

Corresponding to these four points of analysis are few key discoveries:

- Patterns that could be considered to exist at the level of line geometry only, that is, their symmetry is not altered when we consider their color configuration.
- Patterns that present certain symmetry through line geometry, but alters or transforms as color configuration is taken into consideration.
- Patterns that do not alter at the scale of the individual patterned area as the color comes in but the shift occurs in the way the repetition happens in the larger context of the sector and/or the entire floor (Fig. 33.7).

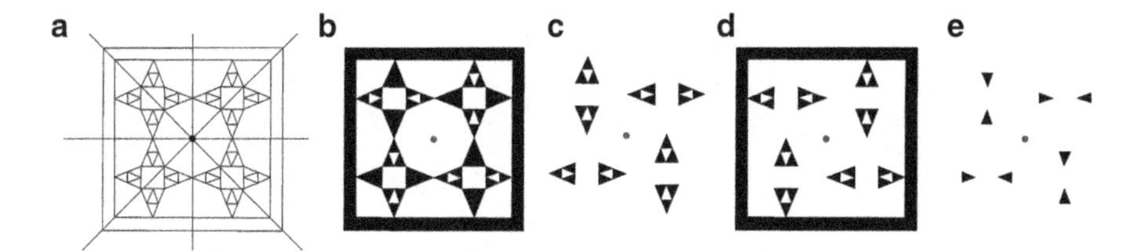

Fig. 33.7 Detail of a pattern. (**a**) line drawing; (**b**) black and white representation of color; (**c**) black color selection; (**d**) red color selection; (**e**) yellow color selection. Image: authors

Analysis as Performance

The above analysis pointed to intriguing geometric phenomena revealed through the separation of colors. Expectations of continuity of a symmetric pattern are defied either through color or through the enlargement of the area of study within the floor. Contrary to expectations of reflection, in most cases, it is possible to visualize the floor patterns generating and repeating through "proper" two dimensional operations only (rotation and translation). These realizations led to questions about both the conceptualization of the floor as a catalyst for imagination and the realization of the floor as an experience. The frame of reference of the study shifted from the "forensic" geometric analysis of a 500 year old marble floor to an experiential, morphogenetic understanding of the floor as a multi-level, geometric and color, transformative phenomenon. The term and concept of "performance," quite similar to George Steiner's pursuit of "understanding that is simultaneously analytical and critical" and an approach of "interpretation as understanding in action" (Steiner 1989: 7–8), helped parallel and simultaneous theorizations about the actual construction of the floor, as well as the formation of its geometric essence. The transformations that occur between the explicit orders of the line geometry and the not-so-explicit orders of color configuration strongly suggest the likelihood of layered geometric patterns at different scales. It seems highly improbable that the constructors had a pile of precisely pre-cut colored geometric marble pieces and then tried to figure out how to assemble them into a composition of the complexity one encounters.

It is also important to point out that without the aid of high-speed digital tools, most of our discoveries about the multi-layered geometry of the floor would have not been visible to the eye. Our analysis as "performance" is very much dependent upon the speed of geometric computation, not unlike the movements discovered when still frames are run at a certain speed to achieve the simulation of movement. This, however, does not imply that the craftsmen who masterly composed this floor did not experience these realities in their own way with their own tools, whether physical, intellectual, or spiritual.

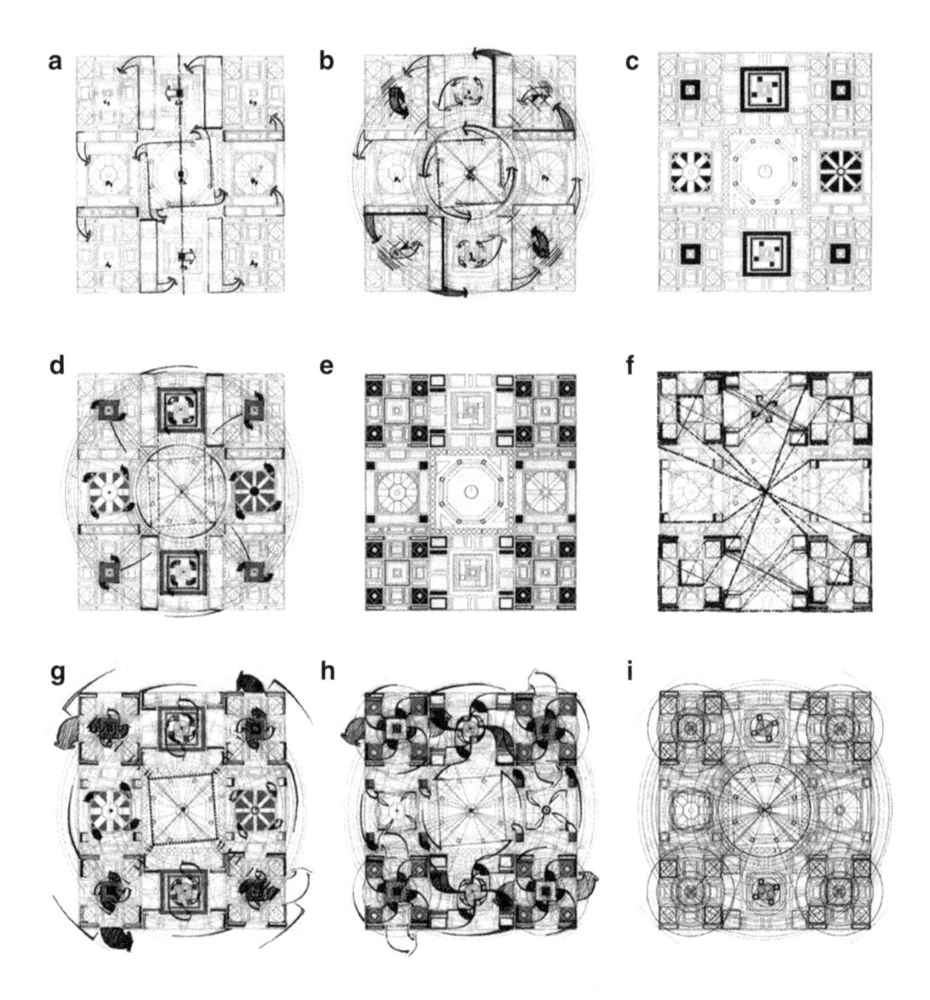

Fig. 33.8 Interpretive diagrams of the morphological aspect of the floor's geometry. Image: authors

Choreography of Symmetry Operations

It has been possible for us to posit a morphogenesis of the floor which can justifiably be called a choreography of symmetry operations leading to the floor as we see it. It is presented here as a series of interpretive drawings which could be viewed in their totality as linked, though each can also be treated as a single finding (Fig. 33.8). The drawings mainly seek to highlight certain generative and operational themes in the floor with their corresponding phases:

1. Rotation (4 phases),
2. Color transformation and complexity (2 phases),
3. Pattern pairing (2 phases), and

4. Synchronization of color and motion (3 phases).

Rotation, Phase 1 (Fig. 33.8a) In the line geometry, the pattern closest to the center of the floor is most evident in suggesting a rotational symmetry. It repeats four times around the central octagonal fountain basin. The geometrical analysis of the motif established an understanding of how the four triangular motifs would relate to the corners of the square that inscribes the central octagon. An anti-clockwise rotation of approximately 4° results in the position of the motifs in the floor. This initial spark of rotation can be seen echoed in the subtle counter-clockwise tilt of the main axis established between sectors A2 and C2. In sector C2, the patterned frames to the left are larger than those to the right. What might seem at a first glance as a form of inaccuracy, repeats again in sector A2 on the opposite side of the floor, only this time the right side patterned frames are larger than those of the left, confirming a 180° rotational relationship around the center of the floor between the two sectors A2 and C2.

Rotation, Phase 2 (Fig. 33.8b) Still looking at the line geometry, the second phase of the operation of rotation could be considered as the 180° rotations of the eight sectors surrounding B2, the central one. The four corner sectors A1, C3, A3, and C1 relate diagonally across. The top central sector A2 relates to the bottom C2, and the left central B1 to the right central B3. This twofold 180° rotation is in clear defiance of axial, reflective, symmetry.

Color Transformation and Complexity, Phase 1 (Fig. 33.8c) There are two ways in which color configuration influences the final group designation of the patterns: transformation on the level of the motif or the pattern unit itself, or transformation only on the level of repetition in the floor sectors. In this first phase of color transformation and complexity, the highlighted patterns transform on the level of the individual patterned area once color configuration is considered. It is intriguing that these patterns are concentrated in the centers of the eight peripheral sectors.

Synchronization of Color and Motion, Phase 1 (Fig. 33.8d) Filtering the color from the previous drawing into the second phase of rotation reveals an intricate relationship between motion and color. Since line geometry always precedes color, this phase could be imagined as a sequential process of motion initiated by the line geometry, resulting in the overall rotation of the floor around its center, which then is followed by the introduction of color into the central patterns of each of the bordering floor sectors.

Color Transformation and Complexity, Phase 2 (Fig. 33.8e) Patterns highlighted in this phase form the second phase of color transformation as the impact of the color configuration is only sensed once the larger context of the floor is considered. Interestingly these patterns appear in the boarders of the eight peripheral sectors.

Pairing, Phase 1 + 2 (Fig. 33.8f) Aside from the apparent presence of the phenomenon of pairing between all the patterns in the floor, its significance becomes more evident through color configuration. More than a few patterns that repeat in groups of four or more pair up into sets of two through color. This phenomenon operates both at the level of a single sector and that of the entire floor. This pairing reinforces the rotational symmetry as the patterns pair diagonally in most cases (180° rotations).

Synchronization of Color and Motion, Phase 2 + Rotation, Phase 3 (Fig. 33.8g) Superimposition of layers reveals a sequence of rotational ripple effect. The initiation of the operation of rotation on the line drawing level, which is most evident in the center of the floor, starts to echo in the bordering sectors as an overall 180° rotation once the color is brought into consideration.

Synchronization of Color and Motion, Phase 3 (Fig. 33.8h) The final phase of the synchronization of color and motion extends the ripple effect mentioned earlier. The third phase of color transformation and complexity integrates into the sequence as the echo of the rotations generated by the central patterns of each of the floor sectors.

Rotation, Phase 4 (Fig. 33.8i) This drawing recapitulates the operation of rotation-like ripples caused in water after the stones sink under the surface. It is possible now to imagine the floor continuously rotating where the motion keeps rippling from the center outwards to the corners. Once there is a sense of rest in our perceptual map another ripple begins from the slight shifts in the heart of the floor and the surrounding sectors and is infused to the spatial understanding of the entire courtyard.

Concluding Comments

The most striking experience in the Sultan Hassan complex is the great space carved out of the heart of this monumental monolith. One is awe-struck at how it softens the harsh Cairo sunlight while illuminating the deepest corners of the four *iwans* of study and supplication (Fig. 33.9). It is the essential space everyone experiences before and after engaging in prayer, arriving at any of the four schools, or visiting the mausoleum. Besides acting as the only access to the four schools, it also forms the interior "public" space for the students of different schools to socialize without leaving the sacred environment of knowledge.

The orientation of the main axis towards the *Kaaba* in Mecca[2] is fundamental to every Islamic prayer and funerary space. The subtle elongation of the courtyard in the direction of Mecca is hardly noticeable from any viewpoint in the courtyard. In

[2] The city of Mecca in Saudi Arabia is the Holy city for Muslims. It houses the *Kaaba* (literally translated as "Cubic"), a stone building within a great court/sanctuary. It is the goal of Islamic pilgrimage and the point toward which Muslims orient themselves towards in prayer. Muslims believe that Prophet Abraham first built it as a landmark for the House of God, for the sole purpose

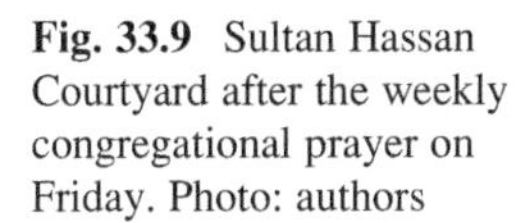

Fig. 33.9 Sultan Hassan Courtyard after the weekly congregational prayer on Friday. Photo: authors

fact it was an unexpected discovery during the documentation process, and later on enticed our curiosity, since the makers' capabilities in achieving precise measurements and geometries are quite evident in other spaces in the building; such as the space of the mausoleum (21×21 m). The floor ambiguously merges the sense of a central fourfold and an axial twofold symmetry. Peculiarly, the courtyard has closely similar quasi-cubic proportions as the *Kaaba*; except the *Kaaba* has a volume about one-eighth of the court's. In other words, the *Kaaba* would approximately occupy a unit volume of an equal $2 \times 2 \times 2$ division of the courtyard's space.

Furthermore, as the geometrical analysis has intriguingly revealed, the monument's main axis rotates slightly as it projects beyond the space of prayer through the plane of the geometric floor (Fig. 33.10). The concept of rotation in the Islamic tradition resonates at different levels. The most prominent is the ritual of circumambulating the *Kaaba* in Mecca (Fig. 33.11). It is conceptually remarkable to realize that the distant orientation towards the *Kaaba* is radial and static, while the closest ring around it in Mecca is concentric and dynamic in a counter-clockwise direction.

In a possibly analogous gesture, the monument's main three-dimensional body imparts an explicit axial symmetry, while the floor pattern generates an implicit rotational field. This concept also manifests itself in one of Islam's most eloquent spiritual paths, Sufism. The courtyard floor of the Sultan Hassan mosque silently mirrors the motion of the heavens, not unlike the whirling dervish who meditates the motion and rhythm of the celestial bodies revolving around a vertical axis while rotating around a distant centre.

The study presented here has revealed that the usual static and restrictive view of geometry is not only unfair but is also potentially limited in helping us see a life beyond a fixed order of points, lines and polygons. Indeed, the tradition speaks of

---

of worshipping of God alone. The first ritual a Muslim would embark on upon arriving at the sanctuary is seven circumambulations (*Tawaf*) around the *Kaaba*.

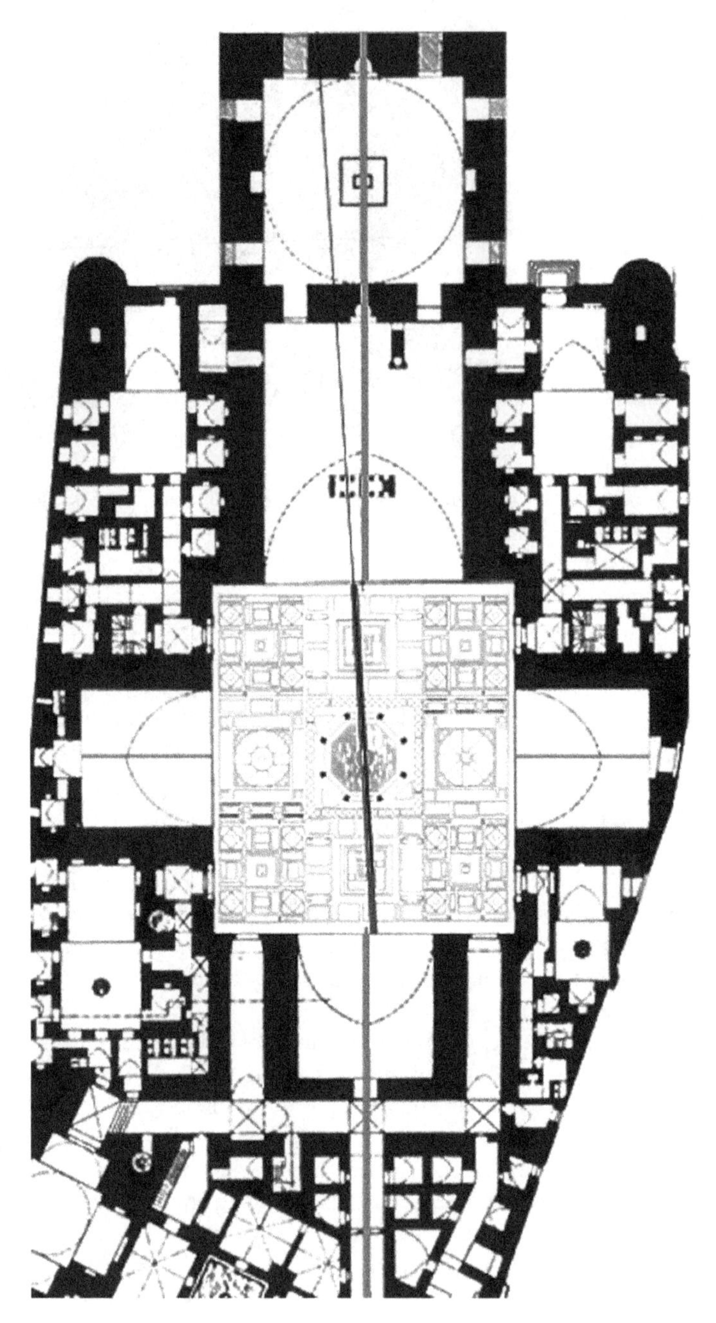

Fig. 33.10 The axial shift of the two dimensional floor from the main three-dimensional body of the edifice. Image: authors

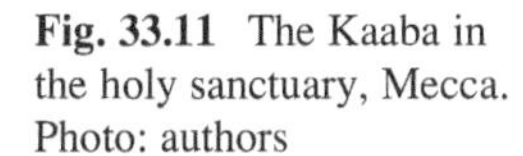

Fig. 33.11 The Kaaba in the holy sanctuary, Mecca. Photo: authors

the different elevations or stations (*maqamat*) of the seeker of divine truth. Form every station one is able to perceive different facets of "reality," where "reality" is composed of multiple layers to which one could relate with different capacities and through different frames of reference at different stages of the ascending journey. To many, the courtyard might seem empty and static, yet as this study has revealed, the void is suggestively inhabited with a mathematical elegance and colorful eloquence of a hidden dance.

The heavens revolve day and night,
Like a potter's wheel,
And every moment the master's wisdom,
Creates a new vessel,
For all that exists,
Comes from one hand,
One workshop.[3]

[3] Mahmoud Shabistari (d. 1320), quoted in Nasr (1978: 105).

Biography Gulzar Haider is a former Professor and Director at the School of Architecture, Carleton University, Canada. He studied architecture and structural engineering as separate professional disciplines at the University of Illinois, Urbana-Champaign and graduated with a B.Arch (1968) and a PhD (1969). He founded the Form Studies Unit at Carleton and introduced the idea of polyhedral chains organized around spatial networks. His work has ranged from structural morphology to architectural history and theory. He was member of the International Commission for Preservation of Islamic Cultural Heritage, Istanbul (1983–1994) and was awarded the Davidson Dunton Research Lectureship in 1999. He is currently the Dean of the School of Architecture at Beaconhouse National University in Lahore, Pakistan.

Muhammad Moussa holds a Bachelor degree in Architecture from Helwan University, Cairo, Egypt (1995), and a Master's degree in Architecture from Carleton University, Ottawa, Canada (2001). His research interests focus on Design and Culture with special emphasis on history of the Middle East, (Ancient, Coptic and Mamluk). He taught Design Studio at Carleton University 1999–2003 as a Sessional Instructor. Besides working to set up his independent design studio, he currently holds a Senior Project Designer position at HHCP Architects in Orlando, Florida.

References

Nasr, S. H. 1978. *An Introduction to Islamic Cosmological Doctrines*. Albany: State University of New York Press.
Steiner, G. 1989. *Real Presences*. Chicago: The University of Chicago Press.
Stevens, P. S. 1980. *Handbook of Regular Patterns*. Cambridge: MIT Press.

Chapter 34
The Fibonacci Sequence and the Palazzo della Signoria in Florence

Maria Teresa Bartoli

Introduction

The Palazzo della Signoria in Florence is the oldest part of the Palazzo Vecchio, built by Arnolfo da Cambio (the architect who began S. Maria del Fiore) between the end of the thirteenth and the beginning of the fourteenth centuries (it was begun in 1298) (Romanini 1980). From its beginning it has been the house of the municipal government. As Florence was one of the biggest towns in Europe, its town hall was exceptionally large; its dimensions were larger than average. Until the dome of S. Maria del Fiore was built by Brunelleschi, the Palace with its extraordinary high tower embodied the Florentine pride and sense of power (Figs. 34.1 and 34.2). Recently, with the help of graduate students who earned their 'Dottorato di Ricerca' in "Surveying and representing architecture and environment," I measured and surveyed the Palazzo della Signoria and now have in hand the results: plans, sections and elevations. Examining the drawings, especially the ground floor plan (Fig. 34.4), I could finally find the secret ratio of the original project. This is the subject of the present chapter.

Since the Middle Ages historians have criticized the irregular form of the palace perimeter. It is a trapezoid, with two non-right angles, which made the interior design of the palace very difficult. I investigated the problem and now believe that this anomaly must be related with the peculiar layout of the last and largest city wall, which dates from the same time.

First published as: Maria Teresa Bartoli, "The Sequence of Fibonacci and the Palazzo della Signoria in Florence", pp. 31–42 in *Nexus V: Architecture and Mathematics*, Kim Williams and Francisco Delgado Cepeda, eds. Fucecchio (Florence): Kim Williams Books, 2004.

M.T. Bartoli (✉)
Dipartimento di Architettura, Università degli studi di Firenze, Florence, Italy
e-mail: mtbarto@unifi.it

K. Williams and M.J. Ostwald (eds.), *Architecture and Mathematics from Antiquity to the Future*, DOI 10.1007/978-3-319-00137-1_34,

Fig. 34.1 Palazzo della Signoria in the Piazza della Signoria, Florence. Photo: author

Fig. 34.2 The tower of the palace in the skyline of the town center. Photo: author

In any case, even given the trapezoidal plan, the rectangle in the trapezoid is a very interesting one. If we define its measurements by means of the *braccio da panno*, the Florentine unit of measure in that time, the dimensions are very significant. From now on I will describe my findings point by point:

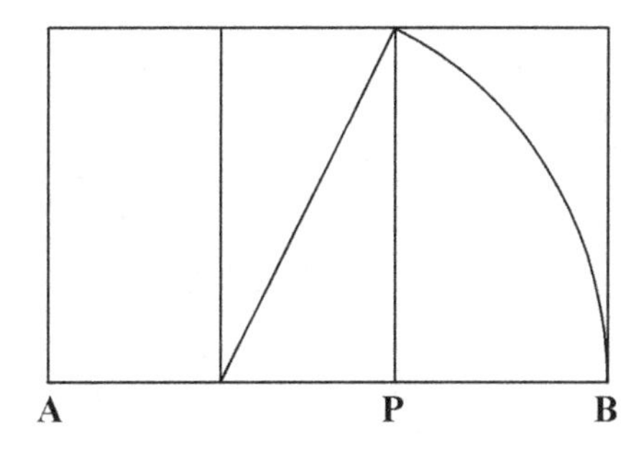

Fig. 34.3 The golden section. $AB : AP = AP : PB \cong 1.618$

1. The rectangle of the palace is 73 *braccia* long and 45 wide: 73/45 = 1.622; a close approximation to the irrational golden number, 1.618039. ... This number is the characteristic ratio of the golden section, a geometrical proportion in which a line AB is divided by P in such a way that AB/AP = AP/PB (Snijders 1985). In a rectangle of sides AB and AP, we can consider a square of side AP and another rectangle of sides AP and PB. This new rectangle is another rectangle of the golden section, divisible into a square and another rectangle, and so on (Fig. 34.3).

   At the beginning of thirteenth century Leonardo Pisano, also known as Fibonacci, in his *Liber Abaci* (Boncompagni 1857–1862), describes the numbers of the sequence that bears his name, obtained by the example of the rate of reproduction of a pair of rabbits. The numbers of the sequence are rational, and, as the sequence increases, the ratio between any two numbers in succession approximates ever more closely the golden number. The sequence is: 0, 1, 1, 2, 3, 5, 8, 13, 21, 34, 55, 89, 144... Mathematicians are familiar with another series, the Lucas series, which is constructed so that every element in the series is equal to the sum of the corresponding term in the Fibonacci series plus the second term following it (for example, the sixth term of the Lucas series—18—is equal to the sixth term of the Fibonacci series—5—plus the eighth—13). These sequences solve in a simple way the problem of finding rational approximations for the irrational golden section.
2. Looking at the palace plans (Fig. 34.4), I realized that the golden section, in the form of the Fibonacci sequence, is the starting point and the basic principle of the whole project. The ground floor consists of a large rectangular hall (the Sala d'armi), a Fibonacci rectangle of six cross-vaulted bays, and an almost-square courtyard; at the second floor, the perimeter of the Sala is divided into two rectangles, according to the golden section (Fig. 34.5).
3. A particular rule is followed in the plan of the tower of the palace. The tower is famous for the fact that, rising over the battlements, it leans out beyond the façade of the palace in a very bold way. A less known fact is that the orientations of its lower façades that lie within the palace are not the same as those of its façades that rise above the palace: the tower turns and changes its position (Fig. 34.6). This anomaly indicates that the building of the palace probably

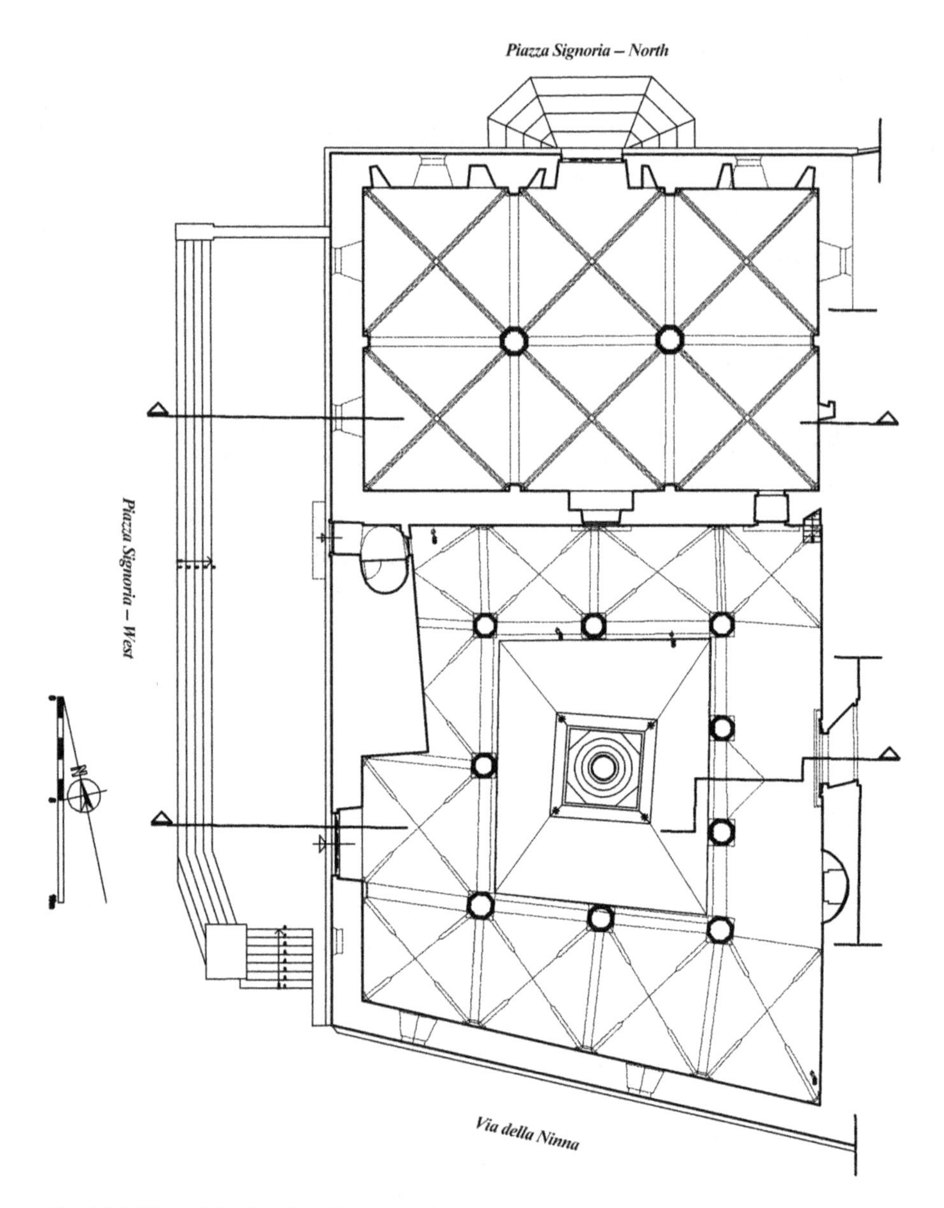

Fig. 34.4 Plan of the first floor. Image: author

started from the tower, then it was decided to change the program (the reason isn't easy to ascertain, but this is a different problem) and so the orientation of the palace was changed, but the inner basement of the tower was preserved. So we must consider the plan of the ground floor, along with the layout of the tower in its upper floors.

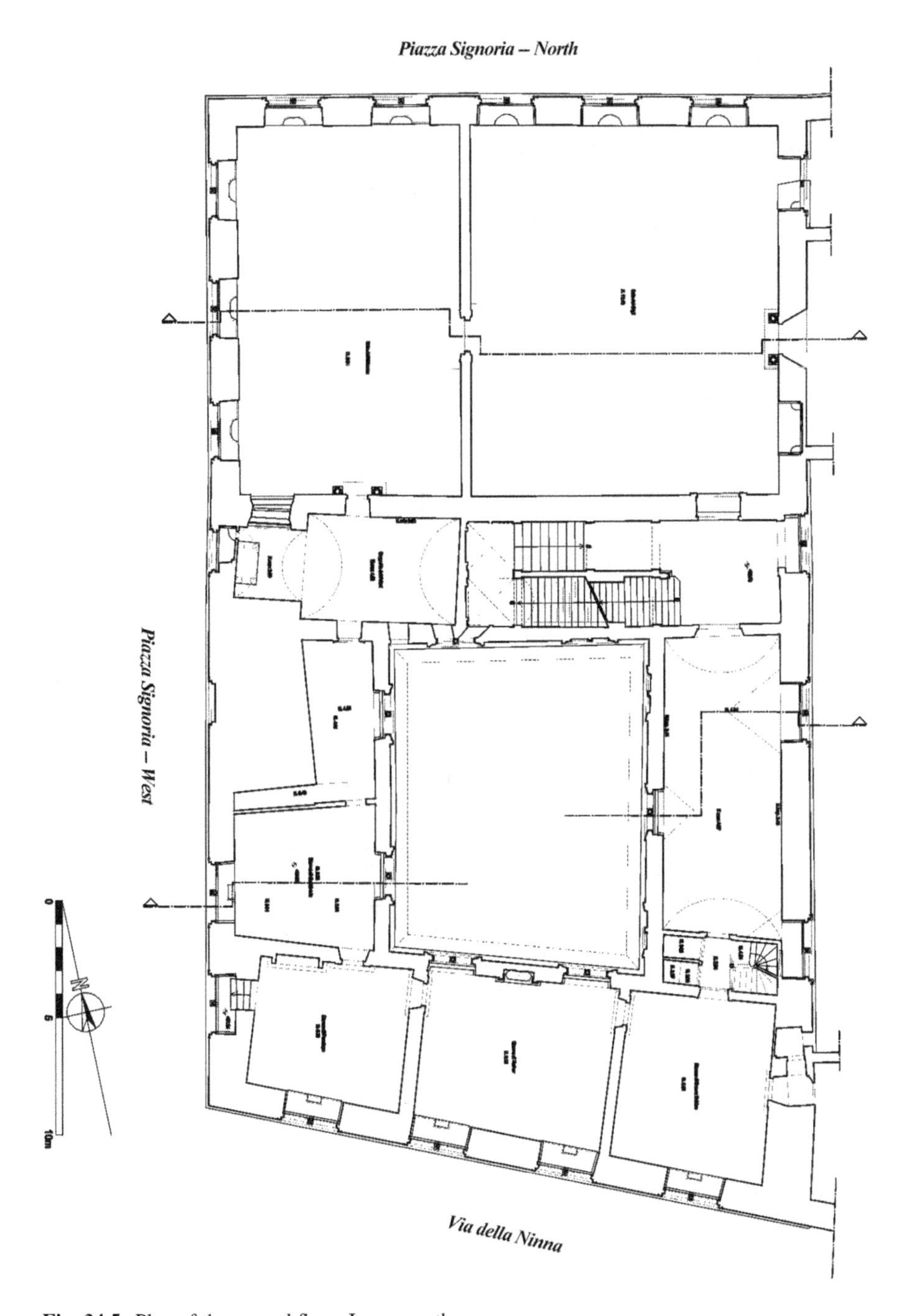

Fig. 34.5 Plan of the second floor. Image: author

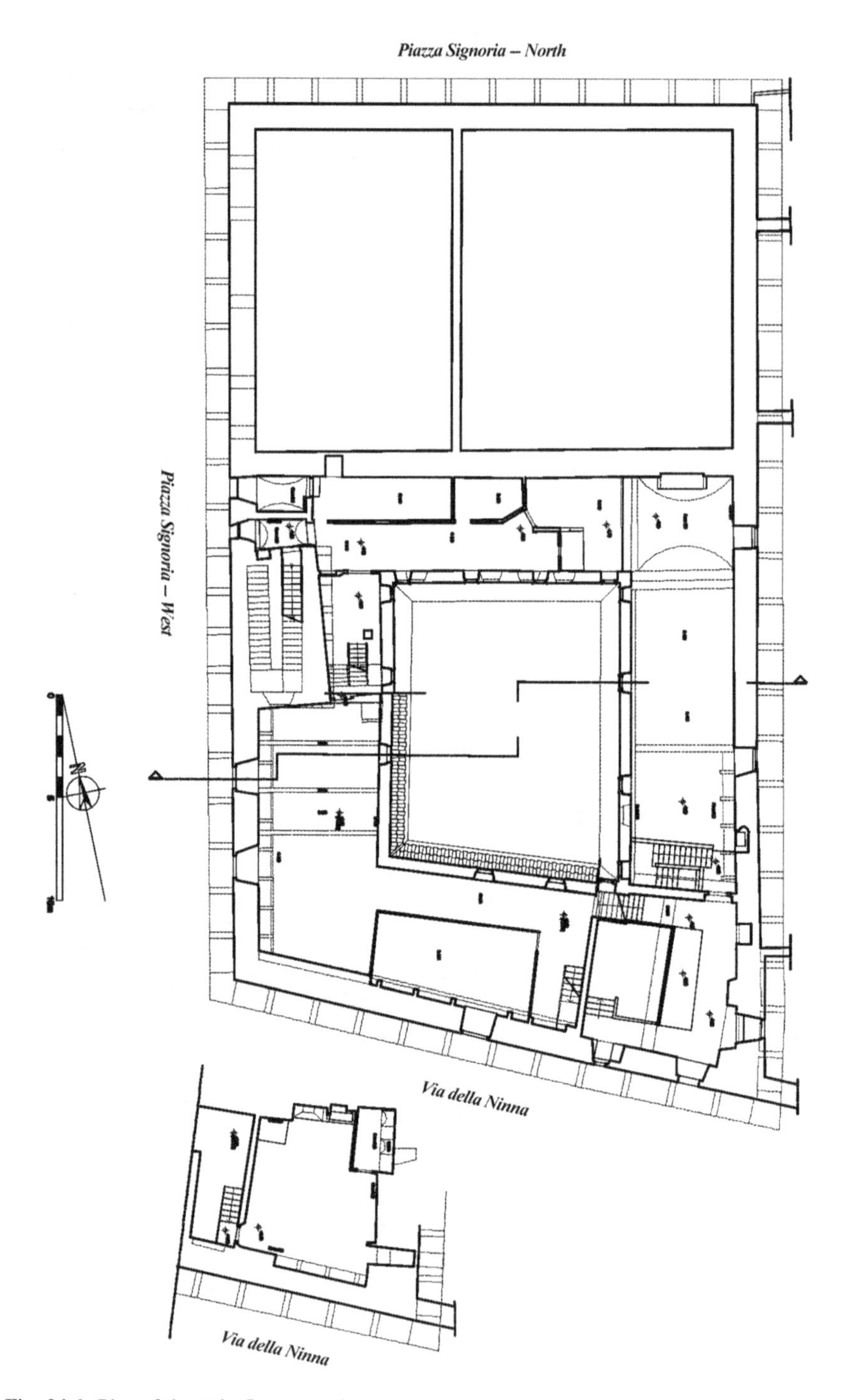

Fig. 34.6 Plan of the attic. Image: author

4. Now, referring to the sketch in Fig. 34.7, I will describe the final results of investigations but without explaining the individual stages, each one being very complicated to reach. Let us start from a couple of numbers in the Fibonacci sequence: 34 and 21. The beginning is a rectangle 68 *braccia* long, 42 *braccia* wide. A grid of 10 × 10 minor rectangles subdivide the major one. Take the 2 × 2 rectangle with one side on the shorter axis and the other one on the longer side of the mayor rectangle: that will be the place of the tower. The line on the right side of the 2 × 2 rectangle is the central axis of a square of 45 × 45 *braccia*, to which is added a rectangle of 28 × 45 *braccia* that can be considered as the addition of two Fibonacci rectangles, 21 × 34 and 7 × 11. Now we have the Fibonacci rectangle 45 × 73. These measurements are the lengths of the two façades of the palace. The total inner length of the west façade is 68 *braccia*. The distance between the outer side of the west façade and the inner side of the east wall is 42 *braccia*.
5. Now we must take into account the non-right angles of the plan. They certainly depend on the intention to have one side parallel to the existing direction of Via della Ninna. The angle between this direction and the Northern wall of the palace is 12°. Starting from the southern extreme of the western façade, we draw a line at such an angle, so marking out the trapezoid. The bisectors of the non-right corners determine the thicknesses of the western and southern walls, which is why their measurements are different and are not whole numbers. That is also why the corner that is the most structurally stressed has the thinnest walls.
6. In the plan of the palace we have a rectangle, a square, and a triangle. In the rectangle is placed the Sala d'Armi, whose ratio, described in the sketch in Fig. 34.7, is very simple. In the square plus the triangle lies the courtyard. The axes of the square become the composition lines of the courtyard. Their divisions determine the location of pillars and arches of bays. Superimposing the sketch in Fig. 34.7 over the ground floor plan, we can verify the congruence of our hypothesis (Fig. 34.8).

Now we go to the elevation, but this requires an introduction. Many years ago I took part in the survey of the Palazzo Strozzi, built at the end of the fifteenth century according to the designs of Giuliano da Sangallo and Antonio Pollaiuolo (Fig. 34.9). The measurements of its elevation seem to derive from the Fibonacci and Lucas sequences: it is exactly 68 *braccia* long and 55 *braccia* high, that is, two rectangles 34 × 55; from the street to under the cornice, it is 48 *braccia* high; it measures 47 *braccia* from the top of the bench (one *braccio* high) to beneath the cornice. The 47 *braccia* are divided into three parts: the lower part is 18 *braccia*, while the sum of the middle part (16 *braccia*), and the upper part (13 *braccia*) is 29. Numbers 18, 29, and 47 are found in the Lucas sequence; 34 and 55 are numbers of Fibonacci (Fig. 34.9). The sketch shows that other details of the elevation, for instance the ground floor windows, take their measurements from the two sequences. We must remember that Filippo Strozzi, who commissioned Palazzo Strozzi, had to demonstrate that it was smaller than the Palazzo della Signoria, in order to obtain the permission of Lorenzo de' Medici.

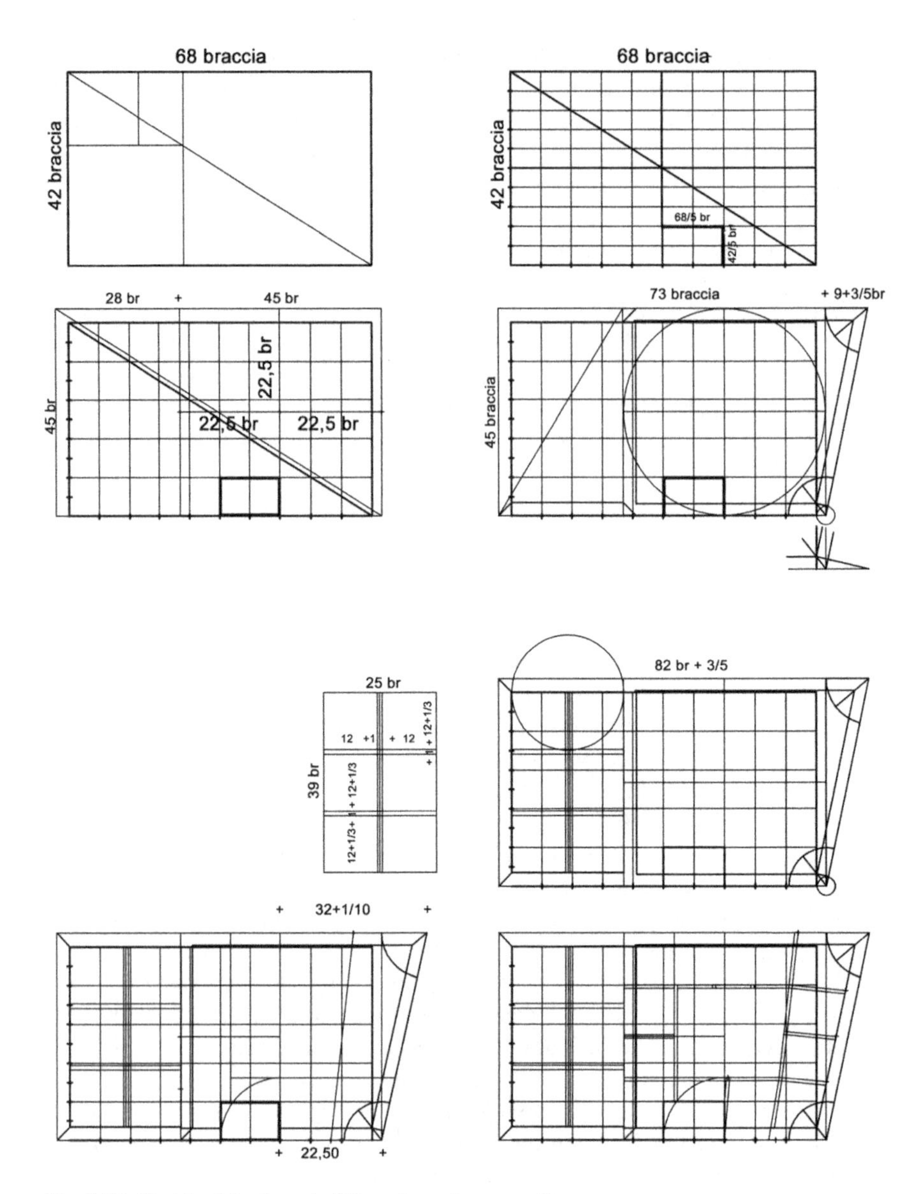

Fig. 34.7 Sketch of the layout of the palace. Image: author

7. Looking at the elevation of the Palazzo della Signoria, we can remark the following numbers (Fig. 34.10): from the level of the entrance to the level of the starting landing of the tower stairs, the height is 45 *braccia*; from the lower landing of this staircase to its top on the first crenellated floor of the tower, the height is 73 *braccia* (219 + 1 steps). These are the same measurements found in the plan. The façade is divided vertically into three parts, on the top of which is placed a crenellated, projecting, gallery. The three parts, the first of 19, the

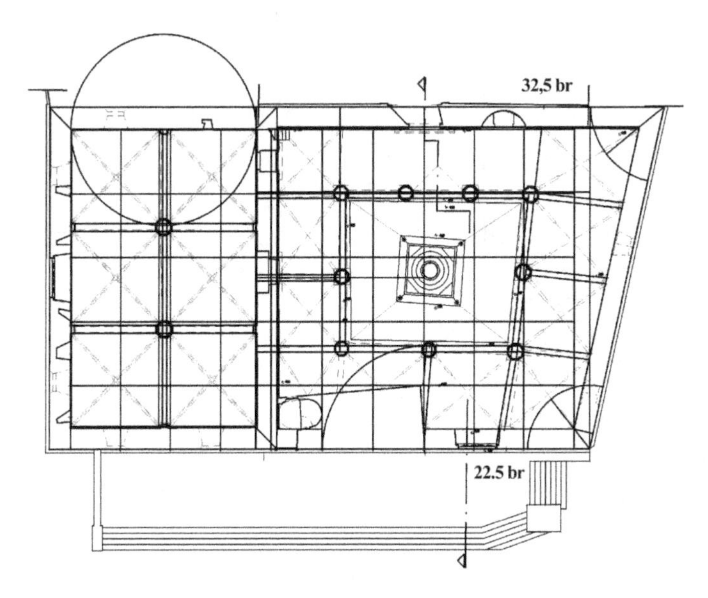

Fig. 34.8 The sketch put on top of the first floor plan. Image: author

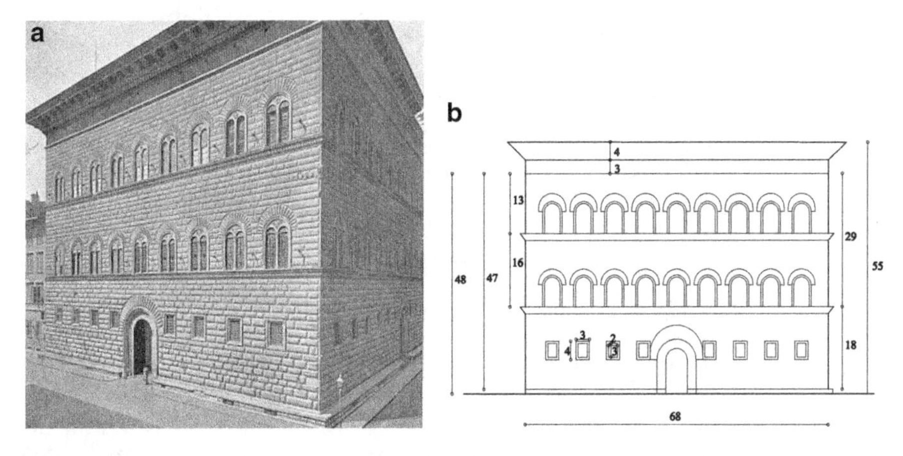

Fig. 34.9 The Palazzo Strozzi (photo, *above*) and the schematic diagram of the façade (*below*). Image: author

second also 19, the third of ten *braccia*, add up to 48 *braccia*. But the first palace façade (of which we have paintings and literary evidence: the fresco on the *Expulsion of the Duke of Athen from Florence* and the *Cronaca* of Giovanni Villani (1991) about political events under the Duke of Athens) had a bench of two stairs above which the height was probably of 47 *braccia*, the sum of 18 + 29, the same numbers that appear in Palazzo Strozzi, but nearly two hundred years earlier. In the Lucas sequence 18, 29, and 47 are followed by 76. The crenellated gallery is a little more than 77 *braccia* long, but it would be very difficult to reach exactly the number 76 because of the projection from the

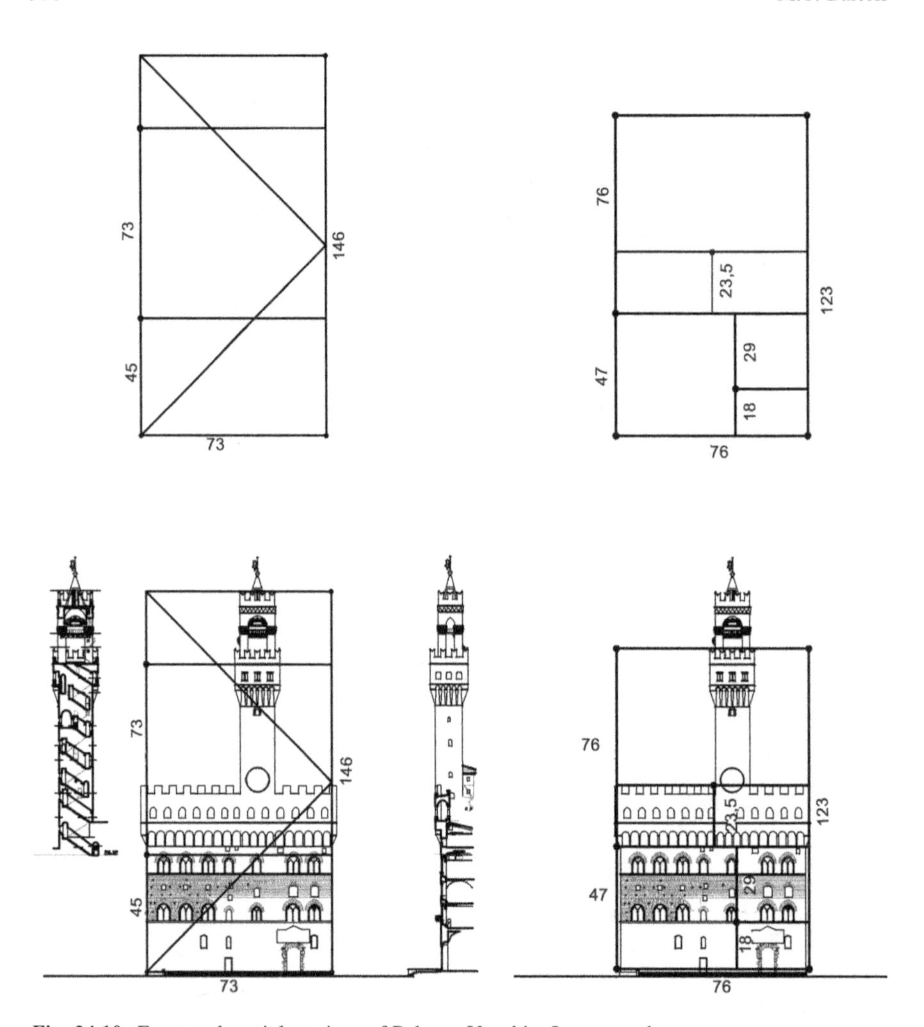

Fig. 34.10 Front and partial sections of Palazzo Vecchio. Image: author

four trapezoidal sides, with non-right angles. Measurement tends toward the number 76, even if they don't reach it exactly; the height of the gallery, 47/2 *braccia*, confirms the intention. The highest points of the tower, on the final embattlements, are 146 *braccia* high on the entrance, that is, 73 $\times$ 2. Thus the western façade is inscribed in two squares, while the northern façade is inscribed in two Fibonacci rectangles.

Conclusions

I hope to have given evidence of the use of the numbers belonging to the Fibonacci and Lucas sequences by the architect of the Palazzo della Signoria. This use had two different aspects. First, it was substantial, and connected with the idea that the palace might possibly have to be enlarged and divided in a progressive and uniform way, preserving the same shape. On the other hand, it was instrumental, connected with the necessity to work out a set of numbers linked by a rule, therefore easy to remember, to control, and to communicate.

According to proceedings of scholasticism, substance and form have the same rule; the scientific paradigm of the sequence gives the solution both for the architectural pattern of a Medieval town hall, and the technical way of arranging its measurements.

Biography Architect Maria Teresa Bartoli has been *Professore Ordinario of* Architectural Surveying since 2002 at the Faculty of Architecture in Florence where she has worked since 1983 first as a *ricercatore* then as *Professore associato* teaching drawing surveying and geometry for architects. She has taken part (either as a team member or as the leader) in the survey of important monuments in Florence such as Palazzo Strozzi Palazzo Medici Palazzo Vecchio the Convents and Churches of Carmine of Santa Maria Novella and of Ognissanti Villa Medici in Careggi the Hospital of San Giovanni di Dio on behalf of the Public Administration or their holders. Her research fields include the history of Renaissance perspective the links between architecture and geometry metrology and history of geometrical paradigms in architecture.

References

Boncompagni, B. 1857-62. *Scritti di Leonardo Pisano, matematico, pubblicati da Baldassarre Boncompagni*. Roma.
Romanini, A. M. 1980. *Arnolfo di Cambio*. Firenze: Sansoni.
Snijders, C. J. 1985. *La sezione aurea*. Padova: Muzio Editore.
Villani, G. 1991. *Nuova Cronica*. G. Porta, ed. Parma: Guanda Editore

Chapter 35
What Geometries in Milan Cathedral?

Elena Marchetti and Luisa Rossi Costa

Introduction

For many years we have promoted connections between mathematics and architecture, therefore it now seems mandatory to analyse with mathematical eyes our own Milan Cathedral, the Duomo, one of the most important and symbolic monuments of our city. In fact, tourists are used to hearing that the Duomo is one of the three symbols of the town, together with *The Last Supper* and *La Scala*.

The Duomo, whose construction started at the end of the fourteenth century, is a surprising monument (Fig. 35.1). Although it is considered Gothic, it is quite different from the traditional Gothic cathedrals found in northern European countries (Brivio and Majo 1980).

At first glance everybody understands that the planning does not follow the canons of the famous French, English or German Gothic monuments, but the result is very harmonious and pleasing. Designers, architects, engineers, artists and construction workers, members of the so-called *Veneranda Fabbrica del Duomo*, which still manages the building today, were probably inspired by both the contemporary culture and by their origins and traditions.

No individual architect or engineer who is credited with having planned the building, and many sculptors and craftsmen worked together during long years of construction; their common local traditions probably resulted in the harmonious elegance of the interior and exterior. In any case, we can confirm that the

First published as: Elena Marchetti and Luisa Rossi Costa, "What Geometries in Milan Cathedral?", pp. 63–76 in *Nexus VI: Architecture and Mathematics,* Sylvie Duvernoy and Orietta Pedemonte, eds. Turin: Kim Williams Books, 2006.

E. Marchetti (✉) • L. Rossi Costa
Department of Mathematics, Politecnico di Milano, Piazza Leonardo da Vinci, 32, 21033 Milan, Italy
e-mail: elena.marchetti@polimi.it; luisa.rossi@polimi.it

K. Williams and M.J. Ostwald (eds.), *Architecture and Mathematics from Antiquity to the Future*, DOI 10.1007/978-3-319-00137-1_35,

Fig. 35.1 Photo: authors

proportions among parts were studied and respected appears evident, for example, in a drawing by Cesare Cesariano (Fig. 35.2).

Year after year, workers combined Gothic forms with traditional Italian shapes. For example, the gable-like Renaissance façade (*a capanna,* typical of the Romanesque style) is in harmony with all the unusual Gothic elements. Nevertheless the monument is popular with everybody!

It would be interesting to know the secret of such harmony! For now we must be content to analyse the structure of some of its geometrical parts, such as the rose windows (Pirina 1986) and the decorated pavement.

After a brief historical description of the Cathedral in sections "A Short Historical Presentation," "A Mathematical Approach in Describing Symmetries" we introduce some mathematical tools suitable for describing symmetries. We especially point out the rose windows and introduce the notion of *group*, as well of cyclic and dihedral groups linked to rotational symmetries (Weyl 1952). The mathematical description is completed with particular examples and virtual reconstructions of recognisable parts of the Cathedral. In section "Analysis of Some Geometrical Decorations" we describe lines and forms found in the pavement and in the windows. We chose to underline these geometrical aspects because they appear only in this artistic context but in other different applied fields (physics, mechanical engineering, etc.) as well (Dedò 1999).

We conclude that Milan Cathedral never ceases to amaze us: it is not only a splendid monument rich in art and elegant architecture, but also a good collection of mathematical examples hidden within its bewildering decorations.

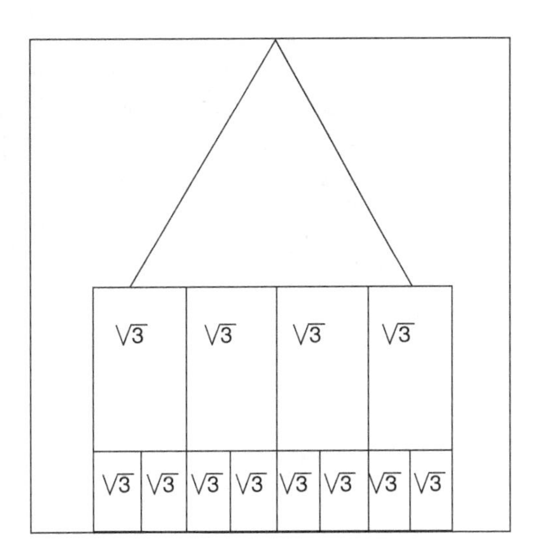

Fig. 35.2 Image: authors

A Short Historical Presentation

The construction of the Duomo went on for very long time, beginning in 1386 as recorded on a tombstone on a wall. Cardinal Carlo Borromeo consecrated the Cathedral in 1577, nearly two centuries after it was begun. The existing façade was completed in the nineteenth century and many details were added in the last century. The very important structural refurbishing of the main pillars was realised in the second half of the twentieth century.

As we said in the introduction, no single architect is credited with having planned the Milan Cathedral but many important figures (included Leonardo da Vinci and Bramante) worked in it, with surprising results. Among others, we mention Simone di Orsenigo, named first general engineer in 1387, and Cesare Cesariano. In 1521 Cesariano, holding to the canons of Vitruvius, designed the *Scenographia* (Fig. 35.3), a vertical section of the Duomo that allows a statue of the Virgin Mary to be placed on the main central spire.

Among the craftsmen we note particularly the sculptors known as *Maestri Campionesi*, who carved the white-pink marble of Candoglia, the main material in the Duomo. The Candoglia quarry is situated in the northwest of Milan, close to the west side of Lake Maggiore. The marble was brought in on barges travelling over rivers, lakes, and canals, such as the famous *Navigli* projected by Leonardo. A special dock was created close to the construction site, in the area where now you find *Via Laghetto* ("small lake"). The Candoglia quarry is actually the property of *La Veneranda Fabbrica* and its marble is employed only for the restoration of the Duomo.

We are still convinced that the *Veneranda Fabbrica*'s workers have to have had a strong professional approach and profound artistic and scientific knowledge, in order to have realized the remarkable aesthetic of the monument.

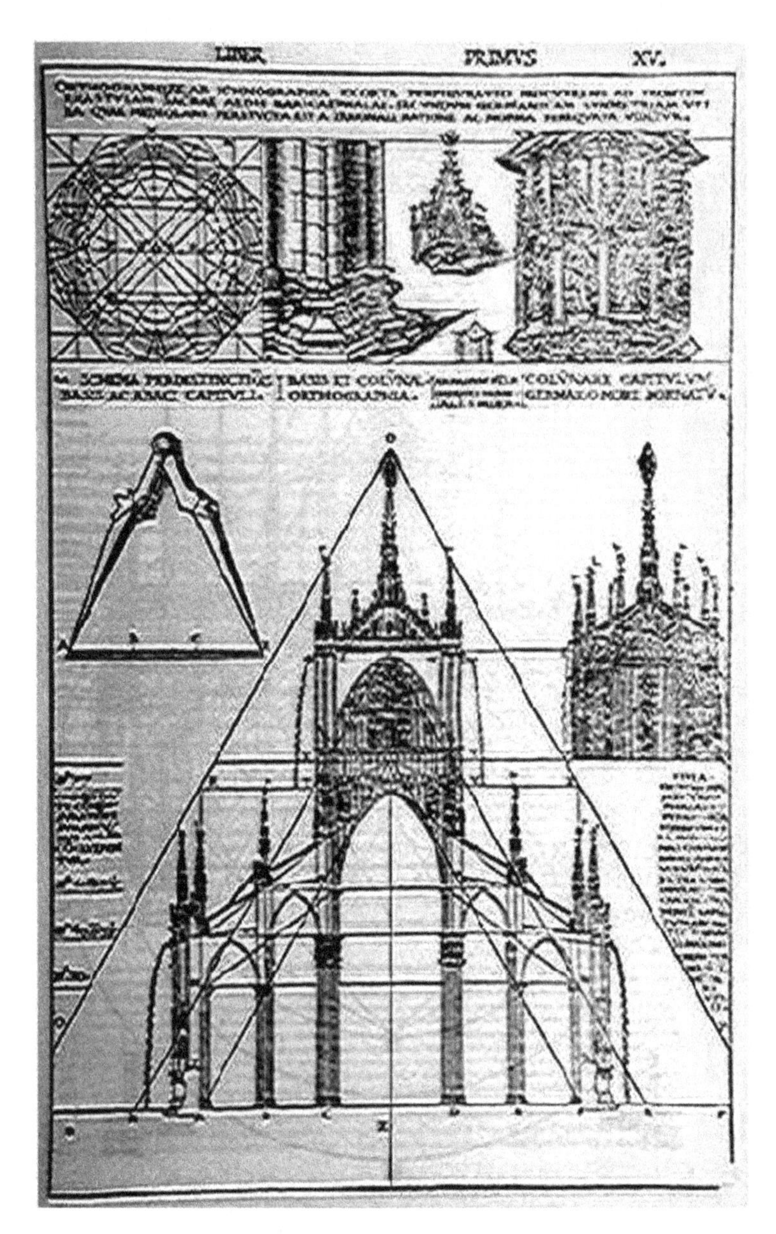

Fig. 35.3

One of the more exciting parts of visiting the Cathedral is to climb the spiral stairs to the roof in a sunny day, to find yourself in a forest of pinnacles with the delighted view of the Alps' snowy peaks. The mathematics that we talk about here provides another key for reading the beautiful sculpture of the Duomo's decoration (Fig. 35.4).

Fig. 35.4 Photo: authors

Construction on the Duomo is still underway even today, and in Milan we have a saying about projects that never end, "It looks like *la fabbrica del Duomo*!"

A Mathematical Approach in Describing Symmetries

In this section we will introduce some of the necessary mathematical tools for analysing symmetries in decorations, with the aim of exploring the Milan Cathedral with mathematical eyes.

Cyclic and Dihedral Figures As well known a general affine transformation of the Oxy Cartesian plane points can be algebraically represented by the following notation

$$\mathbf{v}' = \mathbf{A}\mathbf{v} + \mathbf{h},$$

where $\mathbf{v} = \begin{bmatrix} x \\ y \end{bmatrix}$, $\mathbf{v}' = \begin{bmatrix} x' \\ y' \end{bmatrix}$ are vectors corresponding to the point $P(x,y)$ and its image $P'\ (x',y')$.

As usual $\mathbf{A}(2,2)$ and $\mathbf{h}(2,1)$ are square matrix and column vector respectively, characterizing the transformations.

For our purposes we deal with the following matrices and vectors:

- $\mathbf{A} = \begin{bmatrix} \cos\vartheta & -\sin\vartheta \\ \sin\vartheta & \cos\vartheta \end{bmatrix}$ and $\mathbf{h} = \mathbf{0} = \begin{bmatrix} 0 \\ 0 \end{bmatrix}$ resulting in a rotation (anticlockwise) centred in O of an angle $\vartheta \in [0, 2\pi)$, (in our context often $\vartheta = 2\pi/n$, n positive integer).

The choice $\vartheta = 0$ gives the *identity*.
The choice $\vartheta = \pi$ refers to the symmetry with respect to the centre O.

- $\mathbf{A} = \begin{bmatrix} \cos 2\vartheta & \sin 2\vartheta \\ \sin 2\vartheta & -\cos 2\vartheta \end{bmatrix}$ and $\mathbf{h} = \mathbf{0}$ resulting in a symmetry (or reflection) with respect to a straight line—*symmetry axis*—through O, forming an angle ϑ ($\vartheta \in [0, \pi)$) with the x-axis.

The choice $\vartheta = 0$ or $\vartheta = \frac{\pi}{2}$ refers to the symmetry with respect to x-axis or y-axis, respectively.

- $\mathbf{A} = \begin{bmatrix} \lambda \cos \vartheta & -\lambda \sin \vartheta \\ \lambda \sin \vartheta & \lambda \cos \vartheta \end{bmatrix}$ and $\mathbf{h} = \mathbf{0}$ resulting in rotation and scaling (ϑ angle of rotation, $\lambda > 0$ scaling factor).

The vector $\mathbf{h} \neq \mathbf{0}$ adds a translation to the transformation represented by the matrix $\mathbf{A}$.

For a mathematical description of forms let us characterize plane symmetric figures:

- a figure presents only *cyclic symmetry* when it has no symmetry axis, but rather a centre of symmetry; consequently suitable rotations transpose the figure onto itself (Fig. 35.5);
- a figure presents a *dihedral symmetry* when it has at least one symmetry axis; consequently a reflection, with respect to that axis, transpose the figure onto itself (Fig. 35.6);
- a figure with more than one symmetry axis has a *centre of symmetry* (the intersection of the axes); consequently appropriate reflections and rotations transpose the figure onto itself (Fig. 35.7);
- a *dihedral figure* is also a *cyclic figure* (but not vice-versa) (Loria 1930).

In the following C_n denotes figures presenting only cyclic symmetry, without symmetry axes, and transformed in their self by n rotations around a centre ($\vartheta = 2\pi/n$ rotation angle, n positive integer). In particular C_1 denotes figures having any symmetry at all (Fig. 35.8).

D_n denotes figures having n symmetry axes, that is, dihedral symmetry. Consequently n reflections with respect to the axes transpose the figure onto itself. A D_n figure is also cyclic.

The C_n patterns can be generated by n rotations of a basic motif about the centre. D_n patterns can also be generated by n reflections of a basic motif with respect to the axes.

To operate algebraically it is convenient insert the pattern into an *Oxy* Cartesian system, superimposing the centre of the figure (if present) on the origin O and one of the symmetry axes (if present) on to one of the Cartesian axes.

In this way each point of the figure is represented by a $\mathbf{v}(2,1)$ vector, the rotations and/or reflections are realized by matrices $\mathbf{A}(2,2)$.

Fig. 35.5 Photo: authors

Fig. 35.6 Photo: authors

Dihedral and cyclic figures are aesthetically important: they can be found in furniture and in buildings and are frequently visible in artistic decorations.

In Milan Cathedral many C_n or D_n figures are recognisable. We focus on the rose windows (*rosoni*) present in the big glass windows and on cyclic or dihedral figures in the pavement.

Finite Groups of Symmetry We introduce now the notion of *group* for a succinct description of cyclic and dihedral figures.

A *group* is a set T provided with a binary operation "$\circ$", satisfying the following properties:

- $\forall\, t_1, t_2 \in T \Rightarrow t_1 \circ t_2 \in T$ (the set is closed under the operation $\circ$);
- $\forall\, t_1, t_2, t_3 \in T \Rightarrow t_1 \circ (t_2 \circ t_3) = (t_1 \circ t_2) \circ t_3$ (the operation $\circ$ is associative);

Fig. 35.7 Photo: authors

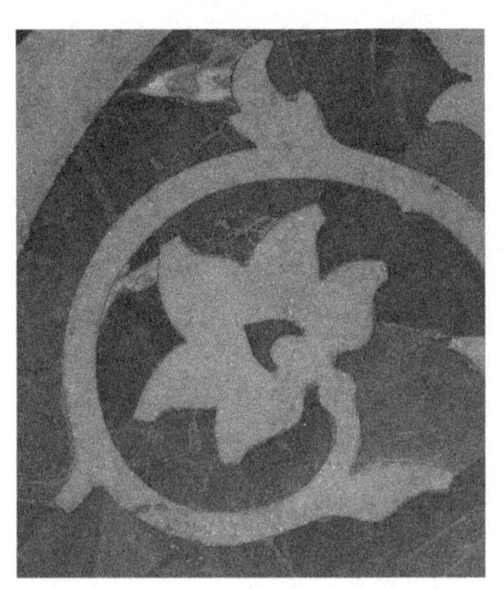

Fig. 35.8 Photo: authors

- $\forall\, t \in T, \exists\ u \in T$ such that $t \circ u = u \circ t = t$ (u is called unit or identity element);
- $\forall\, t \in T, \exists\ t' \in T$ such that $t \circ t' = t' \circ t = u$ (each element t has an inverse t′).

A group with a finite number of elements is said *finite*.

The group T is *commutative* if $t_1 \circ t_2 = t_2 \circ t_1, \quad \forall\, t_1, t_2 \in T$.

The set T of all transformations realizing the symmetries in C_n or D_n figures, forms a finite group, called *finite symmetry group*. In particular considering a pattern in a Oxy Cartesian plane, T can be related to a set of matrices $\mathbf{A}(2,2)$

acting on the vectors $\mathbf{v}(2,1)$ linked to the points of the pattern. The composition law $\circ$ between two elements t_1,t_2 belonging to the group T, has to be intended like the application of the operation t_1 to the result of the operation t_2, that is, $(t_1 \circ t_2)(\mathbf{v}) = t_1(t_2(\mathbf{v}))$.

Following (Budden 1972) we present the structure of some finite groups connected with symmetric forms in Milan Cathedral. We adopt the following conventions:

a. the identity i always appears first, both across and down;
b. the order in which the elements appear shall be the same across as down;
c. to find the product $t_1 \circ t_2$, we take t_1 in row and t_2 in column, i.e., the first operation is given across, the second operation is given down.

Each transformation corresponds to the matrix adequate for the algebraic realisations of the operation itself.

(1) C_1 (Fig. 35.8) is transformed in itself by identity, intended also as rotation r_ϑ, $\vartheta = 0 \pmod{2\pi}$. In the following all the angles are given (mod 2π).

Let us denote with $\mathbf{C_1} = \{i\}$ the set of transformations of C_1, the trivial case with only one element.

$\mathbf{C_1}$	i
i	i

(2) D_1 (Fig. 35.6) is transformed in itself by the identity and by a reflection s_1 with respect to the symmetry axis. Let us denote with $\mathbf{D_1} = \{i, \quad s_1\}$ the set of transformations of D_1 whose compositions can be represented in the following table. We notice that $\mathbf{C_1} \subset \mathbf{D_1}$.

$\mathbf{D_1}$	i	s_1
i	i	s_1
s_1	s_1	i

(3) C_2 (Fig. 35.9) is transformed in itself by the identity and the rotation r_ϑ, with $\vartheta = \pi$. Let us denote with $\mathbf{C_2} = \{i, \ r_\pi\}$ the set of transformations of C_2 whose compositions can be represented in the following table:

$\mathbf{C_2}$	i	r_π
i	i	r_π
r_π	r_π	i

Fig. 35.9 Photo: authors

(4) D_2 (Fig. 35.10) is transformed in itself by the identity, by a rotation r_ϑ with $\vartheta = \pi$, and by two reflections s_1, s_2 with respect to two symmetry axes. Let us denote with $\mathbf{D_2} = \{\, i, \quad r_\pi, \quad s_1, \quad s_2 \,\}$ the set of the four transformations of D_2 whose compositions can be represented in the following table; it appears that the group $\mathbf{D_2}$ is commutative. It is easy to verify that $\mathbf{C_2} \subset \mathbf{D_2}$.

$\mathbf{D_2}$	i	s_1	s_2	r_π
i	i	s_1	s_2	r_π
s_1	s_1	i	r_π	s_2
s_2	s_2	r_π	i	s_1
r_π	r_π	s_2	s_1	i

(5) C_3 (Fig. 35.11) is transformed in itself by the identity and by rotations r_ϑ ($\vartheta = 2\pi/3$ and $\vartheta = 4\pi/3$). Let us denote with $\mathbf{C_3} = \{i, \; r_{2\pi/3}, \; r_{4\pi/3}\}$ the set of transformations of C_3 whose compositions can be represented in the following table:

Fig. 35.10 Photo: authors

Fig. 35.11 Photo: authors

$\mathbf{C_3}$	i	$r_{\frac{2\pi}{3}}$	$r_{\frac{4\pi}{3}}$
i	i	$r_{\frac{2\pi}{3}}$	$r_{\frac{4\pi}{3}}$
$r_{\frac{2\pi}{3}}$	$r_{\frac{2\pi}{3}}$	$r_{\frac{4\pi}{3}}$	i
$r_{\frac{4\pi}{3}}$	$r_{\frac{4\pi}{3}}$	i	$r_{\frac{2\pi}{3}}$

(6) D_3 (Fig. 35.7) is transformed in itself by the identity, by rotations r_ϑ ($\vartheta = 2\pi/3$ and $\vartheta = 4\pi/3$) and by reflections s_1, s_2, s_3 with respect to three symmetry axes. Let us denote with $\mathbf{D_3} = \left\{ i, \quad s_1, s_2 \quad, s_3 \quad, r_{\frac{2\pi}{3}}, r_{\frac{4\pi}{3}} \right\}$ the set of transformations of D_3 whose compositions can be represented in the following table:

$\mathbf{D_3}$	i	s_1	s_2	s_3	$r_{\frac{2\pi}{3}}$	$r_{\frac{4\pi}{3}}$
i	i	s_1	s_2	s_3	$r_{\frac{2\pi}{3}}$	$r_{\frac{4\pi}{3}}$
s_1	s_1	i	$r_{\frac{4\pi}{3}}$	$r_{\frac{2\pi}{3}}$	s_3	s_2
s_2	s_2	$r_{\frac{2\pi}{3}}$	i	$r_{\frac{4\pi}{3}}$	s_1	s_3
s_3	s_3	$r_{\frac{4\pi}{3}}$	$r_{\frac{2\pi}{3}}$	i	s_2	s_1
$r_{\frac{2\pi}{3}}$	$r_{\frac{2\pi}{3}}$	s_2	s_3	s_1	$r_{\frac{4\pi}{3}}$	i
$r_{\frac{4\pi}{3}}$	$r_{\frac{4\pi}{3}}$	s_3	s_1	s_2	i	$r_{\frac{2\pi}{3}}$

It is evident that $\mathbf{C_3} \subset \mathbf{D_3}$.

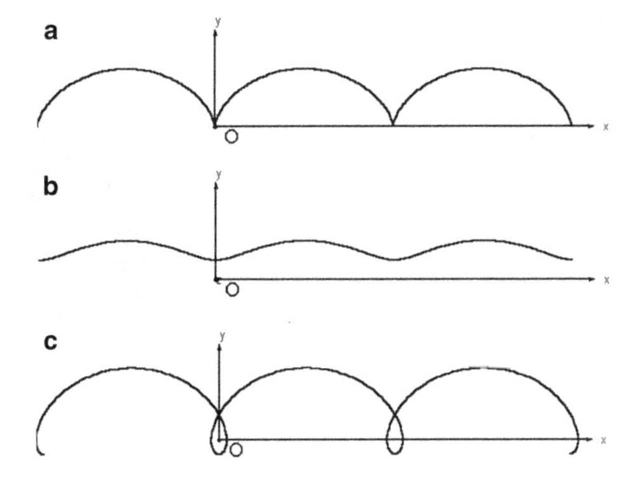

Fig. 35.12 (**a, b, c**). Image: authors

It becomes easy to compose in the same way the tables for any other $\mathbf{C_n}$ or $\mathbf{D_n}$ pattern.

A transformation group is commutative if and only if its table (read like a matrix) is symmetric with respect to the principal diagonal.

Analysis of Some Geometrical Decorations

The Milan Cathedral provides the occasion to discover other mathematical peculiarities: meaningful curves in the inlaid marble pavement decorations or the decorative geometric figures in the windows present properties used even in mechanical engineering.

More precisely we note the cycloid, other related lines such as epicycloids and hypocycloids, and the *Reuleaux* polygons. We identified some of them in the Duomo but all these forms are easily recognisable in many other cathedrals or monuments.

Cycloid, Epicycloids, Hypocycloids The *cycloid* is the curve traced by a point P on the edge of a circle γ (radius r) rolling along a straight line, without slipping or stopping (Fig. 35.12a).

Starting in the sixteenth century many mathematicians and physicians, such as Galilei, Bernoulli, Leibniz and Newton, investigated the numerous peculiarities of the cycloid (Kline 1996). Among other properties we mention that this curve is the solution of the *brachistochrone problem*: that is the cycloid is the curve minimising the travel time of a bead travelling between two points (not on the same vertical line), frictionless and influenced only by the gravity.

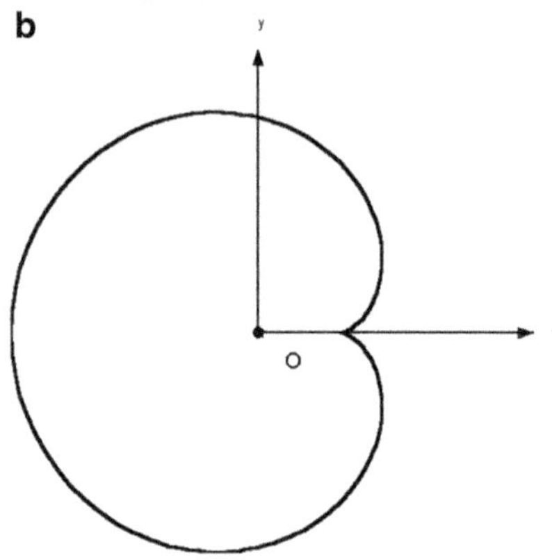

Fig. 35.13 (**a, b**). Photo and Image: authors

The *epicycloids* are curves described by P when the circle γ (radius r) rolls around a circle γ' (radius a) with $r \leq a$. We represent some epicycloids related to the choice of the ratio $q = \frac{a}{r}$, together with pictures of elements of the Duomo in which similar curves are recognizable:

- if $q = 1$ the curve is known as *cardioid* (Fig. 35.13);
- if $q = 2$ the curve is known as *nephroid* (Fig. 35.14);
- if $q = m,\ m = 4, 8, 20$ you can find similar curves in the Duomo decorations (Figs. 35.15, 35.16, 35.17);
- if $q = \frac{3}{2}$, the form is evident in some Duomo windows (Fig. 35.17).

The *hypocycloids* are curves described by P when the circle γ (radius r) rolls inside a circle γ' (radius a) being $r \leq a$ (Fig. 35.18).

We mention some particular cases related to the choice of the ratio $q = \frac{a}{r}$:

- if $q = 1$ the hypocycloid becomes a point;
- if $q = 2$ the curve image is a diameter of γ';
- if $q = 3$ the hypocycloid is known as a *deltoid*;

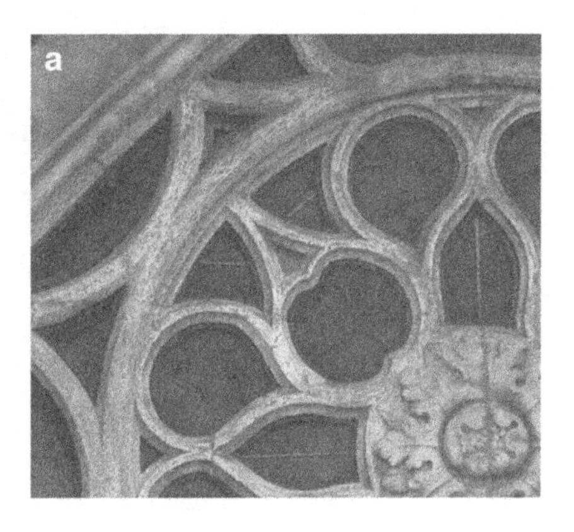

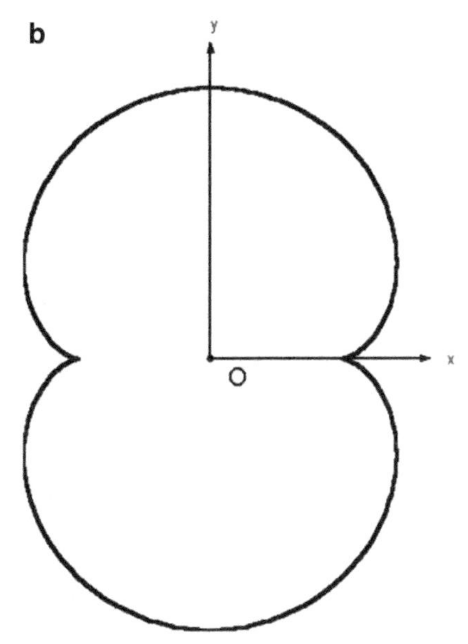

Fig. 35.14 (**a, b**). Photo and Image: authors

- if $q = 4$ the hypocycloid is known as an *astroid,* as seen in the Duomo decoration shown in Fig. 35.19;
  if $q = 5$ the curve is related to the form of a starfish;
  if $q = 8/3$ the curve corresponds to a Duomo rose window shown in Fig. 35.20.

Vector parametric equations of the curves described, with reference to an appropriate *Oxy* Cartesian system, are the following:

$$\mathbf{v}_{cycl} = \begin{bmatrix} r\ (\mathrm{t}\text{-}\sin \mathrm{t}) \\ r\,(1 - \cos t) \end{bmatrix}, \quad t \in R,$$

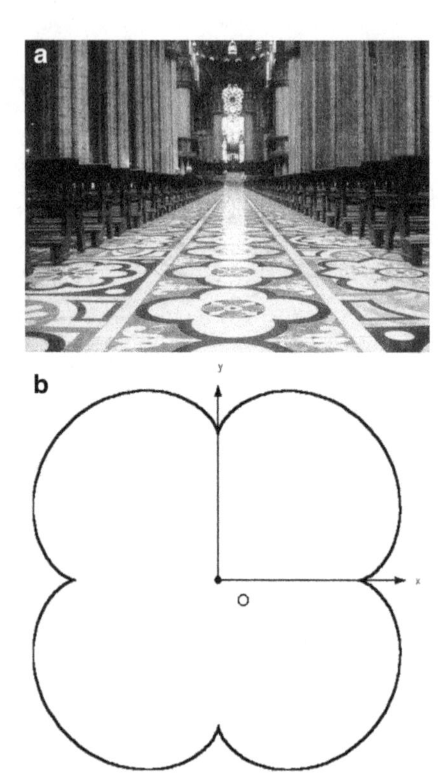

Fig. 35.15 (**a, b**). Photo and Image: authors

$$\mathbf{v}_{epi} = \begin{bmatrix} (a+r)\cos t - r\cos\left(\dfrac{a+r}{r}t\right) \\ (a+r)\sin t - r\sin\left(\dfrac{a+r}{r}t\right) \end{bmatrix}, \quad t \in R, (a \geq r > 0),$$

$$\mathbf{v}_{hypo} = \begin{bmatrix} (a-r)\cos t - r\cos\left(\dfrac{a-r}{r}t\right) \\ (a-r)\sin t - r\sin\left(\dfrac{a-r}{r}t\right) \end{bmatrix}, \quad t \in R, (a \geq r > 0).$$

There are different variations in the cycloid path, if the point P has distance d ($d \neq r$) from the γ centre but it is consistent with the circle; thus a general vector equation:

$$\mathbf{w}_{tro} = \begin{bmatrix} r\,\mathrm{t} - d\sin \mathrm{t} \\ r - d\cos t \end{bmatrix}, \quad t \in R.$$

In Fig. 35.12b–c we show the paths (*trochoids*) traced out by a fixed point closer ($d < r$) or farer ($d > r$) from the centre of the circle respectively.

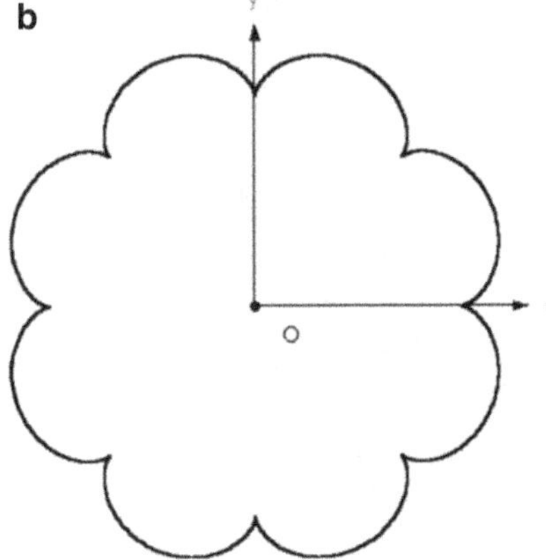

Fig. 35.16 (**a, b**). Photo and Image: authors

Let us consider the circle γ (radius r) rolling outside a circle γ' (radius a, and $q = \frac{a}{r} > 1$). Let us fix a point P inside or outside γ, consistent with it. The trace of P is a curve belonging to the *epitrochoids* family; thus a epitrochoid vector equation, where d is the distance of P from the centre of γ':

$$\mathbf{w}_{epit} = \begin{bmatrix} (a+r)\cos t - d\cos\left(\dfrac{a+r}{r}t\right) \\ (a+r)\sin t - d\sin\left(\dfrac{a+r}{r}t\right) \end{bmatrix}, t \in R, (a \geq r > 0,\ d > 0).$$

In Fig. 35.21 you see one example of epitrochoid corresponding to $q = 2$ and $d < r$.

Starting from the hypocycloids but fixing the point P inside or outside the circle γ, consistent with it, we obtain curves named *hypotrocoids*. A vector equation of the family curves is:

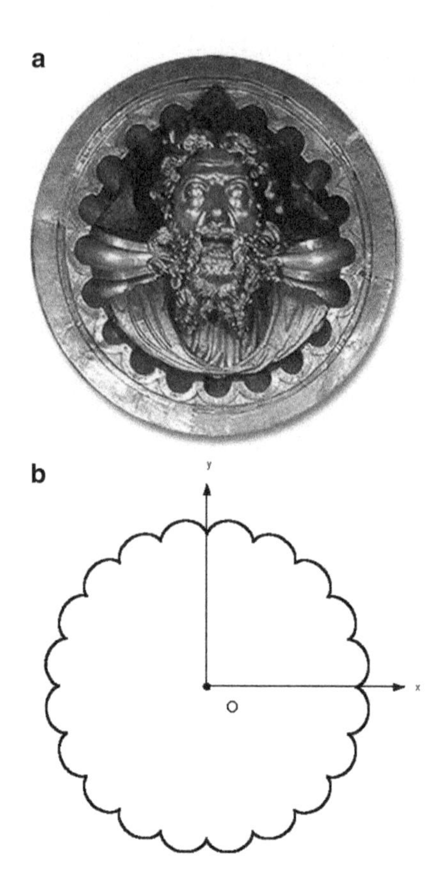

Fig. 35.17 (**a, b**). Photo and Image: authors

$$\mathbf{w}_{hypot} = \begin{bmatrix} (a-r)\cos t + d\cos\left(\dfrac{a-r}{r}t\right) \\ (a-r)\sin t - d\sin\left(\dfrac{a-r}{r}t\right) \end{bmatrix},\ t \in R, (a \geq r > 0,\quad d > 0).$$

For an interesting visualization of all the mentioned curves we refer to the website of Ferréol and Mandonnet (Ferréol and Mandonnet 2005).

The Reuleaux Triangle The circle is the simplest curve of constant *width*: in each direction the maximum of the distance between two points belonging to it is the diameter's length. Consequently a circle can rotate between two parallel straight lines having distance equal to the diameter.

There are many other non-circular curves having constant *width*, just discovered at the time of Leonardo da Vinci and mentioned by Euler (Boyer 1968; Kline 1996): one of these is the *Reuleaux triangle* (Fig. 35.22b).

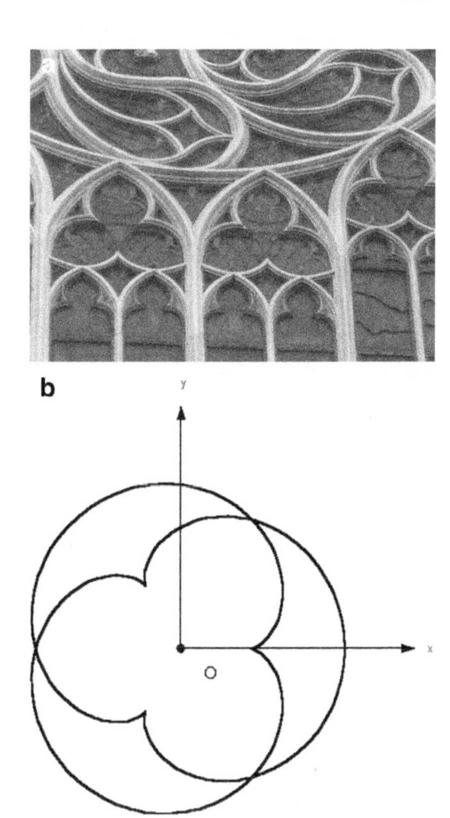

Fig. 35.18 (**a, b**). Photo and Image: authors

Starting from an equilateral triangle *ABC* we replace each side by a circular arc centred in the opposite vertex and radius equal to the side length ℓ (Fig. 35.22a). The three arcs form the Reuleaux triangle of constant width ℓ; the boundary curve has length $\pi\ell$ (the same of the circle having diameter ℓ) and the enclosed area measures $A = \frac{\pi - \sqrt{3}}{2}\ell^2$.

This very elegant shape fits skilfully in art: look the harmony of its insertion in Duomo windows (Fig. 35.23). It has important technical implications as well.

In the *Wankel radial engine* the section of the rolling pistons is a *Reuleaux triangle*. A drill chuck allows us to cut square holes if its section is a *Reuleaux* triangle.

In both cases geometrical properties are crucial (Marr 2000):

- in the Wankel engine the *Reuleaux* piston turns in a specially shaped housing, bordered by an epitrochoid curve (Figs. 35.21 and 35.24) and results in the 4-stroke of an internal combustion engine;
- in the case of a drill chuck we point out that a square, having sides equal to the width of the *Reuleaux* triangle, presents four points of contact with it. This

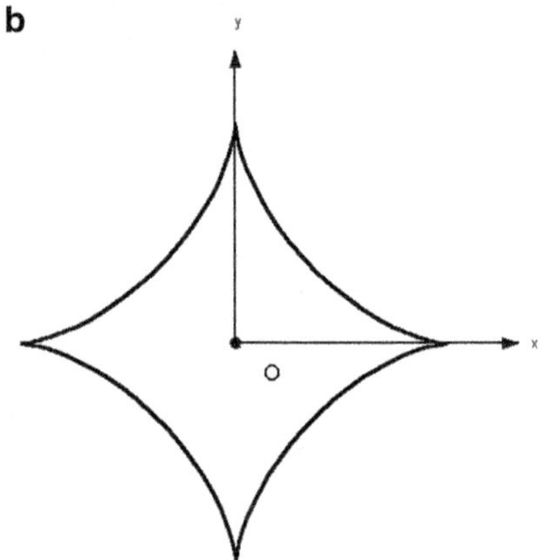

Fig. 35.19 (**a, b**). Photo and Image: authors

property is crucial to realize square holes; in fact during the rotation the three triangle vertices describe approximately the square perimeter (Fig. 35.25).

The *Reuleaux triangle* can be generalized to regular polygons with $(2n + 1)$ sides; the *Reuleaux polygons* have constant width ℓ and $(2n + 1)$ circle-arcs as sides.

The perimeter of each is again $\pi\ell$.

As examples of *Reuleaux polygons* we recall some coins of different countries: the British 20- and 50-pence and the 2-crown coins of the Czech Republic (Fig. 35.26).

Fig. 35.20 (**a, b**). Photo and Image: authors

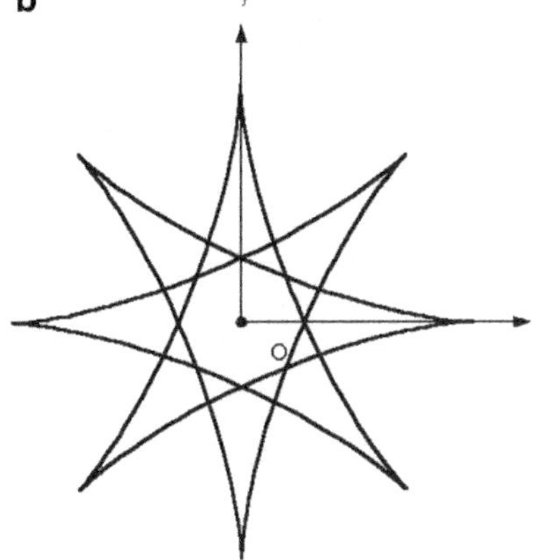

Fig. 35.21 Image: authors

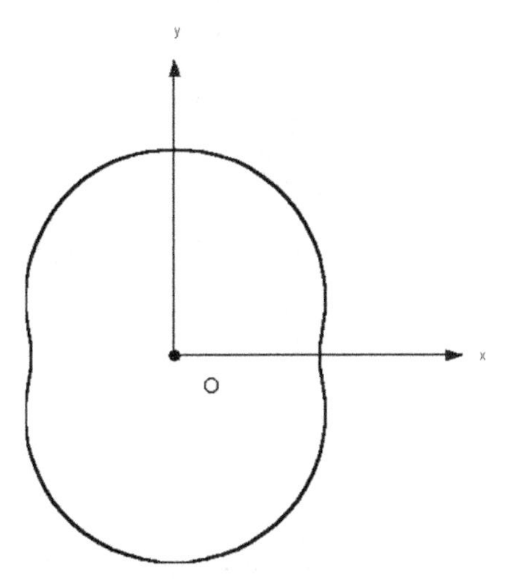

This choice in minting may be related to recognizing coins easily. Like circular pieces they roll in strips of constant width such as are used in coin-operated machines (vending machines, etc.).

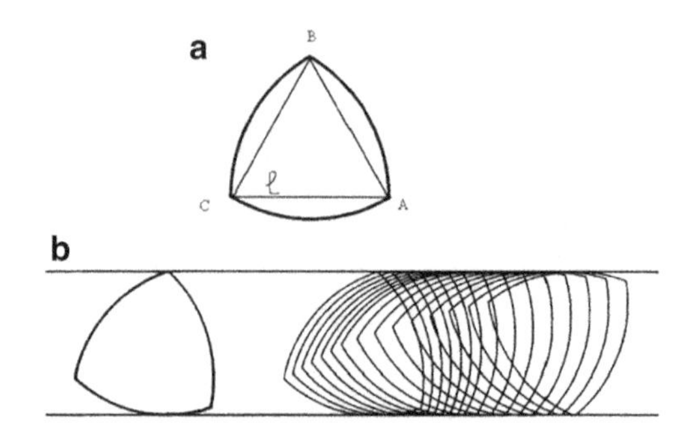

Fig. 35.22 (**a, b**). Image: authors

Fig. 35.23 Photo: authors

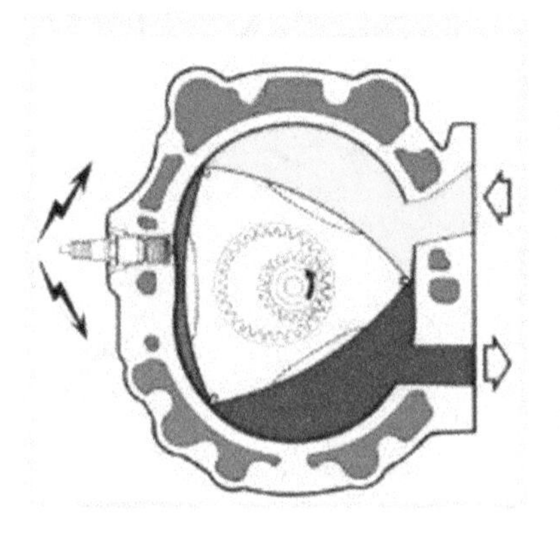

Fig. 35.24

We conclude this section with the mathematical reconstruction of the *Reuleaux triangle*, using the approach described in section "A Mathematical Approach in Describing Symmetries". We consider now the Reuleaux triangle as a D_3 form, generated by suitable rotations or reflections of a part.

Fig. 35.25 Photos and Image: authors

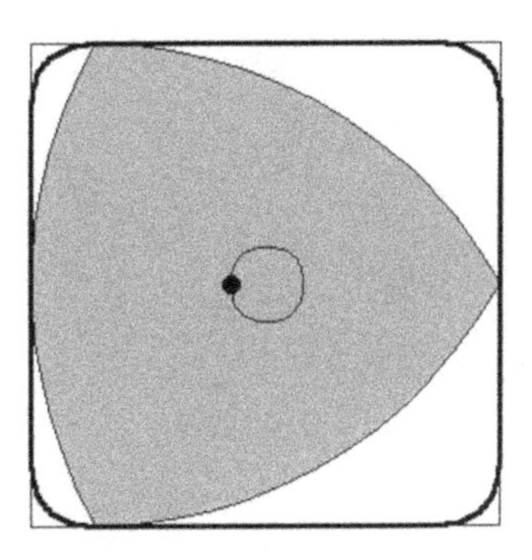

Fig. 35.26

We apply matrix and vector calculus as straightforward procedures to generate all cyclic or dihedral figures.

In the *Oxy* Cartesian plane let us consider the equilateral triangle *ABC* of side ℓ and the corresponding Reuleaux triangle (Fig. 35.27).

Fig. 35.27 Image: authors

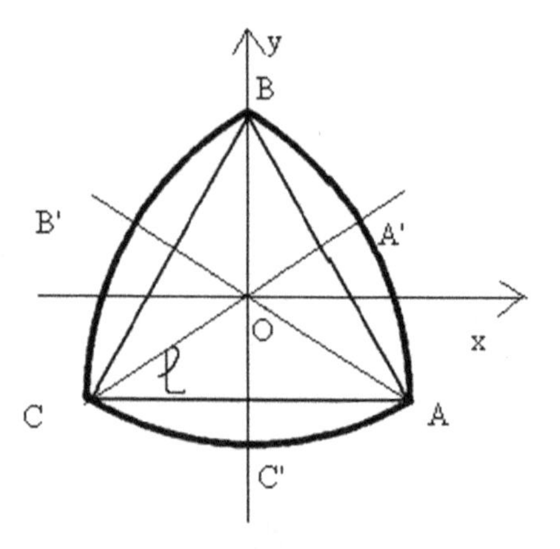

When the vertices are fixed in $A\left(\frac{\ell}{2},\ -\frac{\ell\sqrt{3}}{6}\right)$, $B\left(0, \ell\frac{\sqrt{3}}{3}\right)$, $C\left(-\frac{\ell}{2}, -\frac{\ell\sqrt{3}}{6}\right)$,

then a vector equation of the arc AB is $\mathbf{v}_{AB} = \begin{bmatrix} \ell\left(-\dfrac{1}{2} + \cos t\right) \\ \ell\left(-\dfrac{\sqrt{3}}{6} + \sin t\right) \end{bmatrix}$, $t \in \left[0, \frac{\pi}{3}\right]$.

In the following we consider also the arc AA', obtained with $t \in \left[0, \frac{\pi}{6}\right]$, with A′ as the middle point of the arc AB.

The vector equations of the arcs BC and CA can be obtained in two different ways:

1. starting with the entire arc AB and using rotation matrix

$$\mathbf{A} = \begin{bmatrix} \cos\dfrac{2\pi}{3} & -\sin\dfrac{2\pi}{3} \\ \sin\dfrac{2\pi}{3} & \cos\dfrac{2\pi}{3} \end{bmatrix}$$

so that $\mathbf{v}_{BC} = \mathbf{A}\mathbf{v}_{AB}$ and $\mathbf{v}_{CA} = \mathbf{A}\mathbf{v}_{BC}$;

2. starting with the arc AA' , the matrix $\mathbf{M}_0 = \begin{bmatrix} \cos 2\vartheta & sen2\vartheta \\ sen2\vartheta & -\cos 2\vartheta \end{bmatrix}$, $(\vartheta = \frac{\pi}{6})$ reflects the arc AA' with respect to the axis OA' and gives $\mathbf{v}_{A'B} = \mathbf{M}_0\mathbf{v}_{AA'}$; then other four iterations with the reflection matrices

$$\mathbf{M}_k = \begin{bmatrix} \cos\left(2\vartheta + 2k\frac{\pi}{3}\right) & sen\left(2\vartheta + 2k\frac{\pi}{3}\right) \\ sen\left(2\vartheta + 2k\frac{\pi}{3}\right) & -\cos\left(2\vartheta + 2k\frac{\pi}{3}\right) \end{bmatrix},$$

k = 1,2,3,4, applied to the transformed arcs, give all the boundaries:

$$\mathbf{v}_{BB'} = \mathbf{M}_1\mathbf{v}_{A'B}, \mathbf{v}_{B'C} = \mathbf{M}_2\mathbf{v}_{BB'}, \mathbf{v}_{CC'} = \mathbf{M}_3\mathbf{v}_{B'C}, \mathbf{v}_{C'A} = \mathbf{M}_4\mathbf{v}_{CC'},$$

where B', C' are the middle points of arcs BC, CA respectively.

Conclusions

We have suggested some tools for discovering several architecture and artistic aspects of the Milan Cathedral in a mathematical way.

We especially underlined the symmetry aspects in forms and decorations and described some peculiar curves, so that visitors to the Duomo can follow an interdisciplinary cultural path.

We hope the reader will become aware that mathematical tools are not so difficult to learn and manage. Increasing our mathematical and scientific knowledge can make every tour of a monument, museum or exhibit more interesting. The discovery of mathematical peculiarities both enriches the whole view and provides the opportunity to appreciate individual elements.

There is another geometrical aspect of the Milan Cathedral left to be investigated: the proportions among parts hinted at in the drawings by Cesariano. More secrets of the building's aesthetic canons wait to be revealed.

Biography Elena Marchetti, Associate Professor, has taught mathematics courses to architecture students at the Scuola di Architettura e Società of the Politecnico di Milano since 1988. She has produced numerous publications in Italian and international scientific journals in the area of numerical integration. She has also published many papers about the applications of mathematics, architecture and art. The experience gained through intense years of teaching courses to architecture students led her to collaborate in editing books dedicated to this topic, published with multimedia support packages.

Luisa Rossi Costa, Associate Professor of Mathematical Analysis in the Engineering Schools of the Politecnico di Milano since 1984, developed research projects in numerical and functional Analysis (published in several papers and quoted in books). She also contributed to the creation of the first-level degree in

Engineering via the Internet and of the e-learning platform M@thonline. Since 1998, involved in the relationship between high schools and the Politecnico, she has written many papers on teaching methods related to connections between art, architecture and mathematics to familiarize pupils with mathematics, and has collaborated in editing a book on symmetry, complete with DVD.

References

BOYER, C. B. 1968. *A History of Mathematics*. New York: J. Wiley & Sons.

BRIVIO, E. & MAJO, A. 1980. *Il Duomo di Milano nella storia e nell'arte*. Milan: Arti Grafiche Pizzi.

BUDDEN, F. J. 1972. *The fascination of groups*. Cambridge: Cambridge University Press.

DEDÒ, M. 1999. *Forme: simmetria e topologia*. Bologna: Zanichelli.

FERRÉOL, R. & MANDONNET, J. 2005. *Encyclopédie des Formes Mathématiques Remarquable*. http://www.mathcurve.com/index.htm.

KLINE, M. 1996. *Storia del pensiero matematico*. Vol. II. Turin: G. Einaudi.

LORIA, G. 1930. *Curve piane speciali*. Vol. II. Milan: U. Hoepli.

MARR, A. 2000. *Wankel Rotary Combustion Engine*. http://www.monito.com/wankel/rce.html.

PIRINA, C. 1986. *Le vetrate del Duomo di Milano*. Florence: Le Monnier.

WEYL, H. 1952. *Symmetry*. Princeton: Princeton University Press.

Chapter 36
The Symmetries of the Baptistery and the Leaning Tower of Pisa

David Speiser

The Symmetry of the Leaning Tower

By "symmetry of a tower", I mean here simply the symmetry of the ground plan (*Grundriss*.) The ground plan of the tower of a church is usually a square. Not always but often its covering has the same four-fold symmetry too. Sometimes the tower has only an ordinary roof, the four-fold symmetry is then lost and only a two-fold symmetry survives; we then say that the four-fold symmetry is "broken." Sometimes, to the contrary, the covering shows a more refined eight-fold symmetry. The most famous example of this is the tower of the Cathedral of Freiburg im Breisgau, Germany. In Pisa, the tower of S. Nicola is round below, but its upper levels are octagonal. There are many other, smaller towers with plans following this idea. Occasionally, we find a tower with a six-fold symmetry, but these cases are very rare. Of course, there are also round towers, for example, the ones in Ravenna. And such a round tower is the Leaning Tower. The symmetry of a circle is an infinite one. However, the architectural elements of the tower may reduce this infinite symmetry to a finite one. Before we show how this happens in the case of the Leaning Tower, we shall summarize its structure in a few words (Fig. 36.1).

First published as: David Speiser, "The Symmetries of the Baptistery and the Leaning Tower of Pisa", pp. 135–146 in *Nexus I: Architecture and Mathematics,* ed. Kim Williams, Fucecchio (Florence): Edizioni dell'Erba, 1996. This is a slightly abridged text of an earlier publication: "The Symmetries of the Baptistery and the Leaning Tower of Pisa," *Annali della Scuola Normale Superiore di Pisa,* Classe di Lettere e Filosofia Serie III, vol. XXIV, 2–3 (1994): 511–564, with 7 plates.

D. Speiser (✉)
Bromhubelweg 5, CH-4144 Arlesheim, Switzerland
e-mail: rspeiser@intergga.ch

K. Williams and M.J. Ostwald (eds.), *Architecture and Mathematics from Antiquity to the Future*, DOI 10.1007/978-3-319-00137-1_36,

Fig. 36.1 The Leaning Tower of Pisa. Photo: Author

The massive ground floor consists of a blind arcade. Above the ground level there are five tiers of open arches; a sixth tier is a bit taller. All have twice as many arches as the ground level. On the top there is a somewhat smaller drum (*tamburo*). Thus in all there are eight levels. Obviously the finite symmetry of the tower is determined by the number of pillars and arches. What is this number?

If one counts them, one is stunned to find, that there are neither 16, as one would have guessed at first, nor 12 or maybe 20: there are 15 arches on the ground level, and 30 on the following ones! The drum, finally, has six major and six minor arches, in such a manner that the minor ones are spanned above one, the major one above four arches of the seventh level. This 15-fold and 30-fold symmetry is probably unique; at least, it must be extremely rare.[1] Therefore it demands an explanation, the more so as it is used on such an outstanding building.

In a book about the Leaning Tower Piero Pierotti (1990) demonstrated the importance which the Pisans attached to exact numerical measures: the height of the Tower is exactly 100 Pisan *bracci*, the equivalent of 20 Pisan *pertiche,* while the circumference measures exactly 100 Pisan *piedi.*

These numbers strongly indicate how the Pisans of those times delighted in mathematical relationships and used them in building design: so they knew what they were doing when they constructed a pentadecagonal *campanile.* But it is of fundamental importance to distinguish between significant numerical values such as these and geometric relationships, or in our case, geometrical symmetries. While the former are numerology, the symmetries represent partitions of space or, perhaps, partitions of the plane. The following example illustrates a means of partitioning the plane: in a given square, drawing the diagonals and then a horizontal line through their intersection halves the square. The half-square may in its turn be halved by drawing its diagonals and a second horizontal line (Fig. 36.2a).[2] What might be the use for this construction? Just insert with a colour pencil the following profile (Fig. 36.2b). This profile is not just a mere curiosity noted *post festum*, but it is the basis for the design of the façade of the Duomo of Pisa. That this is true is shown in Fig. 36.3. Obviously (except for the 2, since we twice divided the plane into two equal parts) numbers do not play any role here: the construction is a purely geometric one. Thus in the case of the Leaning Tower and also, as we shall see, the Baptistery, it is not sufficient to speak of 12 or 15 or 20 arches, windows, etc.; these elements express, respectively, a 12-fold, 15-fold, or 20-fold symmetry of an object in space.

The Symmetries of the Exterior of the Baptistery

The visitor who comes from either the Tower or the Duomo to the Baptistery, cannot help feeling at first a certain uneasiness. Partly it must stem from the somewhat strange dome (*cupola*) which begins as a conventional dome-shape, changes to a cone-shape, and is crowned by a final, smaller dome (*cupolina*). This ungainly construction seems neither fish nor fowl. Then too, the spectator will see that, unlike the Tower, the Baptistery is built in two styles, Romanesque and North-Italian Gothic (Fig. 36.4). But the patient spectator will discover deeper discrepancies between the different levels of the Baptistery exterior. These

[1] There are many leaning towers besides the one in Pisa, but to the best of my knowledge, the 15-fold symmetry is never mentioned in any book or scientific paper. Even the *Baedeker*, that faithful companion, fails to mention this fundamental fact. For the only correct reference that I can find, see Carli (1989).

[2] I found this construction in an Italian journal of design shortly after the war. Unfortunately I have forgotten the names of the authors and of the journal, so I cannot give proper credit.

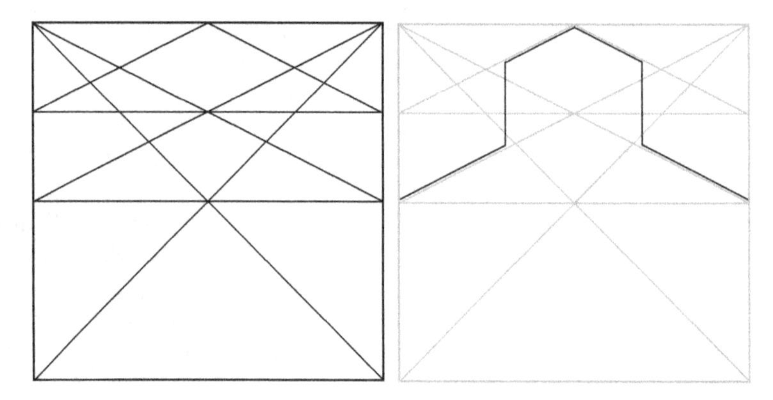

Fig. 36.2 (**a**, left) The partitioning of a square; (**b**, right) The basis for the profile of the façade of the Duomo. Drawing: Kim Williams

Fig. 36.3 The façade of the Duomo of Pisa. Photo: Author

Fig. 36.4 The exterior of the Baptistery of Pisa. Photo: Author

discrepancies are real contradictions which make any grand harmonious effect, like those which impress us of the Duomo and the Tower, impossible. We proceed now to a close analysis of the building.

On the exterior, the Baptistery presents itself as having been constructed, if one includes the *cupola*, in four levels. The second level, counting from the ground floor, is surrounded with a garland of arches. Every pair of arches is topped with a pointed gable. Most of these gables contain a bust in a circular frame, and between gables there is a column-like vertical element. Thus, all counted, there are five elements whose symmetry we shall now investigate one by one.

Of the four levels, the ground floor, in the Romanesque style, is without a doubt the most beautiful and the most impressive. 20 semi-circular arches of equal size rest on 20 pilasters, forming a blind arcade. Four of these arcades contain portals; the remaining 16 are pierced by small arched windows in their upper halves. The walls surface is articulated by 10 white and 9 grey bands. This horizontal subdivision emphasizes the roundness of the wall, thereby stressing the heaviness of the building and, we would like to say, its dignity. The large arches with the portals, as well as the windows, correspond to those of the Duomo, but here at the Baptistery the proportions are more beautiful. One would wish to see the Baptistery completed in this style, but this is not the case.

Ignoring for the moment the second level, we turn to the third. Here too we find a division into 20 equal parts. Each part contains a semi-circular arched window, but the window is split in its middle by a colonnette and filled with Gothic ornaments. Each window is crowned with a pointed Gothic gable. Between gables there are groups of three colonnettes. The gables contain small, round medallions with geometric figures. If we now compare the first level with the third, we see that their symmetry is the same. On the other hand, one cannot be but a little bit disappointed by the quality of the higher one.

When one looks at the cupola, the first impression is hardly that of a cupola at all but of a somewhat unwieldy mass. One gets the feeling that something went wrong during the execution. What really troubles the observer, first subconsciously, but soon consciously, is the new symmetry which is sharply at odds with the first and the third level: the cupola is not divided into 20 but into 12 equal parts. Since these two divisions are not in unison, the cupola does not appear as the artful completion of the entire edifice, but a structure in sharp contradiction to the lower levels.

Let us return to the second level and penetrate, as well as we can, through the surrounding garland to the wall itself. One notes that this level, like the one below, is divided into white and grey horizontal stripes, but less carefully so. The only easily visible elements of the wall are the windows. They are larger than the ones on the ground floor and at least taller if not wider than the ones on the third level. But now a big surprise meets, or I should say, hits, the eye. We find here only 12 windows: thus between the 20-fold symmetry of the ground and third levels, appears a level with 12-fold symmetry; it corresponds to the cupola, but not to the two levels above and below it. Why then is this symmetry clash not as manifest as one would expect it to be?

If one tries to imagine the Baptistery without the garland, one recognizes the painful dissonance between the 20-fold division of the ground and third levels and the 12-fold division of the intermediate level. For instance, the windows of the second level stand in no constructional relation whatsoever to those below or above, and the effect would not only be painful but quite probably a bit comic. Obviously, the garland was added to conceal this situation. But how does the garland accomplish this? For the counting eye the device was simple enough: the garland consists of 60 columns which support 60 arches, and since

$$3 \times 20 = 60 = 5 \times 12$$

the garland stands in a consonant relation to both the levels below and above as well as to the one whose true symmetry it conceals.

The Interior of the Baptistery

The visitor who enters the Baptistery is at once captivated by the upward surge of the 12 interior pillars and columns which surround the innermost space covered by the *cupola*. Above is a gallery supporting 12 pillars, and on top of that finally appears the *cupola* with the 12-fold division. It is a magnificent view indeed. Thus the innermost space is structured 12-fold throughout, although the four pillars mark a square. With respect to the square marked by the four portals, this square is rotated by 45°, so that the portals correspond to the middle arches between the pillars (Fig. 36.5). The octagonal baptismal font, a later addition, responds to this quite cleverly: four of its walls correspond to the portals and four to the pillars, so that between the 12 and the 8 no discrepancy arises, except in the paving design, which reflects the octagonal shape of the font. Everything in the interior, then, is subordinated to the dominant 12-fold symmetry.

Now we may ask, what effect does the symmetry clash between the interior walls have on the interior space? The answer is that the impression made by the central space is so strong and compelling that the contradiction with the exterior wall is hardly felt. Only when one consciously looks for evidence of the 20-fold symmetry of the exterior does the contradiction become evident. The contradiction of symmetry of the windows of the ground level does not really hurt, but it does not help either, let alone develop, the great effect of the central structure. Not quite the same may be said for the windows of the third level, some of which, obviously in conflict with the interior structure and perhaps because of this, are blind.

One would expect the 12 windows of the second level to appear splendidly in the inside, since they are in harmony with it, but, surprisingly, they cannot be seen from the ground floor at all. One sees them only when climbing between the two shells of the outer wall to the gallery. The reason is that they are located right at the floor level of the gallery, so that their compositional effect is lost. At last one then realizes that the interior does not correspond to the exterior at all: in correspondence to the three exterior levels, not counting the *cupola*, there are only *two* interior levels! As we said, the central space is surrounded by two levels of 12 pillars and columns, supporting 12 arches. Behind each arch there is a vaulted bay.[3] Besides the several clashes between contradictory symmetry schemes, this non-correspondence between interior and exterior is the most puzzling aspect of the Baptistery (Fig. 36.6).

[3] For a very careful description of the Baptistery, see Smith (1978).

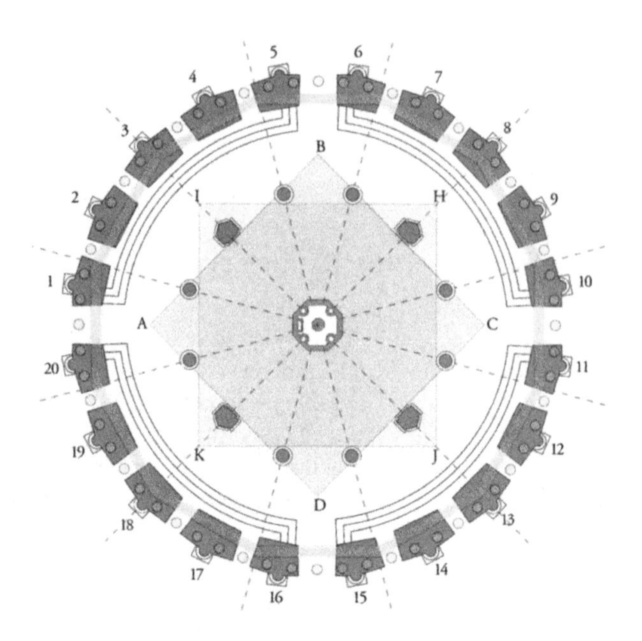

Fig. 36.5 The plan of the Baptistery. Image: Paolo Radicati di Brozolo

We must add one last remark concerning the ground plane. While inside the pavement design simply follows the octagonal symmetry of the baptismal font, the outside presents one more surprise. In the partitioning of the platform on which the Baptistery rests, one would expect either a continuation of the eight-fold symmetry of the interior pavement, or a reflection of the 20-fold symmetry present in the ground level architecture. No: we find again a 12-fold symmetry.

A Proposal for the Phases of the Baptistery's Construction

In light of the many symmetry clashes presented above, two questions come to mind. First, are such changes compatible with what we know from the documents concerning the construction of the Baptistery? Second, is it possible that these plans have been repeatedly revised and fundamentally modified? A complete answer is beyond the scope of this chapter, but we will summarize the argument.

The chronology of the Baptistery construction is complicated by diverse facts. First, the Baptistery has two coverings. The inner cone is partly enclosed by the outer *cupola*, and this doubling presents a riddle. Second, this first covering was preceded by another one, which also may have been a cone or a *cupola*, though the exact history is obscure. But the most striking puzzle remains that there are three exterior levels, and only two interior levels. Figure 36.7 shows schematically the various levels and conflicting symmetry schemes of the exterior, together with the four points of rupture. It seems best to examine the problem that they pose beginning with the latest one, uppermost in our figure.

Fig. 36.6 Section through the Baptistery. Image: From Grassi (1831)

This symmetry rupture, 12-fold versus 20-fold, does not really pose a problem, because for constructional reasons a return to the 12-fold-symmetry was almost unavoidable. The forms of the coverings, especially the cone, follow the 12-fold symmetry of the interior, and thus become, one might say, logically dodecagonal.

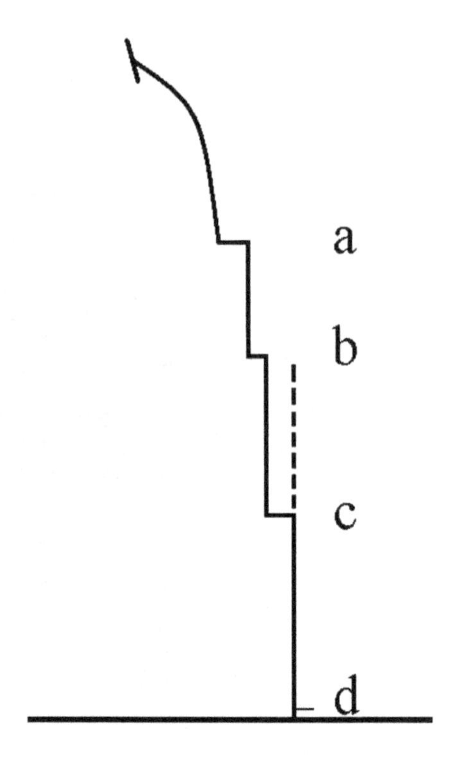

Fig. 36.7 Schematic diagram of the ruptures of symmetry. Drawing: Kim Williams

Automatically we presuppose here that, when the third level was begun, the garland surrounding the second level did not yet exist. In principle this could have been otherwise, the garland might have been built together with the second level, but this seems unlikely. This time the architects did not have their hands tied: they could freely decide whether to go on with the 12-fold symmetry of the second level, or to return to the 20-fold level of the first. As was seen in the Leaning Tower, the Pisans had learned the peculiar relationship between the triangle, the pentagon and their respective families, which could produce the pentadecagon with its family 30, 60, etc. So why not make good use of this relationship? If this hypothesis is correct, sometime after the completion of the second level, two decisions were made, namely, to return to the 20-fold symmetry of the ground level, and to surround the second level (with its 12-fold symmetry) with a garland of 60 arches. 60 is indeed compatible with all three levels.

The situation which confronted the builders after the completion of the ground level seems not unlike the one which a few decades later would confront their successors. Both times, the builders chose a solution that was in contrast with a part of what was already built. The second level could not be built in agreement with both the interior and the ground level. Yet, in one respect the situation was quite different. Unlike the undistinguished second level, the ground level is of a very beautiful and harmonious design. It would be rather extraordinary if the radical changes in the project had not caused the Pisans to quarrel. And indeed there exists a series of documents which testify to a long and bitter feud that pitted the Archbishop, the lord over the Duomo and the Tower, against the Canons to whom the custody of the Baptistery belonged. This fight was about the right to appoint the *operaio*, or head architect of the Baptistery, and it lasted from 1210 to 1221.

The ground level was constructed in wilful defiance of the already existing internal structure and its 12-fold symmetry. Thus it is the most surprising of all ruptures in the Baptistery's history. Those who defied Diotisalvi's plan were under no aesthetic pressure whatsoever to do so; on the contrary, the interior was beautifully conceived, and we may suspect the same of the exterior. There must have been a powerful motive at work for inducing such a change. Moreover, the new ground level stands up to a comparison with the interior with respect to both design and execution, though, to be sure, the aesthetic aim of the new exterior ground level is different. We are forced to look for an artistic event of first rate importance that could have provided the impetus for the artist as well as for his supporters, but we need not look very far. Behind the Duomo the Pisans had begun to build what they wanted to be the world's highest tower, and this tower had a ground plan based on the newest discoveries of the scientific world. Could there be a stronger stimulus for swaying opinion in favour of a new design and overcoming resistance to a break with the old plan? In 1186, a new operaio, Guido (Guidolotto) da Ugone, was appointed and sworn in at a time when the outer wall had already been built to the height of the doors. Thus we are led to affirm that 1185 was the year when a new plan of the Baptistery's exterior was conceived by a new architect.[4]

A Summing Up of the Construction in Chronological Order

In 1185 the original plan for the exterior by Diotisalvi was replaced by one of Guidolotto, who, if he had not himself designed the Leaning Tower, was at least strongly influenced by it. His design made use of a radically new idea based on Pisan geometric knowledge: the construction of the regular pentagon and the regular pentadecagon, which had appeared for the first time in the construction of the Tower. This conception remained unique in the history of European architecture. But Guidolotto's *coup d'état*, though presumably wildly cheered by many, was not to everybody's liking. Circa 1221, the quarrels ended and Diotisalvi's original idea triumphed: the 12-fold symmetry carried the day. The 60-fold garland was added at this time, or a bit later. A third reversal happened circa 1278. The icosagonal symmetry was the victor this time, though the aesthetic quality of the third level was decidedly inferior to the ground level. At about the same time, the 30 gables were added and filled with Giovanni Pisani's sculptures.

[4] The architect of the Tower is not known: could it have been Guido himself?

How and when the *cupola* and cone were added, both with 12-fold symmetry, must be determined by archaeological examination.

Looking today at the Baptistery, we find the greatest artistic impulses and the highest achievements in Diotisalvi's interior and in Guidolotto's exterior ground level. Both are extraordinary conceptions deserving admiration; but they contradict each other, one is tempted to say, fratricidally. The consequence was a continuous striving for reconciliation. But despite its many clashes, the impression the Baptistery gives to the observer, especially when illuminated by the full moon, is an overwhelming one.

Acknowledgment I should like to thank Prof. Luigi A. Radicati di Brozolo, who made this work possible.

Biography David Speiser is Professor Emeritus at the Catholic University of Louvain, where he taught physics and mathematics from 1963 to 1990. His research concerned elementary particles and physical mathematics. From 1990 to 2004, he gave lectures and seminars regularly at the Scuola Normale di Pisa. He was the general editor of the complete works of the mathematicians and physicists of the Bernoulli family from 1980 to 2004. His essays on the relationships between the history of art and the history of science have been collected in *Crossroads: History of Science, History of Art. Essays by David Speiser, Vol. II*, Kim Williams, ed. (Basel: Birkhäuser, 2011).

References

CARLI, E. 1989. Il Campanile. Pp. 219–226 in E. Carli, ed. *Il Duomo di Pisa*, Florence: Nardini.
GRASSI, Ranieri. 1831. *Le frabbriche principali id Pisa ed alcune vedute della stessa città, 24 tavole*. Pisa: Edizioni Ranieri Prosperi.
PIEROTTI, Piero. 1990. *Una torre da non salvare*. Pisa: Pacini Editore.
SMITH, Christine. 1978. *The Baptistery of Pisa*. New York: Garland.

Part V
Theories of Proportion, Symmetry, Periodicity

Chapter 37
Musical Proportions at the Basis of Systems of Architectural Proportion both Ancient and Modern

Jay Kappraff

Introduction

Throughout the history of architecture there has been a quest for a system of proportion that would facilitate the technical and aesthetic requirements of a design. Such a system would have to ensure a repetition of a few key ratios throughout the design, have additive properties that enable the whole to equal the sum of its parts, and be computationally tractable—in other words, to be adaptable to the architect's technical means. The repetition of ratios enables a design to exhibit a sense of unity and harmony of its parts. Additive properties enable the whole to equal the sum of its parts in a variety of different ways, giving the designer flexibility to choose a design that offers the greatest aesthetic appeal while satisfying the practical considerations of the design. Architects and designers are most comfortable within the realm of integers, so any system based on irrational dimensions or incommensurable proportions should also be expressible in terms of integers to make it computationally acceptable.

In his book, *The Theory of Proportion in Architecture* (1958) P. H. Scholfield discusses three systems of architectural proportion that meet these requirements: the system of musical proportions used during the Renaissance developed by Leon Battista Alberti, a system used during Roman times, and the Modulor of the twentieth-century architect, Le Corbusier. All of these systems draw upon

First published as: Jay Kappraff, "Musical Proportions at the Basis of Systems of Architectural Proportion both Ancient and Modern", pp. 115–133 in *Nexus I: Architecture and Mathematics*, ed. Kim Williams, Fucecchio (Florence): Edizioni dell'Erba, 1996.

J. Kappraff (✉)
Department of Mathematics, New Jersey Institute of Technology, University Heights, Newark, NY 07102, USA
e-mail: kappraff@njit.edu

K. Williams and M.J. Ostwald (eds.), *Architecture and Mathematics from Antiquity to the Future*, DOI 10.1007/978-3-319-00137-1_37,

identical mathematical notions already present in the system of musical proportions as we shall show in Sect. 37.2. While the Roman system is based on the irrational numbers $\sqrt{2}$ and $\theta = 1 + \sqrt{2}$, the Modulor is based on the golden mean, $\tau = (1 + \sqrt{5})/2$. Both of these systems can also be approximated arbitrarily closely (asymptotically) by integer series, and these integer series can be used to implement the system with negligible error. We shall demonstrate this in Sect. 37.3 for the Roman system, since the Modulor has been adequately covered elsewhere. We will also show in Sect. 37.4 that at the basis of the Roman system is a geometrical construction discovered in the Renaissance, known as the "law of repetition of ratios." The sacred cut will be shown to lie at the basis of the Roman system. In Sect. 37.5 we shall illustrate, by way of Kim Williams's analysis of the Medici Chapel, that both the law of repetition of ratios and the sacred cut are geometric expressions of the additive properties of the Roman systems and insure the presence of musical proportions in a design. The chapter will conclude with a discussion of Ezra Ehrenkrantz's system of "modular coordination" based on both musical proportions of Alberti and Fibonacci numbers.

The Musical Proportions of the Italian Renaissance

During the Italian Renaissance Leon Battista Alberti and Andrea Palladio developed a system of architectural proportion based on proportions inherent in the musical scale (Scholfield 1958; Wittkower 1971). This movement was a response to the neoplatonic ideas prevalent at the time. Alberti modelled his system on the Pythagorean scale based on the octave, musical fifth, and fourth. To achieve an octave above the fundamental tone, the bridge of a monochord instrument is moved to the midpoint of the string, (i.e., ratio of 1:2 as shown in Fig. 37.1), and the string is plucked. The fifth is obtained by shortening the string by a ratio of 2:3 while the fourth shortens the string by a ratio of 3:4.

All musical proportions of the Pythagorean scale can be expressed as ratios of powers of the prime numbers 2 and 3. For example, the whole tone corresponds to the ratio 8:9. The system of Palladio was based on the Just scale which also included the prime number 5.[1] What is of greater relevance is the manner in which a system of architectural proportion was built from these scales. The first suggestion appears in the "lambda" figure,

found in Plato's Timaeus and referred to as "world soul."

[1] For further details of the musical scale built from these intervals, see Kappraff (1990: 9–12).

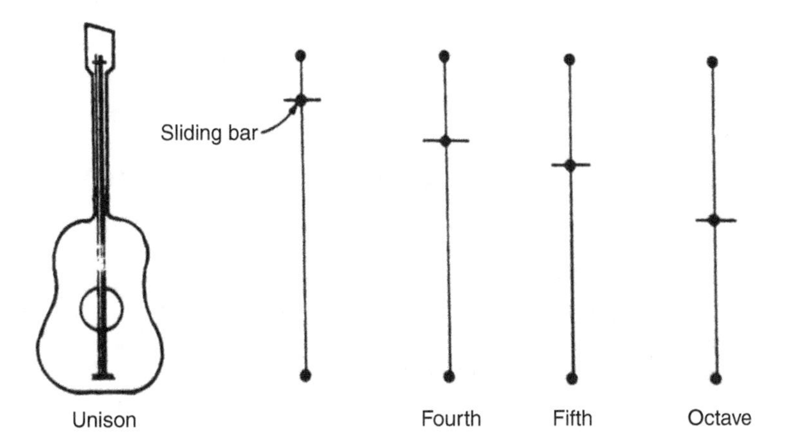

Fig. 37.1 A sliding bridge on a monochord divides the string length representing the fundamental tone into segments corresponding to musical fifth (2:3), fourth (3:4), and octave (1:2). Image: author

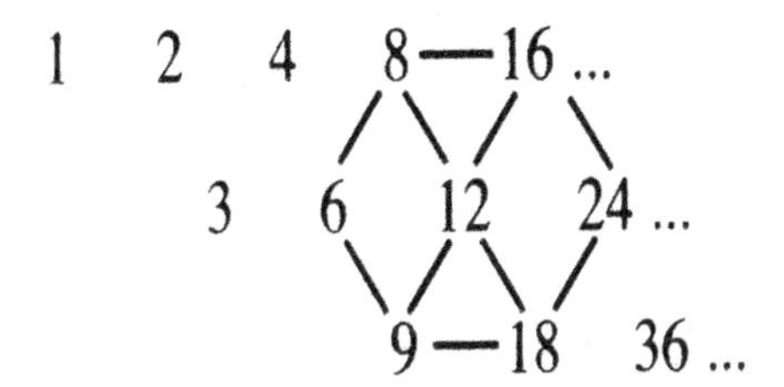

Fig. 37.2 Integer series of Alberti's system of musical proportions

Consider the series,

$$1 \quad 2 \quad 4 \quad 8 \quad 16 \quad \ldots$$

Next, construct additional integer series made up of the arithmetic means of adjacent numbers as shown in Fig. 37.2.

Notice that Plato's lambda appears on the boundary of these series. It is also evident that the ratio of successive terms in the horizontal direction is in the octave ratio, 1:2, while the left-leaning diagonal represents the ratio 2:3 (musical fifth) and the right-leaning diagonal exhibits the ratio 3:4 (musical fourth).

Each number x of these scales is the geometric mean of the numbers y and z to its left and right, i.e. $x = \sqrt{yz}$. By its construction, each number x is the arithmetic mean of the two numbers y,z bracing it from above, i.e., $z - x = x - y$ or $x = 1/2\,(y + z)$. Finally each number x is the harmonic mean of the two numbers y,z bracing it from below, i.e. $(z - x)/z = (x - y)/y$ or $1/x = 1/2(1/y + 1/z)$ which can be rewritten as $x = 2yz/(y + z)$.

As the result of these relationships, any sequence x,u,v,y that includes the arithmetic and harmonic means u,v of its endpoints x,y insures a repetition of ratios as illustrated for the sequence 6, 8, 9, 12. Here, 9:6 = 12:8 and 8:6 = 12:9.

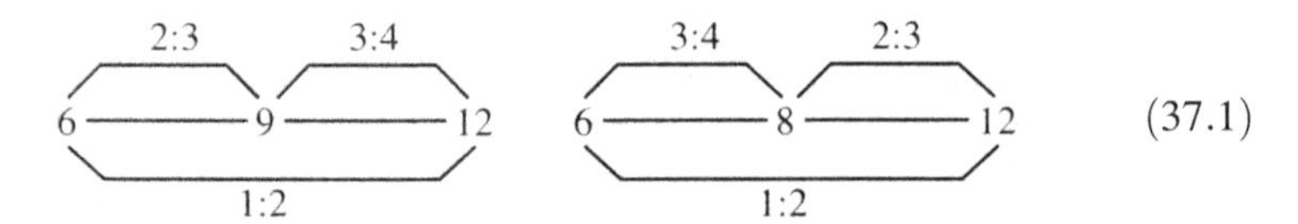

Musically, it is said that "when the musical fifth is inverted in the octave it becomes the musical fourth." Architecturally, any relationship that incorporates musical proportions insures that key ratios repeat in the context of a design.

Alberti used this system by focusing on a hexagon of numbers as shown in Fig. 37.2. The dimensions and subdivisions of the rooms of his buildings had measures given by adjacent numbers within the hexagon. This insured that ratios of these lengths would embody musical ratios. Wittkower has described Alberti's use of musical proportions in the design of S. Maria Novella and other structures of the Renaissance (Wittkower 1971: 33–56).

Although the Renaissance system of Alberti succeeded in creating harmonic relationships in which key proportions were repeated in a design, it did not have the additive properties necessary for a successful system. It is fascinating that a system of proportions used by the Romans and the system of proportionality developed by Le Corbusier, known as the Modulor, both conform to the relationships inherent in the system of musical proportions depicted in Fig. 37.2 with the advantage of having additive properties (Kappraff 1996a, 1996b).

The Roman System of Proportions

The well known integer series,

$$1 \quad 1 \quad 2 \quad 3 \quad 5 \quad 8 \quad 13 \quad 21 \quad \ldots \tag{37.2}$$

in which each term is the sum of the preceding two terms, is known as a Fibonacci series. The ratio of successive terms approaches the golden mean, $\tau = (1 + 5)/2$ in a limiting sense. It is the additive properties of this series that led Le Corbusier to make it the basis of his Modulor series of architectural proportions.

Another integer series possessing additive properties is

$$1 \quad 2 \quad 3 \quad 12 \quad 29 \quad 70 \quad \ldots \tag{37.3}$$

known as Pell's series, in which twice any term in the series added to the previous term results in the next term. The ratio of successive terms from any Pell series such as Series (37.3),

$$2/1 \quad 5/2 \quad 12/5 \quad 29/12 \quad 70/29 \quad \ldots \tag{37.4}$$

approaches the irrational number $\theta = 1 + \sqrt{2}$in a limiting sense. The θ series,

$$\frac{1}{\theta^2} \quad \frac{1}{\theta} \quad 1 \quad \theta \quad \theta^2 \quad \theta^3 \tag{37.5}$$

where $\theta = 1 + \sqrt{2}$, is the only geometric series that is also a Pell series. Thus,

$$\frac{1}{\theta} + 2 = \theta, \quad 1 + 2\theta = \theta^2, \theta + \theta^2 = \theta^3 \tag{37.6}$$

Therefore, a Pell series possesses many additive properties, which is why it was used in ancient Rome as the basis of a system of architectural proportions.

In order to describe the Roman system, first consider a pair of integer Pell series,

$$\begin{array}{cccccccccccc} 1 & & 3 & & 7 & & 17 & & 41 & & & \\ 1 & 2 & & 5 & & 12 & & 29 & & 70 & \ldots \end{array} \tag{37.7}$$

This series has many additive properties, although we shall list only the three fundamental properties from which the others can be derived. Each Pell series certainly has the defining property,

x x x
a b c

Prop. 1: a + 2b = c

Other additive properties are,

d c
x x
x x
a b

Prop. 2: a + b = c, e.g., 2 + 5 = 7,
Prop. 3: a + d = b, e.g., 2 + 3 = 5.

Also, the ratio of diagonally adjacent numbers from Series (37.7) are approximations to the $\sqrt{2}$, i.e., the series

$$1/1 \quad 3/2 \quad 7/5 \quad 17/12 \quad \ldots \tag{37.8}$$

approaches $\sqrt{2}$ in a limiting sense.

Again, these series exhibit the same geometric, arithmetic, and harmonic mean relationships as Fig. 37.2. in an asymptotic sense. Thus each term is the approximate geometric mean of the terms to its right and left, e.g., $5^2 \approx 2 \times 12$. Each term in the second series is the average of the terms that brace it from above, e.g. 5 = 1/2(3 + 7). Each term from the first series is approximately the harmonic mean of the two terms that brace it from below, e.g., 7/21 = 1/3 ≈ 1/2(1/2 + 1/5) = 7/20. Finally, any term in the first series divides the interval bracing it below approximately in the ratio 1: θ, e.g., (41–29)/(70–41) = 12/29 ≈ 1 : θ.

Notice that the double series (37.7) has one problem. The sum of two successive numbers from one series is found in the series above it. Therefore the sum of a pair of numbers in the upper series is not represented in this system. However, we can expand the system by doubling the numbers in the lower series to obtain a third Pell series,

$$\begin{array}{ccccccccccccc} & 2 & & 4 & & 10 & & 24 & & 58 & \ldots \\ 1 & & 3 & & 7 & & 17 & & 41 & \ldots & \\ 1 & 2 & & 5 & & 12 & & 29 & & 70 & \ldots \end{array} \quad (37.9)$$

Notice that the additive properties and the mean relationships continue to hold, in addition to the approximation to the $\sqrt{2}$ gotten by taking the ratio between any term and its diagonally adjacent term from the series above it. This can of course be continued ad infinitum, as indicated by the dots in Series (37.9).

The Roman system of architecture uses the following infinite sequence of θ-series equivalent to Series (37.9):

This system continues to exhibit all additive relationships of Series 9 and the exact mean relationships exhibited by Fig. 37.2. Donald and Carol Watts have studied the ruins of the Garden Houses of Ostia, the port city of the Roman Empire, and found that they are organized entirely by the proportional system of Table 37.1 or its integer approximation, Series (37.9) (Watts and Watts 1986: 132–140).

The Geometry of the Roman System of Proportions

The algebraic properties of the Roman system of proportion can be made palpable by considering the equivalent geometric properties. We shall find that three rectangles of proportions 1: 1 (square—S), 1: $\sqrt{2}$ (square root rectangle—SR) and 1: 1 + $\sqrt{2}$ (Roman rectangle—RR) form an interrelated system. For example, if S is either removed or added to SR, this results in RR, as Fig. 37.3a illustrates. This is equivalent to Properties 2 and 3. That 2S + RR = RR is equivalent to Property I (Fig. 37.3b). Finally, if SR is cut in half it forms two SR at a smaller scale (Fig. 37.3c), while two SR added together form an enlarged SR (see Fig. 37.3d) as predicted by the doubling property of Series (37.9).

The key to understanding the Roman system of proportions is a geometrical construction called the "sacred cut" by the sacred geometer Tons Brunés (Brunés 1967). When the compass point is placed at the vertex of a unit square, and an arc is swept out as shown in Fig. 37.4a, the edge is reduced by a factor of $1/\sqrt{2}$, the sacred cut. If four sacred cuts are drawn from the corners of the square and points of section on the edges are joined, a regular octagon results as shown in Fig. 37.4b. The four sacred cuts also divide the square into four S at the comers, one larger central S, two SR, and two RR (Fig. 37.5). Donald and Carol Watts (1986) have discovered a tapestry preserved from the ruins of the Garden Houses of Ostia organized according to the pattern of Fig. 37.5; other, later designs featuring the sacred cut appear in the Baptistery of Florence (Williams 1994).

The computational properties of the Modulor and the Pell series are also the result of the "law of repetition of ratios," well known in the Renaissance and

Table 37.1 The Roman system of proportions based on $\theta = 1 + \sqrt{2}$

			$2\sqrt{2}$		$2\sqrt{2}\theta$		$2\sqrt{2}\theta^2$		$2\sqrt{2}\theta^3$		$2\sqrt{2}\theta^4$	...
		2		2θ		$2\theta^2$		$2\theta^3$		$2\theta^4$	...	
	$\sqrt{2}$		$\sqrt{2}\theta$		$\sqrt{2}\theta^2$		$\sqrt{2}\theta^3$		$\sqrt{2}\theta^4$	...		
1		θ		θ^2		θ^3		θ^4	...			

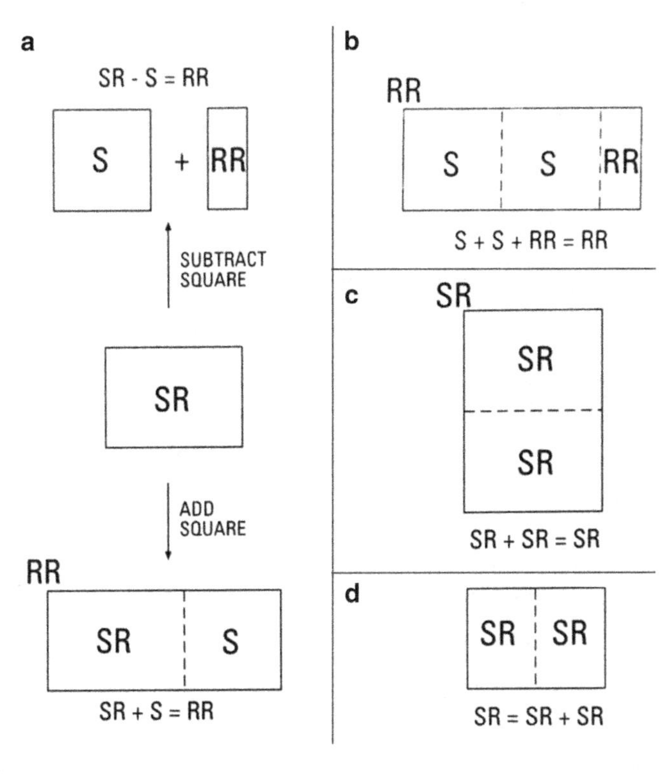

Fig. 37.3 The square (S), square root rectangle (SR), and the Roman rectangle (RR) are interrelated

revived by Jay Hambidge as the key to his concept of dynamic symmetry.[2] To illustrate this law, draw a diagonal to a rectangle and intersect it with another diagonal at right angles as shown in Fig. 37.6a. This subdivides the original rectangle or unit (U), of proportions *a*: *b*, into a rectangle referred to as gnomon (G) and a similar unit of proportions (U) (Fig. 37.6b) *b*: *c*, i.e.,

$$a/b = b/c \text{ and } \mathrm{G} + \mathrm{U} = \mathrm{U} \tag{37.10}$$

This has the effect of reproducing ratios in a rectangle just as the insertion of

[2] For studies of dynamic symmetry, see Hambidge (1967, 1979) and Edwards (1967).

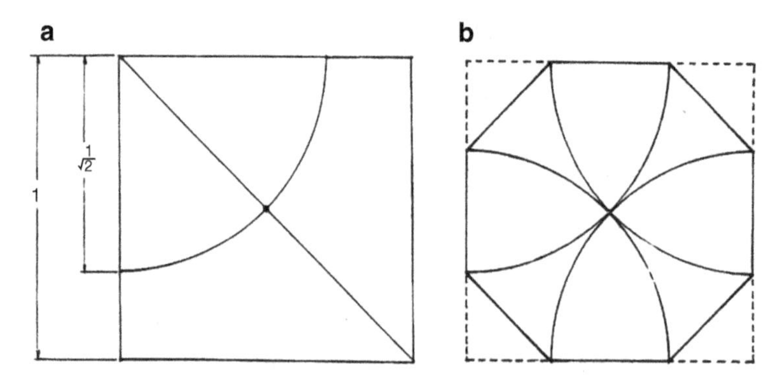

Fig. 37.4 (**a**, *left*) The sacred cut reduces a length by a factor of $1/\sqrt{2}$; (**b**, *right*) four sacred cuts form the vertices of a regular octagon. Image: author

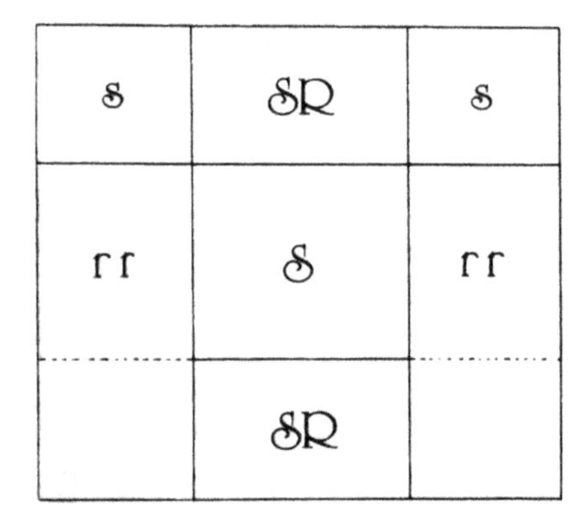

Fig. 37.5 Subdivision of a square by four sacred cuts into squares (S), square root rectangles (SR), and Roman rectangles (RR). Image: author

arithmetic and harmonic means did within the octave for the musical scale (see Series 37.1).

This process can be repeated many times to tile the unit with "whirling gnomons" and one additional unit

$$U = G + G + G + \ldots + U,$$

as shown in Fig. 37.6c. We see that the vertices of the gnomons trace a logarithmic spiral.

In Fig. 37.7 this procedure is applied to a square root rectangle. We see that the gnomon equals the square and SR is subdivided into two SR. However, this construction also possesses a second important geometrical relationship well known to ancient geometers. Notice the upward and downward pointed triangles in Fig. 37.7. It can be shown that for any rectangle, the intersection of such triangles with the diagonals of the rectangle result in a trisection of the length and width of the rectangle (see Fig. 37.8). As a result, the law of repetition of ratios not only

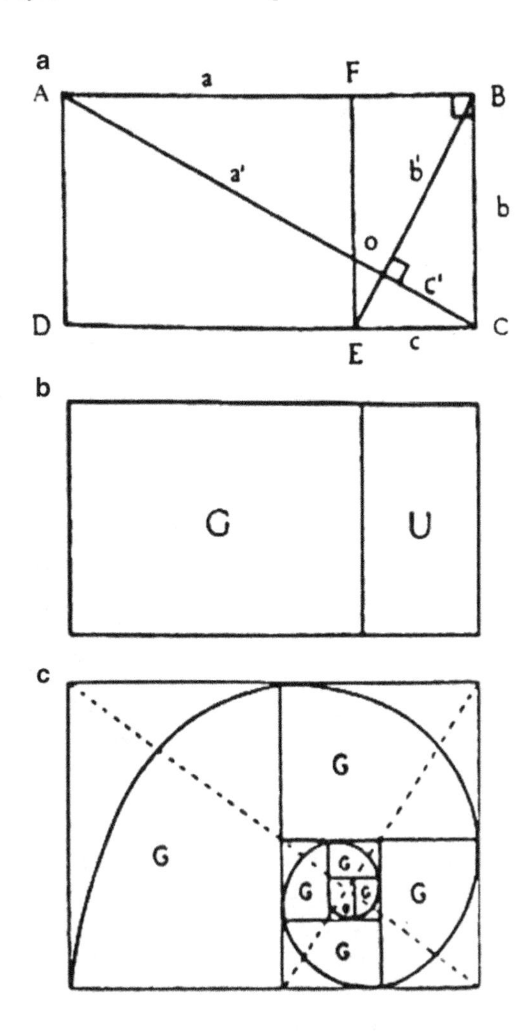

Fig. 37.6 (**a**, **b**) The "law of repetition of ratios" divides a unit rectangle into a gnomon and proportional unit; (**c**) a series of whirling gnomons form a logarithmic spiral. Image: author

bisects but also trisects SR. The beautiful design in Fig. 37.9 captures these relationships.[3] In the next section we shall show that these twin relationships lay at the basis of the architecture of the Medici Chapel as uncovered by Kim Williams.

Relationship Between the Roman System and the System of Musical Proportions

Since the time of the Greeks, there has been a tension in architecture and design between the use of commensurate and incommensurate lengths, i.e., lengths governed by rational or irrational numbers. It was Pythagoras who, it is said, first

[3] This design appears in Edwards, *Design with Dynamic Symmetry*, Fig. 16, p. 25.

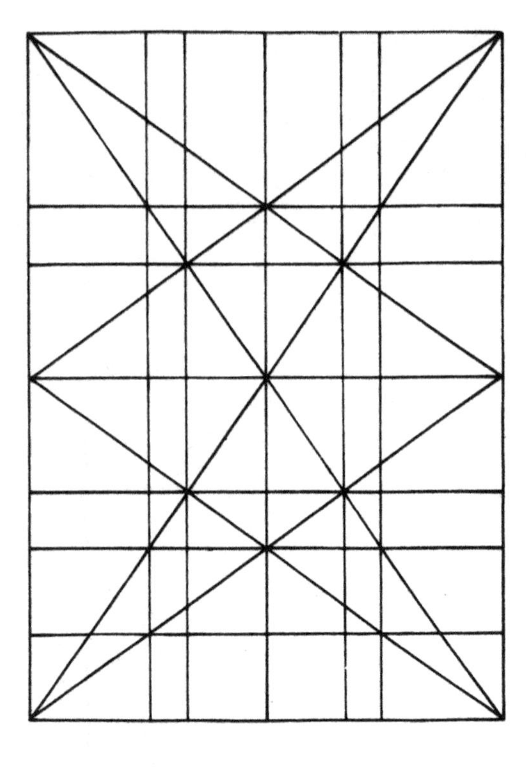

Fig. 37.7 Application of the law of repetition of ratios to a square root rectangle. Image: author

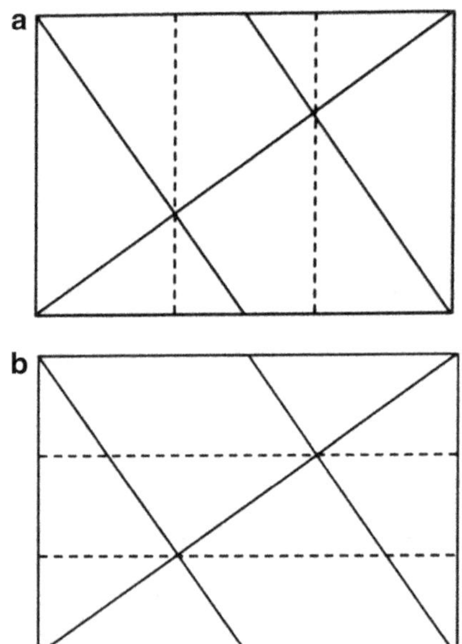

Fig. 37.8 Trisection of (**a**) length; and (**b**) width of a rectangle. Image: author

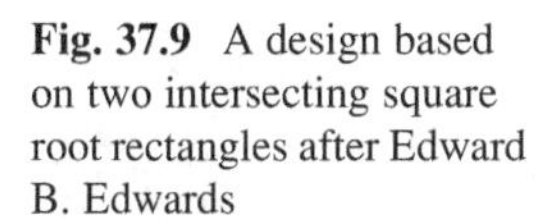

Fig. 37.9 A design based on two intersecting square root rectangles after Edward B. Edwards

discovered that the ratio of the diagonal to the side of a square was incommensurable, i.e., no finite multiple of one equalled a multiple of the other. On the one hand, incommensurable ratios were distressing since they did not fit the model that the Greeks had of number (Kappraff 1990: 16–20). On the other hand, they were easily constructible with compass and straightedge and had interesting geometric properties, some of which have been outlined in this chapter. Although incommensurable measurements were equally incomprehensible from a number theoretic point of view to architects of the Italian Renaissance such as Alberti, Wittkower says:

> Medieval geometry (with its use of incommensurable ratios such as 1:√2 or 1:√5) is no more than a veneer that enables practitioners to achieve commensurable ratios without much ado (Wittkower 1971: 127).

Architect Kim Williams believes that one function of the system of musical proportions may have been to integrate the Roman system of proportions based on the incommensurable ratio √2: 1 with the commensurable ratios at the basis of the musical scale. Williams made these discoveries while surveying the famous Medici Chapel in Florence built by Michaelangelo (Williams 1997) (Fig. 37.10a).

The ground plan of the chapel is a simple square in plan with a rectangular apse, called a *scarsella,* added to the north end as shown in Fig. 37.10b. The sides of the square which form the main space of the chapel measure 11.7 m. The vertical height of the chapel walls measure 11.64 m, suggesting that the main space of the chapel was meant to be a cube. The overall perimeter of chapel and apse fits into a √2 rectangle. Williams recognized that a √2 rectangle is embedded in a cube as the rectangle formed by any pair of opposite edges. Thus the volume of the chapel and the shape of the ground plan are intimately related.

A second √2 rectangle is found in the chapel in the ensemble of the altar and the *scarsella.* Williams makes the important point that the altar protrudes into the

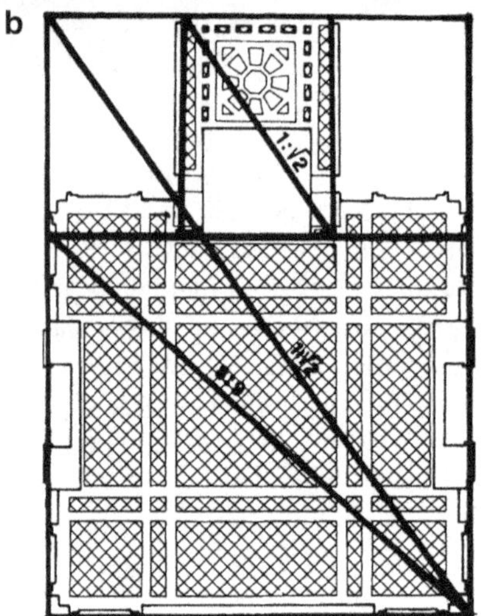

Fig. 37.10 (**a**) Interior of the Medici chapel. Photo: Kim Williams (**b**) Plan of the Medici chapel. Drawing: Kim Williams

Chapel to the extent that the ratio of the distance between the face of the altar and the opposite wall to the width of the chapel is 8:9, the ratio of the musical whole tone. Other dimensions within the chapel were derived from a combination of application of the law of repetition of ratios and the method of trisection illustrated in Fig. 37.8. The method of trisection is itself a means of generating the musical ratios. For example, Fig. 37.11 illustrates how this construction was used by the Renaissance architect Serlio to proportion the portal of a church. Notice the key ratio 1:2 (octave) in the proportion of the door and the ratios 2:3 (fifth) and 1:3 (fifth above an octave) in the positioning of the door.

As we saw in the last section, when the law of repetition of ratios is applied to the 1:$\sqrt{2}$ rectangle, it results in a gnomon equal to the original unit, i.e. the rectangle is divided into two similar $\sqrt{2}$ rectangles. If this construction is repeated, a sequence of $\sqrt{2}$ rectangles is formed with dimensions in the ratio of 1:2, 1:4, 1:8, ... in

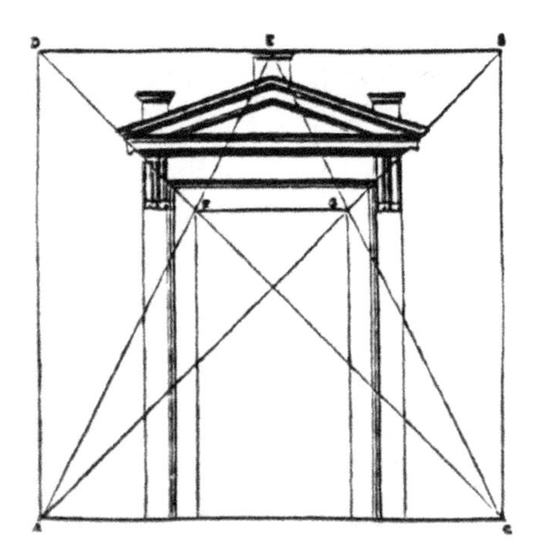

Fig. 37.11 Construction of a door using the method of trisection. From Serlio's first book. Image: Author, after Wittkower (1971: 127, Fig. 10)

comparison to the original. These ratios are the ones generated by the factors of 2 in Plato's lambda.

To demonstrate the geometric system in the Medici chapel, Williams constructed a √2 rectangle with dimensions 27 x 27√2, as shown in Fig. 37.12. By the trisection construction, the intersection point of BJ with diagonal AK is the vertex C of another √2 rectangle with side CD, 2/3 of 27 or 18. The construction is repeated to yield a family of √2 rectangles with short sides 27, 18, 12, 8 ... The law of repetition of ratios creates another sequence of √2 rectangles beginning with long side 27 and proceeding to 18, 12, 8 ... As is evident from Fig. 37.12, the ascending series: 4, 6, 8, 9, 12, 18, 27 inherent in this construction is obtained. These numbers are recognized as being components of the musical proportions of Fig. 37.2 derived from Plato's lambda. Furthermore, geometric series were considered to be proportional systems par excellence, with regard to Renaissance architecture. It was stated by Alberti, "The geometrical mean is very difficult to find by numbers but it is very clear by lines, but of those it is not my business to speak here" (Alberti 1755: 200).

Williams supplies the demonstration that length BC is the geometric means of AB and CD, i.e., $BC = \sqrt{27 \times 18}$. In a similar manner, the zig-zag path AB, BC, CD, DE, EF, FG, GH, ... yields a geometric series with common ratio $\sqrt{\frac{2}{3}}$ and another sequence of √2 rectangles beginning with the rectangle with sides BC: CK = 1:√2.

Choosing a key dimension of the chapel, 3.52 m, which is the overall width of the lateral bays from perimeter wall to the far edge of the pilaster, and using this as the value for side GK of the diagram, Williams found that the other proportional lengths generated in Fig. 37.11 corresponded with other dimensions which appear in the chapel. For example side EM, calculated at 4.31 m, corresponds to the clear width between pilasters of the *scarsella*, which actually measures 4.33 m, with a deviation of only 0.46 %. Making this the long side of a √2 rectangle, its short side, EF, is found to be 3.048, which corresponds to the dimension of half of the rectangle mentioned previously as circumscribing the ensemble of *scarsella* and

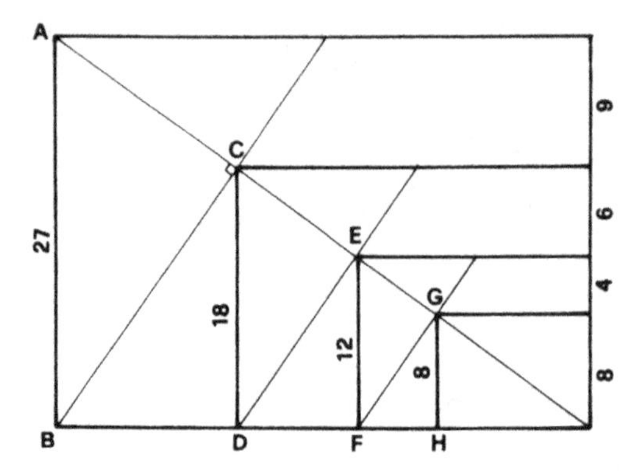

Fig. 37.12 Derivation of the proportions of the Medici Chapel. Drawing: Kim Williams

altar, deviating by only 0.9 % from the measured dimension of 3.02 m. The altar completely fills the other half of the √2 rectangle, and likewise measures 3.02 m. Work by historian Guglielmo De Angelis D'Ossat has revealed proportions in vertical elements of the chapel which, like the ground plan dimensions, may be seen in relation to the repeated trisection of the √2 rectangle. In a diagrammatic analysis of the portal found in each of the lateral bays of the chapel, he points out the ratios 1:2, 1:3, and 2:3. He has also found the ratio 1:2.4, which will be recognized as the proportions of the Roman rectangle (D'Ossat 1984: 189 and Fig. 190). This indicates that all the elements of the chapel were designed with regards to a comprehensive proportional system, and geometric series and musical proportions appear to have been the means of unification for all dimensions of the architecture.

Ehrenkrantz's System of Modular Coordination

In recent times, the architect Ezra Ehrenkrantz created a system of architectural proportion that incorporates aspects of Alberti's and Palladio's systems made up of lengths factorable by the primes 2, 3, and 5, along with the additive properties of Fibonacci series. To picture this system requires a three dimensional coordinate system as shown in Fig. 37.13. As a number moves from left to right, in the X direction, it doubles in value. As a number moves from back to front, in Z direction, the number triples in value. The sum of two numbers in the vertical, or Y direction, equals the next number in the series. Also notice that the upper edge of Plate 1 and the upper left corner points of Plates 1, 2, and 3 of Fig. 37.13 recreate Plato's lambda. The lambda along with the Fibonacci series 1, 2, 3, 5, 8 that comprises the first column of Plate 1 provide the boundary conditions upon which all other numbers of Fig. 37.13 are generated. The Fibonacci series is truncated at 8 because the next number of this series, 13, is a prime number other than 2, 3, and 5.

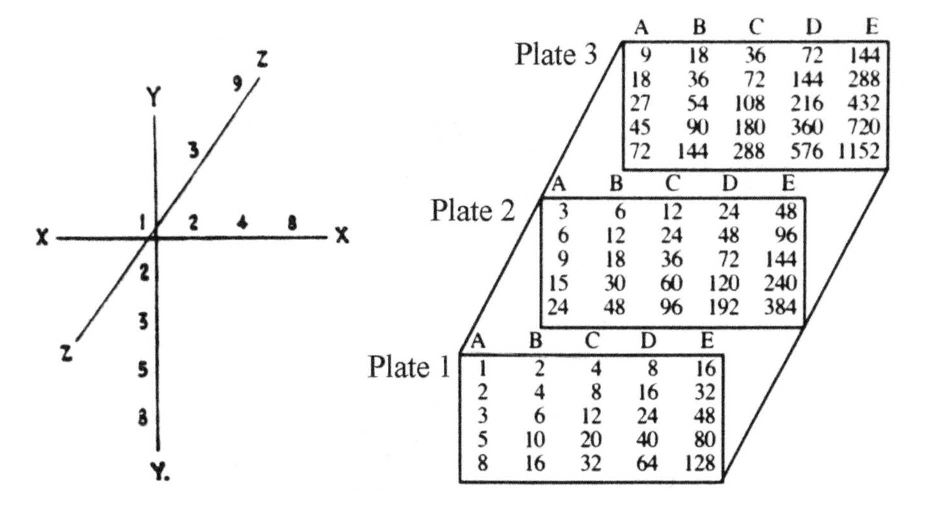

Fig. 37.13 Image: author

This system is successful at providing the architect with standard lengths that insure many possibilities for the subdivision of any length that appears in Fig. 37.14 For lengths up to 144 in., 35 dimensions are available, 20 of which are greater than 2 ft. Of course, dimensions which do not appear in Fig. 37.14, such as 99 in., can be created as the sum of elements from the figure, e.g. $99 = 27 + 45 + 27$. However, compare all the possibilities for creating 96, a number from the figure, with 99 a number not in the figure. Figure 37.15 shows that there are 11 possible summations for 96 as compared to only two for 99.

According to Ehrenkrantz, and referring to Fig. 37.13:

> This system helps to coordinate dimensions which may come from different base modules and therefore be normally considered incompatible. More directly, if one wishes to use multiples of 3 in. with those of 4 in., one can move to the right on the X axis from 3 in. and backwards on the Y axis from 4 in. They intersect at 12 (Plate 2, Column C) and all the numbers below, to the right, or behind 12 are multiples of both 3 in. and 4 in., i.e., foot intervals. Multiples of other base dimensions may be related in a like manner (Ehrenkrantz 1956).

Conclusion

We have discussed a series of relationships inherent in the musical scale well known to the Greeks and even more ancient civilizations, and we showed that they form the basis of three successful systems of proportionality (in addition to the Modulor of Le Corbusier not discussed here): a system used by Alberti and other Renaissance architects, the system of proportions used by the Romans, and a system of modular coordination of Ezra Ehrenkrantz. These systems ensure a repetition of key ratios, and possess properties related to the musical scale, while the latter two systems provide for additive properties that enable designs to be carried out in

Fig. 37.14

Foot Intervals	Dimensions	Foot Intervals	Dimensions
12 ft.	144 in. 360 cm.*	3 ft	36 in. 90 cm.*
11 ft.	135 in. 337.5 cm.		32 in. 80 cm.
	128 in. 320 cm.		30 in. 75 cm.
10 ft.	120 in. 300 cm.		27 in. 67.5 cm.
9 ft.	108 in. 270 cm.	2ft	24 in. 60 cm.
8 ft.	96 in. 240 cm.		20 in. 50 cm.
7 ft.	90 in. 225 cm.		18 in. 45 cm.
	81 in. 202.5 cm.		16 in. 40 cm.
	80 in. 200 cm.		15 in. 37.5 cm.
6 ft.	72 in. 180 cm.	1ft	12 in. 30 cm.
	64 in. 160 cm.		10 in. 25 cm.
5 ft.	60 in. 150 cm.		9 in. 22.5 cm.
	54 in. 135 cm.		8 in. 20 cm.
4 ft.	48 in. 120 cm.		6 in. 15 cm.
	45 in. 112.5 cm.		5 in. 12.5 cm.
	40 in. 100cm.		4 in. 10 cm.
			3 in. 7.5 cm.
			2 in. 5 cm.
			1 in. 2.5 cm.
		0 ft	

Fig. 37.15

99″

27	45	27
36	27	36

96″

48 48
32 32 32
64 32
36 60
30 36 30
24 48 24
36 24 36
24 24 24 24
24 72
24 45 27
40 32 24

which the whole equals the sum of its parts. These systems can also be expressed in terms of integers to facilitate their use.

Biography Jay Kappraff holds a Ph.D. from the Courant Institute of Mathematical Science at New York University. He was associate professor of mathematics at the New Jersey Institute of Technology where he has developed a course in the mathematics of design for architects and computer scientists. Prior to that, he taught at the Cooper Union College in New York City and held the position of aerospace engineer at NASA. He has published numerous articles on such diverse subjects as fractals, phyllotaxy, design science, plasma physics, passive solar heating and aerospace engineering. He has also lectured widely on the relationships between art and science. He is the author of *Connections: The Geometric Bridge Between Art and Science* (1st ed, 1991 2nd ed., 2001). His book, *Mathematics Beyond Measure: A Random Walk Through Nature, Myth and Number,* was published in 2003.

References

ALBERTI, Leon Battista. 1755. *The Ten Books of Architecture*. Rpt. NewYork: Dover Publications, 1986.

BRUNÉS, Tons. 1967. *The Secrets of Ancient Geometry – and Its Use*. Copenhagen: Rhodos.

D'OSSAT, Guglielmo De Angelis. 1984. La Sagrestia Nuova. In Umberto Baldini and Bruno Nardini, eds. *San Lorenzo*. Florence: Nardini Editori.

EDWARDS, Edward B. 1967. *Pattern and Design with Dynamic Symmetry*. Originally published as *Dynamarhythmic Design* (1932). New York: Dover Publications.

EHRENKRANTZ, Ezra. 1956. *Modular Number Pattern*. London: Tiranti.

HAMBIDGE, Jay. 1967. *The Elements of Dynamic* Symmetry (1926). New York: Dover Publications.

___. 1979. *The Fundamental Principles of Dynamic Symmetry as They are Expressed in Nature and Art* Albequerque: Gloucester Art Press.

KAPPRAFF, Jay. 1990. *Connections: the Geometric Bridge between Art and Science*. New York: McGraw–Hill.

___. 1996a. Linking the Musical Proportions of the Renaissance with the Modulor of Le Corbusier and the System of Roman Proportions. *International Journal of Space Structures* **11**, 1 and 2.

___. 1996b. *Mathematics Beyond Measure*: *Essays in Nature, Myth, and Number*. New York: McGraw-Hill.

SCHOLFIELD, P.H. 1958. *The Theory of Proportion in Architecture*. New York: Cambridge University Press.

WATTS, Donald .J. and WATTS, Carol. M. 1986. Roman Apartment Complex. *Scientific American* **255**, 6, (Dec. 1986): 132-140.

WILLIAMS, K. 1994. The Sacred Cut Revisited: The Pavement of the Baptistry of San Giovanni, Florence. *The Mathematical Intelligencer* **16**, 2 (Sp. 1994): 18–24.

___. 1997. Michelangelo's Medici Chapel: The Cube, the Square, and the √2 Rectangle. *Leonardo. Journal of the International Society of Arts, Sciences and Technology* **30,** 2 (1997): 105-112.

WITTKOWER, Rudolf. 1971. *Architectural Principles in the Age of Humanism*. New York: Norton.

Chapter 38
From Renaissance Musical Proportions to Polytonality in Twentieth Century Architecture

Radoslav Zuk

Introduction

The predilection of Renaissance architects and theorists for proportional systems based on consonant musical intervals is well known to scholars of that period. For those who do not have the opportunity to peruse the relevant treatises and to measure exemplary buildings, the writings of Rudolf Wittkower, and especially his seminal *Architectural Principles in the Age of Humanism* (1971), published in the middle of the past century, provide a most helpful guide. In the chapter entitled *The Problem of Harmonic Proportion in Architecture*, Wittkower summarizes the principal reasons for this interest and focuses his attention on the writings and buildings of Leon Battista Alberti and Andrea Palladio, two of the most prominent and influential Renaissance architects, to provide tangible examples. He refers specifically to the Pythagorean monochord, by means of which the relationships between any two consonant sounds can be expressed as whole number ratios, and to the two Platonic sequences of numbers, traditionally shown in the configuration of a *Lambda*, which contains such "consonant" ratios. Wittkower argues that the Renaissance preoccupation with "harmonic" proportions, that is, specific relationships of whole number ratios, stemmed from an attempt to reflect in the built environment the cosmic order which, according to contemporary philosophy, was based on such proportions and revealed in music (1971: 101–154).

An earlier version of this chapter was published as: Radoslav Zuk, "From Renaissance Musical Proportions to Polytonality in Twentieth Century Architecture", pp. 173–188 in *Nexus V: Architecture and Mathematics,* Kim Williams and Francisco Delgado Cepeda, eds. Fucecchio (Florence): Kim Williams Books, 2004.

R. Zuk (✉)
School of Architecture, McGill University, 815 Sherbrooke Street West, Montreal, QC H3A 2K6, Canada
e-mail: radoslav.zuk@mcgill.ca

K. Williams and M.J. Ostwald (eds.), *Architecture and Mathematics from Antiquity to the Future*, DOI 10.1007/978-3-319-00137-1_38,

Towards the end of the chapter Wittkower reviews several instances of the opposition to an objective architectural design approach, implied by the reliance on predetermined proportions, a trend which began in the middle of the seventeenth and continued well into the twentieth century. The subjective attitudes which were engendered by this opposition moved the revolutionary architect, Le Corbusier, to comment that "The Greek, the Egyptian, Michelangelo or Blondel employed regulating lines...The man of to-day employs nothing at all...But he proclaims that he is a free poet and that his instincts suffice..." (Le Corbusier 1983: 70). The subsequent rational attitudes of Modernism, which welcomed Wittkower's findings (Payne 1994: 339), did not last very long, and the free-for-all design trends that were ushered in with Post-Modernism continue to the present day. The subjective and essentially anti-intellectual attitudes in the current design culture are exemplified not only by the fashionable idiosyncratic contemporary architectural production, but also by statements of some of its more prominent exponents. This should not be surprising, as there appears to be little, if any theoretical support for objective design principles in general, and the music-architecture proportional parallels in particular. Among the rational objections raised is the lack of a demonstrable physical proof of optical equivalents to the measurable frequency ratios of consonant acoustic intervals (Forssman 1973: 35). Branko Mitrović (1993) points to another area of skepticism, which centers around the issue of metaphysical or organic interpretations of historical proportional systems.[1] However, he advocates the possibility of an empirical theory of proportion, outside of any system of reference, but rather based on the study of the great works of the past.

This is exactly what Howard and Longair have undertaken in their brilliant study, "Harmonic Proportion and Palladio's *Quattro Libri*" (1982), albeit within the limited sample of the nevertheless impressively prodigious work of Palladio. They examined the length and width dimensions of all the plans illustrated in Book II of Palladio's treatise and found that while not all the examined buildings have plans that correspond fully to musical proportions, there is a significant group of buildings whose plans are based entirely or almost entirely on harmonic numbers. Among them are "... the Villas Emo, Badoer, Barbaro, Malcontenta and Rotonda." In their words, "It can hardly be fortuitous that these are probably the most famous and best loved of all Palladio's villas and palaces" (Howard and Longair 1982: 126–127).

That negative attitudes towards objective theories of architecture have existed not only in the past few centuries, but even in the Renaissance period, is best illustrated by Alberti's statement that

> ... there are some who will ... say that Men are guided by a Variety of Opinions in their Judgment of Beauty and of Buildings; and that the Forms of Structures must vary according

[1] See Mitrović (1993: 67), where he perhaps best summarizes the prevailing attitudes when he states that "The mystic metaphysical argumentation of the theory of proportion can hardly expect to find serious supporters nowadays...," or earlier, "...who would be prepared today to explain the beauty of a numerical relationship by analogy with musical intervals, quoting Pythagorean and Platonist mysticoastrological speculations in support of such a theory?," and is equally skeptical of analogies with natural organisms.

> to every Man's particular Taste and Fancy, and not be tied down to any Rules of Art. A common Thing with the Ignorant, to despise what they do not understand! (Alberti 1955: bk. VI, ch. II, p. 113).

Order

Indeed, it is not easy to grasp something, which is essentially transcendental in its nature—that evasive abstract order that underlies great works of art. It is an order that is sensed rather than consciously perceived and thus lies beyond easy literal interpretations. The great twentieth-century architect Louis I. Kahn (1955: 59) stated: "Order is." He did not continue to explain. An objective analysis of his buildings is necessary to identify the elements of that pervasive order. What such an analysis yields is the distinct coherence and poetry of each of the systems that are integral to a great work of architecture and of their inventive synthesis in an inspiring built form.[2] Among those systems, the system of geometry and of concomitant proportions occupies a central position, as has been amply demonstrated by Klaus-Peter Gast (1998). Earlier in the century a very young Le Corbusier posed the following question: "What is the rule that orders, that connects all things?" (Le Corbusier 1954: 26). Later in life he proclaimed that "...axes, circles, right angles are geometrical truths... Geometry is the language of man" (Le Corbusier 1983: 68). Eventually this quest for order led him to spend several years on developing the well-known system of proportions, the Modulor (Le Corbusier 1954: 36–38). Although not based entirely on consonant whole number ratios, but rather on a combination of one of them with the golden ratio (Zuk 2013: 159), it bears witness to a commitment to objective ordering principles on the part of a creative genius.

The problem with the acceptance of the musical intervals theory may lie not only in the lack of an exactly matching physiological analogy, but also in its frequently narrow interpretation. Limiting the theory to the length to width ratios in isolated spaces or objects renders only the most basic and potentially banal results. After all, many works of Western music contain primarily consonant intervals. They are also those that are often quite ordinary and boring.

Another problem may lie in Wittkower's legacy of linking the theory almost exclusively to Platonic cosmological philosophy. In view of the gigantic advances in scientific knowledge over the past five centuries, this cannot help but generate skepticism. One may argue, however, that it is the sense of the convincing order inherent in the mathematical structures underlying and revealed in the relationships of consonant intervals which has motivated thinkers in various periods of human evolution to formulate philosophical theories based on them. Such theories may then have been used to provide legitimacy to what great artists have discovered to

[2] The systems structure was mentioned already in Zuk (1983: 3) and explored further in Zuk (1999: 65–67).

be intrinsic to the essential quality of their art, and thus to establish sanctioned guidelines for others to follow.

The studies of Howard and Longair suggest that an empirical theory of proportion does not need to exclude the already known proportional systems that may have been legitimized by classical philosophy, but rather may confirm their validity, once a sufficient number of examples is analyzed.

Numerical Relationships

To be meaningful, the comparison between consonant musical intervals and architectural proportions must be extended beyond the parallel between a single frequency ratio of two musical pitches and a length-to-width room ratio. An examination of more complex structures reveals that more comprehensive comparable relationships are inherent in the systems of musical harmony and architectural proportions, respectively. The two Platonic Lambda series of numbers provide a useful starting point, as has been demonstrated by Wittkower.

The series "1, 2, 4, 8" and "1, 3, 9, 27" consist, respectively, of a progression of *geometric means*, $b : a = c : b$, etc. (i.e., 2:1 = 4:2 = 8:4, and 3:1 = 9:3 = 27:9). Wittkower, quoting Francesco Giorgi, who in turn quotes Ficino's Commentary to Plato's *Timaeus*, points out that the two series can be filled in by additional numbers, yielding two additional series "6, 12, 24, 48" and "6, 18, 54, 162" respectively. Both of these series are also *geometric* (i.e., 12:6 = 24:12 = 48:24, and 18:6 = 54:18 = 162:54, respectively). By introducing *arithmetic means,* $b - a = c - b$, and *harmonic means,* $\frac{b-a}{a} = \frac{c-b}{c}$, between the two original series and/or their extensions the same numbers may be obtained (Wittkower 1971: 111). The extended structure may then be shown as follows:

1	2		4		8		16		32		64	...
		3		6		12		24		48		...
					9		18		36		72	...
								27		54		...
											81	...

The numbers following the *arithmetic means* are:
3–2 = 4–3; 6–4 = 8–6; 12–9 = 9–6; 12–8 =16–12; etc.
The numbers following the *harmonic means* are:

$$\frac{4-3}{3} = \frac{6-4}{6}; \quad \frac{8-6}{6} = \frac{12-8}{12}; \quad \frac{12-9}{9} = \frac{18-12}{18} \quad \frac{36-27}{27} = \frac{54-36}{54}; \text{ etc.}$$

The mathematical rigor underlying these relationships is remarkable, irrespective of whether it reflects a "cosmic order" or not. It may therefore provide sufficient rational justification for the employment of these numbers for dimensional relationships in the human environment, with the aim of establishing an abstract order that is in tune with the logical side of the human mind.

Also, it may still be considered significant that specific ratios within this structure constitute the mathematical basis for those musical intervals (literally "distances" between specific musical "pitches," but actually ratios of frequencies of sound), which are perceived by human hearing as being pleasant and therefore referred to as consonant or harmonious. It is a natural phenomenon, rooted in mathematics, to which not only the human ear but also the human eye, as when viewing a Palladian villa, responds with delight.

Consonant Ratios

The consonant intervals referred to as Pythagorean or "perfect" are the frequency ratios of the Unison, 1:1; the Octave, 2:1; the Perfect Fifth, 3:2; and the Perfect Fourth 4:3 (Backus 1969: 116). As can be seen, they are generated by the first four numbers of the two Lambda series, 1, 2, 3, and 4. Moreover, the same ratios of other numbers in the series (e.g., 9, 18, 27 and 36, or 16, 32, 48 and 64) can generate equivalent intervals.

The phenomenon of the similarity of the acoustical effect of those notes whose pitches (frequencies) are at the intervals of octaves (e.g., at double, quadruple, half, quarter, eighth, etc.) of any frequency, has led to the concept of "pitch class." Therefore, all such notes are considered to belong to a specific pitch class. Hence, any interval added to the interval of an octave, or to its double, triple, etc., produces a similar acoustical effect as the interval itself. Thus the ratio 3:1 (3:2 and 2:1, or 3:2 × 2:1, or the interval of a Perfect Fifth + an Octave), is acoustically similar to a Perfect Fifth (3:2) alone. Likewise 6:1 (3:2 and 4:1, or 3:2 × 4:1, or the interval of a Perfect Fifth + two Octaves), is similar to a Perfect Fifth alone. This fact allows the establishing of relationships equivalent to small number ratios between any of the respective numbers in the extended Lambda series.

It can be observed that the ratios of the first four numbers, or their multiples, contained within the extended Lambda series, correspond to the dimensional ratios for the lengths and widths of "areas" (plans of spaces) recommended by Alberti around the middle of the fifteenth century in the first, and thus highly influential, architectural treatise since antiquity, *De Re Aedificatoria* (Alberti 1955: bk. IX, ch. VI, pp. 197–199).

These ratios are: three ratios for "short" areas (1:1, 3:2, and 4:3); three ratios for "intermediate" areas (2:1, 9:4, and 16:9); and three ratios for "long" areas (3:1, 8:3, and 4:1).

As we have seen, 1:1 is the Unison, 3:2 is the interval of the Perfect Fifth, 4:3 is the interval of the Perfect Fourth, 2:1 the interval of the Octave, and 3:1 (3:2 and

2:1) is the interval of the Octave plus the Perfect Fifth, 8:3 (8:6 and 6:3) is the interval of the Octave plus the Perfect Fourth, and 4:1 (4:2 and 2:1) is the interval of two Octaves. These are all "perfect" consonant frequency ratios. Two ratios for "intermediate" spaces are dissonant intervals. 9:4 is the interval of the Major Ninth, and 16:9 the interval of the Minor Seventh. Wittkower tried to explain this apparent inconsistency by a non-musical argument. He suggested that Alberti simply added two consonant intervals to produce these ratios: two Perfect Fifths (9:6 and 6:4—each 3:2) for the 9:4 ratio, and two Perfect Fourths (16:12 and 12:9—each 4:3) for the 16:9 ratio (Wittkower 1971: 116).

But this may have troubled Palladio. He avoided these two ratios when some 120 years later he recommended a similar set of length-to-width ratios for plans of architectural spaces, in Book I of another highly influential treatise, *I Quattro Libri dell' Architettura*, published in Venice in 1570.

Palladio's recommended ratios for rectangular plan shapes are: 1:1, 3:2, 4:3, 2:1, 5:3, and √2:1 (or approximately 45:32). Palladio adds also the shape of the circle to this list (1997: bk I, ch. XXI, p. 57). The first four of these ratios are the same as those recommended by Alberti. They also contain the first four numbers of the Lambda series. In musical terms they are all perfect ratios. The circle (the perfect geometric figure, with a constant diameter) may be seen as the Unison, like the square (1:1), and can be included in this group of ratios. The √2:1, or diagonal of the square: the side of the square (1.414:1), is a dissonant musical frequency ratio and will be dealt with later. The ratio 5:3 will be discussed below.

The complete list of equivalent musical intervals is as follows:

- 1:1 and the circle—the Unison;
- 2:1—the Octave;
- 3:2—the Perfect Fifth;
- 4:3—the Perfect Fourth;
- 5:3—the Major Sixth;
- √2:1 (approximately 45:32)—the Tritone (the Augmented Fourth or the Diminished Fifth).

Wittkower comments astutely that Palladio's inclusion of the 5:3 ratio among his recommended floor plan ratios may have been due to the influence of Ludovico Fogliano of Modena, who in 1529 asserted in his *Musica theorica* that the Major and Minor Thirds, Sixths, Tenths, and Thirteenths are also consonant intervals (Wittkower 1971: 132–133). The frequency ratios producing what is now referred to in musical terminology as "just" consonant intervals are:

- 5:4—the Major Third;
- 6:5—the Minor Third;
- 5:3—the Major Sixth;
- 8:5—the Minor Sixth (Backus 1969: 122).

The Major and Minor Tenths and the Major and Minor Thirteenths are the respective Thirds and Sixths added to Octaves, producing equivalent acoustic effects. This follows the principle of pitch class discussed above in reference to intervals whose ratios are greater than 2:1.

Although Palladio recommends only the 5:3 ratio from among the just intervals, Wittkower states that he used also 5:4, 6:5, and other ratios, which include the number 5 in the designs of his buildings (Wittkower 1971: 132). The √2:1 ratio was and is still technically referred to today as a dissonant interval. So are the ratios 9:8, 9:4 and 16:9, discussed previously.

What may be of interest is to place the number 5 in the mathematical context of the extended Lambda series. As could be seen, these series are made up of geometric, arithmetic and harmonic progressions. Another series based on the progression $c = a + b$, $d = b + c$, etc., starting at 1 yields 1, 2, 3, 5, and 8. This of course is the Fibonacci series, which in higher number ratios approaches the golden ratio. This (Golden) frequency ratio, 1.618:1, is in its effect quite discordant, as it falls (not evenly) between the Major and the Minor Sixth. It is not one of the accepted musical dissonant interval ratios, such as 9:8 or √2:1, which, as will be seen later, form part of distinct musical structures. However, the first five numbers of the Fibonacci series, when combined with the Lambda numbers, can readily yield the just ratios, 5:4, 6:5, 5:3 and 8:5. This can be considered as significant, yet seems to have been overlooked by Wittkower.

The combined series reads then as follows:

1	**2**		4		**8**		16		32		64	...
			5									
		3		6		12		24		48		...
					9		18		36		72	...
								27		54		...
											81	...

The Fibonacci series numbers are shown in bold in the above diagram. The new number is 5 and if it were extended into an arithmetic series (shown also in bold), the beginning of the combined new structure would look as follows:

1	2		4			8				16				32			64	...
				5			**10**		**15**		**20**		**25**		...			
		3			6			12				24				48		...
						9				18				36			72	...
												27				54		...
																	81	...

The significance of the musical implications of the proportional relationships between the following numbers in the modified Lambda series above—1, 2, 3, 4, 5, 6, 8, 9, 12, 15, 16—lies in the fact that they constitute the basis for the system of tonal harmony.

Tonal Harmony

In describing the system of tonal harmony the eminent musicologist, Donald Jay Grout, states:

> That system was the major-minor tonality. . .in the music of the eighteenth and nineteenth centuries: all the harmonies of a composition organized in relation to a triad on the key note or tonic supported primarily by triads on its dominant and subdominant. . . This particular tonal organization had long been foreshadowed in music of the Renaissance, especially that written in the latter half of the sixteenth century. Rameau's *Treatise on Harmony* (1722) completed the theoretical formulation of the system. . . (Grout 1980: 303).

Actually, in spite of the revolutionary twentieth century explorations of alternative harmonic structures, the use of the tonal system or of its modified versions is still very much present today. According to the prominent composer and theorist, Walter Piston, even the "twelve-tone" system, which represents an extreme departure from the tonal system, ". . .as well as almost all music intended to be atonal, that is without a key, comes sooner or later to be heard tonally" (Piston 1962:336).

Grout refers to "triads." A triad is a formation consisting of three notes (frequencies) in a specific relationship of two adjacent intervals of "thirds." (Triads are thus specific chords, a chord being a formation of three or more simultaneously sounding notes). The lowest note (frequency) of a triad is referred to as the "root," the middle note as the "third," and the highest as the "fifth," of the respective chord.

The triad whose root is in the relationship of the intervals of a Perfect Fifth (3:2) and of a Perfect Fourth (4:3), to the roots of two other triads, respectively, becomes the defining chord of a "key" or "tonality." This root note is referred to as the "key-note" or "tonic," the other two as the "dominant" and the "subdominant," respectively. The three triads are also similarly referred to as the Tonic, the Dominant, and the Subdominant of a Tonality. Between them these three triads contain all the seven basic notes (frequencies) of a Key or Tonality. The sequence of ascending frequencies of these notes is referred to as a "diatonic scale." Each of the seven notes is also a potential root for each of the seven triads belonging to the respective tonality. This includes the already familiar principal Tonic, Dominant and Subdominant triads, as well as four secondary triads.

The sequence of notes in an ascending Diatonic Scale helps to explain the designation of musical intervals as, for example, a Fourth or as a Fifth. Since the root of the Tonic triad is also the first note (or "degree") of a Tonality, the nearest higher (frequency) note in the scale is the Second, the one after that, the Third, and thus the Fourth, the Fifth, the Sixth, the Seventh, and then the Octave follow.

The above description is based on frequencies within an Octave. As was already pointed out, similar patterns pertain at double or other exponential frequencies. Hence, the accepted designation of a note frequency by a letter (for example, C, F, A-flat, etc.) is maintained for the double, the half, or the quadruple of that frequency.

As this discussion is meant only to introduce the complex realm of tonal music at the most basic level, the "chromatic" scale, which yields twelve notes in an Octave, as well as the "enharmonic" tonalities, will not be discussed here. The twelve Major and twelve Minor tonalities (and their variants), which are based on these twelve notes and which bring the total of diatonic tonalities to twenty four, will be dealt with briefly below.

Taking into account the various ratios inherent in the modified Lambda series one can observe their striking coincidences with several of the following structuring musical phenomena.

Scales

The degrees of a Diatonic Scale are determined by the intervals between the Tonic (Keynote) and the respective notes, and thus by the relevant frequency ratios. These degrees will be designated here by Roman numerals, I, II, III, IV, etc. It should be noted that some degrees may form either a "major" or a "minor" interval with the Tonic, depending on whether it is in a Major or in one of the variants of a Minor tonality, and the designations + (= Major) and – (= Minor) will be used accordingly. The corresponding degrees, intervals and ratios are shown in Tables 38.1, 38.2, 38.3, and 38.4.

In the Natural Minor Scale, the degrees of the ascending and the descending versions are identical and follow the pattern of the descending Melodic Minor scale shown in Table 38.4.

A diatonic scale can be constructed also with mathematical "means." The importance of considering means here lies in the fact that they deal with three quantities that are combined in very distinct structured patterns. Of crucial importance are the means of the intervals of Fifths, as they generate triads, the basic harmonic structural elements of the tonal system.

Triads

As was already mentioned, the Tonic, the Dominant, and the Subdominant triads contain between them all the seven basic notes of a tonality. The Tonic chord (triad) contains the following three degrees of a tonality, starting from the lowest to the highest (frequency): I, III, V; the Dominant: V, VII, II; and the Subdominant, IV, VI, and I.

Each of the secondary triads contains also three of the seven basic notes of a tonality, at intervals of thirds. The specific notes of each of these chords are determined by the position of the root of the respective chord with respect to the Tonic. Thus the Supertonic triad contains (in ascending frequencies), II, IV, VI; the

Table 38.1 Major scale

Degree	Designation	Interval	Ratio
I	Tonic	Unison	1:1
II+	Supertonic	Major Second	9:8
III+	Mediant	Major Third	5:4
IV	Subdominant	Perfect Fourth	4:3
V	Dominant	Perfect Fifth	3:2
VI+	Submediant	Major Sixth	5:3
VII+	Leading note	Major Seventh	15:8
I	Tonic	Octave	2:1

Table 38.2 Minor scale (Harmonic)

Degree	Designation	Interval	Ratio
I	Tonic	Unison	1:1
II+	Supertonic	Major Second	9:8
III–	Mediant	Minor Third	6:5
IV	Subdominant	Perfect Fourth	4:3
V	Dominant	Perfect Fifth	3:2
VI–	Submediant	Minor Sixth	8:5
VII+	Leading note	Major Seventh	15:8
I	Tonic	Octave	2:1

Table 38.3 Minor scale (Melodic—ascending)

Degree	Designation	Interval	Ratio
I	Tonic	Unison	1:1
II+	Supertonic	Major Second	9:8
III–	Mediant	Minor Third	6:5
IV	Subdominant	Perfect Fourth	4:3
V	Dominant	Perfect Fifth	3:2
VI+	Submediant	Major Sixth	5:3
VII+	Leading note	Major Seventh	15:8
I	Tonic	Octave	2:1

Table 38.4 Minor scale (Melodic-descending)

Degree	Designation	Interval	Ratio
I	Tonic	Unison	1:1
II+	Supertonic	Major Second	9:8
III–	Mediant	Minor Third	6:5
IV	Subdominant	Perfect Fourth	4:3
V	Dominant	Perfect Fifth	3:2
VI–	Submediant	Minor Sixth	8:5
VII–	Subtonic	Minor Seventh	9:5
I	Tonic	Octave	2:1

Mediant triad, III, V, VII; the Submediant triad, VI, I, III; and the Leading Note triad, VII, II, and IV.

Depending on whether the above chords belong to a Major or a Minor Tonality, some of them will be “major,” others “minor,” and others still, “diminished,” or "augmented" (one only) triads. This depends on the interval between the degree of each of the two upper notes in the respective chord and the degree of its root.

The Major triad consists of three notes which are produced by the arithmetic mean of the interval of the Perfect Fifth (3:2): 6:5:4; that is, the interval between the root and the third of the chord is a Major Third (5:4), and between the third and the fifth of the chord, a Minor Third (6:5).

The Minor triad consists of three notes which are produced by the harmonic mean of the interval of the Perfect Fifth (3:2): 15:12:10; that is, the interval between the root and the third of the chord is a Minor Third (6:5), and between the third and the fifth of the chord, a Major Third (5:4).

The Diminished triad consists of three notes which are produced by the geometric mean of the interval of the Diminished Fifth (Augmented Fourth), called also the Tritone, ($\sqrt{2}$:1): approximately 45:38:32; that is, the interval between the root and the third of the chord is a Minor Third (approximately 6:5), and between the third and the fifth of the chord, also a Minor Third (approximately 6:5). The interval of the Tritone is produced by the geometric mean of the Octave (64:45 = 45:32, approximately).

The Augmented triad consists of three notes which are produced by the geometric mean of the interval of the Augmented Fifth (Minor Sixth) (8:5)—approximately 32:25:20; that is, the interval between the root and the third of the chord is a Major Third (5:4), and between the third and the fifth of the chord, also a Major Third (approximately 5:4).

The four types of triads—Major, Minor, Diminished and Augmented—are distributed among the seven degrees of the Major, and of the three variants of the Minor, tonalities as shown in Tables 38.5, 38.6, 38.7, and 38.8.

Inversions

Because of the phenomenon of pitch class, the "third" and the "fifth" (note) of a triad, each belonging to its own respective pitch class, can be disposed at intervals of one, or more, octaves above or below their original position in relation to the root of the respective triad, and still maintain the harmonic character of the resulting chord. Placing these notes an octave, or more, "higher" (at a higher frequency) maintains the original interval, with the interval of one octave, or more, added with respect to the root. Placing them one octave, or more, "lower" and thus below (the frequency of) the root also maintains the harmonic character, or technically "function," of the new chord, but introduces new intervals between the three notes. The latter new positions are called "inversions."

Thus, a First Inversion occurs when the third of a triad is lowered by (the interval of) an octave to become the "bass" (lowest frequency) note of the new chord. The interval between it and the root (now above) is a Minor Sixth (the ratio of 8:5) in a Major triad, and a Major Sixth (the ratio of 5:3) in a Minor triad. When the fifth of the triad is also lowered by an octave, the interval between it and the root (above) is

Table 38.5 Major key (Tonality)

Root degree	Designation	Type	Ratios
I	Tonic	Major	6:5:4
II+	Supertonic	Minor	15:12:10
III+	Mediant	Minor	15:12:10
IV	Subdominant	Major	6:5:4
V	Dominant	Major	6:5:4
VI+	Submediant	Minor	15:12:10
VII+	Leading Note	Diminished	45:38:32

Table 38.6 Minor key—harmonic

Root degree	Designation	Type	Ratios
I	Tonic	Minor	15:12:10
II+	Supertonic	Diminished	45:38:32
III–	Mediant	Augmented	32:25:20
IV	Subdominant	Minor	15:12:10
V	Dominant	Major	6:5:4
VI–	Submediant	Major	6:5:4
VII+	Leading Note	Diminished	45:38:32

Table 38.7 Minor key—melodic (ascending)

Root degree	Designation	Type	Ratios
I	Tonic	Minor	15:12:10
II+	Supertonic	Minor	15:12:10
III–	Mediant	Augmented	32:25:20
IV	Subdominant	Major	6:5:4
V	Dominant	Major	6:5:4
VI+	Submediant	Diminished	45:38:32
VII+	Leading Note	Diminished	45:38:32

Table 38.8 Minor key—Melodic (descending) and natural

Root degree	Designation	Type	Ratios
I	Tonic	Minor	15:12:10
II+	Supertonic	Diminished	45:38:32
III–	Mediant	Major	6:5:4
IV	Subdominant	Minor	15:12:10
V	Dominant	Minor	15:12:10
VI–	Submediant	Major	6:5:4
VII–	Subtonic	Major	6:5:4

a Perfect Fourth (the ratio of 4:3) in both a Major and a Minor triad. The interval ratios within the latter two chords are thus 8:6:5 in a Major chord, and 20:15:12 in a Minor chord, respectively. Similarly, a Second Inversion occurs when the fifth of a triad is lowered by an octave to become the bass note of the new chord. As above, the interval between it and the root is a Perfect Fourth (4:3). When the third of the triad remains in its original position, the interval ratios within the new chords are 5:4:3 in a Major chord, and 24:20:15, in a Minor chord, respectively.

Inversions of a diminished triad (within an octave) produce combinations of related intervals; that is, of an Augmented Fourth ($\sqrt{2}$:1, or approximately 45:32) and a Minor Third (6:5), the latter being the approximate geometric mean of the first. The First Inversion gives the interval ratios of approximately 54:38:32, and the Second Inversion, the interval ratios of approximately 54:45:32 within the resulting chords, whose extreme notes are at the interval of a Major Sixth, or in the ratio of approximately 5:3.

Inversions of an augmented triad (within an octave) produce the same interval relationships as in the original chord. Since the latter consists of two Major Thirds (5:4), the remaining interval within the Octave is a Diminished Fourth (also 5:4), and thus the inversions consist of two acoustically, although not "functionally," equal intervals; that is, a Major Third and a Diminished Fourth, within the resulting chords, whose extreme notes are at the interval of an Augmented Fifth (acoustically equivalent to a Minor Sixth, or in the ratio of 8:5).

Tonal and Polytonal Composition

The great variety of chords and of their inversions within any one tonality offer endless possibilities of their combinations, even within the rather strict rules as regards their structure and sequence. Structured motion from one tonality to another, called "modulation," provides further opportunities of harmonic variety and is a mark of compositional sophistication. The ability of composers to deal creatively with this material and with its underlying structural principles has thus resulted in the extraordinary harmonic richness of the great works of the late Renaissance, Baroque, Classical and the Romantic periods of Western music.

At the beginning of the twentieth century the tonal system of harmony reached, it seemed, the end of its evolution. Various new compositional approaches were initiated, among them that of "polytonality." In this harmonic organization two or more distinct tonalities operate at the same time, hence the terms "bitonality" and "polytonality," respectively. Since each of the concurrent tonal systems maintains its own harmonic structure (series of chords), the ear may be in a position to perceive the inherent order of each, and thus the apparent conflict and the concomitant dissonances can be mitigated, while a new complex composite order emerges.

Several prominent composers have used this technique, such as Darius Milhaud, Bela Bartok, and probably the most influential among them, Igor Stravinsky, who applied it in several of his major works. Examples from two works by this undisputed master may provide illustrations of possible combinations of chords belonging to different tonalities. The first is the notorious—and, for its time (1911), shocking—dissonant sound of two superimposed Major triads, the dissonant interval of the Augmented Fourth ($\sqrt{2}$:1) apart, the lower one in its First Inversion, from the ballet *Petrushka*. The second example can be found at the end of the *Symphony of Psalms* (1931). It shows chords from two Major tonalities whose separating interval is the consonant interval of a Minor Third (6:5).

"Tonal Harmony" in Architecture

Since architecture is a volumetric entity, its geometric definition involves three dimensions. Bringing the length, width and height of individual spaces, of total buildings, and even of smaller architectural components, into optimal harmonic relationships may be thus considered as an essential design challenge. Alberti and Palladio seem to have been aware of that, and made recommendations for determining proportional dimensions for heights of spaces. Among them were the arithmetic, geometric, and harmonic means between the lengths and widths of the plans of such spaces. For square spaces, Palladio recommended the ratio of 4:3 between the height and the length (width) (Alberti 1955: bk. IX, ch. VI, pp. 199–200; Palladio 1997: bk. I, ch. XXIII, p. 58).

Following these recommendations interesting comparisons can be made. Taking Alberti's preferred length-to-widths ratios, distinct length-to-height-to-width proportions can be obtained (Table 38.9), while Palladio's recommended floor plan ratios can yield a substantially different set of spatial proportions (Table 38.10).

As can be seen the length-to-height-to-width (L:H:W) proportions in both charts show some striking parallels to the interval relationships in chords which form the constituent parts of musical structures. Table 38.10 shows also the evolution of Palladio's proportions beyond those of Alberti, in relation to the developments in tonal harmony. Proportions based on Palladio's recommended horizontal ratios, other than those related to the Unison, the Perfect Fourth, and the Octave, are equivalent to triads (or their inversions, in one case) with which either a Major or a Natural Minor tonality could be fully constructed.

"Tonality" and "Polytonality" in Architecture

While direct analogies can be drawn between the frequency ratios of a musical chord and the dimensional ratios of a single architectural space, the differences in the intrinsic nature of the realms of sound and of space may preclude such exact analogies when larger harmonic structures of musical works and the proportional structures of entire buildings are compared.

However, when it is considered that a meaningful architectural experience involves movement from space to space, a broad comparison can be made between such an experience and the experience of a musical composition, where a sequence of varied harmonic events (chords) is revealed to the listener. As in music, where the imaginative and judicious selection of these events and their

Table 38.9 Proportions derived from Alberti's preferred ratios

Plan ratio	Length	Width	Mean	Height	L:H:W	Music equivalent
1:1	1	1	Any mean	1	1:1:1	Unison
3:2	6	4	Arithmetic	5	6:5:4	Major triad
3:2	15	10	Harmonic	12	15:12:10	Minor triad
4:3	8	6	Arithmetic	7	8:7:6	None
2:1	4	2	Arithmetic	3	4:3:2	Perfect Fourth and Perfect Fifth
9:4	9	4	Geometric	6	9:6:4	Two Perfect Fifths
16:9	16	9	Geometric	12	16:12:9	Two Perfect Fourths
3:1	3	1	Arithmetic	2	3:2:1	Perfect Fifth and Octave
8:3	8	3	Fibonacci	5	8:5:3	Minor Sixth and Major Sixth
4:1	4	1	Geometric	2	4:2:1	Two octaves

Table 38.10 Proportions derived from Palladio's preferred ratios

Plan ratio	Length	Width	Mean	Height	L:H:W	Music equivalent
Circle	1	1	Any mean	1	1:1:1	Unison
1:1	3	3	Not a mean	4	3:4:3	Perfect Fourth (special)
3:2	6	4	Arithmetic	5	6:5:4	Major triad
3:2	15	10	Harmonic	12	15:12:10	Minor triad
4:3	8	6	Arithmetic	7	8:7:6	None
2:1	4	2	Arithmetic	3	4:3:2	Perfect Fourth and Perfect Fifth
5:3	5	3	Arithmetic	4	5:4:3	Major triad, second inversion
$\sqrt{2}$:1	45	32	Geometric	12	45:38:32	Diminished triad

structured relationships become a measure of the ultimate success of a work, so in architecture, the specific proportions of the component spaces and their interrelationships (not in a predetermined sequence as in most musical compositions, but in a variety of possible sequences) may determine decisively the inherent quality of a building or group of buildings.

A conjecture can be made that when such spaces are aligned orthogonally, then their proportional relationships may be compared to those of musical chords, which belong essentially to one tonality. Likewise, a building, a group of buildings, or even entire cities, where the architecture is experienced as a series of spatial events revealed in sequences that follow distinct changes in direction, may be considered to be similar to compositions that exhibit a high degree of "modulation," that is, movement to and from other tonalities.[3]

A work of architecture where the spaces and/or volumes intersect at distinct angles, and are therefore experienced simultaneously, may be compared to a polytonal composition where the superimposition of two or more chords belonging to two (or more) respective tonalities is determined by a distinct interval or intervals. In such buildings two or more proportional systems are also superimposed at a distinct angle or angles (Zuk 1986).

[3] An attempt at a possible interpretation of modulation in architecture was made in Zuk (2003).

Assuming that several spaces within an architectural project will follow an orthogonal alignment within one or the other (or more) of its principal underlying "tonal" proportional systems, the "polytonal" proportional superimpositions can be extended to other spaces, and in fact, to an entire building, and even beyond, to the larger urban or rural environment. When each of these systems is of a highly coherent order, and the angle or angles of the superimposition are judiciously chosen, a convincing complex composite geometric order can result.

Since new dimensional ratios are produced in spaces where non-orthogonal juxtapositions of the main proportional systems occur, the character of the resulting totality will depend on both the nature of each of the component proportional systems, as well as on the angle(s) of the super-imposition(s) of these systems.

The geometric shifts in Le Corbusier's Carpenter Center for Visual Arts at Harvard University may be considered to represent a "polytonal" spatial composition. The angled deviation of the principal orthogonal building block from the urban grid and thus from the alignment of the adjoining orthogonal buildings exhibits a "bitonal" juxtaposition of two clearly articulated proportional systems (geometric grids). The straight portions of the two protruding, geometrically distinct, rounded major building components deviate at two different angles from the main block and thus imply two more "tonalities." The various angles of the ramp, even if the latter is relatively small in scale, contribute further to the "polytonal" richness of this twentieth century masterpiece (Sekler 1978: 345–357).

Richard Meier's *Museum für Kunsthandwerk* in Frankfurt offers another striking example of a sophisticated geometric superimposition. Two orthogonal grids intersect within the building at a distinct angle, which also has its reference in the larger urban context. Thus the walls of the interior courtyard, of the main ramp, of the crosswalk and of the café terrace deviate clearly from the prevailing geometry of the main body of the building, which is based on a square. Series of spaces of "bitonal" dissonance result, but the underlying rigor of each of the two principal proportional grids mitigates the "dissonant" and potentially disturbing effects and allows for a complex yet ordered work of architecture to emerge (Klotz and Krase 1985: 124–126).

These and a number of similar twentieth century projects demonstrate that the dynamism of the current aesthetic preference for geometric shifts and collisions in the built environment, which have their justification in the reality of the prevailing complexities in most spheres of human existence, can be revealed in rich and apparently discordant, yet ultimately coherent spatial compositions. Thus, instead of the arbitrary, idiosyncratic, decorative and shallow twists displayed in the configurations of many recent, popularly promoted buildings, equally imaginative, but deeply rooted in the underlying proportional order and therefore inherently profound, geometric structures can result. If other component systems of architectural built form of such a project are also of comparable coherence, then a truly great work of architecture can be achieved.

Biography Radoslav Zuk received a Bachelor of Architecture from McGill University in Montreal, a Master's in Architecture from MIT in Boston, and was recently awarded an honorary doctorate degree from the Ukrainian Academy of Art in Kyiv. He has taught architecture at the University of Manitoba, the University of Toronto, and at McGill University, where he is an Emeritus Professor. He has appeared as a guest lecturer and guest review critic at various universities in Canada, the United States, and Europe. He has designed, among other projects, ten Ukrainian churches in North America and one in Ukraine, in association with or as consultant to architectural firms. He has published on design theory, cultural aspects of architecture, and the relationships between architecture and music. A Fellow of the Royal Architectural Institute of Canada and of several scientific societies, he has received the Royal Architectural Institute of Canada Governor General's Medal for Architecture and Ukraine's State Prize for Architecture.

References

Alberti, Leon Battista. 1955. *Ten Books on Architecture*, Giacomo Leoni, trans. London: Alec Tiranti.

Backus, John. 1969. *The Acoustical Foundations of Music*. New York: W.W. Norton.

Forssman, Erik. 1973. *Visible Harmony*. Stockholm: Sveriges arkitekturmuseum and Konsthogskolans arkitekturskola.

Gast, Klaus-Peter. 1998. *Louis I. Kahn. The Idea of Order*. Basel: Birkhäuser.

Grout, Donald Jay. 1980. *A History of Western Music*. New York: W.W. Norton.

Howard, Deborah and Malcolm Longair. 1982. Harmonic Proportion and Palladio's *Quattro Libri*. *Journal of the Society of Architectural Historians* **41**, 2 (May1982): 116-143.

Kahn, Louis I. 1955. Order and Form. *Perspecta* **3**: 46-63.

Klotz, Heinrich and Waltraud Krase. 1985. *New Museum Buildings*. Stuttgart: Klett-Cotta.

Le Corbusier. 1954. *The Modulor: A Harmonious Measure to the Human Scale Universally applicable to Architecture and Mechanics*. London: Faber and Faber.

___. 1983. *Towards A New Architecture*. Frederick Etchells, trans. New York: Holt, Rinehart and Winston.

Mitrović, Branko. 1993. Objectively Speaking. *Journal of the Society of Architectural Historians* **52**, 1 (March 1993): 59-67.

Palladio, Andrea. 1997. *The Four Books on Architecture*, Robert Tavernor and Richard Schofield, trans. Cambridge, MA: MIT Press.

Payne, Alina. 1994. Rudolph Wittkower and Architectural Principles in the Age of Modernism. *Journal of the Society of Architectural Historians* **53**, 3 (September 1994): 322-342.

Piston, Walter. 1962. *Harmony*. New York: W.W. Norton.

Sekler, Eduard F. 1978. *Le Corbusier at Work*. Cambridge, MA: Harvard University Press.

Wittkower, Rudolf. 1971. *Architectural Principles in the Age of Humanism*. New York: W.W. Norton.

Zuk, Radoslav. 1983. A Music Lesson. *Journal of Architectural Education* **36**, 4 (Spring 1983): 2-6.

___. 1986. Stravinsky, Dissonance and Architecture. *The University of Tennessee Journal of Architecture* **9**: 20-26.

___. 1999. The Ordering Systems of Architecture. Pp. 63-67 in. *Advances in Systems Research and Cybernetics*, Vol. III, George E. Lasker, ed. Windsor: The International Institute for Advanced Studies in Systems Research and Cybernetics.

___. 2003. Modulation in Music and Architecture. Pp. 1-8 in. *Systems Research in the Arts,* Vol. IV, George E. Lasker, Jane Lily and James Rhodes, eds. Windsor: The International Institute for Advanced Studies in Systems Research and Cybernetics.
___. 2013. Three Musical Interpretations of Le Corbusier's Modulor. *Nexus Network Journal* **15**, 1: 155-170.

Chapter 39
Quasi-Periodicity in Islamic Geometric Design

Peter Saltzman

Islamic Geometric Design

It is a commonplace assertion that Islamic cultures share the world's oldest and most sophisticated living tradition of geometric ornamental design. Ever since Jules Goury and Owen Jones completed their monumental book on the Alhambra (1842–1845) and Jules Bourgoin published his classic work on Islamic designs (Bourgoin 1879), Western interest in Islamic geometric design has continued unabated.

Symmetry and Its Discontents

In addition to its aesthetic merit, Islamic geometric design is renowned for its mathematical sophistication, constituting the most highly developed chapter in cultural symmetry studies.[1] Dihedral symmetry groups of high order, all seven frieze groups, all seventeen crystallographic groups of plane isometries, and several non-trivial chromatic symmetry groups may be found in abundance in both eastern and western Islamic countries.[2]

First published as: Peter Saltzman, "Quasi-Periodicity in Islamic Geometric Design". Pp. 153–168 in *Nexus VII: Architecture and Mathematics,* Kim Williams, ed. Turin: Kim Williams Books, 2008.

[1] For a general introduction to the cultural applications of symmetry studies, see Crowe and Washburn (1991).

[2] Good introductions to the symmetries of Islamic geometric designs include the delightful book by Fenoll Hach-Alí and Galindo (2003) and the more comprehensive work by Abas and Salman (1995).

P. Saltzman (✉)
P.O. Box 9003, Berkeley, CA 94709, USA
e-mail: pwsaltzman@leonardcarder.com

K. Williams and M.J. Ostwald (eds.), *Architecture and Mathematics from Antiquity to the Future*, DOI 10.1007/978-3-319-00137-1_39,

Yet the focus of these symmetry studies has little resonance in the scant historical record documenting the techniques of the Islamic masters, and has limited relevance to the aesthetic complexity of Islamic geometric design. In many cases, one can reconstruct a design as a group orbit of a small motif, but this tells us little about the properties of the design other than its symmetry group.[3] The design within a single periodic unit cell is often quite complex, with symmetry playing a subsidiary role. Grünbaum and Shephard have argued against over reliance on group theory to interpret cultural artefacts, insisting that other mathematical measures of order or disorder—and not just symmetries—are necessary to explain the intrinsic features of the artefacts and better reflect the intentions of those who produced them (Grünbaum and Shepherd 1992).

Responding to this challenge, crystallographers Emil Makovicky and Purificacion Fenoll Hach-Alí have published a series of papers in the *Boletín de la Sociedad Española de Mineralogía* over the past decade or more (Makovicky and Fenoll Hach-Alí 1996, 1997, 1999, 2001), applying a variety of crystallographic structural classification principles to the interpretation of Nasrid designs in Spain. Thus, in addition to symmetries, they have developed informative analyses in terms of crystallographic shear, occupancy of Wyckoff positions, rotation of vortex elements and other crystallographic features that together contribute to the development of a more nuanced grammar of Islamic geometric design. Here, however, we will be concerned not with these crystallographic features, but rather with "quasi-crystallographic" features of certain Islamic designs.

The Presentist Fallacy

Before proceeding further, it is worthwhile to issue a caveat concerning the historical significance of mathematical properties that may attach to certain cultural artefacts. In discussing the elaborate symmetries of Islamic designs, for example, it is tempting to impute mathematical knowledge "ahead of its time" to the architects, artists or others with whom they worked. However, we know little about the medieval artists and scientists who were responsible for these masterful designs—although in the tenth-century text of Abu'l-Wafa' al Buzjani on geometric constructions there is mention of regular meetings between mathematicians and artisans concerning the design of geometric ornament (Özdural 2000). Nevertheless, imputing nascent group theory or a nascent theory of quasi-crystals to medieval artists or mathematicians simply cannot be justified on the basis of the historical record.

Ultimately, attempts to find precursors of contemporary mathematical thought in the cultural production of medieval Islam, or any other period for that matter, fall prey to the *presentist fallacy*—the fallacy of reading the present into the past, or, as

[3] Several works analyse the orbit structure of Islamic designs in this manner, including Grünbaum and Shepherd (1986); Abas and Salman (1995); Ostromoukhov (1998).

Butterfield has expressed it in a different context, the fallacy of using the past as "the ratification if not the glorification of the present" (Butterfield 1931). Therefore, in assessing Islamic geometric designs for their mathematical properties, it is important to keep in mind that we are not addressing the *historical question* of what technical knowledge and concerns motivated the construction of those designs, but rather the *aesthetic question* of how the mathematical properties may help to explain the sensible qualities of the designs themselves.

Islamic Dual Designs

Fortunately, one of the few things we *do* know about the historical practice of Islamic geometric design is the widespread use of grid dualization (sometimes referred to as the "polygonal" or "polygons in contact" technique): the use of an underlying polygonal grid from which the design is derived by a stylized variant of topological dualization.[4] This method is as important to Islamic geometric design as linear perspective is to Renaissance painting. Fig. 39.1, showing two designs from the magnificent Topkapi Scroll (Necipoğlü 1995) of the late fifteenth century, suffices to convey the idea.

In each panel, an underlying polygonal grid is first laid down. Similar points (one or more) are then chosen on each of the edges of the polygons, through which "dual" lines are drawn at specified angles of incidence. The dual lines are then continued (not necessarily linearly) until they meet other dual lines of similar origin.

As can be seen from these examples, grid dualization is a highly versatile technique, with three principal design choices: the type of grid, the method of dualization, and the manner in which the design is rendered. Islamic artists used radial grids, lattice grids and a wide variety of other tiling grids. The dual lines can be drawn through just one edge point—usually the midpoint—or through two or more edge points; and their angles of incidence with the edges can be set at various values. The final design may include the underlying grid along with its dual (an "additive" design) or exclude it; and the lines may be rendered by interweaving (alternating "over" and "under" positions along each line), or with the interlinear regions coloured to form a tiling pattern. Grid dualization has enormous aesthetic value: even quite "ordinary" Archimedean or other tilings have duals that appear far more interesting and dynamic than their progenitors.

According to Jay Bonner, the four most common families of eastern Islamic designs were those whose dual lines were drawn through edge midpoints, with angles of incidence chosen to be 36° ("obtuse"), 54° ("middle") or 72° ("acute"),

[4] Discussions of this technique may be found in Hankin (1925, 1934); Wade (1976); Bonner (2003); Kaplan (2005).

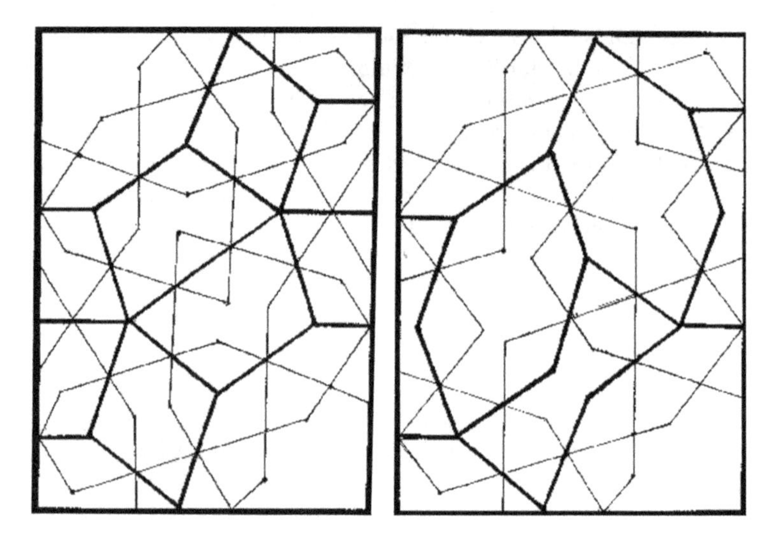

Fig. 39.1 Two dual designs Topkapi Scroll (Necipoğlü 1995). *Thick lines* show the underlying grid. Drawing: author

and those with dual lines drawn through two edge points.[5] Grid dualization (or in Bonner's terminology, the "polygonal technique"),

> is the only method for which there is documented proof that traditional designers used the system widely throughout the Islamic world. The polygonal technique is the only method that allows for the creation of both simple geometric patterns and the most complex compound patterns, often made up of combinations of seemingly irreconcilable symmetries . . . The polygonal technique has the further characteristic of allowing for the creation of all four principal families of Islamic geometric pattern [obtuse, middle, acute and two-point] regularly found throughout the Islamic world (Bonner 2003).

Many writers on Islamic geometric design ignore the grid method, inventing various ad hoc surrogates in its place. For example, Lu and Steinhardt refer to the "direct strapwork method", which they illustrate with a straightedge and compass construction (Lu and Steinhardt 2007a: Figs. 1A–D). They then posit a "paradigm shift" from this direct strap work method to a modular tiling method, whereby a set of five particular *girih* tiles (five of the ten shapes that Bonner includes in what he calls the "5-fold system of geometric pattern generation") decorated with particular dual lines is used to construct a variety of designs. Certainly, it is sometimes useful to develop and construct dual designs in this modular manner, but Lu and Steinhardt's claim that the modular use of these five decorated tiles constituted a "paradigm shift" in medieval Islamic design is unconvincing. Whatever the historical genesis of the *girih* tiles, modular use of decorated tiles is just one facet of dualization, and many other sets of tiles and dual decorations were in constant

[5] An extensive discussion may be found in Bonner (2000).

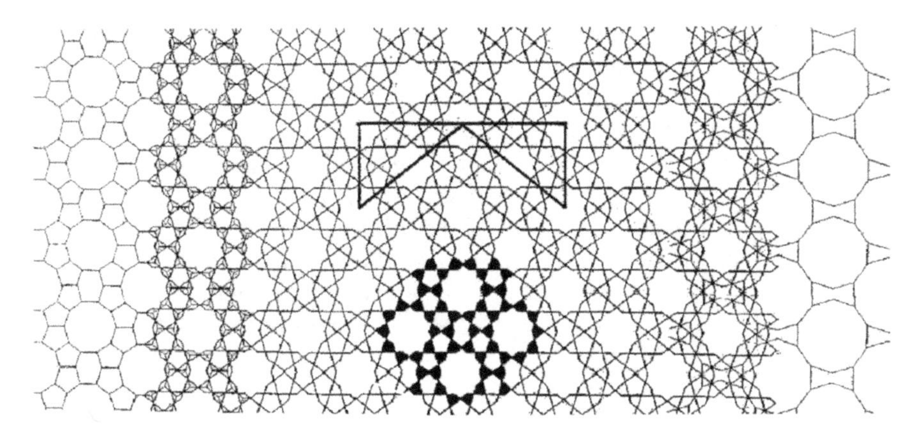

Fig. 39.2 Two different grids which produce the same dual design. The *bracketed triangles* highlight the primary dual design of the spandrel at the Darb-I Imam, Isfahan; the *tiled pattern* below shows one possible rendering of the dual design. Drawings: author, after Kaplan (2005)

use. Moreover, because the same dual design can usually be derived from more than one grid, it is not always clear which tile set was used to generate a given design.

Finally, it should be pointed out that grid dualization is of mathematical interest in its own right. Dual designs inherit many of the symmetry and other properties of the underlying grid, but also introduce certain novel tile shapes, colouring and other combinatorial properties. Bonner (2000) and Kaplan (2005) have noted the fact, already alluded to, that the same dual design can be derived from different grids (the two being connected, in Kaplan's terminology, by a "rosette transform"). For example, the "obtuse" dual of the grid on the left side of Fig. 39.2, consisting of convex pentagons, hexagons and decagons, is the same as the "middle" dual of the grid on the right side of Fig. 39.2, consisting of a decagon packing with non-convex "bowties".

This observation leads to consideration of an equivalence relation on the space of grids (tilings), two grids being "equivalent" if any dual of one is a dual of the other. Equivalent grids behave similarly with respect to symmetry and—to anticipate our main theme—with respect to quasi-periodicity. In fact, grid dualization—in a more modern incarnation due to N. G. de Bruijn—leads directly to quasi-periodicity. These and other mathematical aspects of Islamic grid dualization will be discussed in a sequel to this chapter.

Quasi-Periodicity

Non-Periodic Tilings

One of the frustrations of working with periodic planar grids (i.e., grids with translational symmetries) is the inability to achieve (global) rotational symmetry of orders other than 2, 3, 4 or 6.[6] It is not immediately obvious how to overcome this "crystallographic restriction" with a finite set of tiles, but in 1525, Albrecht Dürer penned an example with just two tiles, a pentagon and a rhomb, achieving global fivefold symmetry through what crystallographers refer to as "pentagonal twinning" (Lück 2000) (Fig. 39.3a). Nearly a century later, Kepler produced another famous example (Fig. 39.3b) (Grünbaum and Shepherd 1987: 52–53, 59). Because they violate the crystallographic restriction, neither of these tilings are periodic.

Prior to Dürer's work, European Renaissance art incorporated only lattice tilings, not pentagonal, non-periodic or other complex tilings. It is certainly conceivable that Dürer derived his interest, and even his tiling, from an Islamic text: there was, after all, an intensive transfer of Arabic scientific works to Europe after the fall of Constantinople to the Ottoman Turks in 1453 (Saliba 2007: 194–195). However, there appear to be no examples of pentagonal twinning in Islamic design, although—as we shall see—Islamic artists did very early develop other—perhaps subtler—examples of fivefold (and tenfold) symmetric designs.

Quasi-Periodic Tilings

In perhaps the most famous example of the "unreasonable effectiveness" of recreational mathematics, in 1974 Roger Penrose constructed a non-periodic tiling of the plane using pentagons, rhombs, pentagrams and partial pentagrams in certain restricted configurations (or satisfying certain "matching rules") (Fig. 39.4); later he found other, essentially equivalent, tilings with fewer tiles (the "kite and dart" and "rhomb" tilings) (Grünbaum and Shepherd 1987: § 10.3, 531–548).

Penrose's tilings, however, are not just non-periodic but are also "quasi-periodic", meaning that any bounded portion of the tiling appears infinitely often in the tiling (and in fact infinitely often in any one of the uncountably many other tilings with the same tile set). Indeed, they have many other properties as well, summed up in the all-encompassing term "quasi-crystalline": the tilings have arbitrarily large bounded fragments with crystallographically forbidden symmetries, they have global "statistical" symmetries, they may be obtained as a projection of slices of higher dimensional lattices, their vertices—considered as complex numbers—possess striking algebraic properties and have "diffractive" Fourier transforms, and more.[7] Thanks largely to the groundbreaking work of N. G. de Bruijn, the class of Penrose tilings has emerged as a mathematical object of great complexity and interest.

After Penrose's discovery, many other families of quasi-periodic tilings of the plane and other spaces were produced, and soon—as with "fractals" and "chaos"—everyone was speaking this new kind of Jourdainian prose. Most famously, in 1985 diffraction

[6] See, for example, Grünbaum and Shepherd (1987: Chap. 1).

[7] Good introductions to this subject may be found in Grünbaum and Shepherd (1987) and Senechal (1995).

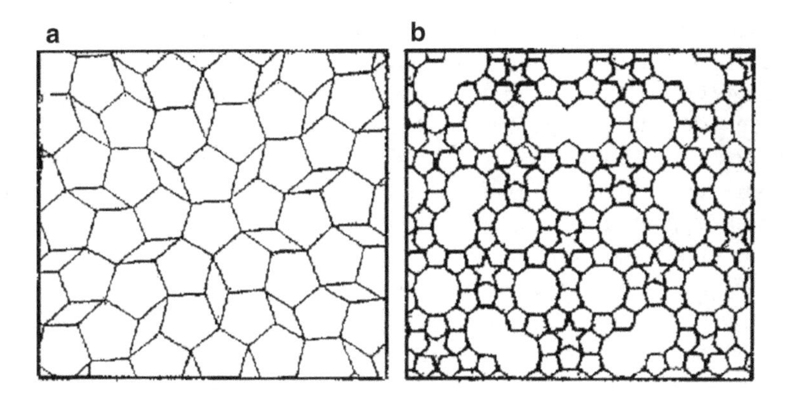

Fig. 39.3 (**a**) Dürer tiling; (**b**) Kepler tiling. Both are non-periodic. Drawing: author

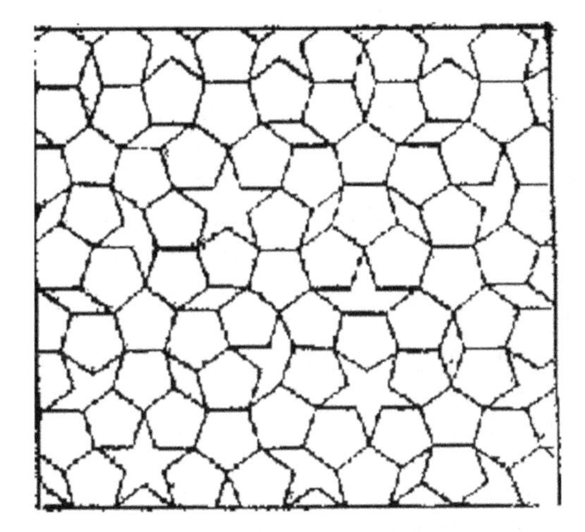

Fig. 39.4 Penrose tiling. Like the Dürer and Kepler tilings, this tiling is non-periodic, but unlike them, it is quasi-periodic. Drawing: author

images of an aluminium manganese alloy produced a quasi-periodic pattern, inaugurating a new era of crystallography and condensed matter physics (Senechal 1995). Of more relevance here is the application of these ideas to aesthetics. Interestingly, in one of his original papers on the subject, Penrose writes about the aesthetic inspiration for his tilings, and proceeds to compare their visual appeal with a design on the window of a mosque in Cairo:

> As one simply stares at the pattern certain regularities seem to jump out. There are a great many regular decagons which tend to overlap at various places ... Things line up in a surprising way. The appeal of this pattern would seem to have something in common with the appeal of the mosque window ... (Penrose 1974).

Indeed, if one believes with Birkhoff that aesthetic value may usefully be related to measures of complexity (though not necessarily in the manner Birkhoff promoted), then the mathematical properties of quasi-periodicity summarized above can be used to argue that quasi-periodic designs do indeed have high "aesthetic value".[8] Quasi-periodic patterns occupy an important niche between highly ordered, periodic patterns and highly complex, random ones, and in E. H. Gombrich's words, give us a sense of the delight that "lies somewhere between boredom and confusion" (Gombrich 1979).[9] As Penrose observed, similar sentiments apply to Islamic geometric designs, and surprisingly "Islamic" looking designs can be produced by dualizing quasi-periodic tilings.[10]

Are certain Islamic designs, then, "quasi-periodic"? As it stands, the question is nonsensical. By definition, quasi-periodicity (infinite repetition of bounded fragments) is a property that pertains only to a tiling of infinite extent. So let us define a bounded design as "quasi-periodic" if, first, it can be derived in some systematic manner from a finite tiling (for example, as a dual design or through other kinds of systematic decorations or erasures), and second, the finite tiling from which it is derived can be extended to a quasi-periodic tiling of the entire plane. This definition is consistent with the use of the term "quasi-periodic" or its cognates in the recent literature on Islamic design.

It must be understood, however, that this or any other definition of "quasi-periodicity" in the context of (bounded) designs has several pitfalls. Most important, it is a fact that *any* finite tiling—even one that can be extended to an infinite quasi-periodic tiling—can also be extended to a *periodic* tiling in a variety of ways (Gähler and Rhyner 1986). Most Islamic geometric designs are, in fact, explicitly embedded in periodic frameworks, and arguably *all* were so intended. The definition given here, however, circumvents this rather sterile issue by focusing on a segment of the design (for example, a unit cell), and calling that segment "quasi-periodic" if its underlying grid is a fragment of an (infinite) quasi-periodic tiling. This definition also accords with our interest in the "aesthetic question" (the extent to which the sensible qualities of Islamic designs may be explained, at least in part, by reference to mathematical properties of quasi-periodicity), rather than in the "historical question" (regarding the intentions or motivations of the artists who created the designs and whether they aimed at periodic or non-periodic patterns).

Inflation Tilings

One of the simplest ways to construct a quasi-periodic tiling is to start with a set of tiles that can be inflated and then subdivided into smaller copies of themselves in

[8] An interesting discussion of Birkhoff measures relevant to these remarks is found in Rigau et al. (2007). A study of the Kolmogorov complexity of finite subsets of tilings of the plane is found in Durand et al. (2008).

[9] Similar views are expressed in Arnheim (1971).

[10] Such designs have been produced by Rigby (2006).

such a manner that the process can be iterated. Subject to a variety of alternative sets of conditions, such inflation rules generate quasi-periodic tilings of the plane. As an example, consider the three tiles (decagon, bowtie and long hexagon) with the inflation rules shown in Fig. 39.5. These elegant inflation rules are due to Lu and Steinhardt, who noted that the subdivisions of the decagon and bowtie are implicit in a design from the Darb-I Imam in Isfahan (Lu and Steinhardt 2007a). The tiles themselves are a subset of the *girih* tiles studied by Lu and Steinhardt, and are a variant of the "M2" tile set introduced by Makovicky (1992: n. 30, Fig. 10 and surrounding text) in analysing an earlier design from Margaha, Iran.[11]

Start with an empty decagon centred at the origin, then inflate and subdivide it as shown in Fig. 39.5. This gives the "level 1" tiles, each of which is then inflated and subdivided to obtain the "level 2" tiles. Note that because the borders of the subdivided tiles consist of symmetric half tiles and are all alike, the second level tiles all line up properly. Also note that the level 2 tiles extend the level 1 tiles, which are still centred at the origin. If this process is now iterated, we have a nested sequence of decagons that increase in size; in the limit we obtain an "inflation tiling" of the entire plane. Since each level of the inflation retains 5-fold symmetry, the same is true of the inflation tiling of the entire plane, which is therefore non-periodic. Quasi-periodicity—infinite repetition of each bounded fragment—follows very naturally from the inflation process itself, as any patch of tiles which appears at some stage will be reproduced at each subsequent stage.[12]

Two Designs from Iran

The Gunbad-i Kabud (Maragha)

We are now in a position to look at two designs that have been cited as the prime examples of quasi-periodicity in Islamic art. In 1992 Emil Makovicky analysed the unit cell design on the walls of the Gunbad-i-Kabud (Blue Tomb), an octagonal tower in Maragha, Iran, dating to the late twelfth century (Makovicky 1992) (Fig. 39.6). The unit cell of the primary design spreads over two walls and repeats four times around the tower. The thin lines in Fig. 39.7 show Makovicky's more recent transcription of the primary design over half of its cell unit (the other half is obtained by reflection through the middle vertical line).[13]

[11] Aspects of the M2 tilings later appeared in high resolution transmission electron microscopy of aluminium cobalt nickel alloys; see Cervellino et al. (2002).

[12] For further discussion of inflation tilings, see Senechal (1995: Chap. 5).

[13] The Gunbad-i Kabud has deteriorated, and thus the original design is obscured in parts. Lu and Steinhardt (2007a: Fig. S6) had pointed out that the lower portion of the original reconstruction given in Makovicky (1992) was incorrect. Emil Makovicky has carefully reconstructed the design based on his inspection of the building (Makovicky 2009).

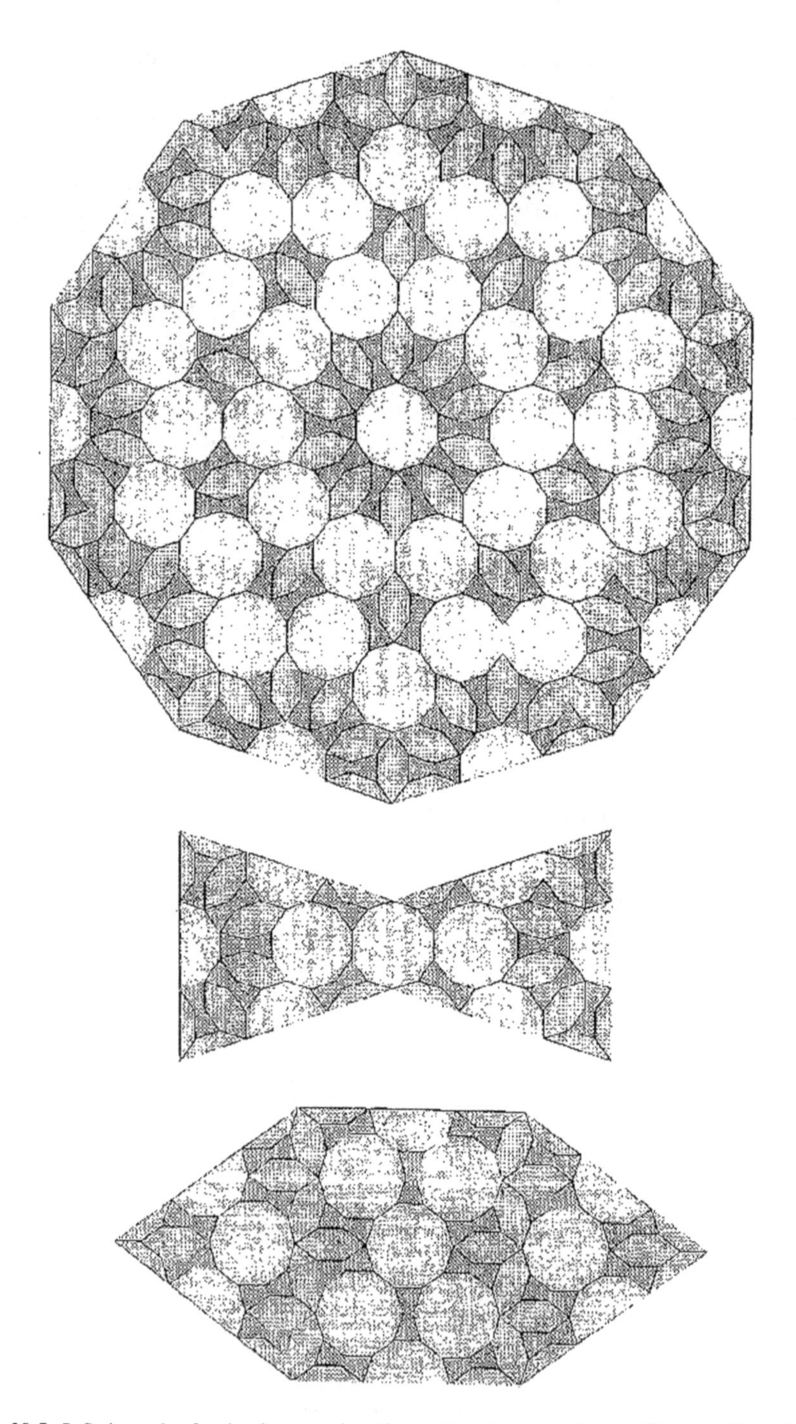

Fig. 39.5 Inflation rules for the decagon, bowtie and long hexagon: Image: Courtesy Peter Lu

Fig. 39.6 The Gunbad-i Kabud, Maragha. Photo: Courtesy Emil Makovicky

There are a number of ways one might test a design like this for quasi-periodicity. Given the five- and ten-fold symmetry elements, one might, for example, suspect that the Penrose tiles themselves could be used to tile the regions between the primary line elements in a way that could be extended to a Penrose tiling of the entire plane.

This is the approach taken, for example, in a recent paper by Arik and Sancak (2007). Such an analysis is incomplete, however, as generally all that can be done in this regard is to reconstruct the design as far as possible with well-matched Penrose tiles. But a well-matched patch of Penrose tiles is no guaranty of extendability to the

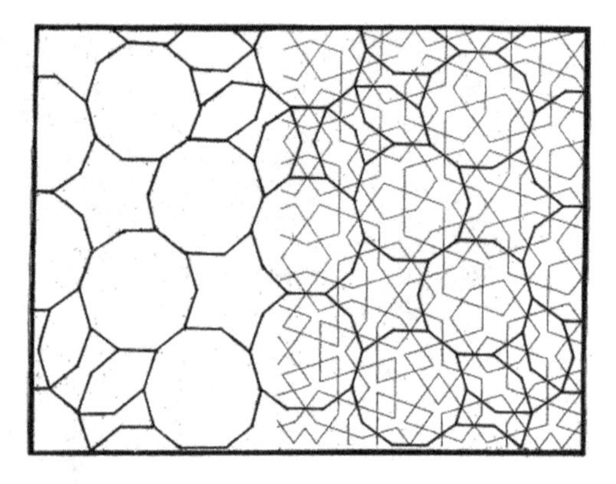

Fig. 39.7 Contours of the underlying grid for the unit cell of the Gunbad-I Kabud, Maragha, with the primary dual design filled in on the *right*. Image: author

entire plane, so without more, this method does not offer a convincing method to establish quasi-periodicity of a design.

A better method—and one implied by the definition of a "quasi-periodic" design here—is taken in Makovicky's paper: find an underlying grid from which the design can be derived, and show that the grid is a fragment of a quasi-periodic tiling. For the grid, Makovicky used the decagons, long hexagons and bowties analysed above, together with five pointed stars. As may be seen from the implied grid in Fig. 39.7, the designers of the Maragha tower decorated these shapes consistently up to rotations of two of the decagons. To show that this grid is a fragment of a quasi-periodic tiling, simply note that it appears in the centre of the subdivided decagon in Fig. 39.5. (The inflation rules in Fig. 39.5 do not aggregate hexagons and bowties into stars, as here, but it is straightforward to incorporate an additional inflation rule for the five-pointed stars.) From the discussion of inflation tilings above, therefore, the grid of Fig. 39.7 is a fragment of a quasi-periodic tiling of the plane, and thus the unit cell of the Gunbad-i Kabud is quasi-periodic in the sense defined here. The fact that the unit cell is repeated four times around the perimeter of the Gunbad-i Kabud, so that the entire design is periodic (with one translational symmetry), does not affect this conclusion.

In addition to the "quasi-periodic" primary design, the Gunbad-i Kabud also features secondary lines within the regions formed by the primary lines. Fig. 39.8 shows a portion of Makovicky's transcription of the complete design with both primary and secondary elements. As may be seen, the secondary design is in effect a two-point dual of the primary design, which itself—as we have seen—is a dual of the implied M2 grid. The result is a masterful example of an "additive" or "double" dual design. The great sophistication and complexity of this design may serve as an appropriate reminder of the fact that Maragha was one of the premier centres of Islamic science: one year after the destruction of Baghdad by the Mongols in 1258, the great astronomer, Nasir al-Din al Tusi—whose key theorem on the "Tusi Couple" was used by Copernicus in *De Revolutionibus*—supervised the construction of the Maragha observatory to which he later brought "the most distinguished company of astronomers ever assembled in one place" (Saliba 2007: 199, 244).

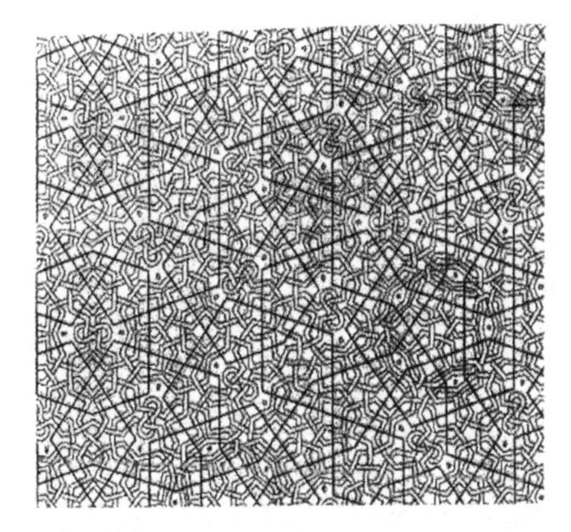

Fig. 39.8 Detail from the *upper right* portion of Fig. 39.7 showing the primary dual design with its (secondary) two-point dual. Image: By permission from Makovicky (2009)

Fig. 39.9 A portion of the right half spandrel at the Darb-I Imam, Isfahan. Image: Courtesy Peter Lu

The Darb-i Imam (Isfahan)

In 2003, Jay Bonner, an architect and leading student and practitioner of Islamic design, published an insightful paper about what he refers to as "self-similar" Islamic designs (2003). In it, he analyses a design found in an arch over a portal at the Darb-i Imam in Isfahan, built in 1453–1454. According to Bonner, this was the work of Sayyid Mahmud-i Naqash, "one of the relatively few architectural ornamentalists in the long history of Islamic art who signed his name to his works." The same design appears also in a spandrel in a different part of the Darb-i Imam, and it is the right half of that spandrel (Fig. 39.9) that Lu and Steinhardt analyse (2007a). For the sake of comparison, I apply Bonner's analysis to the right half spandrel instead of the arch.

Bonner's reconstruction of the Darb-i Imam design proceeds, as did Makovicky's reconstruction of the Maragha design, by imputing an underlying grid—in this case, both a primary and a secondary grid. He first reconstructs the large scale linear design as an obtuse dual of the primary grid on the left side of Fig. 39.2: the two triangles in the centre top of Fig. 39.2 outline the left and right halves of the spandrel with the primary dual lines. Bonner notes that this primary

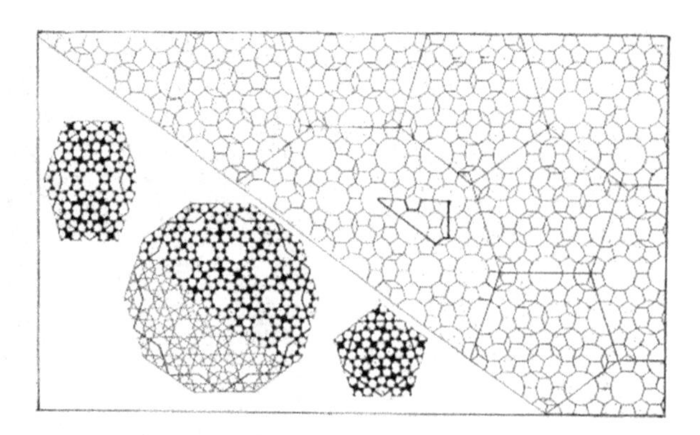

Fig. 39.10 Reconstruction of the primary and secondary grids for the Darb-i Imam spandrel. Individual tiles show the primary (*linear*) and secondary (*tiled*) dual designs; the decagon is partially untiled to show the underlying secondary dual lines. The small region outlined in the central half-decagon is explained in the text. Image: author, after Bonner (2003)

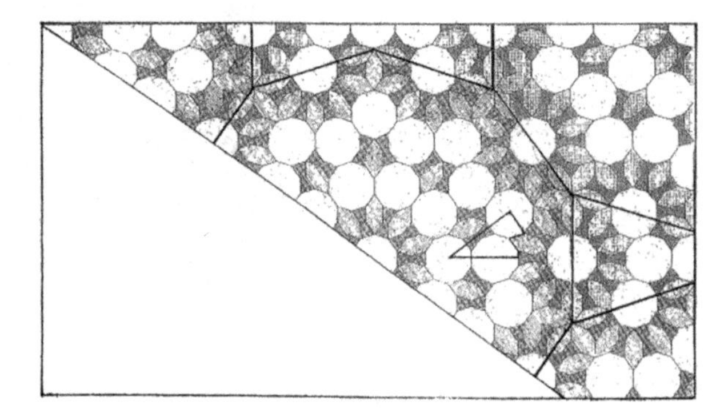

Fig. 39.11 Reconstruction of the primary and secondary grids for the Darb-i Imam spandrel the small region outlined in the central half-decagon is explained in the text. Image: Author, after Lu and Steinhardt (2007a)

grid is the most common fivefold polygonal grid "that can be traced back as far as the year 1000." Next, the primary grid tiles can be subdivided into similar small scale, secondary grid tiles, as shown in Fig. 39.10. The full design then emerges through dualization at both the primary and secondary scales—with the large scale dual design rendered linearly and the small scale dual design rendered in coloured tiles. This procedure reproduces the original design with great accuracy: only a few small tiles in the actual tiling differ from the reconstruction.

Analysing the same design, Lu and Steinhardt impute different, but equivalent, primary and secondary grids—starting with the decagon packing on the right side of Fig. 39.2 rather than the pentagonal tiling on the left (see Fig. 39.11). Furthermore, they show that the secondary grid is a fragment of a quasi-periodic tiling, using the inflation rules discussed above.[14] Thus, with the exception of a few small tiles, the

[14] Although not addressed in Bonner (2003), the subdivisions he uses are in fact part of a set of inflation rules that can also establish the quasi-periodicity of a segment of this design. Those

Darb-i Imam half spandrel design is a fragment of a quasi-periodic tiling of the plane, and so is also quasi-periodic in the sense defined here. Again, the fact that the spandrel design may be embedded in a periodic framework (with two independent translational symmetries)—as is clear from Fig. 39.2—does not affect this conclusion.[15] What is perhaps most compelling about the Darb-i Imam design is the fact that the inflation rules for two of the underlying grid tiles are implicit in the design itself—in effect, a large portion of the design is included in the second level inflation.

It should be emphasized, as Lu and Steinhardt themselves do, that the entire spandrel design is *not* quasi-periodic. Considering the Lu and Steinhardt reconstruction of the right half of the spandrel, the problem is that the large scale grid contains configurations that do not occur in the inflation rules: in particular, the placement of the two large-scale bowties does not occur in any of the inflation rules and cannot occur in the resulting inflation tiling. If, however, the partial bowtie and partial decagon in the lower right corner of the design are removed, what remains is indeed a fragment of the inflation rule for the decagon—as shown in the patch of small tiles outlined in the central half-decagon in Fig. 39.11—and therefore also a fragment of the inflation tiling. The Bonner reconstruction of the design establishes that a different portion of the spandrel is also a quasi-periodic fragment, as shown in the patch of small tiles outlined in the central half-decagon in Fig. 39.10. However, it can be shown that no inflation rules will reproduce the entire design. Lu and Steinhardt attempt to rectify this problem by showing that the entire spandrel can be (approximately) converted to well-matched Penrose tiles, but as already argued, without more this does not establish quasi-periodicity because well-matched Penrose patches are no guarantee of extendability to the entire plane.

Conclusion

As we have seen, the unit cell of the Gunbad-i Kabud design, and a large portion of the Darb-i Imam design, are both quasi-periodic in the sense defined here. In fact, both designs are derived from essentially the same underlying 5-fold symmetric quasi-periodic tiling (M2 or its equivalents.) In both cases, however, the quasi-periodic fragments are embedded in a larger scale periodic framework (in the case of Gunbad-i Kabud, with one translational symmetry) or may be embedded in such a framework (in the case of Darb-i Imam, with two independent translational symmetries).

---

inflation rules, however, require a subdivision rule for the narrow rhomb and one for an extra pentagon, and require more complex matching rules than those used by Lu and Steinhardt.

[15] The large scale dual design from Darb-i Imam appears in multiple guises at many other sites with explicit periodic repetition, though it does not do so in the spandrel studied by Lu and Steinhardt. See Arik and Sancak (2007) and Cromwell (2009).

Invoking our earlier caveat, the conclusion that the Gunbad-i Kabud and Darb-i Imam designs are (in whole or in part) "quasi-periodic" is *not* meant to suggest that medieval Islamic artists or scientists understood or were interested in quasi-periodicity in anything like the sense we define the term today. It is clear, however, that these artists developed a remarkable class of tilings to serve as the underlying grids for complex dual designs that are the hallmark of the Islamic geometric aesthetic.

Although we will not survey other examples here, evidence of quasi-periodicity in Islamic designs from Spain and Morocco has been cited as well (Makovicky et al. 1998).[16] Most interestingly, E. Makovicky, and P. Fenoll Hach-Alí have found evidence of the use of a quite different octagonal quasi-periodic design at the Alhambra (1996). Their analysis of quasi-periodicity in that design proceeds directly from a form of dualization rather than from inflation rules.

Acknowledgments I would like to thank Emil Makovicky, Peter Lu and Peter Cromwell for sharing their work with me and for helpful comments on an earlier draft of this text.

Biography Peter Saltzman is a mathematician and lawyer living in Berkeley, California.

References

Abas, S. J. and A. Salman. 1995. *Symmetries of Islamic Geometrical Patterns*. Singapore: World Scientific.

Arik, M. and M. Sancak 2007. Turkish-Islamic Art and Penrose Tilings. *Balkan Physics Letters* **15**, 3: 1–12.

Arnheim, R. 1971. *Entropy and Art, an Essay on Disorder and Order*. Berkeley: University of California Press.

Bonner, J. F. 2000. *Islamic Geometric Patterns: Their Historical Development and Traditional Methods of Derivation*. New York: Springer Verlag.

___. 2003. Three Traditions of Self-Similarity in Fourteenth and Fifteenth Century Islamic Geometric Ornament. Pp. 1–12 in R. Sarhangi and N. Friedman, eds. *ISAMA/Bridges 2003 Proceedings*. Granada: University of Granada.

Bourgoin, J. 1879. *Les Elements de l'Art Arabe: Le Trait des Entrelacs*. Paris, Firmin-Didot et Cie. English trans. 1971. *Arabic Geometrical Pattern and Design*. New York: Dover.

Butterfield, H. 1931. *The Whig Interpretation of History*. London: Bell and Sons. Rpt. 1968.

Castera, J. 1999. Zellijs, Muqarnas and Quasicrystals. Pp. 99–104 in N. Friedman and J. Barrallo, eds. *ISAMA 99*. San Sebastian: University of the Basque Country.

Cervellino A., T. Haibach, and W. Steurer. 2002. Structure Solution of the Basic Decagonal Al-Co-Ni Phase by the Atomic Surfaces Modelling Method. *Acta Crystallographica* **B58**: 8–33.

Cromwell, P. R. 2009. The Search for Quasi-Periodicity in Islamic 5-Fold Ornament. *The Mathematical Intelligencer* **31**, 1: 36-56.

[16] Castera (1999) has claimed that projections of three-dimensional Islamic "Muqarnas" are quasi-periodic, although Makovicky and Fenoll Hach-Ali (2001) have reached contrary conclusions.

CROWE, D.W. and D. K. WASHBURN. 1991. *Symmetries of Culture: Theory and Practice of Plane Pattern Analysis*. Seattle: University of Washington Press.

DURAND, B., L. LEVIN, and A. SHEN. 2008. Complex Tilings. *Journal of Symbolic Logic* **73**, 2: 593–613.

FENOLL HACH-ALÍ, P. and A. L. GALINDO. 2003. *Science, Beauty and Intuition, Symmetry in the Alhambra*. Granada: University of Granada.

GÄHLER, F. and J. RHYNER1986. Equivalence of the Generalised Grid and Projection Methods for the Construction of Quasiperiodic Tilings. *Journal of Physics A: Mathematical and General* **19**, 2: 267–277.

GOMBRICH, E. H. 1979. *The Sense of Order: A Study in the Psychology of Decorative Art*. London: Phaidon Press.

GOURY, J. and JONES, O. 1842–1845. *Plans, Sections, Elevations and Details of the Alhambra*. London: Owen Jones.

GRÜNBAUM, B. and SHEPHERD, G. C. 1986. Symmetry in Moorish and Other Ornaments. *Computers and Mathematics with Applications* **12B**, 3–4: 641–653.

___. 1987. *Tilings and Patterns*. San Francisco: WH. Freeman.

___. 1992. Interlace Patterns in Islamic and Moorish Art. *Leonardo* **25**, 3-4: 331–339.

HANKIN, E. H. 1925. Examples of Methods of Drawing Geometrical Arabesque Patterns. *The Mathematical Gazette* **12**: 371–373.

HANKIN, E. H. 1934. Some Difficult Saracenic Designs II. *The Mathematical Gazette* **18**: 165–168.

KAPLAN, C. S. 2005. Islamic Star Patterns from Polygons in Contact. Pp. 177–186 in *GI '05: Proceedings of the 2005 Conference on Graphics Interface*. Victoria: British Columbia.

LU, P. J. and STEINHARDT. P. J. 2007a. Decagonal and Quasi-Crystalline Tilings in Medieval Islamic Architecture. *Science* **315**: 1106–1110.

___. 2007b. Response to Comment on "Decagonal and Quasi-Crystalline Tilings in Medieval Islamic Architecture". *Science* **318**: 1383b.

LÜCK, R. 2000. Dürer-Kepler-Penrose, the Development of Pentagon Tilings. *Materials Science and Engineering* **294-296**: 263–267.

MAKOVICKY, E. 1992. 800-Year-Old Pentagonal Tiling From Maragha, Iran, and the New Varieties of Aperiodic Tiling it Inspired. Pp. 67–86 in I. Hargittai, ed. *Fivefold Symmetry*. Singapore: World Scientific.

___. 2009. Another look at the Blue Tomb of Maragha, a site of the first quasicrystalline Islamic pattern. *Symmetry: Culture and Science* **19**: 127-151.

MAKOVICKY. E. and FENOLL HACH-ALÍ. P. 1996. Mirador de Lindaraja: Islamic Ornamental Patterns Based on Quasi-Periodic Octagonal Lattices in Alhambra, Granada and Alcazar, Sevilla, Spain. *Boletín de la Sociedad Española de Mineralogía* **19**: 1–26.

___. 1997. Brick and Marble Ornamental Patterns from the Great Mosque and the Madinat al-Zahra Palace in Cordoba, Spain I. *Boletín de la Sociedad Española de Mineralogía* **20**: 1–40.

___. 1999. Coloured Symmetry in the Mosaics of Alhambra, Granada. *Boletín de la Sociedad Española de Mineralogía* **22**: 143–183.

___. 2001. The Stalactite Dome of the Sala de Dos Hermanas – an Octagonal Tiling? *Boletín de la Sociedad Española de Mineralogía* **24**: 1–21.

MAKOVICKY, E., F. RULL PEREZ, and P. FENOLL HACH-ALÍ. 1998. Decagonal Patterns in the Islamic Ornamental Art of Spain and Morocco. *Boletín de la Sociedad Española de Mineralogía* **21**: 107–127.

NECIPOĞLÜ, G. 1995. *The Topkapi Scroll: Geometry and Ornament in Islamic Architecture*. Santa Monica: Getty Center for the History of Art and Humanities.

OSTROMOUKHOV, V. 1998. Mathematical Tools for Computer-Generated Ornamental Patterns. *Lecture Notes in Computer Science* **1375**: 193–223.

ÖZDURAL, A. 2000. Mathematics and Arts: Connections between Theory and Practice in the Medieval Islamic World. *Historia Mathematica* **27**: 171–201.

Penrose, R. 1974. The Role of Aesthetics in Pure and Applied Mathematical Research. *The Institute of Mathematics and its Applications* **7/8**, 10: 266–271.

Rigau, J., M. Feixas, and M. Sbert. 2007. Conceptualizing Birkhoff's Aesthetic Measure Using Shannon Entropy and Kolmogorov Complexity. Pp. 1–8 in D. W. Cunningham, et al., eds. *Computational Aesthetics in Graphics, Visualization and Imaging*. Banff: Eurographics Association.

Rigby, J. 2006. Creating Penrose-type Islamic Interlacing Patterns. Pp. 41–48 in R. Sarhangi and J. Sharp, eds. *Bridges 2006 Conference Proceedings*. London: University of London

Saliba, G. 2007. *Islamic Science and the Making of the European Renaissance*. Cambridge, Massachusetts: MIT Press.

Senechal, M. 1995. *Quasicrystals and Geometry*. Cambridge: Cambridge University Press.

Wade, D. 1976. *Pattern in Islamic Art*. London: Overlook Press.

Chapter 40
The Universality of the Symmetry Concept

István Hargittai and Magdolna Hargittai

Introduction

The notion of symmetry brings together beauty and usefulness, science and economy, mathematics and music, architecture and human relations, and much more, as has been shown recently with many examples (Hargittai 1986, 1989; Hargittai and Hargittai 1995, 1996). There is a lot of symmetry, for example, in Béla Bartók's music. It is not known, however, whether he consciously applied symmetry or was simply led intuitively to the golden ratio so often present in his music. Bartók himself always refused to discuss the technicalities of his composing and stated merely "We create after Nature." Another unanswerable question is how these symmetries contribute to the appeal of Bartók's music, and how much of this appeal originates from our innate sensitivity to symmetry. This question might be equally asked of symmetries in architectural composition.

The present chapter takes a broad view of the symmetry concept. It demonstrates its breadth and versatility. There are no distinctly different specific symmetries in various disciplines, yet there are discernible differences in emphasis of the application of this concept in different fields. This emphasis changes with time as well. For example, there is a marked emphasis on the presence of symmetry in chemistry, in contrast to physics where the importance of broken symmetries has been stressed during the past decades. Generally though the symmetry concept unites rather than divides the different branches of science, and even helps bridge the gap between what C.P. Snow called "two cultures." Sciences, the humanities,

First published as: Istvàn Hargittai and Magdolna Hargittai, "The Universality of the Symmetry Concept", pp. 81–95 in *Nexus I: Architecture and Mathematics,* ed. Kim Williams, Fucecchio (Florence): Edizioni dell'Erba, 1996.

I. Hargittai (✉) • M. Hargittai
Materials Structure and Modeling Research Group, Budapest University of Technology and Economics, Pf. 91, H-1521, Budapest, Hungary
e-mail: istvan.hargittai@gmail.com; hargittaim@mail.bme.hu

K. Williams and M.J. Ostwald (eds.), *Architecture and Mathematics from Antiquity to the Future*, DOI 10.1007/978-3-319-00137-1_40,

and the arts have all drifted apart over the years and symmetry can provide a connecting link among them. Its benefits are available to us if we free ourselves from the confinements of *geometrical symmetry*.

Everything is rigorous in geometrical symmetry. According to one definition, "symmetry is the property of geometrical figures to repeat their parts" (Shubnikov 1951). Another definition says that "a figure is symmetrical if there is a congruent transformation which leaves it unchanged as a whole, merely permuting its component elements" (Coxeter 1973). In the geometrical sense, symmetry is either present or it is absent. Any question regarding symmetry has a restricted *yes/no* alternative. For the real, material world, however, degrees of symmetry and even gradual symmetry is feasible and applicable. Beyond geometrical definitions there is another, broader meaning to symmetry—one that relates to *harmony* and *proportion,* and ultimately to *beauty*. This aspect involves feeling and subjective judgment and, as a result, is especially difficult to describe in technical terms.

Simple considerations are indispensable in classifying different kinds of symmetry. There are two large classes of symmetry, *point groups* and *space groups*. For point group symmetries there is at least one special point in the object or pattern that differs from all the others. In contrast to this, in space groups, there is no such special point. There are also some terms that are useful in the description of different types of symmetry. Thus, the action that characterizes a particular type of symmetry is called a *symmetry operation*. The tool whereby the operation is performed is called a *symmetry element*.

Point Group Symmetry

The simplest kind of point-group symmetry is *bilateral symmetry*. Bilateral symmetry is present when two halves of the whole are each other's mirror images (Fig. 40.1). This is the most common symmetry and the every-day usage of the term "symmetry" refers to this meaning. The symmetry element is a *mirror plane*, also called a *symmetry plane* or a *reflection plane*. The symmetry operation is *reflection*. Applying a mirror plane to either of the two halves of an object with bilateral symmetry recreates the whole object. Bilateral symmetry is probably the most common symmetry in architecture as well, from simple buildings to larger assemblies (Fig. 40.2a, b).

Another kind of point-group symmetry is *rotational symmetry* (Fig. 40.3). It is present when, by rotating an object around its axis, it appears in the same position two or more times during a full revolution. *Rotation* is the symmetry operation and the *axis of rotation* is the symmetry element. Rotational symmetry may be twofold, threefold, fourfold, etc. It is common that reflection and rotation appear together. The presence of some symmetry elements may generate others and vice versa. If we look at the Eiffel tower from below (Fig. 40.4) we have twice two orthogonal reflection planes which generate a fourfold rotation. The cupolas of many state

Fig. 40.1 The orchid has bilateral symmetry. Photo: authors

capitols and other important buildings have reflectional and rotational symmetry together (Fig. 40.5).

The regular polygons, so basic in architectural design, also have both rotational and reflectional symmetry. Best seen when viewed from above, many buildings have outlines of a regular polygon (Fig. 40.6). The regular polyhedra, also called Platonic solids, all have equal regular polygons as their faces. As H.S.M. Coxeter, professor of mathematics at the University of Toronto, remarked, "the chief reason for studying regular polyhedra is still the same as in the times of the Pythagoreans." Namely, that their symmetrical shapes appeal to one's artistic sense. There are other highly symmetrical polyhedra, called Archimedian polyhedra, whose faces are also regular polygons but not identical ones. Buckminster Fuller's geodesic dome is composed of lightweight bars forming regular polygons. His geodesic dome at the Montreal expo (Fig. 40.7) inspired some chemists who saw that the structure of a newly discovered substance may be the truncated icosahedron. This molecule, C_{60}, called buckminsterfullerene (Fig. 40.8) is characterized, among others, by six axes of fivefold rotation (Hargittai and Hargittai 1994: 100–101). Experimentally discovered in 1985, its great relative stability was predicted already in 1970, based solely on symmetry considerations.

Chirality

A special kind of symmetry relationship is when two objects are related by mirror reflection and the two objects cannot be superposed. Our hands are an excellent

Fig. 40.2 (**a**) The whole assembly of the Blue Mosque in Istanbul, Turkey, has bilateral symmetry. (**b**) The design of St. Peter's Square in Vatican City also shows bilateral symmetry. Photo: authors

Fig. 40.3 This hubcap has sevenfold rotational symmetry. Photo: authors

example, and the term *chiral* derives from the Greek word for hand. Chiral objects have senses and following the hand analogy they are left-handed (L) and right-handed (D). The simplest chiral molecule is a methane derivative in which three of the four hydrogens are replaced by three different atoms, such as, for

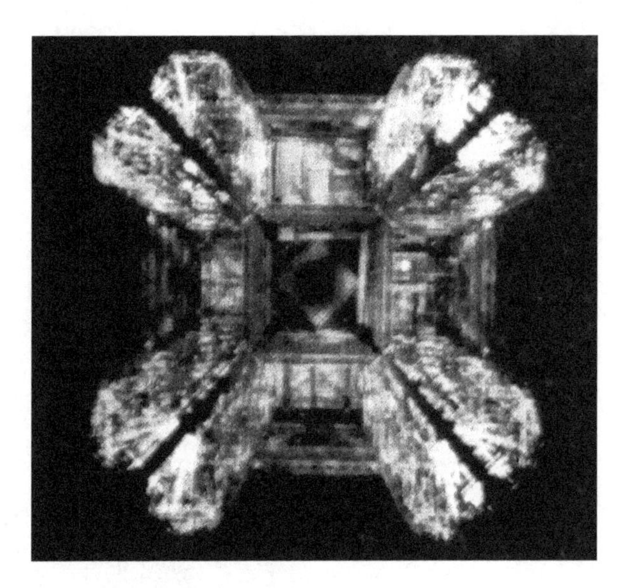

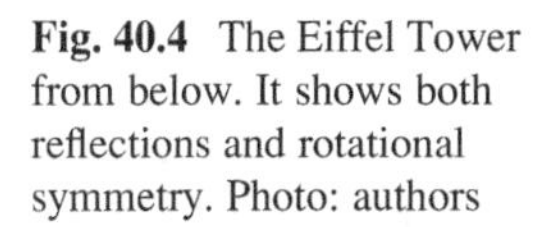

Fig. 40.4 The Eiffel Tower from below. It shows both reflections and rotational symmetry. Photo: authors

example, fluorine (F), chlorine (Cl), and bromine (Br). There may then be a left-handed C(HFClBr) and a right-handed C(HFClBr) molecule which will be each other's mirror images but won't be superposable (Fig. 40.9). A chiral object and its mirror image are called each other's *enantiomorphs*.

The two chiral molecules look the same in every detail; only their senses are different. The distinctions between the twins of a chiral pair have literally vital significance. Only L-amino acids are present in natural proteins and only D-nucleotides are present in natural nucleic acids. This happens in spite of the fact that the energy of both enantiomers is equal and their formation has equal probability in an achiral environment. However, only one of the two occurs in nature, and the particular enantiomers involved in life processes are the same in humans, animals, plants, and microorganisms. The origin of this phenomenon is a great puzzle.

Once a chiral molecule happens to be in a chiral environment, the two chiral isomers will be behaving differently. This different behaviour is manifested sometimes in very dramatic ways. In some cases one isomer is sweet, the other is bitter. In some other cases the drug molecule has an "evil twin." A tragic example was the thalidomide case in the 1950s in Europe, in which the right-handed molecule cured morning sickness and the left-handed one caused birth defects. Other examples include one enantiomer of ethambutol fighting tuberculosis with its evil twin causing blindness, and one enantiomer of naproxen reducing arthritic inflammation with its evil twin poisoning the liver. Ibuprofen is a lucky case in which the twin of the chiral form that provides the curing is converted to the beneficial version by the body.

Even when the twin is harmless, it represents waste and a potential pollutant. Thus, a lot of efforts are directed toward producing enantiomerically pure drugs and pesticides. One of the fascinating possibilities is to produce sweets from chiral

Fig. 40.5 The cupola of the Hungarian Parliament with both reflectional and rotational symmetry. Photo: authors

sugars of the enantiomer that would not be capable of contributing to obesity yet would retain the taste of the other enantiomer.

Chiral symmetry is also frequently found in architectural design either in two- or in three dimensions, as illustrated by Fig. 40.10.

Space Group Symmetry

A different kind of symmetry can be created by simple *repetition* of a basic motif leading us to *space-group symmetries*. The most economical growth and expansion patterns are described by space-groups symmetries. There are three basic cases of space groups, depending on whether the basic motif extends periodically in one

Fig. 40.6 The outline of the Pentagon in Washington, D.C. with its regular *pentagonal shape*. Photo: authors

Fig. 40.7 Buckminster Fuller's Geodesic Dome at the Montreal Expo. Photo: authors

direction only, or in two, or finally, in three. These three cases are described by the so-called *one-dimensional, two-dimensional,* and *three-dimensional* space groups.

Border decorations are examples of one-dimensional space groups. In border decorations a pattern can be generated simply by repeating a motif at equal intervals. This is *translational* symmetry. The symmetry element is *constant translation;* the operation is the *translation* itself. The resulting pattern shows periodicity in one direction. Repetition can be achieved by a simple shift in one direction as can be seen very often in the rows of columns of grandiose buildings (Fig. 40.11) or in the ancient aqueducts of the Romans. Fences are typical examples of one-dimensional space groups (Fig. 40.12), the ease and economy of using the same elements repeatedly makes this obvious. Repetition can also be achieved in

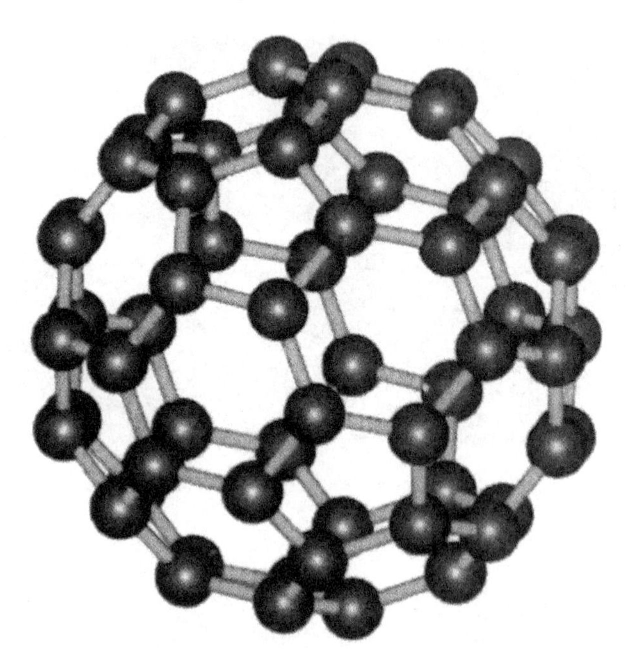

Fig. 40.8 C_{60}, the buckminsterfullerene molecule. Image: authors

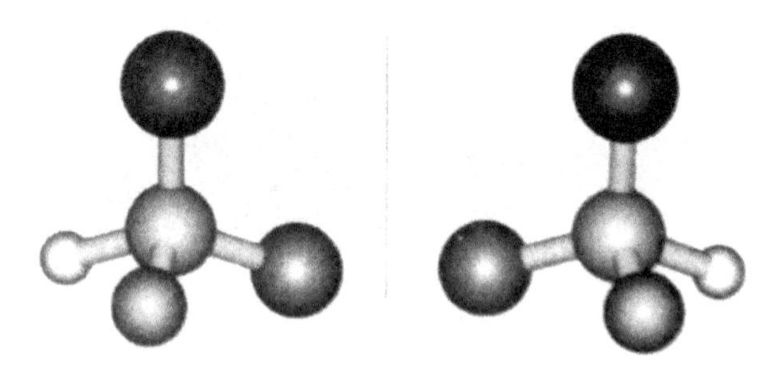

Fig. 40.9 A chiral pair of molecules. Image: authors

Fig. 40.10 Chiral rosettes on a building in Bern, Switzerland. Photo: authors

Fig. 40.11 Colonnade on St. Peter's square in Vatican City. Photo: authors

Fig. 40.12 Repeating pattern of a fence in the Topkapi Palace in Istanbul, Turkey. Photo: authors

Fig. 40.13 Another illustration for one-dimensional space groups: the units turn 90° at every translation in this chain. Photo: authors

other ways, such as by reflection, rotation (Fig. 40.13), or *glide- reflection*. Glide-reflection is another new element that does not occur in point-group symmetries. It means the consecutive application of translation and horizontal reflection. When we walk in wet sand along a straight line we leave behind a pattern of footprints whose symmetry is described by glide-reflection. There is a total of seven possibilities for generating one-dimensional space-group symmetries.

Helices and spirals have also one-dimensional space-group symmetries although their bodies may extend to three dimensions (Hargittai and Pickover 1992). *Helical symmetry* is created by a constant amount of translation accompanied by a constant amount of rotation. In *spiral symmetry*, again, translation is accompanied by rotation but the amount of translation and rotation changes gradually and regularly. An extended spiral staircase has helical symmetry. Well-ordered biological macromolecules also have helical symmetry. Helices are always three-dimensional whereas there are spirals that extend in two dimensions only. Occurrences of spirals may be as diverse as chemical waves and galaxies and snails. Spirals and helices have also been used in various ways in architecture, from ancient times to the present, as Trajan's column in the Forum Romanum (Fig. 40.14) and the spiral ramp of Frank Lloyd Wright's Guggenheim Museum in New York indicate.

Another beautiful example of spiral symmetry is the scattered leaf arrangement around the stems of plants, called *phyllotaxis*. Numbers of the Fibonacci series (1, 1, 2, 3, 5, 8, 13, 21, . . .—each new element is the sum of the two previous elements) characterize the ratios defining the occurrence of every consecutive new leaf in scattered leaf arrangements. Thus, for example, there is a new leaf at each 3/8 parts of the circumference of the stem as we move along the stem in one of the characteristic cases. The pineapple (Fig. 40.15) displays a pattern of spirals that can be thought of as if it were a result of compressed phyllotaxis. Such ratios when involving very large numbers approximate an important irrational number, 0.381966. . ., expressing the so-called *golden ratio*. The golden ratio is created by the golden section in which a given length is divided such that the ratio of the longer part to the whole is the same as the ratio of the shorter part to the longer part. If the whole is 1.00, the lengths of the longer and shorter parts will be 0.618 and 0.382, respectively. This may be the single most important proportion in architecture and in artistic expression. Its relationship to phyllotaxis may have inspired Leonardo da Vinci's description of the scattered leaf arrangement as "more beautiful, more simple, or more direct" than anything humans could devise (Leonardo da Vinci 1939).

Spiral symmetry can also be considered as belonging to the broad concept of *similarity symmetry*. Here pattern generation always involves an increment of a characteristic property (Fig. 40.16).

With two-dimensional space-groups, there is a total of *17* ways to generate different patterns. It is a special case when the planar network covers the plane without gaps and overlaps. Of the regular polygons, only the equilateral triangle, the square, and the regular hexagon are capable of covering the plane without gaps and overlaps. For arbitrary shapes though, there are infinite possibilities. M.C. Escher's periodic drawings and the wall decorations in the Alhambra of Granada, Spain

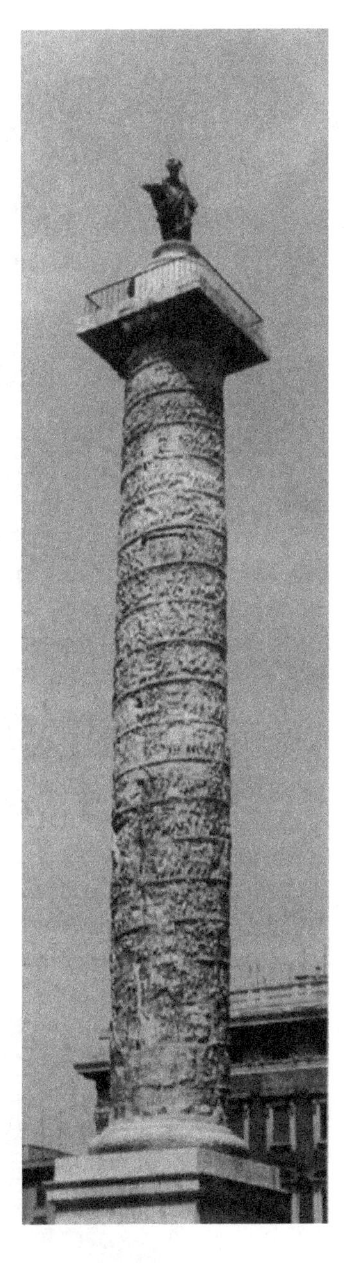

Fig. 40.14 Spiral symmetry of Trajan's column in the Forum Romanum in ancient Rome. Photo: authors

(Fig. 40.17) are famous examples. The façades of buildings, especially those of modern skyscrapers often display symmetries in two dimensions (Fig. 40.18).

Space utilization by periodic arrangements seems to be the underlying principle of the occurrence of three-dimensional space-group symmetries. This is a common arrangement of the building elements in *crystals*. The packing of spheres was first considered as the key to the internal structure of crystals by Johannes Kepler. As he

Fig. 40.15 The pineapple displays a pattern of spirals that can be thought of as if it were a result of compressed phyllotaxis. Photo: authors

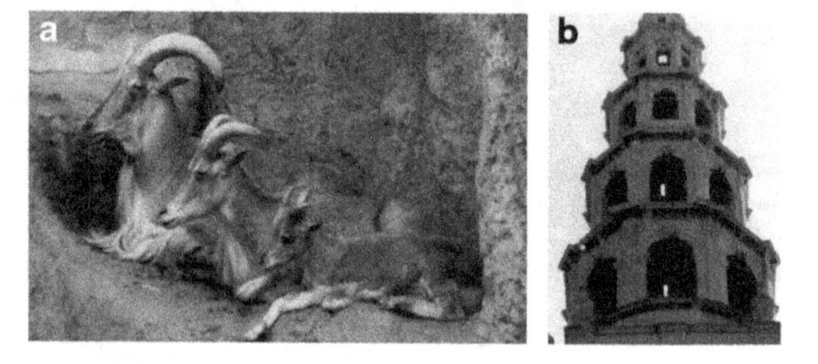

Fig. 40.16 (**a**) Similarity symmetry, the increments being the change in size or the change in age. (**b**) An architectural example of similarity symmetry where the increment is the change in size of the units of the church-tower in London, England. Photo: authors

Fig. 40.17 Two-dimensional space group: decoration from the Alhambra Granada, Spain. Photo: authors

Fig. 40.18 The façades of modern skyscrapers are typical examples of repetitions in two dimensions. Photo: authors

was looking at the exquisitely beautiful hexagonal snowflakes, he made drawings of sphere packing, similar to a pyramid of canon balls (Fig. 40.19).

There are restrictions for the regular and periodic structures, such as the nonavailability of fivefold symmetry in generating them. This can be understood

Fig. 40.19 Random arrangement of canon balls provides much poorer space utilization than their regular arrangement. Photo: authors

easily when we find it impossible to cover the plane without gaps or overlaps with equal-size regular pentagons.

Crystals are advantageous for the determination of the structure of molecules. The great success of X-ray crystallography may have diverted attention from structures of lesser symmetry though of not necessarily lesser importance. The discovery of quasiperiodic crystals [in short, quasicrystals (Hargittai 1990)] by the Israeli scientist Dan Shechtman in 1982 has by now persuaded many scientists that their view of crystals is unnecessary narrow. David Mermin compared abandoning the traditional classification scheme of crystallography, based on periodicity, to abandoning the Ptolemaic view in astronomy, and likened changing it to a new foundation to astronomy's adopting the Copernican view (Mermin 1992).

Recently, even such descriptive fields of biology as zoology have displayed a growing activity in symmetry matters. Not surprisingly, the role of external symmetry is being recognized as decisive in mate selection. Empirical evidence supports the notion relating "animal beauty" to the symmetry of outlook. The degree of left-and-right correspondence of the wings seems to correlate with hormone and pheromone production (Angier 1994: C1).

In view of the fundamental importance of the symmetry concept, it is surprising that even very basic discoveries about it were left to be made in this century. When P.A.M. Dirac was asked about Einstein's most important contributions to physics, he singled out Einstein's "introduction of the concept that space and time are symmetrical" (Yang 1991: 11). An important step was Emmy Noether's recognition that symmetry and conservation are connected. Indeed, the idea that the great conservation laws of physics, like the conservation of energy and momentum, are related to symmetry opened up a wholly new way of thinking for scientists. Realizing that Nature included continuous symmetry in her design physicists started to look for new connections.

It was Dirac who had the prescience to write already in 1949, that "I do not believe that there is any need for physical laws to be invariant under reflections" (Dirac 1949). Yet, even most physicists were surprised by the discovery of the nonconservation of parity in 1957 that brought the Nobel prize in physics to T.D. Lee and C.N. Yang. C.P. Snow called this discovery one of the most astonishing in the whole history of science. Since then broken symmetries have been receiving increasing attention.

There seems to be a difference in approach and emphasis between physicists and chemists in viewing symmetry. It may even be related to the ancient Greek philosophers, stressing the importance of continuum by Aristotle, and of the discreet, by Lucretius and Democritos. From the point of view of continuum, even the ideal crystal may be discussed in terms of broken symmetries. On the other hand, the chemist's approach is succinctly symbolized by Democritos' statement: "Nothing exists except atoms and empty space; everything else is opinion."

Of course, the way symmetry is looked at can vary a great deal. While mathematical symmetry is exact and rigorous, the symmetry we encounter in everyday life is much more relaxed. The vague and fuzzy interpretation of the symmetry concept may also aid scientists to recognize trends, characteristic changes, and patterns. This is getting close to blending fact and fantasy. As Arthur Koestler expressed it, "artists treat facts as stimuli for the imagination, while scientists use their imagination to coordinate facts" (Koestler 1949).

Biography István Hargittai, Ph.D., D.Sc., is Professor Emeritus (active) of the Budapest University of Technology and Economics. He is a member of the Hungarian Academy of Sciences and the Academia Europaea (London) and foreign member of the Norwegian Academy of Science and Letters. He is Dr.h.c. of Moscow State University, the University of North Carolina, and the Russian Academy of Sciences. His recent books include: *Buried Glory: Portraits of Soviet Scientists—Makers of a Superpower* (Oxford University Press, 2013); *Drive and Curiosity: What Fuels the Passion for Science* (Prometheus, 2011); *Judging Edward Teller: A Closer Look at One of the Most Influential Scientists of the Twentieth Century* (Prometheus 2010); *Martians of Science: Five Physicists Who Changed the Twentieth Century* (Oxford University, Press 2006; 2008); *The Road to Stockholm: Nobel Prizes, Science, and Scientists* (Oxford University Press, 2002; 2003).

Magdolna Hargittai, Ph.D., D.Sc., Research Professor at the Budapest University of Technology and Economics. She is a member of the Hungarian Academy of Sciences and the Academia Europaea (London). She is Dr.h.c. of the University of North Carolina. Her recent books (with István Hargittai) include: *Symmetry through the Eyes of a Chemist, 3rd ed.* (Springer, 2009, 2010); *Visual Symmetry* (World Scientific, 2009); *Candid Science I-VI* (Imperial College Press, 2000–2006); *In Our Own Image: Personal Symmetry in Discovery* (Kluwer/Plenum, 2000).

References

Angier, N. 1994. *The New York Times*, February 8, 1994.
Coxeter, H.S.M. 1973. *Regular Polytopes*. 3rd edn. Dover Publications: New York.
Dirac, P.A.M. 1949. Forms of Relativistic Dynamics. *Rev. Mod. Phys.* **21**, 392.

Hargittai, I. ed. 1986 and 1989. *Symmetry: Unifying Human Understanding 1 and 2*. New York and Oxford: Pergamon Press.
___. ed. 1990. *Quasicrystals, Networks, and Molecules of Fivefold Symmetry*. New York: VCH.
___. ed. 1992. *Fivefold Symmetry*. Singapore: World Scientific.
Hargittai, I. and M. Hargittai. 1995. *Symmetry through the Eyes of a Chemist*. 2nd edn. New York: Plenum Press.
___. 1994. *Symmetry: A Unifying Concept*. Bolinas, CA: Shelter Publications. Rpt. New York, Random House, 1996.
Hargittai, I. and C. A. Pickover, eds. 1992. *Spiral Symmetry*. Singapore: World Scientific.
Koestler, A. 1949. *Insight and Outlook*. Macmillan: London.
Leonardo da Vinci. 1939. *The Notebooks. 1508–1518*. Jean Paul Richter trans. Oxford: Oxford University Press.
Mermin, N.D. 1992. Copernican Crystallography. *Phys. Rev. Lett* **68**, 1172 (1992).
Shubnikov, A.V. 1951. *Simmetriya I Antisimmetriya Konechnykh Figure*. Izd. Akad. Nauk SSSR: Moscow.
Yang, C.N. 1991. *The Oscar Klein Memorial Lectures*, Vol 1, G. Ekspong ed. World Scientific: Singapore.

Chapter 41
Contra Divinam Proportionem

Marco Frascari and Livio G. Volpi Ghirardini

An account of the discovery of the role of the Golden Ratio in the gambler's world of probability, and its consequence to us all
(Subtitle of a web page entitled "Golden Ratio selling a gambling method based on the Divine Proportion")

In his *Dictionnaire raisonné de l'architecture française du XIe au XVIe siècle* (1854–1868), under the entry "Proportion," Eugène Viollet-le-Duc asks himself:

> Measuring a hundred times the Parthenon with the difference of few millimetres, what is the use of such a compilation of documents, if we do not know how to derive from them the generating principle of proportions?

What is the generating principle of architectural proportion? The question raised by Viollet-le-Duc is essential, but the clarity of its formulation has been constantly soddened by the mystical and aesthetic clouds of the ϕaithful.

Are architectural proportions metric, numeric, geometric or golden? Which ones among the many in a building are the markers that should be considered reference points for the proportioning of its parts? A golden or divine magnifying glass that distorts rather than clarifies has been applied to everything in the name of aesthetic and mystical impulses. A proportion called the Golden Mean has long been the only explanation for a successive mélange of proportions in all the visual arts. This Golden Mean (also called the Divine Proportion) has been found repeatedly in the pictures of growth patterns embodied in natural events or in the pictures of human products. Since the last century it has so fascinated mathematicians and artists that it is proposed by many as the absolute aesthetic value.

First published as: Marco Frascari and Livio Volpi Ghirardini "Contra Divinam Proportionem", pp. 65–74 in *Nexus II: Architecture and Mathematics*, ed. Kim Williams, Fucecchio (Florence): Edizioni dell'Erba, 1998.
Marco Frascari (1945–2013).

L.G. Volpi Ghirardini (✉)
Accademia Nazionale Virgiliana, Mantova, Italy

Via Chiassi 65, 46100, Mantova, Italy
e-mail: volpighirardini@virgilio.it

K. Williams and M.J. Ostwald (eds.), *Architecture and Mathematics from Antiquity to the Future*, DOI 10.1007/978-3-319-00137-1_41,

By tracing lines onto pictures, this ideal proportion has been found in man-made artefacts and used to mark human achievements. As the acme of this mystically scientific process, pictures of the Parthenon with Golden sections traced on them have been exhibited as demonstrations of the beauty of its man-made, but nature-inspired, rational design. This graphic notion of beauty is so alluring and pervasive that it has been acritically forced upon us as an aesthetic paradigm since grade school. In many school books appear the usual photographs of a nautilus, a pattern of leaf growth and the Parthenon with black or red lines of different thicknesses, all presented as testimonials of the beautiful presence of this perfect ratio in human as well as natural creations, intended to safeguard us from the ugliness of other proportional systems.[1]

What is the *Golden Section*? It is a way to divide a line so as to create an ideal relationship between the parts. A given length is divided such that the ratio of the longer part to the whole is the same as the ratio of the shorter part to the longer part. In other words, "the smaller is to the larger as the larger is to the whole." To explain simply this 'divine concept' for the divination of gambling and for future designs, it is stated that the segment AB is divided so that the ratio of AC to AB is the same as the ratio of CB to AC. If AC is 1.000, then AB becomes 1.618, which is also known as the Golden Mean, or ϕ. The symbol is derived from the initial of Phidias.

The German Apollonian search within the combined sciences of mathematics, philosophy and archaeology lies at the root of the scientific proposal of the Golden Mean as a panacea for explaining the composition of parts and foretelling the aesthetic future of man-made designs. German philosopher Adolf Zeising has made the Golden Mean the only possible principle of a scientific aesthetic and used the Parthenon with the usual diagram traced on it to provide the necessary archaeological authority for his theory of the omnipresence of the aesthetic guarantor ϕ. In 1876, in a ponderous article published in memory of Zeising, mathematician Siegmund Gunter reviewed Zeising's scientific aesthetics in a critical manner, but even he admitted that the presence of ϕ in antique architecture, and notably in the Parthenon, was clear evidence of its being the powerful quintessence of classical aesthetic values. Without any doubt Zeising and Gunter were very skilful at measuring pictures, but it is clear that neither of them had ever measured a building according to tectonic principles (Fig. 41.1).

Vincenzo Scamozzi, in the passage in his treatise on architectural ideals devoted to a discussion of the utility and importance of properly measured drawings, refers to one of Aulus Gellus' *Attic Nights*. In this philosophical story, Gellus depicts an event that took place during a dinner at Frontonis Cornelius's house. A few artisans and an architect present at the dinner are submitting to the host a selection of bathroom designs in order that he might choose the one that he would like to add at his villa. After some consideration, Cornelius makes his selection from among the designs and then asks what the cost of the making of the new bath will be. This appropriate question is used as a premise for raising the philosophical issue on

[1] For a specific discussion of the Golden Mean and architecture, see Ostwald (2001).

Fig. 41.1 The old ϕaithful Parthenon, as probably seen by Adolf Zeising, without the diagram of ϕ on the image. Photo: A. Normand (1851)

which the conversation will be focused during the repast. From the reactions and answers of the architect and the artisans, it appears that the presented set of drawings, although complete in their description of the design, does not allow a precise estimate of the expenditure. A word recurring quite often in the artisans' and the architect's statements is *praeterpropter* (Italian: *pressapoco, circa*; English: about, nearly). The peculiar term *praeterpropter* tickles Cornelius' philosophical inquisitiveness and, after having remarked on its presence in the discourse, he asks a preposterous and precise Apollonian grammarian who is attending the dinner for a definition of this curious term. Immediately this grammarian dismisses *praeterpropter* as a vulgar term used by unscientific artisans. However, Cornelius recalls that the term has been used a few times by Cato, and that Ennius uses it specifically in one of his tragedies. The quote from Ennius' text is *incerte era animus, praeterpropter vita vivitur* (the soul wanders, we live approximately our lives). In the play, Ennius uses this line to point out that he who acts around things is in a state of equipoise between *otium* and *negotium*, between relaxed mental constructions and tense constructive activity. Analogically, we can derive the consideration that architects and builders are just so when they are setting the measures of their designs. *Otium* and *negotium* are the twofold structure of material and existence through which the physical and the metaphysical measures and matrixes of a specific design are elaborated. Mentally conceived, these negotiated measures are the approximate tools for construing and constructing a design. The duality of *otium* and *negotium* relates the ineffable qualities of a design to the storytelling ability of a construction. This polar binomial is the working tool that, through the negotiation of purely and utterly conceived numerical controls, permits the translation of a design into built form or of an idea into a design. These numerical expressions are the resources for the architect's conceiving imagination as well as for the builder's constructive imagination. Even in their approximation, they are very precise tools in the sense that they determine the substance of a design and govern how the design is materially realized.

In metrical terms, every constructive part of building has its geometric order: masonry, in decimetres; wood carpentry, in centimetres; metal works, in millimetres. Every part is exactly approximate. The same is true of the parts conceived by the imagination. Traditionally speaking, a design begins with a tracing of a few lines of a sketch. For instance, one tries to imagine an opening of 3 m by 5 m and traces a few lines. This can be done because most designers know approximately how long 6 m are, but they know much less about 5 m and 3 cm, and they do not know anything at all about 5 m, 3 cm and 5 mm. In designing, we need approximation because it is with the approximation of our body that we know how to measure. Palms, *braccia* or feet are the proper measures. This is because in walking and gesticulating we measure a building. However, our bodies are all deviant from the norm and this humanly precise measuring can be only approximate. We know our palm, but we do not know the palm of the other. Design is the understanding of the other and therefore to approximate is the essential condition of the interpretation of an individual *otium* in the alterity of *negotium*. Furthermore, proportions resulting from designers' *otium* are then negotiated in the material selected for the construction. Approximate measures are tangible and tamable, whereas uncompromising measures are elusive. The Golden Mean is an untamable and intangible measure since, in order for it to be real and efficient, it must be explicitly exact. However, architecture does not permit this categorical exactness, because there are always mitigating factors such as play in the joints and the density of materials. The thickness of a mortar joint, for example, gives dimension to what on paper is an ideal line. The precision of the Golden Mean, an intolerant aesthetic rule, does not acknowledge any play; if such an undesirable factor is discerned, it is called *tolerance*, something that, unfortunately, the ϕ believers must tolerate.

Within a few centuries of each other, two architectural events took place in Milan, each of which dealt with the relationship between the theory and the practice of architecture. Considered together, these events can facilitate the understanding of the positive and negative conditions created when a design is ruled either by the pseudo-precision of ϕ or by an approximation of the rational integration of integers that can be translated into a built form. The first event took place on the construction site of the Cathedral of Milan at an impasse in the design. The second event took place on the grounds of the Milanese Triennale at the particular moment of architectural fervour that had been generated by the need and the forces of the post-war reconstruction in Europe.

The first episode was characterized by the presence of an intolerant French architect. Although he was not advocating ϕ as a universal remedy for the architectural ailments of the Cathedral, the trendy French master builder was setting the *forma mentis* for an understanding of the use of proportion in design based on a rigorous configuration which did not have any direct point of reference with regards to the points of the order ruling the construction of the building. Any point is good for tracing a scientific relationship among the parts of a building. It does not matter if that moulding or cornice is really a point belonging to the configuration of the building erection where a real plumb line or snap line has been applied or must be applied. As for the ϕ believers, any point is good for making the point.

The debate between the French master builder and the Milanese master masons was about the use of geometry. The debate has been presented and discussed by several scholars and the dominant interpretation is that the Parisian master builder advocated geometry as the true science ruling architecture, whereas the Italian master masons did not, instead working by rules of thumb. The famous line of the French master is *Ars sine scientia nihil est* (the art of construction without a geometrical science is nothing). However, the Italian master masons did believe in geometry, not as an intolerant science, but as a system of orderly, constructed and practical associations. As they state in the answer to the French master, they apply an *ordo geometricus*, a pre-instituted unique condition of proportion imposed as a prerequisite for a mystical aesthetic.

This condition of a *pre-posterous* science opposed to a *prosperous* order, a procedure of design based on an elegant ruling of chosen ratios, was the intellectual horizon towards which the advocates of ϕ as an essential condition for proper design would push their agenda during a conference held at the Triennale as well as in an exhibition illustrating a ϕ-biased view of the evolution of historical images showing the use of proportions from the mythological past up to the re-use of the ϕ-proportioned tubular frames conceived by Pagano for the famous Fascist exhibition of the progress of Italian aviation. The aim of this event was the official imposition of ϕ as a design rule on every architect. The resolution, voted on by scholars and designers present at the conference, lost by a vote and the proceedings have been never published (thank G...!).

This preposterous lion's share of ϕ, as we have already pointed out, began with the German Apollonian and mathematical pan-aesthetic. Following an iconophile's logic, the ϕ becomes a sacred measure that is the indispensable means for painterly composition and from that it follows that it is the quintessential universal aesthetic requirement for any human and natural invention. This powerful divining procedure of the human world, which foretells and backtells any human intervention, is the great discovery of Matila prince Ghyka, who combines the two systems by making clear that the science raises the question and the myth gives the answer. Scientific and mystical *otii* can forego the necessary *negotii*.

In recounting the life of Pericles, a forgetful Plutarch does not tell which proportions were preferred by the craftsmen in charge of the renovation of Athens during the second half of the Fifth century BC, nor does he tell what the great Phidias thought of it. We are doubtful of the statement that ϕ originates in a concealed passage of the sixth book of Euclid's well-known *Elements*; that this was written a century after Phidias' time does not help. But for the *ϕaithful*, the Golden Section is a divine principle which has been, is and it always will be.

Plutarch's silence entitles us to make a few observations as to the commensurability of ϕ. First of all, 5:3 and 8:5 are ratios generating a rational number, whereas ϕ generates an irrational number. For a fifth century Athenian, the first two above mentioned ratios were intelligible since they are finite terms, while ϕ, on the contrary, is undefined. It is a vague number that is neither odd nor even, hence neither male nor female: it is not even an hermaphrodite, but an undetermined sexual aberration. For the Greeks, the infinite was undefined, and the indeterminate could not be beautiful, measurable or perfect. This concept of Pythagorean tradition was ordinary wisdom among the agora-goers. Any Athenian could follow the elegant thinking embodied in the paradox of Achilles and the turtle set by such a rational mathematician as Zeno. The same is true for the elegant arts of construction, and the measurable presence of harmony. They cannot be born of antithetical models. Our hypothesis is that all the "discoveries" of the Golden Sections of the golden past are actually based on the ratios 5:3 and 8:5. This is a probable hypothesis, which does depend on few measured millimetres.

In 1674, spending two laborious months on the Acropolis, Carrey produced the first modern set of measured drawings of the Parthenon. At that time, Athens was under Turkish rule, therefore it was before the Venetian Siege, during which a bomb explosion ruined part of the temple. The Turks were judiciously distrustful of foreigners, and if a gate was opened so that Western culture could know the forms of the Parthenon, it was the result of a gift to the ruling Aga: six measures (*braccia*) of Red Venetian brocade, half a dozen four-pound geese and coffee for a total value of 50 gold pieces (*zecchini*). The negotiation that allowed westerners to gaze on the proportions of the Parthenon was forged on two fundamental counting systems that contain the six and the ten (6 measures, 6 geese, 50 gold pieces), their Golden numbers. The ratio between these two Golden numbers—highly regarded by the ancient mathematician and philosophers—is 5:3.[2] Perhaps, in this ratio lies the hidden key for correctly reading the proportions of the Parthenon.

As Sophists and their modern counterpart, the Attorneys, teach us, any reasoning can be turned around. Even where the ratios 5:3 and 8:5 are clearly spelled out, any strict believer in the aesthetic power of ϕ will see in it the Golden Section.

In his *Architectural Principles in the Age of Humanism*, Rudolf Wittkower focused the attention of architects and scholars on how the issue of commensurable entities must have been a substantial problem for the Renaissance artists who were discovering and reinventing the Classical Arts. In *De Re Aedificatoria*, Leon Battista Alberti, the most consequential personality among the architectural theoreticians of the period, defines the 'aureate ratios' in which a proper architect should believe. Alberti's 'aureate ratios' are structured in three groups of three entities each: (a) small areas (1:1, 2:3, 3:4), (b) medium areas

[2] The evil architect Squaronthehypotenus from the Asterix comic book *The Mansions of the Gods* based his design for a typical condominium in a Roman holiday resort in the upper Gallia on the plan of Alberti's San Sebastiano in Mantua; the design is based on the ratio 5:3; see Goscinny and Uderzo (1972: 26). For a correct and recent measured drawings of the plan, see Calzona and Volpi Ghirardini (1994: Dis. 2).

Fig. 41.2 A mind to which Zeising and Ghyka refer frequently is the one of Luca Pacioli. The Italian government dedicated a 500 lira coin to this famous figure. The financial peregrinations of the coin, an instrument for gambling, are always recorded using the Fibonacci double entry system. Photo: authors

(1:2, 4:9, 9:16) and (c) large areas (1:3, 3:8, 1:4) (Alberti 1755: IX iv, 198). The ratios 5:3 and 8:5 do not show up among these three sets of Albertian areas. However, in his practice, Alberti uses the ratio 5:3, primarily to underscore the spaces of the church of San Sebastiano, built in Mantua during the second half of the Quattrocento. First, this forethought can reduce the importance of golden ratios, framing them within a theoretical realm; second, it can establish the ratio 5:3 as a full-fledged reality of the negotiation to be matured and refined within the constructive realm of any architectural plan.

The idea of reading ϕ when it is clearly written 5:3 can be asserted with arguments that are not presumptive. It is well known that during antiquity ratios were used to manifest incommensurable entities. These approximations were judged perfectly acceptable. For instance, in his *Ludi rerum mathematicarum*, Alberti used the ratio 22:7 to define π, but, considering it too roughly estimated, he refused the ratio 7:5 for defining the right angle. However, he then uses the same ratio to substitute Vitruvius proportion of $\sqrt{2}$:1, as Zoubov made it clear by comparing excerpts by the two authors.

During the first conference in the series "Nexus: Relationships Between Architecture and Mathematics" in 1996, the nature of Alberti's proposition of areas and the characteristics belonging to ratios among the measures was demonstrated (Volpi Ghirardini 2015). The areas are born, on the one hand, from the *otium* of a pure mathematical thought, from a *ludus* among numbers embraced by 1 and 4, and, on the other hand, by a *negotium* with the final growth and change of the design translated into a built form. In other words, this is a procedure that takes place within the superior and inferior limits of the Pythagorean *tetraktys*. The ratio 5:3 joins the above-mentioned *ludus*, widening the beginning terms of a unit, that is, enlarging the field from 1 to 5. The logical procedure followed by Alberti has a lucid precedent in the arithmetic of Nicomachus. It is a tribute to the natural number. It is also a tribute to a Greek-Hellenistic mathematics, which is a puissant sibling to geometry. Although related, it is not geometry because of its expression of commensurable entities. The ratio 5:3 cannot represent the incommensurable ϕ, which then is out of the game, pardon, out of the ludus of the negotiations of architectural design.

If someone still harbours doubts that 5:3 can be a guise behind which is concealed the Apollonian ϕ, wandering among edifices to allure our fantasy we come upon another *ab abumdantiam* argument. In the central-plan church of Alberti's San Sebastiano, the numbers 5 and 3 not only zestfully mould the monochord ratio that harmonizes the whole design, but are also the ideal measures of the central room around which the design of the whole edifice takes shape. Indeed, the length of the side of the square central room is equal to the sum of the squares of 5 and 3. Accordingly, the whole is commensurable; as it is said in the Old Testament book *Wisdom*, all is distributed on the base of weight, number and measure [*Wisdom* 11:20]. The irrational numbers do not dwell in this architecture and ϕ is definitely annihilated: take Nicomachus of Gerasa, Livio of Mantua and Marco of Mantua's word for it (Fig. 41.2).

Biography Marco Frascari (1945–2013) was Full Professor of Architecture at the Azrieli School of Architecture & Urbanism at Carleton University in Ottawa. His research interests included theories, histories and practices of architectural representations; theories and histories of architecture in the Veneto; and theories, histories and practices of tectonics and materiality. His last book was *Eleven Exercises in the Art of Architectural Drawing: Slow Food for the Architect's Imagination* (Routledge, 2011).

Livio G. Volpi Ghirardini holds a degree in civil structural engineering. His professional practice is dedicated to the realization of new projects as well as to the restoration of historic monuments. He is a member of the President's Council of the Accademia Nazionale Virgiliana in Mantua. He is the Prefect of the fabric of the Cathedral and the Basilica of S. Andrea in Mantua. This last explains his particular interest in the architecture of Leon Battista Alberti. He is a founding member of the Centro Studi Leon Battista Alberti in Mantua, and is currently serving as its president. Formerly a professor of building science, now he is a contract professor at the Politecnico di Milano, I School of Architecture, and teaches the architecture restoration workshop entitled "Reinforcing of historical buildings".

References

Alberti, Leon Battista. 1755. *The Ten Books of Architecture*. Reprinted 1986 New York: Dover Publications Inc.

Calzona, A. and L. Volpi Ghirardini. 1994. *Il San Sebastiano di Leon Battista Alberti*. Florence: Olschki.

Goscinny, Réne and Albert Uderzo. 1972. *Asterix e il regno degli dei*. Milan: A. Mondadori.

Ostwald, Michael J., 2001. Under Siege: The Golden Mean. *Nexus Network Journal* **2**: 75–83.

Volpi Ghirardini, L. 2015. The Numerable Architecture of Leon Battista Alberti. Pp. 645–661 in Kim Williams and Michael J. Ostwald eds. *Architecture and Mathematics from Antiquity to the Future: Volume I Antiquity to the 1500s*. Cham: Springer International Publishing.

Part VI
From 1400 A.D.–1500 A.D.

Chapter 42
Alberti's Sant'Andrea and the Etruscan Proportion

Michael R. Ytterberg

Introduction

The church of Sant'Andrea in Mantua is a paradox (Fig. 42.1). It is the last of the church designs of Leon Battista Alberti, and though construction was not begun until the year of his death, it is the most complete of his churches, and therefore the one in which his intentions might seem to be clearest in the resulting structure. Yet though the church takes the form of a Latin cross, the evidence suggests that Alberti had intended a basilican plan. In an extant letter, Alberti was explicit about his intentions in at least one respect: his proposal was for a church of the type "known among the ancients as the Etruscan."[1] But the church is not planned like an Etruscan temple as described by Vitruvius. The description Alberti gave in his treatise, *De re aedificatoria*, adhered to the account of Vitruvius in only one way—the presence of the unusual proportion of 5:6.[2] Yet in spite of numerous attempts to discover Alberti's proportional system at Sant'Andrea, no one has convincingly found the presence of the proportion 5:6 in the completed building—until now.[3]

First published as: Michael Ytterberg, "Alberti's Sant'Andrea and the Etruscan Proportion", pp. 201–216 in *Nexus VII: Architecture and Mathematics*, Kim Williams, ed. Turin: Kim Williams Books, 2008.

[1] Alberti sent this letter, accompanied by a sketch, to Ludovico Gonzaga on 20 or 21 October 1470. For a photograph, transcription and translation of the letter, see Johnson (1975: 8, 64, pl. 12).

[2] The manuscript was completed about 1450 but was first published in 1486 after Alberti's death; see Alberti (1988: xvi–xviii). All references to Alberti's *De re aedificatoria* herein are to Alberti (1988).

[3] The following is a partial list of publications which include a proportional analysis of Sant'Andrea: (Sanpaolesi 1961; Krautheimer 1969: 333 ff.; Borsi 1977; 229 ff.; Morolli 1994; Furnari 1995; Tavernor 1998: 169–181; March 1998: 192).

M.R. Ytterberg (✉)
BLT Architects, 1216 Arch Street, Philadelphia, PA 19107, USA
e-mail: mry@blta.com

K. Williams and M.J. Ostwald (eds.), *Architecture and Mathematics from Antiquity to the Future*, DOI 10.1007/978-3-319-00137-1_42,

Fig. 42.1 Sant'Andrea, Mantua, façade. Photo: courtesy Paolo Monti

Alberti and the Etruscan Temple

Why an Etruscan temple? Sant'Andrea is not the cathedral of Mantua, but it nevertheless figured prominently in the efforts of Alberti's patron, Ludovico Gonzaga, to consolidate power over the city of Mantua. The existing building is built on the site of a previous Benedictine abbey church, the last abbot of which died in 1470. At Ludovico's behest, the abbey was abolished in 1472 and refounded as a collegiate church of which Ludovico's son, Cardinal Francesco Gonzaga, was made head (Borsi 1977: 229). The church is in possession of a sacred relic, samples of the Blood of Christ. The Blood is contained in two vials, now on display in the crypt, which were at one time displayed to the pious each Ascension Day. Thus, in Alberti's words, the "principal intention [of the reconstruction] was to have a great space where many people would be able to see the Blood of Christ" (Johnson 1975: 8, 64, pl. 12). It was for this aim that Alberti proposed an Etruscan temple, which he said would be "more capacious" (Johnson 1975: 8, 64, pl. 12) than a competing design. On the basis of the testimony of Virgil, himself a Mantuan, Mantua has traditionally claimed its origin as an Etruscan city. Alberti's strategy, then, at least in part, was to enhance the prestige of the Gonzaga family through this patriotic gesture.

Several historians, the first of which was Richard Krautheimer (1969: 338–339), have commented on the fact that Alberti's description of an Etruscan temple in his treatise bears a much closer relationship to the Basilica of Maxentius than to Vitruvius's description of one (or to actual Etruscan temples, for that matter), and that the nave of Sant'Andrea resembles both the Basilica of Maxentius and Alberti's description (Fig. 42.2). This has led to the current consensus that the existing Latin cross plan was not intended by Alberti.[4] It is possible that the decision to extend the church was made around 1526 according to the designs of Giulio Romano. Giulio may also have been responsible for the first of what seems to have been many changes to the windows in the nave as well as aspects of the interior decoration. Round-headed rectangular windows lit the interior of the Basilica of Maxentius. Robert Tavernor suggests that round-headed rectangular windows and niches once reproduced in the nave of Sant'Andrea the appearance of the façade.[5] Currently, oculi occur in these positions (Alberti was on record as disliking these).[6]

If only the nave was intended, how might Alberti have terminated his design? While the Basilica of Maxentius terminated in the typical semicircular apse, a sixteenth-century drawing shows what appears to be a rectangular chapel at the end of Alberti's nave (Tavernor 1998: 165). That such a configuration was possible is confirmed by Alberti's statement, "The tribunal itself may be rectangular or semicircular" (Alberti 1988: 187).

But why would Alberti identify the Basilica of Maxentius as an Etruscan temple? Since it was vaulted Alberti may not have recognized it as a basilica, which, to him, were all timber-roofed, aisled structures. No other extant monument in Rome or the rest of Italy corresponded to Vitruvius's description of the Etruscan temple with its three cellae side by side facing forward. The Basilica of Maxentius had three sets of chapels facing each other on both sides of a central nave. In the absence of any other evidence this may have motivated Alberti's identification. Perhaps he imagined the Basilica to be a development of the Vitruvian type, and, given his skeptical attitude toward Vitruvius, preferred to describe a building he had actually experienced rather than something whose details he could not corroborate.

[4] Recently, Robert Tavernor analysed the number of bricks said in a contemporary letter to have been stockpiled for the project and concluded that they were sufficient only for the nave of the extant building; see Tavernor (1998: 160–165) and Johnson (1975: 14, 65).

[5] An innovation at Sant'Andrea is the close correspondence of interior and exterior; see Tavernor (1998: 167–168).

[6] Alberti expressed disapproval of oculi in a letter to Matteo de' Pasti, the site architect for another of Alberti's churches, the Tempio Malatestiano Tavernor (1998: 60). Johnson, however, suggested that the original form of the interior elevations of the nave may have included round headed rectangular niches above the doors to the small chapels, similar to the façade, surmounted by the existing oculi. There is evidence for this view in the form of walled-up openings visible from within the western transept piers. See Johnson (1975: 16–17, pl. 17, 18).

Fig. 42.2 Sant'Andrea, Mantua, nave. Photo: courtesy Paolo Monti

Moreover, the Basilica of Maxentius had been misidentified since antiquity; only in the nineteenth century was it accurately identified. A misreading of ancient texts had caused the Basilica to be mistaken for the adjacent Temple of Peace, of which very little remains.[7] The Temple of Peace had been founded by Vespasian to commemorate the suppression of the Jewish revolt of 70 CE and housed plundered treasure from the destroyed Temple of Jerusalem. Since Rome had housed remnants of the Temple of Jerusalem, the Church had additional reason to claim Rome as the successor to Jerusalem, particularly in light of the Muslim occupation of that city. At all times the Temple, whose dimensions and proportions are given in the Bible, has been the model for Christian churches, most commonly as a metaphor but also

[7] For many centuries the Basilica was therefore referred to as the *Templum Pacis* or the *Templum Pacis et Latonae* or simply *Templum Latona*, which is how Alberti knew it. The reference to Latona is another mystery and possibly another case of confusion with an adjacent monument, in this case with the *Arcus Latronis*. Latona was the mother of Apollo and Artemis (Alberti 1988: 22, 370, note 83).

frequently in terms of physical structure.[8] The combination of Etruscan and Jewish references in this single monument—both mistaken—may have made the Basilica of Maxentius an overpoweringly appropriate model for the church of Sant'Andrea.

As mentioned above, the common element in the descriptions of an Etruscan temple by Alberti and Vitruvius is the presence of the proportion 5:6, an unusual proportion because it is not one of the Pythagorean musical consonances upon which so much of ancient and Renaissance proportional theory was based. Alberti's description, which otherwise fits the Basilica of Maxentius very closely, does not initially seem to match in terms of the proportional scheme of the overall plan.[9] But subtracting the vestibule, the main part of Basilica of Maxentius does conform to the proportion of 5:6 (Fig. 42.3). The reality of the situation is that the combination

---

[8] Establishing the Temple of Jerusalem as the model for Sant'Andrea would be most convincing if there were the replication in Sant'Andrea of the dimensions and/or proportions of the inner chamber of the Temple, given in the Bible as 20 cubits wide by 30 cubits high by 60 cubits long, a proportion of 2:3:6 (1 Kings 6.2). This corresponds to Pythagorean musical consonances of a fifth (2:3) and an octave (1:2 = 3:6), ratios condoned by Alberti in his treatise (1988: 305). Numerous observers have measured Sant'Andrea and have found that the width of the nave is 40 Mantuan braccia wide by 60 Mantuan braccia high, a ratio of 2:3, the same as the Temple (Tavernor 1998: 169 f).

The question of the length is more difficult since the extension of the church in the sixteenth century. At least one observer, the local historian Giovanni Cadioli, established in 1763 that the nave of the then Latin cross plan was 120 Mantuan braccia long, see Cadioli (1974: 61). If this figure was the same for Alberti's plan, then the correspondence with the Temple would be perfect, and the ratio between height and length would be 1:2, an octave. Yet the best attempts to reconstruct Alberti's design fail to support this number. Tavernor measured the length of the existing nave as "closer to 115 braccia," based on a photogrammetrical survey made of the church prior to a 1994 exhibition (Tavernor 1998: 171). Based on the same survey, the present study suggests that the length of Alberti's nave was precisely 116 Mantuan braccia long. This result would not seem to fall within an acceptable range of approximation to the Temple for an architect as rigorous as Alberti, so the proportional model for Sant' Andrea must lie elsewhere.

[9] Tavernor has suggested that the disconnect between Alberti's description of the Basilica Maxentius and its proportional scheme may have been a typographical mistake. He suggests that with a simple transposition of numbers, Alberti's proportional scheme can be made to fit the Basilica of Maxentius more exactly. Alberti's text reads, "In plan, their length, divided into six, is one part longer than their width. A portico, serving as the vestibule to the temple, takes up two parts of that length" (Alberti 1988: 197). Tavernor suggests that a closer fit to the Basilica of Maxentius would be obtained if the passage were to read, "In plan, their length, divided into six, is two parts longer than their width. A portico, serving as the vestibule to the temple, takes up one part of that length" (Tavernor 1998: 177). Unfortunately, Tavernor prints a diagram which does not conform to his suggestion. The diagram he publishes divides the Basilica lengthwise into seven parts, not six. But the diagram is correct, for it demonstrates that the main part of the Basilica of Maxentius does conform to the proportion of 5:6. The vestibule is an addition to this proportion, making a total length of seven units, not a subtraction from the overall proportion of six units as Alberti's text suggests. For Alberti's description to be an accurate account of the Basilica of Maxentius, it would have to read, "In plan, their length, divided into six, is one part longer than their width. A portico, serving as the vestibule to the temple, is one part in addition to that length." If this were the case the text would precisely reflect the reality of the plan of the Basilica.

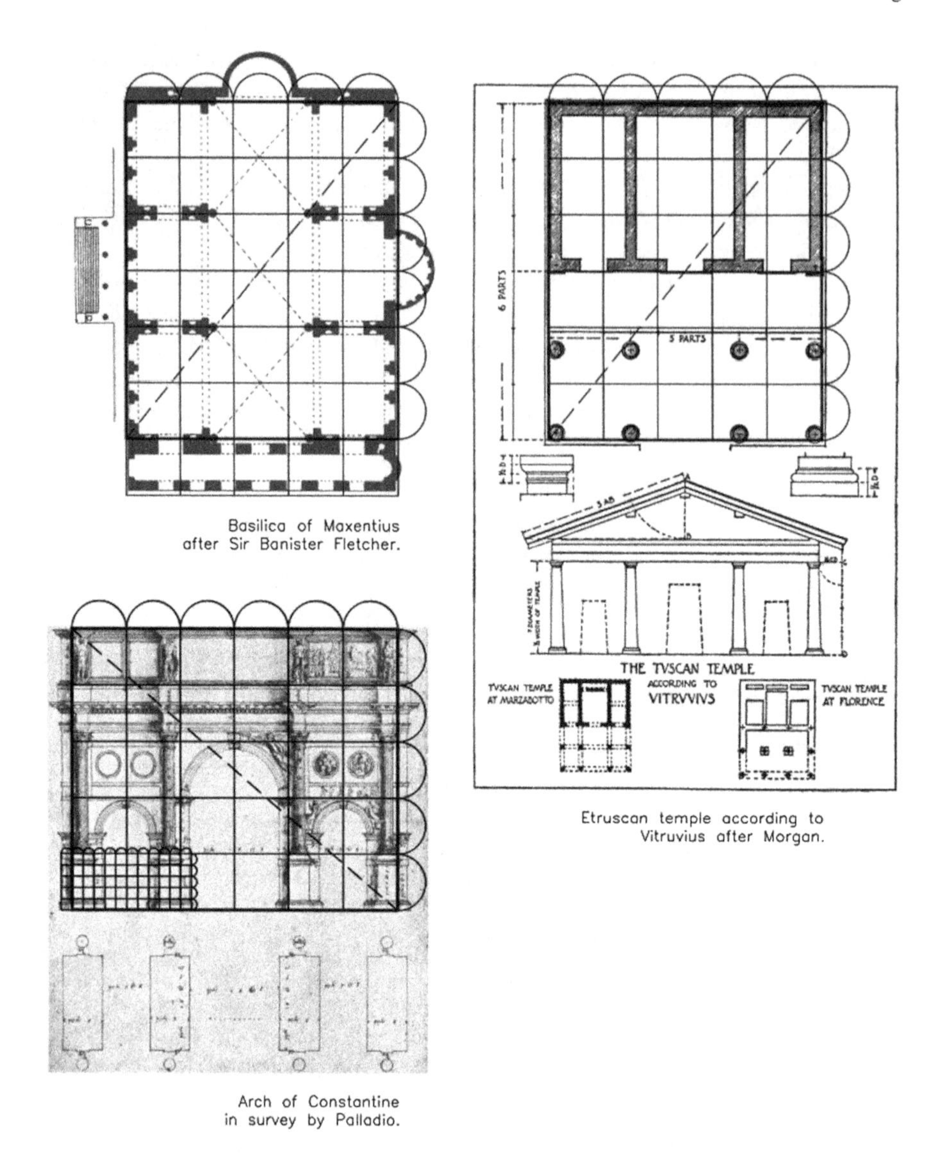

Fig. 42.3 Proportional prototypes: 5:6. Drawing: author

of three chapels on a side and the proportion of 5:6 would have been enough to convince Alberti that the Basilica of Maxentius, known to him by a different name, corresponded more closely to Vitruvius's description of an Etruscan temple than any other surviving example of antique architecture.

Other Prototypes

The Basilica of Maxentius is not the only ancient prototype for Sant'Andrea. The Roman triumphal arch, particularly that of Constantine, was an equally important source for the form of the church. The Blood of Christ, the sacred relic of Sant'Andrea, is both the wine that is drunk at Communion, in remembrance of Him as a community of believers, and the very means to our salvation. It is through Christ's sacrifice and subsequent Resurrection that the pious may defeat death and enter heaven.

It was due to Constantine's efforts that the triumph of Christ came to be associated with the Roman triumph. The specifically Roman column capital, the Composite, originally appeared on triumphal arches and was therefore associated with the triumph of Rome. Seemingly in deference to the Saviour, Constantine's own triumphal arch in Rome, however, has Corinthian capitals.[10]

Alberti had already used the triumphal arch motif for the church façades he had executed at the Tempio Malatestiano at Rimini, where the model was the local Arch of Augustus, and at Santa Maria Novella, where the model was the Arch of Constantine in Rome.[11] At Sant'Andrea the use of the triumphal arch motif for the façade becomes a completely three-dimensional creation for the first time, a building in its own right.[12] Due to the presence of the relic of the Blood of Christ, at Mantua the triumphal arch motif became a specific expression for a particular church in addition to being a standard theme appropriate for every church. Possibly for this reason Alberti continued the triumphal arch motif inside the church onto the walls of the nave. Each major chapel with its two adjacent minor chapels repeats the organization of the façade. This gave Sant'Andrea an unusual degree of correspondence between inside and outside, which was to have important consequences for the future of Renaissance architecture. But in one important way the inside and outside are different: the pilasters on the façade are Corinthian, while those in the interior of the church are Composite.

Neither the façade nor the nave walls conform to the proportional system for a triumphal arch which Alberti provided in his treatise (Alberti 1988: 265–268). Alberti's system does not establish an overall relationship of width to height. The height of the arch depends on the dimensions of the columnar order, which dimensions are not directly controlled by the geometry of the arch as a whole. The illustration that was provided by Cosimo Bartoli in the first illustrated edition

---

[10] See Onians (1988: 59) for a discussion of the meaning of the Composite capital, the Roman triumph, and Christianity.

[11] Alberti's experiments became standard motifs for subsequent Renaissance architecture; see Rudolf Wittkower's classic discussion of Alberti's church façades (1962: 37 ff.). Robert Tavernor (1998: 178) links the Arch of Constantine with the façade of Santa Maria Novella.

[12] One of the most distinctive facts about Sant'Andrea is that the presence of the pre-existing campanile meant that the portico could not be as wide as the church behind, thus accentuating its semi autonomous nature.

of Alberti, published in Florence in 1550 [reproduced in (Alberti 1988: 267)], shows an arch that has a height equal to its width, and this seems to have influenced subsequent authors. If one proportional diagram is printed in any work dealing with Sant'Andrea, it is one that shows that the façade fits within a square. It is true that the width of the portico is equal to the height of the pediment, and this is taken by Tavernor and others as a link to Alberti's preferred proportions for a triumphal arch. In the process, however, a key relationship is obscured. Roman triumphal arches were built to a variety of designs and proportions. Alberti's description of one in his treatise was intended to refer to a typical, not a specific, arch. The Arch of Constantine, to which Alberti's description of an arch otherwise closely adheres, does not fit within a square.

Rather, the Arch of Constantine is controlled by the proportion of 5:6, and its details fit rather neatly into a grid of that proportion (Fig. 42.3). Thus both the appropriateness of the design for a city founded by the Etruscans and the appropriateness of the design to a church that houses the Blood of Christ point to the same proportional system as crucial to the meanings embodied in this particular building. But it still remains to demonstrate the use of this proportion in the building as built.

A Demonstration of the Proportions

This task seems simple enough, but its solution has evaded all those who have attempted it. Not atypical in this regard is the plan diagram reproduced in the beautiful book on Alberti's architecture published by Franco Borsi (1977: 232) (Fig. 42.4).

In this diagram the solution seems simple: the nave has a proportion of 5:6 based on a square module which defines the cells of the major and minor chapels. There are two problems, however. First, this system does not include what would have been the final chapel and is now the crossing pier, suggesting an unlikely asymmetrical elevation of the nave wall in Alberti's building. Second, if one places the plan at the left over the diagram on the right, one discovers the reason for not showing the diagram on top of the plan: they do not correspond. The faint lines of the plan beneath the diagram have been stretched to make the diagram work. More honest, but still unsatisfactory, are all other published attempts to place a diagram over the plan which illustrates an application of the Etruscan proportion.[13]

[13] See the list of publications in n. 4 above.

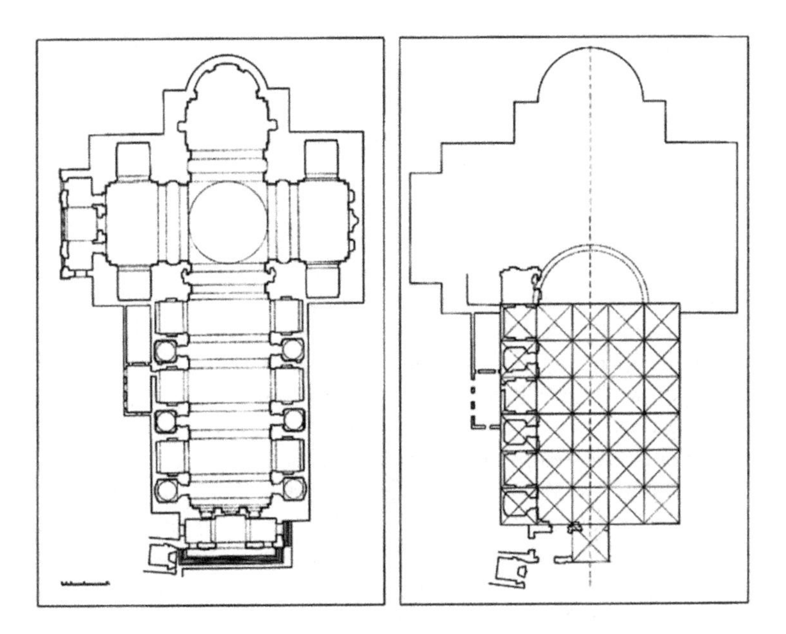

Fig. 42.4 Sant'Andrea, Mantua. Plan and proportional diagram. Drawing: author, after Borsi (1977)

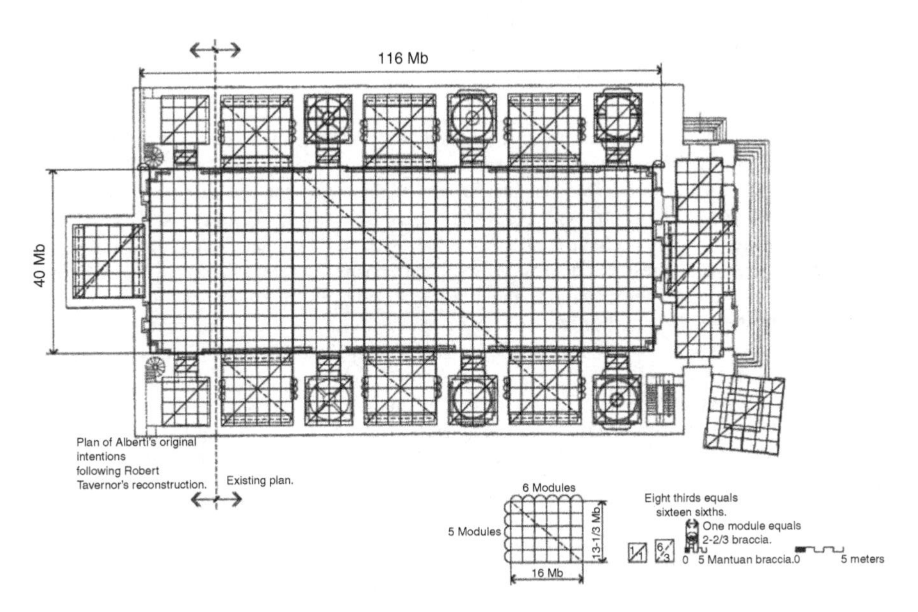

Fig. 42.5 Sant'Andrea, Mantua. Proportional scheme, plan. Drawing: author

The present analysis begins with the plan in its restored state, using the plan restoration published by Robert Tavernor (Fig. 42.5).[14]

It has been immediately noticeable to all observers that the side chapels divide the length of the building such that the void of the major chapels approximately equals the solid mass of wall surrounding the minor chapels. This divides the length of the nave into seven equal spaces, not counting an extra bit at each end caused by the repetition of a pilaster in the corner. However, a square drawn around these chapels, taking as its side the longitudinal modular dimension, produces no apparent relationship to the width of the nave, yet this square is apparently equal in size to the plan of the campanile left over from the Benedictine abbey church. Closer study of the major chapels reveals that the centreline to centreline distance between the pilasters and arches that define the front and rear of the chapels appears to be in the ratio of 5:6 to the width of the chapels. This suggests a modular dimension created by dividing the notional width of the chapels by 6. The width of the nave measures 15 of these proposed modules across. The width of the nave has been given by numerous observers as 40 Mantuan braccia.[15] Dividing by 15 gives the unusual dimension of 2 2/3 braccia as a possible module, a puzzling number. However, multiplication by 6 for the width of the chapels produces the result of 16 braccia, the Vitruvian perfect number, the sum of 10 and 6, the numbers explicit in the human body.[16] In his treatise Alberti does not single out the number 16 as did Vitruvius, but he does call 6 and 10 the "perfect" numbers (Alberti 1988: 304). The dimension of the sides of the existing campanile appears to be 16 braccia, at least notionally, and the starting point for the layout of the new church. Applying

[14] In the case of Sant'Andrea good drawings are available on which to base a proportional study. An exhibition on Alberti's architecture was mounted in 1994 by the Alberti Group, an organization created with the financial assistance of the Olivetti Corporation for the purpose of staging an exhibition. Photogrammetric surveys were made of the major works of Alberti at that time. The surveys that were available for Sant'Andrea were the south and west elevation of the nave and the west façade. These plus a restored plan and west elevation from Tavernor (1998: 142, 185) formed the basis for this study. These drawings show current conditions, of course, and not the original intentions of Alberti. No attempt has been made to restore these elevations or to suggest any disagreement with the restorations published by Tavernor. It should be noted that using another's reconstruction of the original building helps to avoid the trap of devising a plan to fit a proportional system. Verification of key dimensions of the photogrammetrical surveys was possible by comparison with the survey published by E. Ritscher (1899: 2–19, 182–189), republished in Johnson (1975: pl.14, 15, 45, and 79). Every modular dimension in this chapter that can be verified numerically by comparison to Ritscher's work deviates less that a fraction of a percent from the actual given value.

[15] See above, n. 9. A *braccio* (arm, plural *braccia*) is an Italian cubit whose exact length varied from city to city. The Mantuan braccio was equal to .467 m. A stone monument still exists in Mantua which established the official standard for the braccio and other measures. There is a photograph in Rykwert (1979: 76).

[16] The length of the plan unit, 16 Mantuan braccia, can be verified numerically. A close look at Ritscher's (1899) survey and the photogrammetrical survey reveals that the bays of the nave side elevations, which appear to be uniform in terms of their decoration, in fact show a degree of variation, as one might well expect. If we take the average in either case and extend the existing nave by this dimension for the missing final bay, the answer gives a dimension, which, when divided, produces a value of within a fraction of a percent of 16 braccia. See n. 16 above.

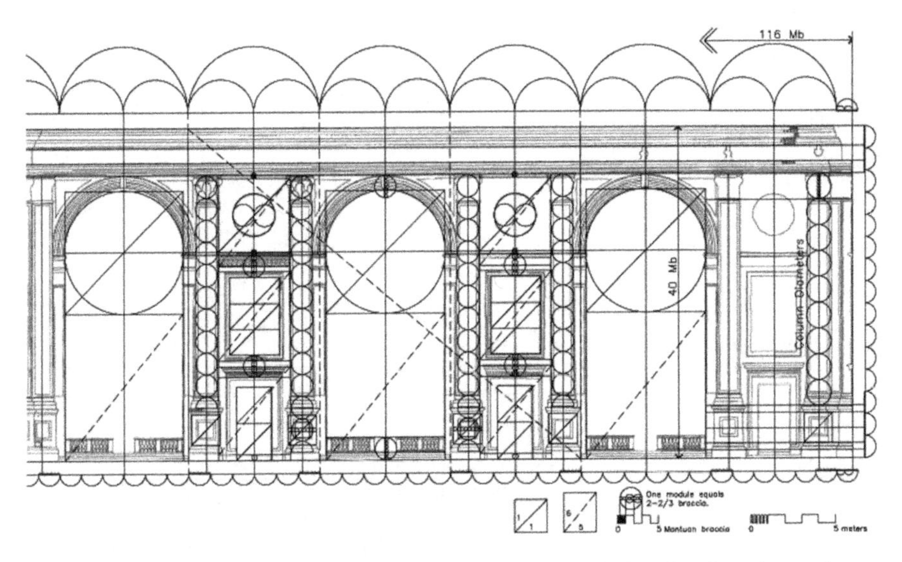

Fig. 42.6 Sant'Andrea, Mantua. Proportional scheme, nave south elevation. Drawing: author

the ratio of 5:6 to the number 16 produces a modular plan unit of 13 1/3 × 16 braccia that defines both the major chapels and the nave itself, which is defined by a grid that is three of these plan modules wide by seven long. The odd fractions produced by the proposed module may have a rational explanation. The module of 2 2/3 braccia equals 8/3, which can be restated as 16/6 braccia.

The proportional scheme of the major chapels is repeated in the central vaulted bay of the façade, with the exception that the rear pilasters at either side of the entry door and their corresponding arch have been truncated at their centre, reinforcing the idea that the centreline dimension has significance.[17] The tribunal at the other end of the nave as restored by Tavernor according to the sixteenth-century sketch and based on the side chapels neatly balances the central bay of the façade. The interior of the porch has the ratio of 1:4, based on a square of four modules. This same four-module square circumscribes the minor chapels.

Turning to the side elevations of the nave, attention is directed first to the triumphal arch motif that has been said to characterize the nave walls when looking at a grouping consisting of one major chapel and the two minor chapels on either side (Fig. 42.6).

The floor of the nave before three contiguous chapels has a dimension of 15 modules by 18 modules, a proportion of 5:6, highlighted on the plan in the centre of the nave. The height of the cornice line of the interior has been given as 40 Mantuan braccia, the same dimension as the width of the nave.[18] This means that

[17] Ritscher gives the width of the central façade bay as 7.1 m versus the width of the typical major chapel as 7.16 m; see Johnson (1975: pl.14, 15, 45, and 79).

[18] Ritscher's measurement is 18.82 m, which is 0.75 % greater than the ideal value of 40 Mantuan braccia, or 18.68 m; see Johnson (1975: pl.14, 15, 45, and 79).

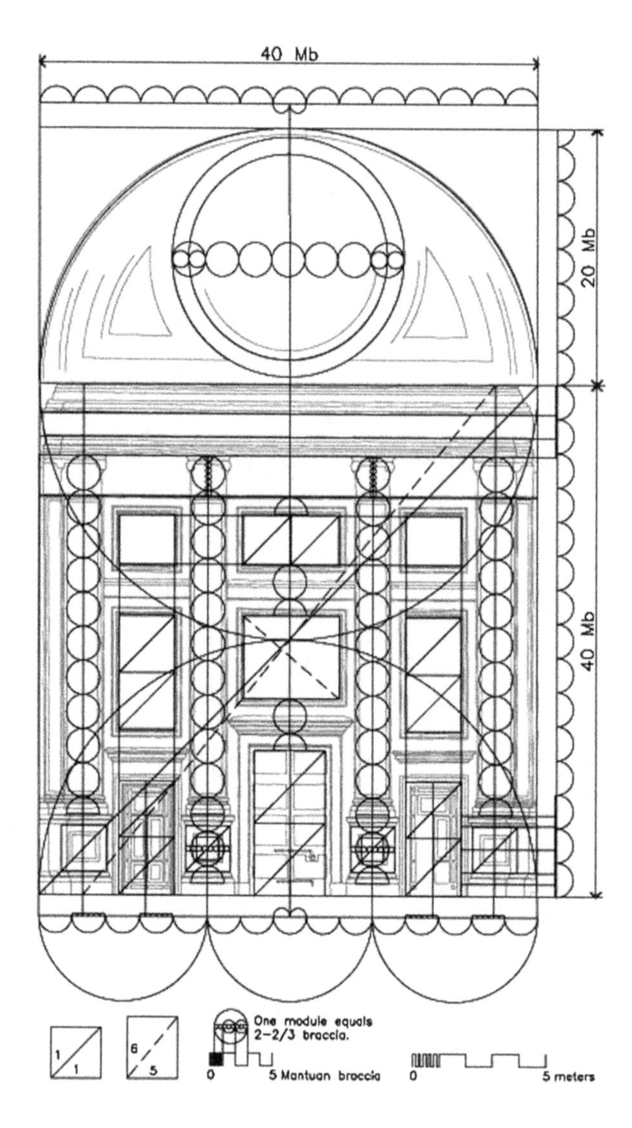

Fig. 42.7 Sant'Andrea, Mantua. Proportional scheme, nave west elevation. Drawing: author

the three bays of the triumphal arch motif also fit within a proportion of 5:6 following the example of the Arch of Constantine. Moreover, the inspiration for the articulation of the plan of the major and minor chapels, where the void of one equals the solid wall mass encasing the next, is now seen explicitly to have been the Arch of Constantine. There the internal dimension between the central two columns equals the out-to-out dimension of the pair of columns on either side. In the case of the Arch of Constantine it is the precisely this arrangement of columns which establishes the proportion of 5:6, and not the mass of masonry beyond to which the columns are attached. In both the Arch and Sant'Andrea, the key dimensions seem to occur at the pedestals.

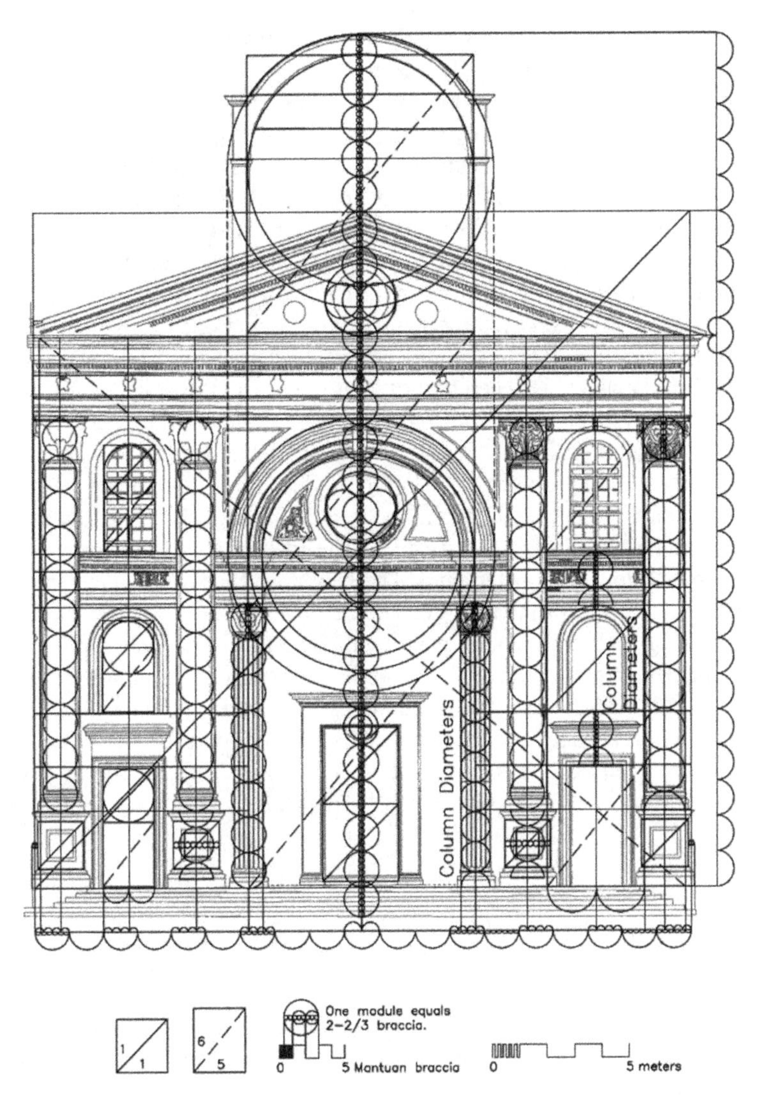

Fig. 42.8 Sant'Andrea, Mantua. Proportional scheme, façade. Drawing: author

Squares as well as 5:6 rectangles occur repeatedly in the details of the side elevations of the nave. The arched openings of the major chapels conform to the proportion of a square on top of a 5:6 rectangle. Squares and 5:6 rectangles also appear on the west end wall of the nave (Fig. 42.7).

Major elements as well as the subdivision of entablatures and other mouldings appear to be dimensioned typically in multiples of either sixths or sixteenths of the module and their factors of three, four, and eight respectively. The west end elevation reveals the 2:3 proportion of the section through the nave.

The height to the apex of the pediment of the west façade is approximately equal to the distance across the façade measured at the dados of the pedestals (Fig. 42.8).

As already mentioned, this is the proportion of the square as commonly observed in writings on Sant'Andrea. But unnoticed by virtually every writer—and undoubtedly more significant—is the fact that the ratio of the height to the top of the entablature to the width across the façade is 5:6, following yet again the example of the Arch of Constantine. Again, squares as well as 5:6 rectangles occur repeatedly in the details and many of the other dimensions of the façade are again multiples of the module or of sixths or sixteenths of the module. The central doorway is rigorously based on a double square, the patterning of which can be further subdivided into a grid of 81-module squares. Perhaps the most spectacular discovery concerns the painted grid which once filled the blank wall spaces of the façade and of which only small traces remain. The module of 2 2/3 Mantuan braccia which has been proposed as the key to the design of Sant'Andrea receives its ultimate validation by precisely dimensioning this grid on a reconstruction drawing by Tavernor, made without knowledge of the proportioning system proposed here (not reproduced here due to lack of space).

The modular system extends to the order of the interior and the major and minor orders of the façade (Figs. 42.6 and 42.8) The interior Composite order is 10 1/2 modules, or 9 implied diameters, tall. The height of the capital equals the column diameter equals 13/12 of the module. The major Corinthian order of the façade is 11 modules, or 9 1/2 implied diameters, tall. The height of the capital equals the column diameter equals 7/6 of the module. The minor Corinthian order of the façade is 1/12 module less than 8 modules, the same less than 9 1/2 implied diameters, tall. The height of the capital is 1/12 module less than the column diameter, which equals 5/6 of the module. From this it would appear that every single dimension of the building can be proportioned according to the modular system.

Alberti's Design Process

In *Architectural Principles in the Age of Humanism*, Rudolf Wittkower described the creation of Renaissance architectural theory through the ideas and architecture of Alberti and Palladio. Wittkower credited to Palladio the creation of an architectural design methodology that unified a systematic proportional procedure with an interest in precedent conceived as *type*. According to Wittkower contrasting Neo-Platonic and Aristotelian doctrines were unified in Palladio's architectural design process through the control by number of a building fabric whose design was determined by the interplay of functional types, based in present patterns of use, and of formal types, found in the Classical past (Wittkower 1962: 68). Clearly, a design process of this type is implicated in the work of Alberti at Sant'Andrea.

The need to house and show the Blood of Christ suitably was met with a scheme which integrated a functional type, the traditional basilica, appropriate for the accommodation of crowds, with three formal prototypes from the ancient world: the Basilica of Maxentius, with its presumed tie to the Temple in Jerusalem; the

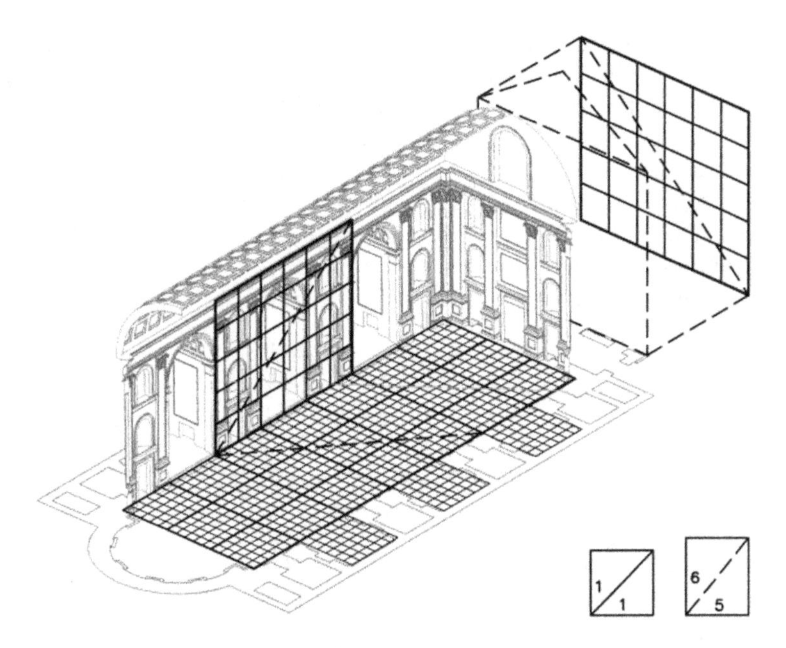

Fig. 42.9 Sant'Andrea, Mantua. Axonometric of 5:6 proportional scheme. Drawing: author

Arch of Constantine, emblematic of the victory over death offered by Christ; and the Etruscan temple, suggestive of civic pride and independence. All three prototypes shared the same generative proportion, 5:6, which was then applied to Sant'Andrea as well. A module was devised which related to the 5:6 proportion through the perfect numbers of 6 and 10, and through this module every detail of the work was related back to the generative ideas which informed the whole.

Figure 42.9 summarizes the relationship of the 5:6 proportion to the fabric of the building.

The axonometric drawing over which the diagram is drawn is taken from Robert Tavernor and illustrates his reconstruction of Alberti's intentions for various details of the interior of the church.[19] A point that has been made as forcibly as possible is that there is no detail that is too small not to be integrated into the overall system, either by Alberti or his many followers, both great and small, in the centuries during which the building was under construction. On this point Alberti was absolutely clear:

> In short, everything should be measured, bonded, and composed by lines and angles, connected, linked and combined—and that not casually, but according to exact and explicit method... so that however much [one] searched, he would not find anything in the entire work inconsistent or incongruous or not contributing its every number and dimension to the splendor and grace (Alberti 1988: 314).

[19] Note that this axonometric drawing shows an apsidal termination to the nave, another possible solution to the question of how Alberti would have completed his design.

Biography Michael R. Ytterberg received undergraduate and graduate degrees in architecture from Rice University and a Ph.D. in the history, theory, and criticism of architecture from the University of Pennsylvania. He teaches urban design and the history of architectural theory at Drexel University in Philadelphia. He is a registered architect in a number of US states and a design principal and member of the executive committee of BLT Architects in Philadelphia. His research interests include Hadrian's Villa, the subject of his Ph.D. dissertation, architectural theory before the eighteenth century, and the changing relationship material culture—and architecture in particular—to the society it serves.

References

Alberti, L. B. 1988. *On the Art of Building in Ten Books*. Trans. J. Rykwert, N. Leach and R. Tavernor. Cambridge, Massachusetts and London: MIT Press.

Borsi, F. 1977. *Leon Battista Alberti*. New York: Harper and Row.

Cadioli, G. 1974. *Descrizione delle Pitture, Sculture ed Architetture che si Osservano nella Città di Mantova e ne' Suoi Contorni* (1763). Bologna: Forni Editore.

Furnari, M. 1995. *Formal Design in Renaissance Architecture*. New York: Rizzoli.

Johnson, E. 1975. *S. Andrea in Mantua: The Building History*. University Park and London: Pennsylvania State University Press.

Krautheimer, R. 1969. *Studies in Early Christian, Medieval, and Renaissance Art*. New York: New York University Press.

March, L. 1998. *Architectonics of Humanism*. Chichester, UK: Academy Editions.

Morolli, G. 1994. I Templa' Albertiani: dal Trattato alle Fabriche. Pp. 106–133 in J. Rykwert and A. Engel, eds. *Leon Battista Alberti*. Milano: Olivetti/Electa.

Onians, J. 1988. *Bearers of Meaning*. Princeton: Princeton University.

Ritscher, E. 1899. Die Kirche S. Andrea in Mantua. *Zeitschrift für Bauwesen*. Vol. **49**: 2–19, 182–189.

Rykwert, J., ed. 1979. Leonis Baptiste Alberti. AD Profiles 21. *Architectural Design* **49**, 5–6.

Sanpaolesi, P. 1961. Il Tracciamento Modulare e Armonico del Sant'Andrea di Mantova. Pp. 95–101 in *Atti del VI Convegno Internazionale di Studi sul Rinascimento. Arte, Pensiero e Cultura a Mantova nel Primo Rinascimento in Rapporto con la Toscana e con il Veneto*. Firenze: G. C. Sansoni.

Tavernor, R. 1998. *On Alberti and the Art of Building*. New Haven: Yale University Press.

Wittkower, R. 1962. *Architectural Principles in the Age of Humanism*. New York and London: W. W. Norton.

Chapter 43
The Numberable Architecture of Leon Battista Alberti as a Universal Sign of Order and Harmony

Livio G. Volpi Ghirardini

I must confess that my interest in the mutual relationships between mathematics and architecture has been tied for many years to practical matters of building and restoration. I began to radically transform my attitude (that is, to comprehend the difference between using mathematics to solve problems and conceiving problems in terms of a mathematical order) when I was charged with the care of the last imposing and complex edifice designed by Leon Battista Alberti, the Basilica of S. Andrea in Mantua. Assuming responsibility for the restoration, and attempting to ensure that even the smallest element which might contain clues helpful in the restoration process would not be disturbed, led me to seek to discover just how much of the project conceived by Alberti, the *father* of modern architecture, is still discernible today.

In order to further the search, it was necessary to return to the theoretical aspect of the project and see how problems involving Alberti's conceptual methods had been treated historically. To my great surprise I discovered that the intimate relationships between number and architecture, even though generally accepted with regards to both the architecture of the Renaissance in general and the architecture of Alberti in particular, had been interpreted in so many different ways that it was possible to schematize them, mathematically speaking, in two opposing schools of thought. The first is the *rational,* rooted in whole numbers and ratios of whole numbers; the second, the *irrational,* admitting the presence of quantities which may not, as Alberti has noted, be expressed in whole numbers.[1]

First published as: Livio Volpi Ghirardini, "The Numberable Architecture of Leon Battista Alberti as a Universal Sign of Order and Harmony", pp. 147–166 in *Nexus I: Architecture and Mathematics,* ed. Kim Williams, Fucecchio (Florence): Edizioni dell'Erba, 1996.

[1] "In establishing dimensions, there are certain natural relationships that cannot be defined as numbers, but that may be obtained through roots and powers" (Alberti 1988: IX.6, p. 307).

L.G. Volpi Ghirardini (✉)
Accademia Nazionale Virgiliana, Mantova, Italy

Via Chiassi, 65, 46100 Mantova, Italy
e-mail: volpighirardini@virgilio.it

K. Williams and M.J. Ostwald (eds.), *Architecture and Mathematics from Antiquity to the Future*, DOI 10.1007/978-3-319-00137-1_43,

The rational school is characterized by an interpretive model elaborated between the 1940s and the 1960s by Rudolf Wittkower (1962), and owes a great debt to the thoughts of Ernst Cassirer (1927), who in his turn followed a path of thought begun by Jacob Burckhardt (1860). It is from this moment that the idea of an Alberti who is *spherical,* champion of the tranquil humanism of Cicero and creator of universal harmonies, began to take hold. Wittkower extended to the whole humanistic phenomenon, and in particular to the architecture of the fifteenth and sixteenth centuries, the basic concepts contained in *De harmonia mundi totius*, a work of esoteric mysticism by the Franciscan Francesco Zorzi printed in Venice in 1525. Into this vision of the period, Leon Battista Alberti was inserted as well; his architecture was held to reflect a supreme harmony and musicality because the divine numbers of Pythagorean and neo-Platonic thought correspond to architectonic measures, and because the same harmonic/musical relationships are found in the architectural elements. Therefore nature, the universe, and God are reflected through a total musicality which takes as its point of departure audible music, and in architecture becomes *crystallized* music.

This hypothesis, which is largely accepted by scholars, has been somewhat forced, however, in the effort to ideally reconstruct Alberti's projects through the application of musical relationships. Of the many scholars who have pursued this trail, I limit myself to mentioning Paul von Naredi Rainer, who has examined the whole of Alberti's work (von Naredi Rainer 1977a: 81–213, 1977b: 178–181, 1982, 1994: 292–299).

On the other side, beginning in the early 1960s, if somewhat timidly at first, a different interpretative model based on Alberti's use of geometric systems of proportions was proposed. The work of Zoubov[2] and Sanpaolesi are particularly representative of this school of thought. Sanpaolesi has justly noted that the decomposition of the facade of S. Maria Novella proposed by Wittkower is so "simple" that it can hardly be considered as representing a compositional method, and additionally, that the elements contained in *De re aedificatoria* are not in themselves sufficient for the "theoretical formulation of a method for the proportioning of the buildings" (Sanpaolesi 1965: 95–101, esp. 95).

The fact that the major part of the monuments designed by Alberti were not completed has certainly facilitated the creation of a multitude of hypotheses (or rather, the multitudinous efforts to rediscover the original design through applications of exact science), and has indirectly allowed opposing interpretative models to coexist. The research into Alberti's design methods has certainly not been helped by Julius von Schlosser's having defined Alberti as a "drawing table architect" (von Schlosser 1929). It is significant that in the buildings themselves

[2]Zoubov, taking as his point of departure the agreement of Vitruvius and Alberti with regards to the dimensions for atria, demonstrates that the latter substitutes $\sqrt{2}$ with the rational value 7/5, and similarly, that the same root is substituted by the rational 10/7 in the Corinthian capital. He attempts through such argument to show that for Alberti the relationships between whole numbers actually represent values for square roots, exactly as the ratio 22/7 was substituted traditionally for the irrational π; see Zoubov (1960: 54–61).

every situation that diverges from the dictates of *De re aedificatoria* is considered either an error of workmanship or an alteration of Albertian thought, as if there must necessarily exist a rigid relationship between cause and effect. In short, there has been an all-too-frequent tendency to attribute to *De re aedificatoria* the traits of a builders' manual which it does not possess.

When scholars have attempted to apply their interpretative models to built architecture, they have often been faced with a very different reality. In various cases attempts have been made to adapt the reality to the model by proposing an ideal Albertian "module" which in some way accommodates the theoretical intentions (Baldini 1989: 155–204). Effectively, the existence of different proportional solutions for the same building, or for isolated elements of a larger whole,[3] demonstrates how difficult it is to convincingly establish the nature of the design process. The explanation of the divergences is relatively simple: it lies in different readings of the original data, in large part owing to the hardships involved in precisely determining the dimensions. Too often one finds drawings in which the dimensions are altered with respect to the reality, and in which there exists a discordance between dimensions related to a single length, as is demonstrated by Table 43.1, which compares values from several surveys of S. Francesco in Rimini[4] (Fig. 43.1). The degree of accuracy obtained by graphic solutions is not usually sufficient to serve as the basis for an analysis; a drawing representing proportional relationships is not credible if it is not supported by numerical values which are clearly expressed, a situation not frequently encountered.

In some cases, however, even precise measurements, though always necessary, are not in themselves sufficient to provide an exact interpretation. For example, the relationship between the larger term and the smaller term of the golden section is 1.618 ..., while the major sixth (5:3) is equal to 1.666 ...: their difference is a matter of hundredths. Similarly, the relationship between the diagonal and the side of a square, $\sqrt{2}:1 = 1.414$, differs by less than 2 % from 7:5 (=1.4). To visualize this, one may reflect that on a drawing with a scale of 1:100, the length of a metre is inferior to the thickness of a thin line. These differences are small, and become completely negligible when one keeps in mind that in actual buildings factors such as settlement, structural damage and restoration may cause translation and/or rotation (out of plumb) and even require the substitution of original elements. When the existence of such factors is verified, it is clear that one may no longer

[3] For example, the geometric representations of the proportions of the façade of S. Andrea share the attempt to circumscribe the elevation of the principle portico with a square. But the square is different each time: Sanpaolesi's version does not include the stairs, while Borsi's version does, though this is achieved by adding only the necessary number of steps; see Sanpaolesi (1975, pl. XX, XXI) and Borsi (1975, p. 234, fig. 244).

[4] The values which appear in this table are taken from von Naredi Rainer (1977a: 170) and (Petrini 1981); my own survey was completed with Arturo Calzona and Bruna Restani in summer 1995.

Table 43.1 Comparative dimensions, façade of San Francesco, Rimini (in cm)

Description	Naredi-R	Petrini	Volpi-G
Total height, lower level (after restoration)	1,396	1,382	–
Height of pedestal	258	227	258.7(l)
Height at the ornamental strip, pedestal	92	97	–
Height of the column, including capital and base	943	965	964.9(r) 975.4(l)
Height of the trabeation	195	190	–
Height of impost, lateral arch	532	526	531.3(r) 529.3(l)
Total width	2967	2972	2964.5
Width, central arch	683	686	683.0(*)
Width, lateral arch	484	484(r) 486(l)	484.5(r) 485.2(l)
Width of piers	332	331(r) 327(l)	328.5(r) 327.7(l)

(r) = right
(l) = left
(*) = the width is not constant: at the opening, 682.2; at the walls, 682.5; at the pedestal, 683.0

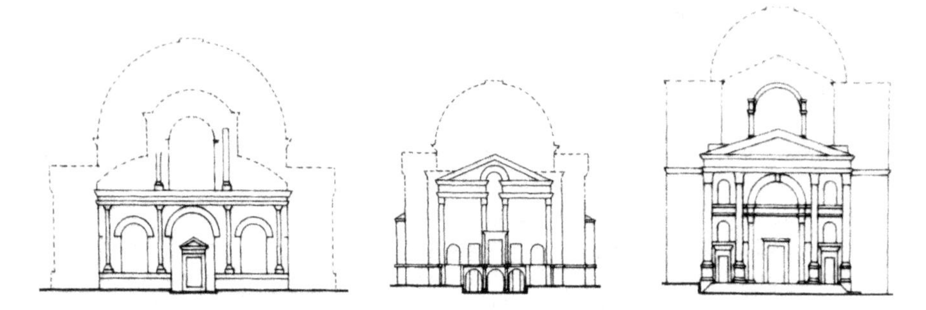

Fig. 43.1 Schematic diagrams of ideal reconstructions of the facades of the churches of San Francesco (Rimini), San Sebastiano (Mantua) and Sant'Andrea (Mantua). Images: author

trust the task of defining the exact relationship between parts to the measurement of individual elements. The solution must be sought, given of course a base of correct dimensional data, within the framework of the whole, in the perfect fit of the parts, which will concatenate only if expressed in a certain way. Elements which cannot be seen as part of a larger framework must remain isolated, incapable of indicating a system of design.

There exists another problem, which regards the identification of a unit of measure. Both the graphic design process and the numeric design process always have a basic unit of length which is carefully chosen. But just as the laws of the cosmos are not made evident if the phenomena in which they are manifest are not correctly understood, if the correct unit of measurement is not adopted for the analysis of the architectural dimensions, then the *number* inherent in them will remain hidden.[5] Further, if the unit of length used as a base is not correctly defined in terms of the original standard of measurement, then the measurements derived from it and the relative proportional relationships proposed remain questionable. It was not coincidental that I dedicated a section of *Il San Sebastiano di Leon Battista Alberti* to the unit of measure, which had never been clearly established (Calzona and Volpi Ghirardini 1994: 229–234).

By now the nature of the arguments I am expounding is evident: they are, to phrase it after Vico, "proper to small intellects, and place in shackles and distress the mind accustomed to the infinite ranges of the various genre ...." In any case, without an examination of the unit of measure, one is forced to remain on the surface of the problem. Unfortunately for us, the total lack of drawings by Alberti relative to his own architecture doesn't aid empirical research.[6] However, in regards to San Sebastiano in Mantua there exists an important document: the sixteenth century drawing by Antonio Labacco representing Alberti's temple ("*di mano di mesere Baptista Alberti*") which offers a fundamental key for its interpretation (Fig. 43.2). The collation of the data furnished by the drawing with that which may be drawn from the built elements which are surely Albertian show that all values of the principle dimensions of the building may be encompassed in a system of a few mathematical relationships.

Labacco's drawing includes a valuable plan with measurements of the upper church, a long annotation of its internal measurements, and a small sketch of the view looking towards the apse. There is not a trace of the actual lower church, more commonly known as the crypt, and neither is there any indication of the stairs (Figs. 43.2 and 43.3). This fact has caused much discussion among scholars, who, in any case, agree that the planimetry and the annotation are derived from a design by Alberti. This is a sufficient base from which to depart in analysing the design method because, to achieve that end, it is immaterial whether or not Labacco has reproduced in its entirety the Albertian model, and that, in its realization, the church is somewhat removed from the design (Fig. 43.4).

Separating ideally the individual members of the building reproduced by Labacco (the central square, the arms of the cross, and the portico), one observes that the respective internal measurements, expressed in *braccia*, form the following three numerical progressions of ratio 3:5:

$$34, \ (34), \ 56\ 2/3 \qquad \text{(a)}$$

representing the side of the central square (the dimension 34 is repeated to signify that the sides are equal), and the height;

$$12, \ 20, \ 33\ 1/3 \qquad \text{(b)}$$

[5] In *De re aedificatoria* (IX, 5), Alberti commented at length on numbers, their significance and their properties, referring to both Vitruvius and the Greek tradition; see Alberti (1988: 303–304).

[6] The drawing attributed to Alberti by Howard Burns cannot be related to any built edifice. See Howard Burns, "A drawing by L.B. Alberti," pp. 45–56, in *Architectural Design Profile 21*, **49**, no. 5–6 (1979).

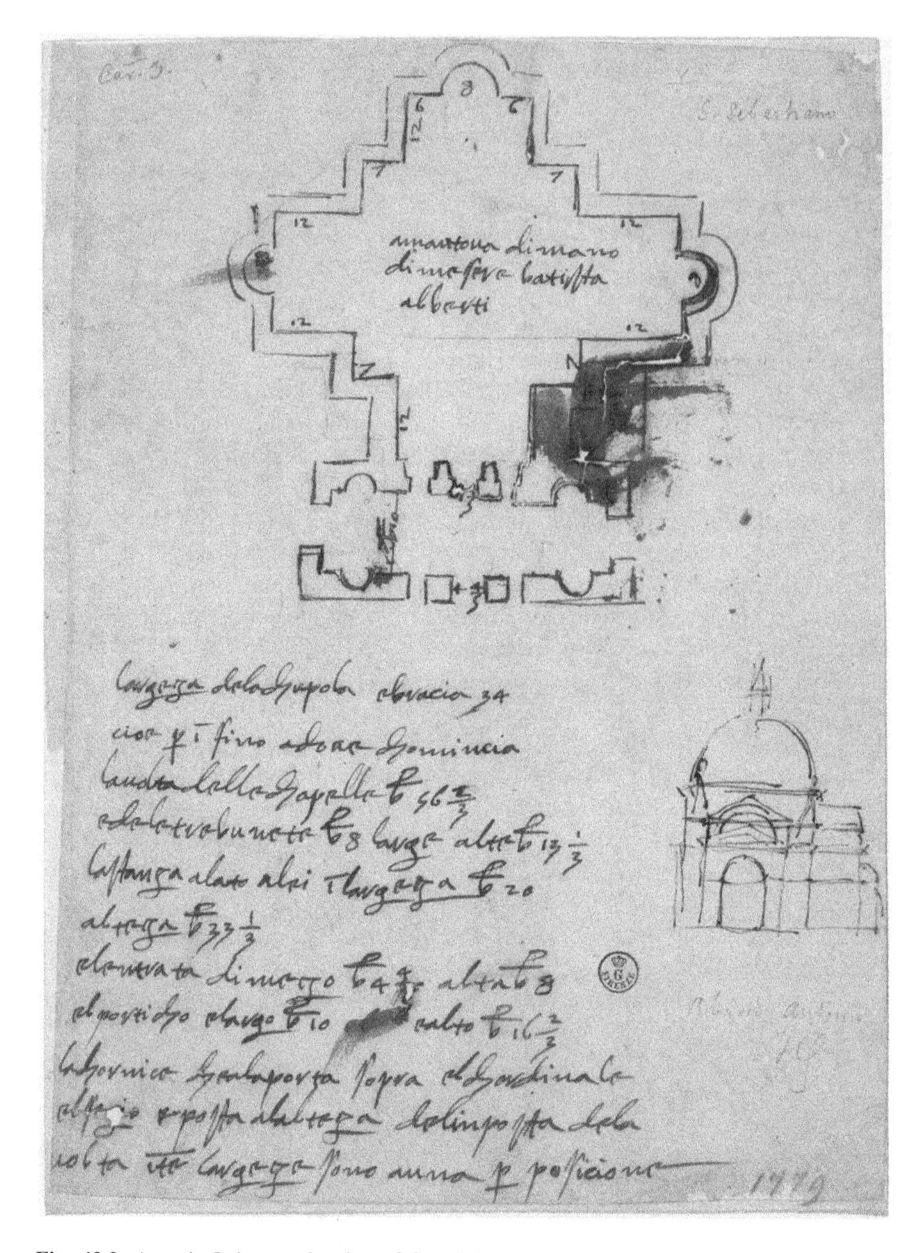

Fig. 43.2 Antonio Labacco, drawing of San Sebastiano. Uffizi, Gabinetto Disegni e Stampe, 1779A, Florence. Image: ©Foto Scala Firenze. Reproduced by permission

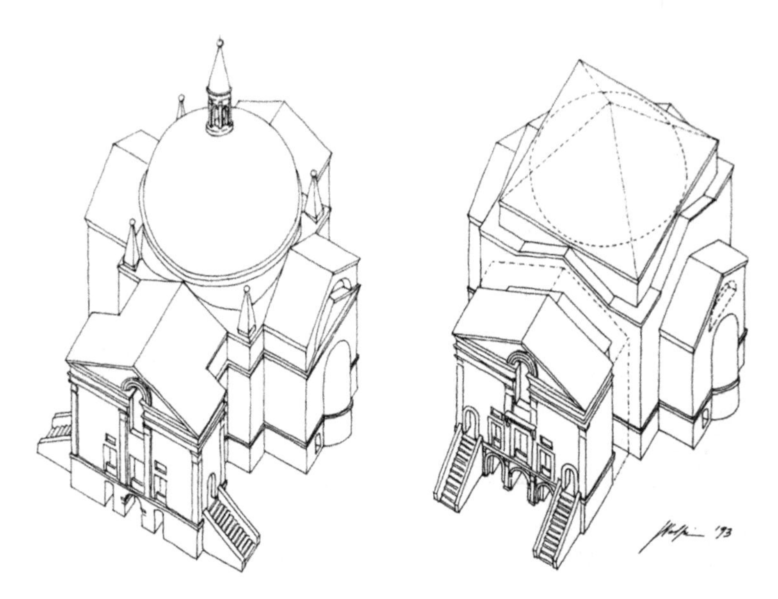

Fig. 43.3 San Sebastiano, longitudinal section and plan constructed with the dimensions indicated by Labacco. Image: author, updated diagram from Calzona and Volpi Ghirardini (1994)

representing the arms of the cross (respectively, the length of sides at their base, and height);

$$10, \quad 16 \quad 2/3 \tag{c}$$

representing the transverse section of the portico (respectively base, length and height) (Fig. 43.5).

The other measurements in Labacco's drawing refer to smaller elements of the church: the main entrance portal (4 4/5 and 8) and the absidioles (8 and 13 1/3). These two architectural elements, besides corresponding to one another with respect to the longitudinal axis of the church, offer a numerical correspondence as well: the height of the portal equals the width of the absidioles. By arranging the dimensions in ascending order, the following sequence is obtained:

$$4 \quad 4/5, \quad 8, \quad 13 \quad 1/3 \tag{d}$$

As were the previous three sequences, this sequence takes its rhythm from the relationship 3:5. Consequently, all of the dimensions of the drawing without exception make up a well-defined group, forming duads or triads in the ratio of 3:5. This means that Alberti's design, as transmitted to us by Labacco, is developed from numerical scales which define the entire architectural space. What remains to be explained, however, is the link between the fundamental dimensions furnished by Labacco for the three elements of the upper church (the portico, arms and square) or rather, the numbers 10, 20, and 34.

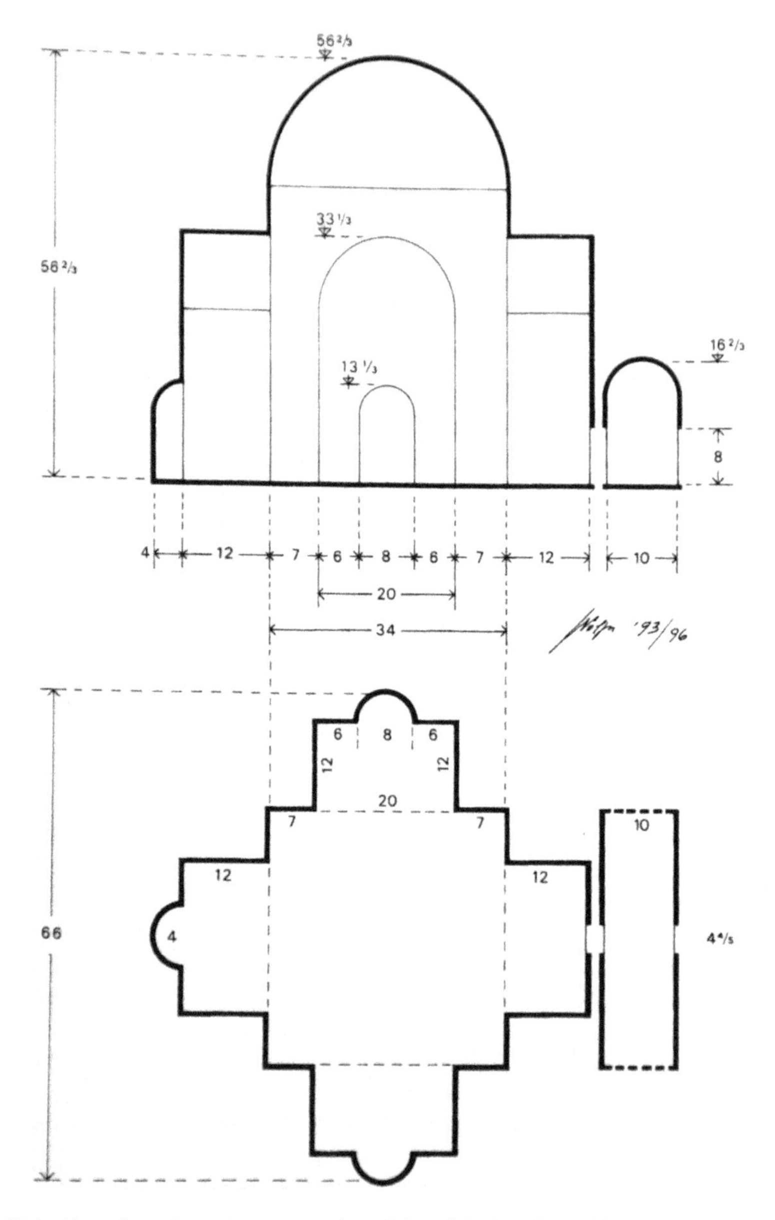

Fig. 43.4 *Above*, hypothetical reconstruction of the original project of San Sebastiano based on Labacco's indications, with the addition of the crypt. *Below*, the model of probable subsequent design modifications. Image: author, from Calzona and Volpi Ghirardini (1994)

By relating the side of the central square, 34, with the square of the numbers which form the recurring relationship, 3 and 5 ($3^2 + 5^2 = 34$) (Calzona and Volpi Ghirardini 1994: 227), I obtain a combination which univocally defines all three fundamental dimensions of the upper church. The values 10, 20, and 34 may be deduced from the series of the squares of the first five numbers, as the progressive sums of odd and even squares (Fig. 43.6).

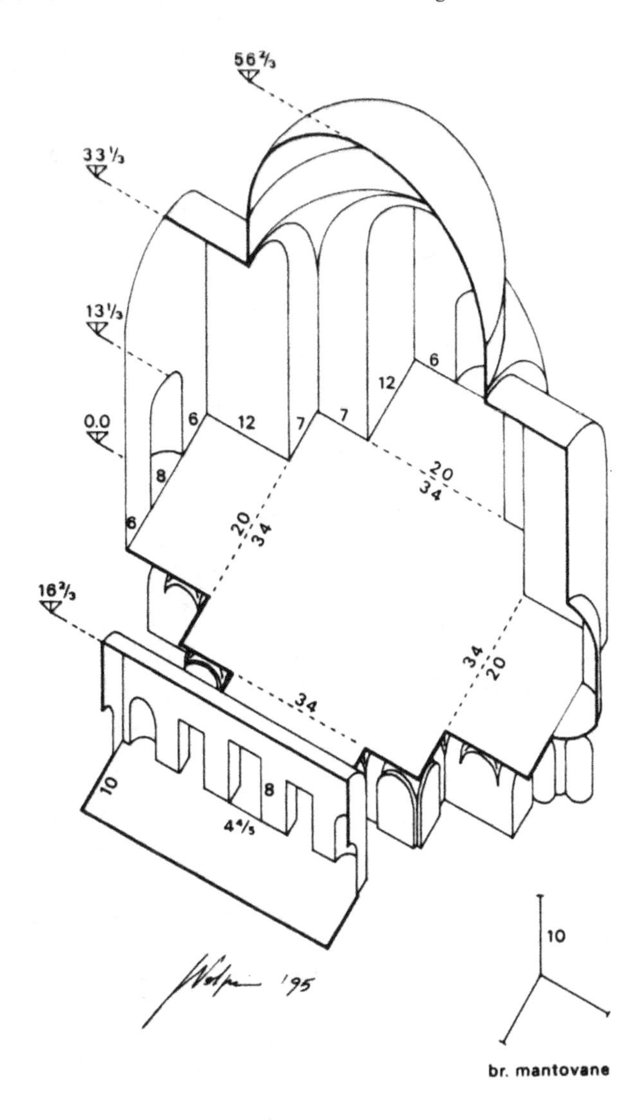

Fig. 43.5 Above, axonometric section of San Sebastiano based on the dimensions furnished by Labacco. Below, numerical progressions which include all Labacco's dimensions, arranged with respect to individual elements of the church. Diagram: author

Central square	34, (34), 56 2/3
Arm	12, 20, 33 1/3
Portico	..., 10, 16 2/3
Entrance and apse	4 4/5, 8, 13 1/3

As I will demonstrate presently, this is the system contained in the mathematical order which governs the design, the logic of which is developed beginning with the first five natural numbers. For now it suffices to link all the numbers noted by Labacco in the following scheme of numeric combinations (Fig. 43.7).

At this point I wish to go beyond the mere interpretation of Labacco's document, because the church, for better or worse, has actually been constructed. By

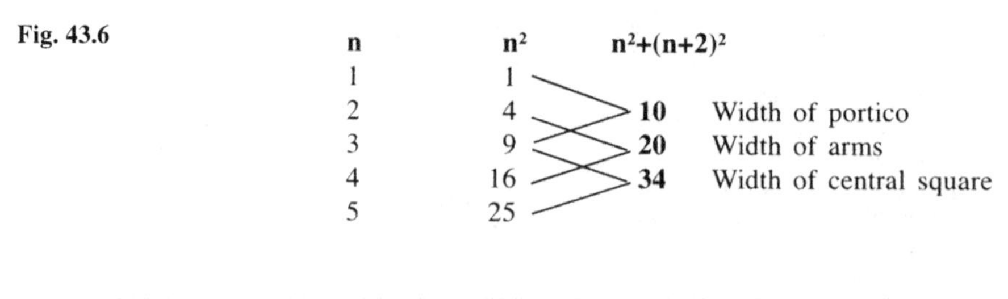

Fig. 43.6

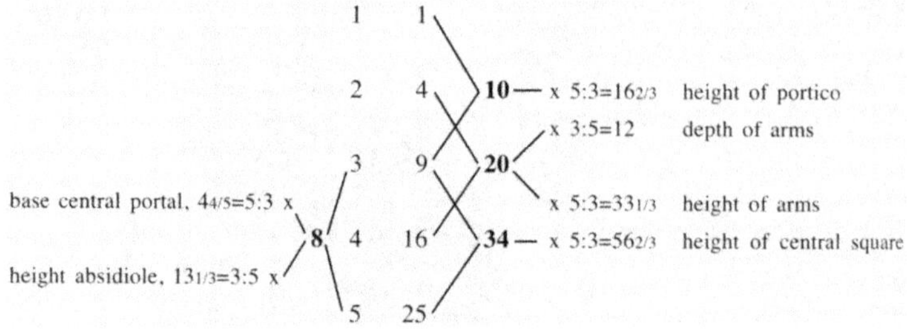

Fig. 43.7

comparing the dimensions of the lower church, not present in the drawing, to those cited by Labacco, it seems that I may legitimately hypothesize that the design process did not make any leaps. This can be easily verified, because if there exists a continuity in the design process, dimensions which relate to each other as those outlined above must be present in the actual construction.

Considering the measures expressed in Mantuan *braccia,* the crypt is generated by addition, from the piers with base 2×2 and by the space covered by the vault, with a base of 5 and height of 8 1/3 (Fig. 43.8). The absidioles make up the last elements which contribute to the form of the interior space, if in a minor way; these have an opening of base 3 and height 7. The sequence formed takes into consideration the essential elements, and does not consider those which give rise to minor proportions.

From the analysis of the crypt and the lower portico, both known to have been realized while Alberti was alive and able to visit the construction site, I am able to obtain the sequence:

$$3,\ 5,\ 8\quad 1/3 \tag{e}$$

of the ratio 3:5, which synthesizes the dimensions of the crypt and other values which integrate relationship (c) (See Table 43.2).

Indeed, the lower portico is 6 *braccia* wide and has a half-width of 16 2/3 *braccia*, values which are justly placed in (c), transforming it from relating only to the section of the superior portico to a progression which is representative of the entire *pronaos*, in that it accounts for all the internal dimensions, with the sole exception of the half-length of the upper portico, of which I shall speak shortly.

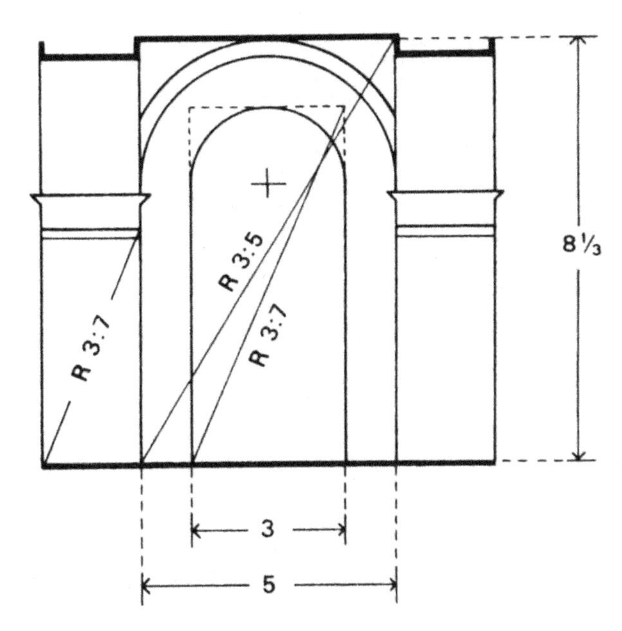

Fig. 43.8 Section of the modular element which is used additively to form the crypt of San Sebastiano. Image: author

When Labacco's data is integrated with the values taken from the actual construction, the following system of four relationships is obtained:

34, (34), 56 2/3	Square	(a)
12, 20,33 1/3	Arm	(b)
6, 10, 16 2/3	Portico	(c)
3, 5, 8 1/3	Crypt	(e)

which is the demonstration that all the dimensions of the interior space of the church are conceived in groups which are internally ordered with regards to the ratio 3:5. Further, relationships (b), (c) and (e) relate to each other as 1:2, therefore representing linear transformations of each other.

The system of relationships discerned may constitute a decent result, but I am able to enter into an even more detailed analysis. The ratio 1:2 serves to link parts of the church which are physically distinct, that is, the crypt, portico and arms, and is also used to double the areas which required a particular length, such as those of the upper and lower porticos. Indeed, Fig. 43.9 illustrates how the numerical value which ties together the dimensions of the interior spaces of the porticos is derived, not from the length, but from the half-length. In particular, the relationships between the base and the half-length of the vestibule of the upper and lower floors is modulated by the squares of the numbers 3, 4, and 5, or 9:25 and 9:16.

Table 43.2 Dimensions of San Sebastiano, Mantua, mentioned in the text

Description	cm	±e	br.m.	Δcm
Crypt				
Side of the base of the cross vault:				
Longitudinal direction	[a]242.2[vm]	6.0		
Transversal direction	[a]236.0[vm]	4.8		
Calculated at level of foundations	235.3[(1)]	1.5	5	+1.8
Height of the cross vault	[a]388.0[vm]	3.0	8 1/3	−1.1
Width of opening, absidioles	[a]138.6[vm]	2.6	3	−1.4
Height of absidioles	[a]328.1[vm]	0.6	7	+1.3
Side of piers, central area	[a]96.1[vm]	2.9	2	+2.7
Lower portico				
Clear width	[a]280.7[vm]	2.3	6	+0.6
Width between walls	[a]174.7[vm]	1.8	8	+1.2
Interior length	1,556.3[vm]	1.8	33 1/3	+0.0
Exterior length (on axis with windows)	1,783.7[vp]	1.0	38 1/4	−2.2
Upper portico				
Base of niches	192.1[vm]	1.3	4 2/25	+1.6
Height of niches	318.2[vm]	0.7	6 4/5	+0.7
Clear width of arches	192.1[vm]	1.3	4 2/25	+1.6
Clear height of arches	318.2[vm]	0.7	6 4/5	+0.7
Width, central portal (with frame)	317.2[vm]	1.3	6 4/5	+0.2
Height, central portal (with cornice)	527.5[(2)]		11 1/3	−1.6
Interior width	469.1[(3)]	1.9	10	+2.2
Interior length	1,660.0[vm]	0.0	35 5/9	−0.1
Space over upper portico				
Interior width	500.0[vv]	4.5	10 2/3	+2.0
Interior length	1,659.0[vv]	4.2	35 5/9	−1.1

All measures were taken manually

±e = max. disparity with respect to vm, or instrument error in the cases of vp and vv

br.m. = theoretic value, Mantuan braccia (1 br.m. = 46.69 cm)

Δcm = difference between real measures and theoretic values (in cm)

[a]Plastered surfaces (as opposed to stone surfaces)

vm = mean value

vp = point value

vv = value in which errors due to deformation, etc. may occur

(1)Value obtained taking the distance between foundation centrelines, the centrelines of the pilasters having been shifted (results from 1995 excavations)

(2) Value obtained by reconstructing a missing piece of cornice

(3) Value obtained by adding the plaster thickness of the church wall to the vm

That I am able to reach these conclusions is due in great part to Labacco, because without his drawing,[7] in the face of a multitude of uncertainties, I would have been

[7] A similar drawing by Aristotile da Sangallo (Lilla, Museo Wicar, carta F.4) is not equally as useful, reporting only plan dimensions without indications of heights. However, it does indicate the width of the openings on either side of the central aperture of the portico, which are not indicated on Labacco's drawing.

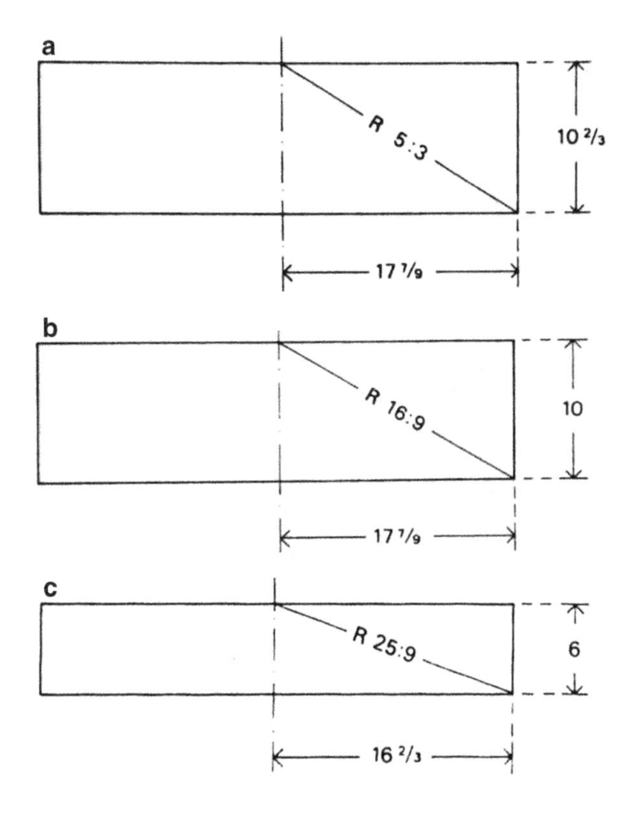

Fig. 43.9 Plan of the three levels of the portico of San Sebastiano. 9a, (*above*), space above the upper portico; 9b, (*centre*) upper portico; 9c, (*below*), lower portico. Diagram with proportional ratios: author

able to attest to only a narrow range of relationships. This was the case in S. Andrea where up to now, beyond the general framework, I have been able to discern with precision the interrelations of the mathematical progressions which govern the dimensions of the central and lateral doors of the portico, which correspond in turn to their physical interrelations.[8] This was also the case in San Francesco, where I discerned the repetition of the same relationships found in San Sebastiano.[9] A full explanation of these cases is beyond the scope of this chapter, the demonstrations having to take into account various alterations of the original structures. I will say that one has only to turn from the dimensions which govern the whole to those of individual elements, such as piers and pilasters, and a whole range of relationships is revealed, diverse but not difficult to interpret. For example, in San Sebastiano the

[8] There exists a significant relationship here as well, 2: 3, which determines the progressions and intervals of the *diapason,* 2:1, linking distant elements. See Volpi Ghirardini (1993).

[9] I refer to the ratio 3:5 in combination with 1:2 (not, however, where these ratios have been found by my predecessors, but rather in the lateral blind arches of the facade).

relationships between 1 and the odd numbers of the first decade completely define the interior elevations of the apses of the crypt.[10]

The above demonstration constitutes a solid base for the study of the Albertian design method. Indeed, the facts demonstrate: (1) that the dimensions of the various architectural elements are determined by finite ratios with the unit of measure; (2) that the numbers which represent the fundamental dimensions are interrelated by mathematical (but not musical) progressions; (3) that the measurements of individual dimensions *(finitio)*,[11] such as width, length, height or distance between elements which correspond spatially, create numerically proportionate triads[12]; (4) that such proportional triads are themselves proportionately interrelated; (5) that the relationships between principle measurements appertain only partially to the set of areas defined in musical terms set out by Alberti.[13]

In regard to this last point, a consideration is in order to clarify that the definition recently proposed by Maria Karvouni (1994: 282–291) for the musical relationships indicated by Alberti is not the only one possible. I may give them a precise univocal mathematical definition. If under a reference series of cardinal numbers I write the two series of the relative squares and *heteromékeis*[14]:

n	1	2	3	4	5	...
n × n	1	4	9	16	25	...
n × (n + 1)	2	6	12	20	30	...

and I limit myself to considering the squares and the *heteromékeis* which may be derived from the first four cardinal numbers,[27] as in the following two series:

n × n	1	4	9	16
n × (n + 1)	2	6	12	

[10] All of my surveys shall be deposited in the "Centro di studi Leon Battista Alberti" in the Accademia Nazionale Virgiliana di Scienze, Lettere e Arti, Mantua.

[11] "For us, the outline is a certain correspondence between the lines that define the dimensions; one dimension being length, another breadth, and the third height" (Alberti 1988: IX.5, p. 305).

[12] *Ternatim autem universos corporis diametros, ut* sic *loquar, coadiugabimus numeris his, qui aut cum ipsis armoniis innati sunt aut sumpti aliunde certa et recta ratione sunt* (Alberti 1966b: IX.5, p. 827).

[13] The dimensions of areas suggested by Alberti are subdivided into three groups [(1:1, 2:3, 3:4), (1:2, 4:9, 9:16) and (1:3, 3:8, 1:4)] which correspond respectively to small, medium and large areas. The ratio 3:5 is not found in these groups; see Alberti (1988: 306).

[14] The term *heteromékeis* is a combination of two Greek roots, *heteros* meaning "different" and *mêkos* meaning "length." In his "Introduction to the arithmetic of Nicomachus," Iamblichus defined the squares as powers of numbers augmented by their own length, while *heteromékeis* are powers of numbers augmented by an unequal length. He writes further of them, "in order to learn which is the most harmonic and natural coupling of the two types of number, that is, of the squares and the *heteromékeis,* which are completely opposite in nature, we need to dispose each of them in parallel rows starting with their own beginning value, that is, the squares beginning with 1 and the *heteromékeis* beginning with 2." See (Iamblichus 1995: 300–302). Nicomachus is described by Alberti as "the greatest mathematician" (Alberti 1988: IX.10, p. 317).

I can affirm that the relationships relative to the Albertian areas are defined by all the possible combinations of the first three squares after unity (which relates to itself)[15] and the first three *heteromékeis,* with the sole exception of the major ratio 4:1. In other words, the ratios obtained from the squares and *heteromekéis* of the first four numbers and contained in the proportions formed of the greater and lesser of the same, coincide with the dimensions of the Albertian areas: not a single relationship more, not a single relationship less.

Alberti described how he mentally prefigured his own architecture using mathematical models, that is, without the instruments necessary for geometrical layout. As much as 10 years before the opening of an actual construction site, he describes his soul as intent on the *res aedificanda*: "I mentally composed and designed some well-ordered buildings, and set out several orders and numbers of columns with various capitals and unusual bases in them, and infused them with a well-balanced and new harmony of cornices and stories."[16] He affirms that, in ordering his thoughts, "nothing satisfies me more than this, nothing absorbs me and involves me as totally, as mathematical research and proofs, especially when I seek to transform them to something useful in daily practice" (Alberti 1966a: 182).

A creative force tends to "fix and stabilize great and inestimable things" (Alberti 1966a: 181). An expressible mathematical order "molds the whole of Nature,"[17] and that same nature aims with all its might towards a harmonic perfection. To the architect falls the task of giving form to an ordered and harmonious complex, which can achieve those qualities only if it contains inherently the intelligence which models nature, "the perfect generator of forms."[18] And because no material image can represent the divine intellect, supreme regulator of the universe,[19] to the temple builder is trusted the difficult choice of meanings and messages to confer on his own creation; but the choice appears to be mandatory if there is to be reflected in the sacred building, as in a game of mirrors, the intelligence of the All, Creator and created, so that the structure and its components, indeed, the very "walls or the

[15] "However, if one is not an actual number, but the wellspring of number..." (Alberti 1988: IX.6, p. 307).

[16] *Composj a mente e coedificai qualche compositissimo edificio, e disposivi più ordini e numeri di colonne con vari capitelli e base inusitate, e collega'vi conveniente e nuova grazia di cornici e tavolati* (Alberti 1966a: 182).

[17] "It is our nature to desire the best, and to cling to it with pleasure ... It has a vast range in which to exercise itself and bloom —it runs through man's entire life and government, it molds the whole of nature" (Alberti 1988: IX.5, p. 302).

[18] "The great experts of antiquity ... have instructed a building is very like an animal, and that Nature must be imitated when [they] delineate it. ... not without reason they declared that Nature, as the perfect generator of forms, should be their model" (Alberti 1988: IX.5, pp. 301–303).

[19] "If no object made by hand could achieve this, they thought it better that each, according to his own powers of imagination, should fashion in his mind an impression of the principal sovereign of all, divine intelligence" (Alberti 1988: VII.17, p. 242).

floors of the temple" all have something of philosophy in them.[20] Nicholas of Cusa, a contemporary of Alberti's, agreeing with his predecessors in this line of thought, affirms: "because one may approach the divine only through symbols, let us resort to mathematical signs as most befitting because of their indisputable certainty."[21]

To conclude, the detailed analysis of the data from San Sebastiano demonstrates a design praxis which is mathematically conceived, elaborated on the basis of interconnected relationships, the fruit of a combinatorial logic of the tightest fit. With respect to some of the explanations offered by *De re aedificatoria*, the actual use of mathematics by Alberti appears more rigorous and complex.

This all leads to the affirmation that the architect's design method is based on a mathematical system, regulated by the harmony present in the variety of relationships evident in nature. Perfect order, harmony, in which number links the visible to the invisible: mathematical order is not only the order of the divine, but also the basis for any order that can be achieved in reality.

Translated from the Italian by Kim Williams

Acknowledgment The author dedicates this work to the memory of Howard Saalman.

Biography Livio G. Volpi Ghirardini holds a degree in civil structural engineering. His professional practice is dedicated to the realization of new projects as well as to the restoration of historic monuments. He is a member of the President's Council of the Accademia Nazionale Virgiliana in Mantua. He is the Prefect of the fabric of the Cathedral and the Basilica of S. Andrea in Mantua. This last explains his particular interest in the architecture of Leon Battista Alberti. He is a founding member of the Centro Studi Leon Battista Alberti in Mantua, and is currently serving as its president. Formerly a professor of building science, he is now a contract professor at the Politecnico di Milano, I School of Architecture, and teaches the architecture restoration workshop entitled "Reinforcing of historical buildings".

References

ALBERTI, Leon Battista. 1966a. Profugiorum ab Aerumna Libri. In Cecil Grayson, ed. *Opere Volgari*, vol. 2. Bari: Laterza.

———. 1966b. *L'architettura. De re aedificatoria*. Giovanni Orlandi, and Paolo Portoghesi, eds. 1966. Milan: Il Polifilo.

[20] "I would have nothing on the walls or the floor of the temple that did not have some quality of philosophy" (Alberti 1988: VII.10, p. 220).

[21] *Hac veterum* via *incedentes, cum ipsis concurrentes dicimus, cum ad divina non nisi per symbola accedendi nobis* via *pateat, quod tunc mathematicalibus signis propter ipsorum incorruptibilem certitudinem convenientius uti poterimus* (Nicholas of Cusa 1913: I, 33, p. 26).

———. 1988. De re aedificatoria. In Joseph Rykwert, Neil Leach and Robert Tavernor, eds. *Leon Battista Alberti: On the Art of Building*. Cambridge, Massachusetts and London: MIT Press.

BALDINI, Gianni. 1989. L'oscuro Linguaggio del Tempio di S. Sebastiano in Mantova. *Mitteilungen des Kunsthistorischen Institutes in Florenz* **XXXIII**, 2 (1989).

Borsi, F. 1975. *Leon Battista Alberti. Opera Completa*. Milan: Electa.

BURCKHARDT, Jacob. 1860. *Die Kultur der Renaissance in Italien*. Basel.

BURNS, Howard. 1979. A drawing by L.B. Alberti. *Architectural Design Profile 21*. **49**, 5–6 (1979).

CALZONA, Arturo and Livio VOLPI GHIRARDINI. 1994. *Il San Sebastiano di Leon Battista Alberti*. Florence: Olschki.

CASSIRER, Ernst. 1927. *Individuum und Kosmos in der Philosophie der Renaissance*. Leipzig: G.B. Teubner.

IAMBLICHUS, 1995. ΠΕΡΙ ΤΗΣ ΝΙΚΟΜΑΧΟΥ ΑΡΙΘΜΗΤΙΚΗΣ ΕΙΣΑΓΩΓΗΣ (*Perì tês Nikomàchou arithmetikés eisapoghés)*. In Francesco Romano, ed. *Giamblico: Il Numero e il Divino*. Milano: Rusconi.

KARVOUNI, Maria. 1994. Il Ruolo della Matematica nel *De re aedificatoria* dell'Alberti. In Joseph Rykwert and Anne Engel, eds. *Leon Battista Alberti*. Milan: Olivetti-Electa.

NICHOLAS OF CUSA. 1913. *De Docta Ignorantia*. In: *Nicolò Cusano:Della Dotta Ignoranza*, P. Rotta, ed. Bari: Laterza.

PETRINI, Gastone. 1981. Ricerche sui Sistemi Proporzionali del Tempio Malatestiano. *Romagna Arte e Storia* **I**, 2: 35-50.

SANPAOLESI, Piero. 1965. II Tracciamento Modulare e Armonico di S. Andrea di Mantova. In Istituto Nazionale di Studi sul Rinascimento ed. *Arte Pensiero e Cultura a Mantova nel Primo Rinascimento in Rapporto con la Toscana e con il Veneto*. Florence: Sansoni.

———. 1975. II Tracciamento Modulare. In Franco Borsi, *Leon Battista Alberti. Opera Completa*. Milan: Electa.

VOLPI GHIRARDINI, Livio. 1993. La 'porta dei sette cieli', Numeri e Geometrie del Portico Principale di Sant'Andrea in Mantova. In Accademia Nazionale Virgiliana, ed. *Atti e Memorie,* n.s. LXI, (1993).

VON NAREDI RAINER, Paul. 1977a. Musikalische Proportionen. Zalenësthetik und Zahlensymbolik im architektonischen Werk L B. Alberti. *Jahrbuch des Kunsthistorischen Institutes der Universität Graz* **12** (1977): 81–213.

———. 1977b. Exkurs zum Problem der Proportionen bei Alberti. *Zeitschrift für Kunstgeschichte,* **40** (1977): 178–181.

———. 1982. *Architektur und Harmonie. Zahl, Mass und Proportion in der abendlandischen Baukunst*. Köln: DuMont.

———. 1994. La Bellezza Numerabile: l'Estetica Architettonica di Leon Battista Alberti. Pp. 292–299. In *Leon Battista Alberti*. Joseph Rykwert and Anne Engel eds. Milan: Olivetti-Electa.

VON SCHLOSSER, Julius. 1929. Ein Künstlerproblem der Renaissance: L.B. Alberti. In *Sitzungsberichte der Akademie der Wissenschaften in Wien* **210**, II (1929).

WITTKOWER, Rudolf. 1962. *Architectural Principles in the Age of Humanism*. London: Academy Editions.

ZOUBOV, V. 1960. Quelques Aspects de la Théorie des Proportions Esthétiques de L.B. Alberti. In *Bibliothèque d'Humanisme et Renaissance. Travaux et documents*, vol. XXII. Genève: Librairie E. Droz.

Chapter 44
Leon Battista Alberti and the Art of Building

Salvatore di Pasquale

Galileo and the Science of Construction

The modern science of construction was born in 1638 with the publication of *the Discorsi e dimostrazioni matematiche intorno a due nuove scienze (Two New Sciences including Centres of Gravity and Force of Percussion)* by Galileo Galilei (1989). This was the last work of the Pisan scientist: composed during the years of his exile in the villa at Arceri, it was an elaboration of notes from his lectures given at the University of Padua at a time when he was in contact with overseers and workers in the shipyard of Venice. The work takes the form of dialogues between three characters over the course of 4 days, first addressing questions about the strength of elements of architectural construction or machinery, and later, the problems of what we today call dynamics. Naturally, today's modern science of construction is very different from that of Galileo, thanks to the contributions of mathematics and physics, just as science is continuing to change today, similarly influenced by the widespread use of computers.

The novelty of Galileo's science lay in the formulation of the problem, posed for the first time as the necessity of knowing what loads result in the breaking of a beam or a column *before* these elements are put in place, so that there exists a reasonable certainty as to the possible causes of structural failures.

It is important to note that this program—as yet limited to very simple elements of architectural composition—required a prior knowledge of the mechanical properties of construction materials and the formulation of hypotheses relative to

First published as: Salvatore Di Pasquale, "Leon Battista Alberti and the Art of Building", pp. 113–125 in *Nexus II: Architecture and Mathematics,* ed. Kim Williams, Fucecchio (Florence): Edizioni dell'Erba, 1998.
Salvatore di Pasquale (1931–2004).

K. Williams and M.J. Ostwald (eds.), *Architecture and Mathematics from Antiquity to the Future*, DOI 10.1007/978-3-319-00137-1_44,

the mathematical modelling of those elements, in order to obtain results that were independent of variations in form:

> ... thus, knowing the strength of a small nail or a small dowel of wood or any other material, I can demonstrate my knowledge of the strength of all the nails, of all the posts, of all iron chains, of all beams, lintels, antennae, masts, in short, of all solids of any material, except in the presence of accidental impediments such as knots, termites, etc.[1]

Galileo formulates the first hypotheses on mechanical proofs, prefiguring what would become, two centuries later, the primary reference, when the definite results were exhibited by J.L. Navier in 1823 in a celebrated course given at the École des ponts et chaussees; another 50 years were to pass before the development of a theory for rigidly-connected post and beam structures and much more complicated structures, such as vaults and domes.

In any case, three points are to be noted: (1) the final layout of *Two New Sciences*, like that of the Paduan notes, took place at a time when the two largest cupolas ever constructed dominated the skylines of Florence and Rome; (2) the science of construction had already produced some well-known failures[2]; (3) some mechanical devices designed and built at a small scale were known not to have worked when ultimately realized.

Galileo takes these negative experiences as a point of departure, explaining why they didn't work and exhibiting his new science. He organizes *Two New Sciences* around the three characters who represent the spirit of the theses being compared: Salviati expresses reason based on experience; Semplicio represents Artistotelian memory; Sagredo is the attentive observer comparing the factual evidence with theoretical knowledge. The confrontation takes place in the Venetian shipyards, where the actions of workers and masters were continually put to the test: the launching of ships, works for lifting and moving weights, cords for pulling, levers and winches used according to a logic that appears natural and harmonic. The harmony, however, doesn't correspond to Sagredo's expectations; to the contrary, he has to admit that actual facts and experiences put his theory in jeopardy. In all probability, the characters of *Two New Sciences* don't represent a single person, but rather the thoughts expressed by those engaged in creating architecture and theorizing about its methodological foundations.

[1] "... sicchè, conosciuta la resistenza di un picciol chiodo, o di una piccola caviglia di legno o di qualsivoglia altra materia, io potrò dimostrativamente sapere le resistenze di tutti i chiodi, di tutti i pali, di tutte le catene di ferro, di tutte le travi, travicelli, antenne, alberi, ed in somma di tutti i solidi di qualsivoglia materia, rimossi però gl'impedimenti accidentari di nodo, tarli, ecc." (Galileo 1990: XVI 241–242).

[2] In the Veneto, the collapse of the vault of J. Sansovino's Marciana library at the end of 1545 was still vividly remembered.

The Use of Proportions

Galileo explicitly criticized the current practice of using small scale models to design buildings and machines, because he believed that they could provide only mechanical information, regardless of scale. The only specific reference to this practice appears in Palladio's *Quattro libri dell'architettura*, in which, after clearly stating in the initial pages that the drawings of constructions he proposed had to be studied according to the unit of measure indicated and their proportions respected, Palladio describes four wooden bridges, adding, "The bridges in these four manners can be constructed as long as necessity requires, by enlarging their parts in proportion."[3] It is difficult to believe that Palladio stood by this statement because the unsuitability of models for this purpose was commonly known; the failure of a model to indicate the real dimensions of an imagined object must have been known to him, because it had already been pointed out by Vitruvius in *De Architettura*, to which Palladio continually refers, as does Alberti in his De Re *Aedificatoria*. Most probably the statement derives from the fact that he was neither an inventor nor a machine builder (that is, of instruments and devices in which the friction between moving parts plays a determining role), nor did he construct architecture of a dimension that would have demonstrated the inconsistency of this claim. Daniele Barbaro, however, with whom Palladio collaborated on a version of *De Architettura*, attributed failures of structures enlarged from models to imperfections in the materials, suggesting that in a perfect world such defects wouldn't exist.[4] But he indisputably indicates the model as an instrument which can furnish information necessary for evaluating conformity to aesthetic standards, determining the number, dimension and type of all parts, and therefore the evaluation of the expense of construction and the technical methods best employed.

A little more than a hundred years later, academic Filippo Baldinucci would define the model in his *Vocabolario Toscano* and describe its use in an unequivocally pre-Galilean vision that testifies to the difficulties arising out of the diffusion of the new science:

> [a model is] a thing that the sculptor or architect makes to present the work to be built; thus the model is sometimes small, sometimes life-size. Models are made of various materials, according to the professors' taste and according to need, that is, of wood, of wax, of clay, of plaster, and others. This is the first and main labour of all the projects, through which the artist arrives at the most beautiful and most perfect. It serves the architect to determine the lengths, widths, heights and bulk; the number, expanse, type and quality of all elements, as

[3] "I ponti di queste quattro maniere si potranno far lunghi quanto richiederà il bisogno, facendo maggiori tutte le parti loro a proporzione" (Palladio 1968: III, 18).

[4] "O quanto deve essere avvertito lo Architetto non solamente rispetto alia forma et ragione che nello animo et mente sua con modi artificiosi rivolge, ma quanto alla materia, i cui difetti sono infiniti, i rimedi pochi et difficili et alcuna fiata niuno, o di niun valore, però è bene che Vitruvio ci propone le maniere difettose, acciochè per lo contrario ci potiamo guardare dagli errori" (Vitruvius 1987: 128–129).

> they should be so that the fabrication is perfect; and also to determine the different trades necessary to carry out the work, and to determine the expenses of accomplishing the work.[5]

The term "bulk" is to be understood as a relation between two of the three dimensions that determine the resistance of beams and columns. It wasn't possible for Baldinucci to go further because the concept of bulk could be stated precisely only after Leonard Euler exactly defined its opposite, slenderness; definite results were achieved only after decades in which experimental results were correlated with hypothetical models. Not all of the Galilean theory was made clear by the *Two New Sciences*, as Descartes critically observed, because of the continual departure from the main argument in order to propose applications of the new science to dogs, whales, ants, ships and giants. When Antonio de Ville, an engineer and military architect who was responsible for the fortifications of Mantua for a 2-year period in the seventeenth century, asked about a bridge to be built over a river of a given width, Galileo responded,

> In order to better declare myself, I take your example of a bridge to span a large trench, say 20 feet, which is powerful enough to support and carry a 1000-pound weight, and not more: try now to see if another bridge spanning a width four times as great and made of the very same material, but with all its members enlarged in a quadruple proportion, will support a weight of 4000 pounds. I say no, and not only that, but I say it might also happen that it can't even support itself but will collapse of its own weight.[6]

Undoubtedly this passage perfectly clarities the operative objects of the new science, but not having inserted them explicitly in the text, as he certainly should have, he forces the reader to reflect on the comprehensive objects that he set out to reach, objects that cannot and should not be limited to architectural or engineering applications, but rather reflect on the manifestations of nature and the lessons it teaches. Thus the force with which emerges Galileo's subtle comparison between a man of normal dimensions and Hercules, and between Hercules and a colossus: it isn't possible that nature, with respect for the rules gleaned from common experience, created beings capable of acting in such a different way without in someway mutating their physical aspect, transforming them into monsters. It is

[5] "...quella cosa che fa lo Scultore o Architetto, per esemplare o mostra di ciò che dee porsi nell' opera da farsi; poichè il modello alcuna volta è minore, alcuna altra è della stessa grandezza. Fannosi i modelli di varie materie, a gusto de' Professori, e secondo il bisogno; cioè di legname, di cera, di terra, di stucco, o d'altro. È il modello prima, e principal fatica di tutta l' opera, e essendo che in essa guastando, e raccomandando, arriva l'Artefice al più bello e al piu perfetto. Serve agli Architetti per istabilire le lunghezze, larghezze, altezze e grossezze: il numero, l' ampiezza, la specie, e la qualità di tutte le cose, come debbono essere; acciò la fabbrica sia perfetta: ed ancora per deliberare sopra le maestranze diverse, delle quali si deve valere nel condurre l'edificio, siccome per ritrovare la spesa che debba farsi in esso" (Baldinucci 1681).

[6] "E per meglio dichiararmi seco, piglio il suo medesimo esempio di un ponte per passare un fosso largo, V. gr., venti piedi, il quale si trovi riuscito esser potente a sostenere e dare il transito a peso di mille libbre, e non più: cercasi ora se per passare un fosso largo quattro volte tanto, un altro ponte, contesto del medesimo legname, ma in tutti i suoi membri accresciuto in quadrupla proporzione, tanto in lunghezza quanto in larghezza ed altezza, sarà potente a reggere il peso di 4000 libbre. Dove io dico di no; e talmente dico di no che potrebbe anco accadere che è non potesse regger sè stesso, ma anche il peso proprio lo fiaccasse" (Galileo 1990: XVI 241–242).

evident that no one has ever seen Hercules or giants and colossi, nor can one reasonably believe that Galileo saw them. But the problem of identifying the subject to whom the message of the new science is addressed remains unsolved, because of the insistence with which Galileo generalizes the problem so that it addresses not only architecture but ship building and machinery as well. One have addressed this in a recent work in which I formulated some hypotheses as to reading and interpretation of Galileo's text that are coherent with the environment in which the ideas matured; here I can add some considerations that have been suggested to me by a rereading of three texts by Alberti and that, to my mind, seem to relate clearly to the new science of Galileo, even while not containing any direct reference.

Leon Battista Alberti

According to Alberti's biographer Girolamo Mancini, "Battista wrote the book on sculpture and on painting, but never used the scalpel and brush to his own glory; instead he wrote on the art of building".[7] In contrast, in the dedication to Lorenzo dei Medici in *De Re Aedificatoria*, Angelo Poliziano presents Alberti as "the man from whom neither the most hidden knowledge nor the most arduous disciplines escapes ... he is also reputed an excellent painter and sculptor. Further, he was expert in all these arts at once as few are in any single art." The fact that no painting or sculpture of Alberti's has come down to us mitigates Poliziano's praise somewhat, but takes away nothing from his synthetic description of an admirable man of multiple activities.

It is with good reason that G.C. Argan, P. Sanpaolesi, C.L. Ragghianti and F. Borsi have all underlined the importance of investigating the relationship between Alberti and Filippo Brunelleschi. In particular, S. Rossi, taking as a point of departure Pevsner's work on the birth of the academies of art in Europe (Pevsner 1982), recognized in the well-known episode of the strike of the workers on the Florentine cupola an important affirmation of the architect as the recognized creator and master of the work; he argues that Alberti was the first, after Cennini, to have consciously and deliberately used design as a means of elevating the artist's situation above that of the craftsman.

While agreeing completely with Rossi, I would like to clarify the relationship, which Alberti had clearly in mind, that exists between the point at which the straight lines that proportionally relates pairs of corresponding points of two similar objects must converge, and the point from which depart the visual rays, "the cusp, that is, the point of the pyramid that is within the eye there where the angle of the quantity is" (Alberti 1975: 1, 7, 20). This passage is of the utmost importance in Albertian thought because the understanding that everything can be determined by the

[7] "Battista dettò i libri della statua e della pittura, ma non riuscì ad adoperare con gloria lo scalpello e i pennelli; invece scrisse sull'arte di edificare..." (Mancini 1882: 365).

prospective representation of a regulating system of points, such as a three-dimensional Cartesian grid, paves the way for the mathematic representation of space and of the objects within it; in order, however, that its legitimacy be established, it is derived directly from the fifth book of Euclid's *Elements*, supported by the theory of proportions. Alberti's use of mathematics is related to the central problem of composition, whether pictorial, sculptural or architectural: to the relationship both between the parts and between the parts and the whole, so that the result is a general harmony. The analogy to music isn't limited to the discernment of certain pleasing sounds, but is mathematically demonstrated by the existence of precise mathematical ratios between the length of chords. This is not the place to go into these ideas, but I permit myself to point out some questions that may not be generally noted and that may inspire further research.

The Galilean critique of the practice of using models is directed at those who were convinced as to the usefulness of models, whether architectural or sculptural, that faithfully reproduced form without paying attention to the differences between the materials used in the model and those used in the actual object. Alberti strenuously defended the use of models; he was the first after Vitruvius to list their advantages (as did Baldinucci) while, however, specifically excluding their use as instruments for determining the strength of a structure or any of its parts. He declares,

> nor has this Design any thing that makes it in its Nature inseparable from Matter; for we see that the same Design is in a Multitude of Buildings, which have all the same Form, and are exactly alike as to the Situation of their Parts and the Disposition of their Lines and Angles; and we can in our Thought and Imagination contrive perfect Forms of Buildings entirely separate from Matter, by settling and regulating in a certain Order, the Disposition and Conjunction of the Lines and Angles. Which being granted, we shall call the design a firm and graceful pre-ordering of the Lines and Angles conceived in the Mind, and contrived by an ingenious Artist (Alberti 1755: I, 1.1–2).

Alberti expresses the concept that the design is fixed in the mind that elaborates it; its form is invariable because it is independent from the material in which it is to be realized.

In actual practice, of course, things are not exactly as Alberti presents them, because forms are determined by numbers and dimensions; one cannot simply ignore the importance of the factor of the weight, except perhaps during the initial formulation of the problem, and only then as long as the scale of the model is not too small with respect to the object to be realized. On the other hand, is it very probable that Alberti was codifying a design practice used in medieval workshops where models, due to their particular architectonic forms, were used to resolve problems of stability, if not of strength. As we shall see shortly, the model of Brunelleschi's design for the Florentine cupola appears to have served this purpose.

Stability and strength contribute equally to the maintenance of a pre-established form and the capacity to resist loads without cracking, constituting even today proof of the validity of a design. Even if in modern mechanics the loss of strength can be viewed as a part of the problem of stability, here we should distinguish the two concepts. Strength and stability are necessary for every structure, but are not necessarily of equal value, in the sense that in any given structure one or the other might be more important. For example, in a Gothic structure, the problem of stability was dominant; in a Roman structure, strength was of the essence. Elements of Roman structure might not be sufficiently strong and crack, without, however, collapsing; a Gothic structure might collapse without necessarily breaking its individual elements (unless they were broken by the fall itself). Today failures of strength can be verified in antique structures for the very reason of their not collapsing, as cracks in the principal beams of many Greek and Roman temples testify, or as in the two great domes of Florence and Rome. In contrast, failures in stability are not often to be found, because they are recorded only in half-forgotten writings seldom consulted by scholars.

Alberti and the *Diffinitore*

If the following argument could be proven, as I believe, Alberti would have had excellent reasons for maintaining that one could ignore material properties in the design of a structure. Brunelleschi's model for the cupola remained in place at the foot of Giotto's bell tower until the first months of 1430 when it was decided to demolish it because it was no longer relevant and because it was so large that it encouraged improper night-time use. If Alberti had seen it, probably he would have had other, more concrete and less idealistic, reasons for writing what he did.

One may recall Alberti's Florentine sojourn in 1434 when he was in the service of Pope Eugene IV. Although some scholars lean towards a later dating for *De statua*, I tend to accept the hypothesis that springs from a direct reading of the small treatise in which Alberti, on the first page, states that he will treat painting in a later moment; since *De pictura* is indisputably dated between 1434 and 1436, it follows that *De statua* preceded *De pictura*. Many scholars from the period of Florentine humanism are also of this opinion, including Mancini, Alberti's biographer. I myself believe this for a reason closely related to Alberti's direct knowledge of the irregularity of the octagon with which Brunelleschi had to contend.

I think it will surprise most readers to learn that the octagon upon which the Florentine cupola is based is very irregular; the lengths of its sides, measured at the last catwalk on the cupola's intrados, varies between 16.98 and 17.60 m, with an average length of 17.32 m. These irregularities had incredible consequences for the builders of the extremely complicated mechanism, that is, the two cupolas and the skeleton structure connecting them, a skeleton which varies greatly in height according to Brunelleschi's design as well as at various ideal levels in horizontal section that result in irregular octagons which are increasingly smaller the higher one goes. A construction project of this complexity could be guided only by one who knew perfectly how to instruct the workers. As Manetti points out, Brunelleschi completely described each step to be executed; when he was absent, work came to a halt (Manetti 1976). Where did Filippo get his dimensions? Though no numerical nor geometric procedure could work because of the enormous

Fig. 44.1 The *diffinitore*. Image: from *De statua*, Cosimo Bartoli, ed. Venice, 1565

difficulties posed by the irregularity of the octagon, a model constructed with the same irregularities might have enormously simplified the problem. It would have been no more and no less than a small sculpture to be replicated at a much larger scale. A *diffinitore*, the instrument Alberti described for sculpting a statue that exactly corresponded to a predetermined model, would have served honourably for measuring cylindrical-polar coordinates from points on the model so that they could be recorded in a notebook and later transferred to the construction site (Fig. 44.1).

We don't know much as to these practical contrivances; it is possible that Alberti had seen something like the tool he describes in *De statua*, perhaps even in Brunelleschi's workshop, a tool that could be used to measure the distances between two points on a model so that they could be transferred to reality using only a simple change of scale. This distance corresponds to the modulus of a vectorial length which, while changing scales, maintains the direction; this is the rational foundation for the use of a model when the independence from materials of model and actual object is postulated, as by Alberti. It is notable that Alberti speaks of it in *De statua*, using as an example the continuous proportions established by corresponding parts of the bodies of Evander, Hercules and a giant to explain when and how classes of size can be defined as proportional. He uses the same example in *De pictura*.[8]

I believe that thinking of the influence of Brunelleschi in the design of a measuring device has a different value than that attributed to the invention of perspective that is the principal argument of *De pictura*, notoriously attributed mostly to Brunelleschi. The problem lies, in fact, in the leap in quality made by Alberti in theorizing about the operations of workshops, and his giving them a precise theoretical and logical structure; in other words, it seems to me that this constitutes the background against which the practical and concrete issues may be examined, constituting the theoretical basis of a discipline. Alberti is much more specific in this regard because he holds that to the things he had written must be attached the thoughts of the philosophers.[9] It is obvious that these are the same themes that Galileo uses to challenge the theory that Alberti elaborated.

A Reevaluation of Alberti and the Science of Construction

I believe it is necessary to restore to Alberti the dimensions of scientist and technician that some scholars of his work have recently obscured, especially with regard for the law of the lever given by Alberti. In the preface to *Ludi Matematici*, Ludovico Geymonat (1980) issued clear warnings as to the absolute necessity of giving works such as Alberti's their full due, because they contain ideas and instruction displayed with a clarity possessed only by those with a full command of the subject. The law of the lever, which Alberti gives in *De Re Aedificatoria*, VI, 7, is completely different from that with which we are familiar and is derived directly from the theory of Archimedes, which was not relative to inclined forces applied to the lever, but rather to a most particular class of weights represented by parallel vectors. Given this restriction, the practical procedure suggested by Alberti is rigorously exact, and was in fact followed up until the end of the nineteenth

[8] "Ma noi, per fare piu chiaro il nostro dire, parleremo in questo più largo. Conviensi intendere qui che cosa sia proporzionale. Diconsi proporzionali quelli triangoli quali con suo lati e angoli abbiano fra sè un ragione che, se un lato di questo triangoio sarà in lunghezza due volte pù che la base e l'altro tre, ogni triangolo simile, o sia maggiore o minore, avendo una medesima convenienza alla sua base, sarà a quello proporzionale; imperò che quale ragione sta da parte a parte nel minore triangolo, quella ancora sta medesima nel maggiore. Adunque tutti i triangoli così fatti saranno fra sè proporzionali" (Alberti 1975: 1, 14).

[9] "...se il cielò, le stelle, il mare e i monti, e tutti gli animali e tutti i corpi divenissero così volendo Iddiò la metà minori, sarebbe che a noi nulla parrebbe da parte alcuna diminuita. Imperò che grande, picciolo, lungo, brieve, alto, basso, largo, stretto, chiaro, scuro, luminoso, tenebroso, e ogni simile cosa, quale può essere e non essere agiunta alle cose, però quelle sogliono i filosofi appellarle accidenti, sono sì fatte che ogni loro cognizione si fa per comparazione..." (Alberti 1975: I, 18).

century. Even critics not eager to review Mach's critiques on the postulates upon which Archimedes based his demonstration have to re-examine their own positions. Vagnetti's attitude is cautious when he declares "the long and intense cultural formation of Alberti does not allow the exclusion of some direct or indirect knowledge of texts then circulating in manuscript form or of an undoubtedly rich oral tradition, with which he could have come into contact during his numerous travels or in the very halls he frequented in Padua and Bologna" (Vagnetti 1972: 183). Geymonat also emphasizes the importance of not ignoring possible ties between the various manifestations of Albertian culture, from literary works to the composition of treatises on Sculpture, Painting and Architecture. P.L. Panza (1994) has adopted this approach by comparing these three treatises with Alberti's literary texts. However, I think there are some remaining grey areas where his train of thought seeks to unite the theoretical foundations of the three arts, probably because this kind of analysis would have required the use of texts about the history of the theories put forth and developed by the *art* of construction, before this became a *science* of construction. Yet I seem to find some germs of this kind of thought expressed by Panofsky, Pevsner and Garin, who intuited something but lacked any possibility of pushing beyond due to the total lack of texts written by a construction technician or by a scholar of the theories upon which the relative knowledge was organized. The fact is that, as far as I know, those who sought to set down this history have taken Galileo as a point of departure, believing that before the Pisan no science was possible. I am not so sure I can state categorically what I believe, but I think that not even terms such as science, art and technique have conveyed the same meaning through the centuries, but rather they have evolved in relationship to their usages and to the particular contexts in which they were used. Without going too far back in time, it is possible to affirm the direct descendance of our culture from the Enlightenment and that the actual organization of knowledge was already at that time characterized by a clear distinction between Art and Science, assigning to Technique the role of realizing the ideas expressed by the first two. In any case, for some decades the boundaries between respective territories have been the object of critical revision aimed at establishing some kind of unity under the common denominator of knowledge.

In this light, the work of Alberti, especially his treatises on the three arts, lends itself to multiple but related analyses which avoid the kind of limiting approaches which betray the spirit of Alberti's work: the unity of knowledge was, in fact, a fundamental principle of humanism. The art of construction as formulated by Alberti is, and could not be other than, the enumeration of codified rules arising out of practice and tested in structural projects that demonstrated their validity by defying the centuries. The scientific demonstration of the validity of Alberti's theories found an instrument of confirmation in the theory of proportion; the accusations against its validity, in few but significant cases, had to wait until failure provided contrary evidence. This is not unlike what occurs in our own age: disasters are occasionally produced by theories that official science holds to be valid.

Translated from the Italian by Kim Williams

Biography Salvatore di Pasquale (1931–2004) was an Italian architect and university professor. He was one of the outstanding twentieth-century Italian scholars of science of construction. His most important theoretical contribution to the science of construction was his book *Scienza delle costruzioni. Introduzione alla progettazione strutturale* (Milan: Tamburini, 1975). One of the principal objects of his studies was Brunelleschi's cupola for the Cathedral of Florence, which culminated in the publication of *Brunelleschi. La costruzione della cupola di Santa Maria del Fiore* (Venice: Marsilio, 2002).

References

ALBERTI, Leon Battista. 1755. *The Ten Books of Architecture*. James Leone, trans. Rpt. New York: Dover Publications, 1986.

———. 1975. *De pictura*. C. Grayson, ed. Bari: Laterza.

BALDINUCCI, F. 1681. *Vocabolario Toscano dell'Arte del Disegno*. Florence: Accademia della Crusca.

GALILEI, Galileo. 1989. *Two New Sciences including Centres of Gravity* and *Forces of Percussion*. Translated with a new introduction and notes by Stillmann Drake, 2nd. edn. Toronto: Wall and Thompson.

———. 1990. *Discorsi e Dimostrazioni Matematiche Intorno a Due Nuove Scienze*, E. Giusti, ed. Turin: Einaudi Editori.

GEYMONAT, Ludovico. 1980. Preface. In Leon Battista Alberti, *Ludi Matematici*, R. Rinaldi, ed. Milan: Guanda.

MANCINI, G. 1882. *Vita di Leon Battista Alberti*. Florence: Sansoni Editore.

MANETTI. A. 1976. *Vita di F. Brunelleschi*. D. de Robertis ed. Milan: II Polifilo.

PALLADIO, Andrea. 1968. *I Quattro Libri dell'Architettura*. Milan: Hoepli.

PANZA, P.L. 1994. *Leon Battista Alberti. Filosofia e Teoria dell'Arte*. Introduction by D. Formaggio. Milan: Guerini.

PEVSNER, Nicholas. 1982. *Le Accademie d'Arte*. Torino: Einaudi Editori.

VAGNETTI, L. 1972. Considerazioni sui Ludi Matematici. *Studi e Documenti di Architettura*. Florence: Teorema.

VITRUVIUS. 1987. *I Dieci Libri di Architettura Commentati da D. Barbaro*. Milan: II Polifilo.

Chapter 45
Verrocchio's Tombslab for Cosimo de' Medici: Designing with a Mathematical Vocabulary

Kim Williams

Introduction

The tombslab by Verrocchio commemorating Cosimo de' Medici, patriarch of the wealthiest of Florentine families, is a relatively small memorial marker laid in the floor of the crossing in the basilica of San Lorenzo in Florence (Fig. 45.1). In spite of its size, it contains interesting lessons on the rich relationships between mathematics and design. Forentine sculptor Andrea del Verrocchio (ca. 1435–1488) was one of the best known artists in Florence at the time, and his workshop was a breeding ground for master artists such as Leonardo da Vinci and Pietro Perugino (Adomo 1991; Bule et al. 1992). As a basis for his composition, Verrocchio used a vocabulary of geometrical figures. The centrepoint in the composition plays a dual role, organizing the figures of the composition and relating the tombslab as a whole to the particular architectural setting in which it was placed. Further, the figures relate to each other through a system of proportions derived from the Pythagorean musical scale. The symbolism of the tombslab is derived from all of these elements: the sacred significance attached to the figures is reinforced by the colours used in the composition, the proportional relationships, and the position of the tombslab in space.

Cosimo de' Medici died in 1464. The tombslab which Verrocchio designed to commemorate him was laid in San Lorenzo in 1467. The slab appears in the pavement in the centre of the crossing; Cosimo's remains are buried in the ground in the crypt below (Baldini and Nardini 1984; Burns 1979). The actual

First published as: Kim Williams, "Verrocchio's Tombslab for Cosimo de' Medici: Designing with a Mathematical Vocabulary", pp. 193–205 in *Nexus I: Architecture and Mathematics*, ed. Kim Williams, Fucecchio (Florence): Edizioni dell'Erba, 1996. Research for this paper was supported by grants from the Anchorage Foundation of Texas, Houston, and the Graham Foundation for Advanced Studies in the Fine Arts, Chicago.

K. Williams (✉)
Kim Williams Books, Corso Regina Margherita, 72, 10153 Turin (Torino), Italy
e-mail: kwb@kimwilliamsbooks.com

K. Williams and M.J. Ostwald (eds.), *Architecture and Mathematics from Antiquity to the Future*, DOI 10.1007/978-3-319-00137-1_45,

Fig. 45.1 Verrocchio's tombslab for Cosimo de' Medici in San Lorenzo. Drawing: author

tomb and the tombslab are connected by a massive pier in the crypt. The central element in the composition of the tombslab is a rectangle of red porphyry. Elongated half-circles are placed on each side of the rectangle. In the top and bottom half-circles are inscription panels[1]; in those on either side are green porphyry *mandorle*. This inner composition is circumscribed by a circle, which is itself circumscribed by a square. In the interstices between the half-circles and the outer circle are small green porphyry roundels. All the geometrical figures are outlined in white marble. An outer square of black marble frames the whole composition. Bronze shields with the Medici symbol, *palle* or balls, in red porphyry, appear in the interstices between the circle and the outer square.[2] Centred on three of the four sides of the outer square are small, square, bronze grilles which provide light for the crypt below. The fourth grille was apparently obliterated when the altar was redesigned in the 1600s[3] (Fig. 45.2).

The tombslab has most intrigued those interested in Renaissance tomb design because it is so different from the kind of tomb markers found at the time (Clearfield 1981). It is devoid of any figural representation of the deceased, for example, such

[1] The inscriptions read, "Here lies Cosimo de' Medici. Publicly declared Father of his Country" and "Lived 75 years, 3 months, and 20 days."

[2] The shields may be a later addition.

[3] In addition to obliterating the fourth grille, the present altar step shaves about 1 cm off the tombslab's upper edge.

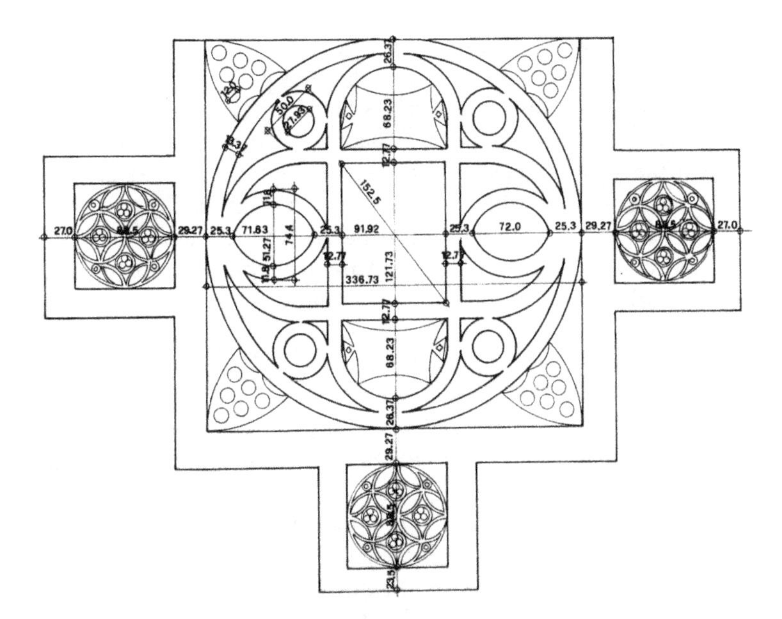

Fig. 45.2 Survey of the tombslab with the key dimensions (all dimensions are in centimetres). Drawing: author

as a bust or portrait. In addition, there are no readily recognized Christian symbols used in the marker. Although Cosimo was the wealthiest and most influential Florentine of his time, his marker is simple to the point of austerity. The simplicity of his tomb marker is unabated in spite of the rich materials used. In part, the restraint demonstrated by the design may be due to the personality of Cosimo: he was known to have avoided overt displays of his wealth and position. To understand fully the significance of the tombslab, however, it is necessary to study the geometrical figures which appear in the composition. My conclusion is that the slab, far from being a terse pagan marker, professes a belief in a divinely-created cosmos. We will first examine the use of the figures as carriers of sacred ideas, and then examine the intriguing relationships suggested by the proportions of the figures.

The Forms of the Tombslab

Verrocchio used five distinct geometrical figures in his composition. These were references to philosophical ideas and readily understood as such by those who were familiar with the neo-Platonic philosophy popular in fifteenth century Florence. In the centre of the panel is a rectangle of red porphyry. The rectangle measures some 92 cm in width and 122 cm in length, with a diagonal of 152.5 cm (see Fig. 45.2 for exact dimensions).[4] The relationship between these dimensions becomes somewhat clearer when they are converted into fifteenth century Florentine units of measure: the rectangle measures 31 soldi in width by 41 soldi in length, with a diagonal equal to some 51 soldi.[5] In other words, the rectangle is formed of two right triangles whose sides relate to each other in the ratio of 3:4:5. I believe that the significance of the 3:4 rectangle in Verrocchio's design is linked to that of the triangles of which it is formed. Triangles in general, as the first plane figures, have always been given special significance. A triangle is formed by three points; Pythagoreans considered three the first "real" number and, therefore, divine. Plato gave a cosmic symbolism to triangles, writing in the *Timaeus* that the world is composed of triangles (Plato 1961: 53c–e). The 3:4:5 triangle, sometimes called an "Egyptian triangle" because it was studied and used by the Egyptians, notably in the Pyramid of Cheops, and sometimes called the "Pythagorean triangle" because it provides a ready proof for the Pythagorean formula $a^2 + b^2 = c^2$, has a special place among triangles. It is the only triangle whose sides form an arithmetic series. In addition, the sum of the lengths of the sides, 3 + 4 + 5, is 12, a particularly significant number. A circle divided into 12 equal segments symbolized the division of the heavens into 12 zodiacal regions. If one imagines the circumference of this circle as a line, then the line can be opened and refolded to form the 3:4:5 triangle (Fig. 45.3). Further, the proportions of the 3:4:5 triangle are related mathematically to the value for the musical whole tone (the mathematics of this proportion will be explained in detail in the next section.) For Neo-Platonists, the musical whole tone was a divine value, symbolic of the Creator, in that it is a number which cannot be divided evenly into two, and is therefore eternal and unchanging.

Thus, the symbolism of the 3:4 centre rectangle, composed of two 3:4:5 triangles, is informed by their symbolizing the cosmos, or creation, as well as their symbolizing the Creator.

The overlapping ovals created by the addition of half-circles to all sides of the rectangle form what is known as a "Solomon's knot." The "knot" has neither beginning nor end, and is therefore symbolic of immortality and eternity. "Solomon's knot" is an ancient pavement motif, and was commonly used in the fifteenth century as a decorative motif. Leonardo da Vinci used it as the basis of a design for a centrally planned church.[6] It was probably used by Verrocchio with much the same intention as it was by Leonardo: to underline the sacred nature of the design by imparting to it the symbolism of immortality and eternity carried by the motif. What differs between Verrocchio's "knot" and that of Leonardo is the

[4] My analysis of the tombslab is based on my own survey. Because accuracy of the values used is critical to the accuracy of the analysis, I follow the recommendations of Howard Saalman (1979), taking each measurement three times and using the average value of the three as the working value. These are the values which appear in Fig. 45.2.

[5] A Florentine braccio was the unit of measure used at the time the tombslab was constructed. 1 braccio = 58.4 cm. It was subdivided into 12 crazie (1 crazia = 4.95 cm) or 20 soldi (1 soldo = 2.92 mm); each soldo was further subdivided into 20 denari; see Zervas (1979).

[6] For an illustration of Leonardo's design, see Pevsner (1972: 202, Fig. 143).

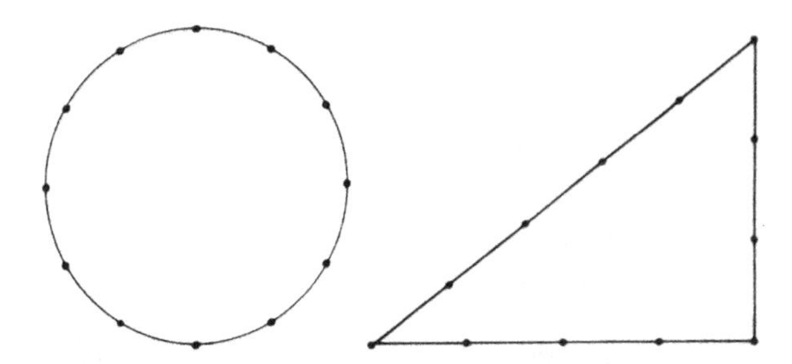

Fig. 45.3 A *circle* divided into 12 segments may be opened and recomposed into a 3:4:5 *triangle*. Drawing: author

proportions of the centre, Verrocchio having used the 3:4 rectangle while Leonardo used a square.

On either side of the central rectangle, in the interstices of the "Solomon's knot," appear two *mandorle*. The *mandorla* ("almond" in Italian) is also known as a *vesica*, or a *vesica pisces* ("fish bladder" in Greek.) The *vesica* is the fish-shaped symbol for Christ, and is linked to the celebration of the Eucharist. This sacrament had a great importance for Cosimo de' Medici: he had expressly requested that his tomb be located at the foot of the altar, so as to symbolize his being present at the celebration of the Eucharist. The *vesica* also represents a zone of intersection between two intersecting circles, one representing the heavens and the other representing earth (in neo-Platonic symbolism the two circles would represent the world of "being" and the world of "becoming"). The *vesica* itself, the intersection, is symbolic of the mediator between the two. For Christians, the mediator is Christ, who was both God and man.

The circle-in-a-square circumscribes the "Solomon's knot." We find references to these two shapes in Plato's *Timaeus* (1961: 54d–55c). The cube and the sphere represented for Plato the earth and the universe respectively, and the two-dimensional forms of square and circle have analogous meanings. As did the geometrical problem of squaring the circle, the circle-in-a-square graphically symbolized the perfecting of the imperfect. It was also a symbol for the cosmos.

Finally, it is important to consider the role of the centre point in the tombslab. Geometrically, a point lacks dimension, possessing only the characteristic of position. In Verrocchio's composition, the centre point serves a fundamental role in organizing the design, and so is an integral part of Verrocchio's language of form. As a reference for both the crossing of the basilica and of the tombslab, it also contributes significantly to the symbolism of the tombslab. The fact that the tombslab appears in a pavement (rather than on a wall, for instance) introduces the element of living man into the composition, by involving not only the deceased commemorated by the slab, but the spectator as well, who, in order to view the memorial, is drawn into the centre of the crossing. The space of the crossing had become increasingly significant in fifteenth century Italian architecture, supplanting the presbytery as the most sacred location within the church. Considering the symbolic content of the design, it is my conclusion that, if the symbolism of the design of Cosimo's marker makes reference to the order of the cosmos, its placement in the centre of the crossing is symbolic of man's central position in that cosmos. This attitude towards man is predominant in fifteenth century humanism.

The colours of the materials used in the tombslab reiterate the symbolism attached to the figures. The red porphyry used for the central rectangle is symbolic of divinity, and also carried connotations of royalty. The colour used for the central rectangle, then, underlines the reference made to the Creator through the 3:4:5 proportions, and may refer to the sovereignty of Cosimo de' Medici as well. In addition, purple was, and is today, a liturgical colour representing sorrow and penitence, used as the colour for Advent and Lent by the Roman Catholic church, and as such is an appropriate colour for a tomb marker. The green porphyry used for the *vesica* shapes refers to resurrection and immortality. As a liturgical colour green symbolizes the predominance of life over death, as the green of new leaves symbolizes winter vanquished by spring. Thus green is apt as a colour for the *vesica*, a shape representing Christ. White is a conventional colour for purity and therefore holiness. The white used to outline the forms of the tombslab alludes to the sacred nature of the message of the memorial.

In terms of the geometrical figures which appear in the design, and the colours in which they are presented, the symbolism of the tombslab is far from pagan. Instead, it is evocative of an indivisible, eternal sovereign Creator of the universe; of the Christ, mediator between heaven and earth, present in the Eucharist; of the desire that the terrestrial achieve the perfection of the celestial.

The Proportions of the Tombslab

Analysis of the panel in terms of abstract geometry presents some difficulties. Geometrical figures are bounded by lines; lines have length but no width. The lines in the tombslab have both length and width. Thus it is not always clear which of the dimensions are to be considered. However, it is clear that the proportions of the forms are interrelated, and this is particularly important with regards to a study of pattern, because the resulting order in the composition integrates the forms. I believe that mathematical proportions related to the musical scale predominate in the composition (as opposed to a system of proportions based on an irrational such as $\sqrt{2}$, ϕ, or π, though irrational proportions can be found in the composition, as we shall see presently).[7] The use of musical proportions implies a "harmonic" treatment of the elements of the composition, and they are found quite frequently in the proportions of elements in Renaissance architecture.

Let us note some of the proportional relationships (please refer to Fig. 45.2). The diameter of the outer circle is equal to twice the long side of the 3:4 rectangle plus

[7] The musical proportions are very closely related to systems of proportion based upon irrational values, so that the use of one does not necessarily preclude the other.

the short side (121.73 + 91.92 + 121.73 = 335.38, a deviation from 336.78 of 0.4 %). The width of the tombslab *vesica* is one-third the diagonal of the 3:4 rectangle (152.5 ÷ 3 = 50.83, a 0.9 % deviation from the actual 51.27). The outer diameter of the small roundels tangent to the "Solomon's knot" is equal to the diagonal of the rectangle. The width of the strips which outline the figures is equal to one quarter of the width of the *vesica* (51.27 ÷ 4 = 12.82, a 0.4 % deviation from 12.77). It is also equal to one-twelfth of the diagonal of the 3:4 rectangle (152.5 ÷ 12 = 12.71, a 0.5 % deviation from 12.77). The outer diameter of the roundels is equal to one-third of the diagonal of the 3:4 rectangle (152.5 ÷ 3 = 50.83, a 1.6 % deviation from 50.0). The outer diameter of the roundels is also equal to two-thirds of the outer width of the *vesica* (74.87 × 2/3 = 49.91, a 0.2 % deviation from 50.0). The doubled border is equal to one-half of the width of the *vesica* (51.27 ÷ 2 = 25.63, a 1.3 % deviation from 25.3, and a 2.8 % deviation from 26.37). We are seeing, then, proportions of 1:2, 1:3, 1:4, 2:3 and 3:4. The diagonal of the rectangle appears to be a key dimension for generating the others.

A search for a geometrical method by which to describe the tombslab has led me to recognize that the composition is based upon three squares *ad quadratum*, the innermost of which is subdivided into modules with musical proportions by extensions of the 3:4 rectangle (Fig. 45.4). The corners of the outermost square come very close to falling exactly on the centrepoints of the grilles (deviating by only 1.5 %).

In an *ad quadratum* composition the $\sqrt{2}$ figures prominently, and it is worthwhile to point out that it is found in other elements of the tombslab. For instance, we noted that while Verrocchio and Leonardo both used the "Solomon's knot," Verrocchio departed from the norm by placing a 3:4 rectangle in the centre instead of a square. A similar variation is found in his use of the *vesica*. The classic *vesica* is formed by two overlapping circles; the circumference of each circle passes through the centre of the other. The width of the *vesica* created is equal to the radius, or one-half of the diameter, of the circles which overlap to create it. It can be circumscribed by a rectangle of proportions $1:\sqrt{3}$. Another way to think of it is as being based upon two equilateral triangles. In contrast, the tombslab *vesica* may be circumscribed by a rectangle of proportions $1:\sqrt{2}$, and the width of the vesica is three-quarters of the diameter of the circles which overlap to create it. It may be thought of as based upon two triangles with sides in the ratios of $1:\sqrt{2}:3$ (Fig. 45.5).

While no documentation exists of the tombslab which would shed light on Verrocchio's actual process for its composition, it is likely that some kind of comprehensive system was used to generate the dimensions of the geometrical figures within the composition. I have been unable in my own work so far to solve the mystery, but two separate systems developed by others are particularly suggestive, and each involves the 3:4:5 triangle. Let us consider first an examination by Jay Kappraff of the geometry of that triangle, which reveals the relationship between that triangle and the Pythagorean musical scale:

> Let the triangle ABC (Fig. 45.6) have lengths in the ratio: $L_1:L_2:L_3 = 3:4:5$. It follows from trigonometry that the angle bisector of angle A cuts the opposite side L_1 in a length $a = L_2/3$, while the angle bisector of angle B cuts its opposite side L_2 in $b = L_1/2$. In other words,

Fig. 45.4 A geometric analysis of the proportions of the tombslab. Drawing: author

tan A/2 = 1/3 and tan B/2 = 1/2. These two fractions have special significance in terms of the Pythagorean musical scale. They form the basis of the Pythagorean tetrachord: 1:1, 1:2, 1:3, and 2:3, respectively the unison, octave, fifth above an octave, and the fifth. Now it follows that 2a:3b:c = 3:4:5. From this it follows that a/b = 8/9, the ratio of a whole tone in the Pythagorean scale (Kapraff 1993: 10).

Dr. Kappraff's study provides one important way to begin to understand just how integrated the dimensions of the tombslab are. This system would provide a justification for the "Solomon's knot" being based upon a 3:4 rectangle rather than a square, and has the further advantage of providing a means of generating both the dimensions of the figures in the tombslab and the resulting musical proportions between them.

Another very interesting geometrical system with regards to the tombslab is the "New Jerusalem" geometry, proposed by John Michell (1988; Kappraff 1991: 4–6) (Fig. 45.7). The point of coincidence between this system and the Verrocchio tombslab lies in the use of the relationships between the 3, 4, 5 and the 11. In the tombslab, each of the three sides of the triangle relate to the diameter of the circle as parts of 11, that is to say, that the three sides of the triangles are related to the diameter of the circle in the following ratios:

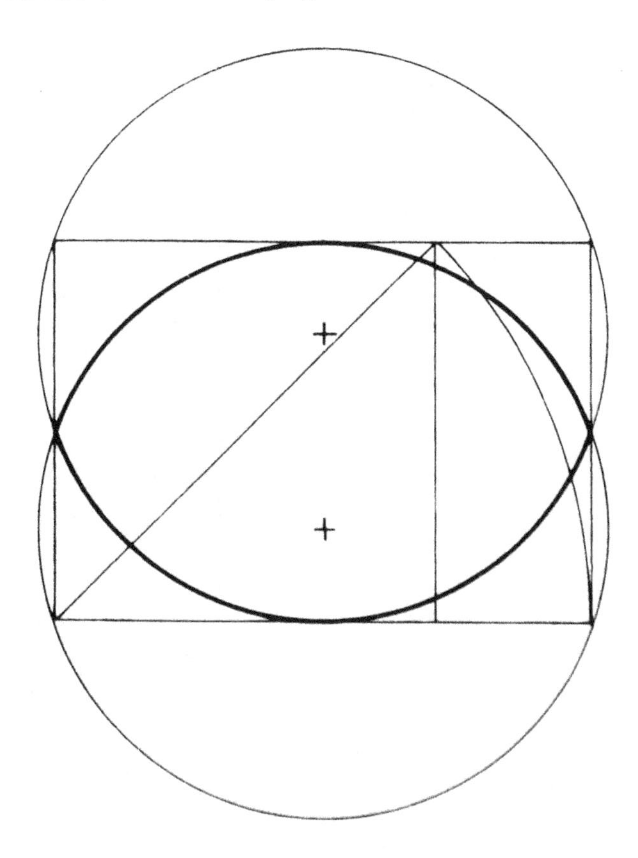

Fig. 45.5 The proportions of the tombslab *vesica*. Drawing: author

$$\frac{91.92}{336.73} :: \frac{3}{11}; \frac{121.73}{336.73} :: \frac{4}{11}; \frac{152.50}{336.73} :: \frac{5}{11}.$$

I have been particularly interested in these proportions because they are found in two other contexts within the basilica of San Lorenzo. First, Filippo Brunelleschi, architect of the basilica, based his plan for the basilica on a module which measures 11 *braccia* to a side. This module is sometimes described in literature on Brunelleschi as a square with a circle inscribed in it. Second, Vasari's pavement design for Michelangelo's New Sacristy in the basilica features a checkerboard design in which the floor plane is subdivided into fields of proportions 3:11, 4:11, 5:11, and 11:11. Turning now to Michell's geometric construction, we see it is based upon a square, the side of which measures 11 units and is created by the addition of two 3:4:5 triangles to a 3 × 3 square. One of Michell's claims for the geometric construction is that a circle which passes through the centre points of the four lateral squares has a circumference equal to the perimeter of the 11 × 11 square, effectively squaring the circle. Some of the similarities between the New Jerusalem construction and the tombslab design are striking, in that we have a circle inscribed in a square, with four smaller squares on each side (analogous to the grilles in the tombslab design) and the recurrent theme of the 3, 4, 5 and

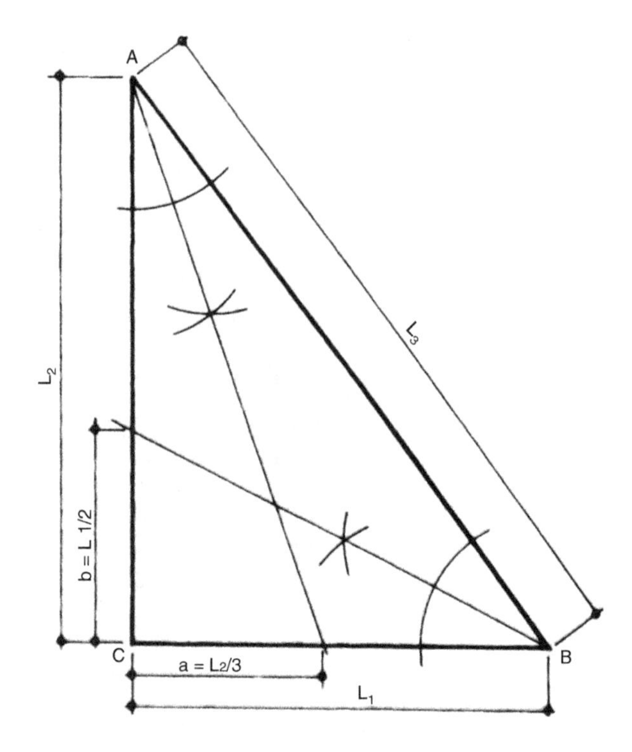

Fig. 45.6 The 3:4:5 triangle is related to the musical 8:9 whole tone. Drawing: author

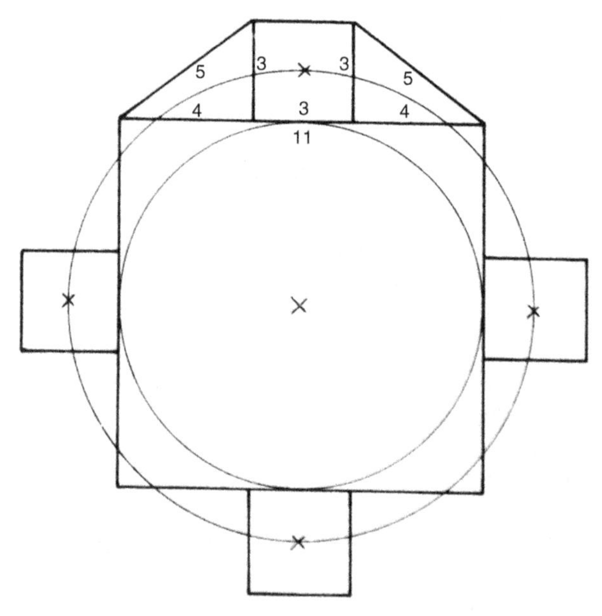

Fig. 45.7 John Michell's "New Jerusalem" geometry, an approximate method for squaring the *circle*. Drawing: author

11 proportions. But extension of the 3 × 3 squares centred on the sides of the New Jerusalem construction won't account for the placement of the 3 × 4 rectangle in the tombslab design, nor am I able to find the generation of the musical proportions within the system.

It may not be possible to develop a system which accounts for all of the properties of Cosimo's funerary marker. At any rate it is clear that mathematics played a major part in Verrocchio's creative process, resulting in a beauty that mathematicians are most apt to appreciate.

Acknowledgments I am indebted to Jay Kappraff for sharing his unpublished paper with me. All images in this chapter are by the author.

Biography Kim Williams was a practicing architect before moving to Italy and dedicating her attention to studies in architecture and mathematics. She is the founder of the conference series "Nexus: Relationships between Architecture and Mathematics" and the founder and editor-in-chief of the *Nexus Network Journal* . She has written extensively on architecture and mathematics for the past 20 years. Her latest publication, with Stephen Wassell and Lionel March, is *The Mathematical Works of Leon Battista Alberti* (Basel: Birkhäuser, 2011).

References

ADOMO, Piero. 1991. *Il Verrocchio*. Florence: Casa Editrice Edam.

BALDINI, Umberto and Bruno NARDINI. 1984. *San Lorenzo*. Florence: Nardini Editore.

BULE, Steven, Alan P. DARR, and Fiorella SUPERBI GIOFFREDI, eds. 1992. *Verrocchio and Late Quattrocento Italian Sculpture*. Florence: Le Lettere.

BURNS, Howard. 1979. S. Lorenzo in Florence Before the Building of the New Sacristy: An Early Plan. *Mitteilung des Kunsthistorisches Institut in Florenz*. **XXIII** (1979): 145–153.

CLEARFIELD, Janis. 1981. The Tomb of Cosimo de' Medici in San Lorenzo. *Rutgers Art Review* **II** (January 1981): 13–30.

KAPPRAFF, Jay. 1991. *Connections:the Geometric Bridge Between Art and Science*. New York: McGraw Hill.

———. 1993. *The Secrets of Ancient Geometry*. Unpublished.

MICHELL, John. 1988. *The Dimensions of Paradise*. San Francisco: Harper and Row.

PEVSNER, Nikolaus. 1972. *An Outline of European Architecture*, 7th ed. Baltimore: Penguin Books.

PLATO. 1961. Timaeus. In *The Collected Dialogues*. Edith Hamilton and Huntington Cairns, eds. Bollingen Series LXXI. Princeton: Princeton University Press.

SAALMAN, Howard. 1979. Designing the Pazzi Chapel: The Problem of Metrical Analysis. *Architectura* **9**, 1: 1–5.

ZERVAS, Diane. 1979. The Florentine Braccio di Panna. *Architectura* **9**, 1: 6–10.

Chapter 46
A New Geometric Analysis of the Pazzi Chapel in Santa Croce, Florence

Mark A. Reynolds

O king, through the country there are royal roads, and roads for common citizens, but in geometry, there is but one road for all

Manaechmus to King Alexander

Introduction

Without original drawings or documented evidence, it is difficult to know with confidence what numbers, measures, and geometric systems have been used in any work of art, architecture, or design. In the Pazzi Chapel, these problems are compounded by the long period of time that transpired in completing the building [construction lasted from 1429 to 1461, and the Chapel was finally finished in 1478 (Fanelli 1980: 3)], and the possibility that there was more than one architect. It is difficult to be sensitive to the artist's or architect's intentions when analysing a work, and to remain objective about these findings. If the geometer does not have command of both the measures and the number relationships—ratios, proportioning and geometric constructions—any sensitivity to the intentions of the creator will remain suspect. There is also the difficulty of knowing with certainty if the builders, tradesmen and labourers were able to carry out the designer's intentions, for many of them had little or no formal education in reading and reckoning. At times, these intentions and goals can be better understood within the context of the artist's or architect's other works, what systems of number and geometry were in use or

First published as: Mark A. Reynolds, "A New Geometric Analysis of the Pazzi Chapel in Santa Croce, Florence", pp. 105–121 in *Nexus III: Architecture and Mathematics*, ed. Kim Williams, Ospedaletto (Pisa): Pacini Editore, 2000.

M.A. Reynolds (✉)
Academy of Art University, San Francisco, CA, USA
e-mail: marart@pacbell.net

K. Williams and M.J. Ostwald (eds.), *Architecture and Mathematics from Antiquity to the Future*, DOI 10.1007/978-3-319-00137-1_46,

known at the time, and through the analyst's knowledge of the subject which is sometimes called "harmonic deconstruction", i.e. geometric analysis. I would like to lay some groundwork regarding my approach to these issues. I find it best to begin, appropriately, with the ground plans of the main hall and the altar space.

Basic Measurements

For this analysis, the measurements of the floor plan and altar space plan were accomplished using a surveyor's tape measure, and a carpenter's metal tape measure, in feet and inches.[1] Three separate measurements were taken: along the central longitudinal axis, running roughly north to south and perpendicular to the longitudinal axis of the Basilica of S. Croce; both floor/wall intersections to the left and right of this longitudinal axis and parallel to it. These calculations were then averaged. This procedure was repeated with the transverse axis and also in the altar space. These measurements were taken from a quarter to an eighth inch fractionally, and calculated from a thousandth to a ten thousandth place decimally.

There were slight discrepancies throughout the floor measurements. Some of the *coccio pesto* and *pietra serena* patterning on the floor was a bit irregular. However, throughout our measurements, I could see that the labourers made adjustments in order to preserve the measurements of the perimeter of the floor. The joints between bands and rectangles were fairly uniform and the irregularities did not affect the overall ratio of length to width. I was actually pleased and surprised to see the amount of uniformity and care in craftsmanship that I observed throughout the chapel.

It is significant to note that the chapel has benches around the perimeter of the floor, which prevent the floor from touching the walls. This fact yields two different ratios, the benches being uniform around the floor perimeter. I have calculated the ratios for both relationships—with and without—the benches.

The longitudinal and transverse axes were measured first, for it is these two axes that determine the ratio of the plan. Most often, in geometric analysis, the floor plan is the critical and primary measure in the structure because it is here that the geometric analysis takes root. These perimeter benches are very nearly a Florentine *braccia*[2] in both height and in depth, averaging 23.825 in. This measurement doubles because there are benches on opposite sides, both on the NS axis as well as the EW axis. In my opinion, when laying out the rectangular plan of the floor, the builder took the measures to the walls, with the benches as part of

[1] Although I indicate the type of measuring systems employed, I suggest that the numbers be viewed as units, and not inches, centimetres or *braccia*. Conversions are not necessary for an analysis.

[2] A *braccio* is an Italian measurement used in the Renaissance. It varied from city to city, with the Florentine *braccio* being the longest. It is about 0.5836. . .m, or 22.7968. . .inches.

the included ratio of the floor's dimensions.[3] In the Pazzi plan, as we shall see, the architect's approach to the design was to place the benches at very nearly the Florentine *braccio* so that the floor plan *inside* the benches would be in the √3 system.

The Floor Plan of the Main Chapel Hall

The following measurements are found in the main space of the chapel:

Longitudinal Axis:

a) to the walls: av. 59.96975... ft.
b) to the benches: av. 55.99895... ft.

Transversal Axis:

a) to the walls: av. 36.02065... ft.
b) to the benches: av. 32.0498... ft.

The two floor plan ratios are:

a) To the walls: 1.66487... to 1
b) To the benches: 1.74724... to 1

These two ratios approximate ratios known and used during the Middle Ages and the Early Renaissance as part of a geometric heritage from the ancients. The first is 3:5, two of the numbers in the Fibonacci Sequence,[4] and the first two in this sequence where the golden section first appears, to the tenth place: 1.6... The percent deviation is: +0.877... % high. The second ratio is $\sqrt{3} = 1.732\ldots$; this is the geometric system that contains the equilateral triangle, the hexagon, and the dodecagon. The √3 system can be seen clearly in the rib and sail design of the cupola, and in the bevel of the altar steps. We will find this same system in other parts of the chapel, on the elevation and the plan. The percent deviation here is −0.107... % low.

[3] I have always contended that the geometric calculations be done on the inside, not outside, of the geometric figure. Thickness of walls does not seem to me to be relevant to the ratio that is seen and experienced. It seems a trivial thing at first, but the ratios derived from the two rectangles are substantially different. The only time equal thicknesses are generated is in the mitred 45° angle.

[4] Keep in mind that Fibonacci, or, Leonardo of Pisa, had just brought to light his summation sequence, and surely it was known in Florence at this time. The other importance to the 3 to 5 ratio is that it was a part of Pythagorean music theory, the major sixth, and later utilized in Palladio's 'Mystic Seven' ratios to be used in the geometric planning for architecture. It is possible the architect of the Pazzi Chapel may have been making a reference to the 6 days of Creation.

The Floor Plan of the Altar Space

As I measured the chapel, I recalled the opinions of those who have said that there is no golden section ratio in the chapel, or that, because no golden section relationship could be found in the chapel, there was no geometric system at all![5] The golden section may not be the epitome of all proportioning, but we still must concede its use if we find it. As we have seen, the measure of the floor plan was in the 3:5 measure of the walls and did not yield the ϕ ratio. This was not to be the case with the altar space. It seemed logical to me, as a geometer, that if this particular relationship were to exist in the chapel, it would be in the floor plan of the main hall or the floor plan of the altar. My hunch proved to be correct; however, ϕ showed up in an arcane geometric construction. I say arcane because it is not a commonly known ratio or geometric construction. It is (Fig. 46.9):

$$\sqrt{5} \div 2 = 1.1180339\ldots \text{to } 1$$

This is the compound rectangle[6] that is composed of the double square and the golden section rectangle. It is the direct development of a line segment divided into mean and extreme ratio using the diagonal of the double square.[7] An additional construction that will also yield this 1.118... to 1 ratio is the double √5. The tangency will be on the long sides of the rectangles.

Here then are the measures:

- The altar space width: 16.354 ft. = 196.25 in.
- The altar space depth: 18.25 ft. = 219 in.
- The ratio of depth to width: 219 in. ÷196.25 in. = 1.1159... to 1
- The percent deviation from the 1.118... (√5 ÷ 2) ratio is: −0.1896...% low.

The Altar

Table 46.1 shows the calculation, ratios and percentage deviations for dimensions of the altar space. These ratios and percent deviations are taken with the comparison to the ideal, which is the double square, √4:1 (or 2:1) ratio in the plan and elevation. The side, sectional, view would then ideally be a square. In substance, the height of the altar is 2.25 in. higher than it is deep.

[5] In particular, see (Guillaume 1990).

[6] See (Hambidge 1967) for a detailed explanation on the subject of simple and compound rectangles.

[7] This construction can be found in several classical and modern sources (among them, Dürer and Hambidge). One of the best uses I have seen is the underlying grid work for *The Resurrection of Christ*, by Piero della Francesca (in Borgo San Sepolcro, Palazzo Communale).

Table 46.1 Key altar dimensions (in inches)

Dimension	Ratio	% Deviation
Height 39.5 (plan)	W:D = 75.5 ÷ 37.25 = 2.2068....:1	+1.34 % high from 2:1
Width: 75.5 (elevation)	W:H = 75.5 ÷ 39.5 = 1.9114....:1	−4.435 % low from 2:1
Depth: 37.25 (section)	H:D = 39.5 ÷ 37.25 = 1.0604...:1	+6.04 % high from 1:1

The Front Elevation

The elevation of the chapel was measured using laser technology, with the *stadio totale* laser. The survey was carried out with the help of geometer Pino Adamo and architect Kim Williams; the data obtained was given to Professor Paul Calter, who crunched and reckoned the numerical data for me to put into the geometric system. It is important to note here that the laser calculations are in the metric system, and these calculations were not converted into feet and inches. My approach to analysis allows me to calculate, measure, and construct with whatever system I have been given. Simply stated, a circle will be a circle regardless of the type of units of measure used for its radius.

From Paul Calter's calculations based on the theodolite measures, we were able to find the following information:

- Height of the chapel, from the ground plane to the intersection of the ball and cross,[8] at the top of the lantern, along the central vertical axis: 27.969 m.
- Width of the chapel, along the base and at the intersection of the cornice and wall: 17.348 m.
- Ratio of height to width: 27.969 m ÷17.348 = 1.61223...

The golden section ratio is $\phi = 1.6180339...:1$. So then, the percent deviation of the height to width ratio from ϕ is −0.3599... % low. Occasionally, one finds the geometric techniques of *ad triangulum* ("from the triangle"), and *ad quadratum* ("from the square") being used in a plan, an elevation[9] or some other primary element of a building. I wanted to see if these constructions were employed in the Pazzi Chapel. I compared our laser measures to the elevation drawings drafted by Cabassi and Tani (1992) and found them to be very nearly identical.[10] Their drawing is indispensable because the portico, believed to be a later addition, has made the original façade impossible to see.[11] By using their survey drawing of the

[8] It is thought-provoking to think about the possibility that the original cross on top of the ball would have taken the elevation from the golden section up to the $\sqrt{3}$ measure!

[9] In 1402, Jean Mignot attempted to settle the debate over the primacy of using a square or a triangle in the elevation of the Milan Cathedral. Mignot declared the triangle to be superior.

[10] As samples, I compared the ratios of the door, a window, and the height to width ratio of chapel façade.

[11] In the Cabassi/Tani drawing, one can still see the original circular oculus that was built into the original façade, and now sealed with mortar and brick.

original wall, I was able to show the geometry of the square and the equilateral triangle unencumbered by the portico's overlapping architectural elements.

Both construction systems use the rotation of the chapel's base width. For *ad quadratum*, the base is rotated to the vertical positions on the left and right sides, and perpendicular to the base. In the case of *ad triangulum*, the apex of the equilateral triangle will be marked where the rotation of the base width towards vertical cuts through the vertical mid-line axis, at 0.866... or, $\sqrt{3} \div 2$, of the height of the square when the side of this square is 1.

In Fig. 46.13, the system of *ad quadratum* is shown. The basic armature of the square is shown with the two diagonals, AM and KZ, and its midlines, NG and PR. All eight half-diagonals are added. With this armature, the fractional parts of 1 assigned to the side of the square can now be located. These fractional parts, 1/2, 1/3, 1/4, 1/5, 1/6 ..., etc., are also a harmonic progression, and can be generated within the grid of the square using only the diagonals and half-diagonals. These fractional parts are indicated. On the left side of the façade, note that K is generating many diagonal types—for example, a thirds diagonal or a fifths diagonal—that yield the fractional/harmonic segments.[12] In the analysis, it can be seen that the architect took advantage of these fractional/harmonic parts to compose some of the key elements in the façade. For example, the fifths (notated f1 and f2) give the altar's width, and the number five is associated with the five wounds of Christ. The widths of the pilasters are generated by using 1/5 and 1/6. The tops of the four windows are on thirds, and the bottoms on tenths. The medallion above the door is in the centre of the square.

In Fig. 46.12, the *ad triangulum* portion of the elevation is shown. Some scholars may argue that the two systems are incommensurable, and I sometimes concur with this point of view. However, in this particular building, there are a few elements generated within the triangular grid work that have not been generated from the square; others result from the two working in concert. I have found that many of the better architects of the time integrated both systems into one cohesive whole. The combination of the two as one grid provides a powerful design tool and a unique structure to work with, as well.

Some of the findings from Paul Calter's calculation of the coordinates points, in addition to the height to width ratio mentioned above, are:

- The drum's ratio, height to width, is, $11.814 \div 3.003 = 3.934\ldots$ to 1. This ratio is nearly $\sqrt{16}$:1 (4:1), the quadruple square. The percentage of deviation is: –1.6483...% low. The 4:1 ratio is one of the proportions used in rendering the male figure, where the figure is 8 heads tall by 2 heads wide at the shoulders, i.e., 4 shoulder widths tall by 1 shoulder width wide.
- The elevation, to *ad triangulum*, 0.866... of the width, is ideally 15.0238; in the Cabassi/Tani elevation, *ad triangulum* cuts the base of the cornice. In the

[12] This geometric function will yield *all* fractional parts, including sevenths, elevenths, thirteenths, and so on.

elevation analysis, the base of the cornice is at 15.948. The percent of deviation is 6.151 % high.

- The front doors to the chapel are nearly the same as the altar, that is, ±2: 1. (Remember, the altar is 2.0268 to 1 in the plan view).

This last ratio is an example of what I originally referred to, namely it is possible to find two solutions to the interpretation of the ratios. The doors are: height (averaged L and R) to width: 4.9505:2.41. The ratio then is: 2.054149:1.

So, we have two choices:

$$\frac{\text{Choice } 1 : \sqrt{4} : 1 \text{ (or } 2 : 1)}{2.05414 - 2 = 0.054149}$$
$$0.054149 \div 2 = 0.02707$$
$$\text{a } 2.70745\ \%\ \text{high deviation}$$

$$\frac{\text{Choice } 2 : \sqrt{\ } \text{ (or } 2.05817 : 1)}{2.05414 - 2 = 0.054149}$$
$$2.05414 - 2.05817 = -0.00402$$
$$-0.00402 \div 2.05817 = -0.0019532$$
$$\text{a} - 0.19575\ \%\ \text{low deviation}$$

Which of the two figures is correct? Because the altar is approximately a double square in plan and elevation (and approximately a double cube in three dimensions), may we assume that there is a symbolic connection between the two elements—doorway and altar? Scriptural texts contain references to the double square and cube. On the other hand, because the builder appears to have used an uncommon ratio involving the golden section, $\sqrt{5} \div 2$, can it be assumed that he would also know another, $\phi\sqrt{\phi} = 2.058\ldots$? By the early 1400s, the golden section had been a staple for artists and builders since the Middle Ages (Bouleau 1963: Chaps. III–IV). Would these ratios be in the builder's toolbox? In this situation, I would have to say that the doors were intended to be a double square. It would make sense to think that the ratio was in the double square because of its spiritual significance. Entering the chapel and sharing in the sacrifice at the altar may have been linked in the architect's intentions. It certainly suggests further research and investigation.

Details and Figures for the Geometric Analysis

The Floor Plan

a) Figure 46.1: Floor plan with the walls, benches and altar space delineated to a simple line drawing.
b) Figure 46.2: Floor plan with Observer 1 (Ol) on the threshold of the chapel floor, and Observer 2 (02) along the inside edge that is marked by the line designating

Fig. 46.1 Floor plan 1. Drawing: author

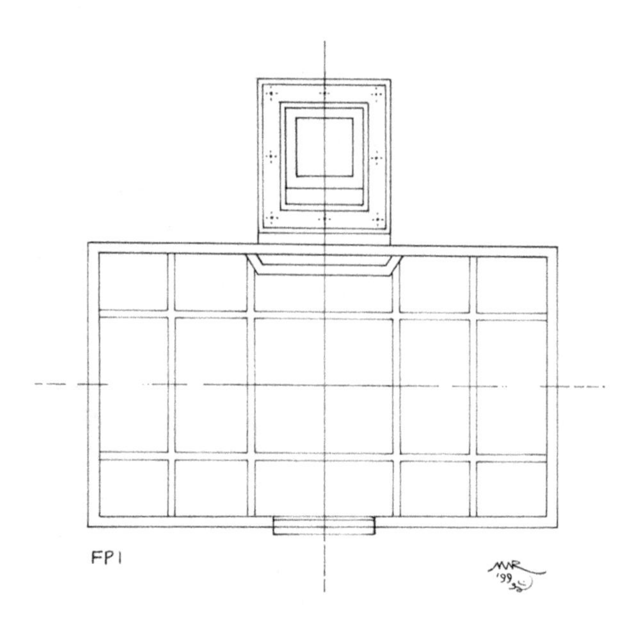

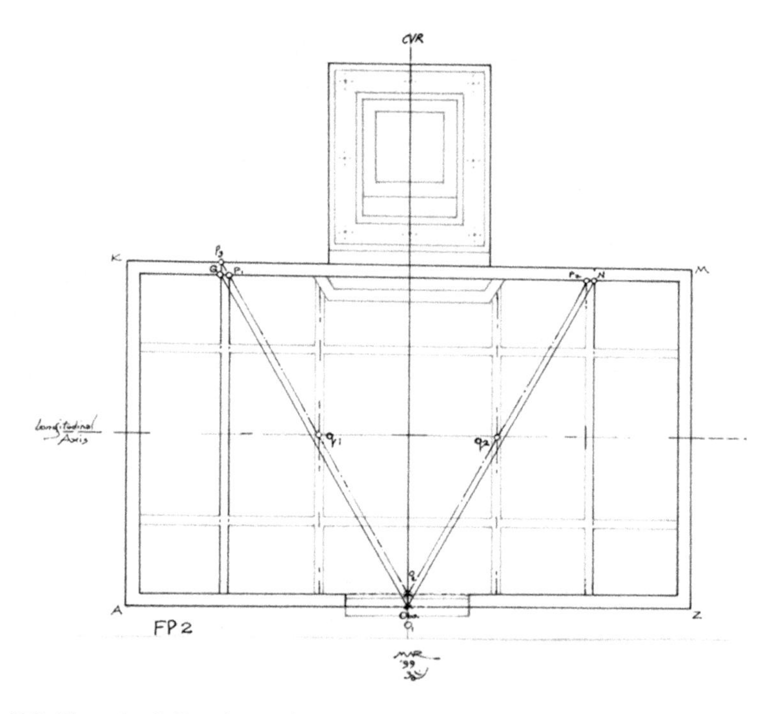

Fig. 46.2 Floor plan 2. Drawing: author

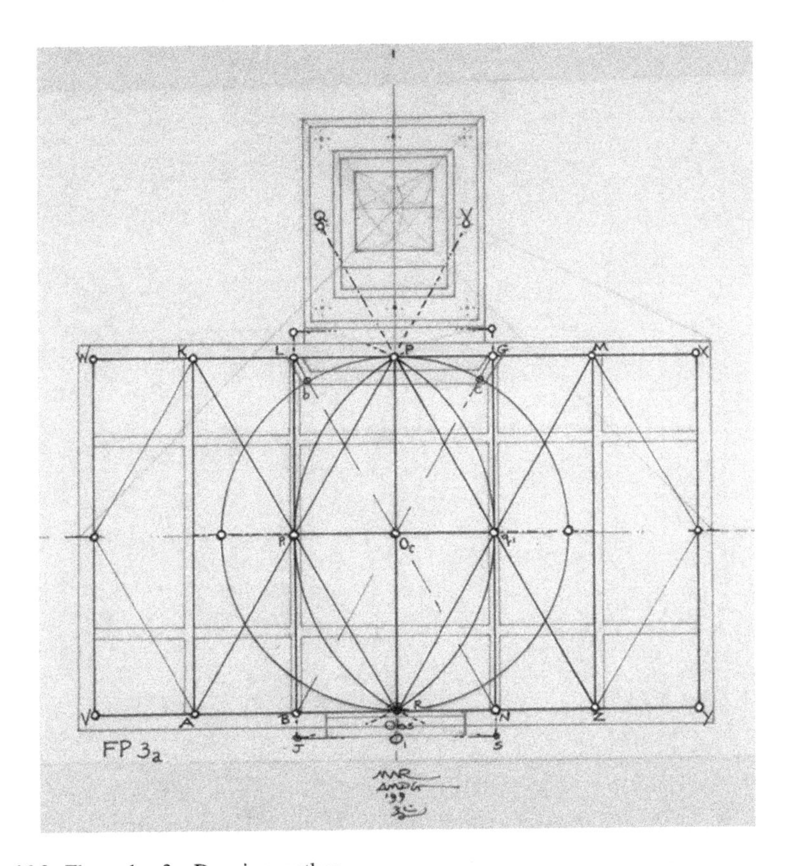

Fig. 46.3 Floor plan 3a. Drawing: author

the bench fronts. A 60° "Cone/Pyramid of Vision"[13] as introduced to Masaccio by Brunelleschi for *The Holy Trinity* painting in Santa Maria Novella, Florence. The lines O1G and O1N, as well as all other lines drawn from the Observer, O, represent lines of sight (named here LOS) from the eye of the Observer. At G and N, the LOS are intercepted by the benches to designate the light bands of stone that crisscross the floor. Points 02P1 and 02P2 from the second Observer mark the other sides of the bands. Note that if 01 is used that the LOS, 01P3, is intercepted by the wall, not the bench, but will still yield the same band edge.

c) Figure 46.3: The Vesica Piscis and the √3 ÷ 2 rectangle,[14] AKMZ. The AKMZ rectangle contains two √3 rectangles, tangent on their long sides. A third √3

[13] It is important here to note that the 60° cone of vision used by the perspectivists of the time equates directly with the use of the *ad triangulum* (from the [equilateral] triangle) system.

[14] That is, the double √3 rectangle, and is either the 0.886... or the 1.154... ratio, depending on whether the long side or short side of the rectangle is used as 1, or unity.

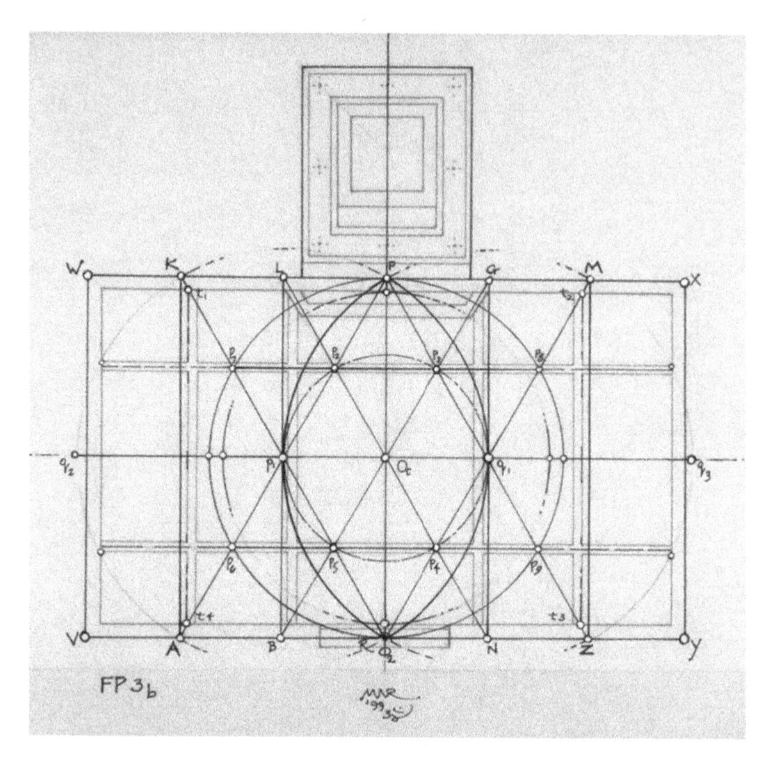

Fig. 46.4 Floor plan 3b. Drawing: author

rectangle, BLGN, frames the Vesica and generates the centre lines of the two bands in the centre of the floor. APZ and KRM are both equilateral triangles. The entire $\sqrt{3}$ system is contained within the *inner* rectangle, VWXY, which is formed by the benches, and is itself a $\sqrt{3}$ rectangle. Points Q and V, where two of the octagonal medallions are located appear to be located on the extended lines of AP and ZP.[15] The lines Lb and Gc are part of the two diagonals, LN and GB, and are 60°. It can be seen that three of the bevels of the altar steps are not at 60°. My measures found that the angles ranged from about 56° to 58.5°. However, one of the four *was* 60°. Here is a prime example of sensing *intentions* of the builder. With the number of findings in the analysis, where *ad triangulum*, equilateral triangles and $\sqrt{3}$ systems have been used, my surmise is that the intention was to cut these steps at that angle, but for whatever reasons, the three others are close. It certainly isn't obvious *in situ*, and it could well be that it passed muster without any authority noticing, or perhaps, even caring.

d) Figure 46.4: This plan is closely related to FP 3a, and shows the second option for the Observer, O2. By placing the Observer back from O1, the observer now

[15] I did not measure or triangulate these two points, so please note that I am only speculating on this. At the same time, I would not be surprised if this were indeed the case.

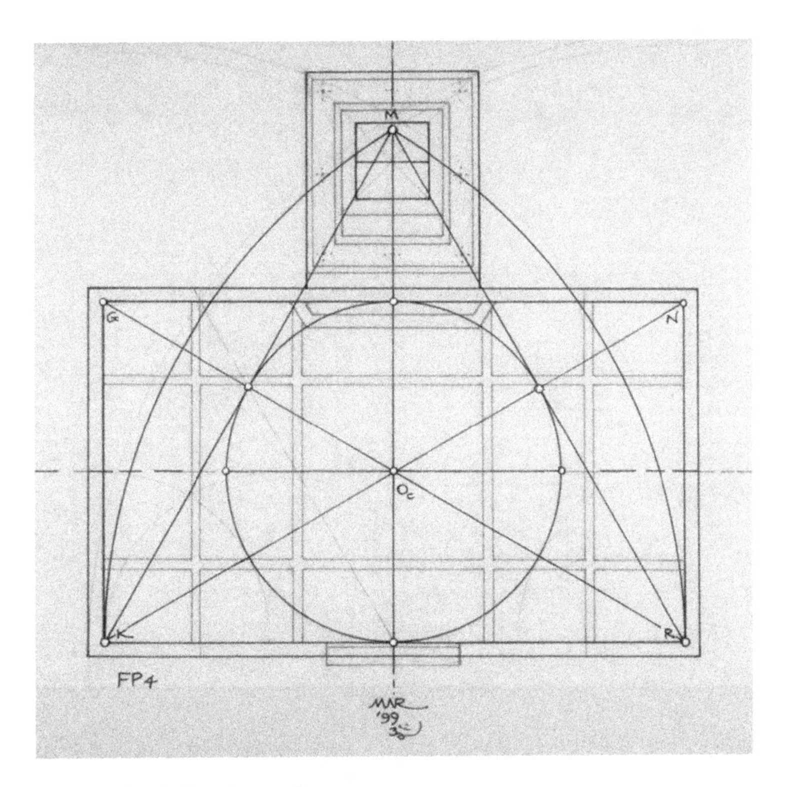

Fig. 46.5 Floor plan 4. Drawing: author

has a somewhat wider 60° cone, as can be seen at points K and M. This makes the double $\sqrt{3}$ rectangle slightly larger, and, consequently, the rectangle, VWXY, aligns with the walls, and not the benches. Rectangle VWXY is *not* a $\sqrt{\int 3}$ rectangle, but is instead approximately a 3 to 5 ratio.[16] Note that lines AP and PZ cut the benches at points t1 through t4, and provide the outer edges of the outer two bands on the floor.

e) Figure 46.5: *ad triangulum* generated by the rabattment of length KR, along the line of the benches. The two lines, KM and RM intersect at apex, M. In the celebration of the Roman Catholic Mass, this point, M, is the area where the Monstrance, which holds the Eucharist, is placed during Mass. Circle *Oc*, drawn tangent to triangle KMR, represents the cupola.
f) Figure 46.6: The *ad quadratum* system. This drawing contains circle *Oc*, and a second circle *n*; these circles generate the Vesica Piscis. It is not needed for *ad quadratum* work, but it is placed here to show the placement of the centre of the longitudinal band at v1 and v2, the band closest to the altar. Circle n also acts as a

[16] The walls would have to be placed at points q2 and q3 in order to be a $\sqrt{3}$.

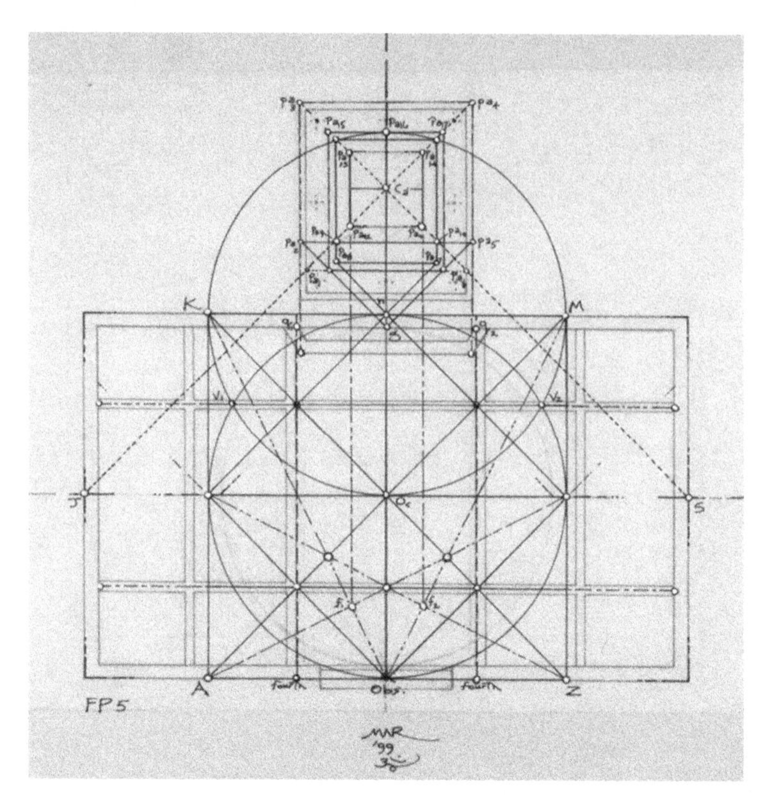

Fig. 46.6 Floor plan 5. Drawing: author

device that links the centre of the chapel floor to the band that frames the altar itself. The master square, AKMZ, has been divided into fractional parts, including fifths at f1 and f2.[17] It is appropriate then to use these fifths to delineate the width of the altar at pa11 and pa12. These two points are harmonically joined where the two diagonals, pa3S and pa4J, cut these fifths. Additionally, points pa11 and pa12 mark the front corners of the altar itself, and are placed by their intersection with the diagonals of a second *ad quadratum* system. This second system of *ad quadratum*, generated in the altar space, starts at points pa3 and pa4, and the 45° diagonal lines are drawn to points J and S, where they intercept the longitudinal axis at the northern and southern walls.

g) Figure 46.7: This last plan drawing is a theory based on other plan drawings, most notably the omnipresent Stegmann and Geymuller (SG) plan drawing and

[17] The number five is synonymous with the five wounds of Christ in the Catholic faith. It would be appropriate to use these fifths to delineate the altar width, as the altar is where the celebrations of Christ's Resurrection take place.

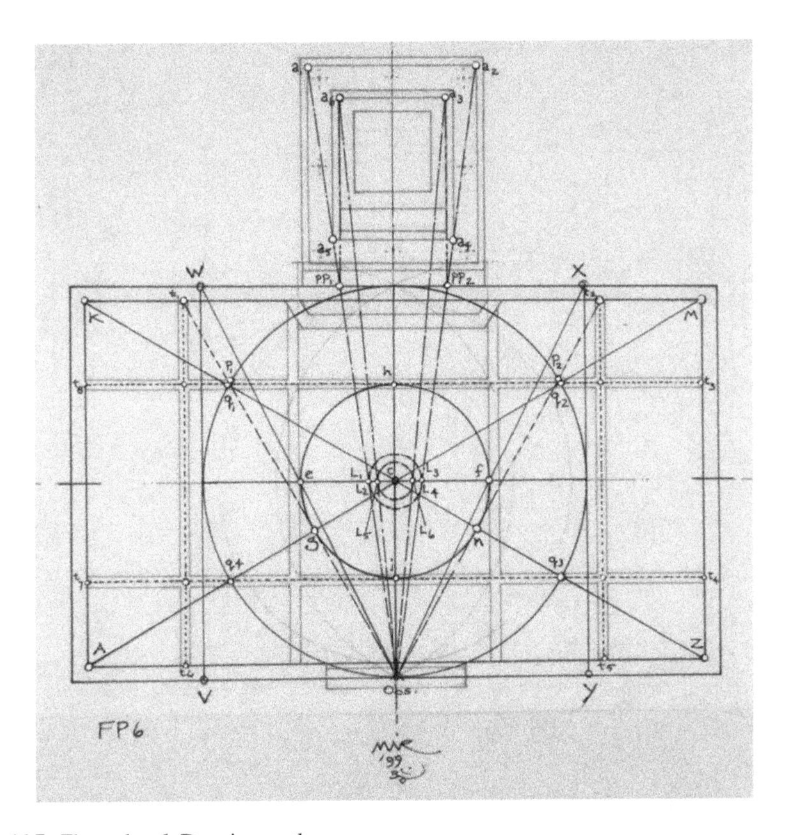

Fig. 46.7 Floor plan 6. Drawing: author

others that borrowed from them. I have not measured the lantern, drum and cupola, so I cannot say with certainty if the following is true. I would welcome the opportunity to either do the survey or have someone send the accurate measures to me. The SG drawing indicates that the ratio of lantern diameter to cupola diameter is 1 to 10. Again, this would make sense if one notes that there are 10 beads to 1, with 5 decades in all, to the rosary used in prayer. If this 10 to 1 ratio is in fact true, then it can be seen that when lines of sight are drawn from the Observer to the corners of the band that surrounds the altar, the LOS intercept the longitudinal axis at this ratio. If the technique of LOS was used, then this 10 to 1 ratio can be generated in this manner.

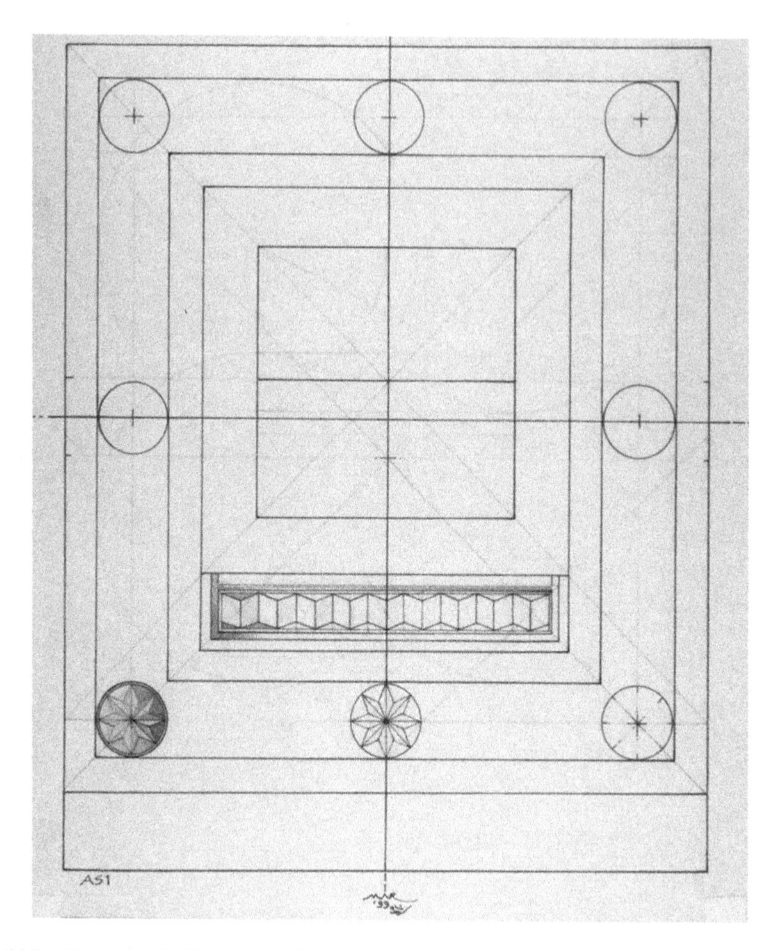

Fig. 46.8 Altar space 1. Drawing: author

The Altar Space

a) Figure 46.8: Line drawing of the altar space area. The plan drawings that I've analysed indicate that this space is square, and it is not. The altar itself is a double cube, and the priest's floor in front of the altar is a double square. As a plan view, these two elements will be a square. The bands around the altar and priest's floor are also in a square format. However, the total space around these items is not a square. There is an intarsia-like stone geometric pattern across the front of the altar between the altar and the chapel hall. It can also be seen that the eight octagonal star medallions are not equidistant from one another. The three in front and back are closer to each other than the ones to the left and right as the altar is faced. The eye can easily be tricked here, as the medallions to the left and

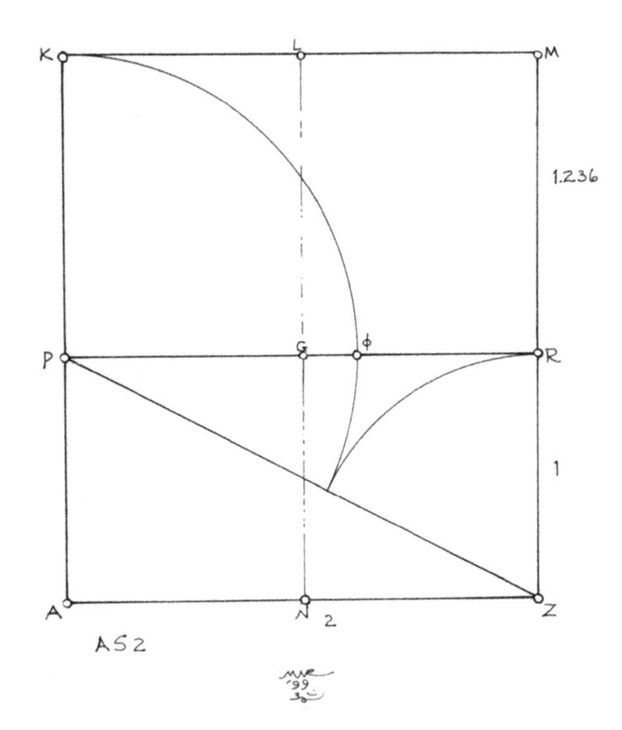

Fig. 46.9 Altar space 2. Drawing: author

right are foreshortened, and appear as though they are all equal. The geometric band in front of the altar is responsible for the non-square floor plan. Drawing Altar Space 2 will discuss this rectangle.

b) Figure 46.9: This rectangle is in fact the "1.1180339...", a ratio that combines the double square and the golden section rectangle; or, if you prefer, two √5 rectangles, tangent on their long sides. It is certainly not a ratio that many scholars and surveyors are aware of, but, at the same time, it is known to some of us working in the fields of geometry and geometric analysis.
c) Figure 46.10: Another way of looking at the "1.118..." is to see that it is really a part of the construction for the golden section rectangle. In this drawing, it can be seen that the bottom part of the square, a double square in its own right, is half, or 0.5, of the square. Subtracting 0.5 from the golden section, 1.618... equals, 1.118....
d) Figure 46.11: Several years ago, I made an extraordinary discovery while working with the 1.118. The golden section is generated from the double square and its diagonal. Instead of stopping at the golden cut, continue around to construct the golden section rectangle. During its rotation, it passes through the midline of the double square extended. At this juncture, the equilateral triangle is generated! This rectangle is quite rich as a compound rectangle, including the double square, the golden section, two √5 rectangles, and the equilateral triangle.

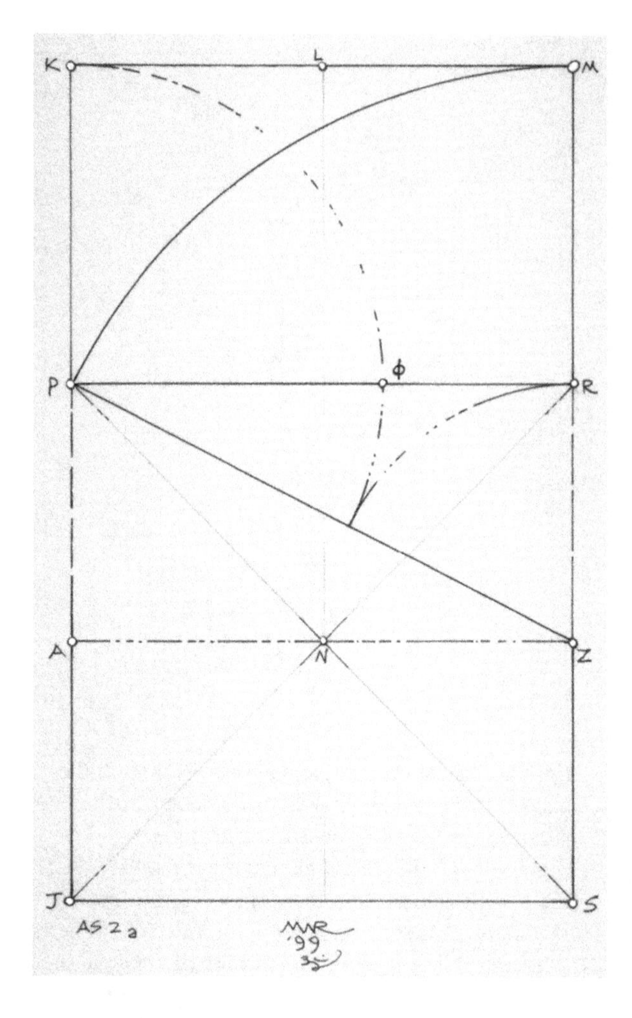

Fig. 46.10 Altar space 2a. Drawing: author

The Front Elevation

a) Figure 46.12: The ad *triangulum* system applied to the façade.
b) Figure 46.13: The *ad quadratum* system applied to the façade. These first two drawings were constructed over a line drawing of the elevation using the survey drawing done by Cabassi and Tani. The drawing from Cabassi and Tani was used for its valuable information concerning the oculus and the original portion of the façade that is attributed to Brunelleschi. The drawings and the harmonic deconstructions speak for themselves.
c) Figure 46.14: This line drawing is a measured line elevation (1 cm = 1 m, although the measuring system can be converted to units of any measure) using the information obtained with the *stadio totale*, and shows the relative

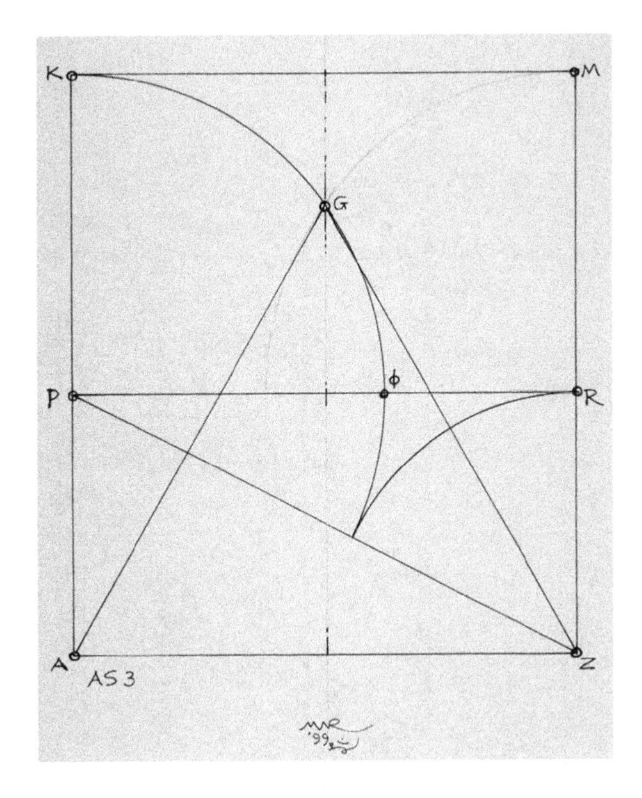

Fig. 46.11 Altar space 3. Drawing: author

magnitudes given by Paul Calter's figures obtained from the measurements done in January 1999 by Kim Williams and Pino Adamo.

The drawing is both convenient and accurate, and it is easy to see the relationships without the "clutter" of the object. For me, the most significant relationship is the height to width ratio, which yields, by 0.006. . ., very nearly the golden section. This is a complex drawing that combines both master grids of *ad triangulum* and *ad quadratum* with the 1.118. . . rectangle of the altar space plan. From this triple compound rectangle, it can be seen that almost every major facet of the elevation is included within the geometry, as there area significant number of eyes (where two or more lines cross) within the grid that align with many of the key elements of the façade. It is an important drawing in that it utilizes *all* the major systems used in the plans of the floor and altar.

Conclusions

The Pazzi Chapel has long been the subject of speculation concerning its builder or builders, the limitations of its construction within the Santa Croce complex, and, most importantly, the geometric systems that are or are not present in the building.

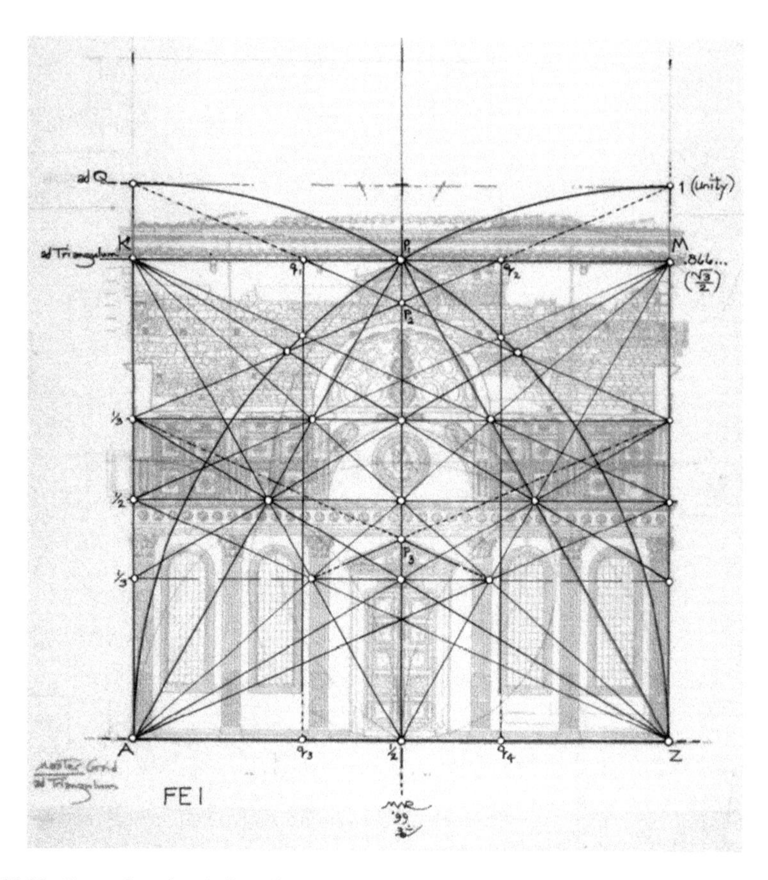

Fig. 46.12 Front elevation 1. Drawing: author

Correct analysis is a difficult task, although it may appear to be an easy one, given the proportioning that is seen and the number of theories that abound on the subject from the annals of art and architectural history. For me, the procedure is always to erase preconceived notions[18] and to approach the work to be analysed as if it were a secret treasure. The goal is to find the geometry within. Find it I did, and, although not surprising, I also found it quite pleasing to behold the variety of geometric systems handled with the deftness of an architectural master.

The Pazzi Chapel is based on the square and its three major divisions: rational whole numbers and fractional parts; square roots and irrationals; the special case of

[18] At the top of the list is: avoid *any* attempts at converting from one system to another. It is foolish at best, and a disaster for sure. Understand that all systems have units and a *unity* that they are based on. All systems work easily when doing geometric analysis. Grid work (structure) and composition come from the circle, square, and the triangle. Know the workings of these and know all that is in the geometry.

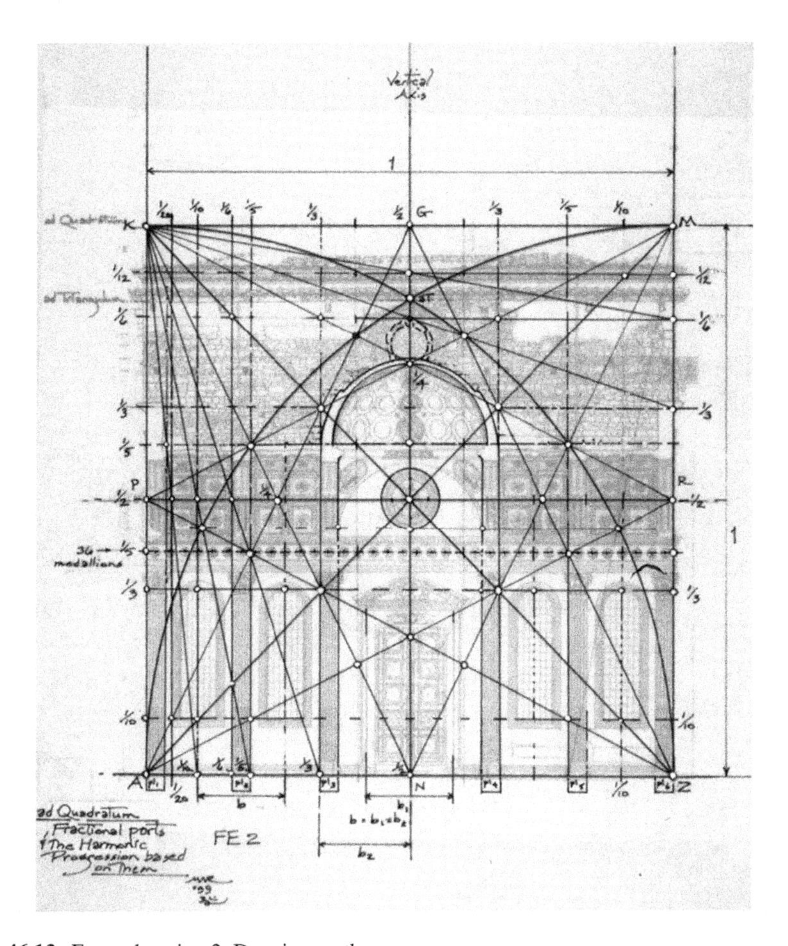

Fig. 46.13 Front elevation 2. Drawing: author

the irrational, the golden section 'family'. The builder also used the square's relationship to the circle and to the equilateral triangle. These three elements exist as the one unifying force in the architecture of the chapel. The artistry was in the integration of these three geometric systems.

It is my hope and intention that the geometric analysis and the drawings based on the measures clearly define for the reader what I have been fortunate to find. I believe that I have found a few geometric relationships that may have evaded others' notice. The human need for order leads to the search for the best and most beautiful ways that it can be achieved. Our tools include geometry and geometric analysis. As we move into a new millennium, we must remember to keep attuned to our collective history. With the state of our technology, there may be a temptation not to.

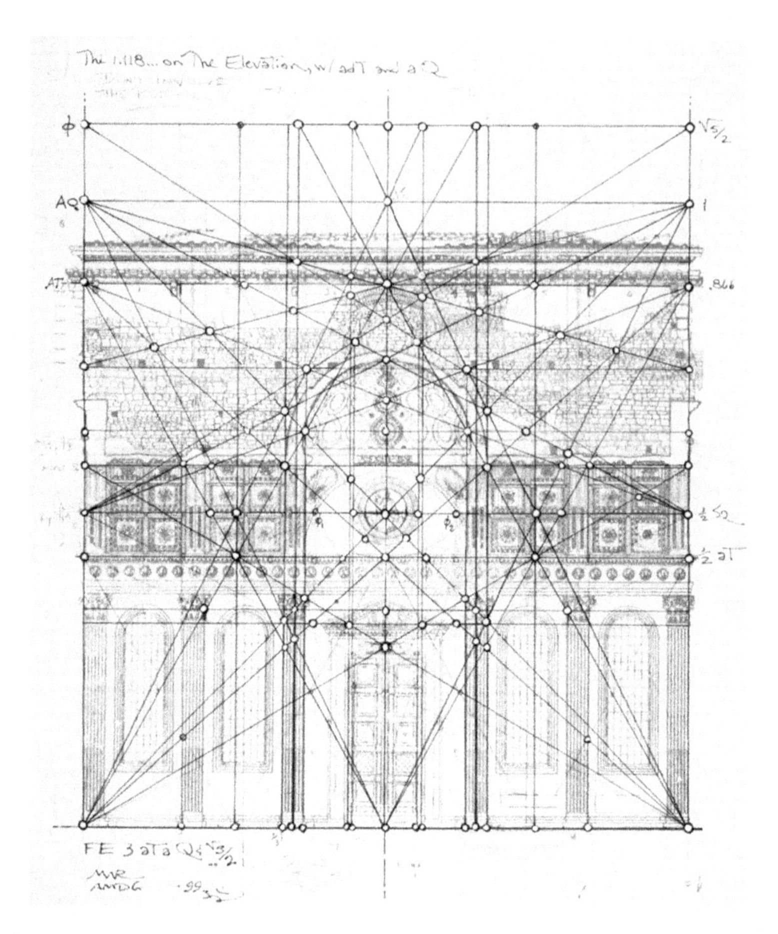

Fig. 46.14 Front elevation 3. Drawing: author

Acknowledgments I would like to thank Kim Williams, whose challenge to take on this project led me to the wonderful people and places of Florence, and whose assistance was indispensable to the completion of the analysis. Pino Adamo's skills with the *stadio laser* totale were inestimable for this project; he saved me many long hours and expenses in scaffolding and labour. Without the critical crunching and reckonings of our raw numerical data by Paul ("Bernoulli") Calter, Professor of Mathematics at Vermont Technical College, this project would be only a shadow of what it is. Stephen Wassell, Chair of the Math Department at Sweet Briar College, deserves special recognition for giving a compassionate ear to me regarding the more exotic qualities of certain geometric systems and Palladio's Means. I would also like to express my deep appreciation to the Franciscans, and to their founder, St. Francis of Assisi, the staff of the Basilica of Santa Croce, and the City and Diocese of Florence, for their permissions and cooperation with this effort. Amy Wilder, an art student at the Academy of Art College in San Francisco, helped me as well, doing much of the initial research for the project. Lastly, I would like to especially thank my family. Most especially, I thank my wonderful wife, Kathy, my Muse-on-Earth, whose help with the measurements, her knowledge of travelling to and through Italy, her incredible sense of order and abilities in organizing, and her faith in me as an artist, were instrumental in making this project the success it was.

Biography Mark Reynolds is a visual artist who works primarily in drawing, printmaking and mixed media. He received his Bachelor's and Master's Degrees in Art and Art Education at Towson University. He was awarded the Andelot Fellowship to do post-graduate work in drawing and printmaking at the University of Delaware. For years Mr. Reynolds has been at work on an extensive body of drawings, paintings and prints that incorporate and explore the ancient science of sacred, or contemplative, geometry. He is widely exhibited, showing his work in group competitions and one-person shows, especially in California. Mark's work is in corporate, public, and private collections. A born teacher, he teaches sacred geometry, linear perspective, drawing, and printmaking to both graduate and undergraduate students in various departments at the Academy of Art University in San Francisco, California. Additionally, Reynolds is a geometer, and his specialities in this field include doing geometric analyses of architecture, paintings, and design.

References

BOULEAU, C. 1963. *The Painter's Secret Geometry*. New York: Harcourt Brace and World.

CABASSI, S. and R. TANI. 1992. *The Portico of the Pazzi Chapel*. Florence: Città di Vita.

FANELLI, G. 1980. *Brunelleschi*. Florence: Scala/Philip Wilson.

GUILLAUME, Jean. 1990. Desaccord Parfait: Ordres et Mesures dans la Chapelle des Pazzi. *Annali*, **2** (1990): 9–23.

HAMBIDGE, Jay. 1967. *Elements of Dynamic Symmetry*. New York: Dover Publications.

Chapter 47
Muqarnas: Construction and Reconstruction

Yvonne Dold-Samplonius and Silvia L. Harmsen

Introduction

Muqarnas is the Arabic word for stalactite vault, an architectural ornament developed around the middle of the tenth century in north eastern Iran and almost simultaneously, but apparently independently, in central North Africa. A muqarnas is a three-dimensional architectural decoration composed of niche-like elements arranged in tiers.[1] In Fig. 47.1 we see an example of an Il Khanid (1256–1336) muqarnas vault: the entrance portal of the shrine of the Holy Bāyazīd at Bastām, in Iran (Pope 1939: 1102).

The two-dimensional projection, or ground plan, of this vault is shown in Fig. 47.6. Like all ground plans, it consists of a small variety of simple geometrical elements.

The first definition of muqarnas is given by the Timurid astronomer and mathematician Ghiyāth al-Dīn al-Kāshī, who ranks among the greatest mathematicians and astronomers in the Islamic world. He was a master calculator of extraordinary ability. His wide application of iterative algorithms, and his sure touch in laying out a computation so that he controlled the maximum error and maintained a running check at all stages, earn him credit as the first modern mathematician. When Ulugh Beg decided to construct the observatory, he invited

First published as: Yvonne Dold-Samplonius and Silvia Harmsen, "Muqarnas, Construction and Reconstruction". Pp. 69–77 in *Nexus V: Architecture and Mathematics,* Kim Williams and Francisco Delgado Cepeda, eds. Fucecchio (Florence): Kim Williams Books, 2004.

[1] For a short introduction, see (Dold-Samplonius 2003).

Y. Dold-Samplonius (✉)
Türkenlouisweg, 1469151, Neckargemünd, Germany
e-mail: dold@math.uni-heidelberg.de

S.L. Harmsen
Erlengrund 23, 68789, St. Leon-Rot, Germany
e-mail: slharmsen@arcor.de

K. Williams and M.J. Ostwald (eds.), *Architecture and Mathematics from Antiquity to the Future*, DOI 10.1007/978-3-319-00137-1_47,

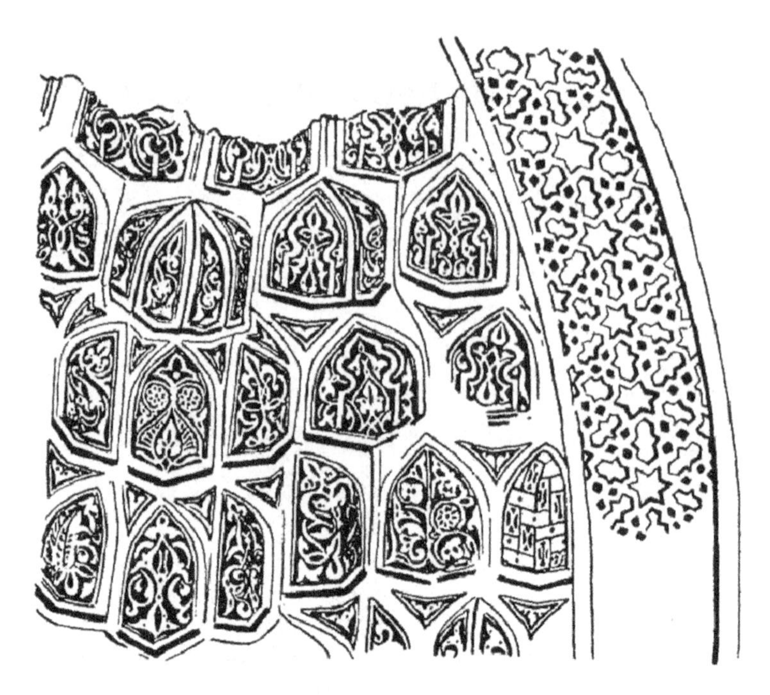

Fig. 47.1 Entrance portal of the shrine at Bastām, Iran. Drawing: authors

al-Kāshī to his court, some time after 1416. Al-Kāshī died outside the Samarqand and observatory in June 1429, probably murdered on the command of Ulugh Beg. Two years earlier he had finished the *Key of Arithmetic*, one of his major works, a veritable encyclopedia of mathematical knowledge. The work is intended for everyday use, as al-Kāshī remarks: "I redacted this book and collected in it all that is needed for him who calculates carefully, avoiding tedious length and annoying brevity." By far the most extensive book is Book IV, *On Measurements*. In the last chapter al-Kāshī approximates the surface area of a muqarnas and gives the following definition:

> The muqarnas is a ceiling like a staircase with facets and a flat roof. Every facet intersects the adjacent one at either a right angle, or half a right angle, or their sum, or another combination of these two. The two facets can be thought of as standing on a plane parallel to the horizon. Above them is built either a flat surface, not parallel to the horizon, or two surfaces, either flat or curved, that constitute their roof. Both facets together with their roof are called one cell. Adjacent cells, which have their bases on one and the same surface parallel to the horizon, are called one tier (Dold 1992: 202).

This last chapter, "Measuring Structures and Buildings", is really written for practical purposes: "The specialists merely spoke about this measuring for the arch and the vault and besides that it was not thought necessary. But I present it among the necessities together with the rest, because it is more often required in measuring buildings than in the rest." Al-Kāshī uses geometry as a tool for his calculations, not for constructions. Besides computing the surface area and volume of arches, vaults, and *qubbas* (cupolas), al-Kāshī establishes here approximate values for such a

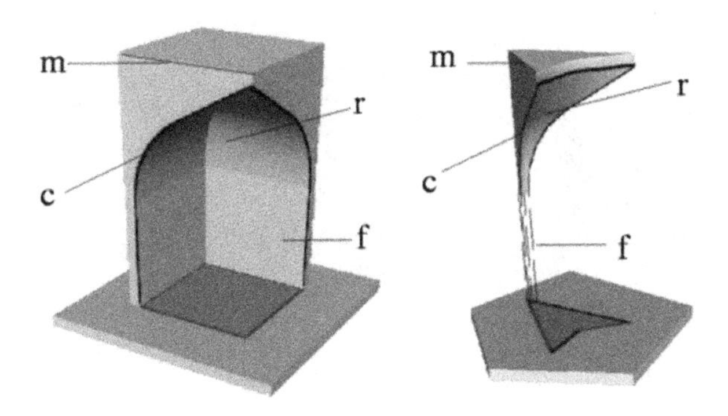

Fig. 47.2 A muqarnas cell standing on a square (*left*) and an intermediate element standing on a "biped" (*right*). A cell consists of two facets, or vertical sides (f), plus their roof (r). The roof (r) is a surface, not parallel to the horizon, or two joined surfaces, either flat or curved. An intermediate element is a surface, or two joint surfaces, connecting the roofs of two adjacent cells. The *curve* (c) is where the cells and elements are put together. The module (m) is defined as the measure of the base of the largest facet, that is, the side of the square; it is the measure-unit of the muqarnas. A tier is a row of cells, with their bases on the same plane surface parallel to the horizon. Rendering: authors

muqarnas surface. He is able to do so, because, although a muqarnas is a complex architectural structure, it is based on relatively simple geometrical elements, as we shall explain below.

Elements of a Muqarnas

The elements of a muqarnas consist of cells and intermediate elements connecting the roofs of two adjacent cells. For a better understanding of the construction of an element, the component parts are shown in Figs. 47.2 and 47.3.

In this Chapter we talk about the design, plane projection and ground plan of a muqarnas. All muqarnas have been designed from a ground plan but most ground plans are now lost. All existing muqarnas have a plane projection which, however, when the muqarnas has partly collapsed, is only partial.

As al-Kāshī explains, the elements stand on simple geometrical figures. This means that the plane projection of an element, or the view from underneath, consists of simple geometrical forms, which al-Kāshī identifies as squares, rhombuses, half-rhombuses, almonds (deltoids), "small bipeds," "jugs," "large bipeds," and "barley-kernels" (which only occur on the upper tier).

In his treatise al-Kāshī shows the plane projection of common elements consisting of simple geometrical forms (Fig. 47.4). These are from left to right: a rhombus and a square, with underneath a "barley-kernel," a "biped", and its complement to a rhombus, an "almond." Other elements like half-squares (cut along the diagonal), half-rhombuses (isosceles triangles with the shorter diagonal

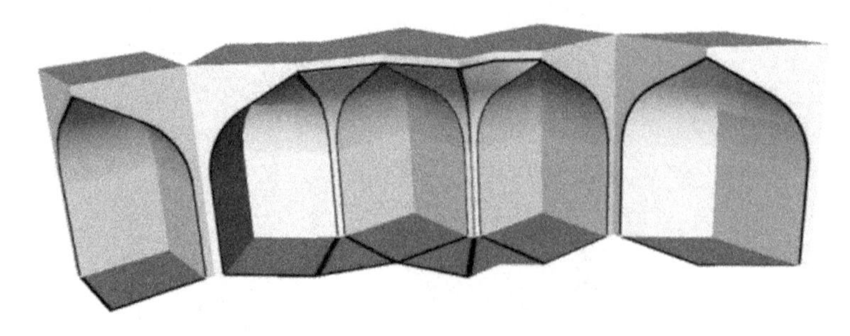

Fig. 47.3 Part of one tier consisting of five cells and three intermediate elements. Rendering: authors

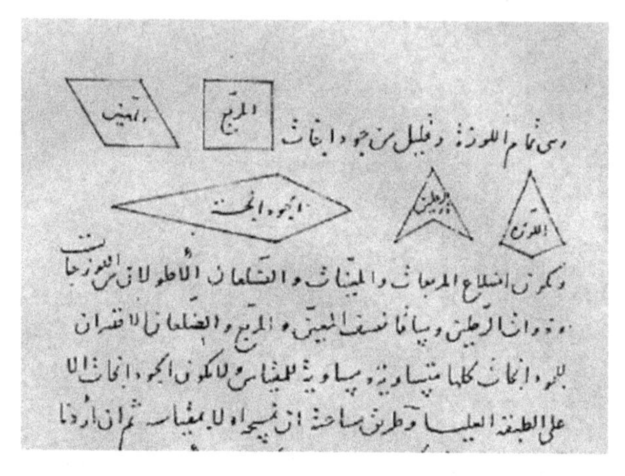

Fig. 47.4 The plane projections of the cells, as given by al-Kāshī in his *Key of Arithmetic*. Malek Library in Tehran, ms. 3180/1

of the rhombus as their base), rectangles, and the "jug" with its complement the "large biped" are only described by al-Kāshī and not drawn.

Figure 47.4 is a reproduction of the oldest extant copy of al-Kāshī's manuscript *Key of Arithmetic*, today conserved in the Malek Library in Tehran (ms. 3180/1). It was copied in the year 1427, the same year in which al-Kāshī finished his manuscript, by al-Kāshī's companion Mo'īn al-Dīn al-Kāshī, who went with al-Kāshī from Kashan to Samarqand around 1420.

With the exception of the intricate Shīrāzī[2] muqarnas, the two-dimensional projection of a muqarnas vault consists of a small amount of simple geometrical elements.

[2] In Timur's time, when building activity exploded, local constructors could manage the simpler buildings. But for the special and more artistic monuments, architects and artisans were imported from the conquered lands, first Khwārizm, then Tabrīz and Shīrāz, and finally India and Syria. It is known that Timur brought in architects from Shīrāz in 1388 and 1393, and that many migrated of

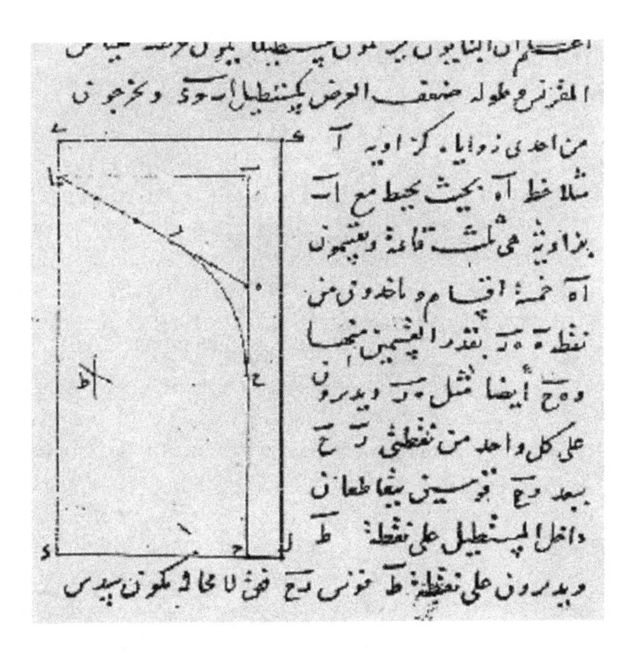

Fig. 47.5 Curve described by al-Kāshī in his *Key of Arithmetic*. Malek library in Tehran, ms. 3180/1

Fitting the Elements

In order to fit together, the muqarnas elements have to be constructed according to the same unit of measure (module) and the curves, on the sides of the elements, have to be the same. This curve (Fig. 47.5) is described by al-Kāshī as the "Method of the Masons", indicating that it is taken from building practice. Al-Kāshī's construction is carefully executed by means of circles, an angle of 30° and exact ratios (Dold-Samplonius, Yvonne 1992–1993: 221–222). In present-day Fez, Morocco, artisans still use more or less the same curve, with the same proportions, drawing it freehand on the wooden beams for their muqarnas construction (Dold-Samplonius 1996). The design of the elements found at Takht-i Sulaiman also follows approximately this same curve.

Some muqarnas are composed of cells with flat roofs, as we have seen in al-Kāshī's definition. In this case we have to fit the facets, or walls, of the cells and construct the flat roofs of the cells above. In this case the normal height of the facets equals the module.

their own free will. The names of several Shīrāz architects have been transmitted, the most famous being Qawām al-Dīn b. Zayn al-Dīn al- Shīrāzī, the only active builder whose surviving structures display a distinctive architectural style. This might well be the reason why the type of muqarnas constructed with many variations, "innumerable possibilities" as al-Kāshī explains, was called *Shīrāzī*.

Muqarnas Design

The earliest known example of a muqarnas ground plan, is an Il Khanid 50 cm. stucco plate from ca. 1270 showing the projection of a quarter muqarnas vault found at the Takht-i Sulaiman, Iran (Harb 1978). There are no known Islamic architectural working drawings from the pre-Mongol era, although we find occasional textual references to plans. After the Mongol conquest of Iran and Central Asia, an abundance of locally produced, inexpensive paper appears to have particularly encouraged architectural drawings on this medium. Rag paper had been introduced to Samarqand by Chinese prisoners of war in 751, and because it was much cheaper than papyrus and parchment, its use had spread throughout the Islamic world after the tenth century. It was not, however, until the Mongols arrived in the 1220s that an extensive paper industry developed in Tabriz and other Iranian towns under Chinese influence. Its extensive use had become essential due to the increasing elaboration of geometric design. Fourteenth-century sources frequently mention architectural drawings produced on either clay tablets or paper. Necipoğlu (1995) describes a late fifteenth-century or early sixteenth-century scroll, now preserved at the Topkapi Museum, Istanbul, which was most likely compiled somewhere in western or central Iran, probably in Tabriz. What we find in the Topkapi scroll, a pattern book from the workshop of a master builder, are patterns to be used for ornaments and as ground plans for muqarnas. The scroll is a high-level design book for architects, builders, and artisans. The Topkapi scroll is the best-preserved example of its kind, with far-reaching implications for the theory and practice of geometric design in Islamic architecture and ornament.

Up until Necipoğlu's discovery of the Topkapi scroll, the earliest known examples of such architectural drawings were a collection of fragmentary post-Timurid design scrolls of sixteenth-century Samarqand paper, preserved at the Uzbek Academy of Sciences in Tashkent. These scrolls almost certainly reflect the sophisticated Timurid drafting methods of the fifteenth century. In 1876, English architect C. Purdon Clarke brought back from Tehran some scrolls and working drawings from the eighteenth and the nineteenth centuries that he had collected following the death of the official state architect, Mirza Akbar; these are now preserved in the Victoria and Albert Museum, London. In 1981, similar material, still in the hands of the master-artisan, was examined by W.K. Chorbaki in two Arab towns (Necipoğlu 1995: 14–15). These scrolls were not only the basic reference manual but also served as a design book. What is evident is that there exists a continuous tradition, from the thirteenth century Takht-i Sulaiman plate to the muqarnas designs still in use in the modern Islamic world.

A few years ago we visited a workshop at Fez, Morocco, where the artisans used a construction plan for a muqarnas on a 1-1 base. The pieces cut out for constructing the muqarnas could actually be put on the draft such that the cross section of the element, that is, the cross section of the wooden beam, exactly matched the figure on the draft.[3] As in the Il Khanid period, 700 years earlier, the plane projection of the elements in the Moroccan plan consists of simple geometrical figures: squares, half-squares, rhombuses, half-rhombuses, rectangles, "almonds," "bipeds" (to use al-Kāshī's terms). Wilber (1955: 73) relates how, at Isfahān, he watched an elderly workman repairing a badly damaged stalactite half-dome of the Şafavid period. On the floor below the damaged elements he had prepared a bed of white plaster, and on this surface he was engaged in incising a half plan of the original stalactite system.

The Timurid scrolls show a decisive switch to the far more complex radial muqarnas, with an increasing variety of polygons and star polygons. The Akbar scrolls are also more elaborate than the twentieth-century Fez drawings. Despite their simplicity, however, the more recent scrolls testify to a relatively unbroken tradition of architectural practice in central Asia from at least the Timurid period onward. The standard patterns compiled in modern Moroccan sketchbooks indicate that the master who drew them repeated inherited formulas rather than inventing new ones.

Figure 47.1 shows part of the entrance portal of the shrine of the Holy Bāyazīd at Bastām, Iran, the whole structure of which, apart from a few Seljuk remains, was constructed in the Il Khanid period. Figure 47.6 shows the plane projection corresponding with that muqarnas structure. The structure is mirrored along the center line. When we look at the right side in the middle, we see three "jugs" connected by two small "bipeds" (outlined in a dark line). This correlates with the second row from below on the left in Fig. 47.1 (Dold-Samplonius 1996: 67–71).

When we study the correlation between the two-dimensional muqarnas ground plan and the three-dimensional muqarnas structure, the question arises: Is this correlation unique? To answer this question we have to take into account two important points:

First, if the height of the muqarnas elements remains the same throughout the whole structure, the structure will be a steep muqarnas, like the Seljuk examples in Anatolia. Hence, when the muqarnas structure has to be inserted into an existing vault, we have to adapt the height of the facets of the elements. In other words, when we want to construct a muqarnas in a vault that is not very steeply pointed, the height of the facets of the elements has to decrease on the higher tiers in such a way that they will fit into the vault. As the height of the facets approaches zero, the remaining part of the vault, that is, the part above the last tier, can then be finished in several ways. In some vaults the original brickwork is left visible; in others the ceiling is plastered and ornamented by painting, or by applying barley-kernels (Fig. 47.4, second row on the left) or by using a combination of these two.

In the Topkapi Scroll we have seen elements based on a semi-regular hexagon on the uppermost tier as in the designs 108/109/110 (Necipoğlu 1995: 344–345). The two sides of the hexagon pointing towards the center of the vault equal the module, whereas the other four sides equal the shorter side of the "biped." The hexagon can

[3] Such a plan, used to construct a muqarnas in present-day Fez, is discussed in (Dold-Samplonius 1996: 71).

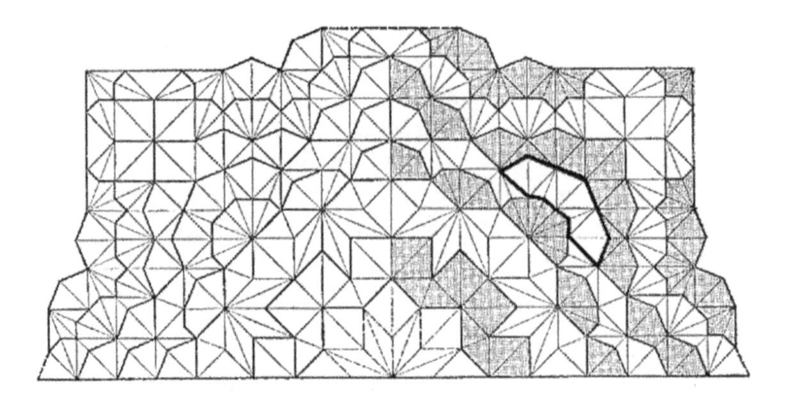

Fig. 47.6 Plane projection of the entrance portal of the shrine at Bastām, Iran. Image: author, after Harb (1978: 47)

be mirrored along the diagonal pointing towards the center of the vault. These elements might well be a badly drawn double "almond" or, maybe, a new element described by al-Kāshī. In these cases also the highest and central part of the vault still has to be decorated.

Second, some designs are sketchy and not worked out in much detail. In the Topkapi Scroll we find several rough designs, for instance designs 96 and 104 (Necipoğlu 1995: 338, 342), where the artist has worked out a small part into detail, probably to avoid confusion. The Timurid Topkapi scroll ranks among the oldest extant designs, but in modern Moroccan designs as well, as seen in the ground plans from Fez, the artist tends to help the artisans by including signs and letters for the required elements in the design.

Based upon the above evidence, we think that the artists and artisans, even while using a standard design from a pattern book, still have some, although restricted, freedom during the construction process. This freedom is necessary when difficulties arise due to irregularities in the building. The artisans repeated endless variations based on old geometric formulas, slightly modifying them by trial and error. For these artists, the muqarnas was, and in a way still is, part of their daily life and culture. For us, outsiders, muqarnas is beautiful geometric art to be studied and admired. We can understand its composition and discover intriguing details but muqarnas forms no part of our cultural identity.

Reconstruction of Muqarnas

There are several interesting questions connected to the principles of constructing muqarnas. While questions about structural integrity and esthetic merit belong to the area of architecture or archaeology, the distinctive properties of a muqarnas design or ground plan are the basis of research to analyze mathematically the

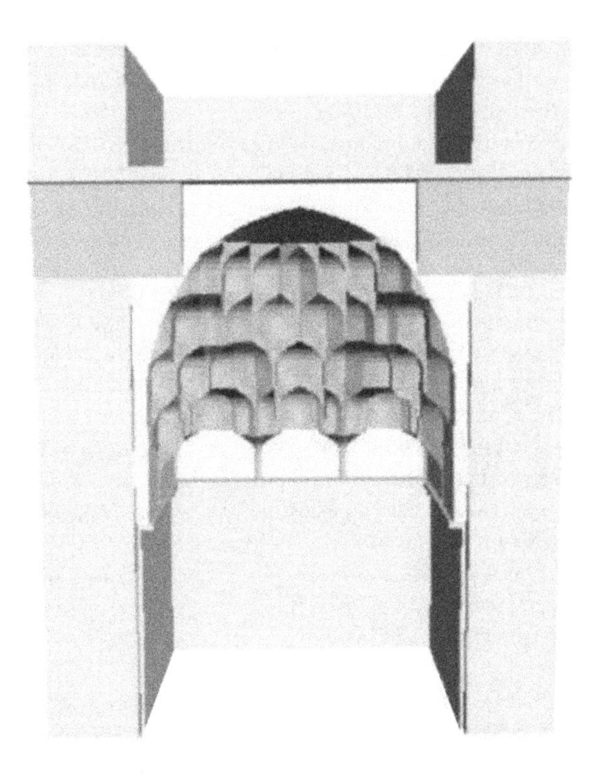

Fig. 47.7 Computer representation of a vault in the friday mosque at Natanz, Iran. Rendering: authors

geometry of muqarnas. Mathematics can help us to decide what the corresponding muqarnas element is for every geometric form in the ground plan, in which tier this element appears, and its orientation.

As many element combinations appear more than once, we are constructing a database of known muqarnas, sorted by period and by region. An algorithm for rebuilding a muqarnas should first calculate which interpretations fit mathematically. Next, these possibile interpretations have to be filtered by comparing them with muqarnas of the same time period and region. The results of the mathematical analysis, together with these two style properties, will then provide the information for building a computer graphics representation of the muqarnas. Having such a tool for computer reconstructions will permit us to make suggestions for reconstructing muqarnas for which only the design is known. If we are dealing with a muqarnas which is partially collapsed, we can make suggestions for reconstruction by calculating which element combinations would fit in the collapsed part of the muqarnas.

An example of a computer representation is given in Fig. 47.7, which shows a computer generated muqarnas vault. The ground plan used for our computer representation is the plane projection of an existing vault. The original vault, bearing the date 1309, is above a niche in the sublevel of the north *iwan* (a roofed or vaulted rectangular room open at one end) in the Friday Mosque at Natanz, Iran. Natanz, a small mountain village on the eastern road between Isfahan, Kashan, and

Qumm, is the site of a famous complex of structures from the Il Khanid period. By comparing our computer representation with the existing vault we can control whether our system functions.

We have developed a video entitled *Magic of Muqarnas* (Dold-Samplonius et al. 2005) which gives an overview of different muqarnas styles. This video explains the construction and reconstruction of muqarnas and also shows our realizations of computer generated muqarnas.

Biography Yvonne Dold-Samplonius studied mathematics and Arabic at the University of Amsterdam, specializing in the History of Islamic mathematics. She wrote her thesis with Prof Bruins, Amsterdam and Prof. Juan Vernet, Barcelona. The academic year 66/67 she spent at Harvard studying under Prof. Murdoch. She has published about forty papers on the History of Mathematics. In recent years her interest has shifted to Mathematics in Islamic Architecture from a historic point of view. Under her supervision two videos concerning this subject, "Qubba for al-Kashi", on arches and vaults, and "Magic of Muqarnas" have been produced at the IWR (Interdisciplinary Center for Scientific Computing), Heidelberg, where she is an associated member. She is an effective member of the International Academy of History of Science.

Silvia Harmsen started studying mathematics in 1996 at the Utrecht University. There she received in 2001 cum laude her diploma in the field of Algebra and Geometry. From 2002 to 2006 she worked on Computer Reconstructions of Muqarnas at the Interdisciplinary Center for Scientific Computing, University of Heidelberg. After finishing her thesis she works as a software developer at SAP AG in Walldorf.

References

Dold-Samplonius, Y. 1992. Practical Arabic Mathematics: Measuring the Muqarnas by al-Kāshī. Centaurus, 35: pp. 193–242.

___. 1996. How al-Kāshī Measures the Muqarnas: A Second Look. Pp. 56-90 in M. Folkerts, ed. Mathematische Probleme im Mittelalter: Der lateinische und arabische Sprachbereich. Vol. X. Wiesbaden: Wolfenbütteler Mittelalter–Studien.

___. 2003. Calculating Surface Areas and Volumes in Islamic Architecture. Pp. 235-265 in The Enterprise of Science in Islam, New Perspectives, J.P. Hogendijk & A.I. Sabra, eds. Dibner Institute Studies in the History of Science and Technology. Cambridge, MA and London: MIT Press.

Dold-Samplonius, Y., Harmsen, S. L., Krömker, S., & Winckler, M. J. 2005. Magic of Muqarnas (Video). Heidelberg: University of Heidelberg. http://www.iwr.uni-heidelberg.de/groups/ngg/Muqarnas/muqarnas-video.php?L=0. Accessed 14 November 2013.

Harb, U. 1978. Ilkhandische Stalaktitengewölbe, Beiträge zu Entwurf und Bautechnik. Archäologische Mitteilungen aus Iran. Vol. IV. Berlin: Dietrich Reiner.

Necipoğlu, G. 1995. The Topkapı Scroll—Geometry and Ornament in Islamic Architecture. Santa Monica: The Getty Center for the History of Art and Humanities, pp. 167-175.

Pope, A. U. 1939. A Survey of Persian Art, from Prehistoric Times to the Present. Vol. II. London and New York: Oxford University Press, p. 1102.

Wilber, D. N. 1955. The Architecture of Islamic Iran: The Il Khanid Period. New York: Greenwood Press.

Index for Volume I[1]

[1] Note: Individual buildings may generally be found listed under one of the following classifications:

- Amphitheatres
- Apartment Buildings
- Archaeological Sites
- Baptisteries
- Basilicas
- Bridges
- Castles
- Cathedrals
- Chapels
- Churches
- Government Buildings
- Houses
- Libraries
- Mausoleums
- Monuments and Memorials
- Mosques
- Museums
- Palaces
- Public Spaces / Squares
- Pyramids
- Shrines
- Temples
- Theatres / Entertainment
- Towers
- Villas

K. Williams and M.J. Ostwald (eds.), *Architecture and Mathematics from Antiquity to the Future*, DOI 10.1007/978-3-319-00137-1,

D

E

Index for Volume II

K. Williams and M.J. Ostwald (eds.), *Architecture and Mathematics from Antiquity to the Future*, DOI 10.1007/978-3-319-00137-1,

R

S

T

U

V

CPSIA information can be obtained
at www.ICGtesting.com
Printed in the USA
LVHW08*1658230918
591096LV00003B/8/P